Endorsed for
Pearson Edexcel
Qualifications

Second Edition

T0351568

Pearson Edexcel GCSE (9–1)

Mathematics

Foundation

Student Book

2

Series editors: Dr Naomi Norman and Katherine Pate

Published by Pearson Education Limited, 80 Strand, London, WC2R 0RL.

www.pearsonschoolsandfecolleges.co.uk

Text © Pearson Education Limited 2020
Project managed and edited by Just Content Ltd
Typeset by PDQ Digital Media Solutions Ltd
Original illustrations © Pearson Education Limited 2020
Cover photo/illustration by © David S. Rose/Shutterstock, © Julias/Shutterstock, © Attitude/Shutterstock, © Abstractor/Shutterstock, © Ozz Design/Shutterstock, © Lazartivan/Getty Images

The rights of Chris Baston, Ian Bettison, Ian Boote, Tony Cushen, Tara Doyle, Kath Hipkiss, Catherine Murphy, Su Nicholson, Naomi Norman, Diane Oliver, Katherine Pate, Jenny Roach, Carol Roberts, Peter Sherran and Robert Ward-Penny to be identified as authors of this work have been asserted by them in accordance with the Copyright, Designs and Patents Act 1988.

First published 2020

24
10 9 8

British Library Cataloguing in Publication Data

A catalogue record for this book is available from the British Library.

ISBN 978 1 292 34638 0

Copyright notice

All rights reserved. No part of this publication may be reproduced in any form or by any means (including photocopying or storing it in any medium by electronic means and whether or not transiently or incidentally to some other use of this publication) without the written permission of the copyright owner, except in accordance with the provisions of the Copyright, Designs and Patents Act 1988 or under the terms of a licence issued by the Copyright Licensing Agency, 5th Floor, Shackleton House, Hay's Galleria, 4 Battlebridge Lane, London SE1 2HX (www.cla.co.uk). Applications for the copyright owner's written permission should be addressed to the publisher.

Printed and bound in Great Britain by Bell and Bain Ltd, Glasgow

A note from the publisher

In order to ensure that this resource offers high-quality support for the associated Pearson qualification, it has been through a review process by the awarding body. This process confirms that this resource fully covers the teaching and learning content of the specification or part of a specification at which it is aimed. It also confirms that it demonstrates an appropriate balance between the development of subject skills, knowledge and understanding, in addition to preparation for assessment.

Endorsement does not cover any guidance on assessment activities or processes (e.g. practice questions or advice on how to answer assessment questions) included in the resource nor does it prescribe any particular approach to the teaching or delivery of a related course.

While the publishers have made every attempt to ensure that advice on the qualification and its assessment is accurate, the official specification and associated assessment guidance materials are the only authoritative source of information and should always be referred to for definitive guidance.

Pearson examiners have not contributed to any sections in this resource relevant to examination papers for which they have responsibility.

Examiners will not use endorsed resources as a source of material for any assessment set by Pearson. Endorsement of a resource does not mean that the resource is required to achieve this Pearson qualification, nor does it mean that it is the only suitable material available to support the qualification, and any resource lists produced by the awarding body shall include this and other appropriate resources.

Pearson has robust editorial processes, including answer and fact checks, to ensure the accuracy of the content in this publication, and every effort is made to ensure this publication is free of errors. We are, however, only human, and occasionally errors do occur. Pearson is not liable for any misunderstandings that arise as a result of errors in this publication, but it is our priority to ensure that the content is accurate. If you spot an error, please do contact us at resourcescorrections@pearson.com so we can make sure it is corrected.

Contents

Pearson Edexcel GCSE (9–1)
Mathematics
Second Edition

Pearson Edexcel GCSE (9–1) Mathematics Second Edition is built around a unique pedagogy that has been created by leading educational researchers and teachers in the UK. This edition has been updated to reflect six sets of live GCSE (9–1) papers, as well as feedback from thousands of teachers and students and a 2-year study into the effectiveness of the course.

The new series features a full range of print and digital resources designed to work seamlessly together so that schools can create the course that works best for their students and teachers.

*Active*Learn service

The *Active*Learn service brings together the full range of planning, teaching, learning and assessment resources.

What's in *Active*Learn for GCSE (9–1) Mathematics?

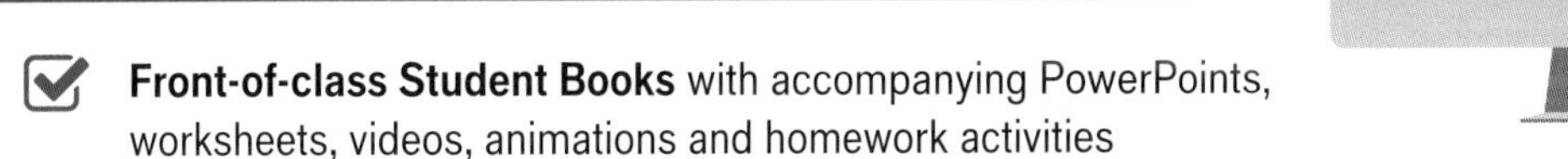

- **Front-of-class Student Books** with accompanying PowerPoints, worksheets, videos, animations and homework activities

- **254 editable and printable homework worksheets**, linked to each Master lesson
- **Online, auto-marked homework activities** with integrated videos and worked examples
- **76 assessments and online markbooks**, including end-of-unit, end-of-term, end-of-year and baseline tests
- **Interactive Scheme of Work** brings everything together, connecting your personalised scheme of work, teaching resources and assessments
- **Individual student access to videos, homework and online textbooks**

Student Books

The Student Books use a mastery approach based around a well-paced and well-sequenced curriculum. They are designed to develop mathematical fluency, while building confidence in problem-solving and reasoning.

The unique unit structure enables every student to acquire a deep and solid understanding of the subject, leaving them well-prepared for their GCSE exams, and future education or employment.

Together with the accompanying online prior knowledge sections, the Student Books cover the entire **Pearson Edexcel GCSE (9–1) Mathematics course**.

The new four-book model means that the Second Edition Student Books now contain even more meaningful practice, while still being a manageable size for use in and outside the classroom.

Foundation tier

Higher tier

Pearson Edexcel GCSE (9–1)
Mathematics Second Edition
Foundation Student Book

2

Building confidence

Pearson's unique unit structure has been shown to build confidence. The **front-of-class** versions of the Student Books include lots of extra features and resources for use on a whiteboard.

Master

Learn fundamental knowledge and skills over a series of lessons.

*Active*Learn **homework**

Links to online homework worksheets and exercises for every lesson.

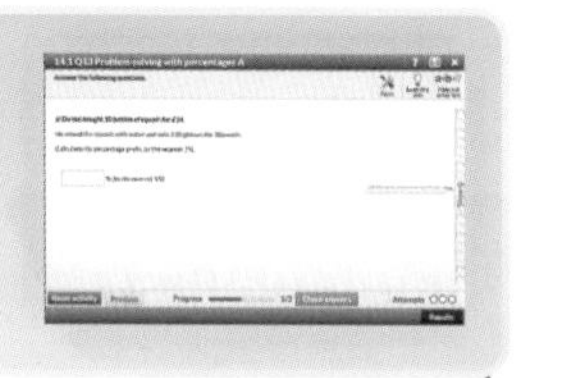

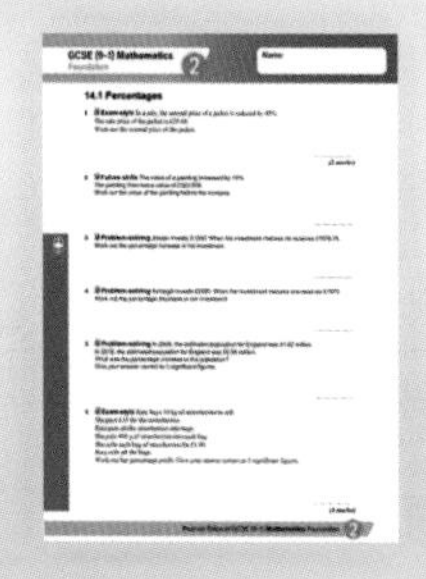

Students can make sure they are ready for each unit by downloading the relevant **Prior knowledge check**. This can be accessed using the QR code in the Contents or via *Active*Learn.

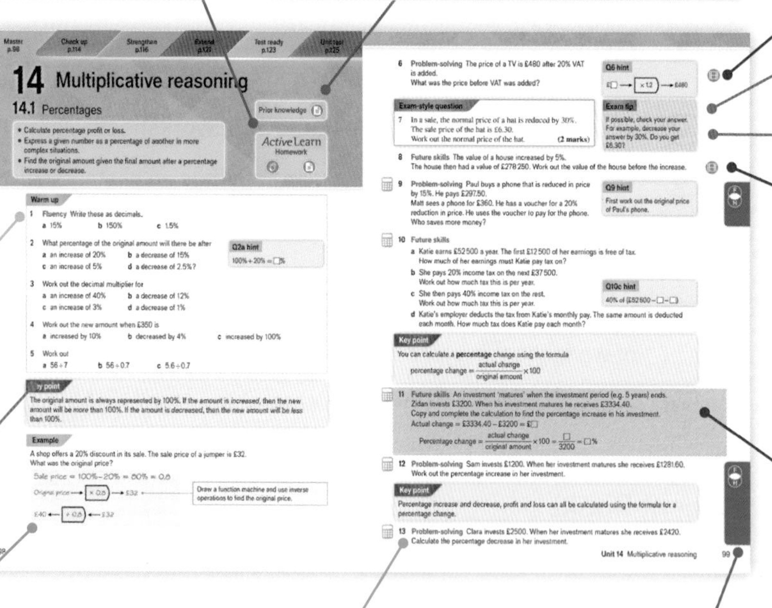

Warm up

Accessible questions designed to develop mathematical fluency.

Key points

Explains key concepts and definitions.

Worked example

Step-by-step worked examples focus on the key concepts.

Problem-solving and **Reasoning** questions are clearly labelled. **Future skills** questions help prepare for life after GCSE. **Reflect** questions encourage reflection on mathematical thinking and understanding.

Crossover content between Foundation and Higher tiers is indicated in side bars.

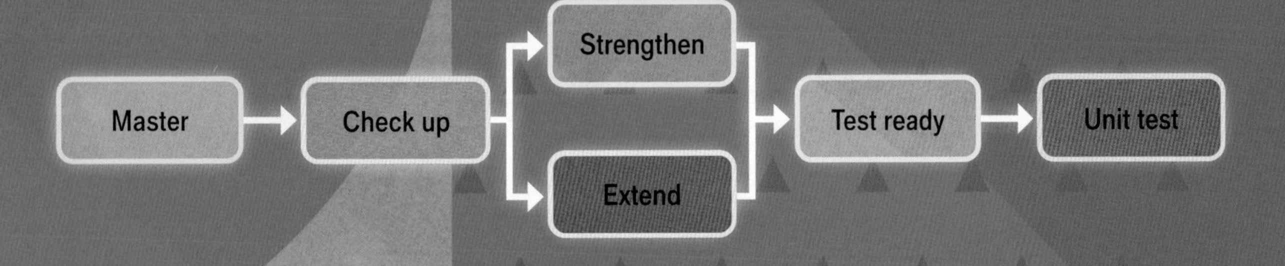

Teaching and learning materials can be downloaded from the blue hotspots.

Exam-style questions are included throughout the books to help students prepare for GCSE exams.

Exam tips point out common errors and help with good exam technique.

Helpful videos walk you step-by-step through answers to similar questions.

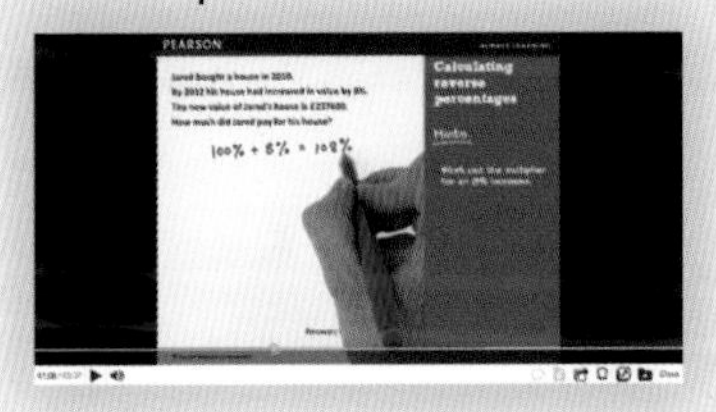

Click on any question to view it full-size, and then click 'Show' to reveal the answer.

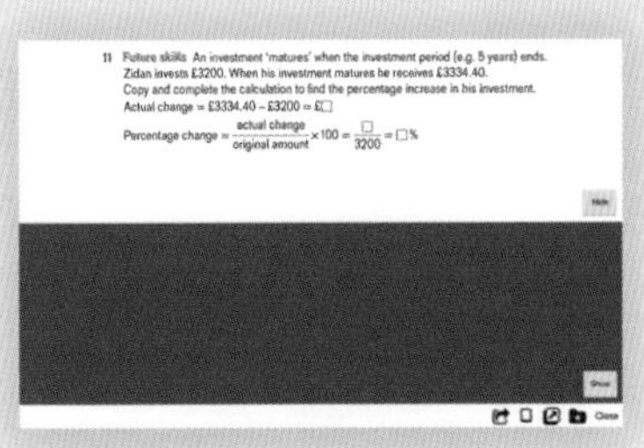

Check up

After the Master lessons, a Check up test helps students decide whether to move on to the Strengthen or Extend section.

Strengthen

Students can choose the topics they need more practice on. There are lots of hints and supporting questions to help.

Extend

Applies and develops maths from the unit in different situations.

Test ready

The **Summary of key points** is used to identify areas that need more practice and **Sample student answers** familiarise students with good exam technique.

Unit test

The exam-style Unit test helps check progress.

Mixed exercises

These sections bring topics together to help practise applying different techniques to a range of questions types, which is required in GCSE exams.

Interactive Scheme of work

The Interactive Scheme of Work makes reordering the course easy. You can view your plan for your year, term or lesson, and access all the related teaching, learning and assessment materials.

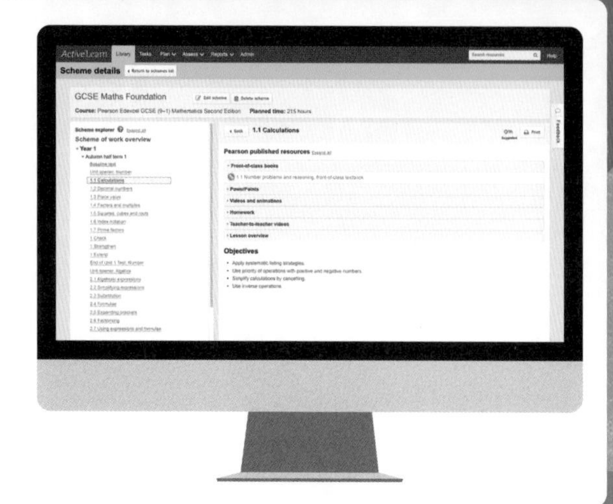

*Active*Learn Progress & Assess

The Progress & Assess service is part of the full *Active*Learn service, or can be bought as a separate subscription. It includes assessments that have been designed to ensure all students have the opportunity to show what they have learned through:

- a 2-tier assessment model
- separate calculator and non-calculator sections
- online markbooks for tracking and reporting
- mapping to indicative 9–1 grades.

Assessment Builder

Create your own classroom assessments from the bank of GCSE (9–1) Mathematics assessment questions by selecting questions on the skills and topics you have covered. Map the results of your custom assessments to indicative 9–1 grades using the custom online markbooks. Assessment Builder is available to purchase as an add-on to the *Active*Learn service or Progress & Assess subscriptions.

Purposeful Practice Books

A new kind of practice book based on cutting-edge approaches to help students make the most of practice.

With more than 4500 questions, our Pearson Edexcel GCSE (9–1) Mathematics Purposeful Practice Books are designed to be used alongside the Student Books and online resources. They:

- use minimal variation to build in small steps, consolidating knowledge and boosting confidence
- focus on strengthening problem-solving skills and strategies
- feature targeted exam practice with questions modified from real GCSE (9–1) papers, and exam guidance from examiner reports and grade indicators informed by **ResultsPlus**.

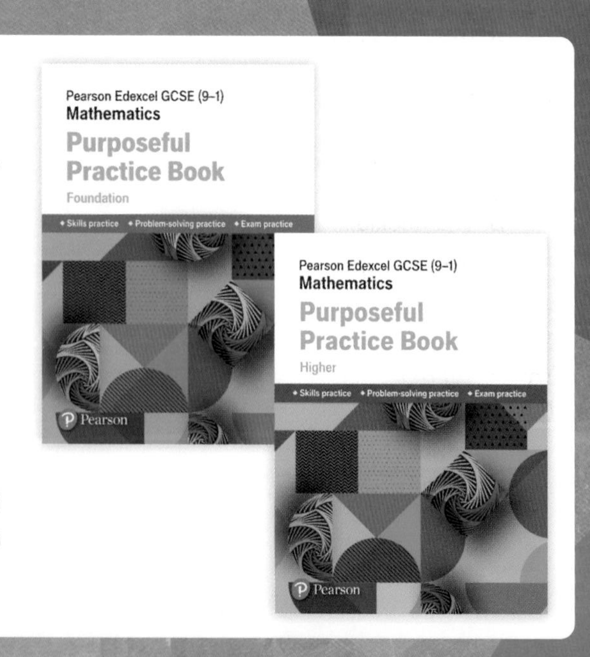

11 Ratio and proportion

Prior knowledge 

11.1 Writing ratios

- Use ratio notation.
- Write a ratio in its simplest form.
- Solve simple problems using ratios.

Warm up

1 **Fluency** Charlie has 3 red beads and twice as many blue beads. How many blue beads does she have?

2 What is the highest common factor of

a 8 and 16 b 35 and 14 c 15, 25 and 100?

3 The bar chart shows the number of teenagers doing some outdoor activities. How many teenagers are doing the activities?

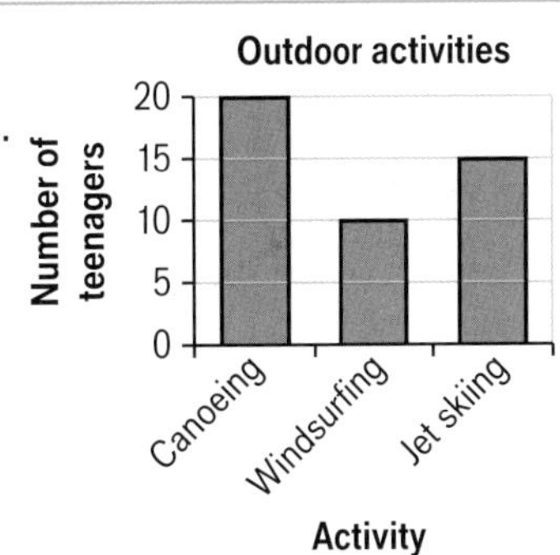

Key point

A **ratio** is a way to compare two or more quantities.

4 Write down the ratio of red tins to yellow tins.

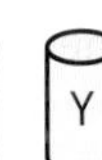

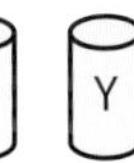

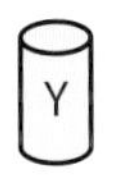

Q4 hint

There are 2 red tins and 5 yellow tins. Write the ratio 'red : yellow' using the numbers.

5 Draw tins to show these ratios of red to yellow.

a 4 : 3 b 3 : 4

c **Reflect** Is the ratio 4 : 3 the same as the ratio 3 : 4?

6 The ratio of purple beads to blue beads on a necklace is 1 : 4.

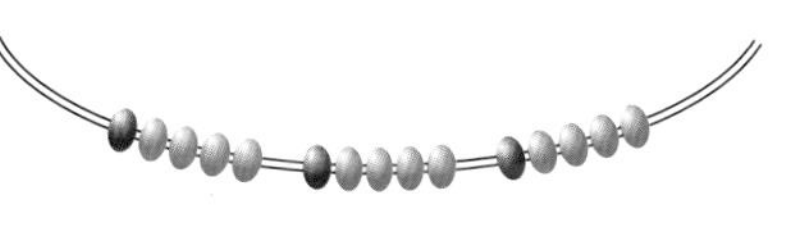

a Copy and complete this sentence.

There is □ purple bead for every □ blue beads.

b The necklace has 30 beads. How many groups of '1 purple, 4 blue' are there?

c How many beads are i purple ii blue?

Key point

You **simplify** a ratio by making the numbers as small as possible (keeping them as integers). Divide the numbers in the ratio by their **highest common factor (HCF)**.

7 Write each ratio in its simplest form.

a 4 : 12 b 16 : 8 c 27 : 9

d 7 : 42 e 15 : 20 f 63 : 28

g 18 : 48 h 12 : 120 i 36 : 45

Q7a hint

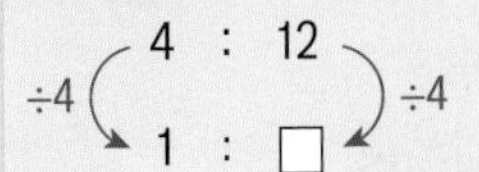

8 **Reasoning** Jon is asked to write the ratio of 450 litres to 15 litres in its simplest form.
Jon writes 3 : 90. What two mistakes has Jon made?

Q8 hint

You do not include units in ratios when the units are the same. Not including litres is not one of Jon's mistakes.

Exam-style question

9 An art shop has 10 tins of pencils.
There are 8 pencils in each tin.
Write as a ratio the number of pencils in three tins to the total number of pencils.
Give your answer in simplified form. **(2 marks)**

Exam tip

Sometimes you must do calculations before finding a ratio.

10 **Problem-solving** The bar charts show the numbers of gold medals and other medals won at a competition by each group from a gym club.

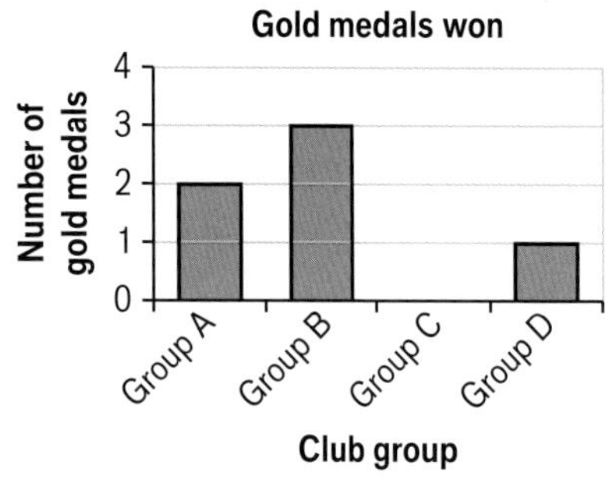

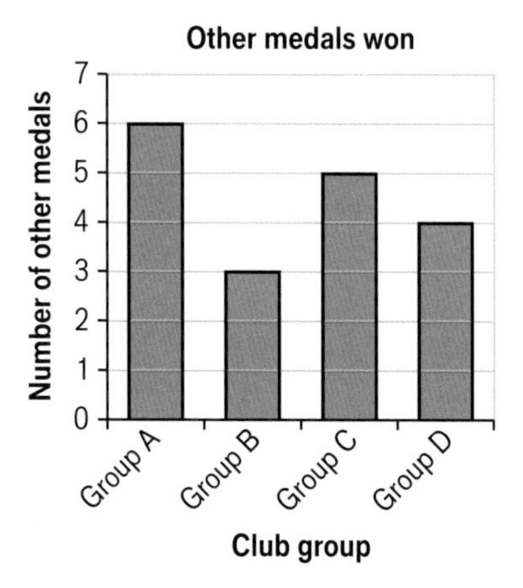

What is the ratio of gold medals won to other medals won?
Give your answer in its simplest form.

11 **Future skills** An after-school club is attended by 32 children. It is run by 4 adults.
The guidelines say that the adult-to-child ratio should be 1 : 8.
Does the club have enough adults? Show your working.

Key point

Ratios are **equivalent** if they have the same simplest form.

12 **Problem-solving** Which of these ratios are equivalent?

A 36 : 16 **B** 135 : 60 **C** 28 : 16 **D** 126 : 56 **E** 49 : 28

13 Write each ratio in its simplest form.

a 20 : 25 : 15 **b** 36 : 24 : 30 **c** 56 : 42 : 35 **d** 16 : 40 : 56

Q13a hint

Divide all three parts by the HCF of 20, 25 and 15.

14 A recipe for shortbread uses 125 g of butter, 55 g of sugar and 180 g of flour. Write the ratio of butter : sugar : flour.
Give your ratio in its simplest form.

Exam-style question

15 There are 96 marbles in a bag.
12 are clay, 30 are china and the rest are glass.
Write the ratio of clay marbles to china marbles to glass marbles.
Give your answer in its simplest form. **(3 marks)**

Exam tip

Make sure you write the numbers in your ratio in the correct order. Check it is fully simplified.

11.2 Using ratios 1

- Solve simple problems using ratios.
- Use ratios involving decimals.

Warm up

1 **Fluency** What are the missing numbers in these equivalent measures?

a 1 m = □ cm b 1 cm = □ mm c 1 m = □ mm d 500 mm = □ m

2 Copy and complete a $2 \times \square = 100$ b $4 \times \square = 120$ c $7 \times \square = 280$

3 Work out a 1.5×10 b 3.71×10 c 9.37×100

4 Find the HCF of each pair of numbers.

a 24 and 30 b 35 and 49 c 64 and 80 d 75 and 325

Example

To make orange paint Maria mixes yellow paint with red paint in the ratio 3 : 1.
She uses 4 tins of red paint. How many tins of yellow paint does she use?

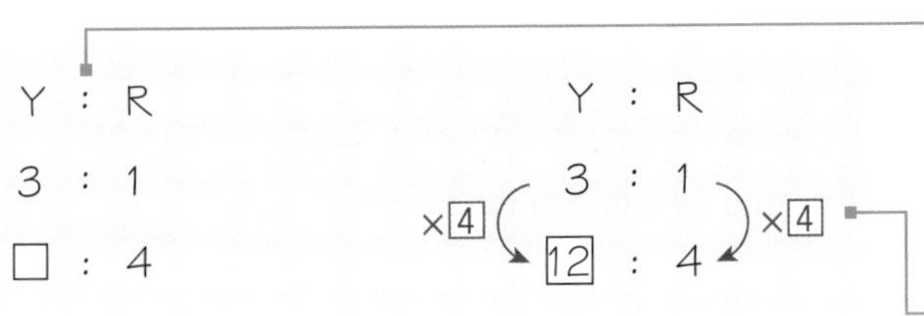

Write down the ratios given. Use Y for yellow and R for red. Write the ratio that is given and any other information given in the question.

Multiply each part by the same number to get an equivalent ratio.

5 **Problem-solving** The ratio of oyster sauce : fish sauce in a stir fry recipe is 2 : 1.
Clare uses 5 tablespoons of fish sauce.
How many tablespoons of oyster sauce does she use?

6 **Problem-solving** The ratio of chick peas : broad beans in a hummus recipe is 9 : 2. Jack uses 180 g of chick peas.
How many grams of broad beans should he use?

Q6 hint

C : B
9 : 2
×□ ×□
180 : □

7 **Reasoning** The ratio of okra : sweet potato in a korma recipe is 15 : 8.
Raj uses 160 g of sweet potato. He has 350 g of okra.
Will he use all of the okra?

Exam-style question

8 Anna splits her monthly net pay into rent money, bill money and money left in the ratio 7 : 1 : 2.
Her rent is £420. How much money does she have left each month? **(2 marks)**

Exam tip

Make sure you work with the correct part of the ratio for the answer that is required.

F
H

9 **Problem-solving** Thomas makes a model of the Eiffel Tower using a ratio of 1:1280.
The height of his model is 250 mm.
What is the height of the Eiffel Tower in metres?

Q9 hint

Model	:	Real
1	:	1280
×□ ↓ □	:	□ ↓ ×□

These are in mm

10 **Reasoning** There are 35 people living in a care home.
At least 1 member of staff is needed for every 4 people in the home.

a Work out the least number of staff needed in the home.

b Another person moves into the home. Does that mean more members of staff are needed? You must give a reason for your answer.

Key point

Ratios in their **simplest form** only have integer parts (whole numbers).

Example

Write 1.5:8 as a whole number ratio in its simplest form.

$$1.5 : 8 \xrightarrow{\times 10} 15 : 80 \xrightarrow{\div 5} 3 : 16$$

1.5 has 1 decimal place so multiply both sides of the ratio by 10 to get a whole number.

The HCF is 5 so divide both sides by 5.

11 Write each ratio as a whole number ratio in its simplest form.

a 0.4:6 **b** 3.5:4.2 **c** 45:13.5 **d** 25.6:46.4

12 **Reflect** What should you multiply numbers in a ratio by if one of the numbers has 2 decimal places?

13 Write each ratio as a whole number ratio in its simplest form.

a 0.25:3.1 **b** 1.4:0.28 **c** 1.62:1.8 **d** 4.8:11.2

14 **Problem-solving** Old computer monitors had a width : height ratio of 4:3.
New computer monitors have a width : height ratio of 16:9.
Is each of these monitors old or new?

a

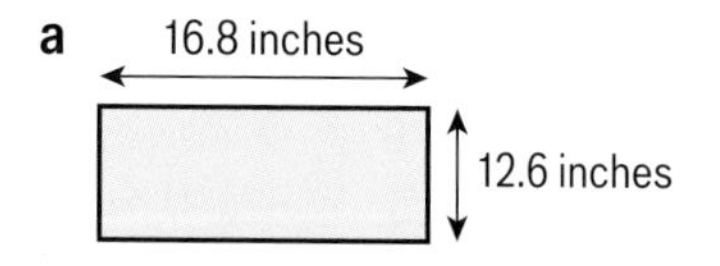

b

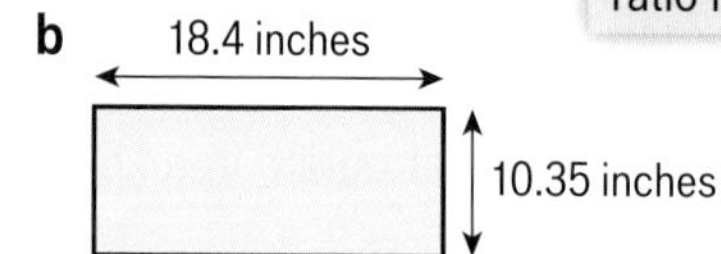

Q14 hint

Write the ratio width : height of each monitor as a whole number ratio in its simplest form.

Exam-style question

15 4 running clubs join together to hold a charity fun run.
The same number of runners from each club take part.
At one of the clubs, the ratio of the number of male runners to the number of female runners that take part is 1:3.
There are 15 female runners from this club that take part.
How many runners in total take part in the fun run?

(4 marks)

Exam tip

Deal with the information one paragraph at a time.

11.3 Ratios and measures

- Write and use ratios for shapes and their enlargements.
- Use ratios to convert between units.

Warm up

1 **Fluency** What is
  a the area of a rectangle with length 7 cm and width 5 cm
  b the volume of a cube with side length 2 cm?

2 Copy and complete these equivalent measures
  a 1 kg = □ g b 1 km = □ m c 1 litre = □ ml
  d 1 hour = □ minutes e 1 minute = □ seconds f 1 hour = □ seconds

3 The diagram shows two rectangles.
Write these ratios in their simplest form.
  a The width of rectangle A to rectangle B.
  b The length of rectangle A to rectangle B.
  c The area of rectangle A to rectangle B.
  d **Reflect** Write a sentence explaining what you notice about your answers to **a**, **b** and **c**.

4 The diagram shows two cubes.
Write these ratios in their simplest form.
  a Height of A to height of B.
  b Area of a face of A to area of a face of B.
  c Volume of A to volume of B.
  d **Reflect** Write a sentence explaining what you notice about your answers to parts **a**, **b** and **c**.

5 **Problem-solving / Reasoning** The ratio of the area of square A to square B is 1:36.
What is the ratio of the length of square A to the length of square B?

6 Write these ratios in their simplest form.
  a 120 mm : 50 cm b 15 minutes : 1 hour
  c 400 g : 1 kg d 2 hours : 30 seconds

Q6 hint
Both parts of the ratio need to be in the same units before you simplify.

7 **Problem-solving** A mug contains 250 ml of coffee and a jug contains 1 litre of coffee.
Write down the ratio of the amount of coffee in the mug to the amount of coffee in the jug.
Give your answer in its simplest form.

Key point

You can use **ratios** to convert between **units**.

8 Write these measures as ratios. The first one is done for you.

a 1 kg = 1000 g kg : g = 1 : 1000

b 1 km = ☐ m

c 1 litre = ☐ ml

d 1 m = ☐ mm

9 Copy and complete these conversions.

a 2.5 km = ☐ m b 34 000 g = ☐ kg

c 3.8 litres = ☐ ml d 7 300 mm = ☐ m

Q9a hint

km : m

1 : ☐

2.5 : ☐

×☐ ×☐

10 1 mile ≈ 1.6 km. Convert

a 5 miles to km b 16 km to miles

c **Problem-solving** The length of the London Marathon is 26.219 miles. How far is this in kilometres?

11 **Reasoning** Lily needs 1500 yards of wool to knit a scarf.
She buys 6 balls of wool. There are 225 m of wool in each ball.
Lily knows that

- 1 yard is 36 inches
- 1 inch is 2.54 centimetres

Does Lily have enough wool to knit the scarf?
You must show all your working.

12 **Future skills** A money exchange has a notice in its window:
£1 buys US$1.38.

a How many dollars (US$) would you get for £500?

b How many pounds would you get for US$483?

c **Reasoning** Which is worth more, £950 or US$1300?
You must show your working.

Q12a hint

£ : $

1 : 1.38

500 : ☐

×500 ×500

Exam-style question

13 Antony went on holiday to Australia.
His flights cost a total of £1760.
Antony stayed for 12 nights in an apartment.
His apartment cost AUD$273 per night.
Antony hired a car for 5 days.
The car hire cost was AUD$136.50 per day.
The exchange rate was AUD$1.82 to £1.

a Work out the total cost of the flights, the apartment and car hire. Give your answer in pounds. **(5 marks)**

b If there were more Australian dollars to £1, what effect would this have on the total cost, in pounds, of Antony's holiday? **(1 mark)**

Exam tip

For part **a**, work out subtotals first, for example the total cost of the apartment. Then correct to the currency asked in the answer.

11.4 Using ratios 2

- Divide a quantity into 2 parts in a given ratio.
- Divide a quantity into 3 parts in a given ratio.
- Solve word problems using ratios.
- Use bar models to help solve ratio problems.

Warm up

1 What is the ratio of blue to white sections in these bars?

a [bar of 5 sections] b [bar of 7 sections]

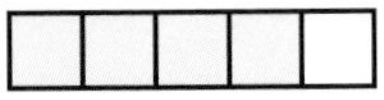

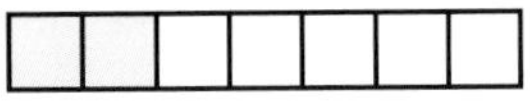

2 Draw a bar to show the blue sections to white sections in the ratio 2 : 3.

3 Work out

a 35 ÷ 7 b 10 ÷ 4 c 4.5 ÷ 3 d 2.8 ÷ 4

4 Round these numbers.

a 12.468 to 2 decimal places b 3.1584 to 3 decimal places

5 Write

a 100 : 350 in its simplest form b two equivalent ratios to 1 : 4

Example

Share £25 in the ratio 3 : 2.

3 + 2 = 5 parts

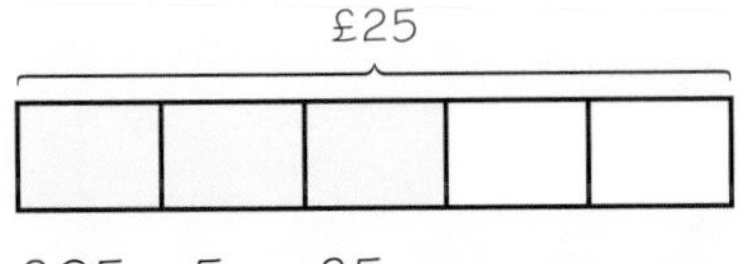

Work out how many parts there are in total.

£25 ÷ 5 = £5 — Work out 1 part.

£5 × 3 = £15
£5 × 2 = £10 — Work out 3 parts and 2 parts.

Answer: £15 : £10

Check: £15 + £10 = £25 ✓ — Check they add up to the correct total.

6 Share these amounts in the ratios given.

a £18 in the ratio 2 : 1 b £42 in the ratio 1 : 6
c £27 in the ratio 4 : 5 d 35 kg in the ratio 2 : 3
e 60 m in the ratio 5 : 7 f 7.5 litres in the ratio 2 : 3

Q6 hint
How many parts are there?

7 **Problem-solving** Before 2012, 10p coins were made from copper and nickel in the ratio 3 : 1.
Each coin had a mass of 6.5 g.
What was the mass of

a copper b nickel?

8 **Problem-solving** Purple paint is mixed from red paint and blue paint in the ratio 5 : 3.
A painter needs 20 litres of purple paint.

a How many litres of red paint should they use?

b How many litres of blue paint should they use?

c **Reflect** What calculation could you do to check your answers to **a** and **b**?

9 Share these amounts in the ratios given.

a £72 in the ratio 2 : 3 : 4

b 100 g in the ratio 2 : 3 : 5

c 360 ml in the ratio 3 : 4 : 5

Q9 hint

How many parts are there?

10 A fruit drink is made from orange, pineapple and apple juice in the ratio 1 : 2 : 4.
Rita wants to make 35 litres of fruit drink. How much of each type of juice does she need?

Exam-style question

11 Mina, Nancy and Oli share £240 in the ratio 3 : 11 : 6.
How much more does Nancy get than Oli? **(3 marks)**

Exam tip

Check your arithmetic carefully. Have you added, subtracted, multiplied and divided correctly?

12 **Reasoning** Stuart is going to make some concrete mix.
He needs to mix cement, sand and gravel in the ratio 1 : 2 : 3 by weight.
Stuart wants to make 150 kg of concrete mix. Stuart has

- 30 kg of cement
- 55 kg of sand
- 70 kg of gravel

Does Stuart have enough cement, sand and gravel to make the concrete mix?
You must show your working.

Q12 hint

Work out how many kg of each Stuart needs to make 150 kg.

13 Share these amounts in the ratios given. Round your answers sensibly.

a £80 in the ratio 2 : 5

b 70 litres in the ratio 2 : 7

c £25 in the ratio 1 : 3 : 4

d **Reflect** How did you round when the ratio was **i** money **ii** litres? Explain.

14 **Reasoning** Bob and Phil buy a dog for £450.
Bob pays £300 and Phil pays £150.
The dog wins a prize at a dog show. The prize is £180.

a Write the amounts Bob and Phil each pay as a ratio in its simplest form.

b Divide their prize money in this ratio.
How much does each of them get?

c **Reflect** Is this a fair way to share the prize money? Explain.

15 **Problem-solving** Andrea and Penny buy a statue for £350.
Andrea pays £140 and Penny pays £210.
They sell the statue 3 years later for £475.
Show how they should share the money fairly.

16 **Problem-solving / Reasoning** Two numbers are in the ratio 1 : 3 and their difference is 12.
What are the numbers?

Q16 hint

Write ratios equivalent to 1 : 3. Which pair of numbers have a difference of 12?

Example

Ethel and Lola share money in the ratio 5 : 2.
Ethel gets £45 more than Lola.
How much does Lola get?

E	E	E	E	E	L	L

Draw a bar to represent the ratio.

E	E	E	E	E
L	L	£45		

Compare the bars for Ethel and Lola and label the difference.

1 part = $45 \div 3$ = £15 — 3 parts represent £45. Work out 1 part.

2 parts = $2 \times$ £15 = £30 — Lola gets 2 parts. Work out 2 parts.

Lola gets £30

Check: 5 parts = $5 \times$ £15 = £75

£75 − £30 = £45 ✓

17 **Problem-solving** Toby and Isy share some money in the ratio 5 : 7.
Isy gets £60 more than Toby.
How much money does Toby get?

18 **Problem-solving** The ratio of Ranjit's age to her dad's age is 2 : 7.
Ranjit's dad is 30 years older than Ranjit.
How old is Ranjit's dad?

19 **Problem-solving** A farmer has sheep, cows and chickens in the ratio 12 : 3 : 2.
The farmer has 150 more sheep than chickens.
How many cows has she got?

20 **Problem-solving / Reasoning** Ed, Kate and Simon each play a game.
Ed's score is six times Kate's score.
Simon's score is half Ed's score.
Write down the ratio of Ed's score to Kate's score to Simon's score.

Q20 hint

Draw 1 part of a bar to represent Kate's score.

Exam-style question

21 £2600 is shared between Advay, Bella, Cara and James.
The ratio of the amount Advay gets to the amount Bella gets is 4 : 1.
Cara gets 1.5 times the amount Advay gets.
James gets half the amount Advay gets.
Work out the amount of money Cara gets. **(4 marks)**

Exam tip

When an exam question includes a ratio, try drawing a bar model to help you to answer it.

11.5 Comparing using ratios

- Compare ratios.
- Write ratios in the form $1:n$ or $n:1$.
- Solve ratio and proportion problems.

Warm up

1 Fluency

a What fraction of this bar is blue?

b What percentage of this bar is white?

2 Work out

a $\frac{3}{9}+\frac{4}{9}$ **b** $1-\frac{2}{5}$ **c** $\frac{2}{5}\times 20$

d $\frac{5}{8}$ of 40 **e** 15% of 40 **f** $\frac{3}{20}$ as a percentage.

3 Simplify these ratios.

a 2.5 : 1.5 **b** 5.5 : 4.5 : 3 **c** 0.36 : 0.4 **d** 3 litres : 500 ml

4 The ratio of staff to managers in a company is 23 : 2.
There are 100 people in the company. How many are managers?

Key point

A **proportion** compares a part with a whole.

5 A bag contains 5 red counters and 3 white counters.

a What is the ratio of red counters : white counters?

b What fraction of the counters are red?

c What fraction of the counters are white?

d **Reflect** Write a sentence explaining what you notice about your answers to part **a**, compared to parts **b** and **c**.

6 Clare and Fiona share a cash prize in the ratio 4 : 3.
What fraction of the prize should

a Clare get

b Fiona get?

c **Reflect** How can you check your answer?

Q6 hint

What should your fractions add to?

7 In a box of mints and toffees, $\frac{3}{4}$ of the sweets are mints and $\frac{1}{4}$ are toffees.

a What is the ratio of mints : toffees?

b There are 8 toffees in the box. How many mints are there?

c How many sweets are in the box altogether?

Q7 hint

M	M	M	T

8 In a canoeing lesson $\frac{2}{7}$ of the group are girls and the rest are boys.

a What is the ratio of girls to boys in the group?

b There are a total of 21 people in the group. Work out the number of boys.

9 **Reasoning** In a band, members are either singers or musicians.
The ratio of musicians to singers in the band is 9 : 4.
Andy says this means $\frac{4}{9}$ of the band are singers.
Is Andy correct? Explain.

10 **Problem-solving** In a gluten-free pizza base, $\frac{7}{12}$ of the flour is rice flour, $\frac{3}{12}$ is potato flour and the rest is tapioca flour.
Work out the ratio of rice flour : potato flour : tapioca flour.

11 **Problem-solving** A shoe shop stocks pairs of shoes, boots and trainers in the ratio 7 : 4 : 9.
What percentage of its stock is trainers?

Q11 hint
Work out the fraction that is trainers first.

Exam-style question

12 Sarah bakes 120 cakes.
$\frac{3}{10}$ of the cakes are carrot cakes.
15% of the cakes are lemon cakes.
The remaining cakes are coffee cakes and chocolate cakes in the ratio 5 : 6.
How many coffee cakes does Sarah bake? **(5 marks)**

Exam tip
When a problem has lots of sentences, read one sentence at a time. For each sentence, try to do a calculation.

Exam-style question

13 The ratio of the number of adults to the number of children staying on a campsite is 7 : 3.
Some people are staying in tents; others are staying in caravans.
$\frac{1}{3}$ of the children are staying in tents.
What percentage of the people staying on the campsite are children that are staying in caravans? **(3 marks)**

Exam tip
Read one sentence at a time.
If it is not possible to do a calculation, then try drawing a picture, like a bar model, to help you to answer it.

14 **Problem-solving** A topping for shortbread uses milk, chocolate, butter and syrup in the ratio 4 : 3.5 : 1.5 : 1.

a What fraction of the topping is milk?

b What percentage of the topping is syrup?

c In 500 g of the topping what are the masses of milk, chocolate, butter and syrup?

d **Reasoning** You have plenty of milk and chocolate, but only 200 g of butter and 100 g of syrup.
What is the maximum amount of topping you can make?

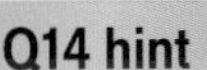

Q14 hint
Write the ratio with whole numbers first.

Key point

You can compare ratios by writing them as **unit ratios.** In a unit ratio, one of the numbers is 1.

15 Copy and complete to write these as unit ratios.

a 12 : 3 (÷3 each side) → □ : 1

b 2 : 7 (÷2 each side) → □ : □

c 3 : 4 (÷4 each side) → □ : □

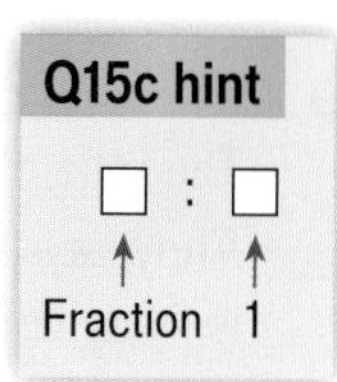

d **Reflect** Which ratios have you written in the form $n:1$? Explain.

16 Write each of these in the form $n:1$.
Give each answer to a maximum of 2 decimal places.

a 3:10 **b** 2:5 **c** 7:4 **d** 16:9 **e** 9:42

F
H

17 **a** Write 3.25:6.5 as a unit ratio in the form $1:n$.

b **Reflect** How else could you write this ratio as a unit ratio?

Exam-style question

18 In an office, $\frac{1}{6}$ of staff wear glasses. The rest do not.
Write down the ratio of the number of staff that wear glasses to the number of staff who do not wear glasses.
Give your answer in the form $n:1$. **(3 marks)**

Example

Molly makes a blackcurrant drink by mixing 30 ml of blackcurrant with 450 ml of water.
Hope makes a blackcurrant drink by mixing 40 ml of blackcurrant with 540 ml of water.
Whose drink is stronger? Explain your answer.

Molly

blackcurrant : water

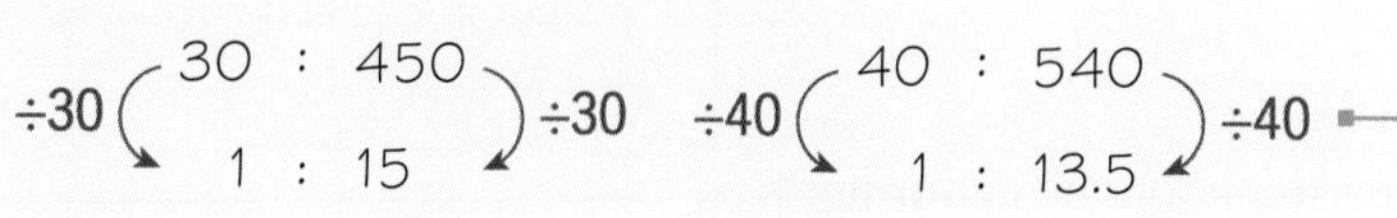

Hope

blackcurrant : water

Simplify to a unit ratio.

Hope's drink is stronger because it uses less water for every millilitre of blackcurrant.

Compare the quantity of water per ml of blackcurrant.

19 **Reasoning** Anna makes orange squash by mixing 50 ml of squash with 850 ml of water.
Jeevan makes orange squash by mixing 60 ml of squash with 1140 ml of water.
Whose squash is stronger? Explain your answer.

20 **Reasoning** Josh makes pink paint by mixing 2 litres of red paint and 500 ml of white paint.
Dexter makes pink paint by mixing 1.5 litres of red paint and 400 ml of white paint.
Whose paint is the darker pink? Explain your answer.

21 **Reasoning** Raj makes concrete using aggregate to cement in the ratio 1930:265.
Sunil makes concrete using aggregate to cement in the ratio 935:175.
Whose concrete has the higher proportion of cement?

11.6 Using proportion

- Use the unitary method to solve proportion problems.
- Solve proportion problems in words.
- Work out which product is better value for money.

Warm up

1 **Fluency** Which is better value if you are shopping?
 a Paying 1.8p per gram or 2.1p per gram? b Getting 3.25 ml for 1p or 2.85 ml for 1p?

2 Work out
 a $350 \div 2$ b $175 \div 5$ c 120×4 d $45 \div 3 \times 2$
 e 2.7×5 f $4.8 \div 3$ g $5 \div \square = 1$ h $7.2 \div \square = 1$

3 Write each of these as a unit ratio in the form n:1. a 4:18 b 11:5 c 14:35

4 A recipe for 6 people uses 900 g of mince.
Copy and complete to work out how much mince is needed for
 a 12 people b 3 people c 9 people

a
P : M
×□ (6 : 900 g) ×□
12 : □

b
P : M
÷□ (6 : 900 g) ÷□
3 : □

c
6 people + 3 people = 9 people
900 + □ = □ g

 d **Reasoning** Show how you could use your answers to parts **a**, **b** and **c** to work out the amounts for **i** 18 people **ii** 15 people.

5 A recipe for 4 people uses 6 eggs. How many eggs are needed for
 a 8 people b 2 people c 6 people d 10 people?

Exam-style question

6 Sam buys 30 forks, 30 spoons and 30 plates.
 Price list
 forks 6 for £2.10
 spoons 15 for £3.60
 plates £1.50 each
What is the total amount of money that Sam spends? **(4 marks)**

Exam tip
After doing all your working, read the question again to make sure you have given the answer that is asked for.

Key point
In the **unitary method**, you find the value of one item before finding the value of more.

7 **Problem-solving** 5 tickets to a theme park cost £125.
How much will 18 tickets cost?

Q7 hint
First find the cost of 1 ticket.

8 **Problem-solving** There are 237 calories in 100 g of apple pie. There are 125 calories in 100 g of custard.
Jaya has 80 g of apple pie and 120 g of custard for dessert.
Work out the total number of calories in Jaya's dessert.

> **Q8 hint**
> First work out the calories in 1 g of apple pie, then 80 g of apple pie.

> **Key point**
> You can use the unitary method to work out which product gives better value for money.

9 **Reasoning** Washing powder comes in two sizes.

a Copy and complete to find the unit ratios.

i

÷□ (1.3 kg : £4.40) ÷□
□ kg : £1

ii

÷□ (2.6 kg : £6) ÷□
□ kg : £1

1.3 kg for £4.40

2.6 kg for £6

b Which gives more grams for £1 and so which size is the better buy?

10 **Reasoning** Jenna can buy her favourite blend of coffee in two different pack sizes: 450 g for £12.60 or 300 g for £8.70.

a Work out the cost of 150 g of coffee for each pack size.

b Which pack size is better value for money?

c **Reflect** There is more than one way to work out the better buy.
Show a different method to work out which pack size is better value for money.

> **Q10c hint**
> Choose a common factor of 300 g and 450 g.

11 **Reasoning** David can buy blackcurrant squash in two sizes: 2 litres for £4.99 or 600 ml for £1.98.
Which is better value for money?

> **Q11 hint**
> Convert both to the same units.

Exam-style question

12 Brenda is going to buy 150 invitations.
Here is some information about the cost of the same invitations in two shops.

> **Stationery for you**
> Pack of 25 invitations £3.19

> **Paper etc.**
> Pack of 10 invitations for £1.95
> Buy 2 packs get 1 pack free

Brenda wants to buy the invitations as cheaply as possible.
Which shop should Brenda buy the 150 invitations from?
You must show how you get your answer. **(4 marks)**

> **Exam tip**
> Read the question, then read any information given in boxes. Then read the question again.

> **Q12 hint**
> For Paper etc.
> 2 packs = 2 × £1.95
> 3 packs is the same price as 2 packs.

11.7 Proportion and graphs

- Recognise and use direct proportion on a graph.
- Understand the link between the unit ratio and the gradient.

Warm up

1 Fluency

a What is the equation of a straight line through the origin (0, 0) with gradient 5?

b What is the gradient of the line $y = 7x$?

2 Draw a line graph for the values in the table.

a What is the gradient of the line?

b What is the equation of the line?

x	0	1	2	3	4	5
y	0	4	8	12	16	20

Key point

When two values are in **direct proportion**, if one value is zero then so is the other. When one value doubles, so does the other.

3 Look at the graph you drew for **Q2**. Are x and y in direct proportion? Explain your answer.

Q3 hint

Does doubling a value of x double the value of y? When x is zero, what is y?

Key point

When two quantities are in direct proportion, plotting them as a graph gives a straight line through the origin. The origin is the point (0, 0) on a graph.

4 The table shows some temperatures in both Celsius and Fahrenheit.

Celsius	5°	10°	20°	25°
Fahrenheit	41°	50°	68°	77°

a Draw a line graph for these values. Put Celsius on the horizontal axis and Fahrenheit on the vertical axis.

b Problem-solving Ice melts at 0 °C. What is this temperature in Fahrenheit?

c Reasoning Are Celsius and Fahrenheit in direct proportion? Explain your answer.

5 Reasoning This graph shows the price of grapes by mass.

a Are price and mass in direct proportion? Explain.

b Work out the gradient of the line.

c How much does 1 kg of grapes cost?

d Reflect How does your answer to part **b** help you to answer part **c**?

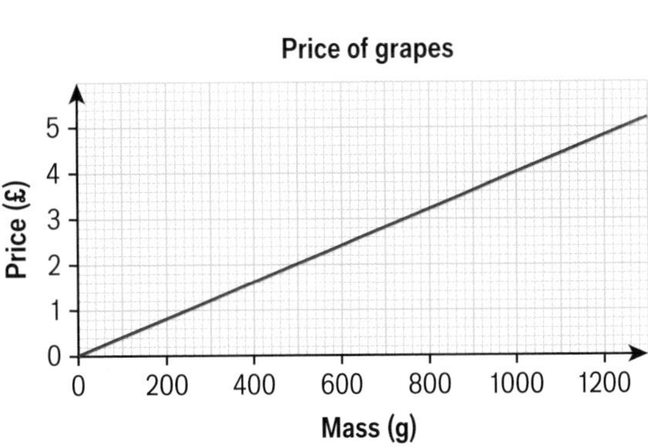

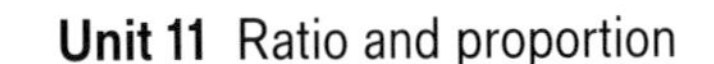

6 **Reasoning** Which of these are in direct proportion? Give reasons for your answers.

- **a** Metres and centimetres
- **b** Age and favourite music type
- **c** Number of swimmers in a pool and time of day
- **d** Euros and pounds

7 **Problem-solving / Reasoning** A plumber charges a callout fee of £50 plus £35 per hour he works.
Is his total charge, C, in direct proportion to the number of hours, h, he works? Explain.

Q7 hint
You could draw a graph.

8 The table shows the amount paid for different numbers of litres of fuel on one day.

Number of litres, n	5	10	20	30	50
Cost, C	£6.60	£13.20	£26.40	£39.60	£66

Q8 hint
$C = \square \times n$

- **a** Draw the graph.
- **b** Work out the equation of the line.
- **c** **Reasoning** Are C and n in direct proportion?
- **d** **Reasoning** Write a formula linking the number of litres, n, and the cost, C.
- **e** Use the formula to work out the cost of 82 litres of fuel on that day.
- **f** **Reflect** Look at the equation of the line and the formula for C and n.
  Write a sentence explaining
  - **i** what is the same
  - **ii** what is different

9 The graph shows the conversion between gallons and pints.

- **a** Write a ratio
  number of gallons : number of pints
- **b** Write the ratio gallons : pints in its simplest form.
- **c** Work out the gradient of the graph.
- **d** Write a formula that links gallons, G, and pints, P.
- **e** **Reflect** What do you notice about the gradient, the unit ratio and the formula?

Conversion graph

10 **Problem-solving** The ratio of miles to km is 5 : 8.

- **a** Copy and complete this table of values for miles and kilometres.

Miles	0	5	10
Kilometres			

- **b** Draw a conversion graph for miles to kilometres.
  Put miles on the x-axis and km on the y-axis.
- **c** Write 5 : 8 as a unit ratio.
- **d** Write a formula linking kilometres (y) and miles (x).

Exam-style question

11 A formula to change between temperature in kelvins (K) and in degrees Celsius (C) is $K = C + 273.15$
Are kelvins and degrees Celsius in direct proportion?
Explain your answer. **(2 marks)**

Exam tip
Writing only 'Yes' or 'No' does not *explain* your answer. You must write 'Yes (or No) because ...

11.8 Proportion problems

- Recognise different types of proportion.
- Solve word problems involving direct and inverse proportion.

Warm up

1 **Fluency** Will 5 builders take more or less time to build a wall than 3 builders?

2 Copy and complete these calculations.

a $12 \times \square = 60$ **b** $600 \times 5 = \square$ **c** $60 \div 8 = \square$

3 Copy and complete

a 5 hours = □ minutes **b** 200 minutes = □ hours □ minutes

c $2\frac{3}{4}$ hours = □ hours □ minutes

Key point

When two values are in **inverse proportion**, one increases at the same rate as the other decreases. For example, as one doubles (×2) the other halves (÷2).

Example

It takes 2 painters 6 days to paint a house.
How many days does it take 3 painters to paint an identical house?

2 people take 6 days,
so 1 person takes $2 \times 6 = 12$ days.
3 people take $12 \div 3 = 4$ days

Work out 1 person first.
It takes 1 person twice (×2) as long as 2 people.

It takes 3 people a third (÷3) as long as 1 person.

4 3 people dig a ditch in 12 hours. How long will it take

a 1 person **b** 5 people?

c **Reflect** What assumption do you have to make to answer parts **a** and **b**?

Q4 hint

Will it take 1 person more or less time than 3 people?

5 **Problem-solving** 2 workers take 5 hours to tile a room. How long will it take

a 4 workers **b** 3 workers?

Give your answers in hours and minutes.

6 **Problem-solving** It takes 5 machines 2 days to complete a harvest.
How long would it take 4 machines?

7 **Reasoning** A park supervisor needs to transplant 420 tree seedlings.
He knows that 1 person can transplant 8 seedlings in an hour.
The supervisor has 7 workers.
Can all the work be done in 7 hours or less? You must show your working.

8 **Reasoning** Which of these are

a in direct proportion **b** in inverse proportion **c** neither?

i Number of windows and time taken to clean them

ii Number of removal men and the time take to empty a house of furniture

iii Number of pages in a book and number of people who have borrowed it from the library

iv Number of newsletters printed and the time it takes to print them

v Number of cooks and the time taken to roast a chicken

Q8 hint

When one value doubles, does the other

- double, $\times 2$ (direct proportion)
- halve, $\div 2$ (indirect proportion)
- stay the same (neither)?

9 It takes 5 hill walkers 1 hour 20 minutes to walk up a hill.

a How long will it take 8 hill walkers?

b **Reflect** Is this a proportion problem? Explain.

10 The mass of 30 identical coins is 150 g.
What is the mass of 800 of these coins? Give your answer in kg.

Exam-style question

11 Machine A and machine B both make engine parts.
Machine A makes 8 parts every 12 minutes.
Machine B makes 10 parts every 15 minutes.
On Friday
machine A makes parts for 4 hours
machine B makes parts for 5 hours
Work out the total number of parts made by the two machines on Friday. **(4 marks)**

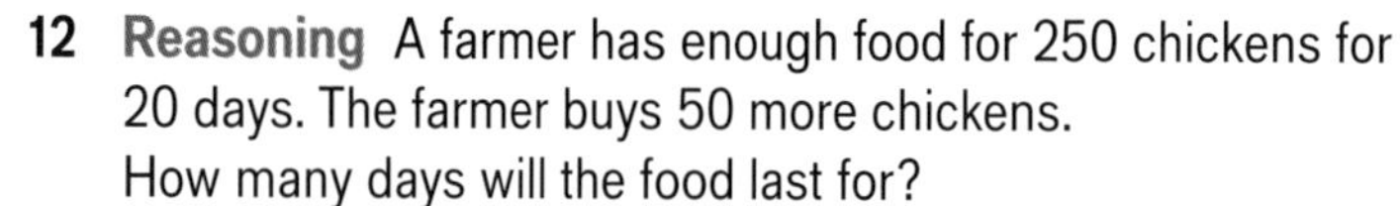

Q11 hint

First work out the number of parts made by machine A in 1 hour, then in 4 hours.

12 **Reasoning** A farmer has enough food for 250 chickens for 20 days. The farmer buys 50 more chickens.
How many days will the food last for?

Q12 hint

How many chickens does the farmer have to feed?

13 **Reasoning** Frank is paid £43.61 for 7 hours of work.
How many hours does he work to earn £112.14?

14 **Reasoning** It takes 5 shredders 4 hours to shred 625 sheets of paper.
Rita gets 3 more shredders.
How long will it take to shred 625 sheets of paper now?

15 **Reasoning** Sarah needs to decorate 20 rooms. She knows that 2 people can decorate a room in 3 hours. She needs all the work to be done in 8 hours.
How many people does Sarah need for the job?

Exam-style question

16 4 tins of wet cat food have a total mass of 1560 g.
3 tins of wet cat food and 2 packets of dried cat food have a total mass of 2890 g.
Work out the total mass of 5 tins of wet cat food and 1 packet of dried cat food. **(4 marks)**

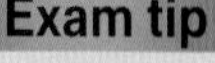

Exam tip

Read the first sentence. What can you work out with this information?

11 Check up

ActiveLearn Homework 

Simple proportion and best buys

1 A recipe for 4 people uses 150 g of sugar. How much sugar is needed for

a 8 people

b 2 people

c 6 people?

2 Helen can buy 300 ml of shampoo for £3.50 or 75 ml of shampoo for £1. Which is better value for money?

Ratio and proportion

3 A necklace has 24 beads. There is 1 red bead for every 3 blue beads. How many beads are

a red   b blue?

4 Write each ratio in its simplest form.

a 24 : 32

b 27 : 18 : 54

c 2.5 : 3.5

5 Copy and complete these conversions.

a 4.2 km = □ m

b 0.05 litres = □ ml

c £200 = □ euros (£1 = 1.23 euros)

d 25 km ≈ □ miles (8 km ≈ 5 miles)

6 Hazel mixes pink paint using red paint and white paint in the ratio 1 : 4. How much white paint should she use with 5 tins of red paint?

7 Iona makes a scale model of a sailing boat using the ratio 1 : 25. Her model is 60 cm long. How long is the real sailing boat? Give your answer in metres.

8 Naadim and Bal share £60 in the ratio 2 : 3.

a How much do they each receive?

b What fraction does Naadim receive?

c Show how you checked that your answer is correct.

9 Write each of these in the form $n:1$.

a 13 : 10

b 7 : 28

c 4.5 : 3

10 The ratio of primary school children to secondary school children in a club is 2:3.
There are 12 primary school children in the club.

a How many secondary school children are there?

b What is the total number of children in the club?

11 Luke makes lemon squash using 40 ml of squash and 380 ml of water.
Jo makes lemon squash using 50 ml of squash and 475 ml of water.
Who has made the stronger squash? Explain your answer.

Proportion, graphs, direct and inverse proportion

12 The graph shows the conversion between US dollars and Barbadian dollars.

a Are US dollars and Barbadian dollars in direct proportion?

b Write a formula linking US dollars and Barbadian dollars.

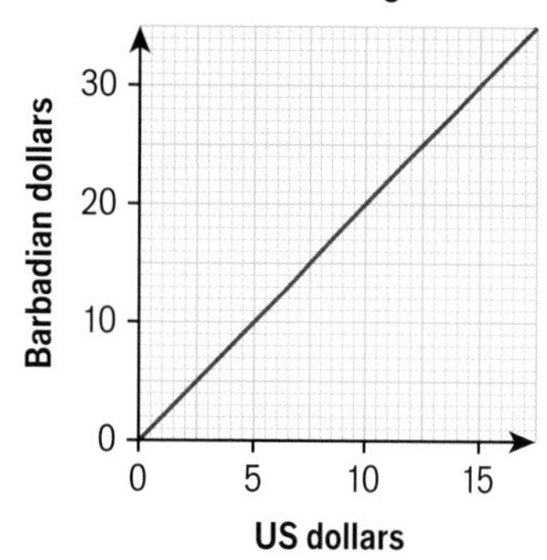

13 It takes 4 people 3 days to lay the cables for broadband.
How long would it take

a 1 person **b** 6 people?

14 The mass of 12 identical empty jam jars is 2.4 kg.
What is the mass of 20 of these jam jars?

15 It takes 3 machines 4 minutes to fill 600 bottles.
How long will it take 8 machines to fill 600 bottles?

16 **Reflect** How sure are you of your answers? Were you mostly

Just guessing ☹ Feeling doubtful 😐 Confident ☺

What next? Use your results to decide whether to strengthen or extend your learning.

Challenge

17 **a** In how many different ways can you divide this rectangle into two parts in the ratio 1:3?

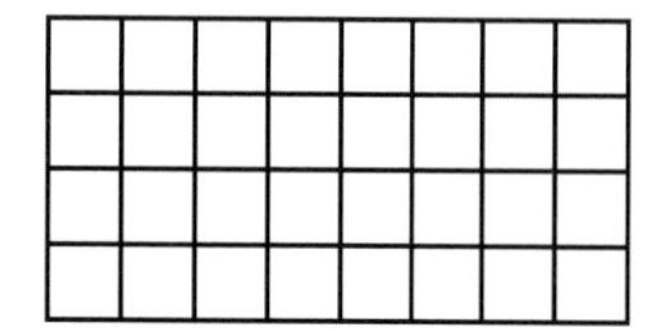

b How many of your answers show a symmetrical pattern?

11 Strengthen

ActiveLearn Homework 

Simple proportion and best buys

1 One bus ticket costs £3. Work out the cost of

a 2 tickets **b** 5 tickets **c** 8 tickets **d** 12 tickets

2 It costs £32 for 4 students to go to the cinema.
How much does it cost

a 1 student **b** 2 students **c** 8 students **d** 10 students?

3 **Reasoning** 200 g of lotion in jar A costs £2.50.
300 g of lotion in jar B costs £3.60.

a What is the highest common factor of 200 g and 300 g?

b Work out how much you pay for the number of grams in your answer to part **a** for

i jar A **ii** jar B

c Use your answer to part **b** to decide whether jar A or jar B is better value for money.

Ratio and proportion

1 Andie makes a bracelet with green beads and yellow beads.
She uses 1 green bead for every 2 yellow beads.
The bracelet has 12 beads altogether.
How many beads are

a green **b** yellow?

Q1 hint

2 Copy and complete to write each ratio in its simplest form.

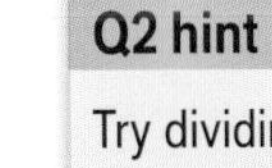

c

10 : 15 : 35
÷□ ÷□ ÷□
□ : □ : □

Q2 hint

Try dividing all the numbers in the ratio by 2. If this does not give whole numbers, try dividing by 3, and so on.

3 Keep multiplying each of the numbers in these ratios by 10 until you get a ratio with whole numbers.
Then write the ratio in its simplest form.

a 0.6 : 0.8

b 0.5 : 3.5

c 1.2 : 4

d 1.5 : 1.25

e 1.05 : 1.35

f 0.04 : 1

Q3a hint

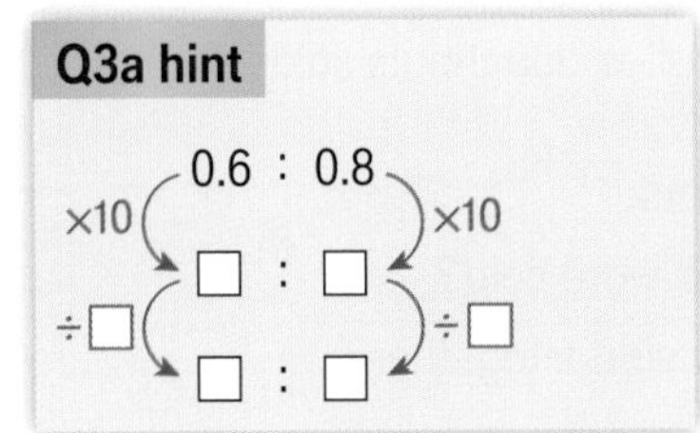

Q3d hint

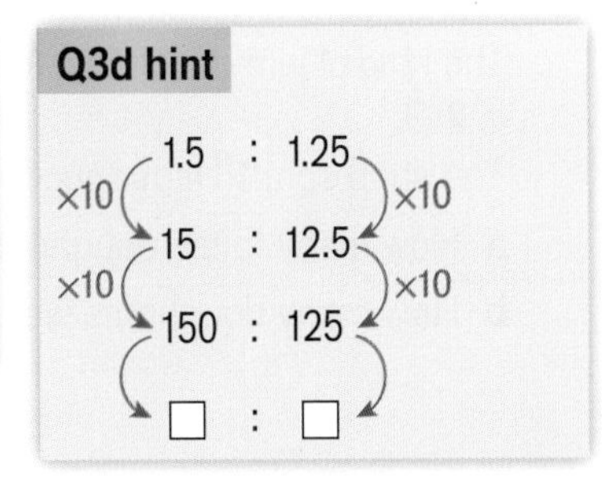

4 Complete these conversions.

a 3.8 m = □ cm

b 2.5 litres = □ ml

c £100 = □ euros (£1 = 1.23 euros)

d 24 pints = □ gallons (1 gallon = 8 pints)

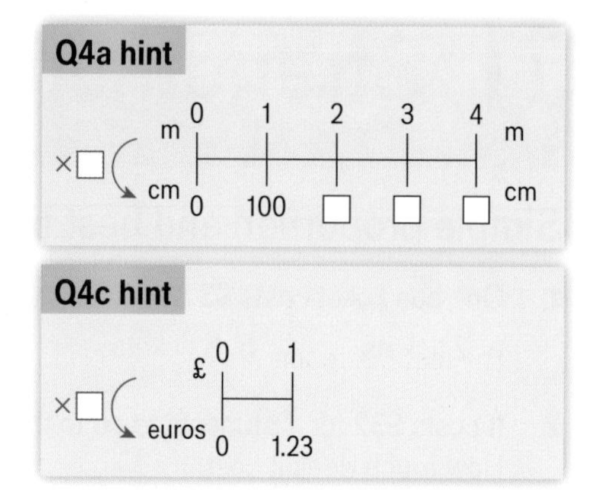

5 Jim mixes red paint and blue paint in the ratio 1 : 5.

R	B	B	B	B	B

He uses 3 tins of red paint.
Blue paint comes in the same size tins as red paint.
How much blue paint should he use?

Q5 hint

R	B	B	B	B	B

×□ ×□

R	R	R

6 The ratio of butter to sugar in a recipe is 2 : 1.
Ian uses 6 ounces of butter.
How many ounces of sugar should he use?

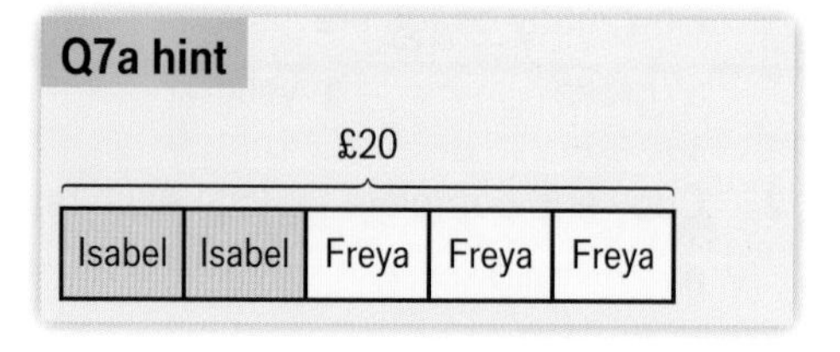

7 Isabel and Freya share £20 in the ratio 2 : 3.

a How much do they each receive?

b Add your answers for Isabel and Freya's shares together to check your answer.

Q7a hint

£20

Isabel	Isabel	Freya	Freya	Freya

8 A piece of cloth is cut into three pieces in the ratio 1 : 2 : 5.
The piece of cloth is 240 cm long.

a How long is each piece?

b Show how you have checked your answer.

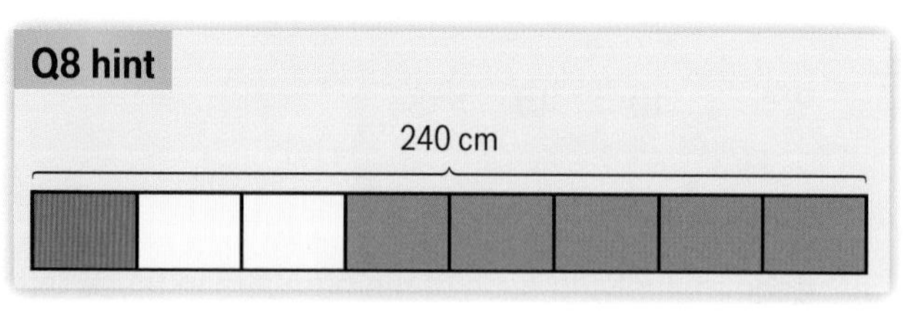

9 A shop sells jam doughnuts and toffee doughnuts.
The ratio of jam doughnuts to toffee doughnuts sold is 2 : 3.
The shop sold 18 jam doughnuts.

a How many toffee doughnuts were sold?

b How many doughnuts were sold altogether?

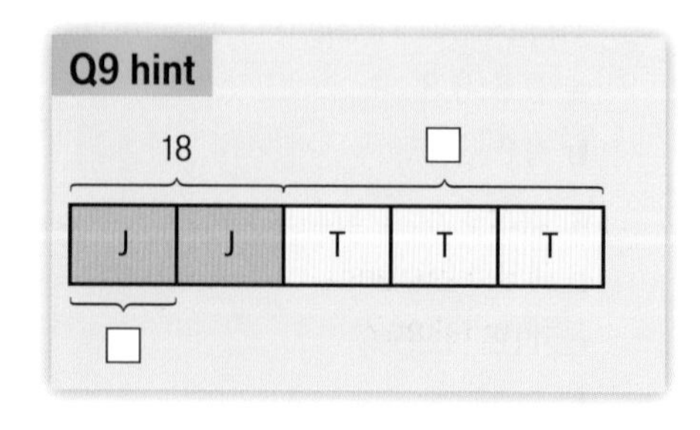

10 The ratio of quad bikes to go-karts at an activity centre is 3 : 7.

a Copy and complete the bar model to show the ratio.

q	q	

b What fraction of the vehicles are go-karts?

c What fraction of the vehicles are quad bikes?

11 $\frac{3}{4}$ of the bikes at a cycle shop are road bikes.

The rest are mountain bikes.

a Copy and complete the bar model to show the fractions of road bikes and mountain bikes.

r	r	

b What is the ratio of road bikes : mountain bikes?

c There are 27 road bikes. How many mountain bikes are there?

d How many bikes are there in the cycle shop altogether?

12 Match each ratio to one of the unit ratios in the box.

a 2 : 10 **b** 7 : 10 **c** 7 : 2 **d** 2 : 5

A 1:2.5 **B** 1:5 **C** 3.5:1 **D** 0.7:1

13 **Reasoning** Amar makes lemon squash using 20 ml of squash and 180 ml of water.
Ben makes lemon squash using 50 ml of squash and 500 ml of water.

a Copy and complete the unit ratios for Amar's squash and Ben's squash.

Amar
squash : water
20 : 180
÷□ ÷□
1 : □

Ben
squash : water
50 : 500
÷□ ÷□
1 : □

b Who has made the stronger squash? Explain your answer.

Q13b hint
Who has less water for 1 litre of squash?

Proportion, graphs, direct and inverse proportion

1 The sketch graph shows an object moving at a constant speed.

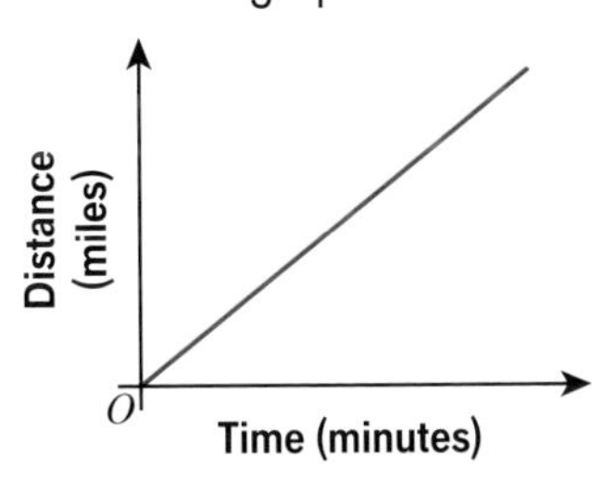

a Is this graph a straight line?

b Does the graph go though (0, 0)?

c Is the distance travelled in direct proportion to the time taken?

Q1c hint
Did you answer 'Yes' to parts **a** and **b**?

2 The ratio of kilograms to pounds is 1:2.2.

a Copy and complete the table of values for the ratio.

Kilograms	0	1	2	3
Pounds	0	2.2	☐	☐

b Write a formula that shows the relationship between kilograms (k) and pounds (p).

Q2b hint

$p = \square \times k$

3 A shop sells material for bridesmaid dresses at £4 per metre.

a Copy and complete the table of values for the cost of the material.

Length, l (m)	0	1	2	3	4	5	6	7	8	9	10
Cost, c (£)	☐	☐	☐	☐	☐	☐	☐	☐	☐	☐	☐

b Draw a graph to show the cost of the material.

c Write a formula that shows the relationship between length of material (l), and cost (c).

d Is the price of material proportional to the length bought?

e What is the cost of 20 m of material?

Q3b hint

4 3 m of pipe weighs 375 g.

a Will 1 m of pipe weigh more or less than 3 m?

b Will 2 m of pipe weigh more or less than 1 m?

c Copy and complete the table.

Length of pipe	More or less	Mass
3 m	–	375 g
1 m		
2 m		

5 It takes 3 people 4 hours to lay some pipes.

a Will 1 person take more or less time than 3 people?

b Will 2 people take more or less time than 3 people?

c Copy and complete the table.

Number of people	More or less	Hours
3	–	4
1		
2		

Q4 and Q5 hint

If more, then multiply. If less, then divide.

11 Extend

1 Which of these sketch graphs show one variable in direct proportion to another? Explain your answer.

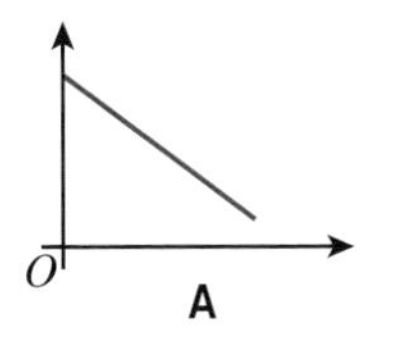

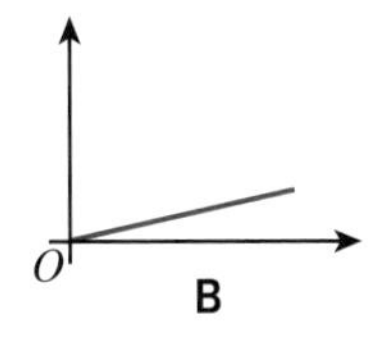

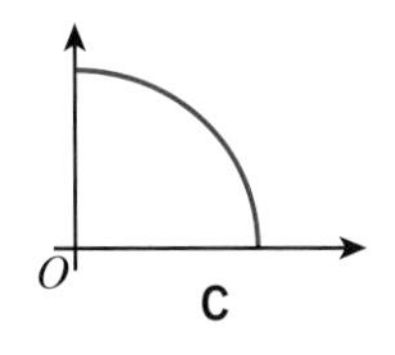

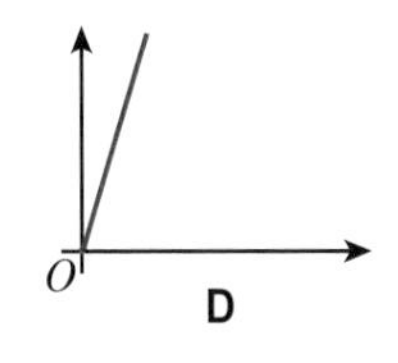

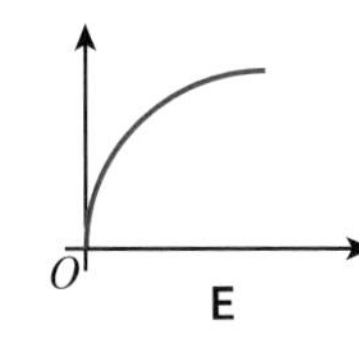

2 Superfoods and Healthy4You sell boxes of the same type of porridge oats.
Each store has a special offer.
Which box of porridge oats is the better value for money?
You must show your working.

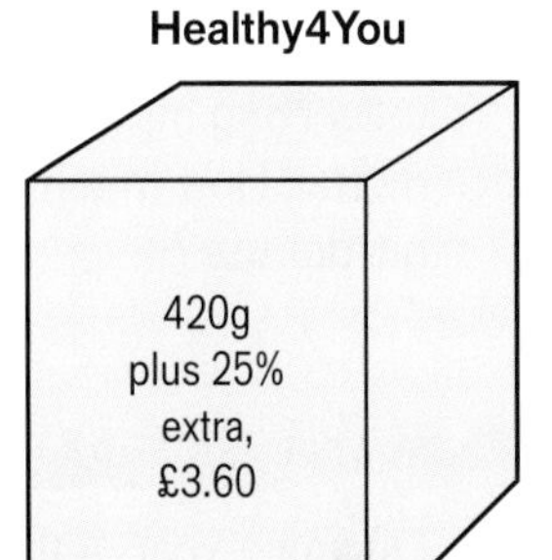

Exam-style question

3 Here is a list of ingredients for making fruit scones.

Makes 8 fruit scones
200 grams flour
60 grams margarine
40 grams sugar
60 grams dried fruit
160 ml milk

Wendy wants to make 30 fruit scones.
Wendy has 180 grams of sugar and 700 grams of flour.
She has plenty of the other ingredients.
What is the greatest number of fruit scones she can make?
You must show all your working. **(3 marks)**

Exam tip

Write down all the steps of your working, so you can gain marks for method even if you get the wrong answer.

4 **Reasoning** 3 painters take 12 hours to paint a house.

a How long would it take 4 painters to paint the house?

b The charge for the work is £18 per painter per hour. What is the total cost?

Exam-style question

5 Larry buys a bicycle for £1100 plus VAT at 20%.
Larry pays a deposit for the bicycle.
He then pays the rest of the cost in 12 equal payments of £88.75 per month.
Find the ratio of the deposit Larry pays to the total of the 12 equal payments.
Give your answer in its simplest form. **(5 marks)**

Exam tip

Check your arithmetic at each step of solving the problem.

Q5 hint

Start by working out
£1100 + VAT.
↑
20% of £1100

6 **Problem-solving** The sizes of the angles in a quadrilateral are in the ratio 2 : 4 : 5 : 7. What size is the largest angle?

7 **Reasoning** Sunil and Karl shared some money in the ratio 3 : 5.
Sunil gave $\frac{1}{6}$ of his money to Lucas. Lucas got £12.50.
How much money was shared originally by Sunil and Karl?

8 **Problem-solving** A box contains 480 counters.
There are twice as many red counters as yellow counters.
There are three times as many blue counters as red counters.
The rest of the counters are green.
The box contains 50 yellow counters.
How many green counters are in the box?

Q8 hint
Start with the 50 yellow counters. What can you work out next?

9 **Problem-solving** The chart shows the proportions of sand and cement needed to mix mortar for brickwork. A builder uses 78 kg of cement to make some mortar. How much sand should the builder use?

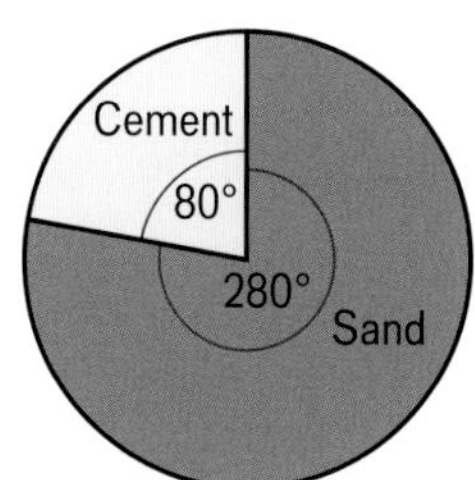

Q9 hint
Write down the ratio of sand : cement.

Exam-style question

10 During a year, a theatre has performances of plays (P), musicals (M) and dance shows (D).
In a year, the ratio of the number of plays (P) to the number of musicals (M) is 7 : 4 and the number of musicals (M) to the number of dance shows (D) is 3 : 2.
There are 8 dance shows (D) in the year.
How many performances are there altogether in a year? **(4 marks)**

Q10 hint

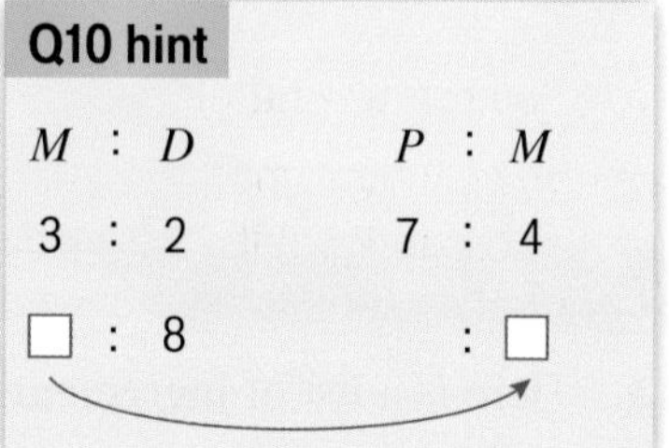
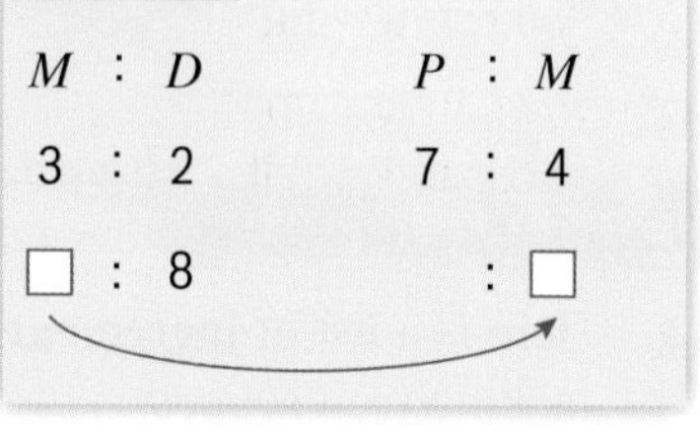

11 **Problem-solving** 3 boys did some gardening. They earned £129.60.
They shared the money in the ratio of the number of hours they each worked.
Ollie worked for $9\frac{1}{2}$ hours, Sam worked for $8\frac{1}{4}$ hours and Peter worked for $6\frac{1}{4}$ hours.
How much money did each boy receive?

Exam-style question

12 Dan is 3 years older than Abbie. Lucy is twice Dan's age.
The sum of their three ages is 45.
Find the ratio of Abbie's age to Dan's age to Lucy's age. **(4 marks)**

Q12 hint
Let Dan's age be x. How can you write Abbie's age and Lucy's age? What is the value of x?

Exam-style question

13 In London, 1 litre of petrol costs 125.2p.
In Chicago, 1 US gallon of petrol costs \$2.99.
1 US gallon = 3.785 litres
£1 = \$1.38
In which city is petrol better value for money?
You must show your working. **(3 marks)**

Q13 hint
Start with London.
For 1 litre
p : £ □ □ £ : \$ □ □
Then work out the cost in US for 1 gallon (3.785 litres).

Exam tip
When there is more than one conversion in a question, work with one at a time.

11 Test ready

Summary of key points

To revise for the test:

- Read each key point, find a question on it in the mastery lesson, and check you can work out the answer.
- If you cannot, try some other questions from the mastery lesson or ask for help.

Key points

1	A **ratio** is a way to compare two or more quantities.	→ 11.1
2	You **simplify** a ratio by making the numbers as small as possible. Divide the numbers in the ratio by their **highest common factor** (**HCF**).	→ 11.1
3	Ratios are **equivalent** if they have the same **simplest form**.	→ 11.1
4	Ratios in their **simplest form** only have integer parts (whole numbers).	→ 11.2
5	You can use **ratios** to convert between units.	→ 11.3
6	To share an amount in a given ratio, first work out how many parts there are in total. Then work out the amount in 1 part to find the amount in each part.	→ 11.4
7	A **proportion** compares a part with the whole.	→ 11.5
8	You can compare ratios by writing them as **unit ratios**. In a unit ratio, one of the numbers is 1.	→ 11.5
9	Unit ratios are often represented as $1:n$ or $n:1$, where n is any number.	→ 11.5
10	In the **unitary method**, you find the value of one item before finding the value of more.	→ 11.6
11	You can use the unitary method to work out which product gives better value for money.	→ 11.6
12	When two values are in **direct proportion**, if one value is zero then so is the other. When one value doubles, so does the other.	→ 11.7
13	When two quantities are in direct proportion, plotting them as a graph gives a straight line through the origin.	→ 11.7
14	When two values are in **inverse proportion**, one increases at the same rate as the other decreases. For example, as one doubles (×2) the other halves (÷2).	→ 11.8

Sample student answers

Exam-style question

1 Antony is going to use these instructions to make pink lemonade.

Mix 2 parts of cranberry juice with 3 parts of lemonade.

Antony thinks he has 200 ml of cranberry juice and 150 ml of lemonade.

a If Antony is correct, what is the greatest amount of pink lemonade he can make? **(2 marks)**

b Antony has 200 ml of cranberry juice and 120 ml of lemonade. Does this affect the amount of pink lemonade he can make? Give a reason for your answer. **(1 mark)**

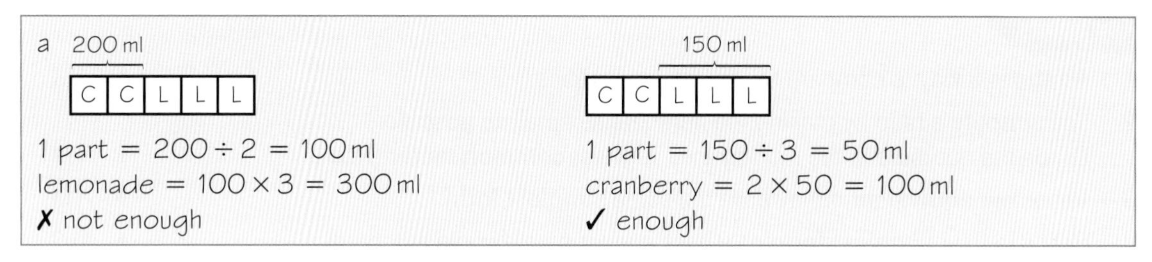

a What else must the student do to provide a complete answer for part **a**?

b How can you use the student's models to answer part **b**?

Exam-style question

2 A shop sells packs of blue pens, packs of gold pens and packs of red pens. There are

5 pens in each pack of blue pens.

6 pens in each pack of gold pens.

3 pens in each pack of red pens.

On Saturday,

number of packs of blue pens : number of packs of gold pens : number of packs of red pens = 2 : 1 : 4

A total of 112 pens were sold. Work out the number of red pens sold. **(4 marks)**

Student 1

Packs

B G R

2 : 1 : 4

blue pens	gold pens	red pens	
2 × 5	1 × 6	4 × 3	
10	6	12	Total = 28
20	12	24	Total = 56
30	18	36	Total = 84
40	24	48	Total = 112 pens

48 red pens were sold.

Student 2

blue gold red

2 : 1 : 4 = 7 parts

1 part = 112 ÷ 7 $7\overline{)11^{4}2}$ = 16

red pens = 4 parts

= 4 × 16

= 64

64 red pens were sold

One student shared the packs of blue pens, gold pens and red pens in the ratio 10 : 6 : 12.
The other student shared the number of blue pens, gold pens and red pens in the ratio 2 : 1 : 4.

a Read the question again. Which student understood the ratio correctly?

b Which student's answer was correct?

11 Unit test

1 At the 2014 Commonwealth Games, Australia won 42 silver medals and India won 30 silver metals.
Write the ratio of the number of silver medals won by Australia to the number of silver medals won by India. Give your ratio in its simplest form. **(2 marks)**

2 1 kg of cheddar cheese costs £6.40. How much does 100 g cost? **(2 marks)**

3 There are 45 lemons and limes in a box.
The ratio of the number of lemons to the number of limes is 1 : 4.
Work out the number of limes in the box. **(3 marks)**

4 Greg changed £475 into euros. The exchange rate was £1 = 1.2 euros.
How many euros did he get? **(2 marks)**

5 The number of hotel staff working in housekeeping, catering or the office is in the ratio 6 : 4 : 3.
There are 24 housekeeping staff. How many staff work in the office? **(3 marks)**

6 At Mini Mart, 4 tubs of ice cream cost £9.40.
At Dave's Deli, 3 tubs of the same ice cream cost £7.50.
At which shop is ice cream better value for money? **(3 marks)**

7 A school with 900 students has 150 computers.
Write the ratio of the number of students to the number of computers in the form $n:1$. **(1 mark)**

8 The ratio of the amount of water that bucket A holds to bucket B is 8 : 5. Bucket A holds 15 litres more than bucket B. How much water does bucket B hold? **(3 marks)**

9 Sam, Jack and Ali share £45 in the ratio 2 : 3 : 4.

a What fraction does Sam get? **(2 marks)**

b How much does Ali get? **(2 marks)**

10 James, Isaac and Lucas share £30 in the ratio of their ages.
James is 10 years old, Isaac is 8 years old and Lucas is 7 years old.
Isaac gives a third of his share to his Dad.
How much money does Isaac have now? **(4 marks)**

11 Here are some of the ingredients needed to make a beef pie for 6 people.

Beef pie (serves 6 people)
180 g flour
320 g beef
4 eggs
160 ml milk

a How much beef is needed to make a beef pie for 15 people? **(2 marks)**

b Half as much butter as flour is needed.
Work out how much butter is needed to make a beef pie for 4 people. **(3 marks)**

c If you have 5 eggs and plenty of the other ingredients, can you make a beef pie for 8 people? **(3 marks)**

12 The graph shows a car moving at a constant speed.

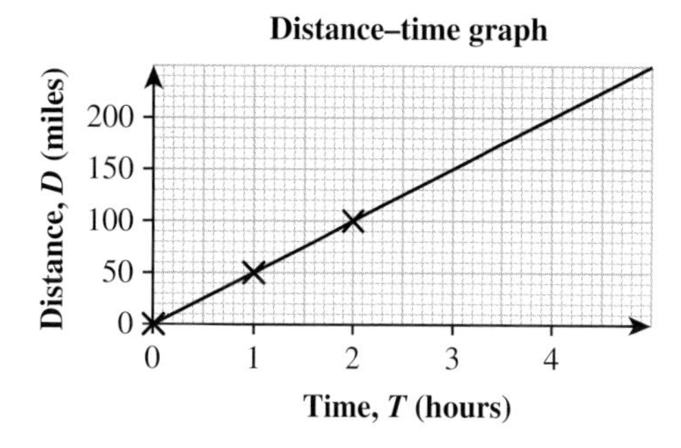

a Does the graph show distance and time in direct proportion?
Explain your answer. **(1 mark)**

b Write a formula linking distance (D) and time (T). **(1 mark)**

13 It takes a teacher 1.5 minutes to mark one spelling test.
How long will it take the teacher to mark 30 students' spelling tests? **(1 mark)**

14 3 machines can print a batch of leaflets in 2 hours.
How long would it take 4 machines to print the same batch of leaflets? **(3 marks)**

15 Anna, Bob and Sally each earn the same monthly salary.
Anna saves $\frac{3}{10}$ of her salary and spends the rest.
Bob spends 75% of his salary and saves the rest.
The amount of salary Sally saves : the amount of salary she spends = 2 : 3.
Work out who saves the most of their salary each month.
You must show how you get your answer. **(4 marks)**

(TOTAL: 45 marks)

16 Challenge

a The sizes of the interior angles of some polygons are in these ratios.

Triangle	Quadrilateral	Pentagon
1 : 2 : 3	1 : 2 : 3 : 4	1 : 2 : 3 : 4 : 5

Write the size of each angle in each polygon.

b Use the same pattern to write the ratio of the sizes of the interior angles and the size of each angle in an octagon.

17 Reflect For each statement, **A**, **B** and **C**, choose a score:
1 – strongly disagree; 2 – disagree; 3 – agree; 4 – strongly agree
A I always try hard in mathematics.
B Doing mathematics never makes me worried.
C I am good at mathematics.
For any statement you scored less than 3, write down two things you could do so that you agree more strongly in the future.

12 Right-angled triangles

12.1 Pythagoras' theorem 1

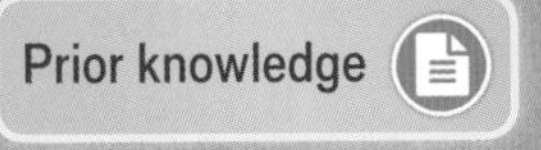

- Understand Pythagoras' theorem.
- Calculate the length of the hypotenuse in a right-angled triangle.
- Solve problems using Pythagoras' theorem.

ActiveLearn Homework

Warm up

1 **Fluency** Work out

a 6^2 **b** $\sqrt{49}$ **c** the square root of 81

2 What is the area of this square?

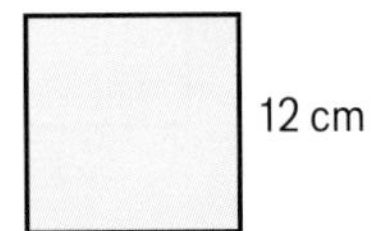

3 $x = 7$ and $y = 24$. Work out

a $x^2 + y^2$ **b** $\sqrt{x^2 + y^2}$

4 Find the positive solution of each equation.

a $c^2 = 65$ **b** $c^2 = 11^2 + 15^2$

Give your answers correct to 3 significant figures.

Key point

In a right-angled triangle the longest side is opposite the right angle.
It is called the **hypotenuse**.

5 **a** Draw each triangle on centimetre square paper.

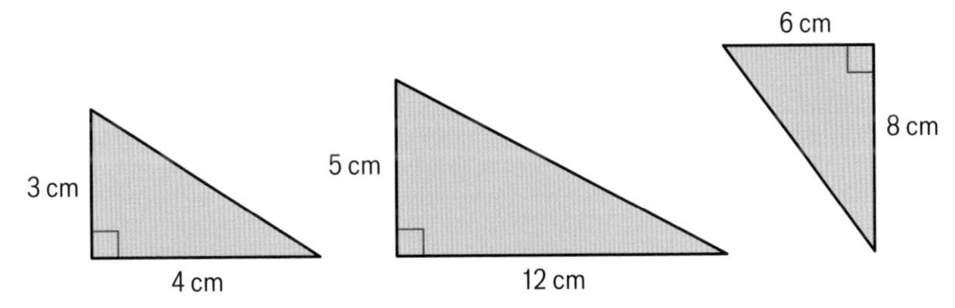

b **i** Measure the length of the unknown side. **ii** What is the length of the hypotenuse?

6 Write the length of the hypotenuse for each of these triangles.

a

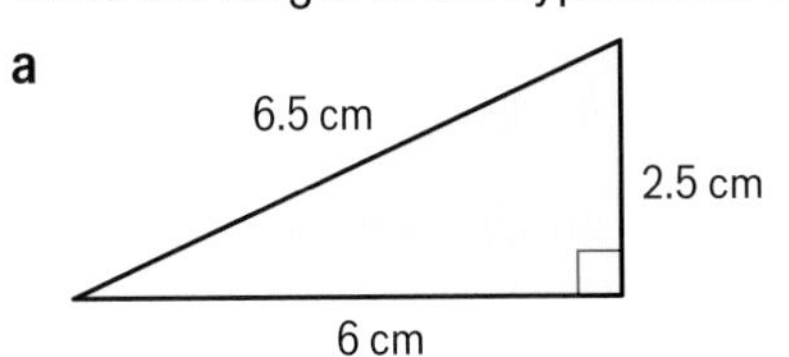

b

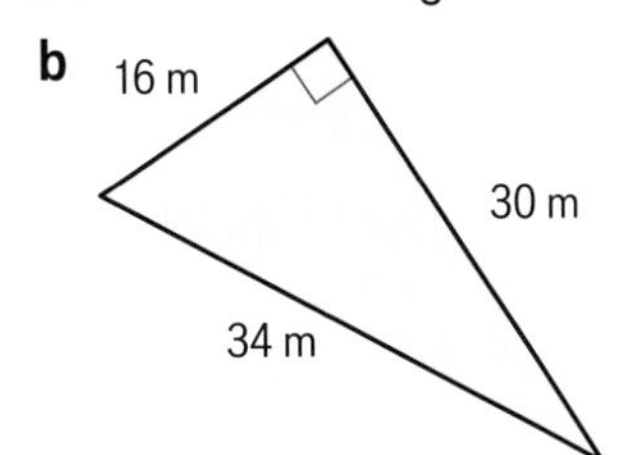

c

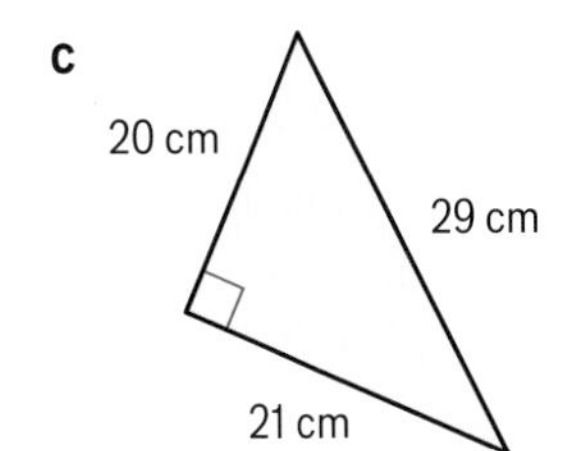

7 A square is drawn on each side of a right-angled triangle.

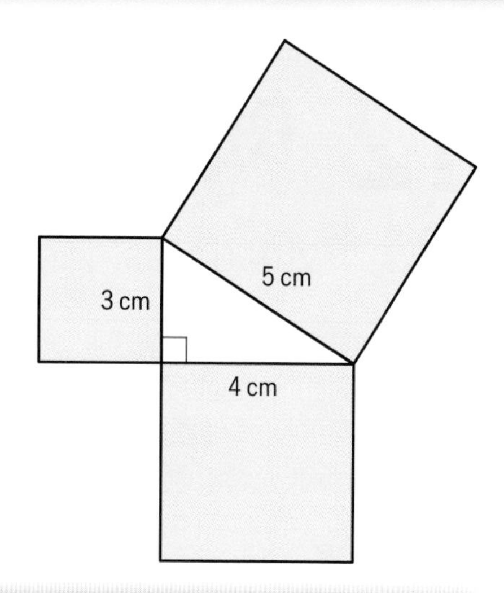

a Check that the measurements match your answer to **Q5a**.

b Find the area of the square on the hypotenuse.

c Find the sum of the areas of the squares on the other two sides.

d Copy and complete.
$\square^2 = 3^2 + 4^2$

e Reasoning Copy and complete.
The square on the hypotenuse has the ______ area as the sum of the areas of the ______ on the other two sides.

Key point

Pythagoras' theorem shows the relationship between the lengths of the three sides of a right-angled triangle.

$c^2 = a^2 + b^2$

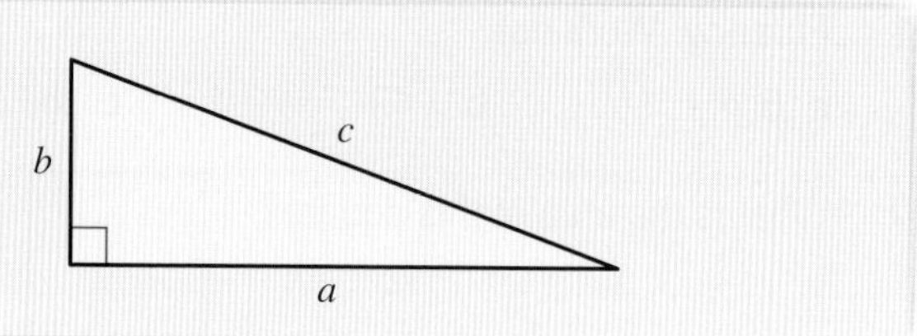

Example

Calculate the length of the hypotenuse, x.
Give your answer correct to 3 significant figures.

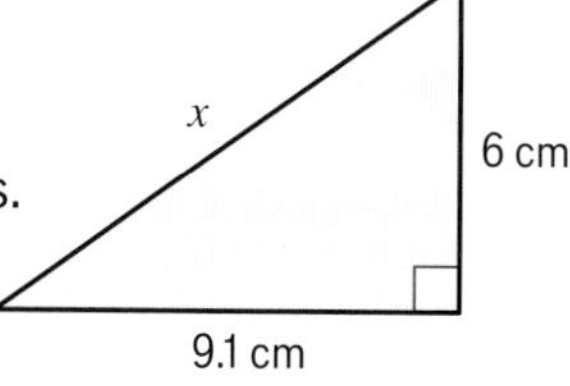

Sketch the triangle. Label the hypotenuse c and the other two sides a and b.

$a = 6$, $b = 9.1$, $c = x$ — Write $a = \square$, $b = \square$, $c = \square$

$c^2 = a^2 + b^2$

$x^2 = 6^2 + 9.1^2$ — Substitute the values of a, b and c into the formula for Pythagoras' theorem, $c^2 = a^2 + b^2$.

$x^2 = 118.81$

$x = \sqrt{118.81}$ — Use a calculator to find the square root of x^2.

$x = 10.9$ cm (3 s.f.) — Give your answer correct to 3 s.f. and include the units.

8 a i Sketch this right-angled triangle.

ii Label the side that is 28 cm as a, and the side that is 45 cm as b.

iii Calculate the length of the hypotenuse, x.

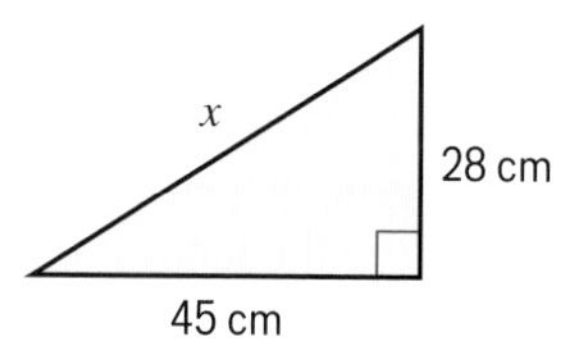

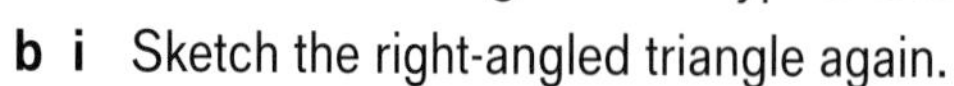

b i Sketch the right-angled triangle again.

ii Label the side that is 45 cm as a, and the side that is 28 cm as b.

iii Calculate the length of the hypotenuse, x.

c Reflect Does it matter which side is a and which side is b? Explain.

9 Calculate the length of the hypotenuse, x, in each right-angled triangle.
Use the correct units (mm, cm, m or km) in your answer.
Give your answers correct to 3 significant figures.

Q9 hint

Do not round before finding the square root.

a

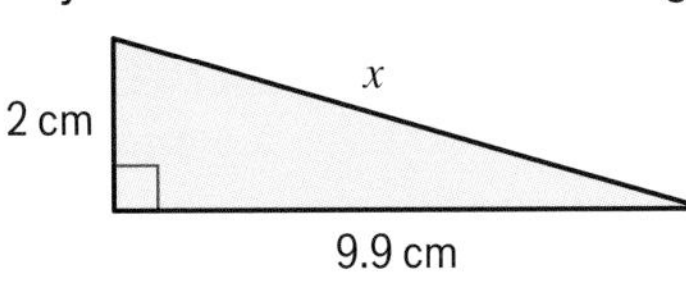

b

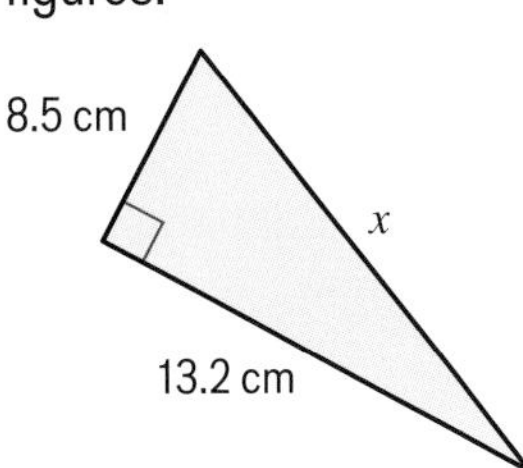

c

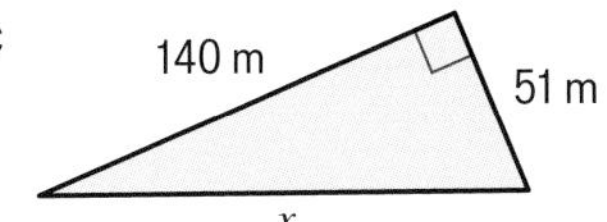

d

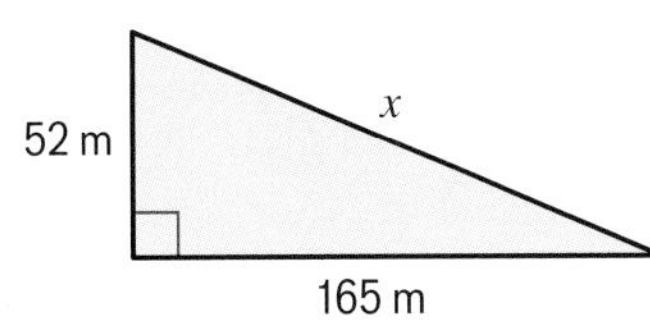

e

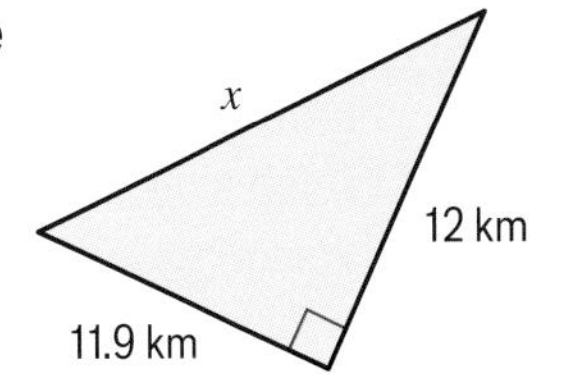

10 Calculate the length of PR in each right-angled triangle.
Give your answers correct to 2 decimal places.

a

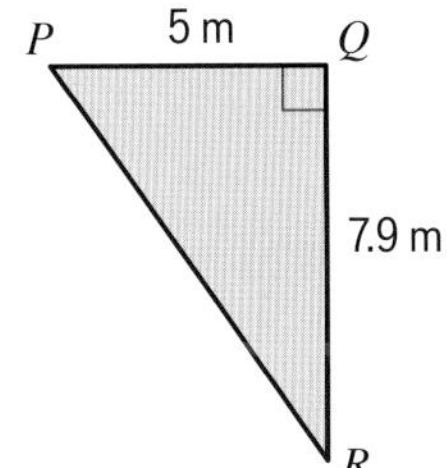

b

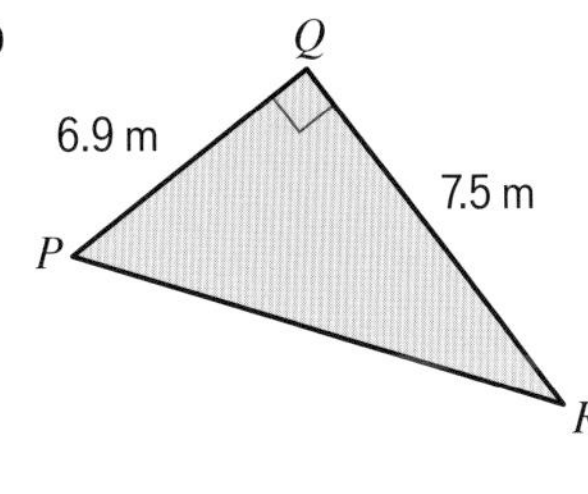

c

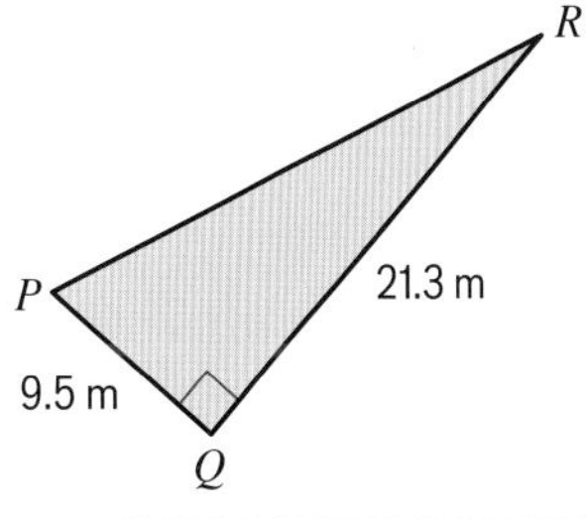

Exam-style question

11 A boat's sail is a right-angled triangle.
The height of the sail is 7.6 m.
The base of the sail is 2.8 m.
Work out the length of the longest edge of the sail.
Give your answer correct to 2 decimal places. **(3 marks)**

Exam tip

Sketch the right-angled triangle.
Label the information you know.
Label the length you want to find x.

12 **Reasoning** Caroline is working out the length BC for this right-angled triangle.
She writes

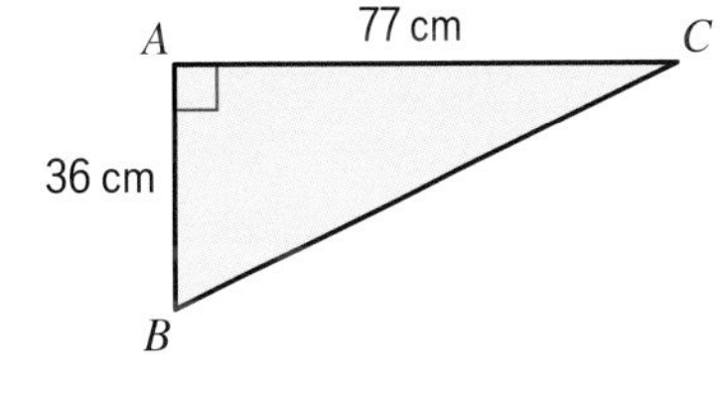

$a = 36, b = 77, c = BC$
$BC^2 = 36 + 77$
$BC^2 = 113$
$BC = \sqrt{113}$
$BC = 10.6$ cm (1 decimal place)

a **Reflect** Look at the diagram of the triangle.
How do you know that BC cannot be length 10.6 cm?

b Look at Caroline's working. What mistake has she made?

c Work out the correct length of BC.

12.2 Pythagoras' theorem 2

- Calculate the length of a line segment AB.
- Calculate the length of a shorter side in a right-angled triangle.
- Solve problems using Pythagoras' theorem.

Warm up

1 **Fluency** Here is a right-angled triangle.
Which side is the hypotenuse?
Explain how you know.

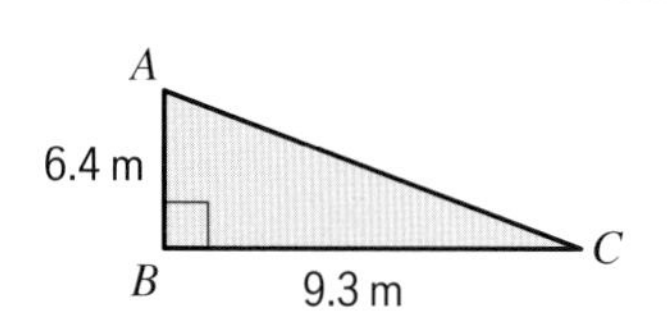

2 Calculate the length of AC for the right-angled triangle in **Q1**.
Give your answer correct to 3 significant figures.

3 Solve these equations.

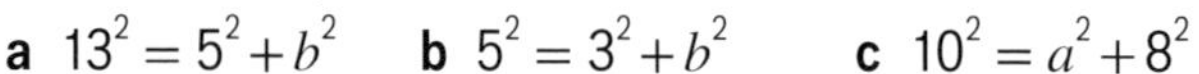

a $13^2 = 5^2 + b^2$ **b** $5^2 = 3^2 + b^2$ **c** $10^2 = a^2 + 8^2$

Q3a hint

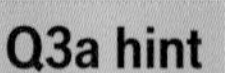

$13^2 - 5^2 = 5^2 + b^2 - 5^2$

4 **Problem-solving**

a Plot the points $A(1, 2)$ and $B(5, 4)$ on a centimetre square grid.

b Draw the right-angled triangle with AB as the hypotenuse.

c Work out the length in centimetres of AB.
Give your answer correct to 3 significant figures.

Q4 hint

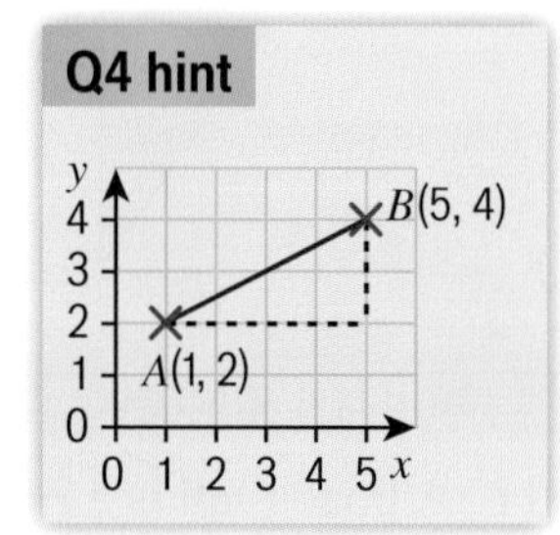

5 **Problem-solving** The points P and Q are plotted on a centimetre square grid. For each set of points, calculate the length of PQ.

a $P(3, 7)$ and $Q(6, 3)$ **b** $P(-5, 2)$ and $Q(7, 7)$ **c** $P(-7, 5)$ and $Q(8, -3)$

Key point

You can use Pythagoras' theorem to work out the length of a shorter side in a right-angled triangle.

Example

Work out the length of the unknown side in this right-angled triangle.
Give your answer correct to 2 decimal places.

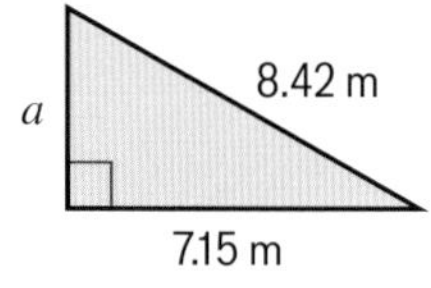

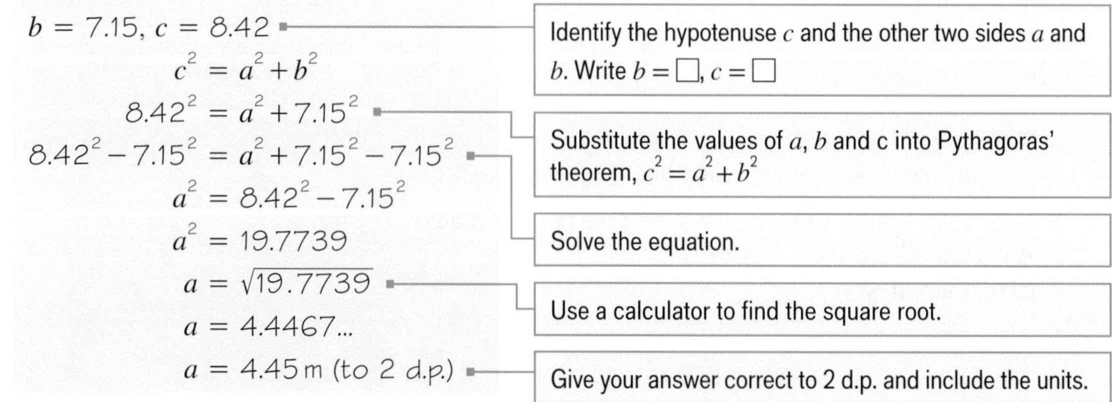

$b = 7.15,\ c = 8.42$ — Identify the hypotenuse c and the other two sides a and b. Write $b = \square$, $c = \square$

$c^2 = a^2 + b^2$

$8.42^2 = a^2 + 7.15^2$ — Substitute the values of a, b and c into Pythagoras' theorem, $c^2 = a^2 + b^2$

$8.42^2 - 7.15^2 = a^2 + 7.15^2 - 7.15^2$ — Solve the equation.

$a^2 = 8.42^2 - 7.15^2$

$a^2 = 19.7739$

$a = \sqrt{19.7739}$ — Use a calculator to find the square root.

$a = 4.4467...$

$a = 4.45$ m (to 2 d.p.) — Give your answer correct to 2 d.p. and include the units.

6 Work out the length of the unknown side in each right-angled triangle.
Give your answers to an appropriate degree of accuracy.
State the degree of accuracy after each answer, e.g. 2 d.p.

a

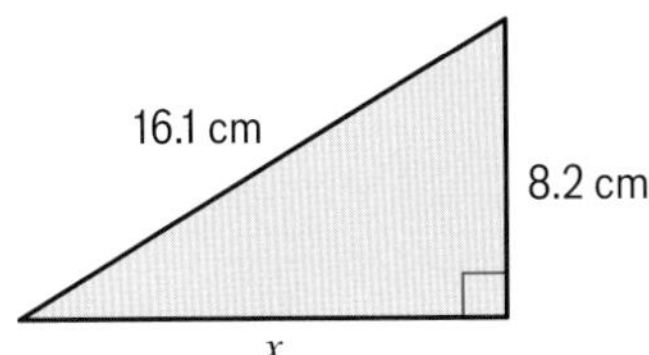

b

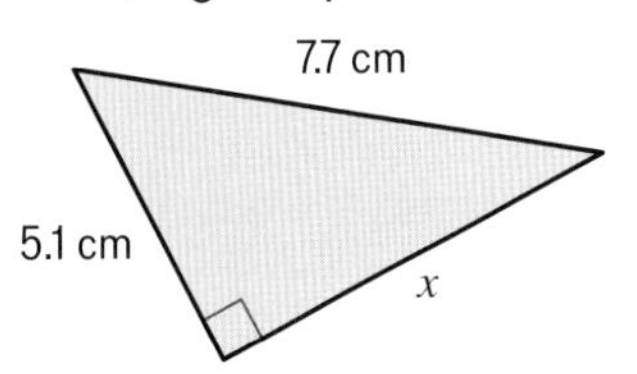

c

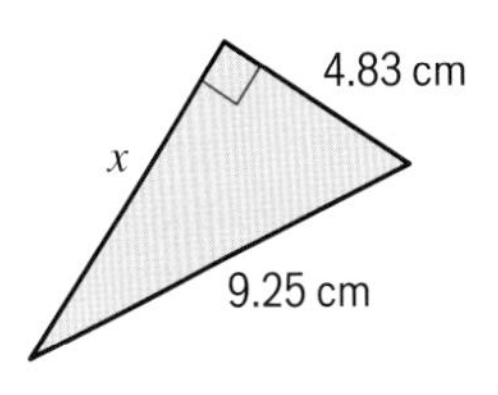

d **Reflect** How do the values in the question help you decide on a suitable degree of accuracy for the answers?

7 **Reasoning**

a Which of these triangles have the same length sides?

A

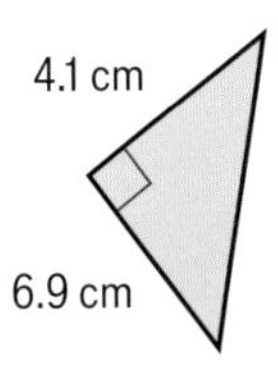

B

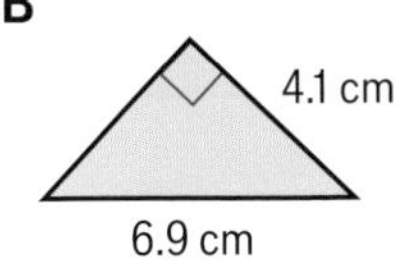

C

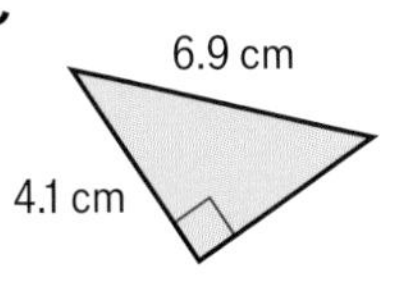

D

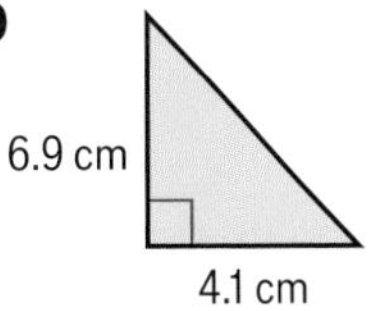

Diagram NOT accurately drawn

b Work out the length of the unknown side for each right-angled triangle.
Give your answers correct to 1 decimal place.

Exam-style question

8 Work out the perimeter of triangle PQR.

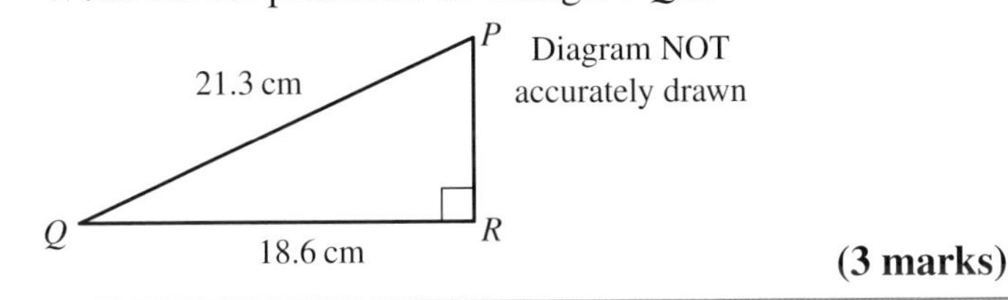

Diagram NOT accurately drawn

(3 marks)

Exam tip

Sometimes exam questions require you to use Pythagoras' theorem to find a length. Then, you must use this length to work out another length.

9 **Problem-solving / Reasoning** Find the lengths of the unknown sides in these diagrams.

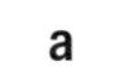

a

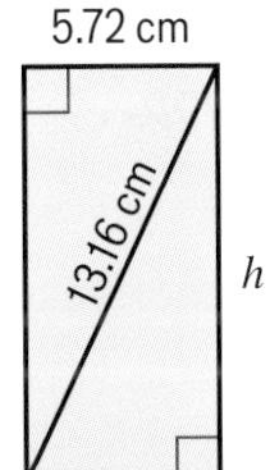

b

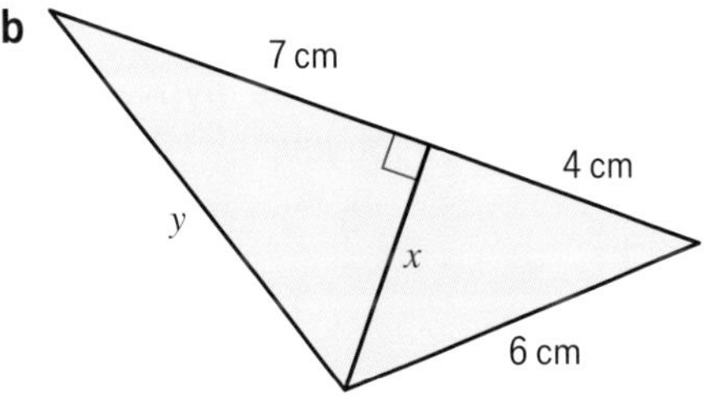

Key point

A triangle with sides a, b and c, where c is the longest side, is right-angled only if $c^2 = a^2 + b^2$.

10 **Problem-solving** Here are the lengths of sides of triangles.
Which of these triangles are right-angled triangles?

a 5 cm, 6 cm, 8 cm

b 6 km, 8 km, 10 km

c 4 m, 5 m, 6 m

Q10 hint

If the triangle is right-angled, the longest side will be the hypotenuse.

12.3 Trigonometry: the sine ratio 1

- Understand and recall the sine ratio in right-angled triangles.
- Use the sine ratio to calculate the length of a side in a right-angled triangle.
- Use the sine ratio to solve problems.

Warm up

1 **Fluency** ABC is a triangle.
$AB = 4\text{ cm}$ and $AC = 8\text{ cm}$.

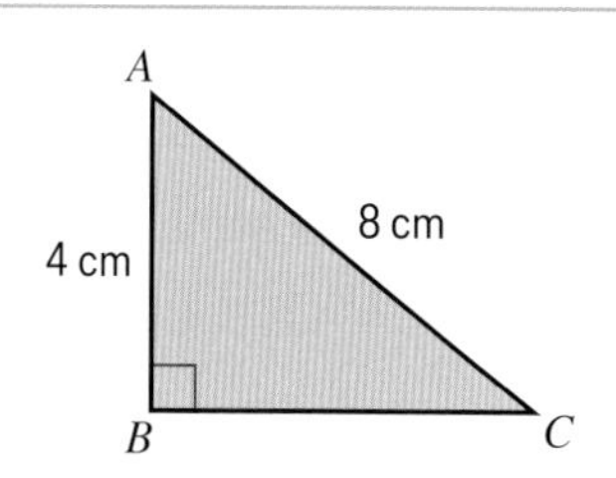

a What type of triangle is ABC?

b What is $\frac{AB}{AC}$ as

i a fraction in its simplest form **ii** a decimal

iii a ratio in its simplest form?

Q1b iii hint

$AB : AC$

2 Use a protractor to draw an angle of 30°.

3 Solve the equation $0.417 = \frac{x}{3}$.

Key point

The side opposite the right angle is called the **hypotenuse**.
The side opposite the angle θ is called the **opposite**.
The side next to the angle θ is called the **adjacent**.
θ is the Greek letter 'theta'. It is often used in maths to mark an unknown angle.

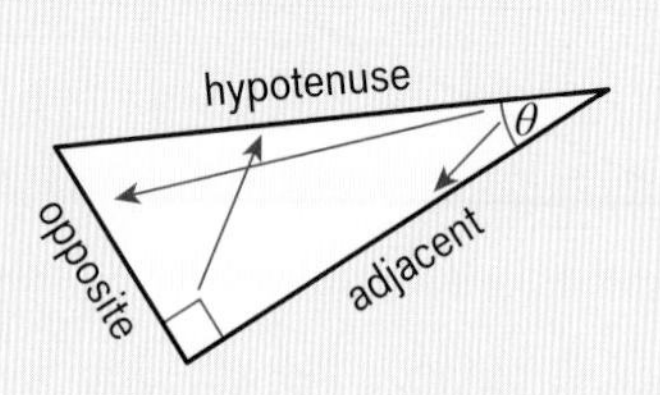

4 Copy these right-angled triangles.
For each triangle, label the **hyp**otenuse '**hyp**' and the side **opp**osite the angle θ '**opp**'.
The first one has been started for you.

a

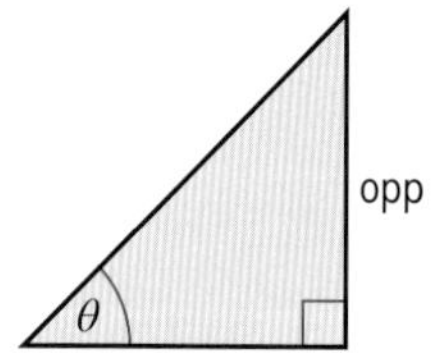

b

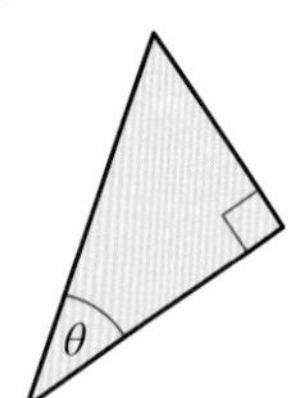

c

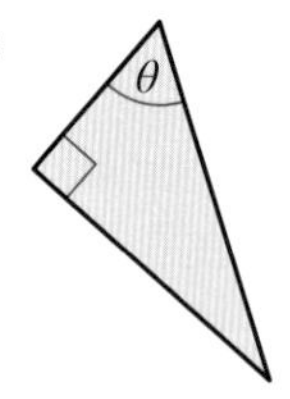

5 **Reasoning** **a** Use a ruler and protractor to draw these triangles accurately.

6 cm
30°

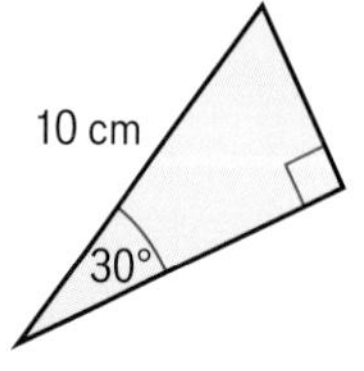

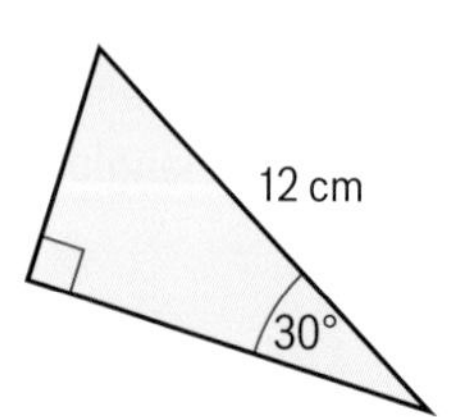

Q5a hint

Draw the hypotenuse first. Then draw a line at 30° to the hypotenuse. Complete the right-angled triangle.

b Label the hypotenuse (hyp) and the opposite side (opp).

6 **a** For each triangle you drew in **Q5a**

i measure the opposite side

ii write the fraction $\frac{\text{opposite}}{\text{hypotenuse}}$ and convert it to a decimal

b What do you notice?

Key point

In a right-angled triangle, the **sine** of an angle is the ratio of the **opp**osite side to the **hyp**otenuse.
The sine of angle θ is written as $\sin\theta$.

$$\sin\theta = \frac{\text{opp}}{\text{hyp}}$$

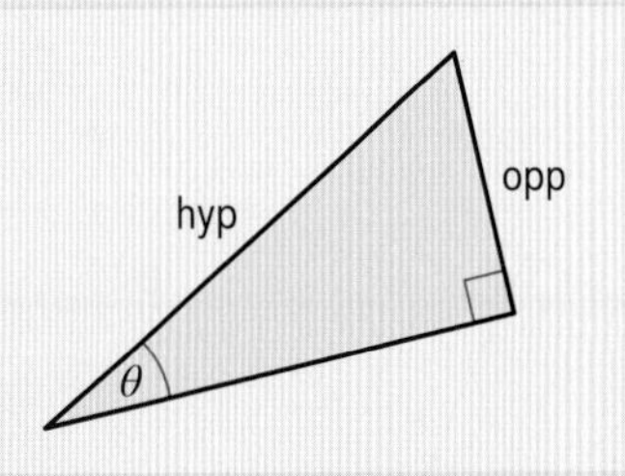

7 Use your calculator to find, correct to 3 decimal places

a $\sin 50°$ **b** $\sin 38°$ **c** $\sin 46.8°$

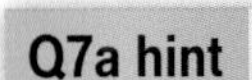

On your calculator enter    

8 **Reasoning** Use your answers to **Q6** to write the value of $\sin 30°$. Check using a calculator.

9 Write $\sin\theta$ as a fraction for each triangle.

a

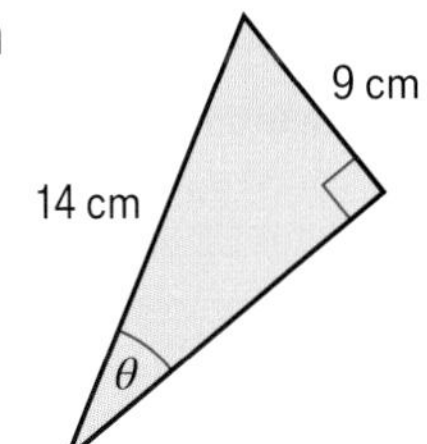

b

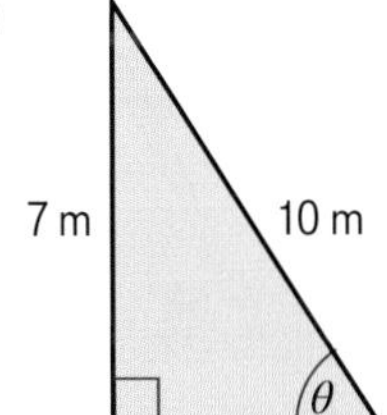

c

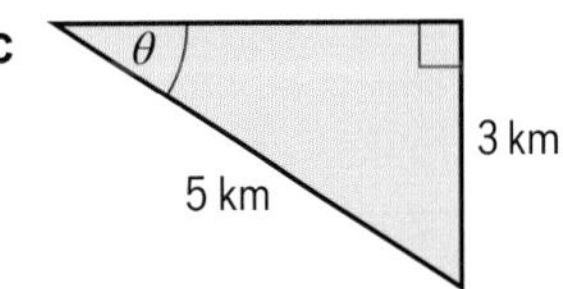

Q9 hint

$\sin\theta = \frac{\text{opp}}{\text{hyp}}$

Example

Use the sine ratio to work out the value of x.

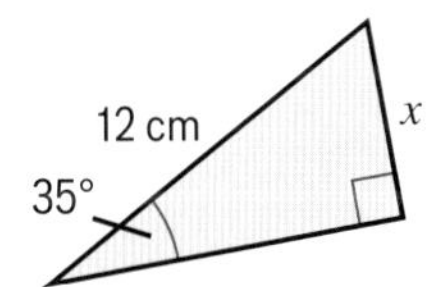

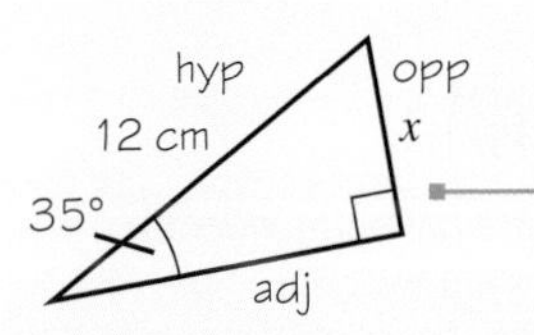

Sketch the triangle. Label the hypotenuse, and the opposite and adjacent sides.

$\theta = 35°$
opposite $= x$
hypotenuse $= 12$ cm

Identify the information given: angle, opposite and hypotenuse.

$\sin(\text{angle}) = \frac{\text{opp}}{\text{hyp}}$

Write the sine ratio.

$\sin 35° = \frac{x}{12}$

Substitute the sides and angle into the sine ratio.

$x = 12 \times \sin 35°$

Multiply both sides by 12. Use your calculator to work out $12 \times \sin 35°$.

$x = 6.8829...$
$x = 6.9$ cm (to 1 d.p.)

Round your answer to 1 decimal place.

10 Use the sine ratio to find the value of x in each triangle.
Give your answers correct to 1 decimal place.

a
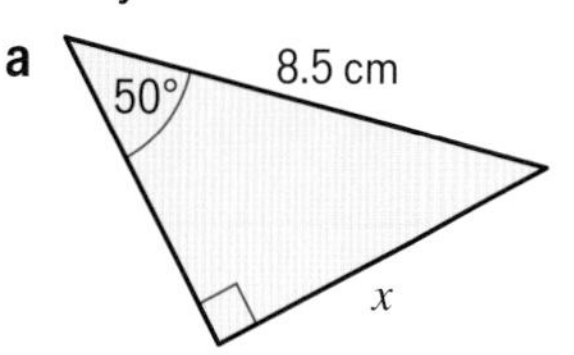

b
62°
x
9.8 cm

c
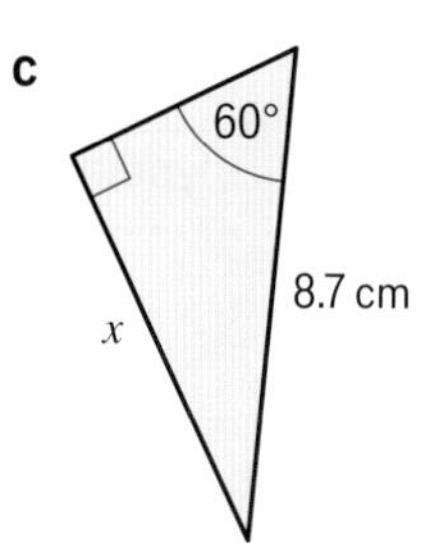

d
71°
x
16 cm

e
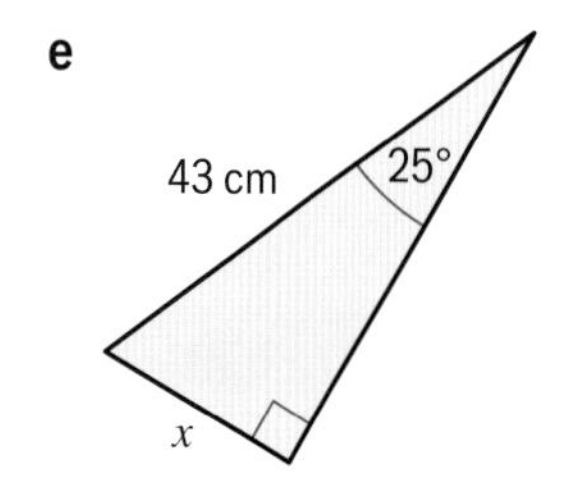

f
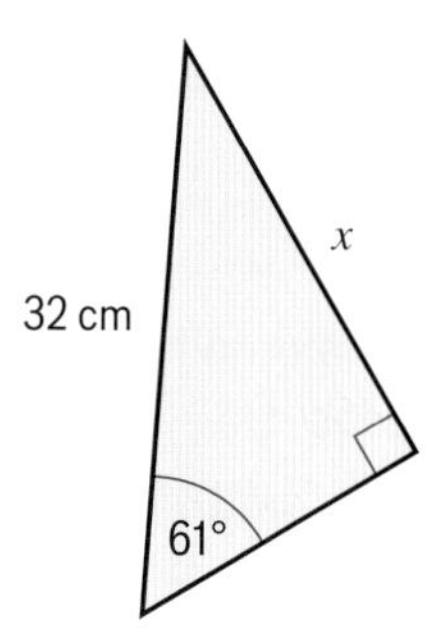

11 Problem-solving
A ladder of length 3 m leans against a vertical wall.
The ladder makes an angle of 68° with the horizontal ground.
How far is the top of the ladder from the ground?
Give your answer correct to 2 decimal places.

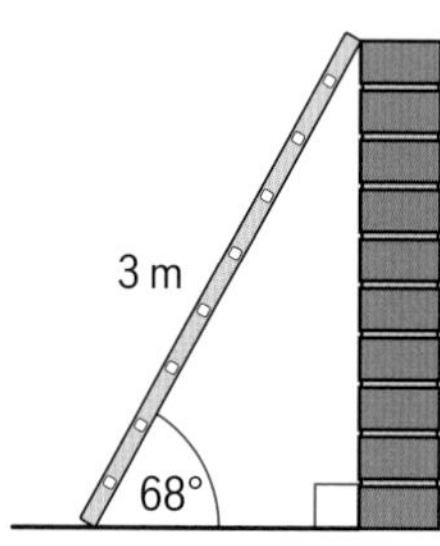

Q11 hint
Sketch the diagram. Label the distance you are trying to find as x.

12 Problem-solving A wheelchair ramp of length 2.15 m leads to the top of a step.
The ramp makes an angle of 4° with the horizontal ground.
What is the vertical height of the step?
Give your answer in centimetres.

Q12 hint
Sketch the right-angled triangle.

Exam-style question

13 A vertical flagpole is 7.5 m tall. The top of the flagpole is to be secured to the ground by a rope.
The rope will make an angle of 60° with the horizontal ground.
Will a rope of length 8 m reach the top of the flagpole?
(4 marks)

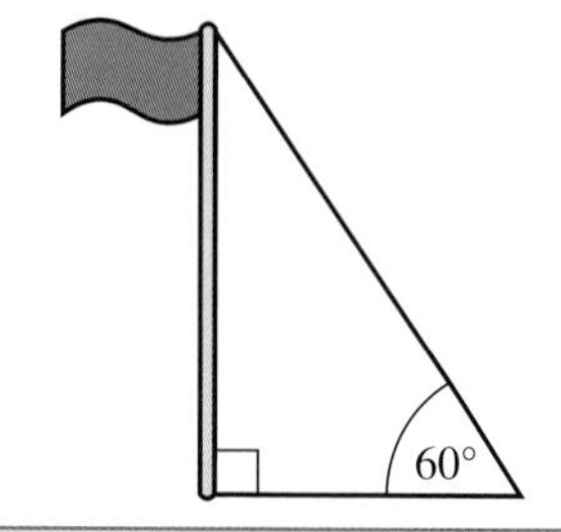

Exam tip
Marks are given for working, so clearly show every step of your working. Remember to write a sentence answering the question.

12.4 Trigonometry: the sine ratio 2

- Use the sine ratio to calculate an angle in a right-angled triangle.
- Use the sine ratio to solve problems.

Warm up

1 **Fluency** Use the diagram to write $\sin 22.6°$ as a fraction.

12 m, 22.6°, 5 m, 13 m

2 Use your calculator to find $\sin 32°$.

3 Find angle θ. Each one is a multiple of 10°.

a $\sin\theta = 0.642\,787\,609\,7$

b $\sin\theta = 0.939\,692\,620\,8$

c $\sin\theta = 0.173\,648\,177\,7$

Q3 hint

Use the sin key on your calculator. You only need to try angles such as 10°, 20°, 30°, ...

Key point

$\sin^{-1}$ is the inverse of sine.
When you know the value of $\sin\theta$, you can use $\sin^{-1}$ on a calculator to find θ.
On most calculators, you will need to use a SHIFT key or 2nd key to find $\sin^{-1}$.

4 Copy and complete these diagrams by using $\sin^{-1}$ on your calculator to find the angle.

a
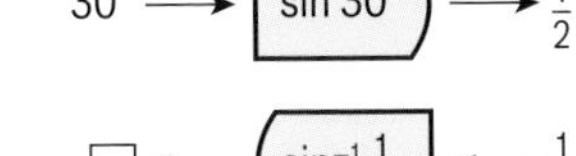

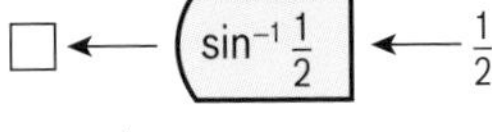

b

5 Use $\sin^{-1}$ on your calculator to check your answers to **Q3**.

6 Use $\sin^{-1}$ on your calculator to find the value of θ correct to 0.1°.
(This means correct to 1 decimal place.)

a $\sin\theta = 0.345$ b $\sin\theta = 0.8241$ c $\sin\theta = 0.8672$

d $\sin\theta = \frac{11}{12}$ e $\sin\theta = \frac{5}{8}$ f $\sin\theta = \frac{3.7}{5.9}$

Q6d hint

Enter $\sin^{-1}\left(\frac{11}{12}\right)$

7 **Reasoning** Annie and Joe are told that $\sin\theta = \frac{7}{11}$. They are asked to find θ.

Annie writes:

$7 \div 11 = 0.64$ (2 d.p.)
$\sin\theta = 0.64$
$\theta = \sin^{-1} 0.64$
$= 39.8$ (1 d.p.)

Joe writes:

$\sin\theta = \frac{7}{11}$
$\theta = \sin^{-1}\left(\frac{7}{11}\right)$
$\theta = 39.5211...$
$\theta = 39.5$ (1 d.p.)

a Write a sentence explaining what is different about Annie's working and Joe's working.

b Whose answer is correct? Explain.

Q7b hint

When have Annie and Joe rounded their answers?

Example

Use the sine ratio to find the size of angle x.

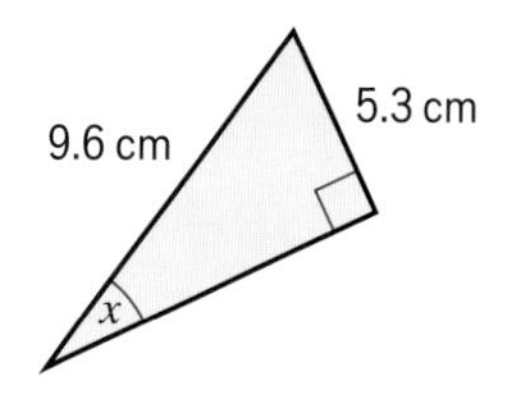

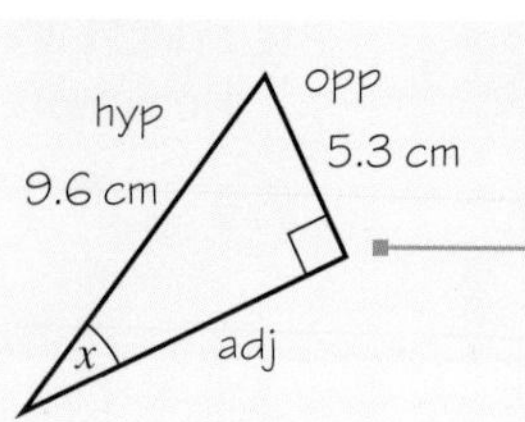

Sketch the triangle. Label the hypotenuse, and the opposite and adjacent sides.

angle = x
opposite = 5.3 cm
hypotenuse = 9.6 cm

Identify the information given: angle, opposite and hypotenuse.

$$\sin(\text{angle}) = \frac{\text{opp}}{\text{hyp}}$$

Write the sine ratio.

$$\sin x = \frac{5.3}{9.6}$$

Substitute the sides and angle into the sine ratio.

$$x = \sin^{-1}\left(\frac{5.3}{9.6}\right)$$

Use $\sin^{-1}$ to find the angle.

$x = 33.5100...$

$x = 33.5°$ (to 1 d.p.)

Round your answer to 1 decimal place.

8 Use the sine ratio to find the size of angle θ in each triangle.

a

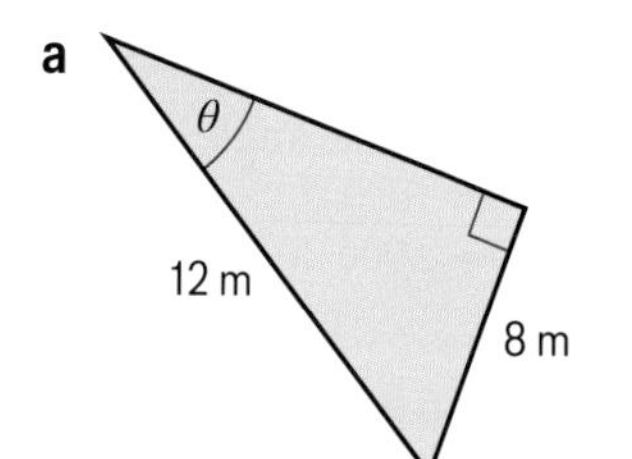

b

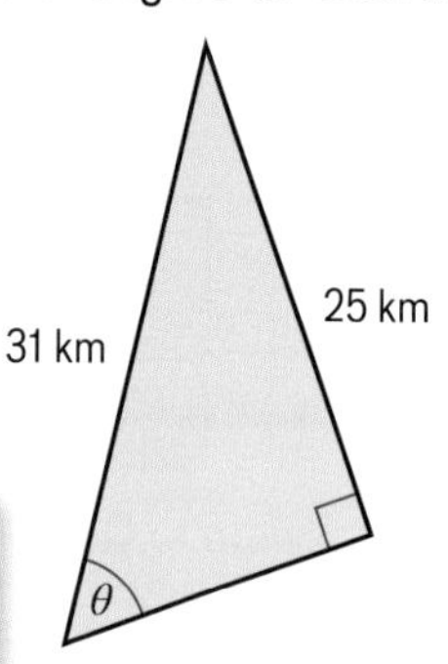

c

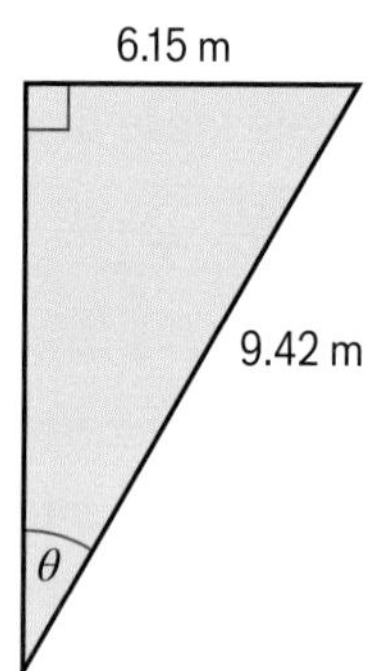

Q8 hint

State the degree of accuracy after your answer. Angles are normally given to the nearest 0.1°.

F
H

9 Problem-solving
The track for the Lisbon tram rises 1 m for every 7.47 m travelled.
What angle does the track make with the horizontal?

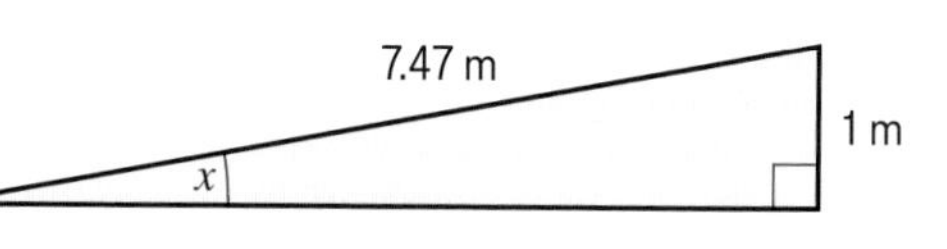

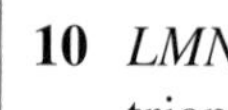

Exam-style question

10 LMN is a right-angled triangle.
$MN = 8.5$ cm
$LN = 11$ cm
Calculate the size of the angle marked x.
Give your answer correct to 1 decimal place.

(2 marks)

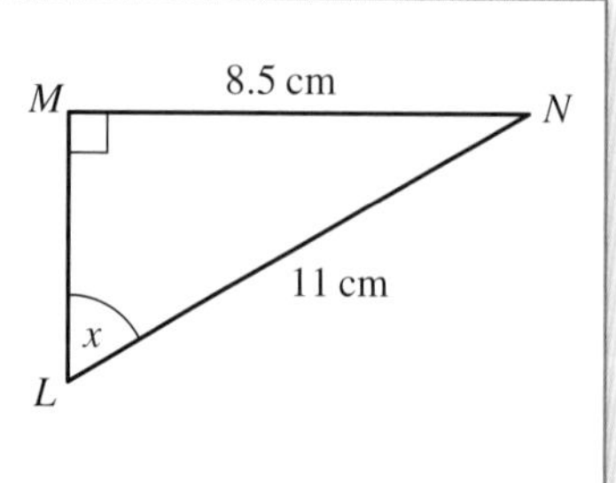

Exam tip

Make sure your calculator is in degree mode. Use all the figures on your calculator display for answers to steps in your working. Round only when you reach your final answer.

12.5 Trigonometry: the cosine ratio

- Understand and recall the cosine ratio in right-angled triangles.
- Use the cosine ratio to calculate the length of a side in a right-angled triangle.
- Use the cosine ratio to calculate an angle in a right-angled triangle.
- Use the cosine ratio to solve problems.

Warm up

1 **Fluency** Name the hypotenuse and the adjacent side in each triangle.

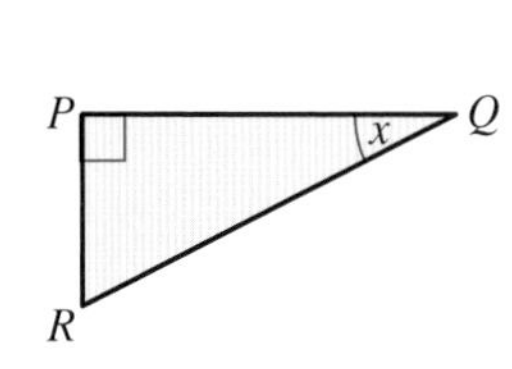

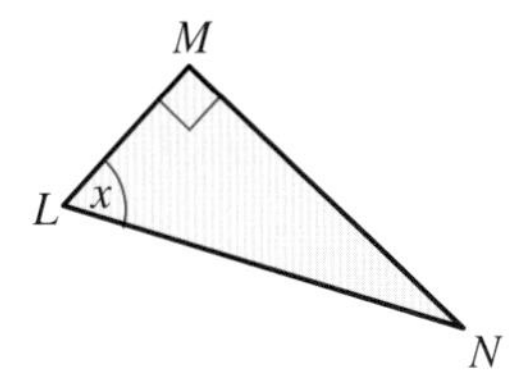

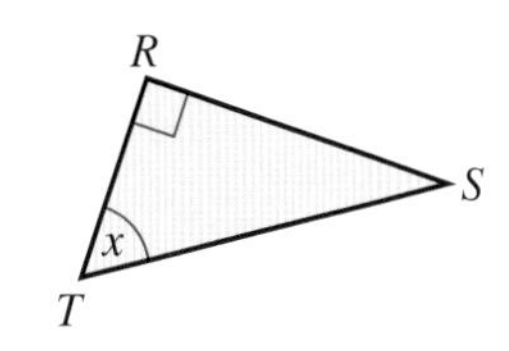

2

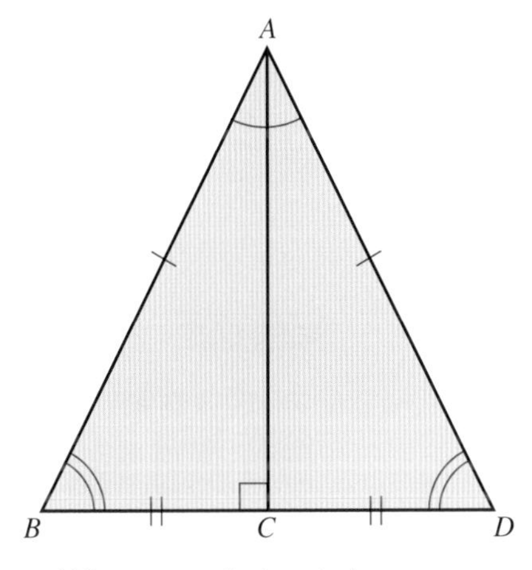

a What type of triangle is

i ABC **ii** ACD **iii** ABD?

b Copy and complete.

$BC = \square \times BD$

3

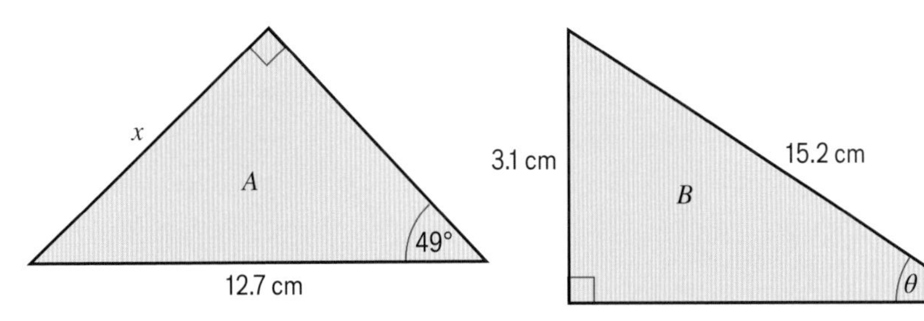

Use the sine ratio to work out

a the value of x in triangle A

b the size of angle θ in triangle B

Give your answers correct to 1 decimal place.

4 **Reasoning**

a Use a ruler and protractor to draw these triangles accurately.

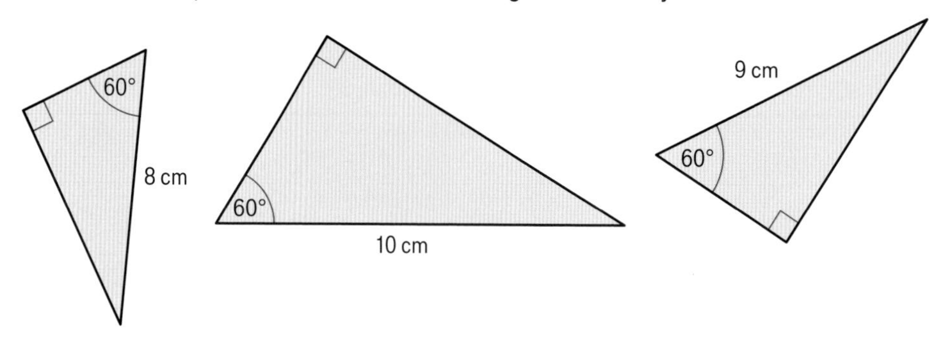

b For each triangle

i label the hypotenuse (hyp) and the adjacent side (adj)

ii measure the adjacent side

iii write the fraction $\frac{\text{adjacent}}{\text{hypotenuse}}$ and convert it to a decimal

c What do you notice?

Key point

In a right-angled triangle, the **cosine** of an angle is the ratio of the **adj**acent side to the **hyp**otenuse.
The cosine of angle θ is written as $\cos\theta$.

$$\cos\theta = \frac{\text{adj}}{\text{hyp}}$$

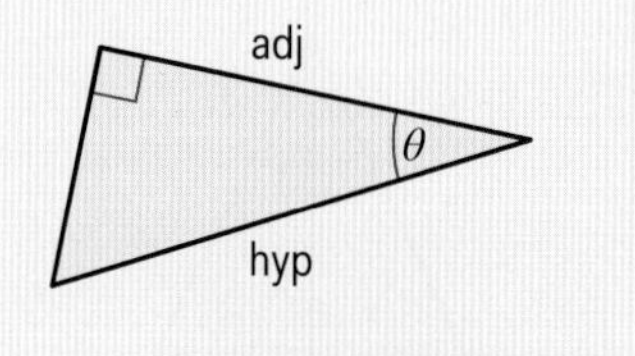

5 Use your calculator to find, correct to 3 decimal places

a $\cos 37°$

b $\cos 82°$

c $\cos 5.8°$

Q5a hint

On your calculator enter

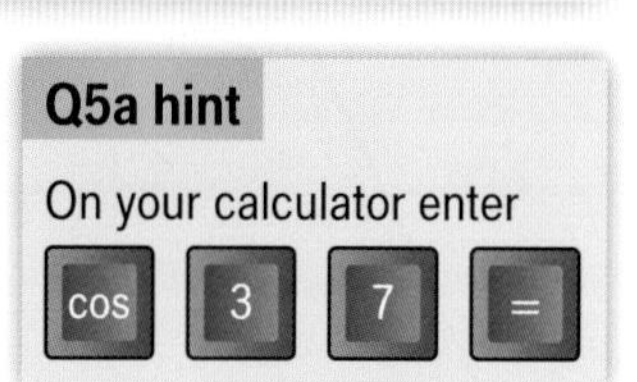

6 Write $\cos\theta$ as a fraction for each triangle.

a

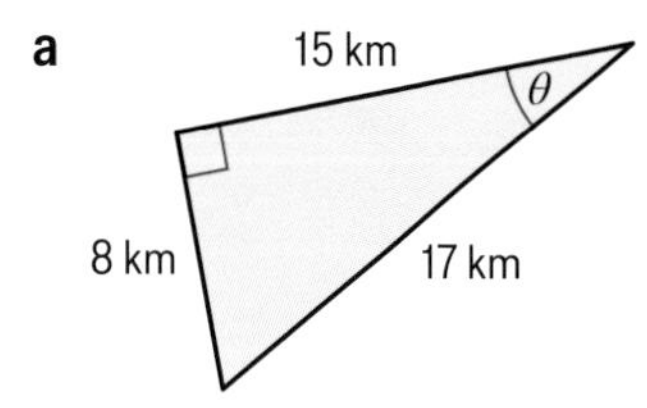

b

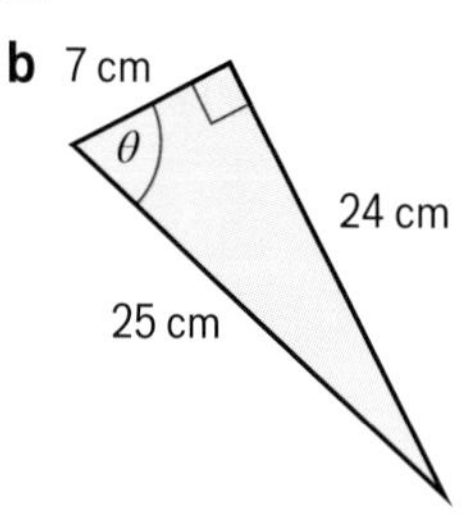

c

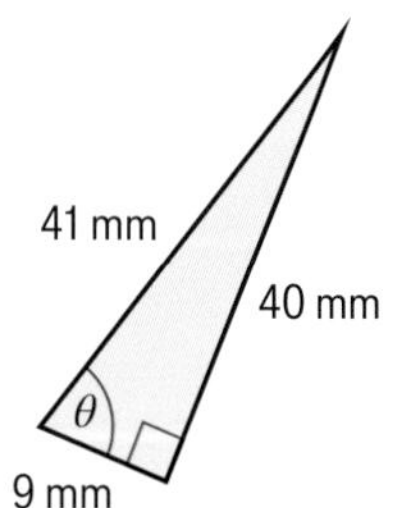

7 **Reasoning** Use this triangle to explain why $\cos 60° = \sin 30°$.

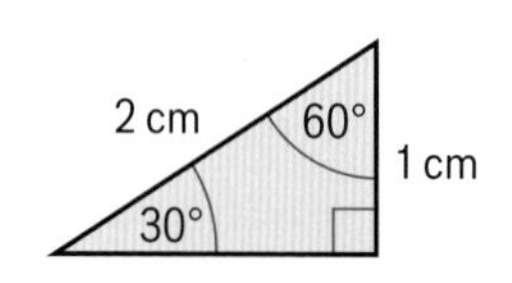

Example

Use the cosine ratio to work out the value of x.

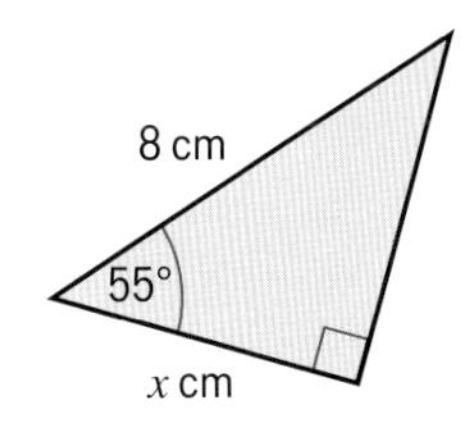

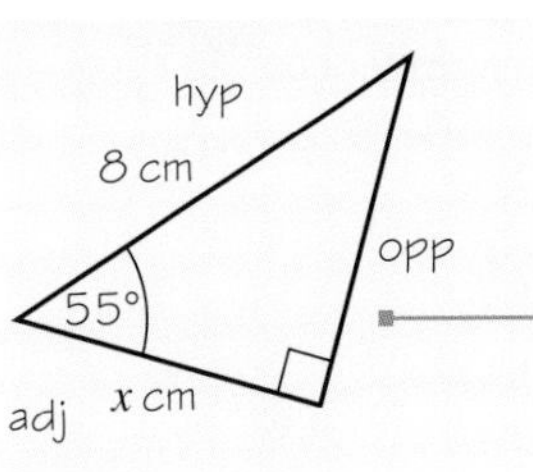

Sketch the triangle. Label the hypotenuse, and the opposite and adjacent sides.

angle = 55°
adjacent = x
hypotenuse = 8 cm

Identify the information given: angle, adjacent and hypotenuse.

$\cos(\text{angle}) = \frac{\text{adj}}{\text{hyp}}$ — Write the cosine ratio.

$\cos 55° = \frac{x}{8}$ — Substitute the sides and angle into the cosine ratio.

$x = 8 \times \cos 55°$ — Multiply both sides by 8. Use your calculator to work out $8 \times \cos 55°$.

$x = 4.5886...$

$x = 4.6$ cm (to 1 d.p.) — Round your answer to 1 decimal place.

8 Find the value of x in each triangle. Give your answers correct to 1 decimal place.

a

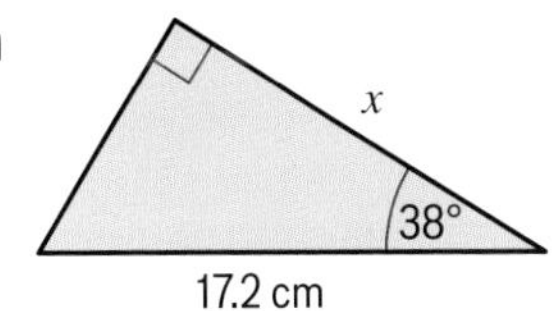

b

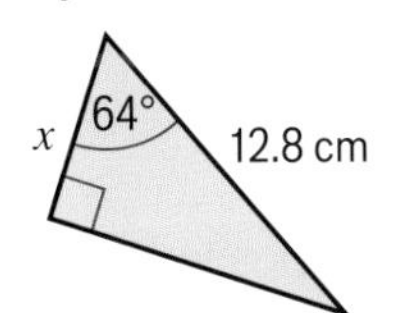

c

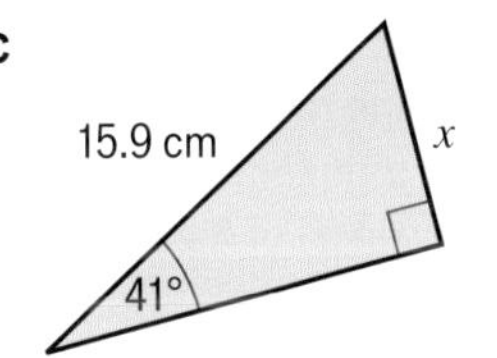

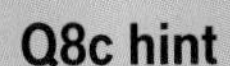

Q8c hint

The cosine ratio is not the correct trigonometric ratio to use to find x in this triangle.
What other trigonometric ratio do you know?

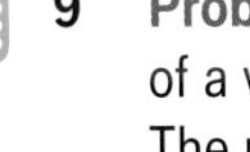

9 **Problem-solving** A tree surgeon ties a 10 m rope to the top of a vertical tree trunk. He pulls it tight and pegs it to the ground. The rope makes an angle of 55° with the horizontal ground. What is the distance of the peg from the base of the tree trunk? Give your answer correct to 2 decimal places.

Q9 hint

Sketch the diagram. Label the distance you are trying to find as x.

Key point

cos^{-1} is the inverse of cosine.
When you know the value of $\cos\theta$, you can use $\cos^{-1}$ on a calculator to find θ.

10 Copy and complete these diagrams by using $\cos^{-1}$ on your calculator to find the value of θ correct to 0.1°.

a $\theta \rightarrow$ [$\cos\theta$] $\rightarrow$ 0.362

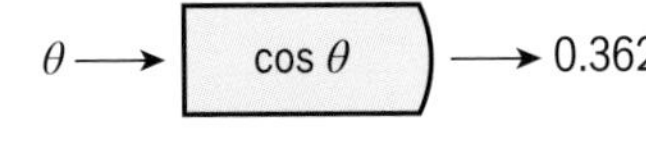

□ $\leftarrow$ [$\cos^{-1} 0.362$] $\leftarrow$ 0.362

b $\theta \rightarrow$ [$\cos\theta$] $\rightarrow$ 0.6735

□ $\leftarrow$ [$\cos^{-1} 0.6735$] $\leftarrow$ 0.6735

11 Use your calculator to find the value of θ correct to 0.1°.

a $\cos\theta = 0.5729$

b $\cos\theta = \frac{5}{8}$

c $\cos\theta = \frac{24}{25}$

d $\cos\theta = \frac{9.8}{15.6}$

Q11b hint

Enter $\cos^{-1}\left(\frac{5}{8}\right)$

12 Copy and complete these diagrams. The first one has been done for you.

a

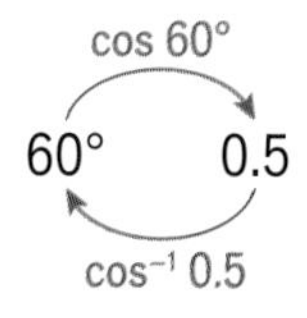

b

cos 28°

28° □

cos⁻¹ □

c

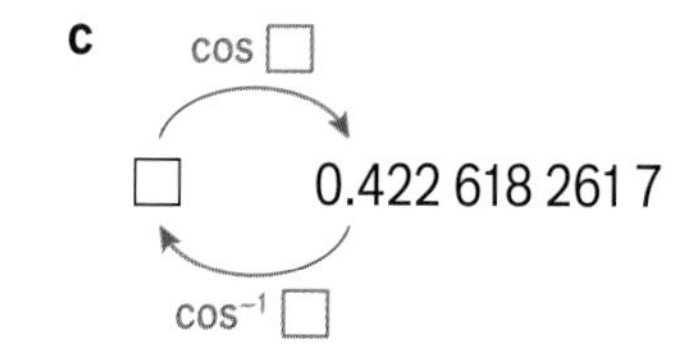

13 Sketch these triangles.

i

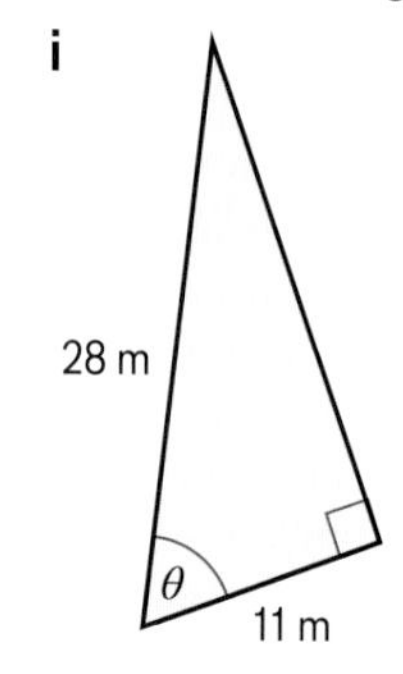

ii

8.37 m, θ, 3.81 m

iii

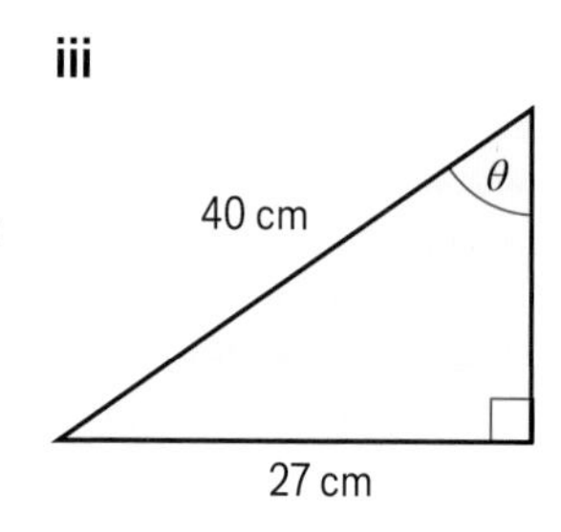

For each triangle

a label the hypotenuse (hyp) and the opposite (opp) and adjacent (adj) sides

b write the trigonometric ratio involving the two sides you are given

c substitute the vales you are given into the trigonometric ratio

d use the inverse trigonometric ratio to find the size of the angle θ.
Give your answer correct to the nearest 0.1°.

Q13 hint

The cosine ratio is not the correct trigonometric ratio to use to find θ in all of these triangles.

14

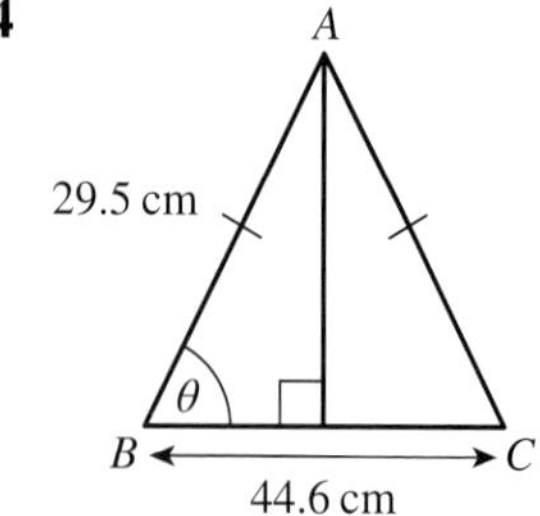

ABC is an isosceles triangle.

$AB = 29.5\,\text{cm}$

$BC = 44.6\,\text{cm}$

Work out the size of angle θ.

Give your answer correct to 1 decimal place. **(3 marks)**

Exam tip

In exam questions you sometimes have to work out an unknown length first to help you to find the answer you are asked for.

12.6 Trigonometry: the tangent ratio

- Understand and recall the tangent ratio in right-angled triangles.
- Use the tangent ratio to calculate the length of a side in a right-angled triangle.
- Use the tangent ratio to calculate an angle in a right-angled triangle.
- Solve problems using an angle of elevation or angle of depression.

Warm up

1 **Fluency** Name the opposite and adjacent sides in each triangle.

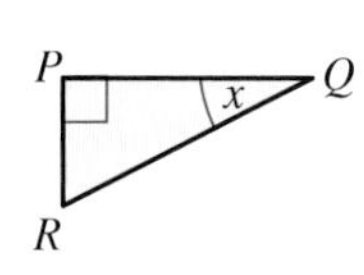

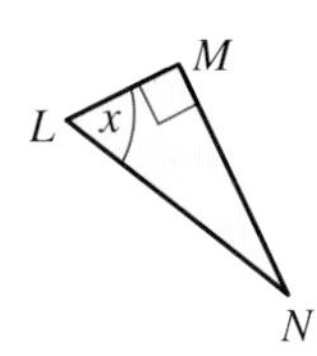

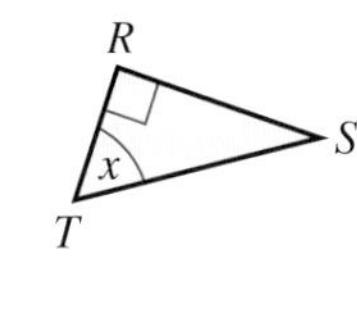

2 Write the fraction $\frac{\text{opp}}{\text{adj}}$ for each triangle.

a

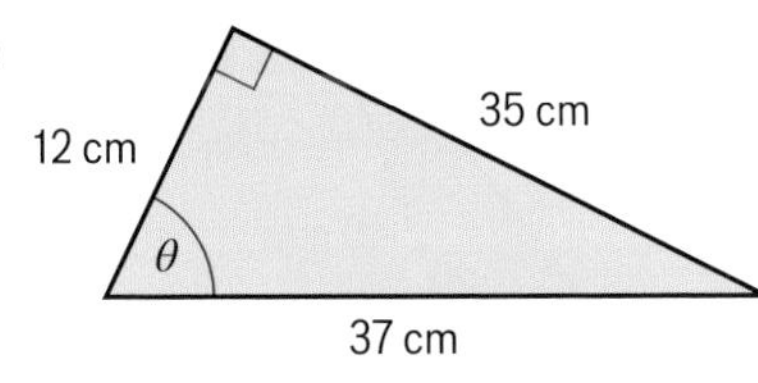

b

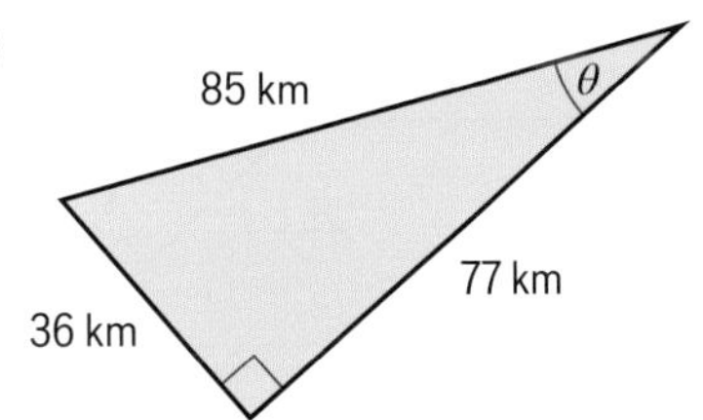

3 Use your calculator to work out

a $\sqrt{3}$ **b** $\frac{\sqrt{3}}{7}$ **c** $\frac{7}{\sqrt{3}}$

Give your answers correct to 3 decimal places.

4 **Reasoning**

a Use a ruler and protractor to draw these triangles accurately.

31°
10 cm

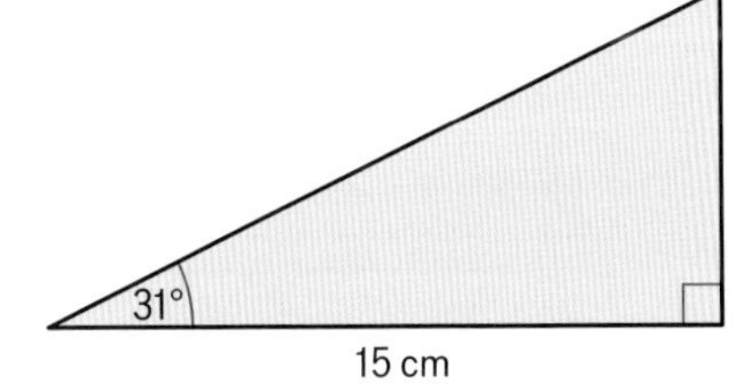

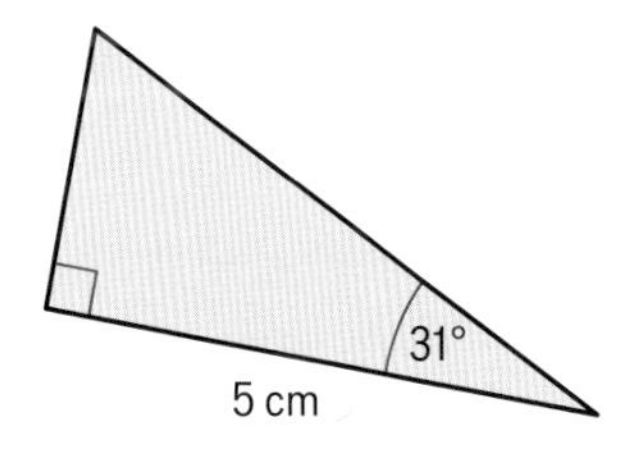

b For each triangle

i label the opposite side (opp) and the adjacent side (adj)

ii measure the opposite side

iii write the fraction $\frac{\text{opposite}}{\text{adjacent}}$ and convert it to a decimal (correct to 1 decimal place)

c What do you notice?

Key point

In a right-angled triangle, the **tangent** of an angle is the ratio of the **opp**osite side to the **adj**acent side.

The tangent of angle θ is written as $\tan\theta$.

$$\tan\theta = \frac{\text{opp}}{\text{adj}}$$

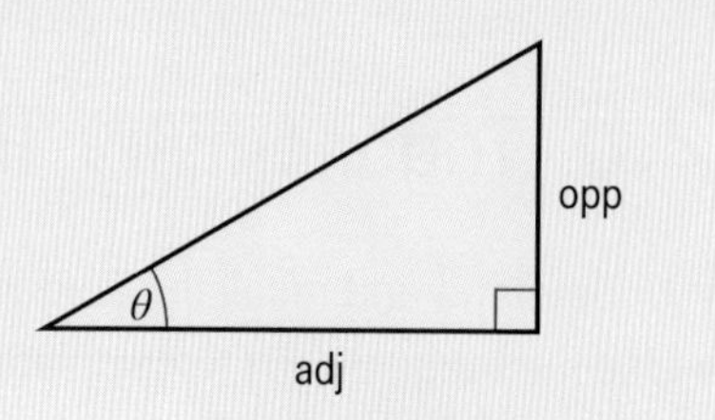

5 Use your calculator to find, correct to 3 decimal places

a $\tan 25°$

b $\tan 50°$

c $\tan 75°$

Q5a hint

On your calculator enter

tan 2 5 =

6 **Reasoning** Use the triangles you drew in **Q4** to write the value of

a $\tan 31°$

b $\tan 59°$

Check using a calculator.

Q6b hint

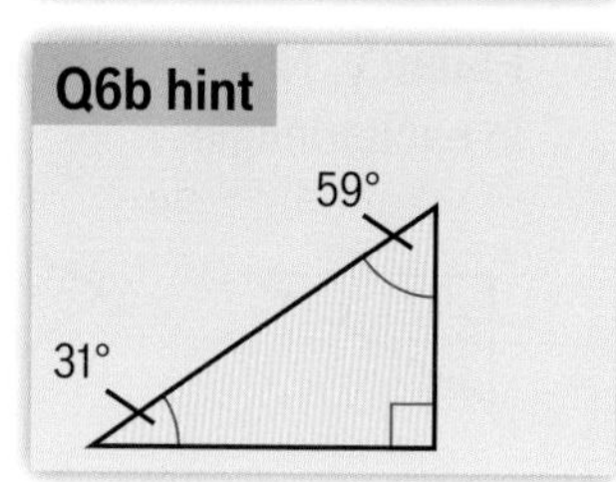

7 Write $\tan\theta$ as a fraction for each triangle.

a

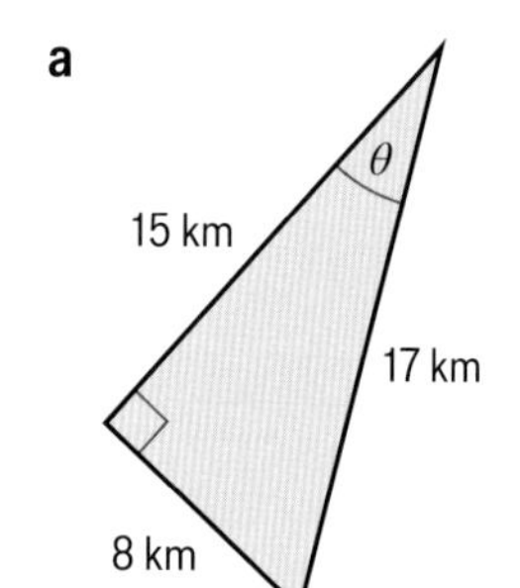

b

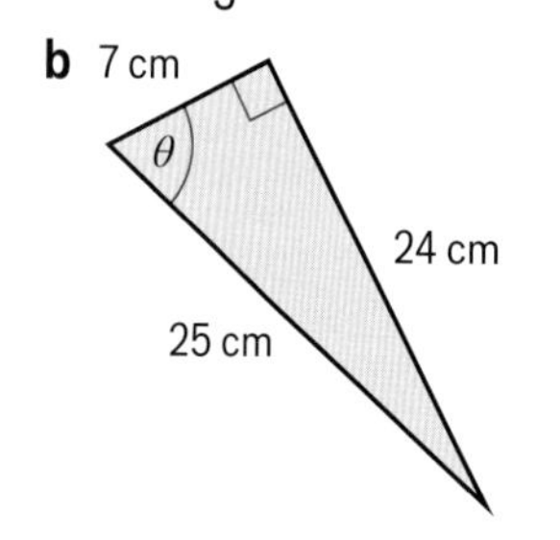

c

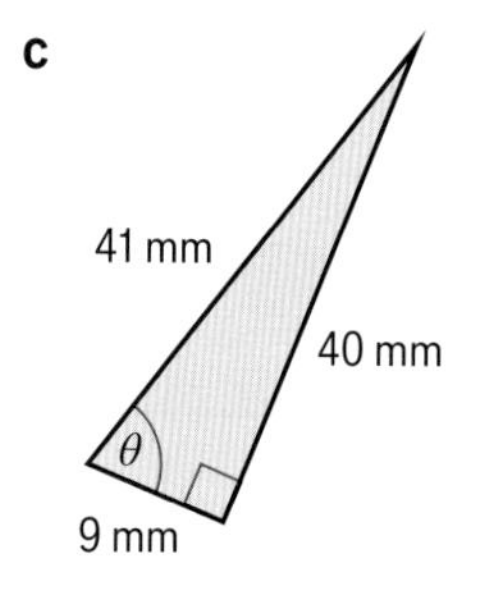

Example

Use the tangent ratio to work out the value of x.

8.7 cm

34°

x cm

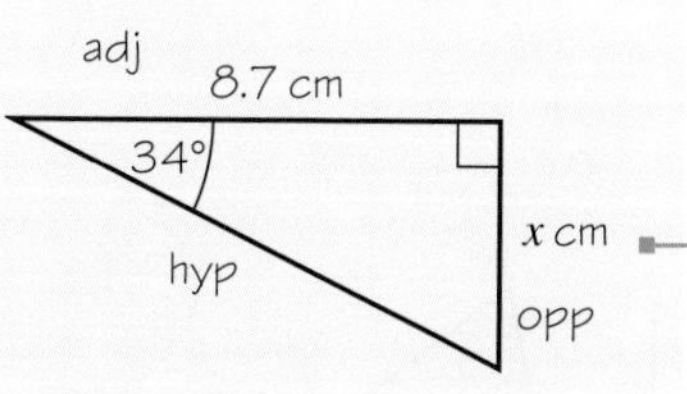

Sketch the triangle. Label the hypotenuse, and the opposite and adjacent sides.

angle = 34°

opposite = x cm

adjacent = 8.7 cm

Identify the information given: angle, opposite and hypotenuse.

$\tan(\text{angle}) = \frac{\text{opp}}{\text{adj}}$ — Write the tangent ratio.

$\tan 34° = \frac{x}{8.7}$ — Substitute the sides and angle into the tangent ratio.

$x = 8.7 \times \tan 34°$ — Multiply both sides by 8.7. Use your calculator to work out $8.7 \times \tan 34°$.

$x = 5.8682...$

$x = 5.9$ (to 1 d.p.) — The final answer will be $x = \square$ without units as the unknown side is labelled x cm, not simply x.

8 Use the tangent ratio to find the value of x in each triangle.
Give your answers correct to 1 decimal place.

a

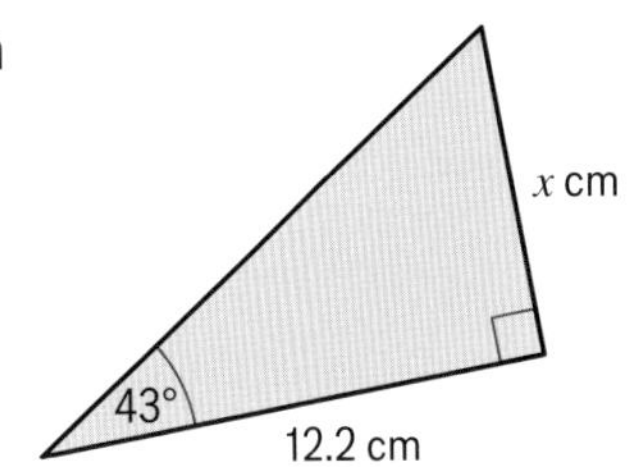

b

x cm

63.5 cm

29°

c

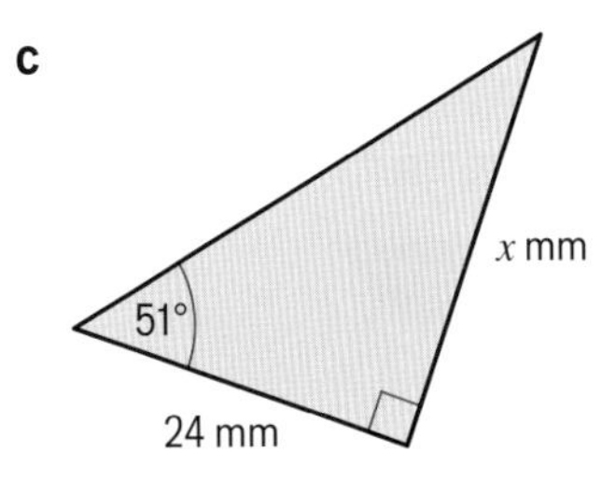

9 Problem-solving
A vertical flagpole is secured to the ground by a rope.
The end of the rope is 4 m from the base of the flagpole.
The rope makes an angle of 50° with the horizontal ground.
At what height does the rope attach to the flagpole?

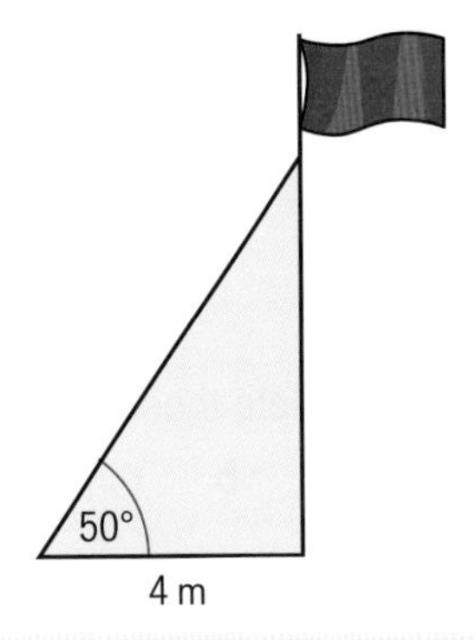

Key point

The angle of **elevation** is the angle measured upwards from the horizontal.
The angle of **depression** is the angle measured downwards from the horizontal.

10 For each diagram, does it show an angle of elevation or an angle of depression?

Q10 hint

Start at the horizontal line (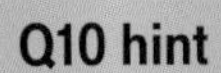). Does the angle turn upwards or downwards?

a

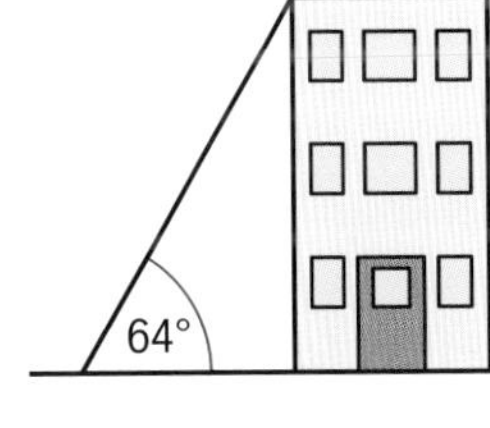

b

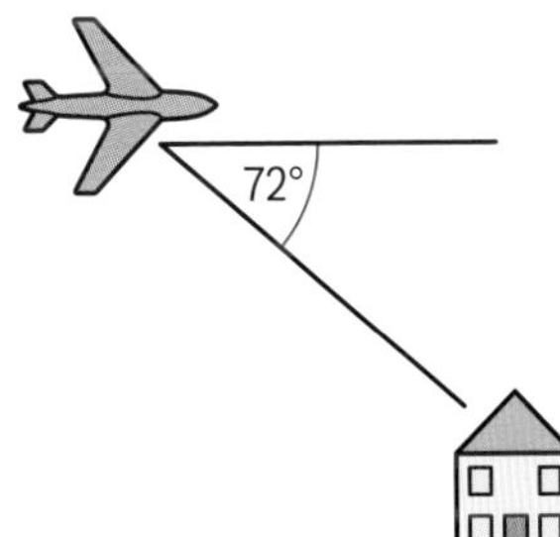

c

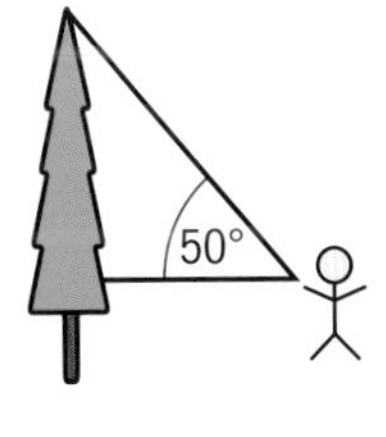

11 Problem-solving
The angle of elevation of a cliff top, C, from a small boat, B, out at sea is 36°.
The boat is 100 m from the bottom of the cliffs.

a Copy the diagram and label it with this information.

b How high are the cliffs? Give your answer correct to the nearest metre.

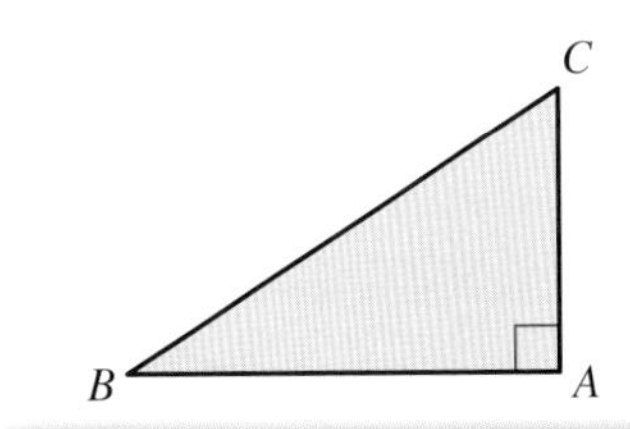

Q11 hint

Assume that the surface of the sea is horizontal and that the cliffs are vertical.

12 Problem-solving
From the top of a tower in York Minster, Georgia can see a boat on the river.
The angle of depression from Georgia to the boat is 10°.
The boat is 400 m away. How high is the tower? Give your answer correct to the nearest metre.

Q12 hint

Sketch the right-angled triangle.

Key point

$\tan^{-1}\theta$ is the inverse of tangent.
When you know the value of $\tan\theta$, you can use $\tan^{-1}$ on a calculator to find θ.

13 Use $\tan^{-1}$ on your calculator to find the value of θ correct to 0.1°.

a $\tan\theta = 0.853$

b $\tan\theta = 1.725$

c $\tan\theta = \frac{7}{8}$

d $\tan\theta = \frac{30}{17}$

Q13a hint

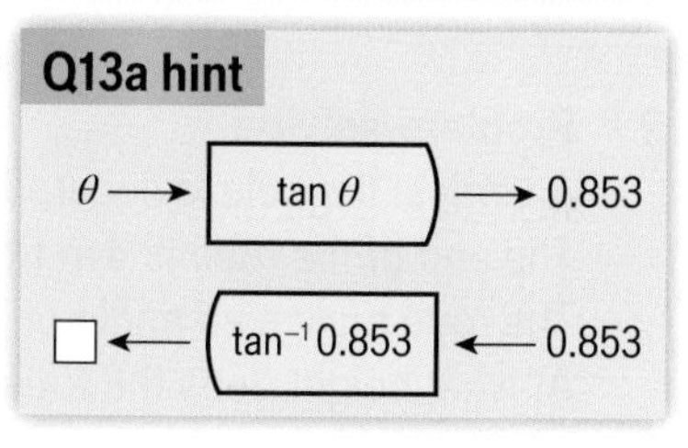

14 Copy and complete these diagrams. The first one has been done for you.

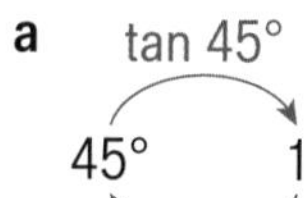

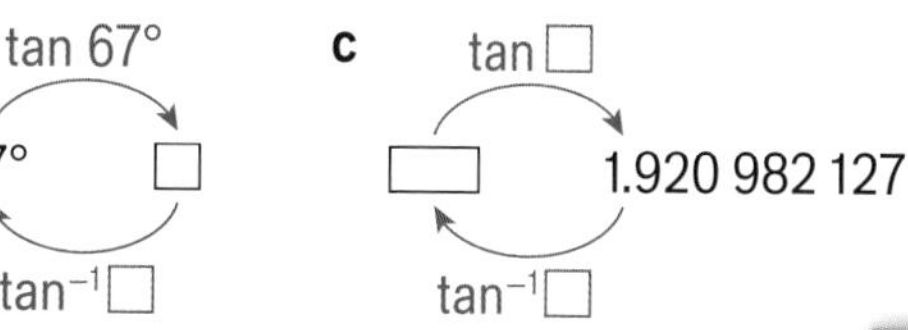

15 Use the tangent ratio to calculate the size of the unknown angle in each of these triangles.

Q15 hint

Unknown angles are not always marked with θ.

a

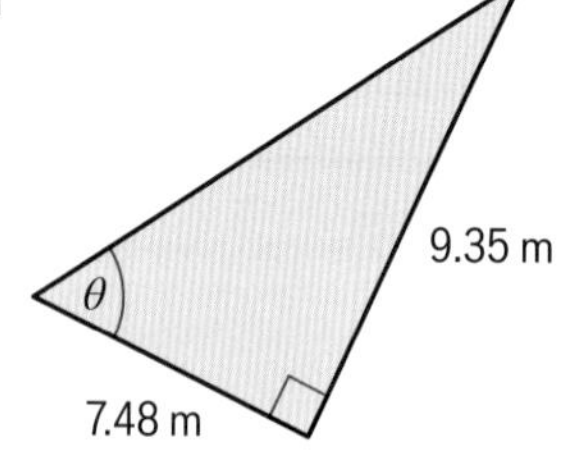

b

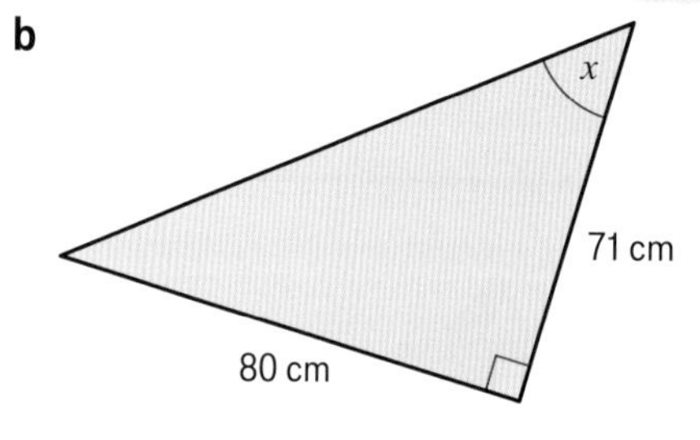

c

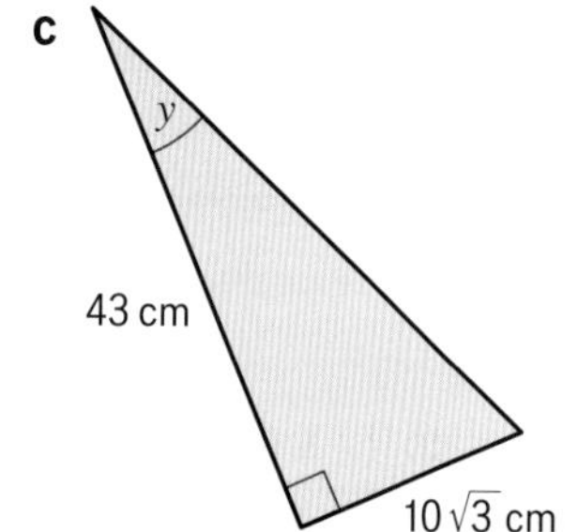

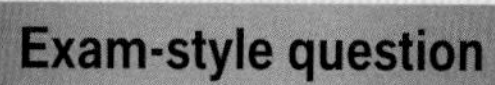

Exam-style question

16

Diagram NOT accurately drawn

ABC and *BCD* are right-angled triangles.
$AB = 3$ cm $BD = 8$ cm angle $BAC = 49°$
Work out the size of angle *BCD*.

(6 marks)

Exam tip

When a shape is made up of two or more smaller shapes, it can help to sketch the smaller shapes separately, for example

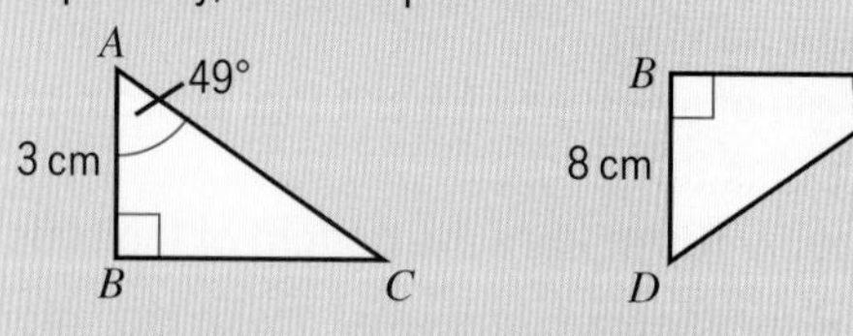

Q16 hint

Use triangle *ABC* to work out length *BC* first.

12.7 Finding lengths and angles using trigonometry

- Understand and recall trigonometric ratios in right-angled triangles.
- Use trigonometric ratios to solve problems.
- Know the exact values of the sine, cosine and tangent of some angles.

*Active*Learn Homework

Warm up

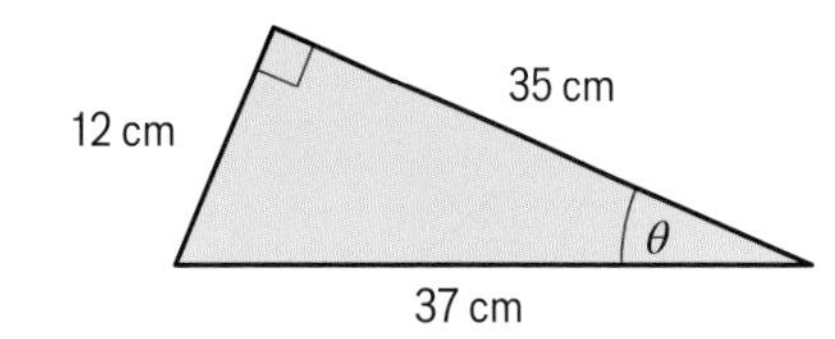

1 **Fluency** For this triangle, match each of $\sin\theta$, $\cos\theta$ and $\tan\theta$ to one of these fractions.

A $\frac{12}{35}$ **B** $\frac{12}{37}$ **C** $\frac{35}{37}$

2 Solve $0.765 = \frac{17.3}{x}$. Give the value of x to 3 significant figures.

3 Sally uses Pythagoras' theorem to work out the length of the hypotenuse in this right-angled triangle. Copy and complete her working.

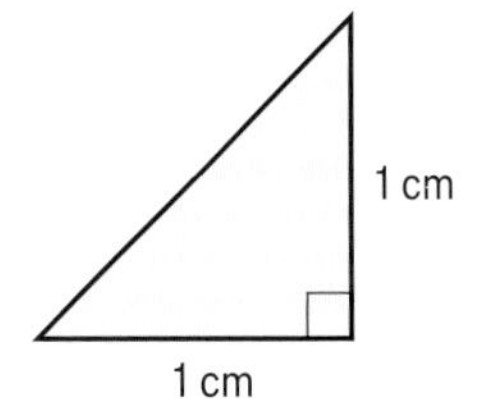

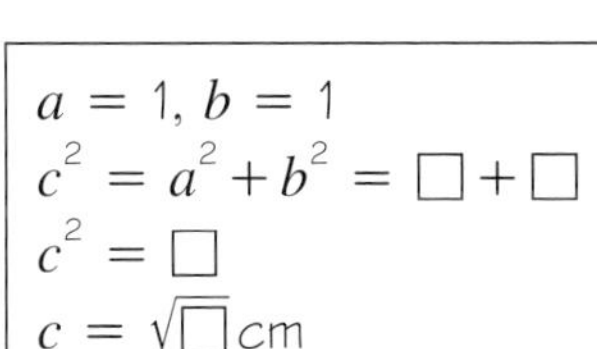

$a = 1,\ b = 1$
$c^2 = a^2 + b^2 = \square + \square$
$c^2 = \square$
$c = \sqrt{\square}$ cm

Key point

You need to be able to choose the correct trigonometric ratio to solve a problem.

Example

Calculate the value of x. Give your answer correct to 3 significant figures.

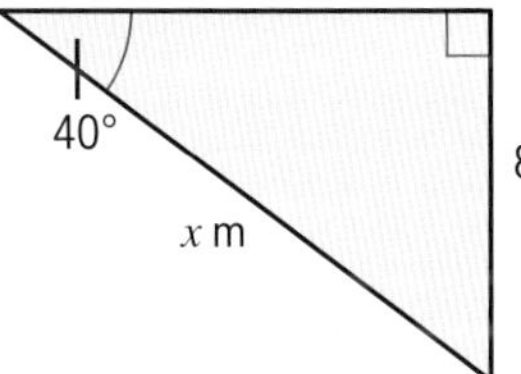
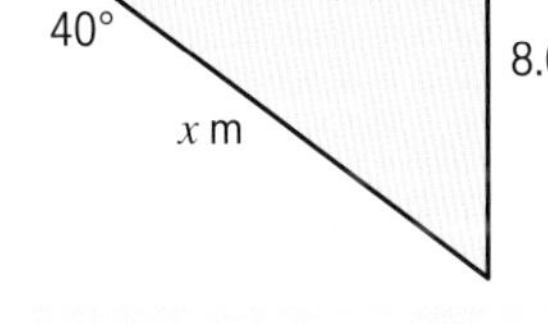

angle = 40° opposite = 8.63 m hypotenuse = x m — Identify the information given.

$\sin(\text{angle}) = \frac{\text{opp}}{\text{hyp}}$ — Decide on the ratio (sin, cos, tan) you need to use. You are given 'opp' and 'hyp' so use the sine ratio.

$\sin 40° = \frac{8.63}{x}$ — Substitute the sides and angle into the sine ratio.

$x \times \sin 40° = \frac{8.63}{x} \times x$ — Multiply both sides by x.

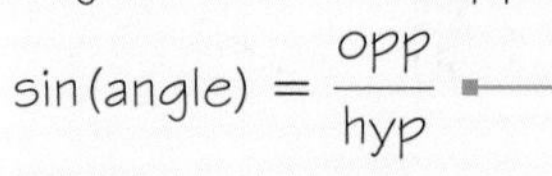

$x \times \sin 40° = 8.63$

$x = \frac{8.63}{\sin 40°} = 13.4258... = 13.4$ (to 3 s.f.) — Divide both sides by sin 40°. Then round to 3 significant figures.

4 Calculate the length of AB in each triangle.
Give your answers correct to 2 decimal places.

a
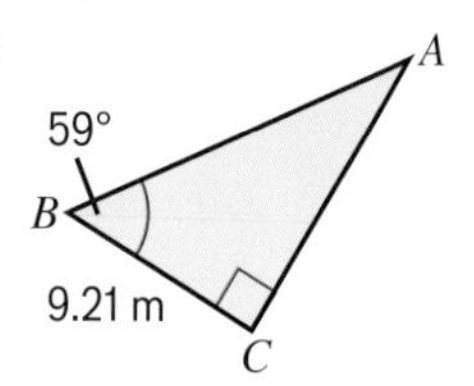

b
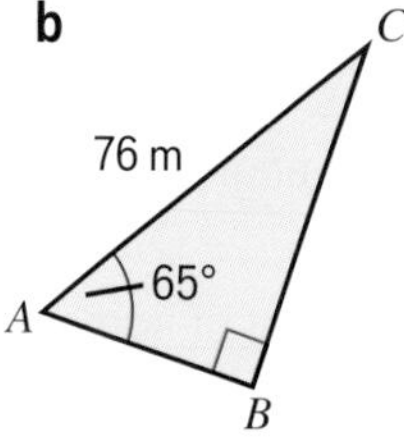

c **d**
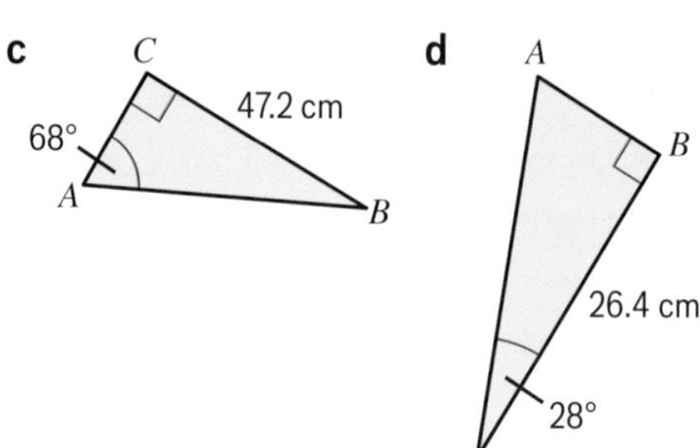

5 Calculate the size of angle x in each of these triangles.
Give your answers correct to the nearest 0.1°.

a
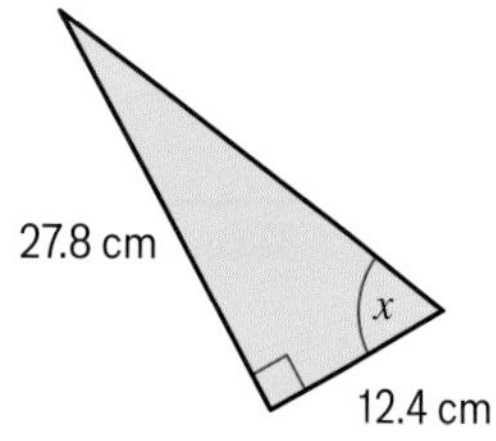

b
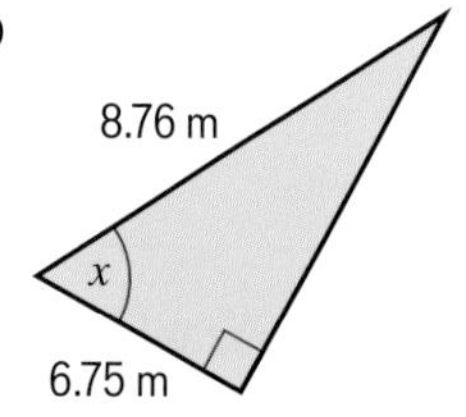

c
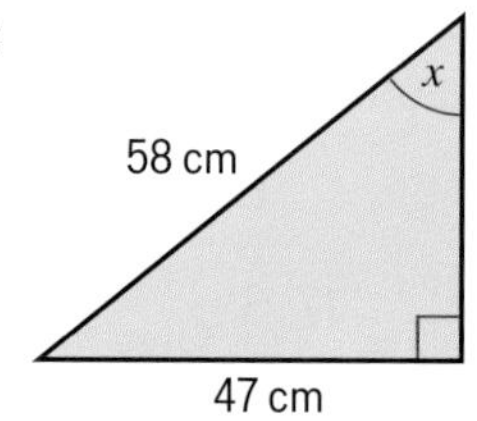

6 **Reflect** Some people use the 'word' sohcahtoa (pronounced 'soak a toe uh') to remember the sine, cosine and tangent ratios.

$\sin\theta = \frac{\text{opp}}{\text{hyp}}$ $\cos\theta = \frac{\text{adj}}{\text{hyp}}$ $\tan\theta = \frac{\text{opp}}{\text{adj}}$

Others use
'Some Old Horse Caught Another Horse Taking Oats Away' or
'Some Of Her Children Are Having Trouble Over Algebra'.
Which do you prefer? Do you have one of your own?

Exam-style question

7 PQR is a right-angled triangle.

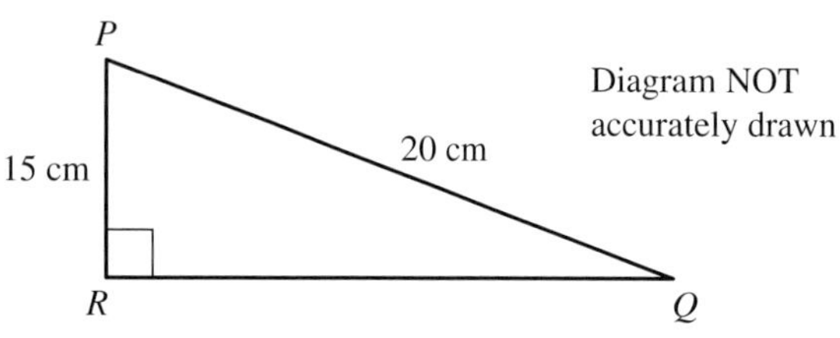

a Work out the size of angle RPQ.
Give your answer correct to 1 decimal place. **(2 marks)**

b The length of PQ is reduced by 4 cm.
The length of PR is reduced by 3 cm.
Angle PRQ is still 90°.
Will the value of $\cos RPQ$ increase, decrease or stay the same?
You must give a reason for your answer. **(1 mark)**

Q7 hint

Simplify the cosine ratio $\frac{\square}{\square}$ for the triangle in part **a** and the triangle in part **b**.

8 **Problem-solving** The diagram shows a zip wire going between two vertical posts..
The zip wire makes an angle of 4° with the horizontal.
How long is the zip wire?

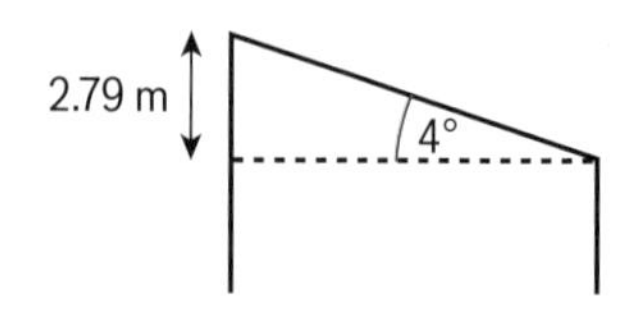

9 **Problem-solving** Rectangle $ABCD$ is 14.1 cm wide.
Angle BAC is 36°.
Work out the length of the diagonal AC.

10 ABC is a right-angled triangle.

a Copy and complete these ratios.

i $\tan 45° = \square$ **ii** $\sin 45° = \frac{\square}{\sqrt{\square}}$

b Write the ratio cos 45°.

c **Reflect** What do you notice about the ratios sin 45° and cos 45°?

11 PQR is a right-angled triangle.

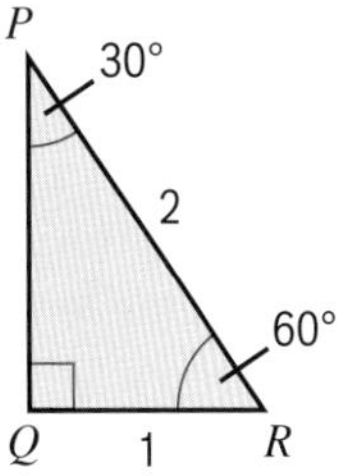

a Use Pythagoras' theorem to show that $PQ = \sqrt{3}$.

b Write these ratios.

i sin 30° **ii** sin 60° **iii** cos 30° **iv** cos 60°

c **Reflect** What do you notice about the ratios

i sin 30° and cos 60° **ii** sin 60° and cos 30°?

12 **Reasoning** Use your answers to **Q10** and **Q11** to find the lengths of the sides marked with letters.

a

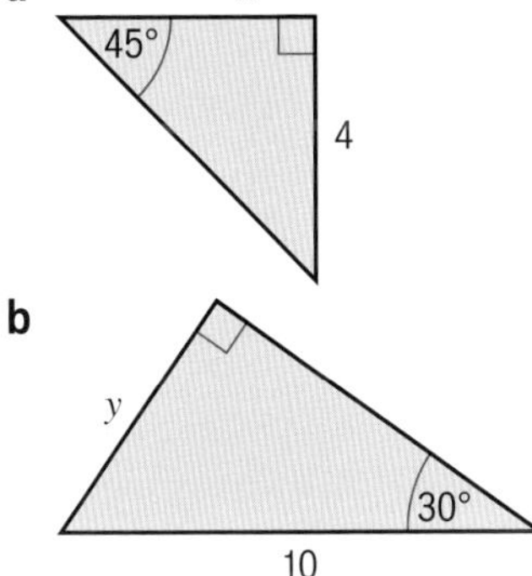

b

c

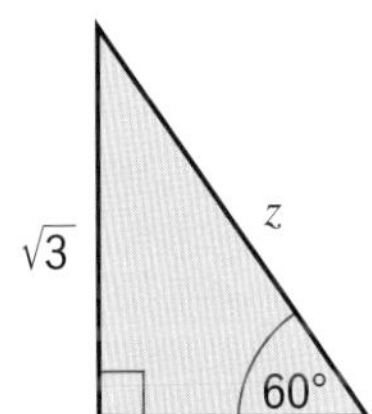

Q12a hint

From **Q10a i**, $\tan 45° = \square$
In this question, $\tan 45° = \frac{\square}{\square}$
Use this information to work out x.

Key point

You need to know these exact values.

$\sin 0° = 0$	$\sin 30° = \frac{1}{2}$	$\sin 45° = \frac{1}{\sqrt{2}}$	$\sin 60° = \frac{\sqrt{3}}{2}$	$\sin 90° = 1$
$\cos 0° = 1$	$\cos 30° = \frac{\sqrt{3}}{2}$	$\cos 45° = \frac{1}{\sqrt{2}}$	$\cos 60° = \frac{1}{2}$	$\cos 90° = 0$
$\tan 0° = 0$	$\tan 30° = \frac{1}{\sqrt{3}}$	$\tan 45° = 1$	$\tan 60° = \sqrt{3}$	

13 **Reasoning**
Work out all the angles in triangle ABC.

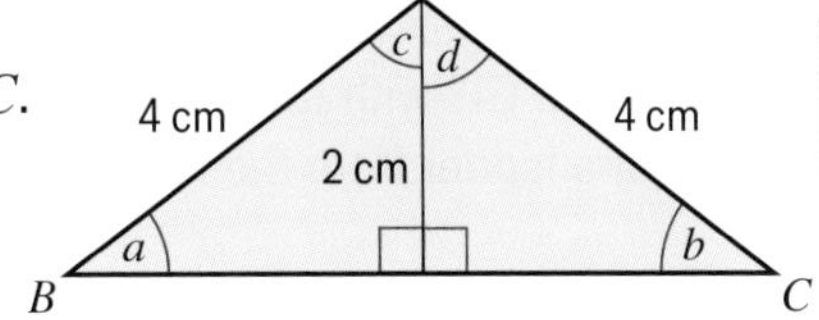

Q13 hint

Simplify your ratios.

12 Check up

ActiveLearn Homework 

Pythagoras' theorem

1 Calculate the value of x in each of these right-angled triangles.

a

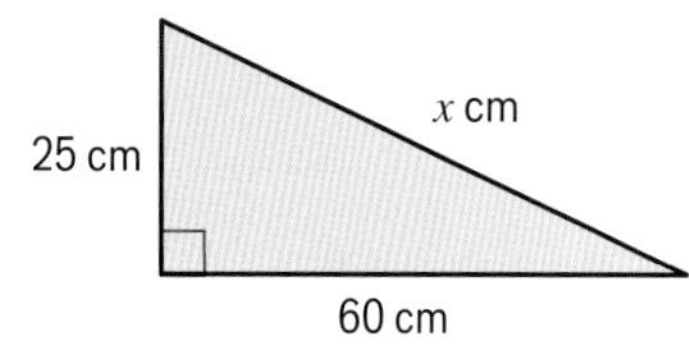

b

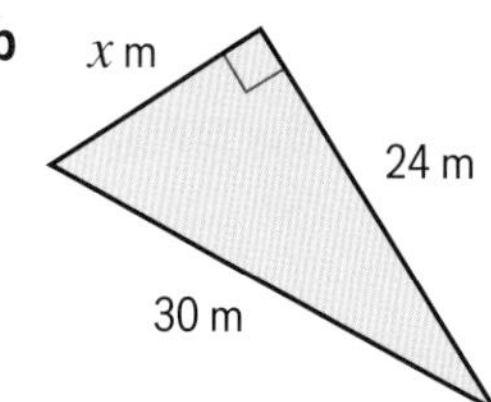

2 The points $A(-3, 2)$ and $B(4, -6)$ are plotted on a centimetre square grid. Work out the length of AB.

3 Tom thinks that this triangle is right-angled.

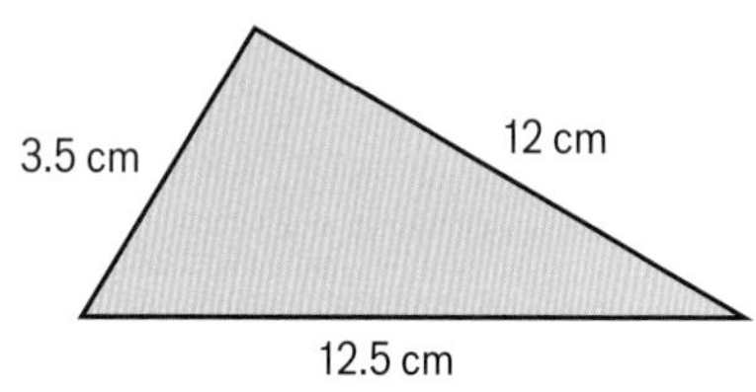

Is Tom correct? Explain your answer.

Finding lengths using trigonometry

4 **a** Use the sine ratio to find x.

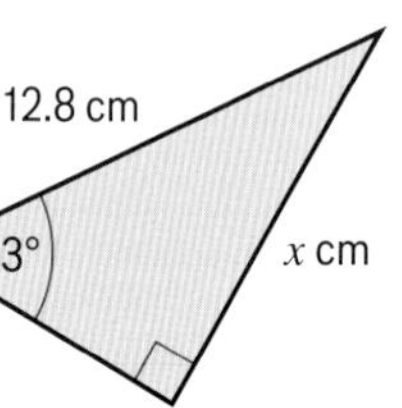

b Use the tangent ratio to find x.

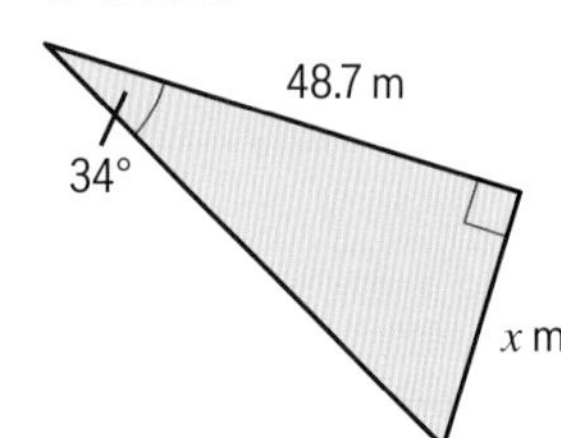

c Use the cosine ratio to find x.

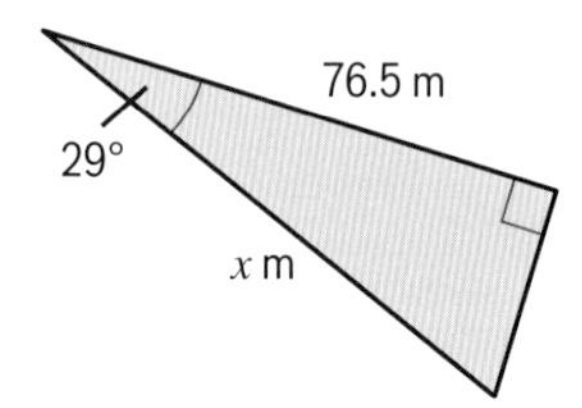

5 A kite flies on the end of a 5 m string.
The string is tied to a stone on the ground.
The string makes an angle of 50° with the horizontal ground.
What is the vertical height of the kite above the ground?

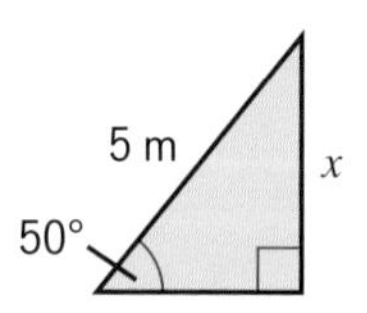

6 A ladder of length 4.5 m leans against a vertical wall.
The ladder makes an angle of 61° with the horizontal ground.
How far is the bottom of the ladder from the base of the wall?

Finding angles using trigonometry

7 **a** Use the sine ratio to find θ.

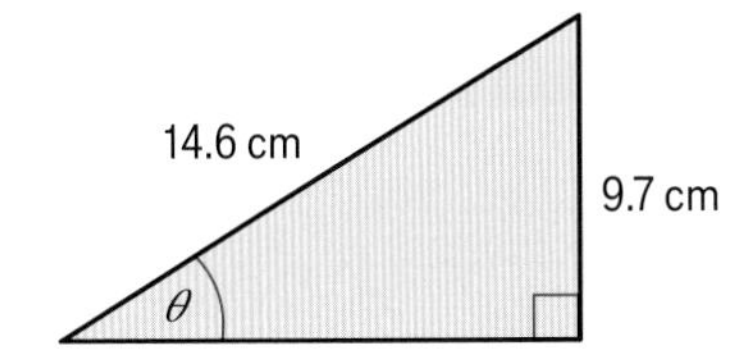

b Use the cosine ratio to find θ.

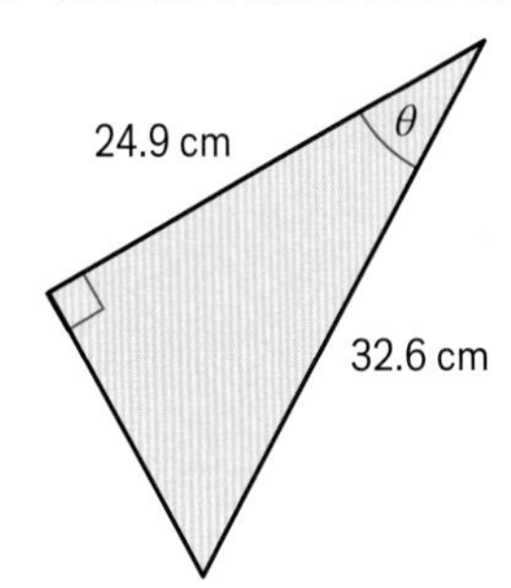

8 Calculate the size of angle x in this triangle.
Give your answer correct to 1 decimal place.

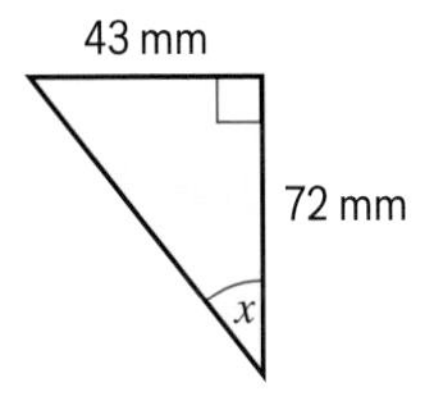

9 Write the exact value of

a $\tan 45°$

b $\sin 30°$

c $\cos 30°$

d $\sin 0°$

10 **Reflect** How sure are you of your answers? Were you mostly

Just guessing ☹ Feeling doubtful 😐 Confident ☺

What next? Use your results to decide whether to strengthen or extend your learning.

Challenge

11 Draw a square of side 5 cm.
Draw the diagonal.
Calculate $\tan\theta$.
Explain why $\tan\theta = 1$ for any square.

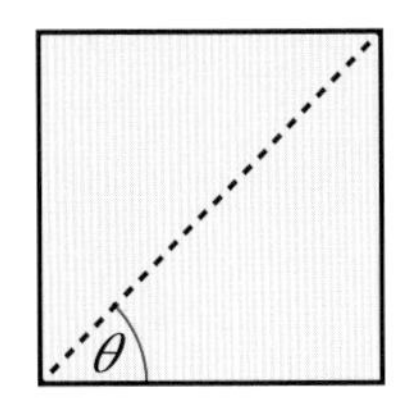

12 Strengthen

ActiveLearn Homework 

Pythagoras' theorem

1 The hypotenuse is the longest side and does not touch the right angle. Write the length of the hypotenuse for each of these triangles.

a
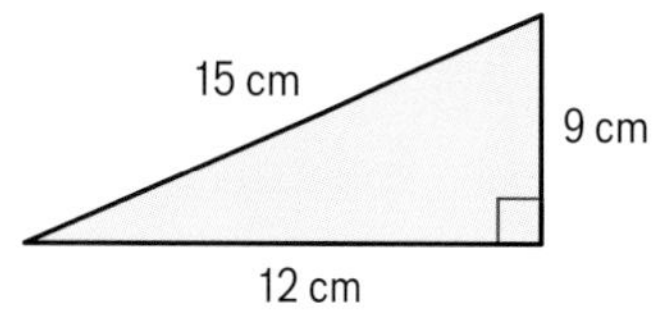

b
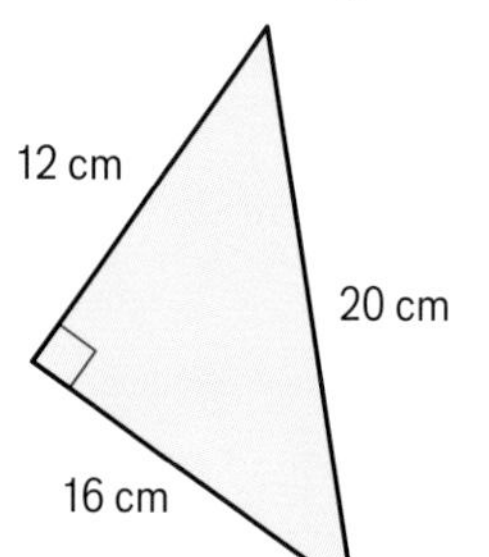

c
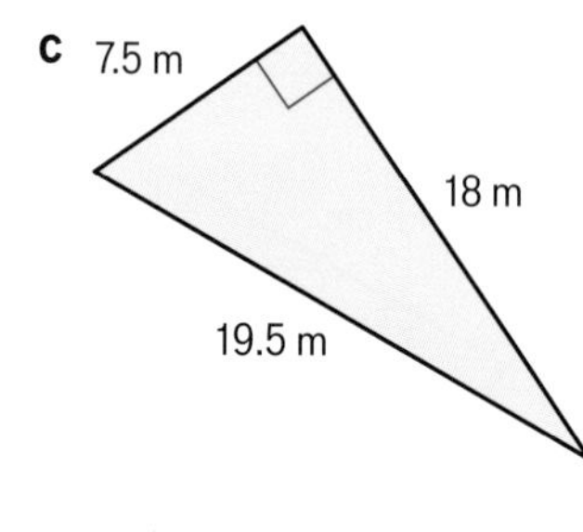

2 A square is drawn on each side of a right-angled triangle. Copy and complete these statements.

a The coloured square is on the __________ of the right-angled triangle.

b The square on the 6 cm side has area
6 cm × ____ = ____ cm^2

c The square on the 5 cm side has area
____ × ____ = ____

d The coloured square has area
____ + ____ = ____

Q2d hint

Add the areas of the smaller squares to get the area of the coloured square.

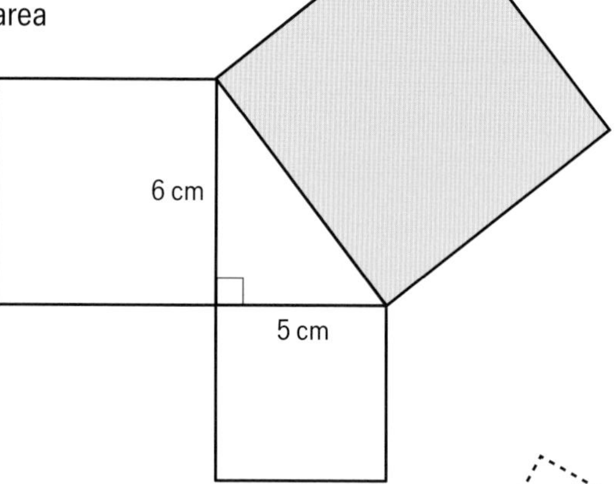

3 The hypotenuse of this right-angled triangle is labelled c. Copy and complete these steps to find the value of c.

$c^2 = 8^2 + \square^2$
$c^2 = \square$
$c = \sqrt{\square}$
$c = \square$ km

Q3 hint

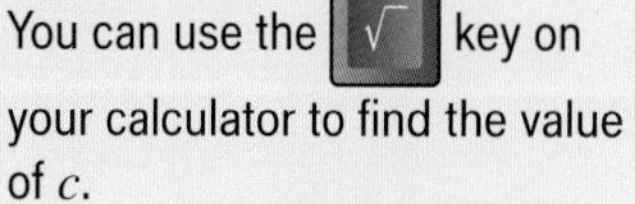

You can use the √ key on your calculator to find the value of c.

8 km, c km, 15 km

4 Use the method in **Q3** to calculate the value of c in each of these right-angled triangles.

a

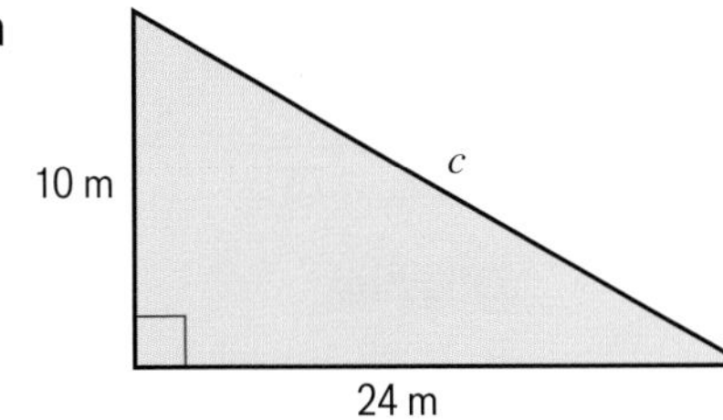

b

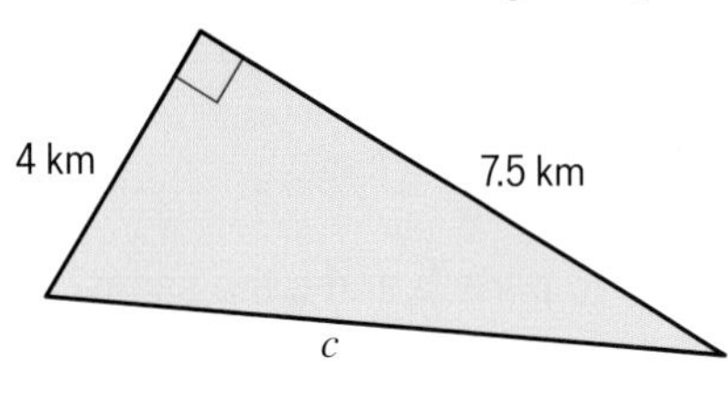

5 The right-angled triangle ABC is drawn on a centimetre square grid.

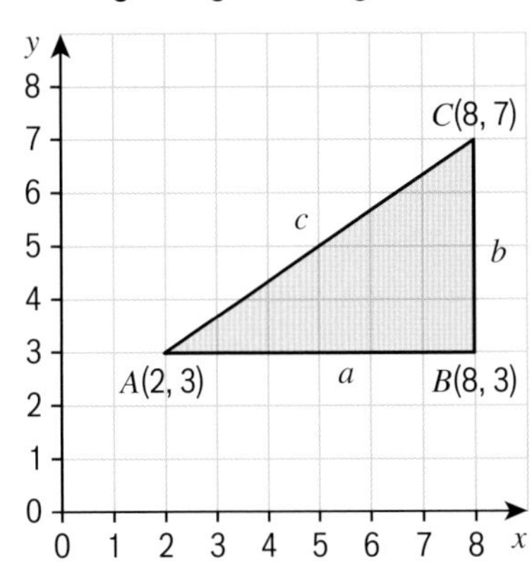

a Work out the lengths of AB (a) and BC (b).

b Use Pythagoras' theorem to find the length of AC (c).

6 The points $A(5, -3)$ and $B(-1, 8)$ are plotted on a centimetre square grid.
Work out the length of AB.

Q6 hint

First draw a sketch. Label the ends of the line and draw a right-angled triangle.

7 Sketch these right-angled triangles.

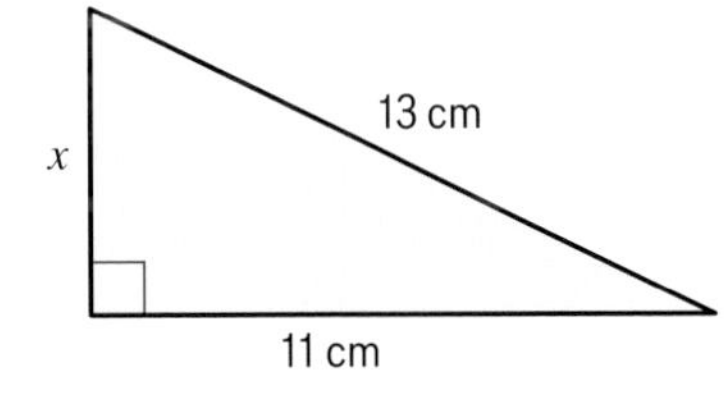

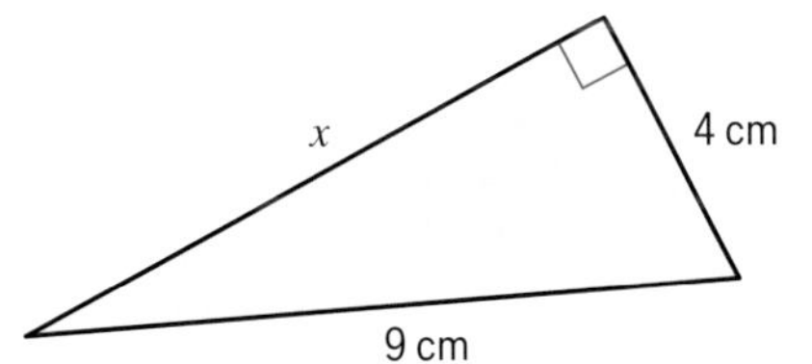

For each triangle

a label the hypotenuse c and the other two sides a and b

b substitute the values of a, b and c into Pythagoras' theorem, $c^2 = a^2 + b^2$

c solve the equation to work out the length of the unknown side.
Give your answers correct to 3 significant figures.

8 Sketch this triangle.

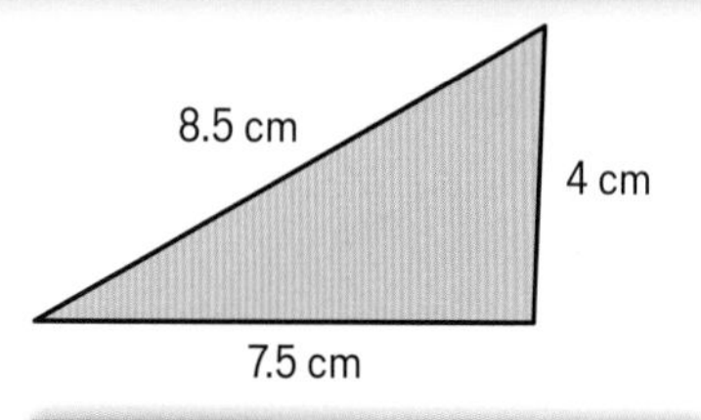

a Label the longest side c and the other two sides a and b.

b Work out c^2.

c Work out $a^2 + b^2$.

d Are your answers to parts **b** and **c** the same?

e Is this a right-angled triangle?

Q8e hint

If your answers to parts **b** and **c** are the same, you can say that the triangle is right-angled. If your answers to parts **b** and **c** are not the same, you can say that the triangle is not right-angled.

Finding lengths using trigonometry

1 Sketch each triangle.

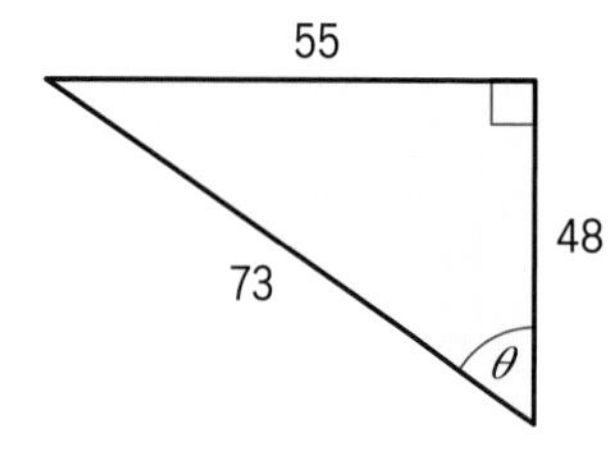

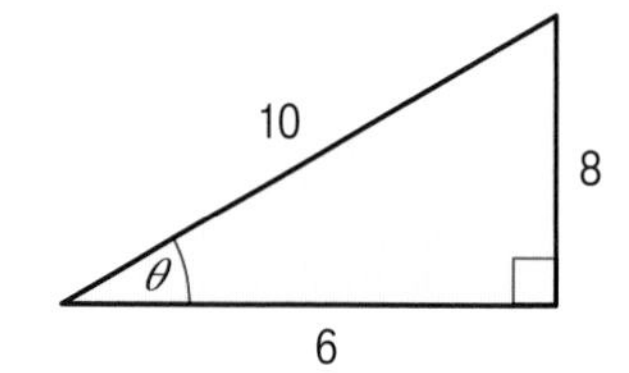

On each triangle

a label 'hyp' on the hypotenuse (the longest side)

b label 'adj' on the adjacent side (this is the side that touches the angle θ but is not the hypotenuse)

c label 'opp' on the opposite side (this is the side that does not touch the angle θ).

d write as fractions

i $\sin\theta$

ii $\cos\theta$

iii $\tan\theta$

Q1d hint

For the trigonometric ratios use

soh **cah** **toa**

$\sin\theta = \frac{\text{opp}}{\text{hyp}}$ $\cos\theta = \frac{\text{adj}}{\text{hyp}}$ $\tan\theta = \frac{\text{opp}}{\text{adj}}$

2 Sketch each triangle.

a

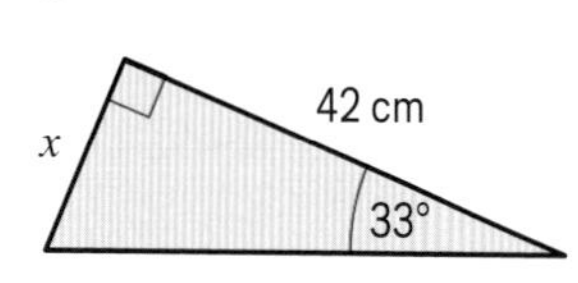

b

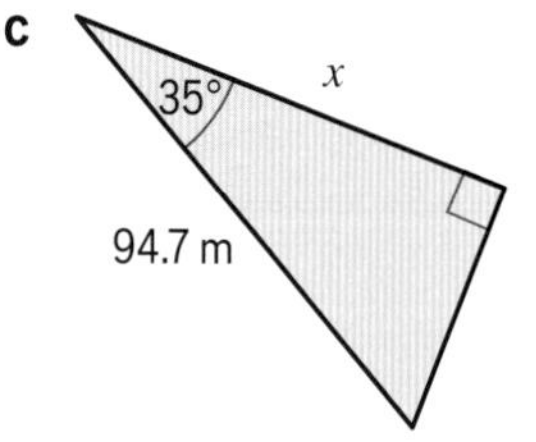

c

35°
x
94.7 m

For each triangle

i copy and complete $\theta = \square°$

ii label any length that is given (as a number or x) as 'hyp', 'adj' or 'opp'

iii write down the trigonometric ratio (sin, cos or tan) that includes the labels you used in part **b**

iv write an equation

$$\_\_\_\theta = \frac{\square}{\square}$$

v rearrange your equation to make x the subject

vi use your calculator to work out x correct to 3 significant figures

Q2v hint

$x = \ldots$

3 Follow the steps in **Q2** to work out the length of AC in each of these triangles.
State your chosen degree of accuracy after your answer.

a

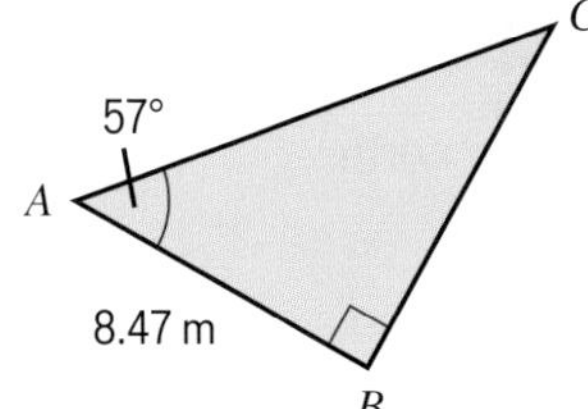

b

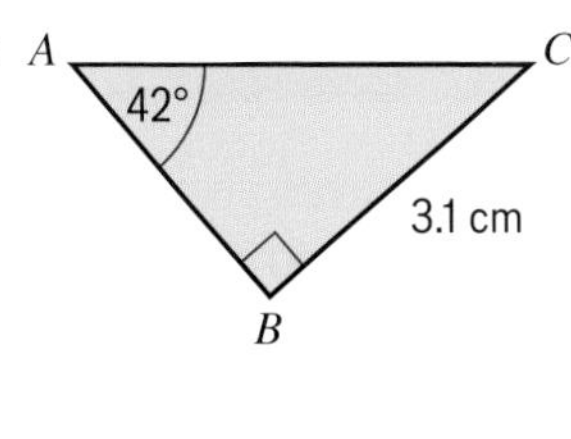

Q3 hint

When rearranging your equation, first multiply both sides by AC.

Finding angles using trigonometry

1 Use the inverse functions on your calculator to copy and complete.
Give each value of θ correct to 0.1°.

a $\sin\theta = \frac{1}{5}$
$\sin^{-1}\left(\frac{1}{5}\right) = \square$
So $\theta = \square$

b $\cos\theta = 0.2$
$\cos^{-1}0.2 = \square$
So $\theta = \square$

c $\tan\theta = \frac{2}{3}$
$\tan^{-1}\left(\frac{2}{3}\right) = \square$
So $\theta = \square$

2 Sketch these triangles.

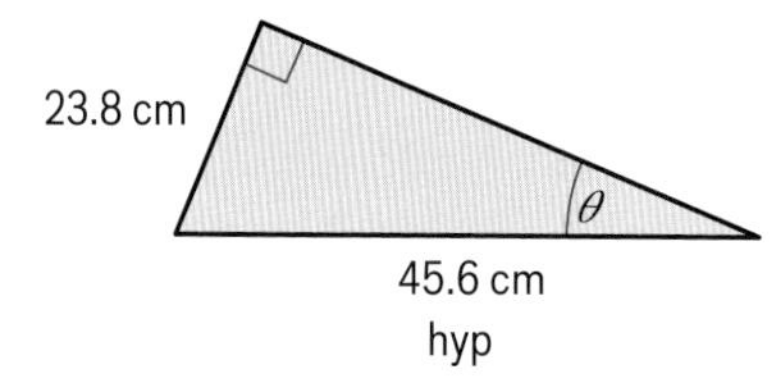

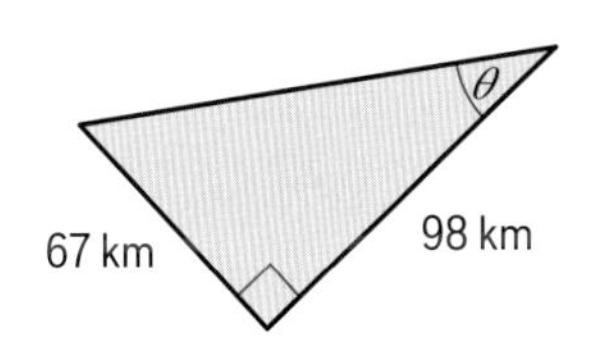

For each triangle

a label any length that is given as 'hyp', 'opp' or 'adj'

b write down the trigonometric ratio (sin, cos or tan) that includes the labels you used in part **a**

c substitute the given lengths into the trigonometric ratio

$\_\_\_\theta = \frac{\square}{\square}$

d use the relevant inverse ratio ($\sin^{-1}$, $\cos^{-1}$ or $\tan^{-1}$) on your calculator to work out the value of θ correct to 0.1°

Q2b hint

Which trigonometric ratio uses the labels 'hyp' and 'opp'? ... 'adj' and 'opp'?

3 Use your calculator to find these values as fractions or surds ($\sqrt{\square}$).
Copy and complete the table.

Angle	0°	30°	45°	60°	90°
sin					
cos					
tan					

Q3 hint

Use the $S\Leftrightarrow D$ button on your calculator to switch your answer between decimal form and surd or fraction form.

12 Extend

1 Problem-solving An aeroplane leaves an airport and flies 50 km due north.
It then flies due east a further 120 km to reach its destination.
Calculate the direct distance from the airport to the destination.
Give your answer correct to 3 significant figures.

Exam-style question

2 A ladder leans against a vertical wall.
The ladder stands on horizontal ground.
The length of the ladder is 7 m.
The bottom of the ladder is 2.5 m from the bottom of the wall.
A ladder is safe to use when the angle between the ladder and the horizontal ground is less than 75°.
Is the ladder safe to use?
You must show all your working. **(3 marks)**

Exam tip

After you have shown all your calculation, remember to answer the question that was asked. Write 'Yes, the ladder is safe to use' or 'No, the ladder is not safe to use'.

3 A rectangle's height is 2.9 cm.
This is 1.6 cm shorter than its diagonal.
Work out the area of the rectangle.

4 Problem-solving Calculate the length of AC.
Give your answer correct to 1 decimal place.

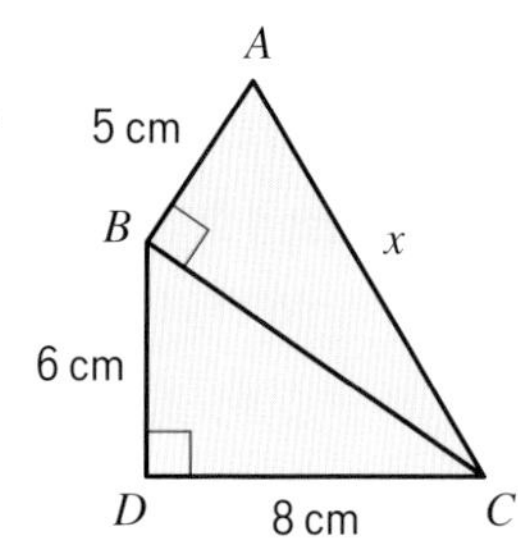

Exam-style question

5 Use your calculator to work out

$$\sqrt{\frac{\sin 15° + \sin 65°}{\cos 15° - \cos 65°}}$$

a Write down all the figures on your calculator display. **(2 marks)**

b Write your answer to part **a** correct to 2 decimal places. **(1 mark)**

Exam tip

When working with $\frac{\text{calculation 1}}{\text{calculation 2}}$ work out
- the numerator
- the denominator
- numerator ÷ denominator.

6 Reasoning Copy and complete these statements.

a As angle x increases from 0° to 90°, $\sin x$ __________ .

b As angle x increases from 0° to 90, $\cos x$ __________ .

Q6 hint

Write down the values of $\sin x$ when $x = 0, ..., 90°$.
Write a statement about how the value of $\sin x$ changes.

7 Reasoning

a Show that angle ABC is a right angle.

b Calculate the size of angle BCA.

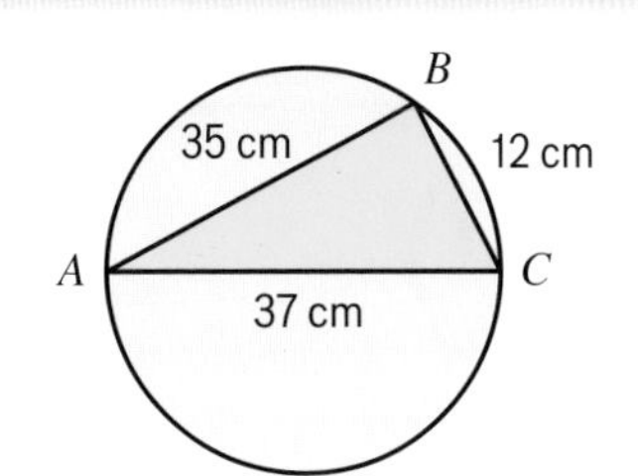

8 Problem-solving / Reasoning

The diagram shows a triangular prism.
Calculate the surface area of the prism.

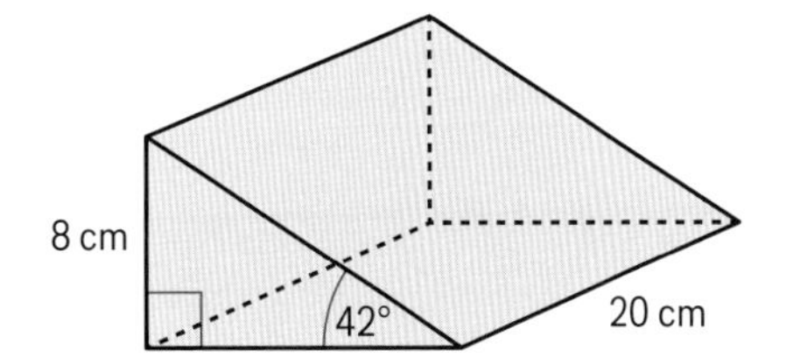

Exam-style question

9 This rectangular frame is made from 5 straight steel rods.
The mass of the rods is 0.25 kg per metre.
Work out the total mass of the steel rods used in the frame. **(5 marks)**

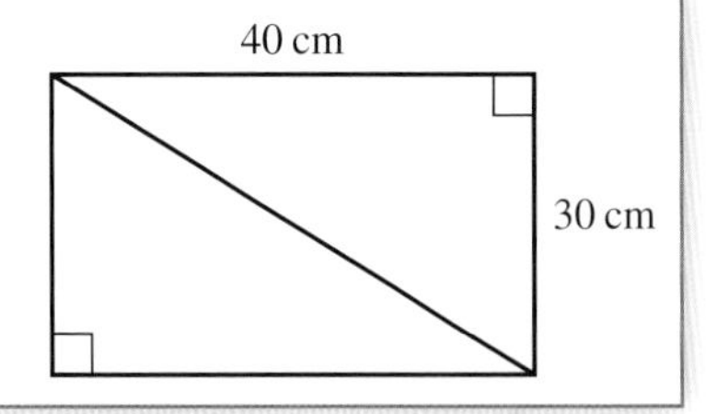

Exam-style question

10 $PQRS$ is a trapezium.

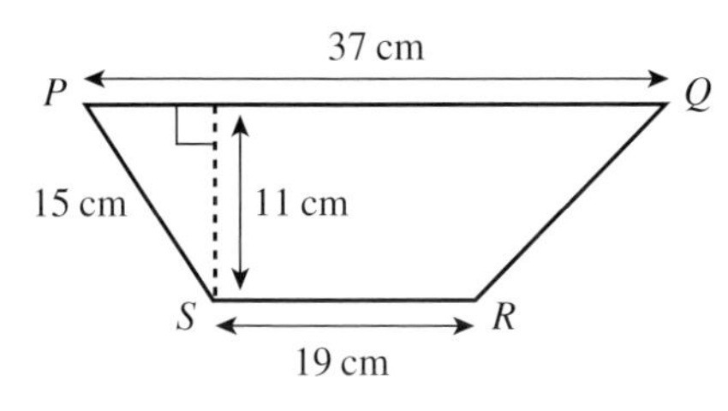

Work out the size of angle PQR.
Give your answer correct to 1 decimal place. **(5 marks)**

Exam tip

Look for right-angled triangles in shapes. Can you use Pythagoras' theorem or trigonometry to find useful unknown lengths or angles?

Q10 hint

Sketch the trapezium. Draw in another right-angled triangle that includes angle PQR.

11 An isosceles triangle has a base of 12 cm, and two other sides, each of length 9 cm.
What are the angles in the isosceles triangle?

12 An isosceles trapezium has an area of 70 cm^2 and parallel sides with lengths 6.2 cm and 13.8 cm.
What is its perimeter?

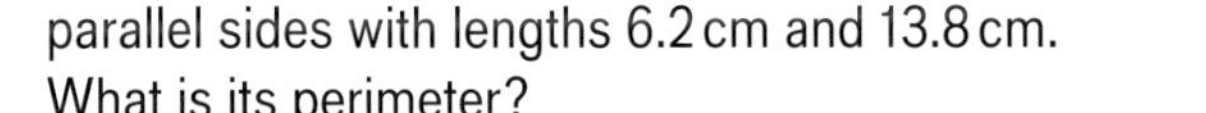

Q12 hint

Area of a trapezium $= \frac{1}{2}(a+b)h$
Divide the trapezium into a rectangle and two right-angled triangles to help you.

13 A right-angled triangle has a height of 3 cm.
Its base is 80% of the length of its hypotenuse.
Work out the length of its base.

Q13 hint

Write 80% as a decimal.
Sketch the triangle. Label the hypotenuse x and label the base in terms of x.

12 Test ready

Summary of key points

To revise for the test:

- Read each key point, find a question on it in the mastery lesson, and check you can work out the answer.
- If you cannot, try some other questions from the mastery lesson or ask for help.

Key points

1 In a right-angled triangle, the longest side is opposite the right angle. It is called the **hypotenuse**. → 12.1

2 **Pythagoras' theorem** shows the relationship between the lengths of the three sides of a right-angled triangle.
$c^2 = a^2 + b^2$ → 12.1

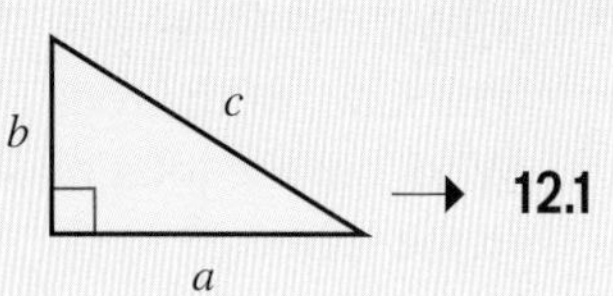

3 You can use Pythagoras' theorem to work out the length of a shorter side in a right-angled triangle. → 12.2

4 A triangle with sides a, b and c, where c is the longest side, is right-angled only if $c^2 = a^2 + b^2$. → 12.2

5 The side opposite the right angle is called the **hypotenuse**.
The side opposite the angle θ is called the **opposite**.
The side next to the angle θ is called the **adjacent**.
θ is the Greek letter 'theta'.
It is often used in maths to mark an unknown angle. → 12.3

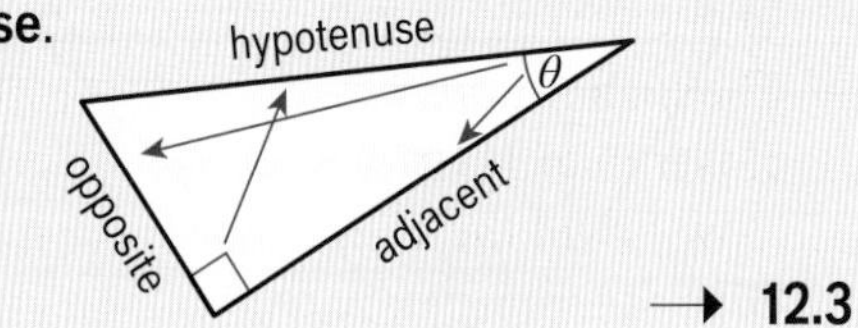

6 In a right-angled triangle, the **sine** of an angle is the ratio of the **opp**osite side to the **hyp**otenuse.
The sine of angle θ is written as **sin** $\boldsymbol{\theta}$.

$$\sin\theta = \frac{\text{opp}}{\text{hyp}}$$
→ 12.3

7 In a right-angled triangle, the **cosine** of an angle is the ratio of the **adj**acent side to the **hyp**otenuse.
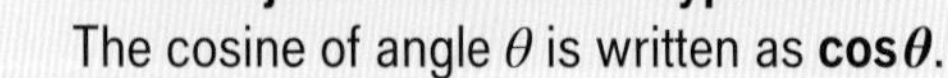
The cosine of angle θ is written as **cos** $\boldsymbol{\theta}$.

$$\cos\theta = \frac{\text{adj}}{\text{hyp}}$$

→ 12.5

8 In a right-angled triangle, the **tangent** of an angle is the ratio of the **opp**osite side to the **adj**acent side.
The tangent of angle θ is written as **tan** $\boldsymbol{\theta}$.

$$\tan\theta = \frac{\text{opp}}{\text{adj}}$$
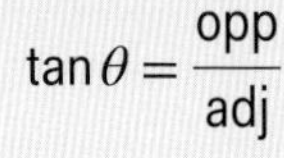

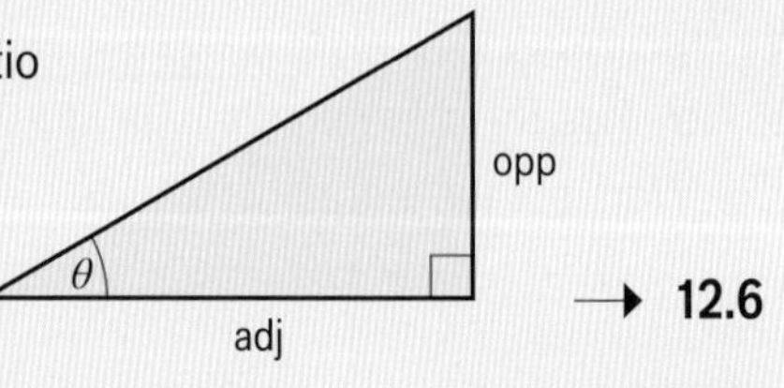

→ 12.6

9 You can use $\sin^{-1}$, $\cos^{-1}$ or $\tan^{-1}$ on a calculator to find the size of an angle. → 12.4, 12.5, 12.6

Key points

10 The angle of **elevation** is the angle measured upwards from the horizontal. → **12.6**

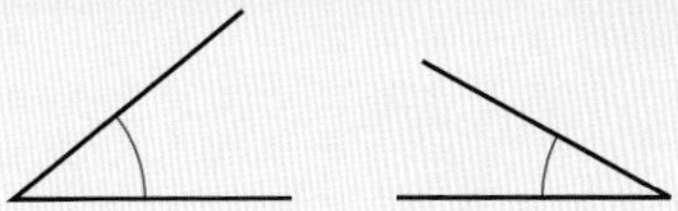

11 The angle of **depression** is the angle measured downwards from the horizontal. → **12.6**

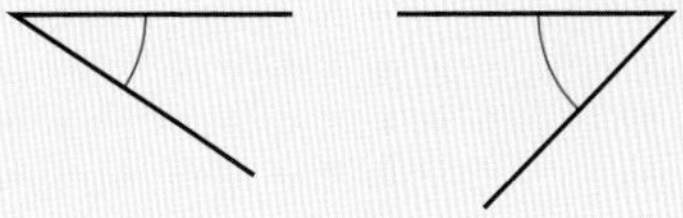

12 You need to know these ratios and be able to choose the one you need to solve a problem.

$\sin(\text{angle}) = \frac{\text{opp}}{\text{hyp}}$ $\quad\cos(\text{angle}) = \frac{\text{adj}}{\text{hyp}}$ $\quad\tan(\text{angle}) = \frac{\text{opp}}{\text{adj}}$ → **12.7**

13 You need to know these exact values.

$\sin 0° = 0$ $\quad\sin 30° = \frac{1}{2}$ $\quad\sin 45° = \frac{1}{\sqrt{2}}$ $\quad\sin 60° = \frac{\sqrt{3}}{2}$ $\quad\sin 90° = 1$

$\cos 0° = 1$ $\quad\cos 30° = \frac{\sqrt{3}}{2}$ $\quad\cos 45° = \frac{1}{\sqrt{2}}$ $\quad\cos 60° = \frac{1}{2}$ $\quad\cos 90° = 0$

$\tan 0° = 0$ $\quad\tan 30° = \frac{1}{\sqrt{3}}$ $\quad\tan 45° = 1$ $\quad\tan 60° = \sqrt{3}$ → **12.7**

Sample student answers

Exam-style question

1 A ladder is 4.8 m long.
The ladder is placed on horizontal ground, resting against a vertical wall.
The instructions for using the ladder say that the bottom of the ladder must *not* be closer than 1.2 m from the bottom of the wall.
How far up the wall can the ladder reach?
Give your answer correct to 1 decimal place. **(3 marks)**

Student A

$c = 4.8\text{ m}$ $\quad b = 1.2\text{ m}$

$c^2 = a^2 + b^2$

$4.8^2 = a^2 + 1.2^2$

$23.04 = a^2 + 1.44$

$23.04 - 1.44 = a^2$

$a^2 = 21.6$

$a = 4.6\text{ m}$

The ladder can reach to 4.6 m up the wall.

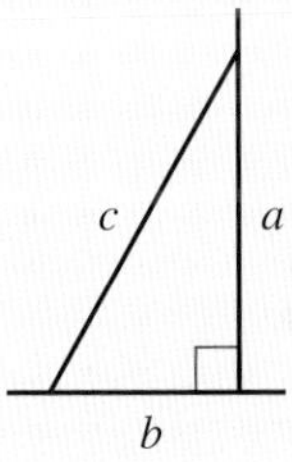

Student B

$c^2 = a^2 + b^2$

$c^2 = 4.8^2 + 1.2^2$

$c^2 = 23.04 + 1.44$

$c^2 = 24.48$

$c = \sqrt{24.48}$

$c = 4.9\text{ m}$

The ladder reaches 4.9 m.

a Which student has the correct answer?

b How has the sketch and the labelling helped Student A?

Exam-style question

2 ABC is a right-angled triangle.

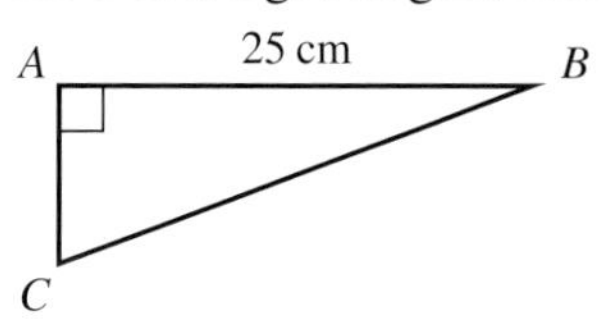

$AB = 25\,\text{cm}$
Angle $A = 90°$
Size of angle B : size of angle $C = 1 : 4$
Work out the length of BC.
Give your answer correct to 3 significant figures. **(4 marks)**

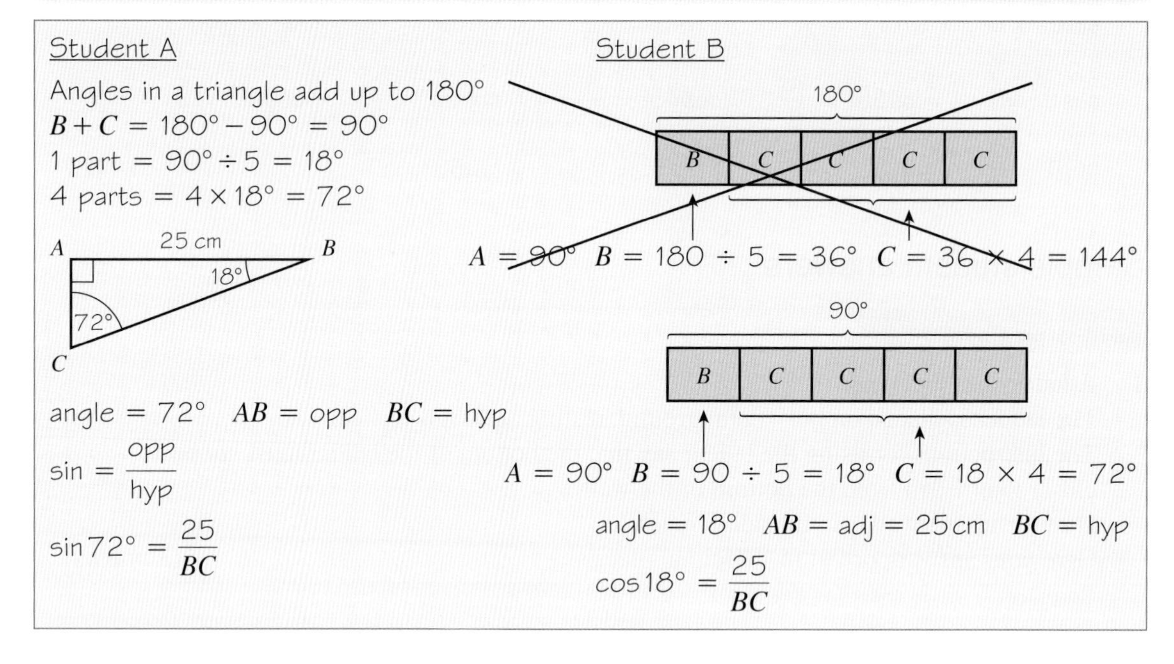

a Student B first of all works out
$A = 90°$ $B = 36°$ $C = 144°$
Then they cross it out.
Explain how you think Student B knows this is incorrect.

Only the first part of each student's working is shown.
Student A uses the sine ratio. Student B uses the cosine ratio.

b Show that they will both get the same answer.

c Explain why both students' workings (one using the sine ratio and the other using the cosine ratio) are correct.

12 Unit test

1 The diagram shows a right-angled triangle.
Calculate the value of x.
Give your answer correct to 3 significant figures. **(2 marks)**

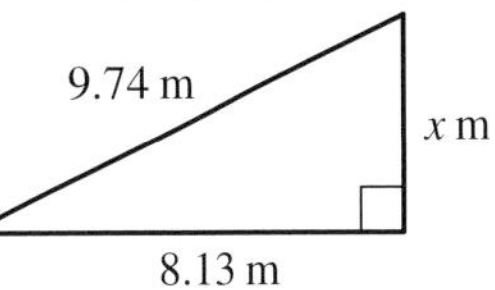

2 $PQRS$ is a square with side 7 cm.
Work out the length of the diagonal QS in $PQRS$.
Give your answer correct to 2 decimal places. **(3 marks)**

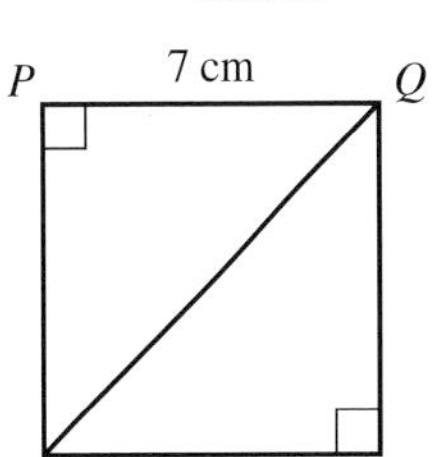

3 Karen thinks that triangle LMN is a right-angled triangle.

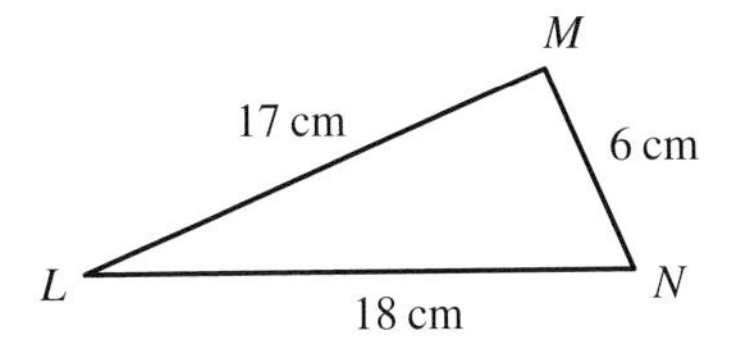

Is Karen correct? Explain your answer. **(3 marks)**

4 Use the tangent ratio to find
a angle x **(1 mark)**
b length y **(2 marks)**

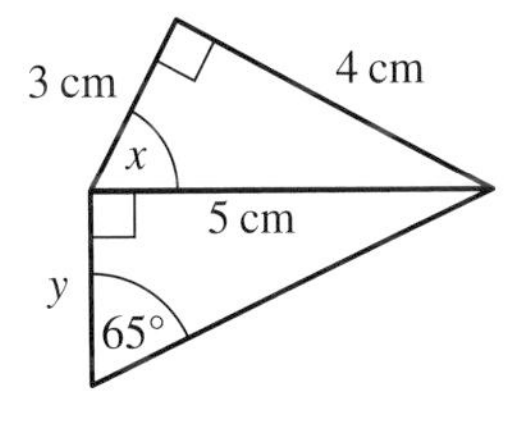

5 Work out the length XY in each triangle.

a **(3 marks)**

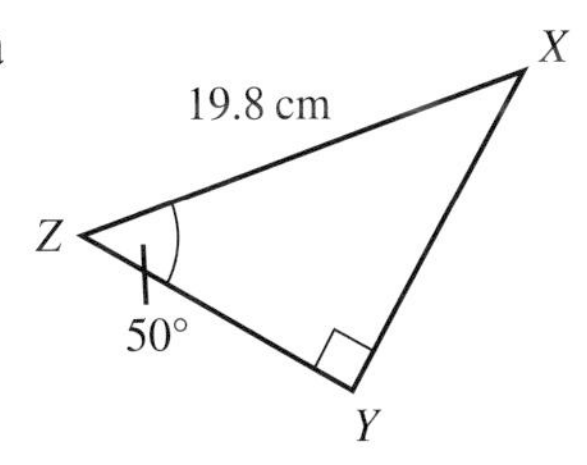

b **(3 marks)**

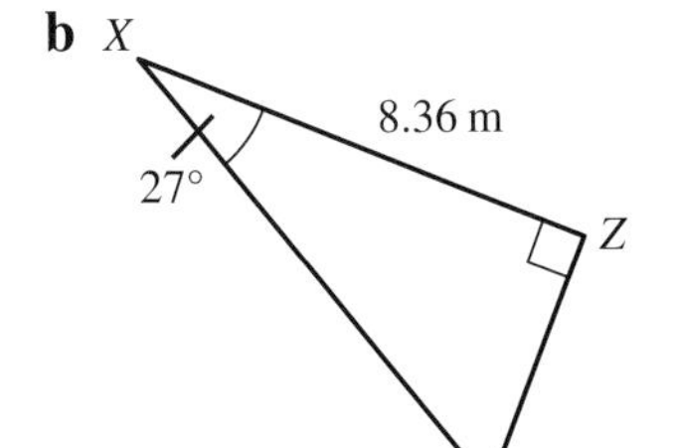

6 Calculate the size of angle θ in this triangle.

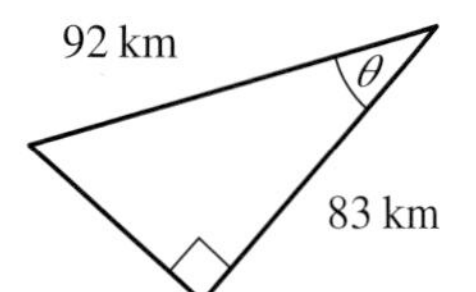

(4 marks)

7 The diagram shows the positions A, B and C of Alice, Bob and Charlie at the side of a canal.
Charlie is directly opposite Bob.
Alice and Bob are 8 m apart on the same side of the canal.
The direction of Charlie from Alice makes an angle of 41° with the side of the canal.
Calculate the width of the canal.
Give your answer correct to 3 significant figures.

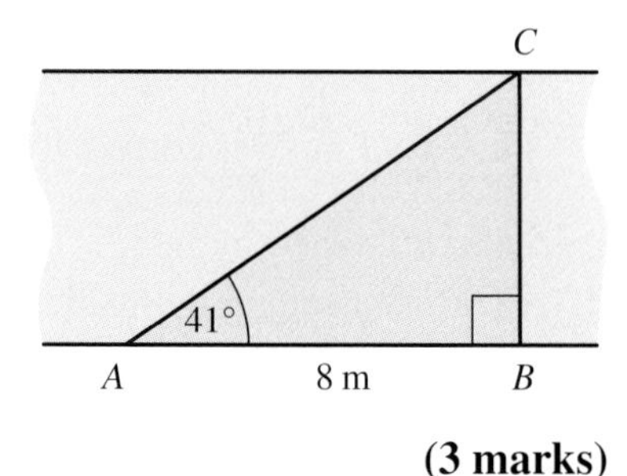

(3 marks)

8 A ramp of length 2 m is used to load the back of a van.
One end of the ramp rests on the horizontal ground.
The other end rests on the back of the van, at a height of 45 cm.
What angle does the ramp make with the horizontal ground? **(4 marks)**

9 Work out the height of this isosceles triangle.

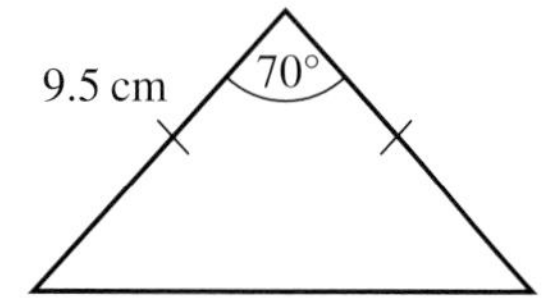

(4 marks)

10 Without using a calculator, work out $\tan 45° - \sin 30° + \cos 90°$. **(3 marks)**

(TOTAL: 35 marks)

11 Challenge This octagon is made from isosceles triangles.
Calculate its area.

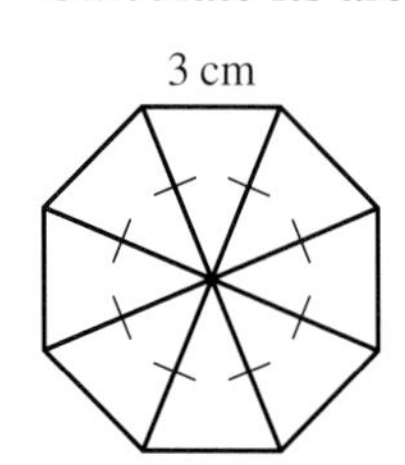

Q11 hint

Work out one angle in the centre of the octagon. Then work out the height of a triangle.

12 Reflect You are asked a question about this right-angled triangle.
Which of these would you use when asked to find length AB

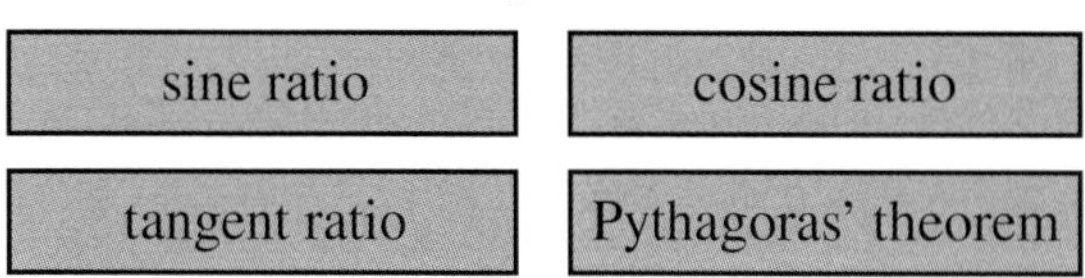

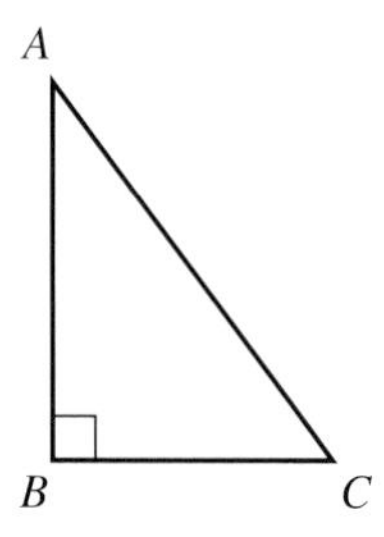

when given

a length AC and angle BAC

b length AC and angle ACB

c lengths AC and BC

d length BC and angle BAC

e length AC only, but also told that ABC is an isosceles right-angled triangle?

13 Probability

13.1 Calculating probability

- Calculate probabilities from equally likely events.
- Calculate probabilities of mutually exclusive and exhaustive events.
- Solve probability problems.

Prior knowledge

ActiveLearn Homework

Warm up

1 **Fluency** **a** What is the probability of this fair spinner landing on green as

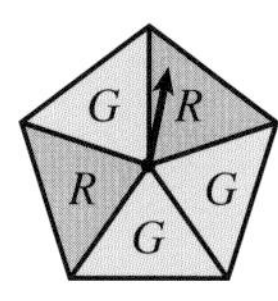

i a fraction **ii** a decimal **iii** a percentage?

b Are the events 'red' and 'green' equally likely?

2 Work out

a $\frac{3}{13} + \frac{10}{13}$ **b** $1 - \frac{1}{4}$ **c** $1 - \frac{3}{7}$ **d** $1 - 0.7$

Key point

$$\text{Probability} = \frac{\text{number of successful outcomes}}{\text{total number of possible outcomes}}$$

3 For this regular 7-sided fair spinner

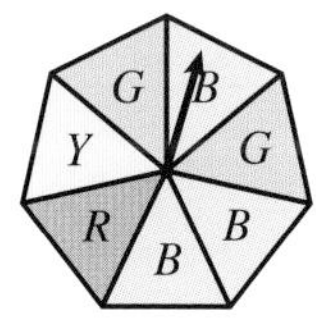

a write down the probability of the spinner landing on

i blue **ii** blue or green **iii** purple

b Which colour is twice as likely as red?

Exam-style question

4 An ordinary fair dice is thrown once.

a Copy the probability scale below. Mark with a cross (×) the probability that the dice lands on an even number. **(1 mark)**

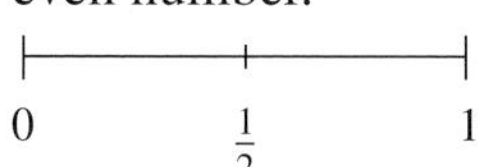

b Write down the probability that the dice lands on a number less than 3. **(1 mark)**

Exam tip

Write your answer to part **b** as a fraction. You do not need simplify it.

5 The letters from the word *PROBABILITY* are written on cards and placed in a bag.
Ella picks one card at random from the bag.
Work out

a $P(R)$
b $P(B)$
c $P(O \text{ or } A)$

Q5 hint

$P(R)$ means probability of picking R.

Exam-style question

6 There are only 4 red counters, 5 blue counters and 2 green counters in a bag.
One counter is taken at random from the bag.
Write down the probability that this counter is blue. **(2 marks)**

7 A bag contains 4 purple balls, 3 yellow balls and 5 green balls.
Steven takes one ball from the bag at random.
Work out

a P(green)
b P(purple)
c P(yellow)
d P(not yellow)
e P(yellow) + P(not yellow)

Key point

If the probability of event A happening is $P(A)$, then the probability of it not happening is $1 - P(A)$.

8 A bag contains coloured counters. The probability of picking at random a blue counter is $\frac{4}{7}$.
What is the probability of picking a counter that is *not* blue?

Exam-style question

9 The probability that a new cooker has a fault is 0.025.
What is the probability that a new cooker does not have a fault? **(1 mark)**

Exam tip

Calculate the answer carefully, and check your calculation.

Key point

Events are **mutually exclusive** when they cannot happen at the same time.

10 **Reasoning** A fair six-sided dice is rolled.

Are the following pairs of events mutually exclusive?

a Rolling an even number and rolling an odd number
b Rolling an even number and rolling a prime number
c Rolling a factor of six and rolling an odd number

Q10a hint

Can you roll a number that is both odd and even at the same time?

11 Are the pairs of events in **Q10** exhaustive?

Key point

The probabilities of an **exhaustive** set of **mutually exclusive** events sum to 1.

Example

The table shows the probabilities of taking different coloured marbles at random from a bag.

Colour	blue	green	silver
Probability	0.3		0.5

One marble is taken from the bag.

Work out the probability that the marble will be green.

$P(\text{green}) = 1 - 0.3 - 0.5$ — The probabilities add up to 1.

$= 0.2$

Exam-style question

12 Here is a four-sided spinner.

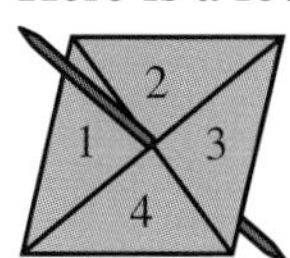

The table shows the probabilities that when the spinner is spun it will land on 1, on 2 or on 4.

Number	1	2	3	4
Probability	0.1	0.3		0.2

The spinner is spun once.

a Work out the probability that the spinner will land on 3. **(1 mark)**

b Which number is the spinner least likely to land on? **(1 mark)**

c Work out the probability that the spinner will land on 1 or 4. **(1 mark)**

Exam tip

The answer to part **b** is a number not a probability.

13 The probability that Stratworth Town win a football match is $\frac{1}{3}$.
The probability that they draw the match is $\frac{1}{6}$.
Work out the probability that they lose the match.

Key point

Estimate for predicted number of outcomes = probability × number of trials.

14 When you roll a fair dice, the probability that it lands on 1 is $\frac{1}{6}$.

Work out an estimate for the number of times a fair dice will land on 1 in

a 30 rolls

b 60 rolls

c 120 rolls

15 The probability that a new fridge is faulty is 0.02.
A shop sells 500 of these fridges.
Work out an estimate for the number of these fridges that are faulty.

Exam-style question

16 The table shows the probabilities that a biased dice will land on 1, on 2, on 3, on 4 and on 5.

Number on dice	1	2	3	4	5	6
Probability	0.15	0.08	0.17	0.2	0.21	

Lyra rolls the biased dice 300 times.

Work out an estimate for the number of times it will land on 5 or on 6. **(3 marks)**

Exam tip

Use a calculator. This question was on a calculator paper.

17 There are only red counters, blue counters and green counters in a bag. The table shows the probability of taking at random a red counter, a blue counter and a green counter from the bag.

Colour	red	blue	green
Probability	0.2	0.3	0.5

a Are there more blue counters than red counters in the bag? Explain how you know.

There are 8 red counters in the bag.

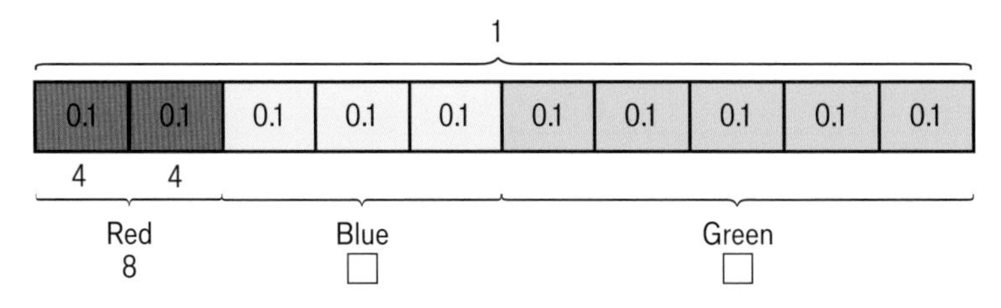

b Use the bar model to work out the number of

i blue counters

ii green counters

c Work out the total number of counters in the bag.

Exam-style question

18 There are only white cubes, green cubes and yellow cubes in a box. The table shows the probability of taking at random a white cube from the box.

Colour	white	green	yellow
Probability	0.6		

The number of green cubes in the box is the same as the number of yellow cubes in the box.

a Copy and complete the table. **(2 marks)**

There are 9 white cubes in the box.

b Work out the total number of cubes in the box. **(2 marks)**

13.2 Two events

- Work out probabilities from sample space diagrams.
- Draw and use sample space diagrams to solve probability problems.

Warm up

1 **Fluency** **a** What are the possible outcomes of flipping a coin?
b A fair coin is flipped. What is P(head)?

2 Tina is buying a sandwich.
She can choose one type of bread and one filling.

Bread	Filling
Roll	Egg
Wrap	Cheese
	Lettuce
	Prawn

a Write down all the possible combinations Tina can choose.
b How many possible combinations are there?

3 In **Q2** Tina picks the bread and filling at random.
Work out
a P(R, L) **b** P(W, E) **c** P(the sandwich filling is prawn)

4 A bag contains 3 colours of counters, in two sizes.

Colour	Size
Red	Large
Blue	Small
Yellow	

A counter is taken from the bag at random.
a Copy and complete this list of possible outcomes: (R, L) (B, L) ...
b How many possible outcomes are there?
c Work out
i P(Y, S) **ii** P(the counter is large) **iii** P(the counter is red)

5 A bag contains one brick of each colour
Green Blue Red Yellow
Molly takes two bricks at random from the bag.
a Write down all the different possible combinations of two bricks.
Note: (red, green) is the same as (green, red).
b Work out the probability that
i Molly takes the red and yellow bricks
ii one of the bricks Molly takes is blue

Key point

A **sample space** diagram shows all the possible outcomes.
You can use it to find a theoretical probability, based on equally likely outcomes.

Example

A fair dice is rolled and a fair coin is flipped.

a Draw a sample space diagram.

b Write the probability of getting a tail on the coin and an odd number on the dice.

c Write the probability of getting a head on the coin or an even number on the dice.

a

	1	2	3	4	5	6
Head	H, 1	H, 2	H, 3	H, 4	H, 5	H, 6
Tail	T, 1	T, 2	T, 3	T, 4	T, 5	T, 6

Draw a two-way table to show the possible outcomes.

b $P(T, \text{odd}) = \frac{3}{12}$

There are 12 equally likely outcomes. Three are tail and an odd number: 'T, 1', 'T, 3' and 'T, 5'

c $\frac{9}{12}$

6 outcomes are 'H, □' and 3 outcomes are 'T, even number'
$3 + 6 = 9$

6 Two fair coins are flipped.

a Copy and complete the two-way table to show the possible outcomes.

	Head	Tail
Head		
Tail		

b How many possible outcomes are there?

c Work out

i P(two heads) **ii** P(one head and one tail)

iii P(at least one head)

d Reflect Show how you can use P(no heads) to work out P(at least one head).

Key point

When there are 3 ways of making the first choice and 3 ways of making the second, there are $3 \times 3 = 9$ ways of choosing two objects.

2nd choice			
3			
2			
1			
	1	2	3

1st choice

7 A coin is flipped and this spinner is spun.

P B G R Y

a How many different possible outcomes do you think there are?

b Copy and complete the two-way table showing the possible outcomes, to check your answer to part **a**.

	Blue	Red	Yellow	Green	Pink
Heads	H, B	H, R			
Tails	T, B				

8 Two fair spinners are spun.

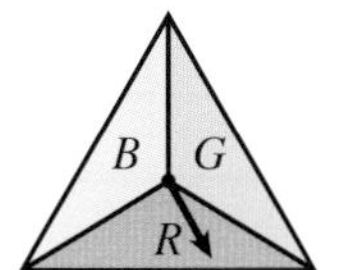

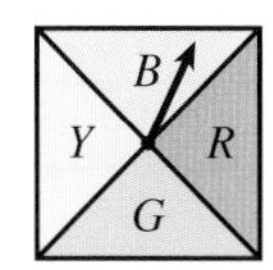

a Copy and complete the two-way table showing the possible outcomes.

	Red	**Green**		
Red	R, R	R, G		
Green	G, R			
Blue				

Work out

b P(R, R) **c** P(at least one green) **d** P(Y, Y)

e **Reasoning** Mischa says, 'The probability of getting at least one blue is $\frac{7}{24}$ because there are 24 letters in the table and 7 of them are B.'
Explain why Mischa is wrong.

9 Two 3-sided fair spinners are spun.

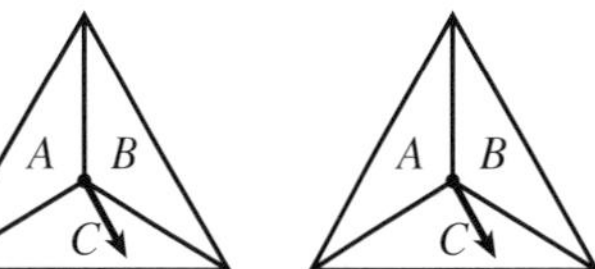

a Draw a sample space diagram to show all the possible outcomes.

b Find the probability that both spinners land on the same letter.

c Calculate an estimate for the number of times both spinners will land on the same letter in 30 spins.

10 Emily is deciding what to wear. She has a choice of three different skirts and three different tops.

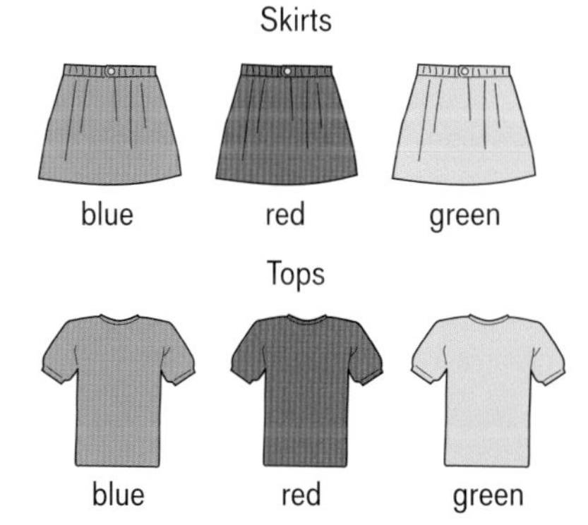

a Draw a sample space diagram showing Emily's options.

b How many combinations has she to choose from?

c Emily chooses a skirt and top at random.
Write the probability that she wears

i only red

ii a green top and a blue skirt

iii some blue

d Emily also has two possible scarves to wear with her outfit. How many different combinations does she have now?

11 **a** Copy and complete this sample space diagram for the possible outcomes of throwing a fair dice twice.

b Work out

i P(4, 2)

ii P(1st throw is 3)

iii P(both numbers are the same)

iv P(one of the numbers is 3)

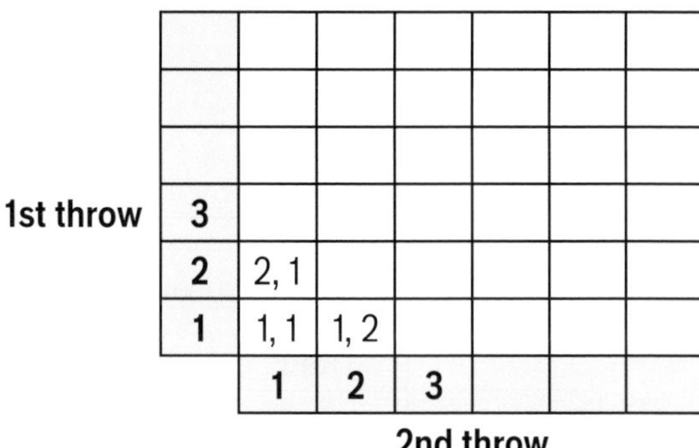

F
H

Exam-style question

12 Kate throws an ordinary six-sided fair dice twice.
She says,
'The probability of getting 1 on both throws is $\frac{2}{6}$.'
Is Kate correct? You must explain your answer. **(1 mark)**

Exam tip

Show working and write a sentence to explain your answer.

13 A fair dice is numbered 2, 4, 6, 8, 10 and 12. Another fair dice is numbered 1, 3, 5, 7, 9 and 11. The dice are rolled together, and the scores added.

a Copy and complete the sample space diagram showing the possible total scores.

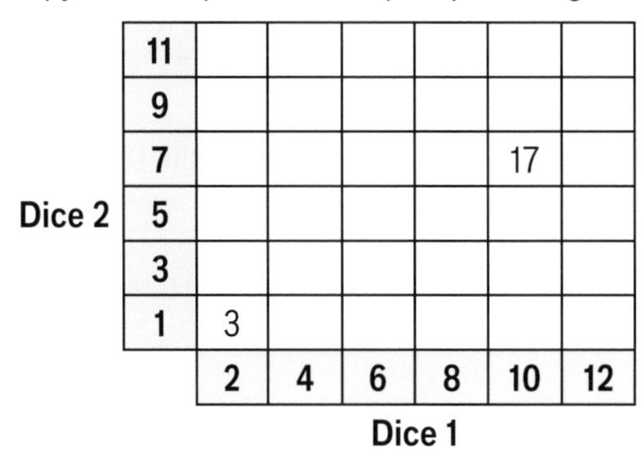

b Find the probability of scoring

i a total of 7

ii more than 12

iii an even number

iv a multiple of 3

c Which is more likely, scoring a multiple of 3 or a total of 7?

14 **Problem-solving/Reasoning** Contestants spin two fair numbered wheels and multiply the two numbers they score together. If the overall score is odd, they win a prize.

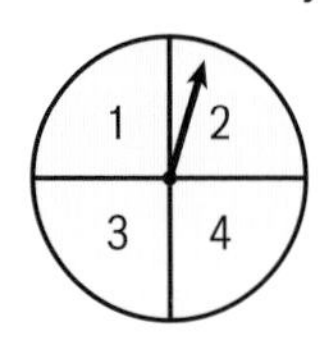

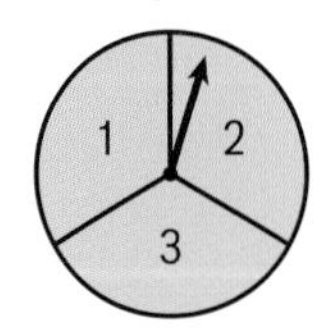

Mr Dixon says, 'Winning and losing are equally likely.'
Is he correct?

15 **Problem-solving** Two ordinary fair dice are rolled.
Work out the probability that the difference between their two scores is greater than 1.

13.3 Experimental probability

- Estimate and interpret probabilities based on experimental data.
- Make predictions from experimental data.

Warm up

1 **Fluency** **a** What is the theoretical probability of rolling 3 on a fair dice?

b A fair dice is rolled 60 times. How many 3s do you expect?

2 Bella recorded the type and colour of vehicles passing the school gate. Here are her results.

	Red	Silver	White	Blue
Car	5	10	3	2
Van	2	8	15	0
Bus	17	2	1	5

a How many cars passed the gate?

b How many vehicles passed the gate?

c What proportion of the vans were silver?

Key point

You can estimate the probability of an event from the results of an experiment or survey.

$$\text{Estimated probability} = \frac{\text{frequency of event}}{\text{total frequency}}$$

This **estimated probability** is also called the **experimental probability**.

3 Eric rolled a six-sided dice 20 times.

a Copy and complete the frequency table to record his results.

6, 3, 4, 6, 2, 3, 3, 4, 6, 4, 3, 2, 3, 2, 3, 1, 3, 5, 6, 5

Outcome	1	2	3	4	5	6
Frequency						

b What is the total frequency?

c What is the frequency of 6?

d Find the estimated probability of 6.

e Eric rolls the dice 100 times.
How many times would you expect it to land on 6?

f **Reflect** Think carefully about what ‘frequency’ tells you. Write a definition in your own words.

Key point

The **relative frequency** of an event is also an estimate of the probability.

$$\text{Relative frequency} = \frac{\text{number of successful trials}}{\text{total number of trials}}$$

4 Stephan drops a drawing pin 60 times and records whether it lands 'point up' or 'point down'. He records his results in a table.

	Point up	Point down
Frequency	35	25

a The relative frequency of 'point up' is $\frac{35}{60}$. Write the relative frequency of 'point down'.

b Reasoning Stephan states that the probability of a drawing pin landing 'point up' is $\frac{7}{12}$. Is he correct? Show working to explain.

5 Jack spins this spinner. He records his results in a table.

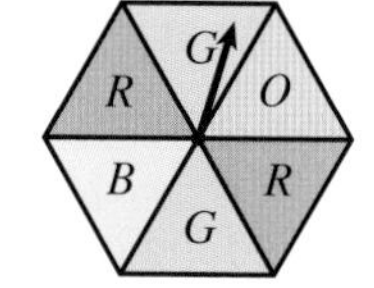

	Red	Blue	Green	Orange
Frequency	7	3	6	2

a How many trials did Jack do?

b What is the relative frequency of spinning red?

Key point

The more times you repeat an experiment, the more accurate the experimental probability.

6 Reasoning Esme tests the statement 'Toast usually lands butter side down', by dropping a piece of toast 20 times.
The toast lands butter side down 12 times.
Esme concludes that the statement is true.
Do you agree? How could she get a more accurate estimate of the probability?

Q6 hint

Show your working. Write a sentence or two comparing your working to Esme's conclusion.

7 Anjenna and Belinda both spin a wheel.
Their results are recorded in the table.

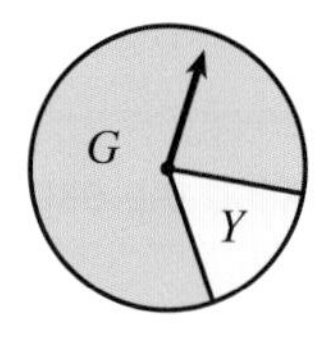

	Green	Yellow
Anjenna	72	28
Belinda	120	40

a How many times did Anjenna spin the wheel?

b Write the estimated probability of green from Anjenna's results.

c Whose results will give the better estimate for the probability that the spinner will land on green? Give a reason for your answer.

d Altogether Anjenna and Belinda spin the wheel 260 times.
What is the estimated probability of green for their combined results?

e Which is the best estimate for P(green)? Explain.

8 Arthur repeatedly rolls a standard four-sided dice and notes the number that is rolled.

	1	2	3	4
Frequency	14	18	12	16

a What is the experimental probability of Arthur rolling a 2?

b What is the theoretical probability of Arthur rolling a 2 on a fair 4-sided dice?

c Reasoning Do you think the dice is fair? Give a reason for your answer.

9 Two dice are rolled 120 times. The difference in the scores is recorded as being 0 or not 0.

	0	Not 0
Frequency	42	78

a What is the experimental probability of a difference of 0?

b What is the theoretical probability of a difference of 0?

c How many 0 differences would you predict from 120 rolls of the dice?

d **Reasoning** Do you think the dice are biased? Explain your answer.

Exam-style question

10 a Tony throws a biased dice 100 times.
The table shows his results.
He throws the dice once more.
Find an estimate for the probability that he will get a 5. **(1 mark)**

Score	Frequency
1	25
2	12
3	15
4	11
5	20
6	17

b Tony rolls the biased dice 200 more times.
He gets 5 in 37 of these rolls.
Work out the best estimate of getting 5. **(2 marks)**

Exam tip

Show your working. If you calculate your answer incorrectly, you may still get marks for method.

11 The table shows some information about Year 11 students.

	Staying at 6th form	Going to college	Total
Male	116	59	175
Female	88	65	153
Total	204	124	328

a Work out the probability that a student picked at random from all the Year 11 students is

i male

ii a female going to college

b A male student is picked at random.
What is the probability that the student is staying at 6th form?

Q11a hint

$\frac{\text{total number of males}}{\text{total number of students}}$

Q11b hint

$\frac{\text{number of males staying}}{\text{total number of males}}$

12 **Problem-solving** This table shows students' ice cream choices.

	Chocolate	Vanilla	Toffee
Year 10	14	9	16
Year 11	8	5	3

Work out the probability that

a a student picked at random chose vanilla

b a Year 11 student picked at random chose toffee

13 **Problem-solving** Look back at the table of vehicles passing in **Q2**.

a Estimate the probability that the next vehicle to pass the gate is

i a red car **ii** a blue bus

b Estimate the probability that the next silver vehicle to pass the gate is a car.

c Estimate the probability that the next bus to pass the gate is red.

13.4 Venn diagrams

- Understand the language of sets and Venn diagrams.
- Use Venn diagrams to solve probability problems.

Warm up

1 **Fluency** What is an integer?

2 Write down the first five

a prime numbers **b** multiples of three **c** square numbers

Key point

Curly brackets { } show a set of values.
$\in$ Means is an 'element of'.
5 $\in$ {odd numbers} means '5 is in the set of odd numbers'.
An element is a 'member' of a set. Elements are often numbers, but could be letters, items of clothing or even body parts.
$\mathscr{E}$ means the universal set – all the elements being considered.

3 Set A = {odd numbers less than 10}
Set B = {square numbers less than 10}

a List the numbers in each set. A = {1, 3, ...}, B = {1, 4, ...}

b Write 'true' or 'false' for each statement

i $6 \in A$

ii $9 \in B$

iii $11 \in A$

c Which numbers are in both sets?

d Which numbers are in set A only?

e Which numbers are in either A or B (or both)?

f Which numbers less than 10 are not in A or in B?

4 The Venn diagram shows two sets X and Y, and $\mathscr{E}$, the set of all numbers being considered.

a Copy and complete these sets.

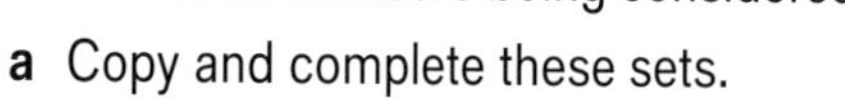

X = {2, 4, □, □, □, □}
Y = {3, □, □, □}
$\mathscr{E}$ = {1, 2, 3, ...}

$\mathscr{E}$ | X: 2, 4, 8, 10 | X and Y: 6, 12 | Y: 3, 9 | outside: 1, 5, 7, 11

b Match each set to its description.

{integers 1 to 12}

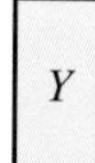

{multiples of 2 up to 12}

{multiples of 3 up to 12}

c A number is chosen at random from the universal set $\mathscr{E}$.
What is the probability that the number is in set X?

Key point

$A \cap B$ means A intersection B.
This is the elements that are in A *and* in B.

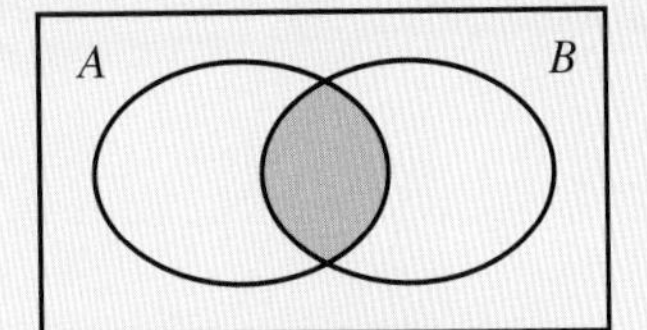

$A \cup B$ means A union B.
This is the elements that are in A *or* in B *or* both.

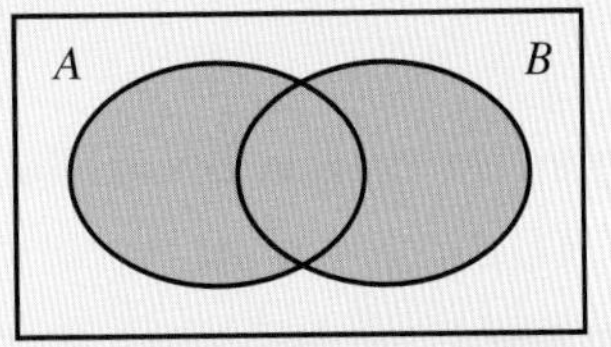

A' means the elements not in A.

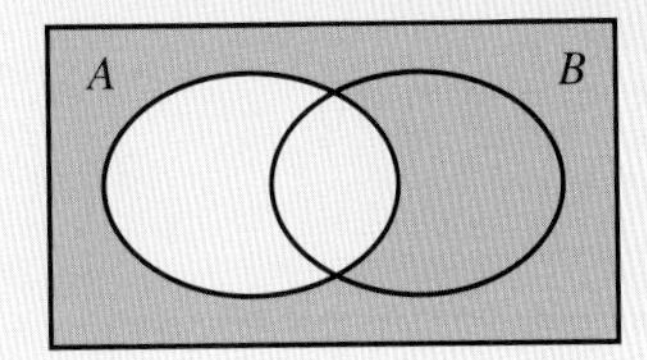

Example

$\mathscr{E} = \{\text{integers from 1 to 10}\}$
$A = \{2, 6, 10\}$
$B = \{2, 3, 5, 7\}$
Draw a Venn diagram to represent this information.

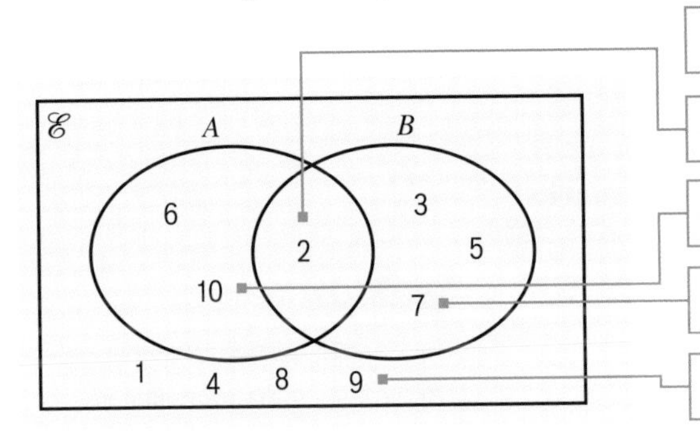

1 Label $\mathscr{E}$ and the sets A and B.

2 Write in the elements in A **and** B.

3 Write in the elements in A but not in B.

4 Write in the elements in B but not in A.

5 Write in the elements of $\mathscr{E}$ not in A or B.

5 **a** List the elements of each set.
$\mathscr{E} = \{\text{integers from 1 to 15}\}$
$A = \{\text{even numbers from 1 to 15}\}$
$B = \{\text{multiples of 3 from 1 to 15}\}$

b Copy the Venn diagram.
Which elements are in A **and** B?
Write these elements in the overlap of A and B.

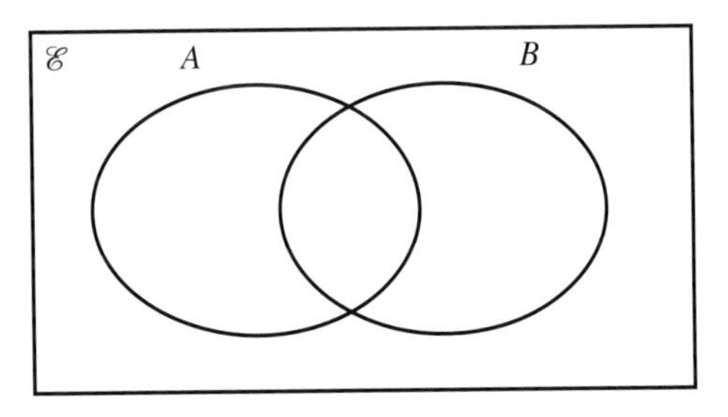

c Write in the other elements of

i A

ii B

d Write the numbers that are in $\mathscr{E}$ but not in A or B outside.

e **Reflect** How can you check you have included all the numbers in the diagram?

6 For the Venn diagram in **Q4**, copy and complete these sets.

a $X \cap Y = \{6, ...\}$

b $X \cup Y = \{2, 3, ...\}$

c $X' = \{1, ...\}$

d $Y' = \{1, 2, ...\}$

e $X' \cap Y = \{...\}$

Q6e hint

X Y

7 For the Venn diagram you drew in **Q5**, list the members of

a $A \cap B$ b $A \cup B$ c A' d B' e $A' \cap B$ f $A \cap B'$

Exam-style question

8 $\mathscr{E} = \{1, 2, 3, 4, 5, 6, 7, 8, 9, 10\}$
$A = \{1, 4, 7, 9, 10\}$
$B = \{4, 6, 10\}$

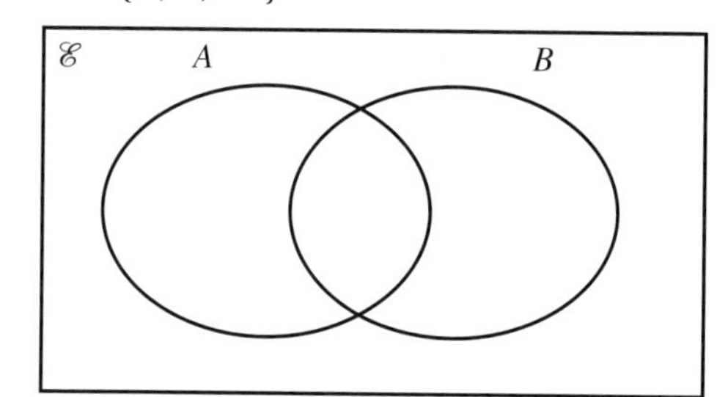

a Complete the Venn diagram to represent this information. **(3 marks)**

A number is chosen at random from the universal set $\mathscr{E}$.

b Find the probability that the number is in the set $A \cap B$. **(2 marks)**

Exam tip

Check that all members of $\mathscr{E}$ are in your diagram exactly once.

Exam-style question

9 A = {multiples of 3 between 10 and 20}
B = {even numbers less than 20}

a List the members of $A \cup B$. **(2 marks)**

b Describe the members of $A \cap B$. **(1 mark)**

Exam tip

The question doesn't ask you to draw a Venn diagram. But you can draw one if it helps you to answer the question.

10 $\mathscr{E}$ = {odd numbers between 0 and 20}
$A = \{3, 5, 7, 11, 17\}$
$B = \{3, 5, 13, 15\}$
$C = \{5, 9, 15\}$

a Copy and complete the Venn diagram for this information.

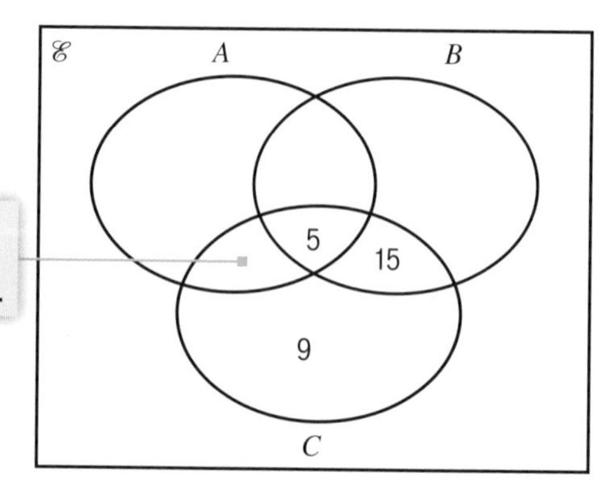

Q10a hint

There is no element in this section, so leave it empty.

b Write down the elements of $A \cap B$.

11 **Reasoning** Jessica asks 25 people in her school if they have a mobile phone or tablet. The Venn diagram shows her results.

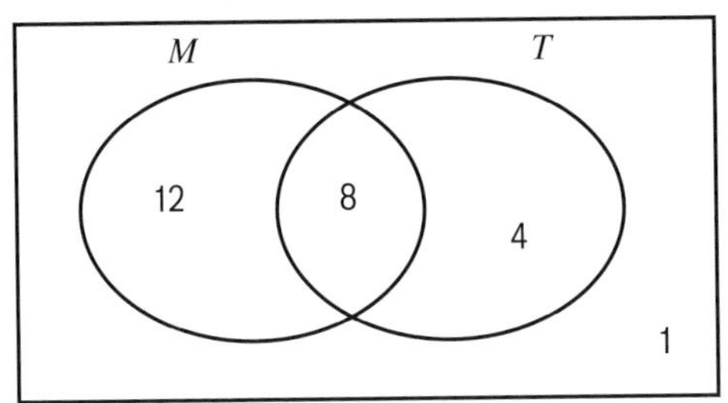

a How many people have a tablet.

b How many people have either a mobile phone or a tablet?

c What is the probability that a person chosen at random has a tablet and a mobile phone?

d What is the probability that a person chosen at random just has a tablet?

e Jessica's friend Amy says, 'The Venn diagram shows that 12 people have mobile phones'. Explain why Amy is not correct.

12 The Venn diagram shows the numbers of students studying French and Spanish.

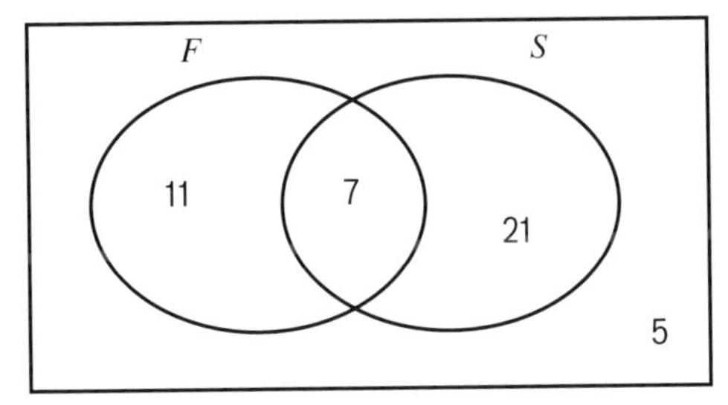

a Work out the total number of students represented in the diagram.

b One student is chosen at random. What is the probability that this student

i studies French

ii studies Spanish but not French?

13 Adrian is doing a survey of music tastes. He asks 30 people if they like rock (R) or pop (P). 17 said they like both rock and pop, 8 said they like neither and 3 said they like only pop music.

a Draw a Venn diagram to show Adrian's findings.

b What is the probability that a person chosen at random likes

i only pop

ii rock?

Master p.65 | Check up p.87 | Strengthen p.89 | Extend p.92 | Test ready p.94 | Unit test p.96

13.5 Tree diagrams

- Solve problems using frequency trees and tree diagrams.
- Work out probabilities using tree diagrams.
- Understand independent events.

ActiveLearn Homework

Warm up

1 **Fluency** Work out

a $\frac{1}{3}\times\frac{1}{3}$ **b** $\frac{1}{2}\times\frac{1}{2}$ **c** $\frac{2}{9}+\frac{4}{9}$

2 A bag contains 10 coloured sweets. There are 7 red sweets and 3 green sweets. 1 sweet is taken from the bag at random. Write

a the probability of taking a red sweet

b the probability of taking a sweet that is not red

Key point

A **frequency tree** shows the number of options for different choices.

3 **Reasoning** The frequency tree shows people's orders in a café.

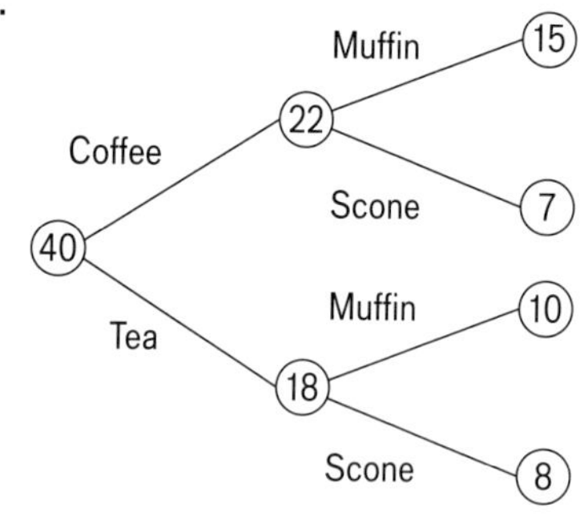

a How many people ordered something in the café?

b How many people ordered

i tea

ii coffee

iii coffee and scone

iv tea and muffin

v muffin

vi scone?

c **Reflect** What do you notice about the numbers on each pair of branches?

Exam-style question

4 30 people work in an office.
18 of these people are female.
7 of the males are aged 40 or over.
15 of the 30 people are under 40.

Copy and complete the frequency tree for this information. **(3 marks)**

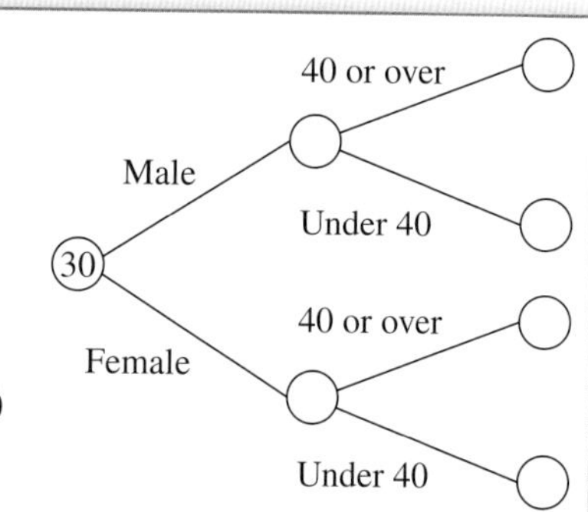

Exam tip

Check that each pair of numbers adds to the correct total.

5 **Reasoning** There are 100 flats in a block. Each flat has 1 bedroom or 2 bedrooms.
30 of the 100 flats have a balcony. 20 of the flats with a balcony have 2 bedrooms.
60 of the 100 flats have 2 bedrooms.

a Copy and complete the frequency tree to show this information.

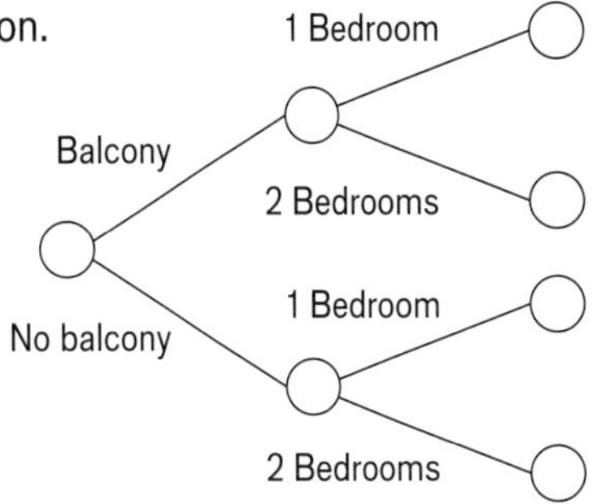

b One flat is selected at random.
Find the probability that it has

i a balcony **ii** exactly 1 bedroom

c One of the flats with no balcony is selected at random.
Find the probability that it has 2 bedrooms.

d **Reflect** Draw a two-way table to show the information about the flats.
Which do you find easier for finding probabilities – the frequency tree or the table?
Explain why.

Exam-style question

6 50 people are asked if they like chocolate.
36 of these people are women.
42 of the people like chocolate.
3 of the men do not like chocolate.

a Use this information to complete the frequency tree. **(3 marks)**

Women
Like chocolate
Do **not** like chocolate
Men
Like chocolate
Do **not** like chocolate

One of the people who likes chocolate is chosen at random.

b Find the probability that this person is a woman. **(2 marks)**

Exam tip

Complete the frequency tree diagram carefully – you will need it to answer the rest of the question.

Key point

Two events are **independent** when the results of one do not affect the results of the other.

7 In a box of chocolates, there are 2 hard centres and 5 soft centres.

a What is the probability of picking at random a soft centre?

David picks a chocolate at random. It is a soft centre and he eats it.

b How many chocolates are left in the box?

c How many soft centres are left in the box?

He picks another chocolate at random.

d What is the probability that he picks a soft centre this time?

e Are the two events 'picking a soft centre first time' and 'picking a soft centre second time' independent? Write a sentence to explain.

8 Which of these pairs are independent?

a Rolling a 5 on a dice and then rolling another 5

b Getting full marks in a physics test and then getting full marks in a biology test

c Getting a head on a coin on the first flip and getting a head on a coin on the second flip

d Picking a counter from a bag, keeping it and then picking a second counter from the same bag

Key point

A **probability tree diagram** shows the probabilities of different outcomes.

Example

There are 3 blue and 4 green counters in a bag.
Steven picks one counter at random, notes its colour and replaces it.
He then chooses a second counter and notes its colour.

a Draw a tree diagram to show this.

b Work out the probability that Steven picks 2 blue counters.

c Work out the probability that Steven picks 1 blue counter and 1 green counter.

a

$\frac{3}{7}$ B: $\frac{3}{7}$ B, $\frac{4}{7}$ G

$\frac{4}{7}$ G: $\frac{3}{7}$ B, $\frac{4}{7}$ G

In a probability tree diagram, the branches are labelled with the probabilities.
$P(B) = \frac{3}{7}$ and $P(G) = \frac{4}{7}$

These two events are independent so the probabilities for the second choice are the same as for the first choice.

b $P(B, B) = \frac{3}{7} \times \frac{3}{7} = \frac{9}{49}$

When events are independent, multiply the probabilities together along the branches of the tree to work out the probability of the final outcome.

c $P(B, G) = \frac{3}{7} \times \frac{4}{7} = \frac{12}{49}$

$P(G, B) = \frac{4}{7} \times \frac{3}{7} = \frac{12}{49}$

Total: $\frac{12}{49} + \frac{12}{49} = \frac{24}{49}$

There are two ways of getting 1 green, 1 blue. Look at the events 'B, G' and 'G, B'. Work out each probability and add them together.

9 The tree diagram shows the probabilities that Mrs Johnson has to stop at two sets of traffic lights.

a What is the probability she stops at the first set?

b What is the probability that she stops at both sets?

c What is the probability that she does not stop at either set?

d Work out the probability that she stops at just one set?

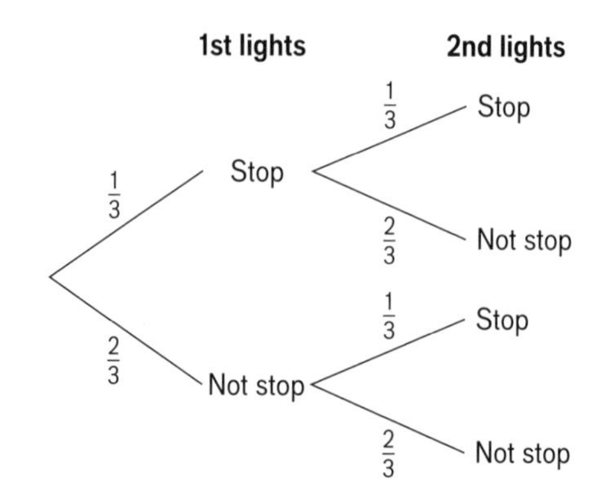

10 On each day, the probability that the bus is late is $\frac{1}{4}$.

a Write the probability that the bus is not late.

b Copy and complete the tree diagram.

c Work out the probability that the bus is not late two days running.

d Work out the probability that the bus is late on the first day and not late on the second.

e **Reflect** Explain how you can check the probabilities on your diagram are correct.

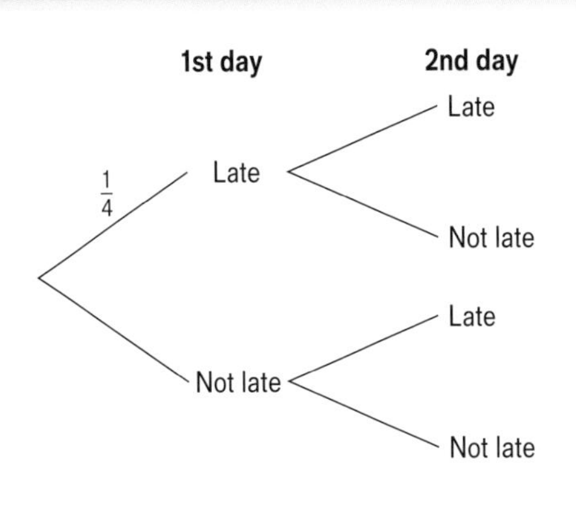

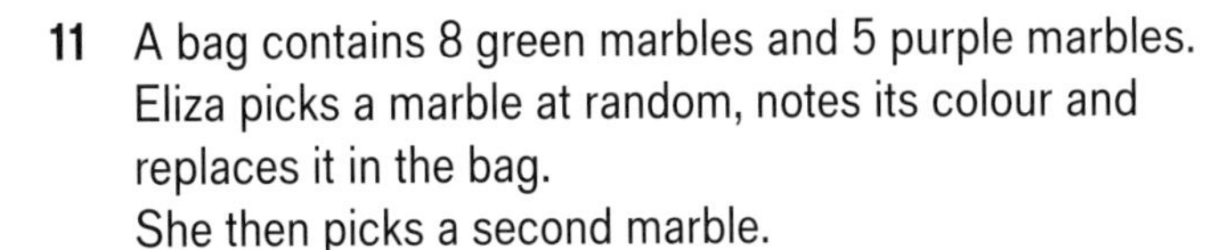

11 A bag contains 8 green marbles and 5 purple marbles.
Eliza picks a marble at random, notes its colour and replaces it in the bag.
She then picks a second marble.

a Copy and complete the tree diagram.

b Work out

i the probability that Eliza picks one marble of each colour

ii the probability that Eliza picks at least one green

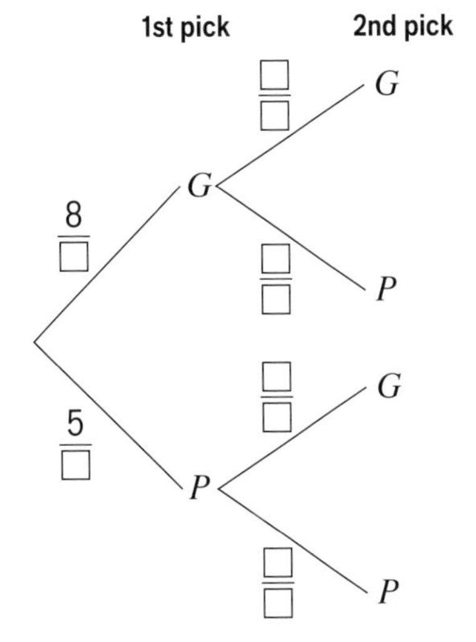

12 Archie rolls a fair black dice once and a fair red dice once.

a Copy and complete the probability tree diagram to show the outcomes.
Label clearly the branches of the probability tree diagram.

b Calculate the probability that Archie gets a 6 on both the red dice and the black dice.

c Calculate the probability that Archie gets at least one six.

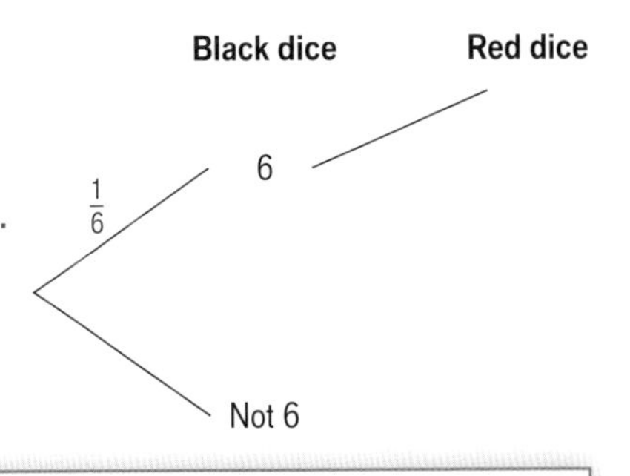

Exam-style question

13 A spinner can land on red or blue.
When the spinner is spun once, the probability it will land on red is 0.75.
The spinner is spun twice.
Evie draws this probability tree diagram.
The diagram is not correct.
Write down two things that are wrong with the probability tree diagram. **(2 marks)**

1st spin | 2nd spin
0.75 red: 0.75 red, 0.15 blue
0.15 blue: 0.15 red, 0.75 blue

Master p.65 | Check up p.87 | Strengthen p.89 | Extend p.92 | Test ready p.94 | Unit test p.96

13.6 More tree diagrams

- Understand when events are not independent.
- Solve probability problems involving events that are not independent.

Warm up

1 **Fluency** Work out

a $\frac{2}{3}\times\frac{1}{2}$ **b** $\frac{2}{7}\times\frac{5}{6}$ **c** $\frac{5}{64}+\frac{21}{64}$ **d** $1-\frac{23}{64}$

2 A bag contains 8 coloured sweets.
There are 5 red sweets and 3 green sweets.
Fiona picks a red sweet, changes her mind and puts it back in the bag.
She then chooses another sweet at random.

a Copy and complete the tree diagram.

b Work out the probability that she picks 2 green sweets.

c Work out the probability that she picks at least one red sweet.

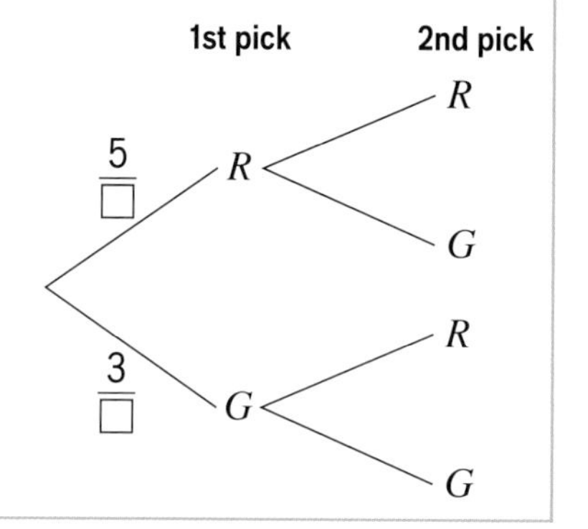

3 A bag contains 6 orange and 4 pink sweets.

a Debbie picks a sweet at random. What is P(pink)?

Debbie's sweet is pink. She eats it.

b How many sweets are left in the bag?

c How many pink sweets are left in the bag?

d Now what is

i P(pink)

ii P(orange)?

Key point

When the outcome of one event changes the possible outcomes of the next event, the two events are not independent. The second event is **dependent** on the first.

4 Are these pairs of events dependent or independent?

a Taking a sweet, eating it, then taking another sweet

b Picking a blue sock at random from a drawer, putting it on, then picking another sock

c Rolling an even number on a fair dice and rolling a second even number

d Rolling a 6, then rolling another 6

Example

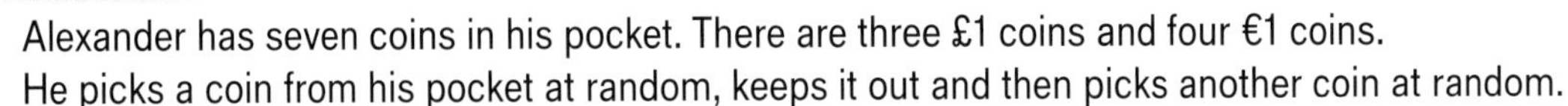

Alexander has seven coins in his pocket. There are three £1 coins and four €1 coins.
He picks a coin from his pocket at random, keeps it out and then picks another coin at random.

a Draw a tree diagram to show this.

b Work out the probability that he picks two £1 coins.

c Work out the probability that he picks one of each type of coin.

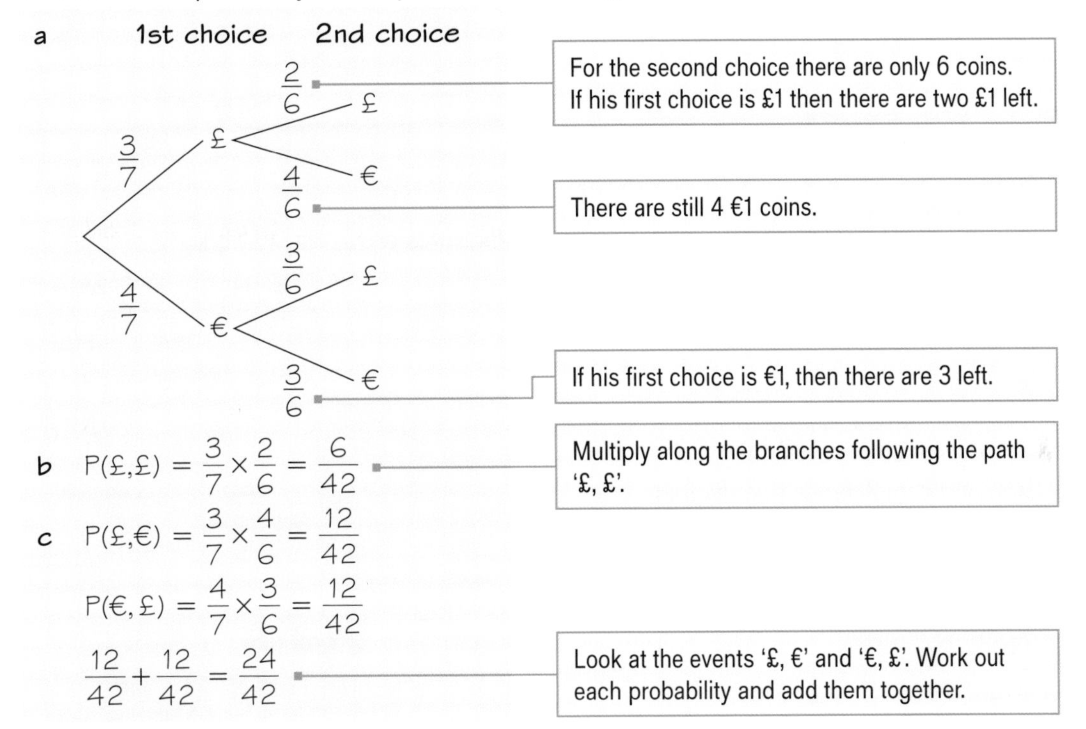

b $P(£,£) = \frac{3}{7} \times \frac{2}{6} = \frac{6}{42}$

Multiply along the branches following the path '£, £'.

c $P(£,€) = \frac{3}{7} \times \frac{4}{6} = \frac{12}{42}$

$P(€,£) = \frac{4}{7} \times \frac{3}{6} = \frac{12}{42}$

$\frac{12}{42} + \frac{12}{42} = \frac{24}{42}$

Look at the events '£, €' and '€, £'. Work out each probability and add them together.

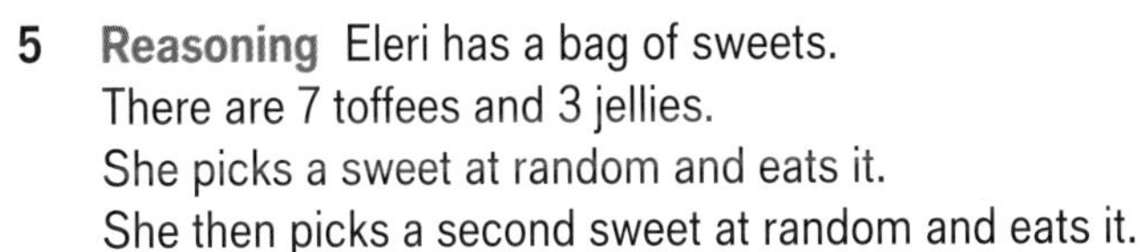

5 **Reasoning** Eleri has a bag of sweets.
There are 7 toffees and 3 jellies.
She picks a sweet at random and eats it.
She then picks a second sweet at random and eats it.
Use the tree diagram to find the probability that Eleri eats

a 2 toffees

b 1 of each type of sweet

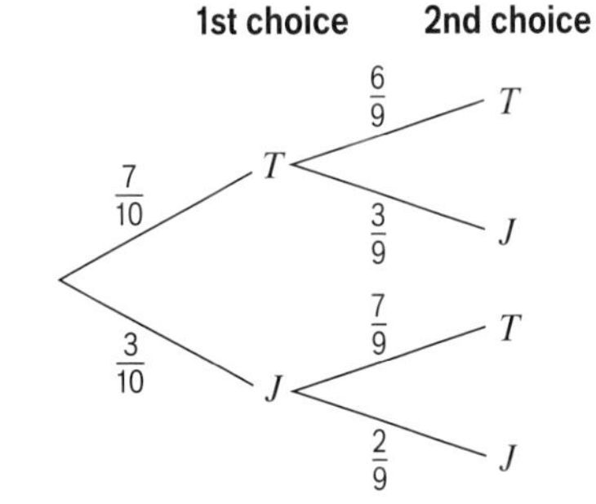

6 **Reasoning** June has these numbered cards.

She picks 2 cards at random without replacing them.
She records whether the number on each card is odd or even.

a Copy and complete the tree diagram.

b Find the probability that she picks

i 2 even numbered cards

ii 1 even numbered card and 1 odd numbered card

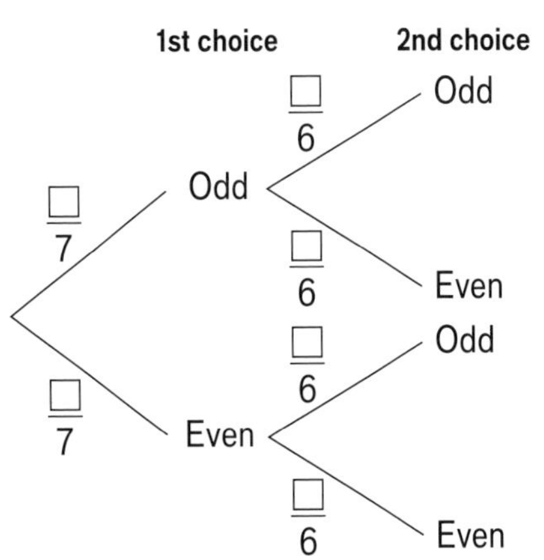

7 **Reasoning** Yasmin has a box of sweets containing 3 toffees (T) and 8 mints (M).

a Copy and complete the tree diagram.

b Work out the probability that Yasmin takes at random

i no mints

ii at least one mint

c **Reflect** Write down what you notice about your answers to parts **bi** and **bii**.

1st sweet 2nd sweet

$\frac{3}{\square}$ T $\frac{\square}{10}$ T

M

M T

M

8 **Reasoning** Mr Brown drives through two sets of traffic lights on his way to work.
The probability that he stops at the first set of lights is 0.4.
If he stops at the first set, the probability he stops at the second set is 0.6.
If he does not stop at the first set, the probability that he stops at the second set is 0.25.

a Copy and complete the tree diagram.

b Find the probability that Mr Brown does not stop at either set of lights.

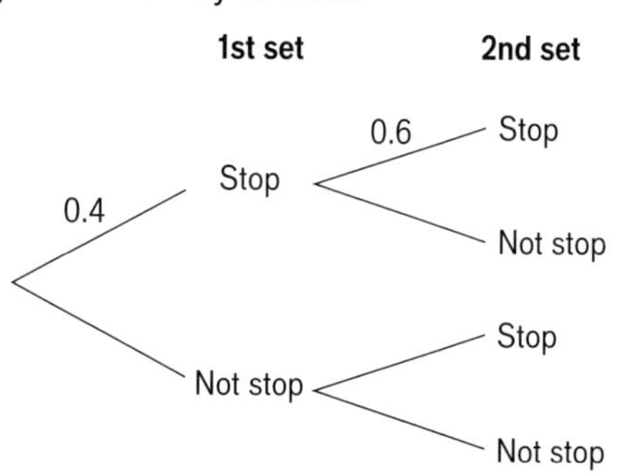

Exam-style question

9 There are 8 socks in a drawer.
5 of the socks are white. 3 of the socks are grey.
Maya takes two socks from the drawer at random.

a Complete the probability tree diagram.

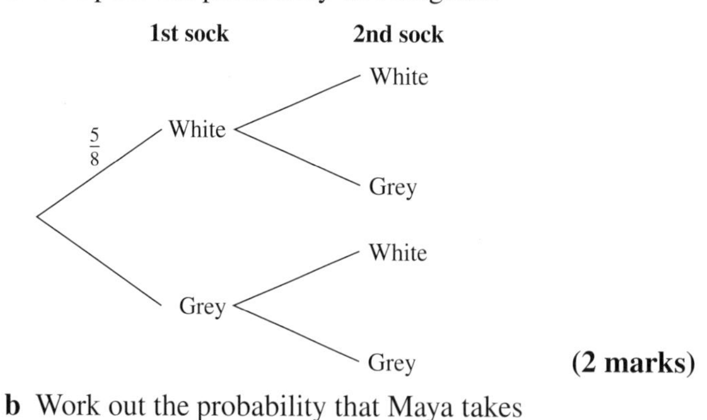

(2 marks)

b Work out the probability that Maya takes two socks of the same colour. **(3 marks)**

Exam tip

Make sure you label each calculation clearly, for example P(white, white) or P(W, W).

13 Check up

ActiveLearn Homework 

Calculating probabilities

1 The probability of getting a 4 on a spinner is $\frac{1}{4}$. Write the probability of not getting a 4.

2 The table shows the probabilities of certain outcomes on a spinner with 3 colours.
Work out P(blue).

Colour	red	yellow	blue
Probability	0.3	0.2	

3 There are 38 counters in a bag.
20 of the counters are blue.
The rest of the counters are red.
One of the counters is taken at random.
Find the probability that the counter is red.

Experimental probability

4 Bill and Fred do an experiment with a spinner.
The table shows their results.

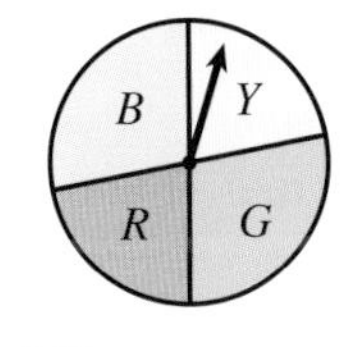

	Yellow	**Blue**	**Green**	**Red**
Bill	32	25	22	31
Fred	18	15	17	20

a Estimate the experimental probability of spinning blue using Bill's results.

b Estimate the experimental probability of spinning blue using Bill and Fred's results.

c Which experimental probability is the more accurate?

5 Luca flips a coin 200 times. He gets 'heads' 80 times.

a Write the experimental probability of getting a 'head'.

b Write the theoretical probability of getting a 'head'.

c How many 'heads' would you predict in 200 flips of a fair coin?

d Do you think Luca's coin is fair? Explain.

Probability diagrams

6 Anna and Beth each have a set of cards.
Flo takes one card at random from each set and adds the numbers.

Anna

Beth

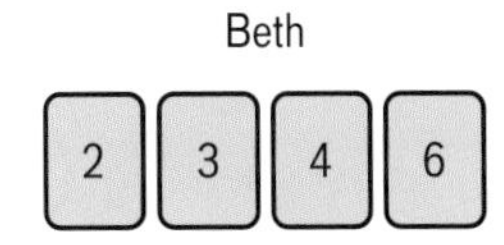

a Draw a sample space diagram showing the possible total scores.

b How many possible outcomes are there?

c What is the probability that Flo takes cards with

i a total of 7 **ii** a total greater than 5?

7 A bag contains two 10p coins and two 20p coins.
Another bag contains one 10p coin and three 20p coins.
Olivia takes one coin at random from each bag.
What is the probability of her getting 40p?

8 Zainab asks students in her class whether they like football or tennis.
The Venn diagram shows the results.

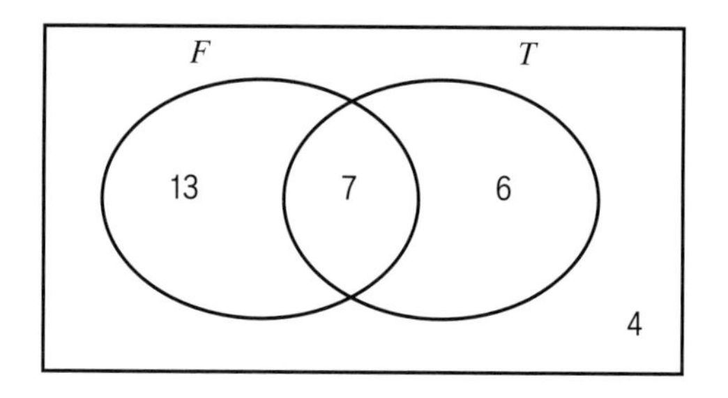

a How many people are in Zainab's class?

b How many people like football?

c What is the probability that a person chosen at random likes football but not tennis?

9 **a** Copy and complete the Venn diagram for these sets.

$\mathscr{E}$ = {integers less than 10}
A = {odd numbers less than 10}
B = {factors of 8}

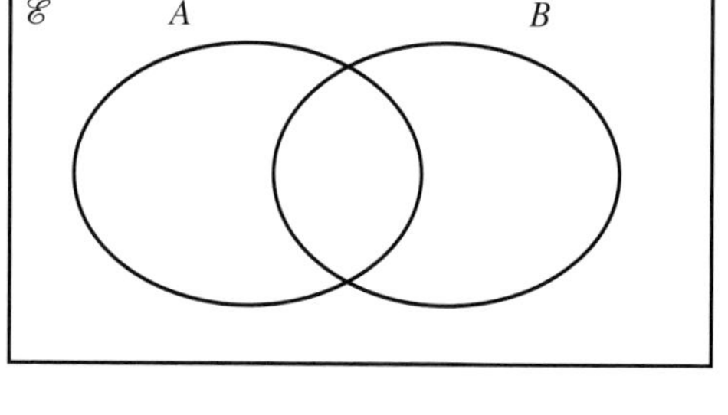

b List the elements of these sets.

i $A \cap B$ **ii** $A \cup B$ **iii** A'

10 60 students are either left-handed or right-handed.
35 of the 60 students are in Year 10.
7 of the 11 left-handed students are in Year 11.

a Complete the frequency tree for this information.

One student is picked at random.

b Find the probability that this student is in Year 11 and right-handed.

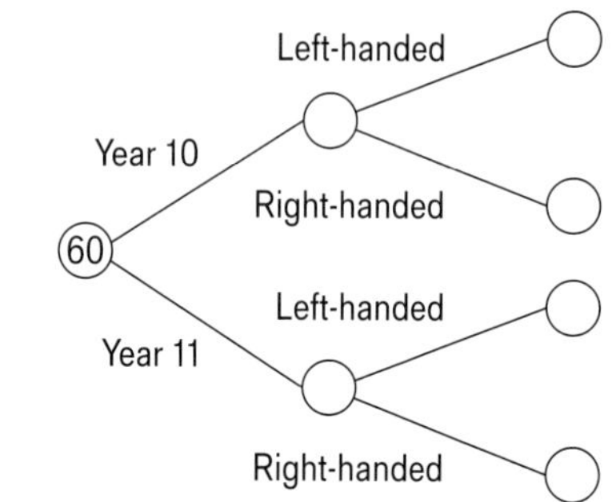

Dependent events

11 A box contains 7 necklaces and 3 bracelets.
Pip chooses an item from the box at random, puts it on and then chooses another item.

a Copy and complete the tree diagram.

b Find the probability that Pip chooses two necklaces.

c Work out the probability that she chooses one necklace and one bracelet.

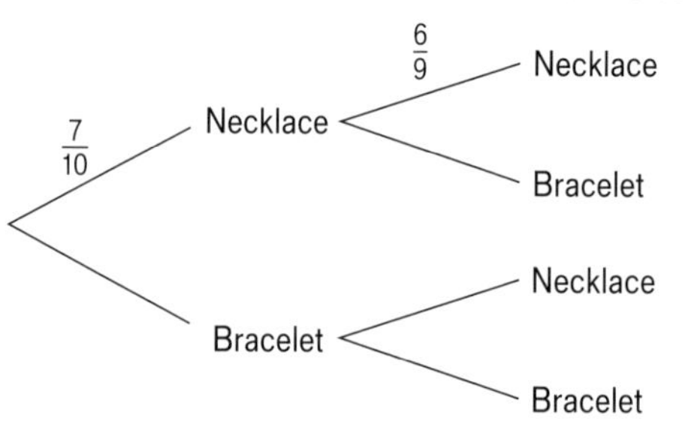
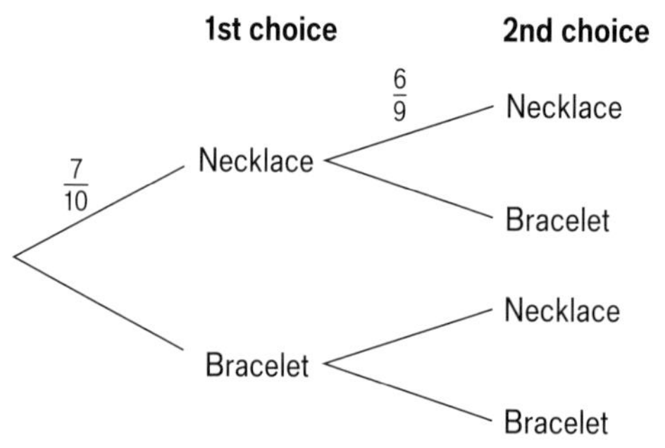

12 **Reflect** How sure are you of your answers? Were you mostly

Just guessing ☹ Feeling doubtful 😐 Confident ☺

What next? Use your results to decide whether to strengthen or extend your learning.

Challenge

13 A fairground game involves spinning two fair wheels.
You win if your total score is less than 7.

a Is the game fair? Explain.

b The fairground game operator suggests changing the rules so that you win if the difference in the scores is less than 3. Is the game fair now? Explain.

13 Strengthen

ActiveLearn Homework

Calculating probabilities

1 The probability of getting a 6 on a fair dice is $\frac{1}{6}$.
Write the probability of not getting a 6.

Q1 hint

P(not 6) = P(6) $= \frac{1}{6}$

1	2	3	4	5	6

2 The table shows the probabilities of trains arriving early, late or on time.

Arrival	Early	Late	On time
Probability	0.1		0.6

Use the bar model to work out the probability of a train arriving late.

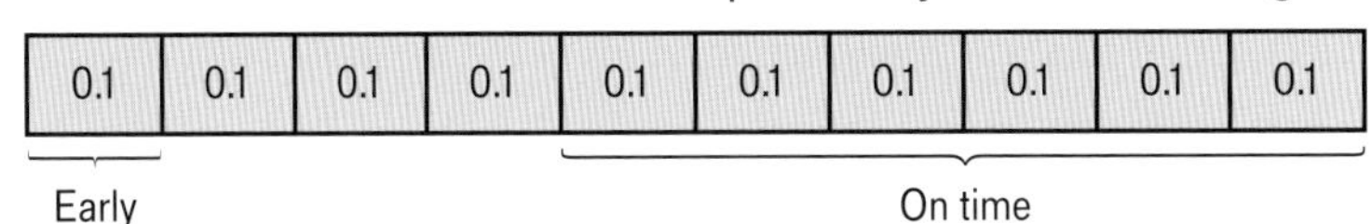

0.1	0.1	0.1	0.1	0.1	0.1	0.1	0.1	0.1	0.1

Early On time

3 There are 30 counters in a bag.
12 of the counters are red.
The rest of the counters are blue.

a How many of the counters are blue?

b Work out i P(red) ii P(blue)

Experimental probability

1 Maddie rolls a dice and records the number of times she gets a 2.
Freya does the same with a different dice.

	Maddie	**Freya**
Number of rolls	60	90
Number of 2s	12	30

a What is the theoretical probability of rolling a 2 on a fair dice?

b What is the total number of rolls for
i Maddie ii Freya?

c Write the experimental probability of
i Maddie getting a 2 ii Freya getting a 2

d Compare the experimental probabilities with the theoretical probability.
Whose dice is more likely to be fair?

2 Seb flips a coin 10 times and gets 7 heads.

a Explain why he thinks the coin might be biased.

He flips the same coin 200 times and gets 102 heads.

b Explain why he now thinks the coin is fair.

c Which is his most accurate estimate of the experimental probability of getting a head? Explain.

Q2c hint

More _____ → more accurate estimate.

Probability diagrams

1 Rachel takes one card at random from each set.

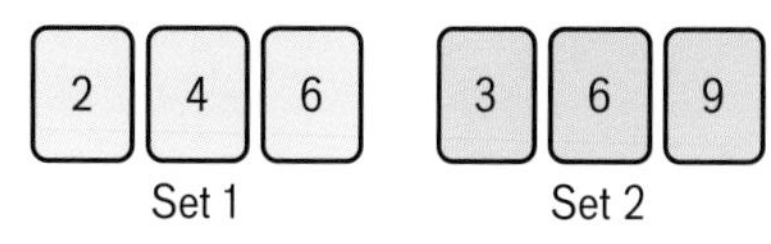

		Set 1		
		2	**4**	**6**
Set 2	**3**	(2, 3)		
	6			
	9			

a Copy and complete the sample space diagram to show all the possible outcomes.

b How many possible outcomes are there?

c Work out the probability of

i taking (4, 6)

ii both numbers being at least 4

iii taking two even numbers

d Rachel now adds the scores together.
Draw a new sample space diagram to show the totals for adding the scores.

e What is the probability of getting a total greater than 8?

2 Bag 1 contains one 10p coin and one 50p coin.
Bag 2 contains one 10p coin, one 20p coin and one 50p coin.
Jo takes one coin at random from each bag.

a Draw a sample space diagram to show the possible outcomes.

b Ring all the outcomes that give a total of 60p.

c Write down the probability that Jo takes 60p.

3 Francis asks 25 students in his class if they play music or sports outside school.
The Venn diagram shows the results.

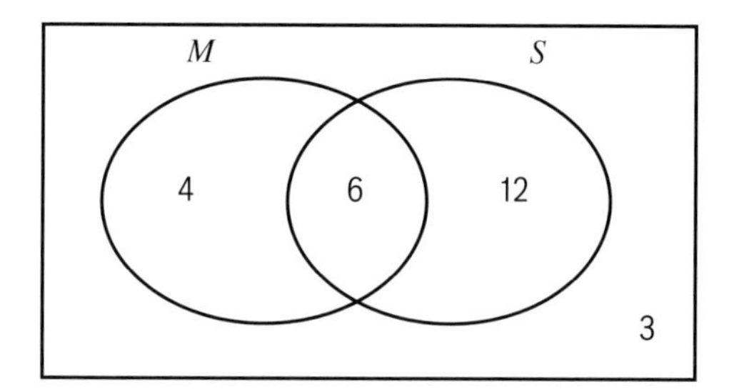

a How many students play music or sport or both?

b Work out the probability that a person picked at random

i plays music or sport or both

ii plays music and sport

iii doesn't play music

Q3a i hint

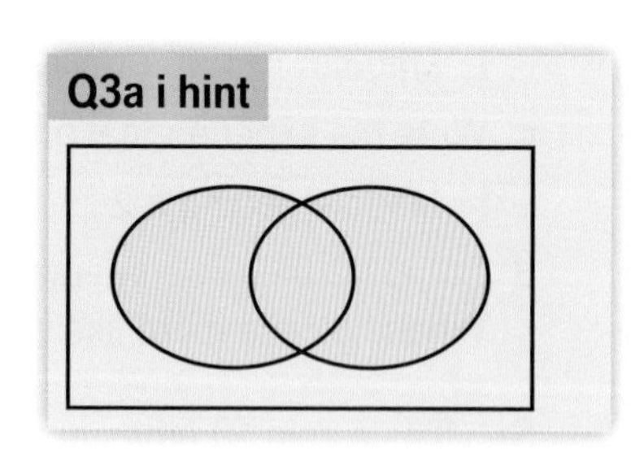

4 Match each label to a diagram.

a $A \cap B$ **b** $\mathscr{E}$ **c** $A \cup B$

i

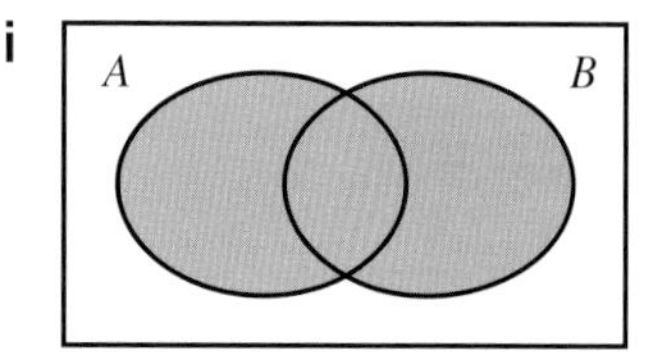

ii

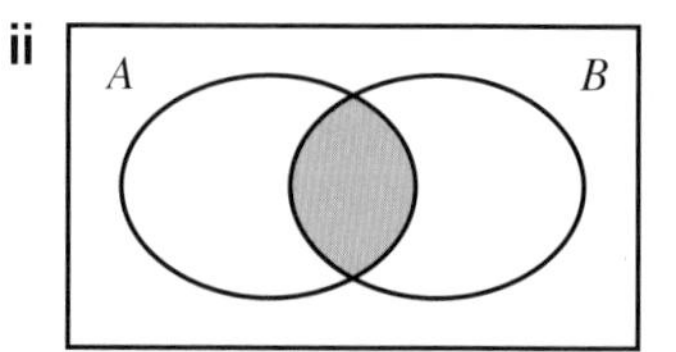

iii

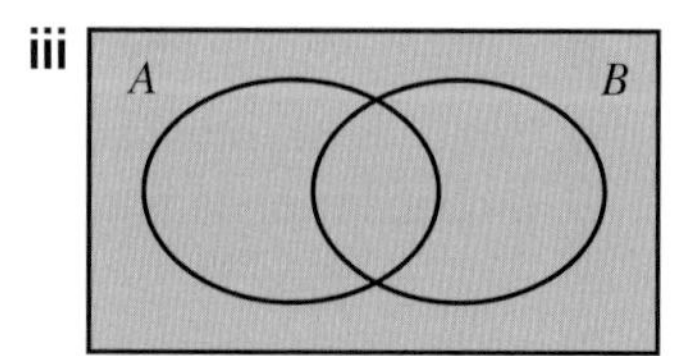

5 Look at this Venn diagram.

List the numbers

a in A **b** in B

c in $A \cup B$ **d** in $A \cap B$

e in $\mathscr{E}$.

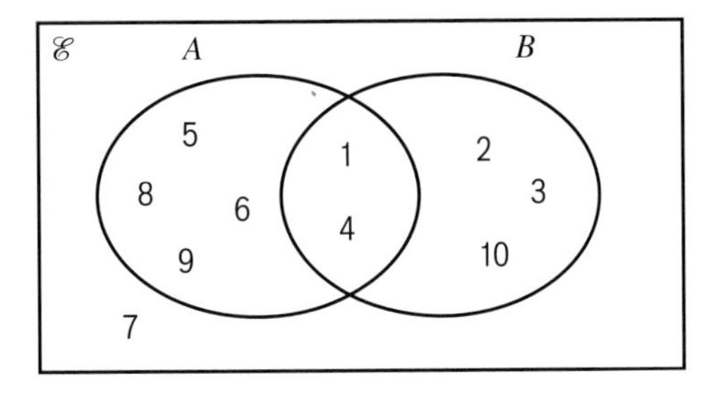

6 Copy and complete the part of the frequency tree for each statement.

a 50 students study Spanish or German.
20 of these students are male.

b 15 of the male students study Spanish.

c 35 students study Spanish.

d Work out and write in the remaining two numbers.

Male
Female
Spanish
German
Spanish
German

e One student is picked at random.
What is the probability that this student is a male who studies German?

$$\frac{\text{number of males who study German}}{\text{number of students}} = \frac{\square}{\square}$$

Dependent events

1 Geeta's pack of sweets contains 7 jelly sweets and 3 fruit sweets.
She takes one sweet at random, eats it and then takes a second sweet at random.

a If her first sweet is jelly, the probability that her second sweet is jelly is $\frac{6}{9}$. Explain why.

b If her first sweet is fruit, the probability that her second sweet is fruit is $\frac{2}{9}$. Explain why.

c Copy and complete the tree diagram.

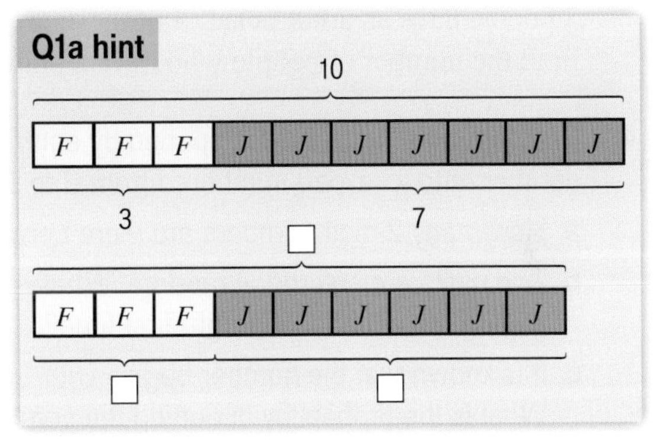

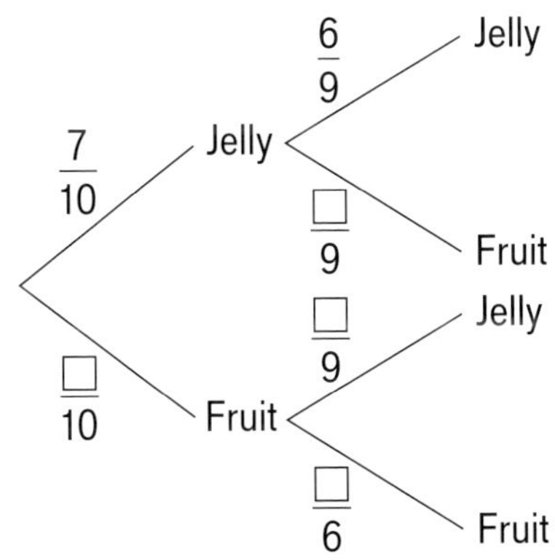

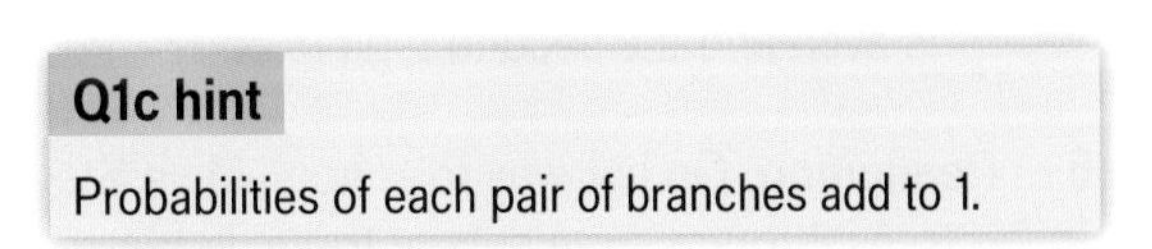

d Work out the probability that Geeta eats 2 jelly sweets.

e **i** Copy and complete this list of all the possible combinations of sweets.

(J, J) $(J, \square)$ $(F, \square)$ $(F, \square)$

ii Which combinations have 1 jelly sweet and 1 fruit sweet?

iii Use the tree diagram to find the probability of each combination from part **e ii**.

iv Add your results from part **e iii** to find the probability that Geeta eats 1 jelly sweet and 1 fruit sweet.

13 Extend

1 **Problem-solving** Lynn has a bag containing 7 blue counters and 8 red counters.

 a Lynn picks a counter at random. Write the probability that the counter is red.

 b Lynn adds blue counters to the bag so the probability of picking a red counter is now $\frac{1}{3}$.
 How many blue counters did she add?

Q1b hint

$\frac{8}{\square} = \frac{1}{3}$

2 **Reasoning** The table shows the probabilities of picking coloured counters from a box.

	Blue	Green	Red	Yellow
Probability	0.1	0.15	0.5	0.25

 a Calculate the expected results for 10 counters.

 b Explain why there must be more than 10 counters in the box.

 c What is the smallest number of counters that could be in the box?

Exam-style question

3 There are 20 people on a bus. Each person has an adult ticket or a child ticket or a bus pass.
There are 14 females on the bus.
8 of the females have a child ticket. 2 of the males have a bus pass.
5 people have an adult ticket. 3 of these are male.
Find the number of people who have a child ticket. **(4 marks)**

4 **Problem-solving** A safe is opened by entering a 2-digit number onto a keypad numbered from 0 to 9.

 a How many 2-digit numbers are there between 00 and 99?

 b Two digits are entered at random. What is the probability of getting the combination right first time?

 c It is known that the number begins with 1 or 2.
 What is the probability of getting the combination right?

 d How many possible combinations are there where the first digit is even and the second digit is odd?

5 **Reasoning** On her way to school, Miss Stevens passes through three sets of traffic lights.
The probability that she stops at any of the sets of lights is 0.4.

 a Copy and complete the tree diagram.

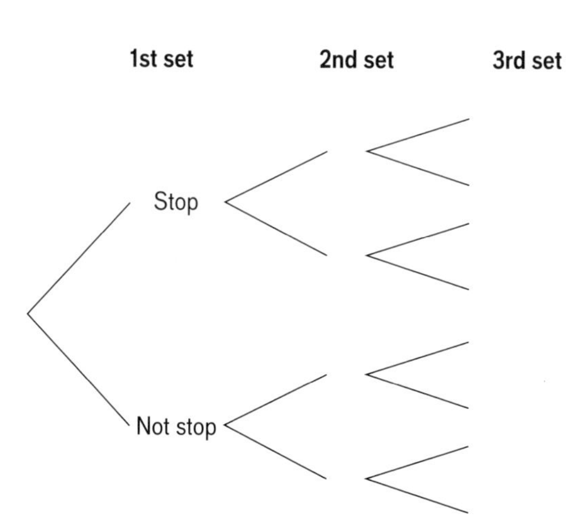

 b Work out the probability that Miss Stevens stops at

  i all three sets of lights

  ii none of the sets of lights

  iii just one set of lights

  iv two sets of lights, using your answers to parts **b i** to **iii**

Exam-style question

6 There are some counters in a bag.
The counters are red or blue or green or yellow.
Ed is going to take a counter at random from the bag.
The table shows each of the probabilities that the counter will be red or will be blue.

Colour	red	blue	green	yellow
Probability	0.15	0.55		

There are 12 red counters in the bag.

The probability that the counter Ed takes will be green is twice the probability that the counter will be yellow.

a Work out the number of green counters in the bag. **(4 marks)**

A coin is taken at random from a bag of coins.
The probability that the coin is silver is 0.5.
There must be an even number of coins in the bag.

b Explain why. **(1 mark)**

Exam tip

If you use diagrams on your working or explanation, make sure they are clear and easy to read.

Exam-style question

7 $\mathscr{E}$ = {even numbers between 11 and 31}
A = {12, 14, 16, 18, 20}
B = {12, 18, 24, 30}
C = {16, 18, 20, 24, 26}

a Complete the Venn diagram for this information.

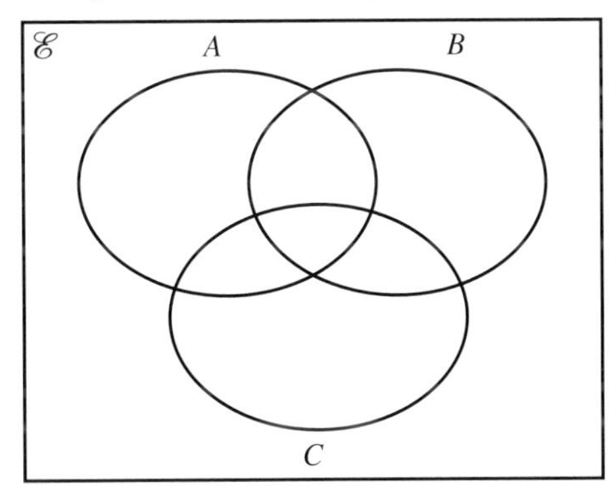

(4 marks)

A number is chosen at random from $\mathscr{E}$.

b Find the probability that the number is a member of $B \cap C$. **(2 marks)**

Exam tip

Shade the part of the diagram you need.

Exam-style question

8 A biased coin is flipped 200 times.
The experimental probability of landing 'Heads' is $\frac{7}{10}$.
Work out the estimate for the probability that the coin will land 'Heads' the first time and 'Tails' the second time. **(2 marks)**

13 Test ready

Summary of key points

To revise for the test:

- Read each key point, find a question on it in the mastery lesson, and check you can work out the answer.
- If you cannot, try some other questions from the mastery lesson or ask for help.

Key points

1. Probability $= \dfrac{\text{number of successful outcomes}}{\text{total number of possible outcomes}}$ → **13.1**
2. If the probability of an event happening is P, the probability of it not happening is 1 – P. → **13.1**
3. Events are **mutually exclusive** when they cannot happen at the same time. → **13.1**
4. The probabilities of an **exhaustive** set of **mutually exclusive** events sum to 1. → **13.1**
5. Estimate for predicted number of outcomes = probability × number of trials. → **13.1**
6. A **sample space diagram** shows all the possible outcomes.
   You can use it to find a theoretical probability, based on equally likely outcomes. → **13.2**
7. When there are 3 ways of making the first choice and 3 ways of making the second, there are 3 × 3 = 9 ways of choosing two objects.

2nd choice			
3			
2			
1			
	1	2	3

1st choice → **13.2**

8. You can estimate the probability of an event from the results of an experiment or survey.

   Estimated probability $= \dfrac{\text{frequency of event}}{\text{total frequency}}$

   Estimated probability is also called **experimental probability.** → **13.3**
9. The **relative frequency** of an event is also an estimate of the probability.

   Relative frequency $= \dfrac{\text{number of 'successful' trials}}{\text{total number of trials}}$ → **13.3**
10. The more times you repeat an experiment, the more accurate the experimental probability. → **13.3**
11. Curly brackets { } show a set of values.
    ∈ means is an 'element of'.
    5 ∈ {odd numbers} means '5 is in the set of odd numbers'.
    An element is a 'member' of a set. Elements are often numbers, but could be letters, items of clothing or even body parts.
    $\mathscr{E}$ means the universal set – all the elements being considered. → **13.4**
12. $A \cap B$ means A intersection B.
    This is the elements that are in A *and* in B.

A B

Key points

$A \cup B$ means A union B.
This is the elements that are in A *or* in B *or* both.

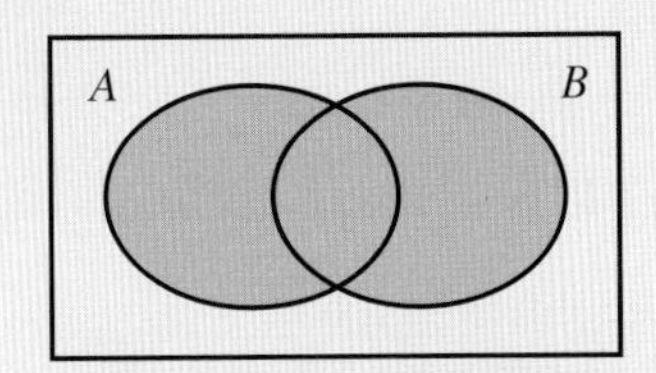

A' means the elements not in A.

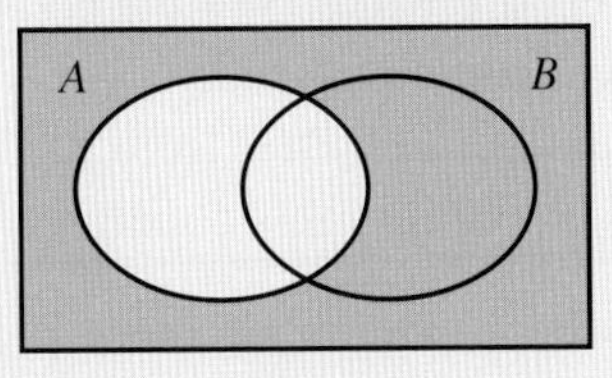

→ 13.4

13 A **frequency tree** shows the number of options for different choices. → 13.5

14 Two events are **independent** when the results of one do not affect the results of the other. → 13.5

15 A **probability tree diagram** shows the probabilities of different outcomes. → 13.5

16 When the outcome of one event changes the possible outcomes of the next event, the two events are not independent. The second event is **dependent** on the first. → 13.6

Sample student answer

Exam-style question

Leah has 2 bags of sweets.
In the first bag there are 3 red sweets and 4 yellow sweets.
In the second bag there are 5 red sweets and 1 yellow sweet.
Leah takes at random a sweet from each bag.
The tree diagram shows the probabilities.

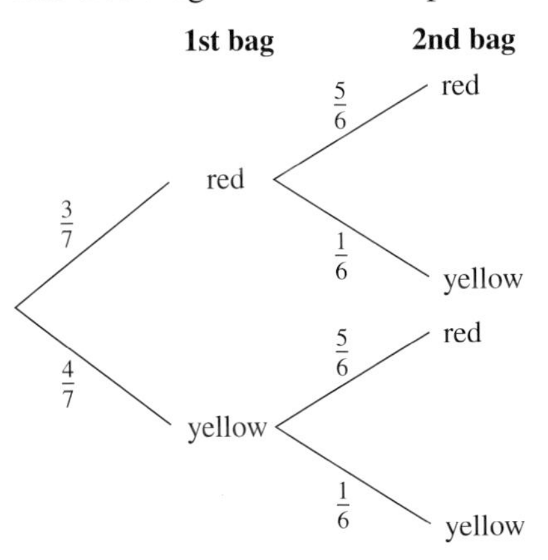

Work out the probability that Leah takes two yellow sweets. **(2 marks)**

$$\frac{4}{7} + \frac{1}{6} \rightarrow \frac{5}{13}$$

This student's answer is not correct.

a What mistakes has the student made?

b Work out the correct answer.

13 Unit test

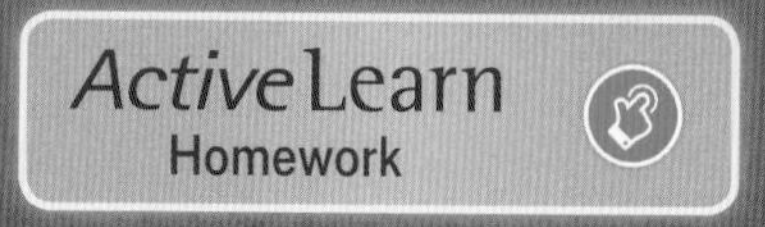

1 There are 15 counters in a bag.
2 of the counters are white. 3 of the counters are pink. 6 of the counters are yellow.
The rest of the counters are red.
Zoe takes at random a counter from the bag.
Show that the probability this counter is red or white is $\frac{2}{5}$. **(3 marks)**

2 Richard spins these two fair spinners and records the total score.
Work out the probability that the total score is an odd number. **(2 marks)**

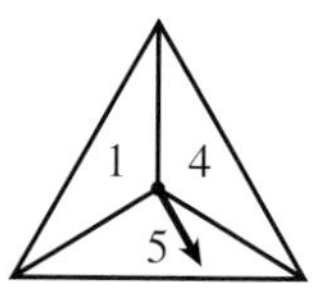

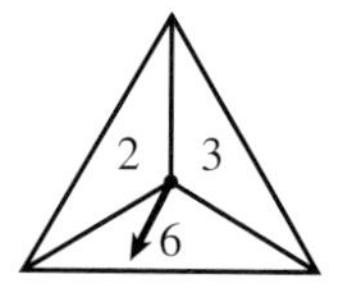

3 There are 32 students in Pat's class.
20 of the students are female. 10 of the male students have short hair.
14 of the students with long hair are female.

a Use the information to complete the two-way table.

	Male	Female	Total
Long hair			
Short hair			
Total			

(3 marks)

One of the students is chosen at random.

b Write down the probability that this student is male with short hair. **(1 mark)**

4 Ben has two bags.
In the first bag there are 3 red balls and 5 blue balls.
In the second bag there are 4 red balls and 2 blue balls.
Ben takes at random a ball from the first bag.
He then takes at random a ball from the second bag.

a Complete the probability tree diagram. **(2 marks)**

b Work out the probability that Ben takes two blue balls. **(2 marks)**

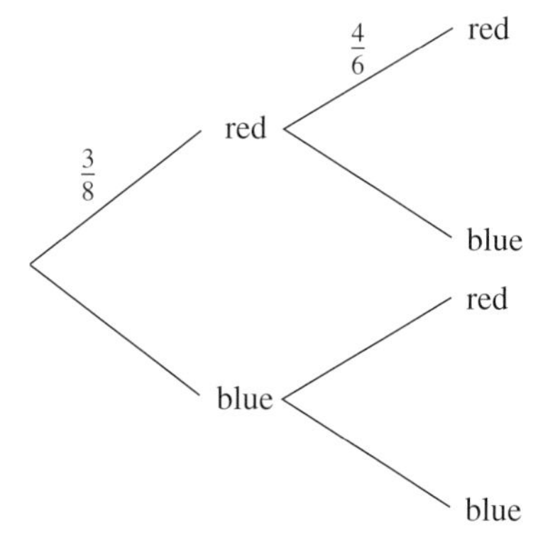

5 A bag contains green, blue and purple balls.
The probability of picking at random a blue ball is $\frac{1}{4}$.
The probability of picking at random a green ball is $\frac{1}{3}$.

a Write the probability of picking a purple ball. **(1 mark)**

b There are 6 blue balls. How many balls are there in total? **(1 mark)**

c Sam adds 12 balls to the bags so that
P(purple) $= \frac{1}{2}$ and P(blue) $= \frac{1}{4}$
How many balls of each colour does he add? **(3 marks)**

6 
$\mathscr{E}$ = {integers from 1 to 20}
A = {odd numbers between 0 and 10}
B = {1, 4, 9, 16}

a Complete the Venn diagram to represent this information. **(4 marks)**

A number is chosen at random from the universal set, $\mathscr{E}$.

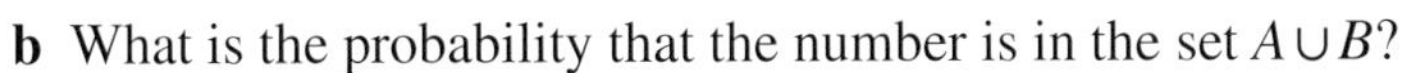
b What is the probability that the number is in the set $A \cup B$? **(2 marks)**

7 
Dr Goode can either walk or drive to work.
The probability that he drives is $\frac{3}{5}$.
If he drives, the probability that he is late is $\frac{1}{3}$.
If he walks, the probability that he is late is $\frac{2}{3}$.

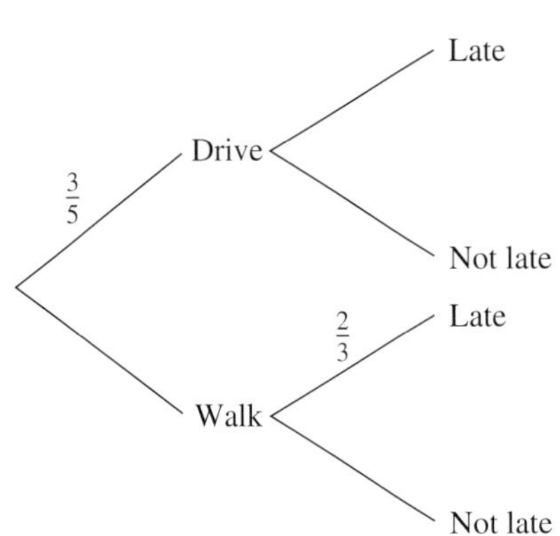

a Copy and complete the tree diagram. **(3 marks)**

b Find the probability that Dr Goode drives and is late. **(1 mark)**

c Work out the probability that Dr Goode is not late. **(2 marks)**

8 Anu and Beth are experimenting with a spinner. Their results are shown in the table.

	Red	Blue	Green
Anu	17	6	7
Beth	36	11	13

a Work out Anu's experimental probability of getting red. **(1 mark)**

b Work out the best estimate for P(red). **(2 marks)**

c The spinner has 10 sections. How many are likely to be green? **(2 marks)**

9 There are 120 students in Year 11.
Each student will stay at 6th form, go to college or get an apprenticeship.
There are 70 boys in Year 11.
15 of the boys get an apprenticeship.
18 of the girls go to college.
20 of the 60 students who stay at 6th form are girls.
One student is selected at random.
Find the probability that this student goes to college. **(5 marks)**

(TOTAL: 35 marks)

10 Challenge Layla rolls two unbiased dice. She adds the scores.
Which is more likely – a total score greater than 7 or less than 7? Explain your answer.

11 Reflect Choose **A**, **B** or **C** to complete each statement about probability.

In this unit, I did…	**A** well	**B** OK	**C** not very well
I think probability is …	**A** easy	**B** OK	**C** hard
When I think about doing probability, I feel…	**A** confident	**B** OK	**C** unsure

Did you answer mostly **A**s and **B**s? Are you surprised by how you feel about probability? Why?
Did you answer mostly **C**s? Find the three questions in this unit that you found the hardest.
Ask someone to explain them to you. Then complete the statements above again.

14 Multiplicative reasoning

14.1 Percentages

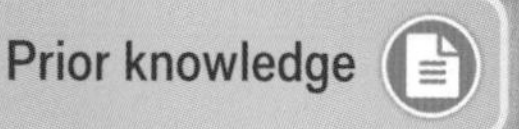

- Calculate percentage profit or loss.
- Express a given number as a percentage of another in more complex situations.
- Find the original amount given the final amount after a percentage increase or decrease.

*Active*Learn
Homework

Warm up

1 **Fluency** Write these as decimals.

a 15% **b** 150% **c** 1.5%

2 What percentage of the original amount will there be after

a an increase of 20% **b** a decrease of 15%

c an increase of 5% **d** a decrease of 2.5%?

Q2a hint

100% + 20% = □%

3 Work out the decimal multiplier for

a an increase of 40% **b** a decrease of 12%

c an increase of 3% **d** a decrease of 1%

4 Work out the new amount when £350 is

a increased by 10% **b** decreased by 4% **c** increased by 100%

5 Work out

a $56 \div 7$ **b** $56 \div 0.7$ **c** $5.6 \div 0.7$

Key point

The original amount is always represented by 100%. If the amount is *increased*, then the new amount will be *more* than 100%. If the amount is *decreased*, then the new amount will be *less* than 100%.

Example

A shop offers a 20% discount in its sale. The sale price of a jumper is £32.
What was the original price?

Sale price = 100% − 20% = 80% = 0.8

Original price → [× 0.8] → £32

Draw a function machine and use inverse operations to find the original price.

£40 ← [÷ 0.8] ← £32

Original price was £40.

6 **Problem-solving** The price of a TV is £480 after 20% VAT is added.
What was the price before VAT was added?

Q6 hint

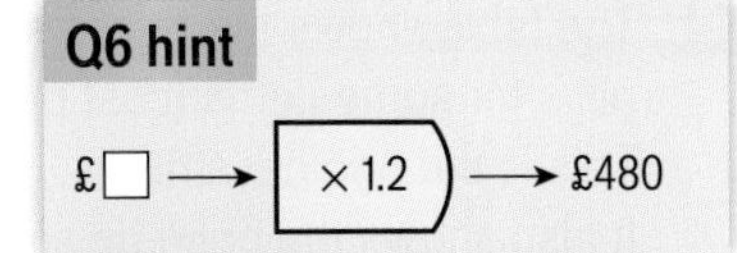

Exam-style question

7 In a sale, the normal price of a hat is reduced by 30%.
The sale price of the hat is £6.30.
Work out the normal price of the hat. **(2 marks)**

Exam tip

If possible, check your answer. For example, decrease your answer by 30%. Do you get £6.30?

8 **Future skills** The value of a house increased by 5%.
The house then had a value of £278 250. Work out the value of the house before the increase.

9 **Problem-solving** Paul buys a phone that is reduced in price by 15%. He pays £297.50.
Matt sees a phone for £360. He has a voucher for a 20% reduction in price. He uses the voucher to pay for the phone.
Who saves more money?

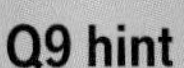

Q9 hint

First work out the original price of Paul's phone.

10 **Future skills**

a Katie earns £52 500 a year. The first £12 500 of her earnings is free of tax.
How much of her earnings must Katie pay tax on?

b She pays 20% income tax on the next £37 500.
Work out how much tax this is per year.

c She then pays 40% income tax on the rest.
Work out how much tax this is per year.

Q10c hint

40% of (£52 500 – □ – □)

d Katie's employer deducts the tax from Katie's monthly pay. The same amount is deducted each month. How much tax does Katie pay each month?

Key point

You can calculate a **percentage** change using the formula

$$\text{percentage change} = \frac{\text{actual change}}{\text{original amount}} \times 100$$

11 **Future skills** An investment 'matures' when the investment period (e.g. 5 years) ends.
Zidan invests £3200. When his investment matures he receives £3334.40.
Copy and complete the calculation to find the percentage increase in his investment.
Actual change = £3334.40 – £3200 = £□

$$\text{Percentage change} = \frac{\text{actual change}}{\text{original amount}} \times 100 = \frac{\square}{3200} = \square\,\%$$

12 **Problem-solving** Sam invests £1200. When her investment matures she receives £1281.60.
Work out the percentage increase in her investment.

Key point

Percentage increase and decrease, profit and loss can all be calculated using the formula for a percentage change.

13 **Problem-solving** Clara invests £2500. When her investment matures she receives £2420.
Calculate the percentage decrease in her investment.

14 The table shows the prices a shopkeeper pays for some items (cost price).
This is the original amount.
It also shows the price he sells them for (selling price).

Item	Cost price	Selling price	Actual profit	Percentage profit
ring	£5	£8		
bracelet	£12	£18		
necklace	£20	£30		
watch	£18	£25		

a Copy and complete the table to work out the percentage profit on each item.

b **Reflect** Is the item with the greatest actual profit the item with the greatest percentage profit?

15 **Future skills** Kellie bought a car for £15 000. A year later she sold it for £13 900.
Work out her percentage loss. Give your answer to 1 decimal place (1 d.p.).

16 **Problem-solving** In 2008, the Office for National Statistics estimated that the UK population was 61 823 800. In 2018, it estimated that the UK population was 66 435 600.
What was the percentage increase in the population?
Give your answer to 3 significant figures (3 s.f.).

17 **Problem-solving** Satvir buys 30 T-shirts for £4 each.
She sells $\frac{1}{2}$ of them for £6 each, $\frac{1}{3}$ of them for £5 each and the rest for £4.50 each.
What is her percentage profit?

18 **Problem-solving / Reasoning** Last year Maria paid £240 for her home insurance.
This year she has to pay £564 for her home insurance.

a **Reflect** Write a sentence explaining how you can tell by looking at the costs for last year and this year that the percentage increase is greater than 100%.

b Work out the percentage increase in the cost of Maria's home insurance.

Exam-style question

19 Paul buys 6 kg of tomatoes to sell.
He pays £15 for the tomatoes.
Paul puts all the tomatoes into bags.
He puts 400 g of tomatoes into each bag.
He sells each bag of tomatoes for £2.20.
Paul sells all the bags.
Work out his percentage profit. **(4 marks)**

Exam tip

Write down the information you need to work out the final answer. For example, to work out the percentage profit, you need the original (cost) price of all the tomatoes and the amount Paul made by selling all the bags of tomatoes.

20 **Problem-solving / Reasoning**
The table shows information about visitor numbers to a cinema in 2018 and in 2019.

Year	Total number of visitors	Ratio of children to adults	Price of child ticket	Price of adult ticket
2018	13 275	2 : 1	£5.60	£8.60
2019	10 410	3 : 2	£7	£10

a Work out the percentage change in the total number of visitors from 2018 to 2019.
Give your answer to 1 decimal place (1 d.p.).

b Does your answer to part **a** show a percentage increase or decrease?

c How many children visited the cinema in 2018?

d Work out the percentage change in the amount of money taken in ticket sales from 2018 to 2019.
Give your answer to 1 d.p.

Q20d hint

What information do you need? How can you use the figures in the table?

14.2 Growth and decay

- Find an amount after repeated percentage changes.
- Solve growth and decay problems.

Warm up

1 **Fluency** **a** $5 \times 5 \times 5 = 5^{\square}$ **b** $1.5 \times 1.5 \times 1.5 = 1.5^{\square}$

c $5 \times 5 \times 5 \times 5 = 5^{\square}$ **d** $1.5 \times 1.5 \times 1.5 \times 1.5 \times 1.5 = 1.5^{\square}$

2 Work out the decimal multiplier for

a an increase of **i** 30% **ii** 3% **iii** 3.5%

b an decrease of **i** 20% **ii** 2% **iii** 2.5%

3 Sunir bought a car for £6000. It lost 30% of its value in the first year.

a What is the multiplier to find the value of the car at the end of the first year?

b What was the value of the car at the end of the first year?

It lost 10% of its value in the second year.

c What is the multiplier to find the value of the car at the end of the second year?

d Use the value you found in part **b** and the multiplier you found in part **c** to find the value of the car at the end of the second year.

e Copy and complete the diagram to show the single decimal number that the original value of the car can be multiplied by to find its value at the end of the two years.

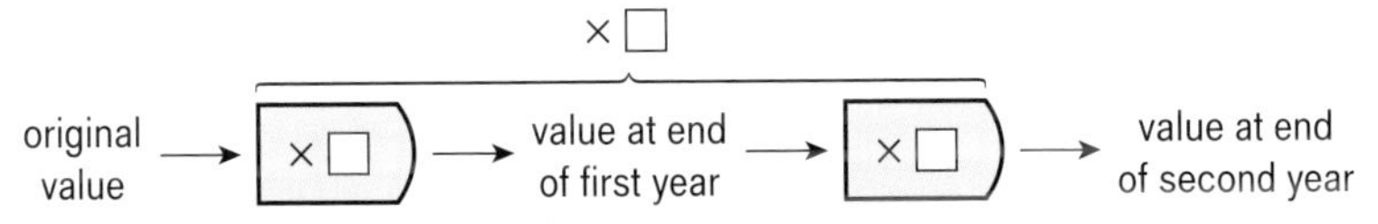

f Use your diagram to check your answer to part **d**.

Q3a hint
Decrease by 30%.

4 Work out the decimal multiplier that represents

a an increase of 40% **b** an increase of 1.5%

c an increase of 40% followed by an increase of 1.5%

5 **Future skills** Munir has a job with an annual (yearly) salary of £25 000. At the end of the first year, he is given a salary increase of 2.5%. At the end of the second year, he is given an increase of 3%.

a Write the single number, as a decimal, that Munir's original salary can be multiplied by to find his salary at the end of the 2 years.

b Work out Munir's salary at the end of 2 years.

6 **Future skills** Crista buys a flat for £120 000. In the first year, the value of the flat increased by 3%.

a What was the value of the flat at the end of the first year?

In the second year, the value of the flat decreased by 2%.

b What was the value of the flat at the end of the second year?

c Write the single decimal number that the original value of the flat can be multiplied by to find its value at the end of two years.

7 Work out the decimal multiplier that represents

a an increase of 11%

b a decrease of 7%

c an increase of 11% followed by a decrease of 7%

d a decrease of 7% followed by another decrease of 7%

8 **Problem-solving** Three years ago, a bird survey counted 1500 tawny owls. Last year, a bird survey of the same area counted 8% fewer tawny owls.
This year, a bird survey of the same area counted 10% fewer tawny owls.
How many tawny owls were counted this year?

Q8 hint

8% fewer means a decrease of 8%.

9 **Reasoning** Mia says, 'An increase of 20% followed by a decrease of 8% is the same as an increase of 12%.'
Is Mia correct? Show working to explain your answer.

10 Copy and complete the calculation for the decimal multiplier that represents

a an increase of 10% for

i 2 years = □×□ ii 3 years = □×□×□ iii 4 years = □×□×□×□

b a decrease of 10% for

i 2 years = □×□ ii 3 years = □×□×□ iii 4 years = □×□×□×□

11 **Reasoning** Ruth is working out the current value of her motorbike.
She bought it 3 years ago for £5200. It has decreased in value by 10% each year.
Here is Ruth's working: $5200 \times (0.9)^3$.

a **Reflect** Write a sentence explaining why Ruth uses the multiplier $(0.9)^3$.

b Work out the current value of Ruth's motorbike.

Key point

Banks and building societies pay **compound interest**. At the end of the first year, interest is paid on the money in the account. The interest is added to the amount in the account. At the end of the second year, interest is paid on the original amount in the account and on the interest earned in the first year, and so on.

Example

£2000 is invested for 2 years at 5% per annum compound interest.
Work out

Hint

Per annum means per year.

a the total amount in the account after 2 years

b the total interest earned over the 2 years

a 100% + 5% = 105% so the multiplier is 1.05 — Work out the multiplier for an increase of 5%.
After 1 year: $£2000 \times 1.05$
After 2 years: $£2000 \times 1.05 \times 1.05$
$= £2000 \times 1.05^2$ — Multiply the original amount by 1.05^2 to find the amount in the account after 2 years.
= £2205 — Use a calculator.
The total amount in the account after 2 years is £2205.

b £2205 − £2000 = £205 — Subtract the original amount to find the interest.
The total interest earned over the 2 years is £205.

12 **Future skills** £2500 is invested for 2 years at 4% per annum compound interest.
Work out the total interest earned over the 2 years.

13 **Future skills** £3000 is invested for 3 years at 2% per annum compound interest.
Work out the total interest earned over the 3 years.

Exam-style question

14 Antony invests £20 000 in a savings account for 3 years.
The account pays compound interest at a rate of 1.5% per annum.
Calculate the total amount of interest Antony will receive at the end of 3 years. **(3 marks)**

Exam tip

Do not round during your working. Only round your final answer.

Exam-style question

15 Pounds Bank has two types of account.
Both pay compound interest.

Cash saver	**Shares saver**
Interest	Interest
1.2% per annum	1.9% per annum

Annie invests £3100 in the cash saver account.
Jo invests £2950 in the shares saver account.

a Work out who will get more interest by the end of 3 years.
You must show your working. **(4 marks)**

b In the third year, the rate of interest for the shares account is changed to 2%.
Does this affect who will get more interest by the end of 3 years?
Give a reason for your answer. **(1 mark)**

Exam tip

Always think carefully about your multiplier when working out compound interest.

16 **Future skills** 2000 bacteria are put into a Petri dish.
The number of bacteria increases by 30% every hour.
How many bacteria will be in the dish after 8 hours?

Q16 hint

- after 1 hour $= 2000 \times 1.3$
- after 2 hours $= 2000 \times 1.3^{\square}$

17 **Problem-solving** The level of activity of a radioactive source decreases by 5% per hour.
The activity is 1400 counts per second.
What will it be 10 hours later?

Q17 hint

Counts per second after

- 1 hour $= 1400 \times 0.95$
- 2 hours $= 1400 \times 0.95^{\square}$

18 **Problem-solving** The level of activity of a sample containing a radioactive isotope is 120 000 counts per minute.
The half-life is the time taken for the count rate to fall to half its starting value.
The half-life is 2 days. What will the count rate be after 10 days?

Q18 hint

What is the multiplier for a half-life?

19 **Problem-solving** In 2014 a fast-food chain has 160 outlets in the UK.
The number of outlets increases at a rate of 8% each year.

a How many outlets will it have in 2020?

b **Reflect** What is an appropriate degree of accuracy for this question?

14.3 Compound measures

- Solve problems involving compound measures.

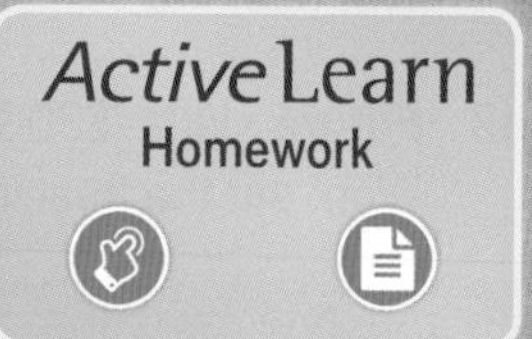

Warm up

1 Fluency Work out these rates.

a Payment of £80 for an 8-hour day is a rate of £□ per hour.

b Travelling 240 km on 12 litres of petrol is a rate of □ km per □.

2 $a = \frac{b}{c}$

Find

a a when $b = 12$ and $c = 3$ **b** b when $a = 12$ and $c = 3$ **c** c when $a = 12$ and $b = 3$

3 A cuboid has dimensions 2 m by 3 m by 4 m. Work out its volume.

4 Copy and complete.

a 1 kg = □ g **b** 1 m^3 = □ cm^3

Key point

Compound measures combine measures of two different quantities.
For example, rate of pay is a measure of money and time.

5 Future skills George works a basic 35-hour week. He is paid 'time and a half' for each hour he works on a Saturday and 'double time' for each hour he works on a Sunday.
His basic hourly rate of pay is £16.50.

a Work out George's rate of pay for Saturday = 1.5 × □ = □

b Work out George's rate of pay for Sunday = 2 × □ = □

c How much is George paid when he works a basic 35-hour week, plus 4 hours on Saturday and 3 hours on Sunday?

6 Problem-solving Water is leaking from a water butt at the rate of 3 litres per hour.

a Work out how much water leaks from the water butt in

i 20 minutes **ii** 10 minutes

b Reflect How can you use your answer to **Q6 a ii** to work out how much water leaks from the water butt in 50 minutes?

Initially there are 120 litres of water in the water butt.

c Work out how long it takes for all the water to leak from the water butt.

Q6 hint

litres		minutes
3	:	60
□	:	20

÷□ (litres), ÷□ (minutes); 1 hour = 60 minutes

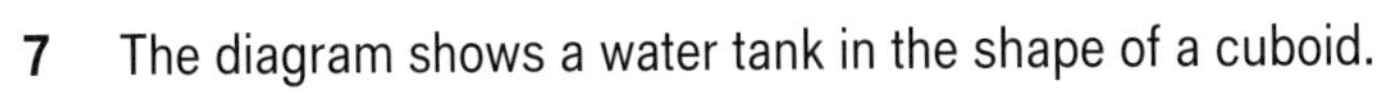

7 The diagram shows a water tank in the shape of a cuboid.

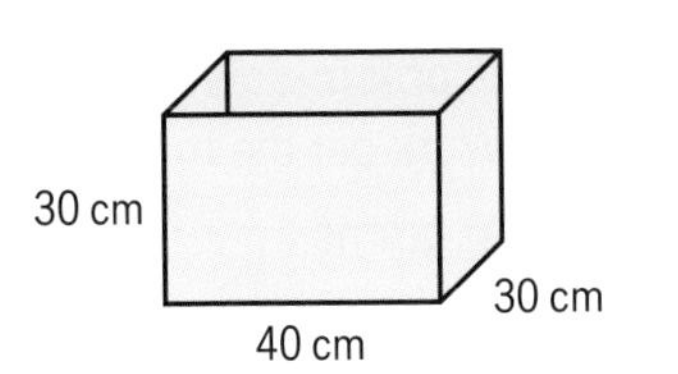

a Work out the volume of the water tank.

Water is poured into the tank at a rate of 5 litres per minute.
1 litre = $1000\,cm^3$

b Work out the time it takes to fill the water tank completely.
Give your answer in minutes.

8 **Problem-solving** A car travels 300 km and uses 20 litres of petrol.

a Work out the average rate of petrol usage in kilometres per litre.

b **Reflect** Write a sentence explaining why the question asks for 'average' rate rather than exact rate.

Q8 hint

Kilometres per litre means the number of kilometres travelled on each 1 litre of fuel.

Key point

Density is a compound measure. It is the **mass** of substance contained in a certain **volume**.

$$\text{density} = \frac{\text{mass}}{\text{volume}} \text{ or } D = \frac{M}{V}$$

Density is usually measured in g/cm^3. To calculate density in g/cm^3, you need to know the mass in grams (g) and the volume in cubic centimetres (cm^3). It can also be measured in kg/m^3.

9 A cast iron nail has a mass of 10.8 g. Its volume is $1.5cm^3$.
Copy and complete to work out its density in g/cm^3.

$$\text{density} = \frac{\text{mass}}{\text{volume}} = \frac{\square}{\square}$$

10 **Problem-solving** A sample of bronze has a mass of 2 kg.
Its volume is $250\,cm^3$.
What is its density in g/cm^3?

Q10 hint

2 kg = $\square$g

11 **Problem-solving** A cubic metre of concrete has a mass of 2400 kg.
What is the density of the concrete in g/cm^3?

Q11 hint

$1\,m^3 = \square cm^3$
2400 kg = $\square$g

12 **Reasoning** $5\,cm^3$ of gold has a mass of 96.6 g.
$5\,cm^3$ of platinum has a mass of 107.25 g.

a Work out the density of each metal.

b Which metal is more dense? Explain your answer.

13 **Problem-solving** This block of wood has a mass of 288 g.
What is its density?

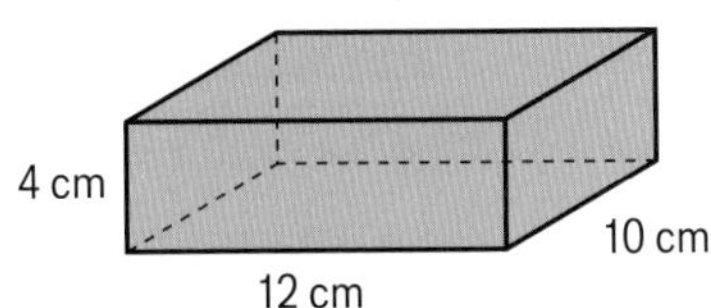

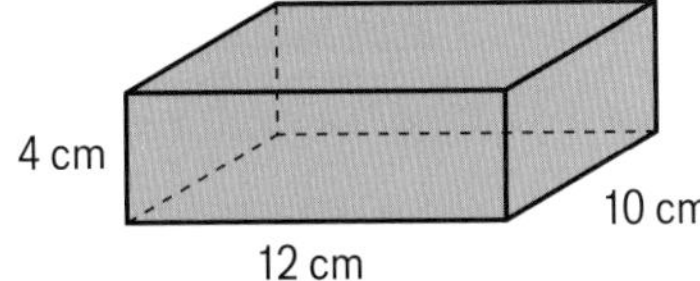

Q13 hint

Volume of the block of wood
$\square cm^3$

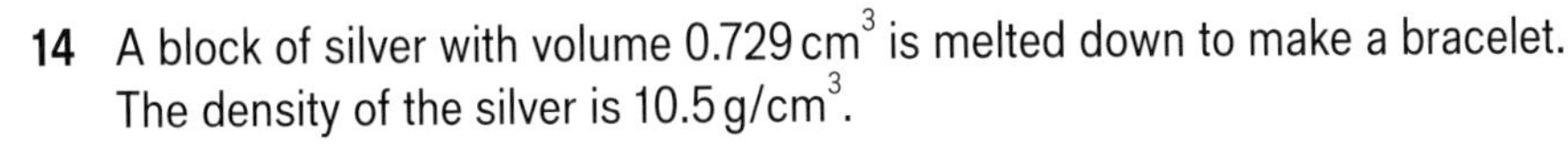

14 A block of silver with volume $0.729\,cm^3$ is melted down to make a bracelet.
The density of the silver is $10.5\,g/cm^3$.

a Substitute the values you know into $D = \frac{M}{V}$.

b Rearrange the formula to work out the mass of the silver bracelet.

15 Problem-solving An iron bar has volume 650 cm^3 and density 7.87 g/cm^3.
Work out the mass of the iron. Give your answer in

a grams **b** kilograms

16 Problem-solving The density of copper is 8.96 g/cm^3. The density of zinc is 7.13 g/cm^3.

a A metal-worker has 6.5 cm^3 of copper and 3.5 cm^3 of zinc.

i What mass of copper does he have?

ii What mass of zinc does he have?

b The metal-worker uses the copper and zinc to make brass.
Work out the density of the brass.

17 Iron has a density of 7.87 g/cm^3.
The mass of a piece of iron is 5.4 kg.

a Substitute the values you know into $D = \frac{M}{V}$.

b Rearrange the formula to work out the volume of the piece of iron. Give your answer correct to 3 significant figures.

Exam-style question

18 A ball of clay has a mass of 1.3 kg.
The density of the clay is 1.746 g/cm^3.
Work out the volume of the ball of clay.
Give you answer correct to 3 significant figures.

(3 marks)

Exam tip

You must know the formula

density $= \frac{\text{mass}}{\text{volume}} = \frac{\square}{\square}$

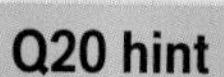

Key point

Pressure is a compound measure. It is the **force** applied over an **area**.
To calculate pressure, you need to know the force in newtons (N) and the area in square metres (m^2).

pressure $= \frac{\text{force}}{\text{area}}$ or $P = \frac{F}{A}$

Pressure is usually measured in N/m^2.

19 Problem-solving A force of 45 N is applied to an area of 2.6 m^2.
Work out the pressure in N/m^2.

20 A force is applied to an area of 4.5 m^2.
It produces a pressure of 20 N/m^2. Work out the force in N.

Q20 hint

Substitute the values you know into $P = \frac{F}{A}$.
Rearrange to find F.

21 Problem-solving A rectangular metal sheet of length 80 cm and width 50 cm exerts a force of 8 N on a table.
Work out the pressure in N/m^2.

Exam-style question

22 A force of 90 newtons acts on an area of 25 cm^2.
The force is increased by 30 newtons.
The area is doubled.
Noel says 'The pressure decreases by a third.'
Is Noel correct?
You must show how you get your answer.

(3 marks)

14.4 Distance, speed and time

- Convert between metric measures of speed.
- Calculate average speed, distance and time.
- Use formulae to calculate speed and acceleration.

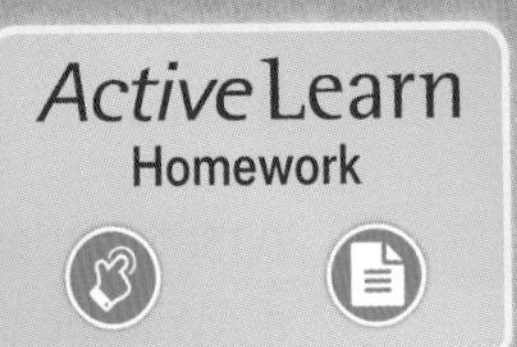

Warm up

1 **Fluency** A cyclist travels 15 km in 1 hour.

a What is the cyclist's average speed in km/h?

b How far does the cyclist travel in metres?

2 $d = st$

a Find d when $s = 10$ and $t = 0.3$.

b Find s when $d = 400$ and $t = 80$.

c Find t when $d = 150$ and $s = 50$.

3 Write these times in hours (as decimals).

a 30 minutes b 15 minutes

c 75 minutes d 6 minutes

Q3 hint

$\frac{\square}{60} = \square$

4 Write these times in hours and minutes.

a 0.2 hours

b 1.75 hours

c 3.4 hours

Q4 hint

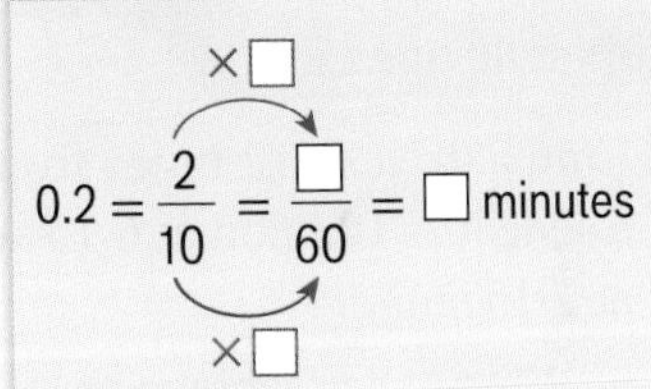

Key point

Speed is a compound measure. It is a measure of distance travelled and time taken.
You can calculate speed using the formula

$$\text{speed} = \frac{\text{distance}}{\text{time}} \text{ or } S = \frac{D}{T}$$

Speed is often measured in metres per second (m/s), kilometres per hour (km/h) or miles per hour (mph).

5 **Problem-solving** A train travels 426 km in 3 hours.
Copy and complete to work out its average speed.

$$\text{speed} = \frac{\text{distance}}{\text{time}} = \frac{\square}{\square}$$
$$= \square \text{ km/h}$$

6 **Problem-solving** A car travels 108 miles in 2 hours 15 minutes.

a Write the time as a decimal.

b Work out the average speed in mph.

7 **Problem-solving** Work out the average speed for these journeys.

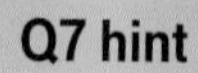

Q7 hint First write time as a decimal.

a A man cycles 45 km in 3 hours 45 minutes.

b A train travels 200 miles in 2 hours 6 minutes.

8 A motorcycle travels 45 minutes at an average speed of 46 mph.

a Write the time as a decimal.

b Substitute the values you know into $S = \frac{D}{T}$.

c Rearrange the formula to work out the distance travelled by the motorcyclist.

9 **Problem-solving** Work out the distance travelled for these journeys.

a An aeroplane travels for 3 hours 20 minutes at an average speed of 600 mph.

b A tram travels for 50 minutes at an average speed of 21 km/h.

10 Tina jogs 12 km at an average speed of 10 km/h.

a Substitute the values you know into $S = \frac{D}{T}$.

b Rearrange the formula to work out the time taken for Tina's jog.

c Convert your decimal answer in part **b** to hours and minutes.

11 **Problem-solving** Work out the time taken for these journeys.

a A spider climbs a 3.2 m wall at an average speed of 8 cm/s.

b A swallow flies 0.5 km at an average speed of 11 m/s.

Exam-style question

12 Emma drives 412 miles from London to Glasgow.
She leaves London at 6.15 am and arrives in Glasgow at 2.15 pm.

a What is her average speed? **(2 marks)**

Emma drives on from Glasgow to Inverness at an average speed of 48 mph.
She leaves Glasgow at 4.50 pm and arrives in Inverness at 8.20 pm.

b What distance does Emma drive between Glasgow and Inverness? **(2 marks)**

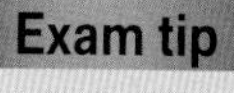

Exam tip You must know the formula

$\text{speed} = \frac{\text{distance}}{\text{time}} = \frac{\square}{\square}$

13 Paul lives 504 metres from the tram station.
It takes him 6 minutes to walk to the station.
What is his average speed in m/s?

Q13 hint The question asks for the speed in m/s so convert time to seconds.

14 A snail moves 20 cm at an average speed of 1 mm/s.
Work out the time it takes.

Q14 hint The speed is in mm/s so convert the distance to millimetres.

15 A horse gallops an average speed of 45 km/h.

a How far does the horse gallop in 1 hour?

b **Reasoning** Would the horse gallop more or less distance in 1 minute?

c How far will the horse gallop in 1 minute? Give your answer in m.

d How far will the horse gallop in 1 second?

Q15c hint

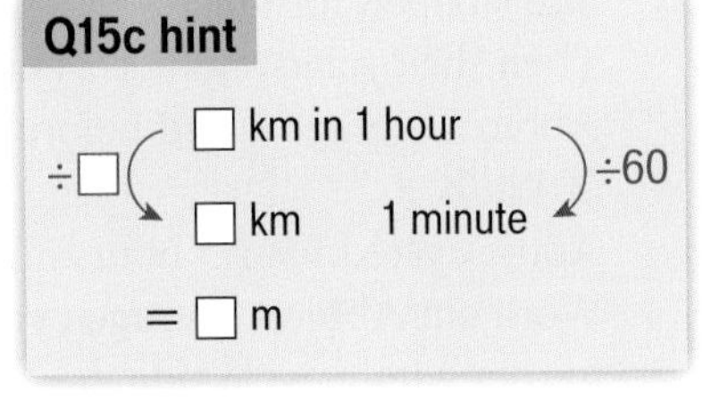

16 Convert these speeds from metres per second (m/s) to metres per hour (m/h).

a 1 m/s

b 12 m/s

c 8 m/s

Q16a hint

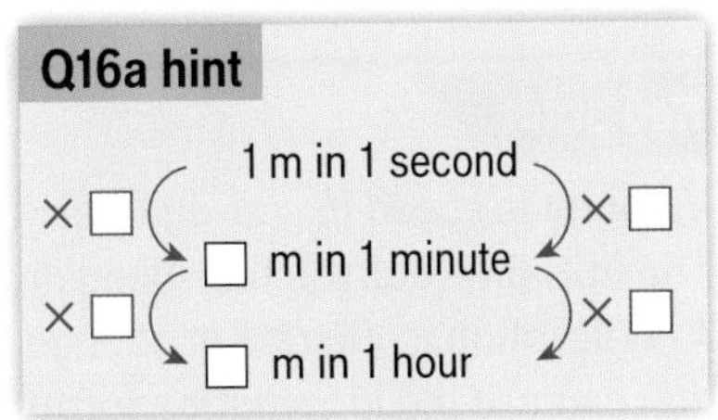

17 Convert these speeds from metres per second (m/s) to kilometres per hour (km/h).

a 5 m/s **b** 18 m/s **c** 30 m/s

Q17 hint

Convert m/s to m/h then
□ m in 1 hour = □ km in 1 hour

18 **Problem-solving** Convert these speeds from kilometres per hour (km/h) to metres per second (m/s).

a 54 km/h

b 72 km/h

c 9 km/h

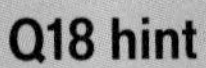

Q18 hint

Start with
54 km in 1 hour = □ m in 1 hour.

19 **Problem-solving / Reasoning** A Formula 1 racing car has a top speed of 350 km/h. A peregrine falcon (the fastest bird) has a top speed of 108 m/s. Which is faster? Explain your answer.

20 **Problem-solving** A plane travels 570 miles per hour.

a Work out an estimate for the number of miles the plane travels in 1 minute.

b Work out an estimate for the number of seconds the plane takes to travel 1 mile.

Q20 hint

The question says work out an 'estimate' so round 570 miles to the nearest 100 miles.

Key point

$$\text{Average speed} = \frac{\text{total distance}}{\text{total time}}$$

21 Karl travels the first 35 kilometres of his journey in 45 minutes.

He then travels the last 65 km in $1\frac{1}{2}$ hours.

a Work out the total distance.

b Work out the total time.

c Work out Karl's average speed for the whole journey in km/h.

Exam-style question

22 Abbie cycled 40 km.
She stopped for lunch.
Then she cycled another 24 km.
Abbie's average speed before lunch was 20 km/h.
Abbie cycled for 1.5 hours after lunch.
Work out Abbie's average cycling speed for her whole cycle ride. **(4 marks)**

Exam tip

When a question includes a lot of information, it can help to put it in a table. Then fill in the gaps with the information you require to answer the question. For example

	Distance	Time	Speed
Before lunch			
After lunch			

Key point

Velocity is speed in a given direction. It is often measured in m/s.
The **initial velocity** is the speed in a given direction at the start of the motion. It may be zero.
Acceleration is the rate of change of velocity. It is often measured in m/s^2.
You can use the **kinematics formulae** for calculations with moving objects.

$v = u + at \qquad s = ut + \frac{1}{2}at^2 \qquad v^2 = u^2 + 2as$

a is the constant **acceleration**, u is the initial **velocity**, v is the final velocity, t is the time taken and s is the displacement from the position when $t = 0$.

23 Use the formula $v = u + at$ to work out

a v when
  i $u = 3$, $a = 2$ and $t = 4$
  ii $u = 4$, $a = 0.5$ and $t = 10$

b u when
  i $v = 5$, $a = 1$ and $t = 3$
  ii $v = 4$, $a = 0.5$ and $t = 6$

c a when
  i $v = 7$, $u = 1$ and $t = 2$
  ii $v = 4$, $u = 0.2$ and $t = 5$

Q23b, c hint

First substitute into the formula, then solve.

24 Copy and complete the table using the formula $s = ut + \frac{1}{2}at^2$.

s (m)	u (m/s)	a (m/s^2)	t (s)
10		2	1
8	2		1
15		3	2

25 Use the formula $v^2 = u^2 + 2as$ to work out

a v when $u = 8$, $a = 2$ and $s = 9$

b u when $v = 6$, $a = 1$ and $s = 5.5$

Q25a hint

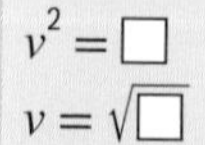

$v^2 = \square$

$v = \sqrt{\square}$

26 A car starts from rest (so $u = 0$) and accelerates at 5 m/s^2 for 200 m.
Work out the final velocity in m/s.

Q26 hint

You are given a, u and s, and need to find v. t is not given so use $v^2 = u^2 + 2as$

14.5 Direct and inverse proportion

- Use ratio and proportion in measures and conversions.
- Use inverse proportion.

Warm up

1 **Fluency** Which of these graphs shows direct proportion?

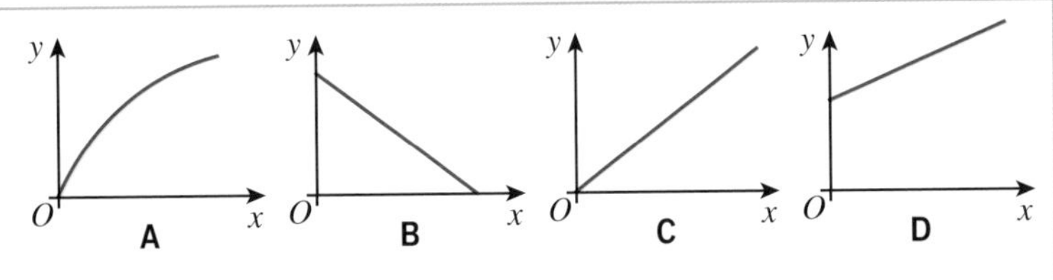

2 Solve these equations to work out the value of n.
Write your answers as a decimal.

a $4n = 5$ **b** $5n = 4$ **c** $6n = 3$

3 Write each ratio as a unit ratio.

a 3 : 12 **b** 6 : 2 **c** 11 : 2 **d** 3 : 5

4 The ratio of kilograms to pounds is 1 : 2.2.

a Copy and complete this table of values for kilograms and pounds.

Kilograms	0	5	10
Pounds			

b Draw a conversion graph for kilograms to pounds.
Put kilograms on the x-axis and pounds on the y-axis.

c Are kilograms and pounds in direct proportion? Explain.

d Write a formula linking kilograms (x) and pounds (y).

Q4d hint
$y = \square x$

5 The table shows the pressure for different forces on an area of $4\,m^2$.

Force, F (N)	4	10	16	20
Pressure, P (N/m^2)	1	2.5	4	5

a Write the ratio force : pressure for each force.

b Write each ratio from part **a** in its simplest form. What do you notice?

c Write a formula linking the values of F and P in the table.

d **Reflect** How can the unit ratio help you write the formula?

Q5c hint
$F : P$
$\square : \square$
$F = \square P$

Key point

When two variables are in **direct proportion**, pairs of values are in the same ratio.

6 For each ratio, write a formula connecting the variables.

a $x : y$
3 : 1

b $s : t$
1 : 6

c $p : q$
1 : 3.5

d $F : m$
5 : 2

e $a : b$
4 : 5

f $r : s$
4 : 7

Q6d–f hint
Write each ratio as a unit ratio

7 **Reasoning** The table shows the distance (d) travelled by a car over a period of time (t).

Distance, d (miles)	8	16	24	32	40
Time, t (minutes)	10	20	30	40	50

a Write the ratios for the pairs of values in the table.

Q7a hint
$d : t$

b Is d in direct proportion to t? Explain your answer.

c Write a formula that shows the relationship between distance (d) and time (t).

d Use your formula to work out the distance travelled after 25 minutes.

Key point

When
- y varies as x
- y varies directly as x
- y is in **direct proportion** to x

you can write $y \propto x$

$y \propto x$ means 'y is proportional to x'.

When $y \propto x$, then $y = kx$, where k is the **constant of proportionality**.

This means the value of k is constant (stays the same) when x and y vary.

Example

The number of dollars, D, is directly proportional to the number of pounds, P.

a One day, \$1 = £0.64. Write a formula linking D and P.

b Use your formula to convert £15 to dollars.

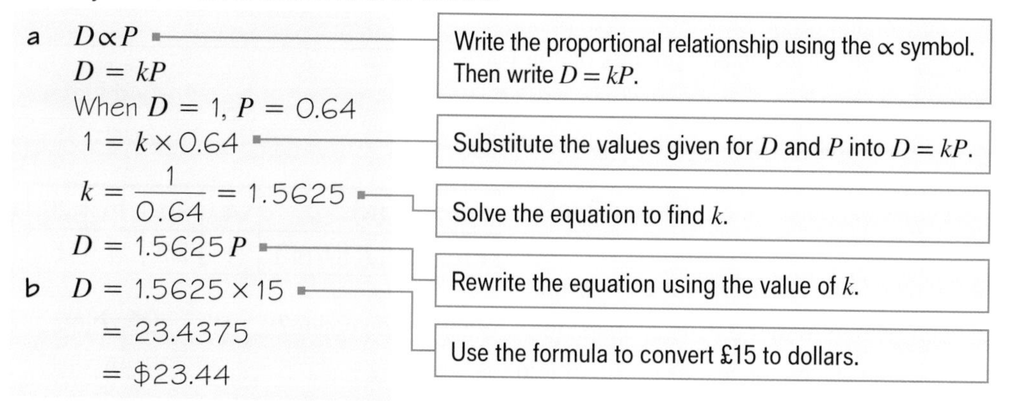

a $D \propto P$ — Write the proportional relationship using the $\propto$ symbol. Then write $D = kP$.

$D = kP$

When $D = 1$, $P = 0.64$

$1 = k \times 0.64$ — Substitute the values given for D and P into $D = kP$.

$k = \dfrac{1}{0.64} = 1.5625$ — Solve the equation to find k.

$D = 1.5625P$ — Rewrite the equation using the value of k.

b $D = 1.5625 \times 15$ — Use the formula to convert £15 to dollars.

$= 23.4375$

$= \$23.44$

8 **a** Copy and complete each proportional relationship.

i F is proportional to m: $F \propto \square$ ii V is proportional to x: $V \propto \square$

b Write each proportional relationship using k, the constant of proportionality.

9 x is proportional to t. When $x = 6$, $t = 4$.

a Write the proportional relationship for x and t.

b Use the given values of x and t to find k.

c Find x when $t = 10$.

Q9 hint
$x \propto$
$x =$

10 r is directly proportional to s. When $r = 5$, $s = 4$.

a Write a formula for r in terms of s.

b Find r when $s = 7$.

Q10a hint
Start $r \propto$
$r =$

11 The number of pounds, P, is directly proportional to the number of euros, E.

a One day €1 = £0.80. Write a formula connecting P and E.

b Use your formula to convert €80 to pounds.

Key point

When two variables X and Y are in **inverse proportion**,

$X \propto \frac{1}{Y}$ $\quad X = \frac{k}{Y}$ $\quad Y = \frac{k}{X}$ $\quad XY = k$ (constant)

Exam-style question

12 Vic's pay is directly proportional to the hours he works.
When Vic works 7.5 hours in a day, his pay is £165.
How much pay does Vic receive for a month when he works 172.5 hours? **(2 marks)**

Exam tip

When a question states that two quantities are in direct proportion, write the proportional relationship and then a formula.

13 **Reasoning** X and Y are in inverse proportion. Copy and complete the table.

X	10	20	15		
Y	15			60	12

Q13 hint

XY = constant

14 Does each equation represent direct proportion, inverse proportion or neither?

a $y = 2x$ **b** $y = \frac{4}{x}$ **c** $x + y = 7$

d $xy = 5$ **e** $\frac{y}{x} = 8$ **f** $2 = \frac{y}{x}$

Q14c hint

Rearrange the equation to $y = \square$

15 **a** Copy and complete each proportional relationship using the $\propto$ symbol.

i y is inversely proportional to x: $y \propto \frac{1}{\square}$

ii P is inversely proportional to A: $P \propto \frac{1}{\square}$

b Write each proportional relationship using k, the constant of proportionality.

16 y is inversely proportional to x. When y is 6, x is 10.

a Write the proportional relationship for y and x.

b Use the given values of x and y to find k.

c Find y when $x = 15$.

17 **Problem-solving** s is inversely proportional to t. When $t = 0.4$, $s = 20$.
Calculate the value of s when $t = 0.5$.

Q17 hint

Start $s \propto$

$s =$

Exam-style question

18 In an electrical circuit, the current, I, is inversely proportional to the resistance, R.
When the resistance in the circuit is 12 ohms, the current is 8 amperes.
Find the current in amperes when the resistance is 6.4 ohms. **(3 marks)**

Exam tip

Some of the marks will be awarded for method. Therefore, make sure you show your working clearly.

14 Check up

ActiveLearn Homework

Percentages

1 Louise buys a pack of 24 bottles of a fruit drink for £12.49. She sells all of them for 95p each. Work out her percentage profit. Give your answer to 3 significant figures.

2 A dishwasher originally cost £300. It is reduced to £250 in a sale. What is the percentage decrease in price?

3 £2500 is invested for 2 years at 3.5% compound interest. Work out the total amount in the account after 2 years.

4 The number of bees in a beehive decreases by 2% each year. There are 6500 bees at the beginning of 2019. How many will there be at the end of 2025?

Compound measures

5 Sarah works a 30-hour week, Monday to Friday. Her hourly pay is £10.20.
She is paid 'time and a quarter' for each hour she works on a Saturday,
and 'time and a half' for each hour she works on a Sunday.
Sarah works her 30-hour week, plus 3 hours on Saturday and 4 hours on Sunday.
How much is she paid?

6 A bottle of water of weight 19.6 N rests on a table.
The area of the base of the bottle in contact with the table is $7\,\text{cm}^2$.
What pressure in N/cm^2 does the bottle of water exert on the table?

$$\text{pressure} = \frac{\text{force}}{\text{area}}$$

7 These two metal blocks each have a volume of $0.5\,\text{m}^3$.

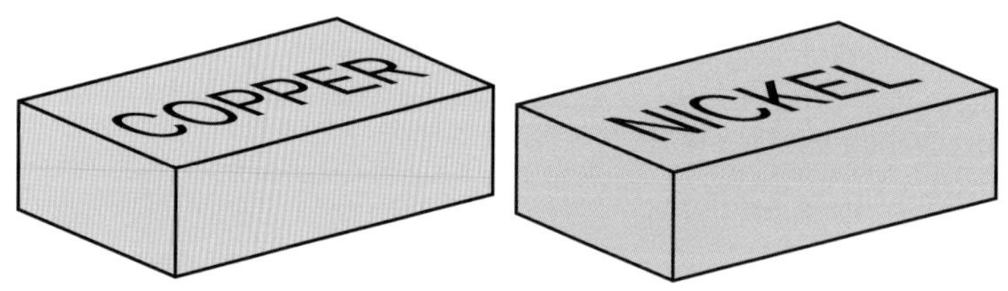

The density of the copper block is 8900 kg per m^3.
The density of the nickel block is 8800 kg per m^3.
Calculate the difference between the masses of the blocks.

Distance, speed and time

8 **a** Steffan drives 35 km in 45 minutes. What is his average speed in km/h?

b Michelle walks 10 km at an average speed of 3.2 km per hour. How long does it take her?

9 A car starts from rest and accelerates uniformly for 8 seconds for a distance of 200 m.
Work out the acceleration of the car.

Kinematics formulae
$v = u + at$
$s = ut + \frac{1}{2}at^2$
$v^2 = u^2 + 2as$

10 The fastest recorded speed of Usain Bolt is 12.4 m/s.
The fastest speed of a great white shark is 40 km/h.
Which is the faster speed? Explain your answer.

11 Amy drives on the motorway at an average speed of 64 mph for 2.5 hours.
Then she drives on country roads for a distance of 17 miles for 30 minutes.
Work out Amy's average speed for the whole journey.

Direct and inverse proportion

12 The table shows the amounts received when British pounds are changed to euros.

a Copy and complete the table.

Pounds (£)	Euros (€)
150	189
400	504
320	
	352.80

b Is the number of pounds in direct proportion to the number of euros?
Explain your answer.

c Write a formula connecting pounds to euros.

13 f is directly proportional to g, so $f = kg$ for some value of k.

a When $f = 8$, $g = 2$. Find the value of k.

b Work out f when $g = 1.5$.

14 y is inversely proportional to x. When $x = 5$, $y = 9$.

a Write a formula for y in terms of x.

b Find y when $x = 15$.

15 **Reflect** How sure are you of your answers? Were you mostly
Just guessing ☹ Feeling doubtful 😐 Confident 
What next? Use your results to decide whether to strengthen or extend your learning.

Challenge

16 Emma says, 'I find it difficult to remember the percentage change formula.
Is it actual change divided by original amount or actual change divided by final amount?'
Matt says, 'I know that if I invest £10 and get £15 in return, I've made £5. This is 50% of £10.
So there has been a 50% change. That helps me to remember the formula.'
Matt sketches a diagram.

£10 | £5
50%

Try both of Emma's suggestions for the formula using Matt's numbers.
Which formula is correct?
What do you think of Matt's strategy? Do you have another way of remembering the formula?
If so, what is it?

14 Strengthen

ActiveLearn Homework 

Percentages

1 Copy and complete the working to find the percentage profit made on each item.

original price = □
actual profit = □
percentage profit = $\frac{\text{actual profit}}{\text{original price}} \times 100 = \frac{\square}{\square} \times 100 = \square\%$

a Bought for £12, sold for £15
b Bought for £25, sold for £32.50
c Bought for £140, sold for £260

2 Follow the same method as in **Q1** to work out the percentage loss made on each item.

a Bought for £16, sold for £12
b Bought for £450, sold for £360
c Bought for £60, sold for £42

Q2 hint
actual loss = buying price − selling price

3 Steve bought a box of 12 pineapples for £5.
He sold all of them for £2.50 each.

a How much did Steve receive for selling them all?
b Follow the same method as in **Q1** to work out his percentage profit.

4 A shop has a sale with 10% off the normal prices.
Copy and complete the working to find the normal price of a coat with a sale price of £45.

100% − 10% = 90%

÷□ 90% = £45 ÷ 90
1% = £0.50
×□ 100% = □ ×□

5 The price of a house increases by 20% to £360 000.
Copy and complete the working to find the price of the house before the increase.

100% + 20% = 120%

÷□ 120% = £360 000 ÷ 120
1% = £3000
×□ 100% = □ ×□

6 Work out the original price of each of these items.

a Discount 15%, sale price £51

b Discount 25%, sale price £66

c Discount 40%, sale price £96

Q6a hint

□% = £51

÷ □ ↓ ÷ □

1% = □

7 Work out the original value of each of these before the increase.

a Increase of 15% to 460 g

b Increase of 30% to 390 km

c Increase of 45% to 580 litres

8 Saima invests £700 for 2 years at 3.8% per annum compound interest.
Copy and complete the table to work out the total amount in the account after 2 years.

Q8 hint

Write your final answer to money calculations rounded to the nearest penny. Only round your final answer.

Year	Amount at start of year	Interest at end of year	Amount at start of year + interest = total at end of year
1	£700	$700 \times 0.038 =$	
2		$\square \times 0.038 =$	

9 £750 is invested for 2 years at 4.3% per annum.
Draw a table to work out

a the total amount in the account after 2 years

b the total interest earned

10 Write the multiplier for each of these percentage *increases* and *decreases* as a decimal number.

a 15%

b 8%

c 2.6%

d 21%

e 7%

f 4.5%

g 35%

h 11.2%

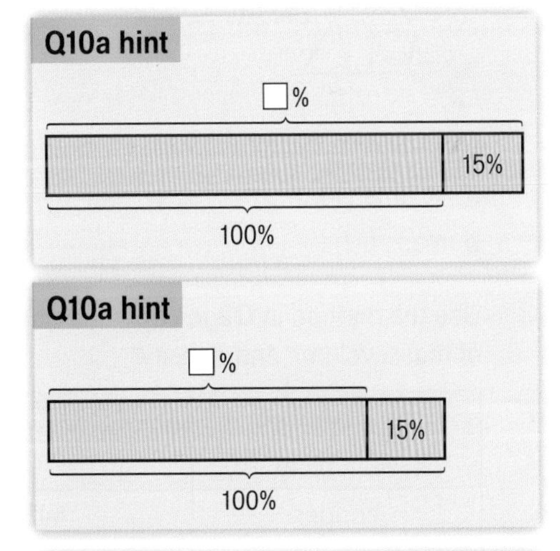

11 A population of rabbits increases by 20% each month.

a Write the multiplier for the percentage increase in rabbits.

At the beginning of January the population is 15 rabbits.

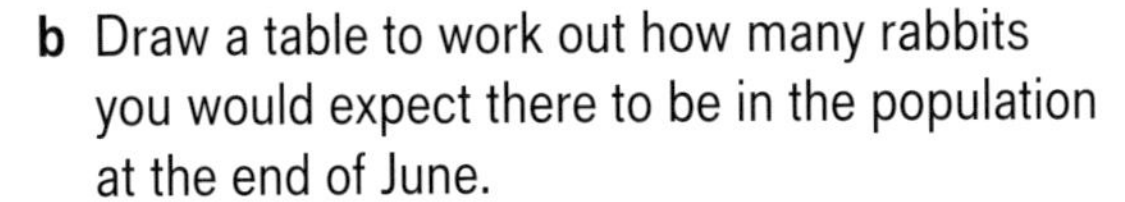

b Draw a table to work out how many rabbits you would expect there to be in the population at the end of June.

Q11b hint

Month	Number at the start of the month	Number at the end of the month
January		

12 The winter Arctic ice area is decreasing at a rate of 4.2% every 10 years.

a Write the multiplier for the percentage decrease in winter Arctic ice.

In 1979 the area was 16 million square metres.

b What is the winter area expected to be in 2029?

Q12b hint

Draw a table.

Compound measures

1 Jamal works a basic 20-hour week. His hourly rate of pay is £6.60. Jamal also works overtime.

a He is paid double for each hour he works on Sunday.
Work out his Sunday hourly rate of pay.

b He is paid 'time and a quarter' for each hour he works on a Saturday.
Copy and complete his Saturday hourly rate of pay.
$\square \times 1.25 = \square$

c He is paid time and a half for each hour he works in the evenings.
Copy and complete his evening hourly rate of pay.
$\square \times \square = \square$

d How much is Jamal paid for a week when he works a basic 20-hour week, plus 4 hours on Saturday, 3 hours on Sunday and 10 hours in the evenings.

Q1d hint

Work out how much Jamal earns for the basic 20 hour week.
Then use your answers to **a, b** and **c** to work out how much he earns for his work on Saturday, Sunday and in the evenings.

2 You can use this triangle to represent the formula for density.

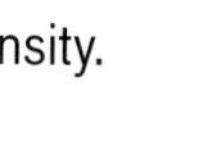

Cover the quantity you want to find.

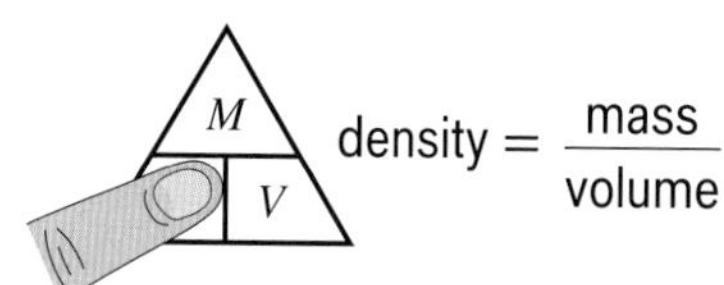

$$\text{density} = \frac{\text{mass}}{\text{volume}}$$

Use this method to write the formula for

a volume (V)

b mass (M)

Q2b hint

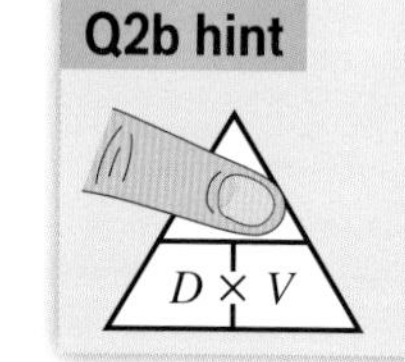

3 Use the method in **Q2** to copy and complete this table of mass, volume and density.

Metal	Mass (g)	Volume (cm^3)	Density (g/cm^3)
aluminium		10	2.70
copper	448		8.96
zinc	427.8	60	

4 $\text{pressure} = \frac{\text{force}}{\text{area}}$

Copy and complete this table of force, area and pressure.

Force (N)	Area (cm^2)	Pressure (N/cm^2)
60	15	
	20	11
45		9

Q4 hint

You can use this triangle.

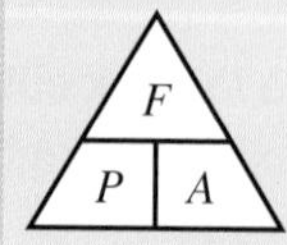

Distance, speed and time

1 You can use this triangle to represent the formula for speed.
Cover the quantity you want to find.
Use this method to write the formula for

a time (T)

b distance (D)

Q1b hint

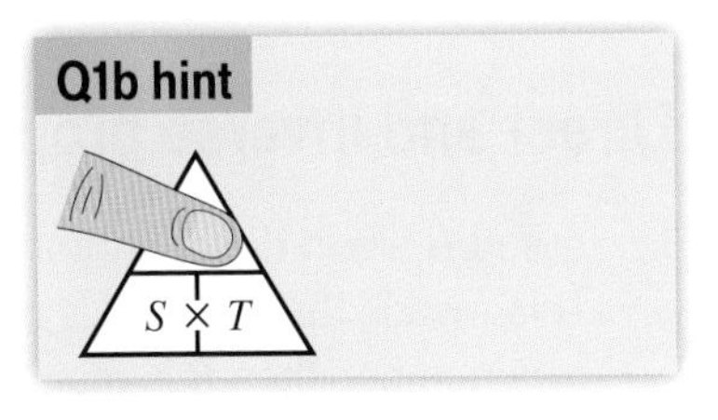

2 Use the method in **Q1** to copy and complete this table of distance, time and speed.

Distance (miles)	Time (hours)	Speed (mph)
	4	45
	2.5	58
150	3	
120	1.25	
45		30
154		56

3 $\text{Average speed} = \dfrac{\text{total distance}}{\text{total time}}$

Ali drives 40 miles in 50 minutes.
Then she drives 32 miles in 40 minutes.

a What is the total distance Ali drives?

b What is the total time Ali takes? Write your answer as decimal hours.

c What is Ali's average speed for the whole journey?

4 Is the answer to each of these conversions going to be higher or lower than the number given?

a 30 km/h in m/h

b 4000 m/h in m/min

c 80 m/min in m/s

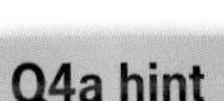

Q4a hint

If you travel 30 km in 1 hour, do you travel further or less far than 30 m in 1 hour?

5 Copy and complete this table to convert from km/h to m/s.

km/h	m/h	m/min	m/s
9			
			5
12			
			8

Q5 hint

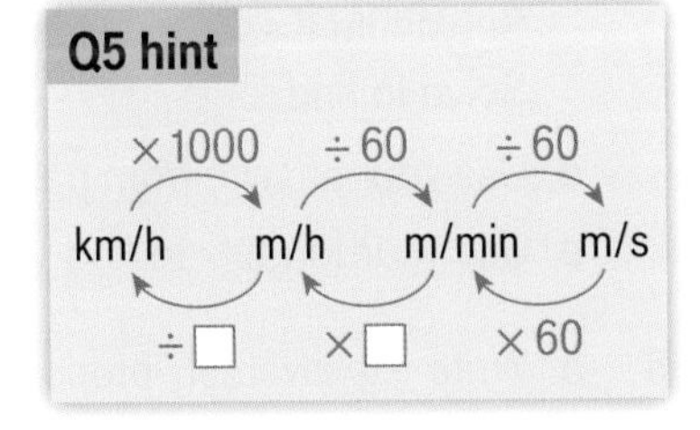

6 Write down the kinematics formula that does not include

a s

b t

c v

Kinematics formulae

$v = u + at$

$s = ut + \frac{1}{2}at^2$

$v^2 = u^2 + 2as$

7 Choose the correct kinematics formula to work out each of these.

a Find a when $s = 24$, $t = 3$ and $u = 5$.

b Find u when $s = 12$, $v = 8$ and $a = 2$.

c Find v when $u = 4$, $a = 3$ and $t = 7$.

Q7a hint

You are not given a value for v, or asked to find it.
Look for the formula that does not include v.

Direct and inverse proportion

1 A and B are in direct proportion.
This means that when A is multiplied by a number, then B is multiplied by the same number.
Work out the values of W, X, Y and Z.

A	B
4	7
8	W
X	42
16	Y
Z	35

Q1 hint

$\times\square$ (4 : 7 → 8 : W) $\times\square$

Now use your value of W to work out X.

$\times\square$ (8 : W → X : 42) $\times\square$

2 A and B are in inverse proportion.
This means that when A is multiplied by a number, then B is divided by the same number, or vice versa.
Work out the values of W, X, Y and Z.

A	B
4	12
24	W
8	X
Y	3
Z	30

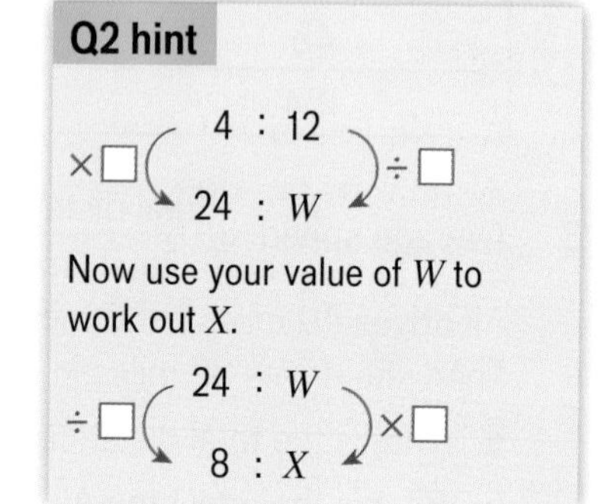

3 Match the equivalent statements.

y is proportional to x | $y = kx$ | $y \propto \frac{1}{x}$

$y = \frac{k}{x}$ | $y \propto x$ | y is inversely proportional to x

4 **a** Write y is directly proportional to x using $\propto$.

b Write a formula showing that y is directly proportional to x.

c When $x = 7$, $y = 21$.
Substitute these values into your formula from part **b**.
Solve to find k.

d Use your answer from part **b** to rewrite your formula as $y = \square x$.

e Use your formula to find y when $x = 4$.

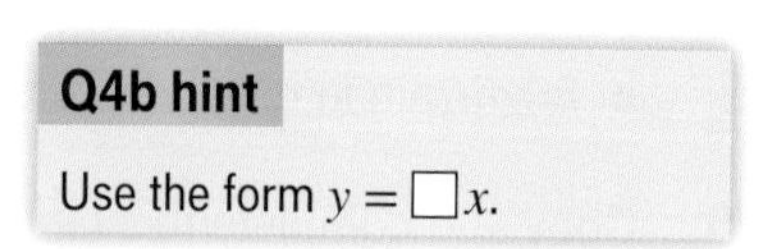

5 **a** Write y is inversely proportional to x using $\propto$.

b Write a formula showing that y is inversely proportional to x.

c When $x = 5$, $y = 20$.
Substitute these values into your formula from part **b**.
Solve to find k.

d Use your answers from part **b** and **c** to rewrite your formula as $y = \frac{\square}{x}$

e Use your formula to find y when $x = 4$.

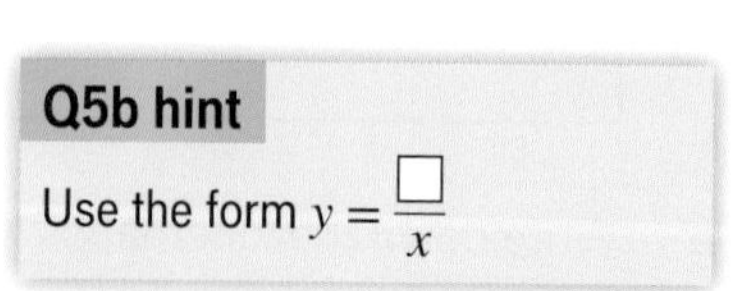

14 Extend

1 **Reasoning** Stevie's manager says she can have either a 2% pay rise this year and then a 1.5% pay rise next year, or a 2.5% pay rise this year only. Which is the better offer? Explain.

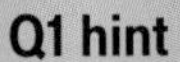

Q1 hint
You could decide on a salary to work with.

2 An object starts from rest and travels for a distance of 15 m with an acceleration of 5 m/s^2.
Work out its final velocity.

Q2 hint
Use a kinematics formula.
u = initial velocity = 0 as the object starts from rest.

3 A 25 g block of aluminium has volume 9260 mm^3.
Calculate its density in g/cm^3.

Q3 hint
What units do you need for volume?

Exam-style question

4 A train line is 201.5 miles long.
A train travels along the line at an average of 62 mph.
It is replaced with a new train.
The new train is advertised as being able to travel the length of the line in 3 hours 3 minutes.

a How many minutes faster is the new train than the old train? **(3 marks)**

b The new train actually travels at 63 mph.
How does this affect your answer to part **a**? **(1 mark)**

5 A shop is having a sale. Each day, prices are reduced by 20% of the price on the previous day.
Before the start of the sale, the price of a fridge is £550.

a Work out the price of the fridge on the third day of the sale.

On the first day of the sale, the price of a table is £200.

b Work out the price of the table before the start of the sale.

6 **Problem-solving** A car is travelling at 17.88 m/s.

a Work out an estimate for the number of minutes the car takes to travel 36 km.

b Is your answer to part **a** an underestimate or an overestimate?
Give a reason for your answer.

7 The extension E of a string (in mm) is directly proportional to the mass m on the string in grams.
A mass of 250 g gives an extension of 12 mm.

a Find the extension of the string for a mass of 600 g.

b Find the mass that extends the string by 36 mm.

Q7 hint
Write a formula for E in terms of m.

8 y is inversely proportional to x.
When $x = 2$, $y = 36$.
Work out

a the value of y when $x = 12$

b the value of x when $y = 9$

9 The volume V (m^3) of a gas is inversely proportional to the pressure P (N/m^2).
When $P = 600\ N/m^2$, $V = 5\ m^3$.

a Write a formula for V in terms of P.

b Find the volume when the pressure is $300\ N/m^2$.

c Calculate the pressure on $7.5\ m^3$ of the gas.

Exam-style question

10 The diagram shows a prism.

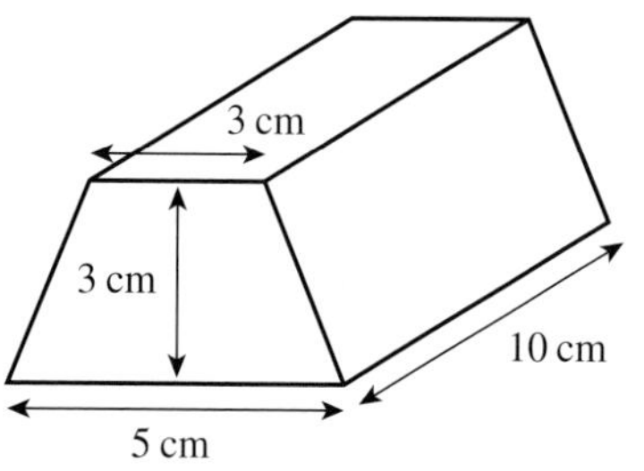

Diagram NOT accurately drawn

The cross-section is a trapezium.

$$\text{pressure} = \frac{\text{force}}{\text{area}}$$

a Work out the force needed for the base of the prism to exert a pressure of $0.5\ N/cm^2$. **(3 marks)**

The prism is made out of gold.
Gold has a density of 19.3 grams per cm^3.

b Work out the mass of the prism.
Give your answer in kilograms. **(2 marks)**

Exam tip

Sometimes you must work out some missing information before finding the answer.
For example, to find the mass, given the density, first you must work out the volume.

Exam-style question

11 Jed wants to invest £2000 for 3 years in the same bank.

The Rapid Bank	The Eco Bank
Compound interest	Compound interest
1.4% for the first year 1% for each extra year	1.5% for the first year 0.5% for each extra year

At the end of 3 years, Jed wants to have as much money as possible.
Which bank should he invest his £2000 in?
You must show your working. **(4 marks)**

Exam tip

You must make clear which working is for which bank and make sure you state which is better and why.

12 Caspar drove from Winchester to Oxford.
Caspar's average speed from Winchester to Newbury was 100 km/h.
Caspar's average speed from Newbury to Oxford was 80 km/h.
Caspar says that the average speed from Winchester to Oxford can be found by working out the mean of 100 km/h and 80 km/h.
If Caspar is correct, what does this tell you about the two parts of Caspar's journey?

14 Test ready

Summary of key points

To revise for the test:

- Read each key point, find a question on it in the mastery lesson, and check you can work out the answer.
- If you cannot, try some other questions from the mastery lesson or ask for help.

Key points

1 The original amount is always 100%. If the amount is *increased,* then the new amount will be *more* than 100%. If the amount is *decreased,* then the new amount will be *less* than 100%. → **14.1**

2 You can calculate a **percentage** change using the formula

$$\text{percentage change} = \frac{\text{actual change}}{\text{original amount}} \times 100$$ → **14.1**

3 Percentage increase and decrease, profit and loss can be calculated using the formula for percentage change. → **14.1**

4 Banks and building societies pay **compound interest**. At the end of the first year, interest is paid on the money in the account. The interest is added to the amount in the account. At the end of the second year, interest is paid on the original amount in the account *and* on the interest earned in the first year, and so on. → **14.2**

5 **Compound measures** combine measures of two different quantities. → **14.3**

6 **Density** is a compound measure. It is the mass of substance contained in a certain volume. It is usually measured in grams per cubic centimetre (g/cm^3). It can also be measured in kg/m^3.

$$\text{density} = \frac{\text{mass}}{\text{volume}} \text{ or } D = \frac{M}{V}$$ → **14.3**

7 **Pressure** is a compound measure. It is the force applied over an area. It is measured in newtons per square metre (N/m^2).

$$\text{pressure} = \frac{\text{force}}{\text{area}} \text{ or } P = \frac{F}{A}$$ → **14.3**

8 **Speed** is a compound measure. It is a measure of distance travelled and time taken. Speed is often measured in metres per second (m/s), kilometres per hour (km/h) or miles per hour (mph).

$$\text{speed} = \frac{\text{distance}}{\text{time}} \text{ or } S = \frac{D}{T} \qquad \text{average speed} = \frac{\text{total distance}}{\text{total time}}$$ → **14.4**

9 **Velocity** is speed in a given direction. It is often measured in m/s. The **initial velocity** is the speed in a given direction at the start of the motion. It may be zero. **Acceleration** is the rate of change of velocity. It is often measured in m/s^2. → **14.4**

10 You can use the **kinematics formulae** for calculations with moving objects.

$$v = u + at \qquad s = ut + \tfrac{1}{2}at^2 \qquad v^2 = u^2 + 2as$$

where a is a constant acceleration, u is the initial velocity, v is the final velocity, t is the time taken and s is the displacement from the position when $t = 0$. → **14.4**

11 When two variables are in **direct proportion**, pairs of values are in the same ratio. → **14.5**

Key points

12 You can write $y \propto x$ when y is **directly proportional** to x → 14.5

13 When $y \propto x$, then $y = kx$, where k is the **constant of proportionality.**
This means that the value of k is constant (stays the same) when x and y vary. → 14.5

14 When two variables X and Y are in **inverse proportion**,
$X \propto \frac{1}{Y}$ $X = \frac{k}{Y}$ $Y = \frac{k}{X}$ This means that $XY = k$ (constant). → 14.5

Sample student answers

Exam-style question

1 Mr and Mrs Fernandez sold their house for £660 800.
They made a profit of 12% on the price they paid for the house.
Calculate how much they paid for the house. **(3 marks)**

Student A

12% of 660800 = 0.12 × 660800
= 79 296

```
  6 6 0 8 0 0
-   7 9 2 9 6
  5 8 1 5 0 4
```

They paid £581504 for the house.

Student B

100 + 12 = 112%

Price → × 1.12 → 660800

590000 ← ÷1.12 ← 660800

They paid £590000 for the house.

a Which student gave the correct answer? Explain what the other student did wrong.

b What check could the students have done to make sure their answers were correct?

Exam-style question

2 Paul completed a go kart race in 1 minute 45 seconds. The race was 2100 m in length.
Paul assumes that his average speed is the same for each race.

a Using this assumption, work out how long Paul should take to complete an 800 m race. **(3 marks)**

b Paul's average speed increases the further he goes.
How does this affect your answer to part **a**? **(1 mark)**

a Time = 1 minute 45 seconds = 1.75 Distance = 2100 m

$\text{Speed} = \frac{\text{Distance}}{\text{Time}} = \frac{2100}{1.75} = 1200 \text{ m/s}$ Distance = 800 m

$\text{Time} = \frac{\text{Distance}}{\text{Speed}} = \frac{800}{1200} = 0.67 \text{ s (2 decimal places)}$

b If Paul's average speed increases, his time will be faster.

a The student uses the correct method for part **a** but gets the answer wrong.

i What mistake has the student made?

ii Why should this student realise the final answer does not make sense?

b Is the student's answer to part **b** acceptable? Explain.

c What is the correct answer to part **b**?

14 Unit test

1 Michael's hourly rate of pay increases from £15.40 to £18.00.
What is the percentage increase in his rate of pay?
Give your answer correct to 1 decimal place. **(3 marks)**

2 Jamie got a pay rise of 5%. His new pay was £1785 per month.
Work out his pay per month before the pay rise. **(2 marks)**

3 A metal cube of side 20 cm exerts a force of 6 N on a table.

$$\text{pressure} = \frac{\text{force}}{\text{area}}$$

Work out the pressure in N/m^2. **(4 marks)**

4 Polly goes for a jog.
She jogs 18 km in 2.5 hours.

a Work out Polly's average speed.
Give your answer in kilometres per hour. **(2 marks)**

Suvi says that Polly jogs 2 metres every second.

b Is Suvi correct?
Show all your working. **(2 marks)**

5 Dom cycles 44.5 km at an average speed of 15.5 km/h.
He then cycles 500 m at an average speed of 25 km/h.
What is his average speed for the whole journey? **(5 marks)**

6 Dan buys a house for £250 000. He sells it for £274 500.

a Work out Dan's percentage profit. **(3 marks)**

Dan invests £274 500 for 2 years at 1.8% per year compound interest.

b Work out the value of the investment at the end of 2 years. **(3 marks)**

7 Claire invests £600 in a savings account for 3 years.
The account pays compound interest at an annual rate of 1.7%.

a Work out the total amount of money in Claire's account at the end of 3 years. **(3 marks)**

Katie invests £600 in a savings account for 3 years.
The account pays compound interest at an annual rate of 2.2% for the first two years only.
Then it pays compound interest at an annual rate of 1.5%.

b How much more is in Katie's account than Claire's account at the end of 3 years? **(4 marks)**

8 The time, T seconds, it takes a pan of water to boil is directly proportional to the volume of water, v cm^3, in the pan.
When $v = 100cm^3$, $T = 40$ seconds.
How long will it take to boil 120 cm^3 of water? **(5 marks)**

9 In electrical circuits, the current I is inversely proportional to the resistance R.
In one circuit, the current is 9 amperes when the resistance is 14 ohms.
Find the current in amperes when the resistance is 12 ohms. **(5 marks)**

10 The wavelength of a wave is inversely proportional to its frequency.

a A radio wave of wavelength 1000 metres has a frequency of 300 kHz.
A different radio wave has a frequency of 600 kHz.
What is its wavelength? **(2 marks)**

b Calculate the frequency of a radio wave with wavelength 842 metres. **(3 marks)**

11 The density of lemon oils is 0.85 g/cm^3.
The density of honey is 1.4 g/cm^3.
A mixture contains 1.7 g of lemon oil and 7 g of honey.
Work out the density of the lemon oil and honey mixture. **(4 marks)**

(TOTAL: 50 marks)

12 Challenge £650 is invested at 3.4% compound interest.

a Copy and complete the table.

Year	Amount at start of year	Multiplier in index form	Total amount at end of year
1	£650	1.034	
2		$1.034^{\square}$	
3		$1.034^{\square}$	
4			
5			

b What is the multiplier for year 10?

c What is the multiplier for year n?

d Write a formula for the amount in the account at the end of year n, when £P is invested at r%.

13 Reflect This unit is called 'multiplicative reasoning'.
'Multiplicative' means involving multiplication or division.
'Reasoning' is being able to explain why you have used multiplication or division in this unit.

a List three ways you have used multiplication or division in this unit.

b Why is it good to reason in mathematics?

Mixed exercise 4

ActiveLearn Homework 

Exam-style question

1 PQR is a straight line.

The length PQ is 7 times the length QR.
$PR = 200$ cm
Work out the length PQ. **(3 marks)**

2 Reasoning Karl goes to the supermarket on his drive home from work.
He leaves work at 5 pm and arrives at the supermarket at 5.15 pm.
Karl is in the supermarket for 20 minutes. He then drives 10 miles to his house.
The speed limit on the road is 30 mph. Karl does not exceed the speed limit.
Can Karl arrive home before 6 pm?
Give reasons for your answer.

Exam-style question

3 The stem and leaf diagram shows the marks gained by 20 students in a maths test.

5	8
6	1 2 4 4 6 8 8 9
7	0 3 5 6 6 6 7
8	1 2 8 9

Key: 5 | 8 means 58

The pass mark for the maths test is 65.
One of the students is chosen at random.
Ceris writes, 'The probability that this student passed the maths test is $\frac{4}{5}$.'
Ceris is wrong. Explain why. **(2 marks)**

4 Problem-solving The diagram shows a fair 16-sided spinner.
Copy the spinner and label each section A, B, C, D so that

$P(A) = \frac{1}{8}$
$P(B) = 2P(A)$
$P(C) = P(D)$

Exam-style question

5 The ratio of the cost of one jar of a supermarket's own coffee to the cost of the same size jar of branded coffee in a supermarket is 2 : 3.
Complete the table of costs.

	2 jars	6 jars	8 jars	11 jars
Supermarket's own coffee	£7			
Branded coffee				

(3 marks)

6 **Problem-solving** Two 4-sided fair spinners are spun. Here is spinner A.

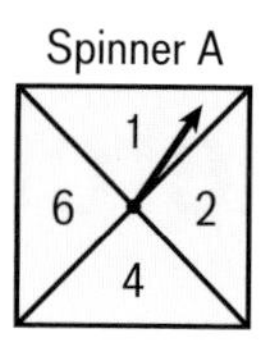

The score is the difference between the two numbers. Evie has lost spinner B but has this partially completed sample space diagram.

		Spinner A			
		1	**2**	**4**	**6**
Spinner B		0			
			3		
		4			1
					0

All the numbers on spinner B are positive.
What are the four missing numbers on spinner B?

Exam-style question

7 A spinner can land on black, white, pink or orange.
The table shows the probabilities of the spinner landing on black or orange.

Colour	black	white	pink	orange
Probability	0.2			0.48

The ratio of the probability of landing on white to the probability of landing on pink = 5 : 3.
Copy and complete the table. **(4 marks)**

8 **Problem-solving** Sian wants to invest £8000 for 3 years.
A bank offers these two savings accounts.

Account A	**Account B**
Simple interest at 5% per annum	Compound interest at 4.8% per annum

Sian wants to earn as much interest as possible.
In which account should she invest her money?
You must show your working.

9 **Reasoning** Ben is driving on a motorway.
He sees this road sign showing the distance to Junction 16 and the expected time it will take.

Junction 16
20 miles
18 minutes

The speed limit on the motorway is 70 mph.
Ben thinks he will have to drive faster than the speed limit to reach junction 16 in 18 minutes.
Is he right? You must show your working.

10 Reasoning Here are the ingredients needed to make 24 flapjacks.

Flapjacks
ingredients to make 24 flapjacks
225 g of butter
225 g sugar
75 g syrup
275 g porridge oats

Ed has
600 g butter
575 g sugar
200 g syrup
650 g porridge oats

Does Ed have enough ingredients to make 60 flapjacks? Give reasons for your answer.

Exam-style question

11 The diagram shows a right-angled triangle.

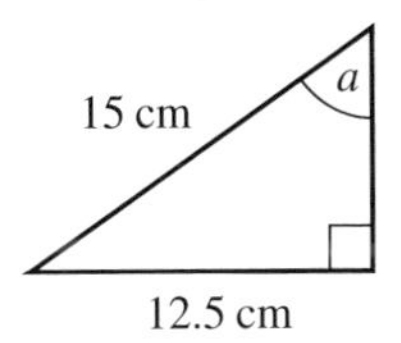

Show that $a = 56.4°$ (to 1 decimal place). **(2 marks)**

12 Reasoning The diagram shows a trapezium.

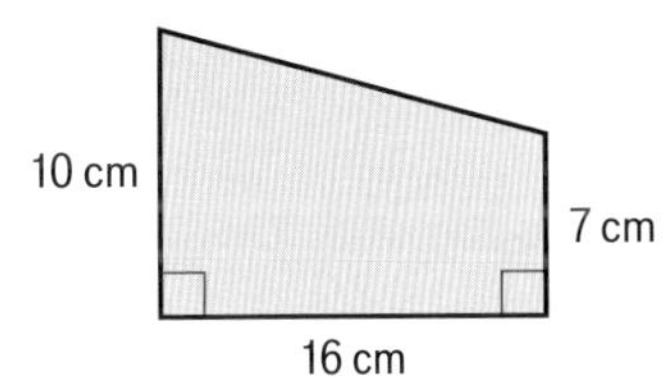

Dina says, 'The perimeter of the trapezium is greater than 50 cm.'
Is she correct? Give reasons for your answer.

13 Problem-solving In a school, Year 9 students must choose either geography or history or both.
227 students study only geography.
47 students study only history.
Three times as many students choose geography as choose history.

a Copy and complete the Venn diagram.

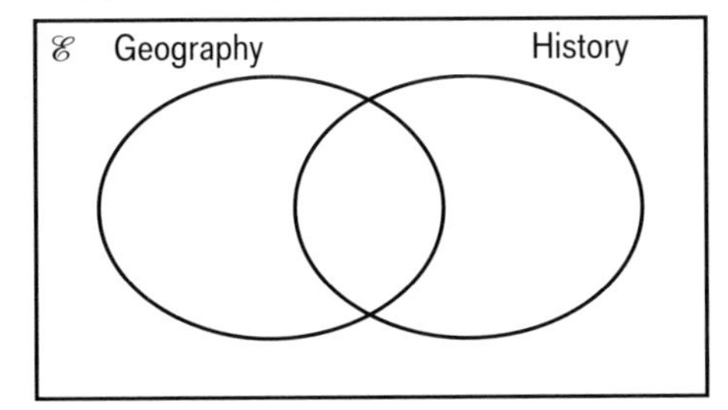

A Year 9 student is chosen at random.

b What is the probability that this student studies both geography and history?

Exam-style question

14 Angus and Khaled have a total of 180 collectors' cards.
The ratio of the number of Angus's cards to the number of Khaled's cards is 5 : 7.
Khaled gives some cards to Angus.
The ratio of the number of Angus's cards to the number of Khaled's cards is now 3 : 2.
How many cards does Khaled give to Angus?
You must show all your working. **(4 marks)**

15 **Problem-solving** The mass of a solid X is 4 times the mass of solid Y.
The volume of solid X is double the volume of solid Y.
How many times greater is the density of solid X than the density of solid Y?

16 **Problem-solving** Josie takes her family out for a meal.
She has a voucher for 15% off the total bill. Josie pays £46.41.
How much would the meal have cost without the voucher?

Exam-style question

17 Waleed buys some furniture for £6500 plus VAT at 20%.
Waleed pays a deposit for the furniture.
He then pays the rest of the cost in 12 equal payments of £406.25.
Find the ratio of the deposit Waleed pays to the total of the 12 equal payments.
Give your answer in its simplest form. **(4 marks)**

18 **Reasoning** Which of these triangles are right-angled?
You must give reasons for your answers.

a

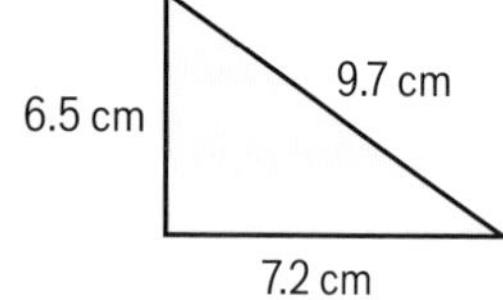

b

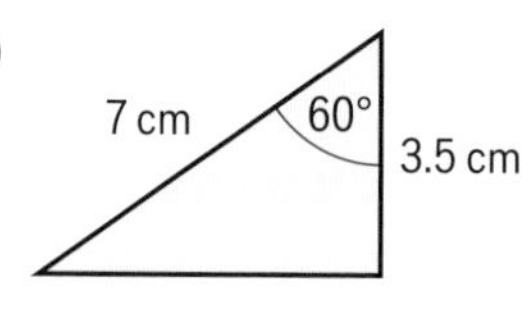

c

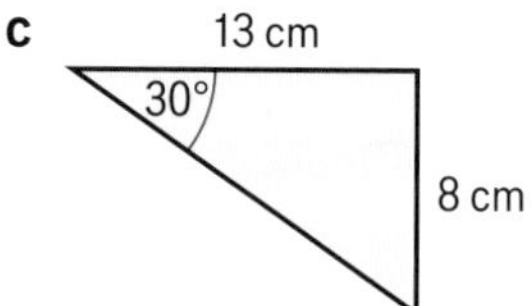

19 **Problem-solving** The diagram shows right-angled triangle ABC.
$x : y = 2 : 1$
Calculate the sizes of angles x and y.
Give your answers correct to 3 significant figures.

A
D
C
B
23 cm
20 cm
y
x

20 **Problem-solving** George has a box of chocolates.
There are 5 milk chocolates and 3 dark chocolates in the box.
George takes 2 chocolates from the box at random.

a Copy and complete the probability tree diagram.

b Work out the probability that George takes 2 milk chocolates from the box.

First chocolate **Second chocolate**

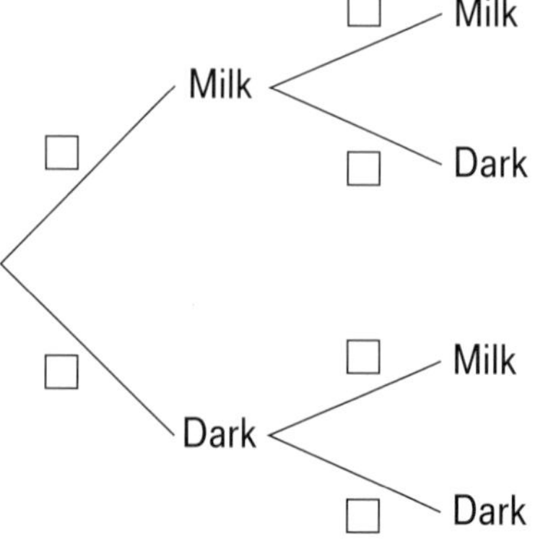

15 Constructions, loci and bearings

15.1 3D solids

- Recognise 3D shapes and their properties.
- Describe 3D shapes using the correct mathematical words.
- Understand the 2D shapes that make up 3D objects.

ActiveLearn Homework

Warm up

1 **Fluency** What are the names of these two-dimensional shapes?

a
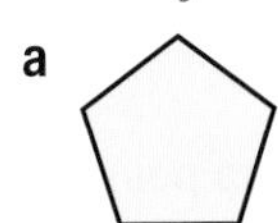

b
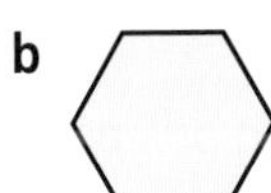

c
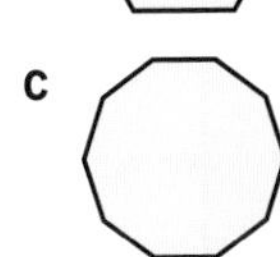

2 Draw these shapes accurately on squared paper.

a A square of side length 3 cm

b A rectangle 4 cm by 6.5 cm

Key point

The flat surfaces of 3D shapes are called **faces**, the lines where two faces meet are called edges and the corners at which the **edges** meet are called **vertices.** The singular of vertices is **vertex**.

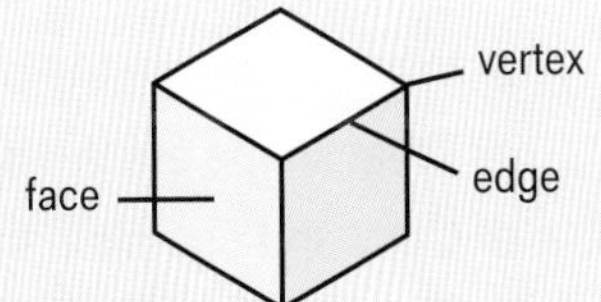

3 Here are two ordinary dice.

a How many faces does each dice have?

b How many edges does each dice have?

edge

face

Exam-style question

4 Here is a 3D shape.

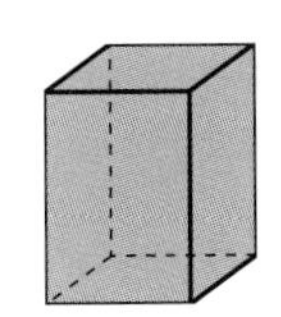

a Write down the name of this 3D shape. **(1 mark)**

b Write down the number of vertices of this 3D shape. **(1 mark)**

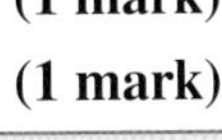

Key point

A **dimension** is the size of something in a particular direction.
Length, width, height and diameter are all dimensions of shapes or solid objects.

5 Here is a cuboid.

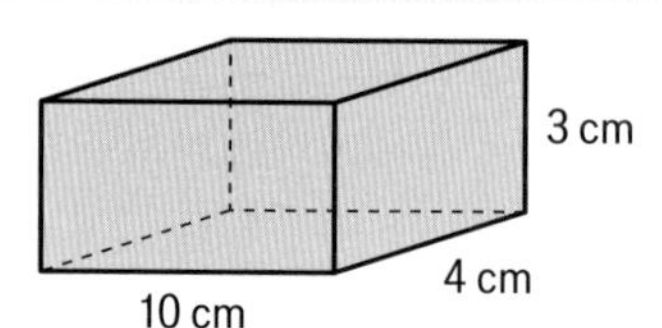

a What is the name of the shape of its faces?

b Write down the dimensions of each of the faces.

6 Here are some three-dimensional solids.

a 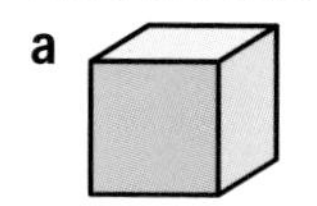 **b** 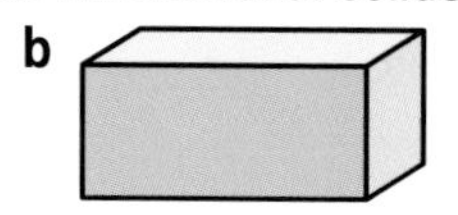 **c** 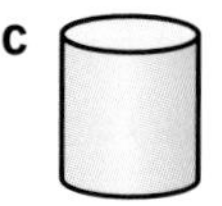 **d** 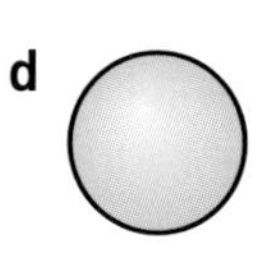 **e** 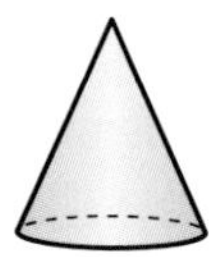

cylinder	sphere	cube	cuboid	cone

Match the correct name to each solid.

Key point

Pyramids have a base that can be any shape and sloping triangular sides that meet at a point.
A **right prism** has two parallel faces. All the other faces are perpendicular to these.

7 **Reasoning** Here are some more three-dimensional solids, this time pyramids and prisms.

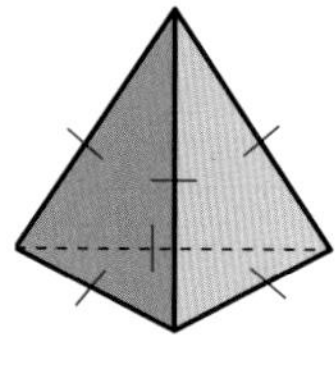

tetrahedron

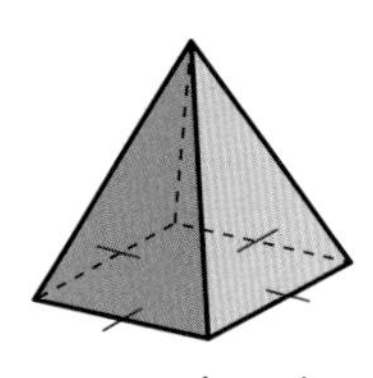
square-based pyramid

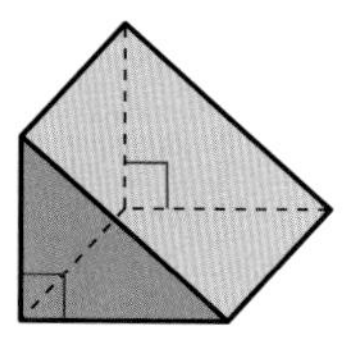
triangular prism

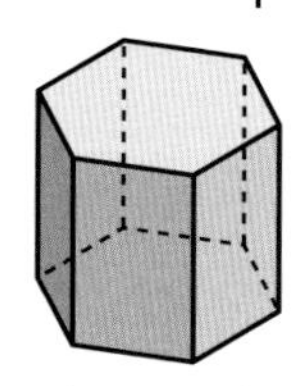
hexagonal prism

When the shape is a triangle, state clearly which type.

a What shape is each face of the tetrahedron?

b What shape are the parallel faces of the triangular prism?

c What shape are the non-parallel faces of the triangular prism?

d Copy and complete this table.

	Number of faces	Number of edges	Number of vertices
Cube	6	12	8
Tetrahedron	4		
Square-based pyramid			
Triangular prism			
Hexagonal prism			

e Add together the number of faces and vertices and compare this with the number of edges.
Write a rule connecting the number of faces (F), vertices (V) and edges (E).

8 Reasoning Here is another prism.
Cathy says, 'It has twice as many edges as it has faces.'
Is she correct?
Explain your answer.

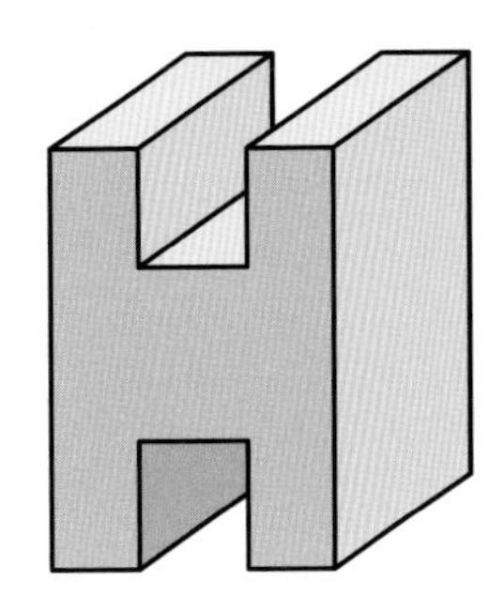

Example

A triangular prism has ends that are equilateral triangles with side lengths 4 cm.
It is 12 cm long. Sketch the prism.

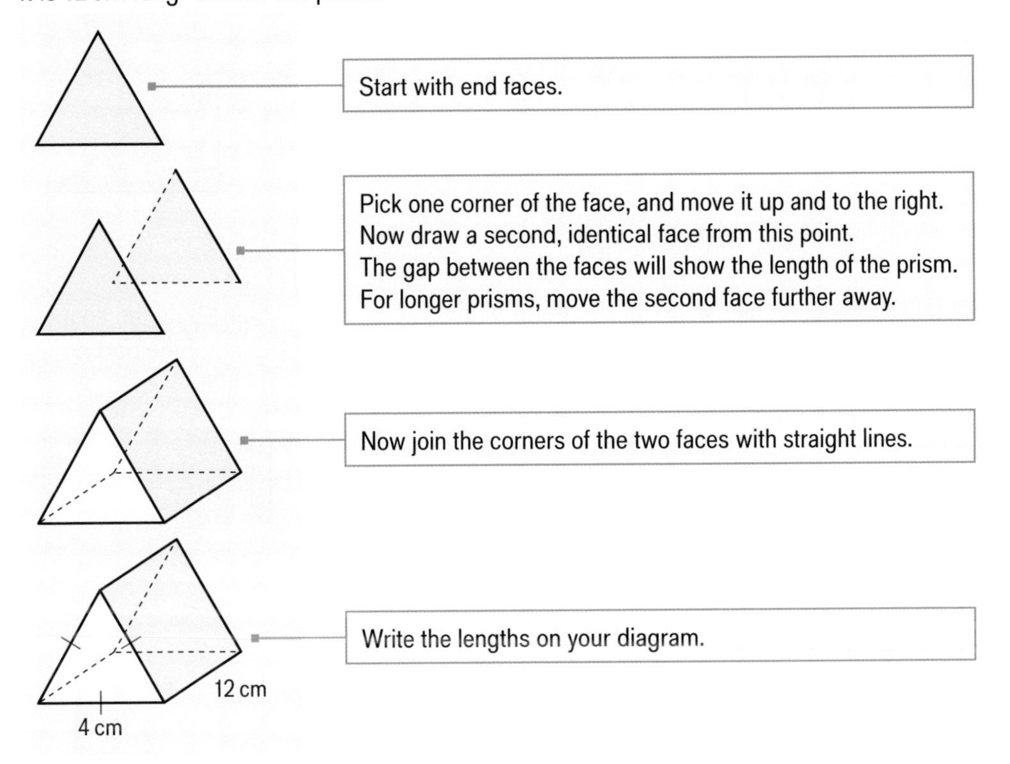

9 A triangular prism has ends that are isosceles triangles with side lengths 5 cm, 5 cm and 3 cm.
It is 10 cm long. Sketch this prism.

10 Reasoning The ends of a cuboid are square faces of side 10 cm.
The length of the cuboid is 6 cm. Sketch this cuboid.

11 Reasoning The parallel faces of a right prism are identical isosceles trapezia.

a What shape are the other faces of this prism?

b How many faces does this prism have?

15.2 Plans and elevations

- Identify and sketch planes of symmetry of 3D shapes.
- Draw and interpret plans and elevations of 3D shapes.

Warm up

1 **Fluency** Draw these lines accurately using a ruler.

a 7.5 cm **b** 47 mm **c** 6.9 cm **d** 24 mm

2 **Fluency** How many lines of symmetry does each of these shapes have?

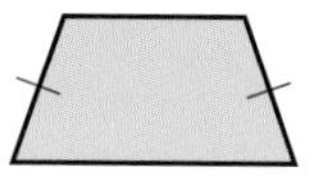
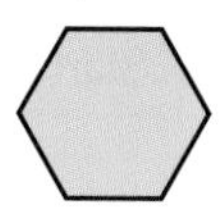
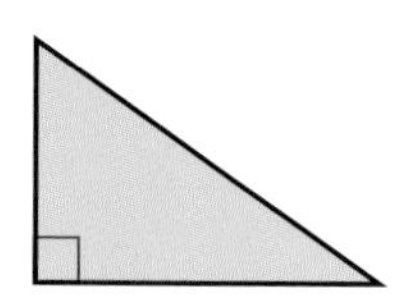

Key point

A **plane** is a flat (2D) surface. A solid shape has a **plane of symmetry** when a plane cuts the shape in half so that the part of the shape on one side of the plane is an identical reflection of the part on the other side of the plane. The planes of symmetry for this cuboid are shown in blue.

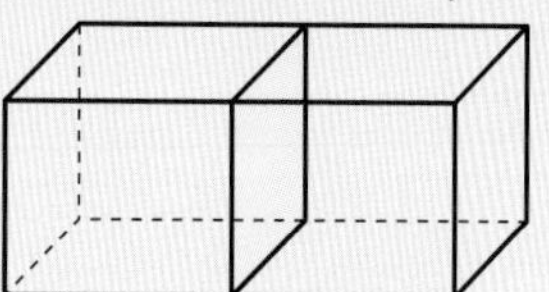
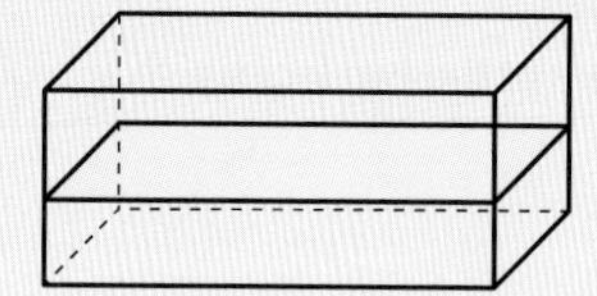
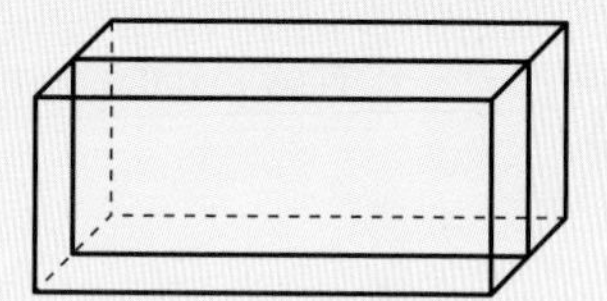

3 Here are two right prisms and two pyramids.

a
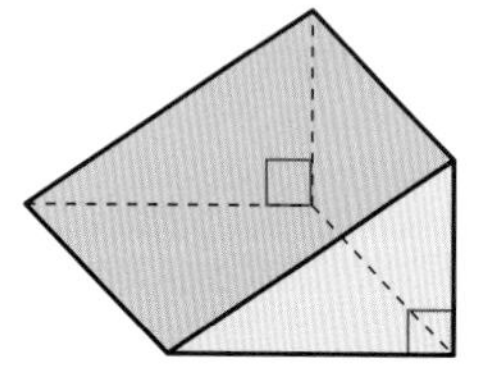
b
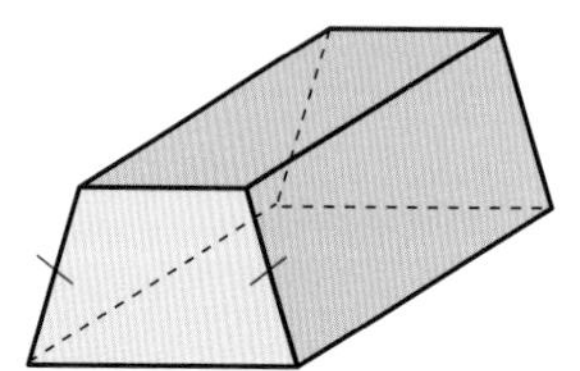
c
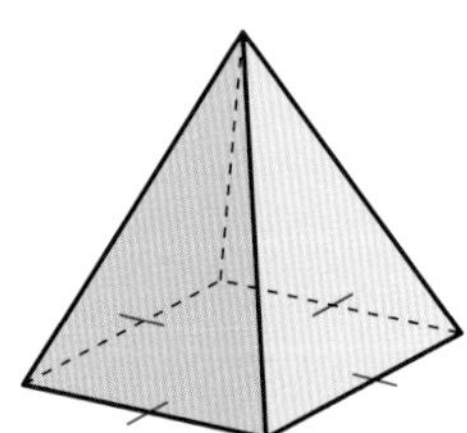
d
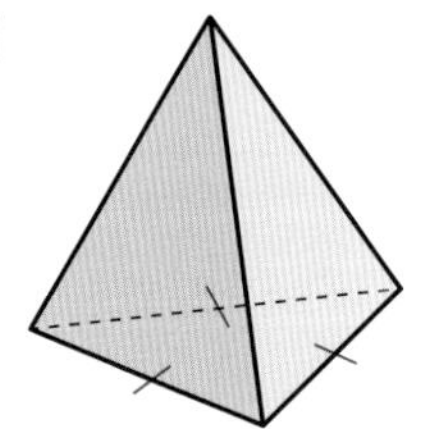

Copy the diagrams and draw in all the planes of symmetry for each shape.

4 This diagram represents a shed (not drawn to scale).
The shed is in the shape of a prism.
Each end face of the shed is an equilateral triangle on top of a square.
The shed has two planes of symmetry.
Copy the diagram and draw in the two planes of symmetry.

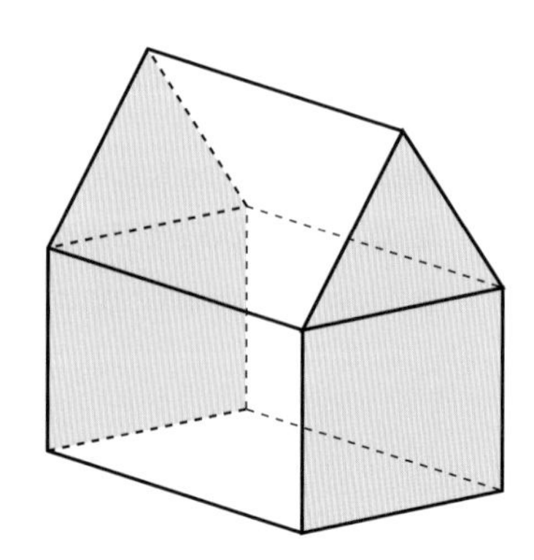

5 Reasoning Here is an 8th birthday cake for twins.
It is made from two identical circular cakes.
In how many ways can the cake be cut to give each twin exactly half of the cake?

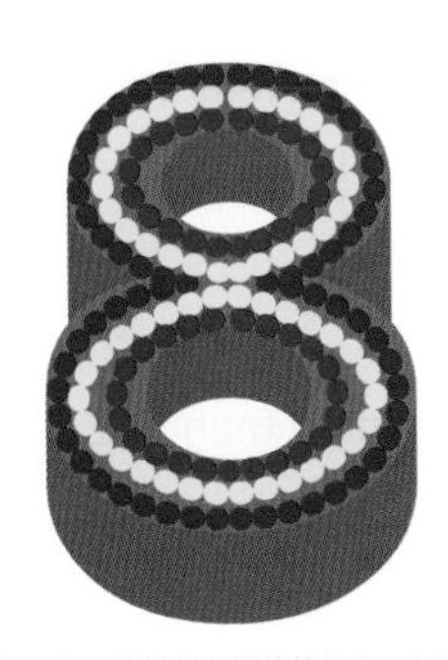

Key point

The **plan** is the view from above an object.
The **front elevation** is the view of the front of an object.
The **side elevation** is the view of the side of an object.

Example

On squared paper, draw the plan, the front elevation and the side elevation of this cuboid.

plan
3 cm
side elevation
4 cm
6.5 cm
front elevation

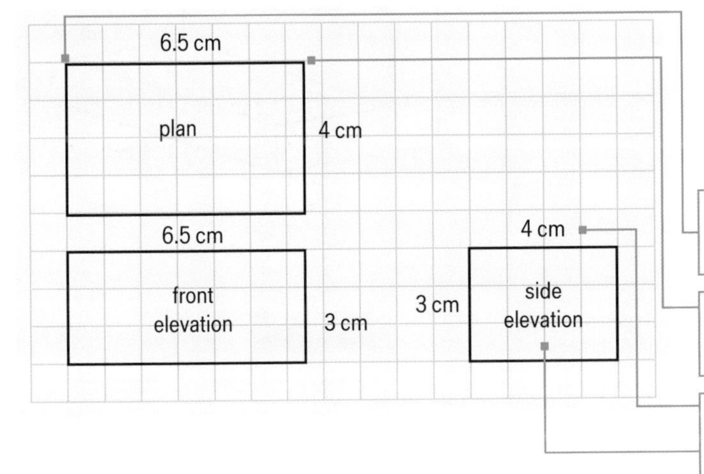

Start each view on the corner of a grid square.

You can still draw the view even if the lengths are not whole numbers.

Remember to label the lengths, including units, and say which view you have drawn.

6 Draw the plan view, the front elevation and side elevation of these cuboids on squared paper.

a

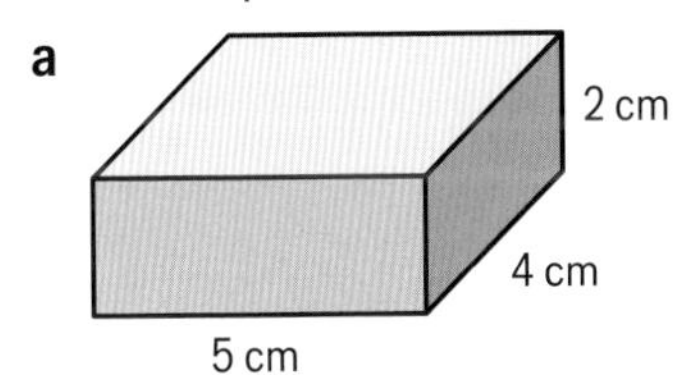

b

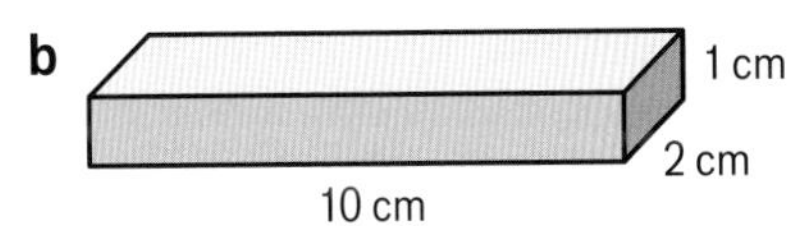

c

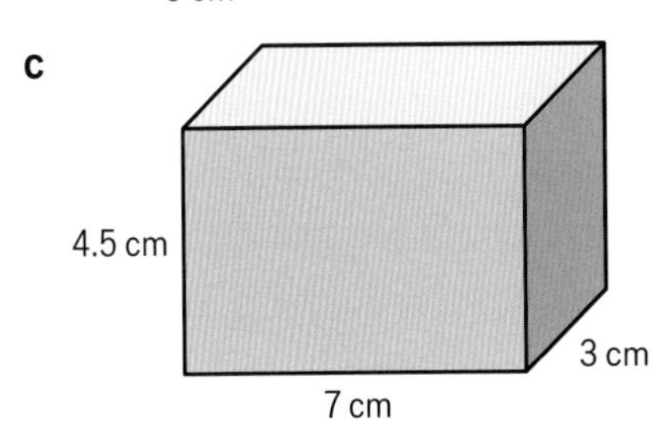

Q6 hint

If you are using 5 mm paper, remember that two squares = 1 cm.

7 **a** **Reasoning** Here are the plan views of some solids. Name the solid each could be.

i  ii iii iv

b Is there more than one possible answer for each plan?

8 **Problem-solving** On squared paper, draw the plan, front and side elevations of this cylinder.

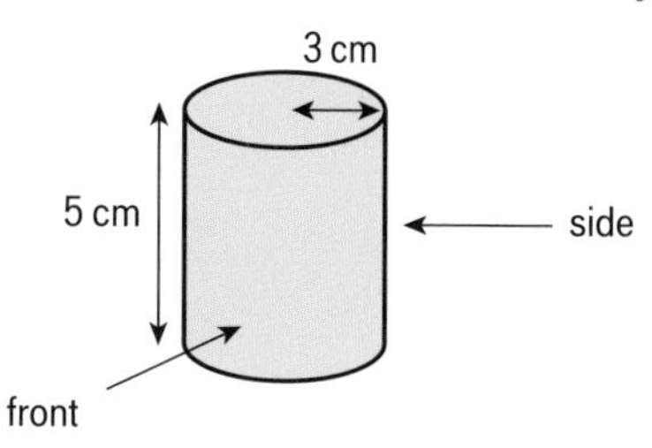

9 On squared paper, draw the plan, front elevation (shown by the arrow) and side elevation of this prism.

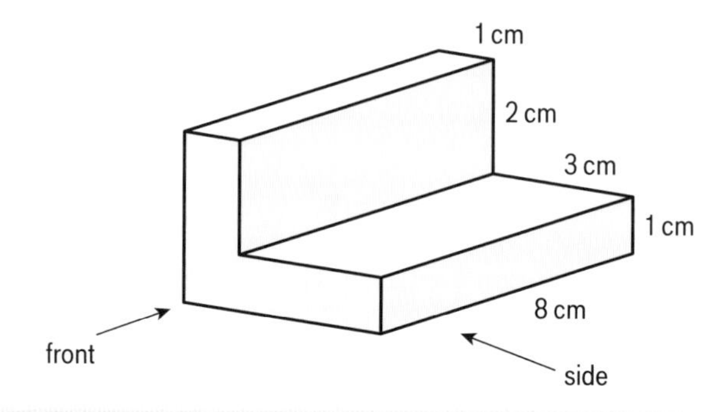

Exam-style question

10 This diagram shows a prism with a cross-section in the shape of a trapezium.
On a centimetre grid, draw the front elevation and the plan of the prism. **(3 marks)**

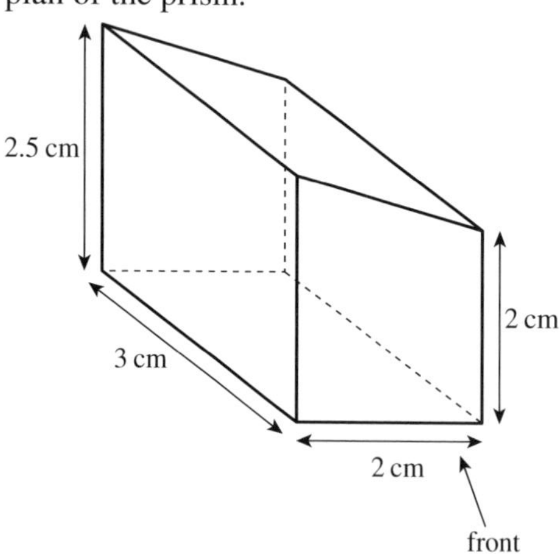

Exam tip

Make sure your lines show up clearly on the grid.

11 Here are the plan, front elevation and side elevation views of a cuboid, drawn on squared paper.

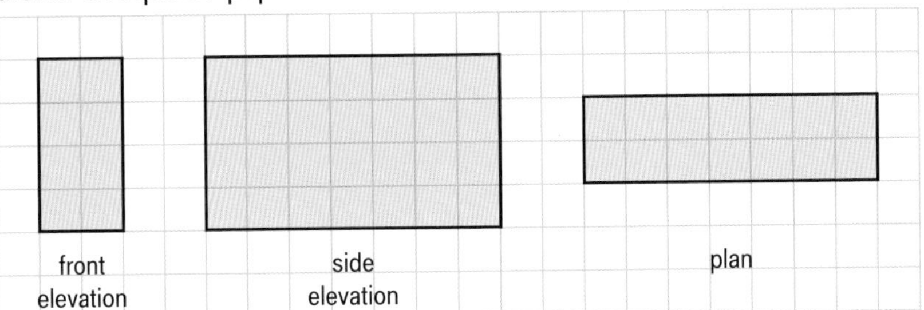

Sketch the cuboid and label its dimensions.

12 Here are the plan, front elevation and side elevation views of a solid shape, drawn on squared paper.

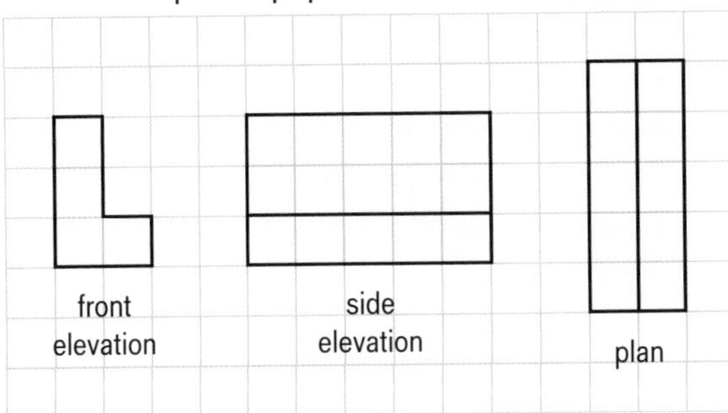

Draw a sketch of the solid. Label its dimensions.

Exam-style question

13 Here are the plan, front elevation and side elevation views of a solid shape drawn on squared paper.

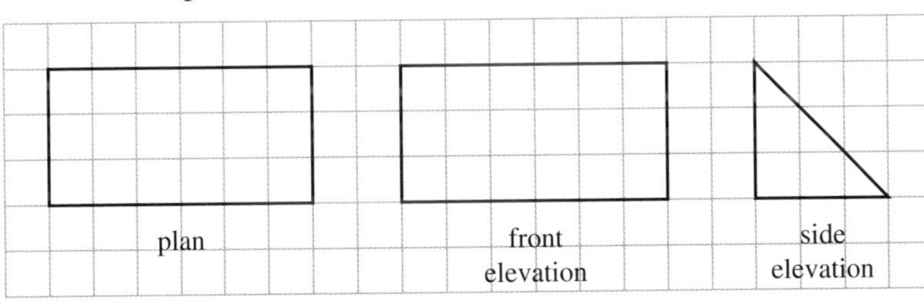

Draw a sketch of the solid shape.
Give the dimensions of the solid on your sketch. **(2 marks)**

Exam tip

Use a pencil and a ruler to ensure your shape is drawn neatly.

15.3 Accurate drawings 1

- Make accurate drawings of triangles using a ruler, protractor and compasses.
- Identify SSS, ASA, SAS and RHS triangles as unique from a given description.
- Identify congruent triangles.

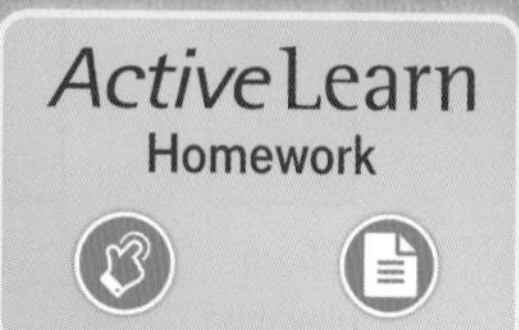

Warm up

1 **Fluency** Which of these shapes are congruent?

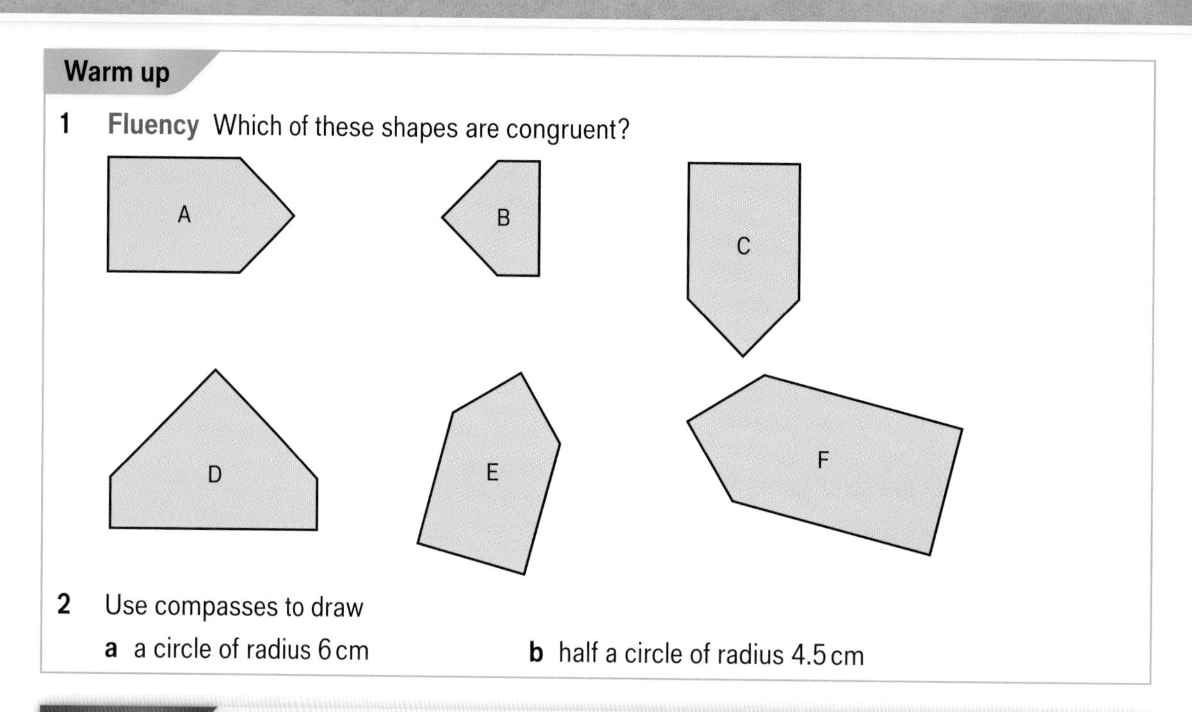

2 Use compasses to draw

a a circle of radius 6 cm

b half a circle of radius 4.5 cm

Key point

You can draw an accurate diagram of a triangle with a ruler and protractor if you know three measurements. In an **ASA** triangle, you are given an **A**ngle, a **S**ide length and another **A**ngle.

3 Make an accurate drawing of this triangle using a ruler and protractor.

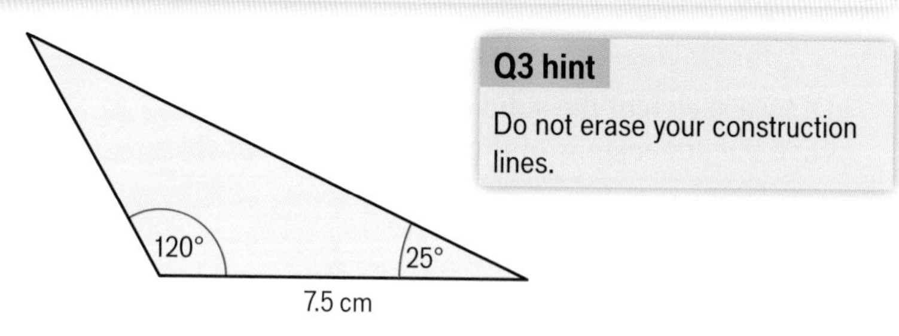

Q3 hint

Do not erase your construction lines.

4 Make an accurate drawing of triangle PQR using a ruler and protractor.

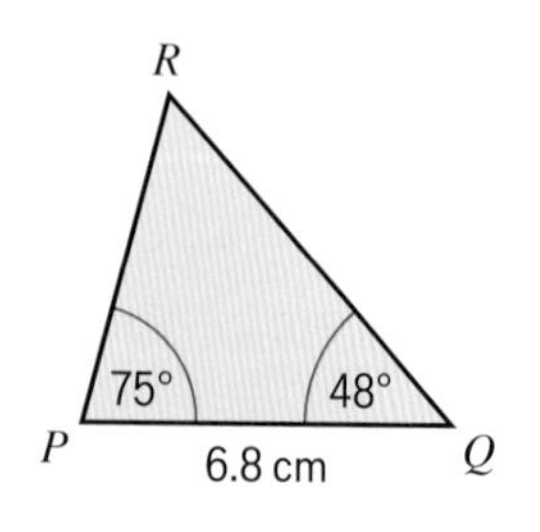

Key point

In an **SAS** triangle you are given two **S**ide lengths and the **A**ngle in between.

5 Make an accurate drawing of triangle XYZ using a ruler and protractor.

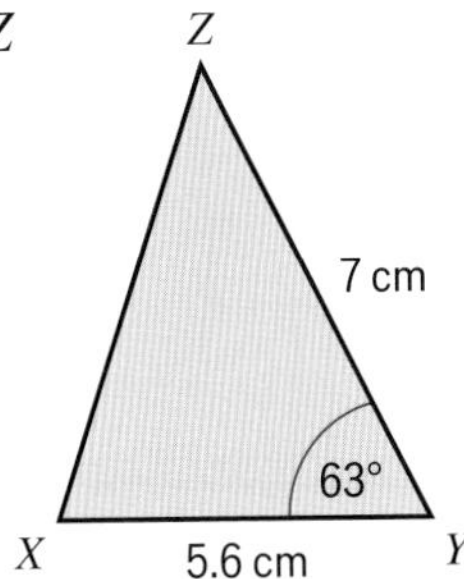

Q5 hint

Start by drawing and labelling the line XY, and then measure the angle XYZ at point Y using a protractor.

Exam-style question

6 The diagram shows a sketch of triangle ABC.

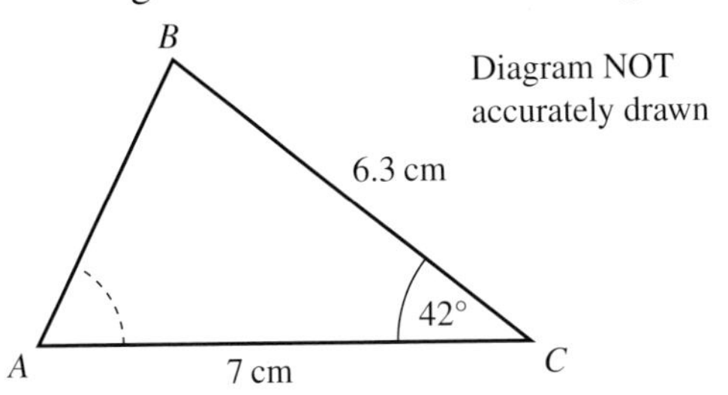

a Make an accurate drawing of triangle ABC. **(2 marks)**

b Measure the size of angle BAC on your diagram. **(1 mark)**

7 **a** **Reasoning** Which of these five triangles are ASA triangles and which are SAS triangles?

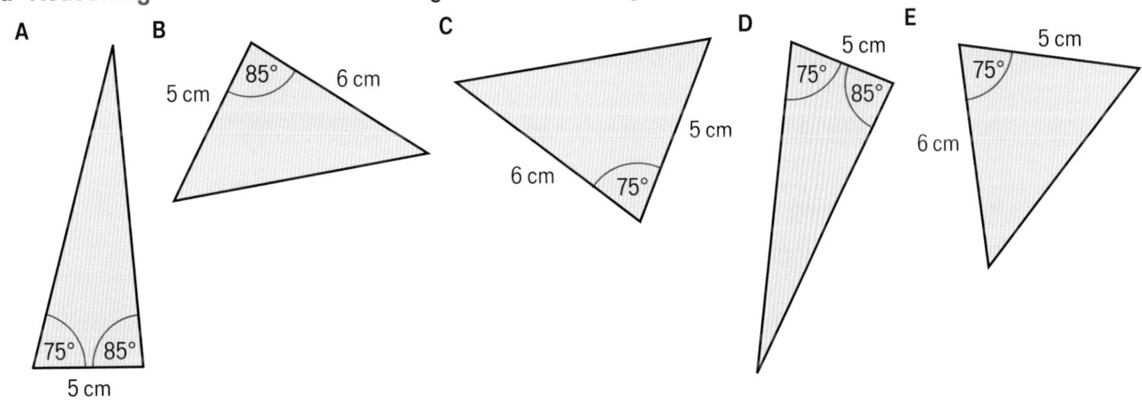

b Make accurate drawings of all the triangles using a ruler and protractor.

c Which pairs of triangles are congruent?

d **Reflect** For each of these statements, write down whether it is always, sometimes or never true.

i Two ASA triangles are congruent if the given side length and two angle sizes are identical in each triangle.

ii Two SAS triangles are congruent if the given angle size and two side lengths are identical in each triangle.

Key point

In an **SSS** triangle, you are given all three **S**ide lengths but none of the angles.

Example

Construct a triangle with sides 7 cm, 5 cm and 10 cm.

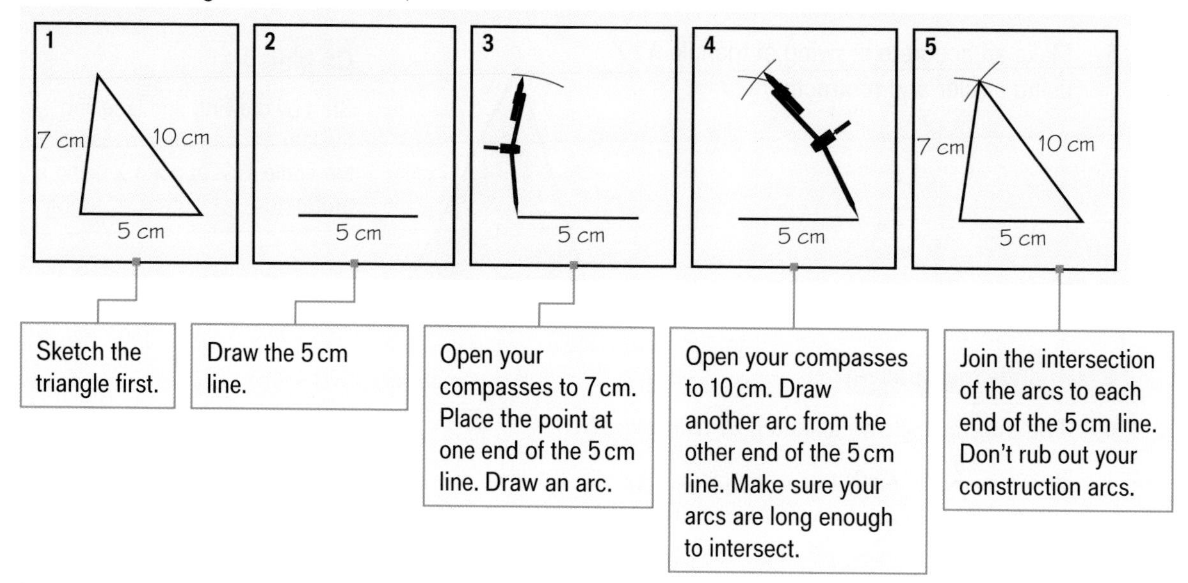

Key point

To **construct** means to draw accurately using a ruler and compasses.

8 Construct each triangle using a ruler and pair of compasses.

a

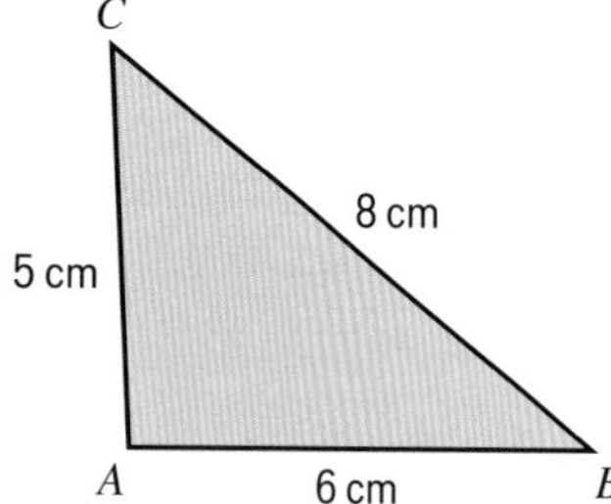

b

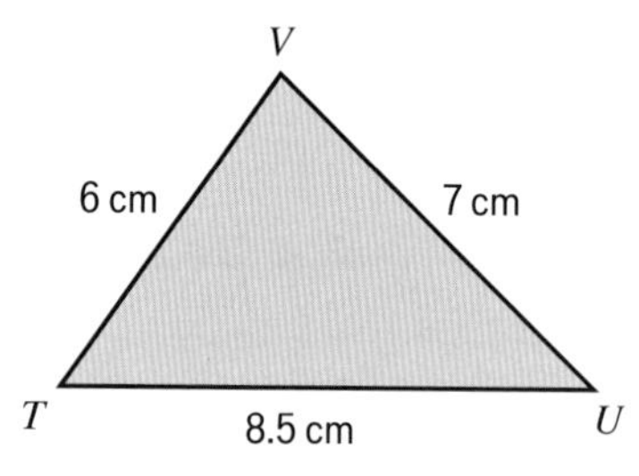

9 **a** **Problem-solving** Construct an equilateral triangle with side lengths 5 cm.

b **Reflect** What angle have you constructed here?

10 **Problem-solving** Construct a triangle with sides of length 6 cm, 3 cm and 8 cm.

Key point

In an **RHS** triangle, you are given the **R**ight angle, the **H**ypotenuse length and another **S**ide length.
The **hypotenuse** is the longest side of a right-angled triangle.

11 **a** Make an accurate drawing of triangle VWX.
Use a protractor to draw the right angle.

b Find the length of the side VW.

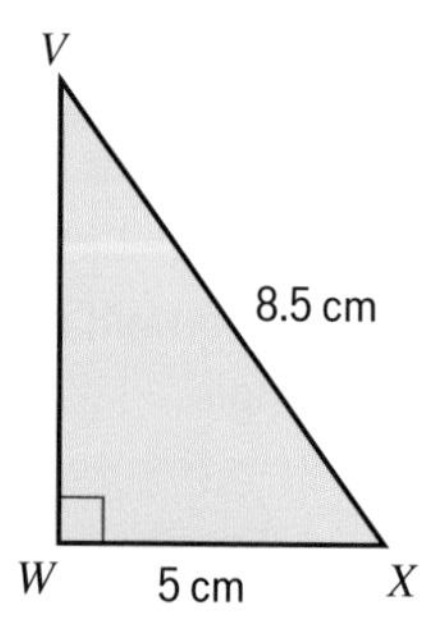

12 a Reasoning Here are some sketches of RHS triangles and SSS triangles.
Make accurate drawings of all the triangles.

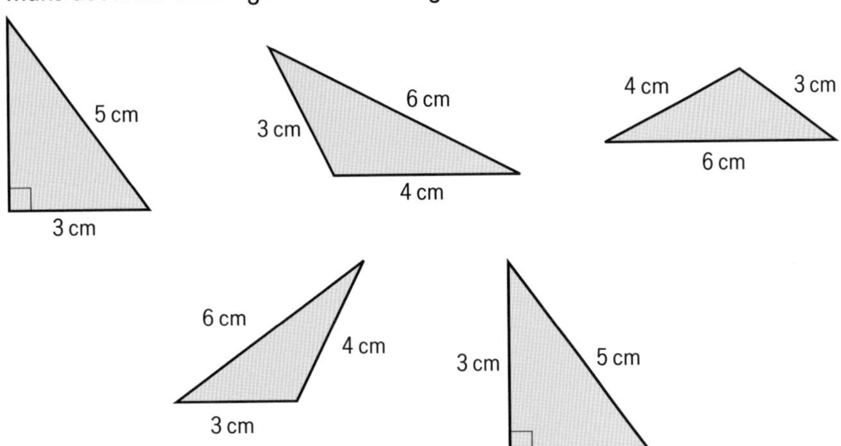

b Reflect For each statement, write down whether it is always, sometimes or never true.

i Two RHS triangles are congruent if one given side length and hypotenuse length are identical in each triangle.

ii Two SSS triangles are congruent if the three given side lengths are identical in each triangle.

13 Reasoning For each set of triangles, decide which two are congruent.
For each pair, choose the correct description from the box.

ASA	SSS	RHS	SAS

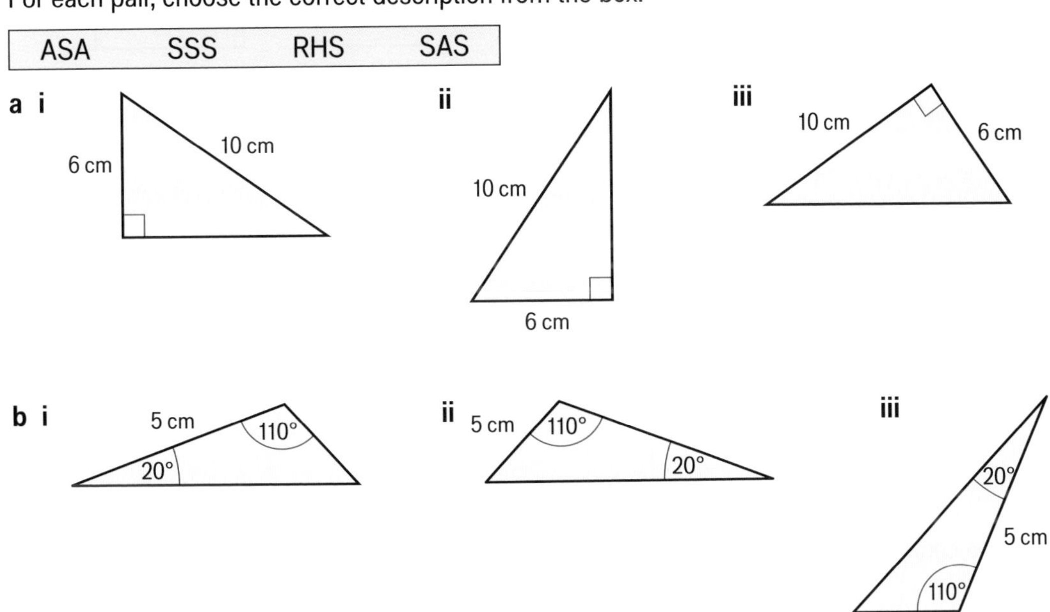

15.4 Scale drawings and maps

- Draw diagrams to scale.
- Use scales on maps and diagrams to work out lengths and distances.
- Solve problems involving scales.

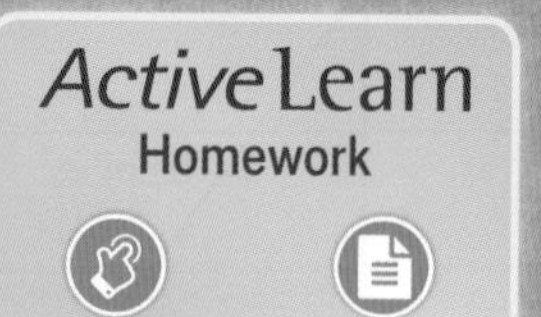

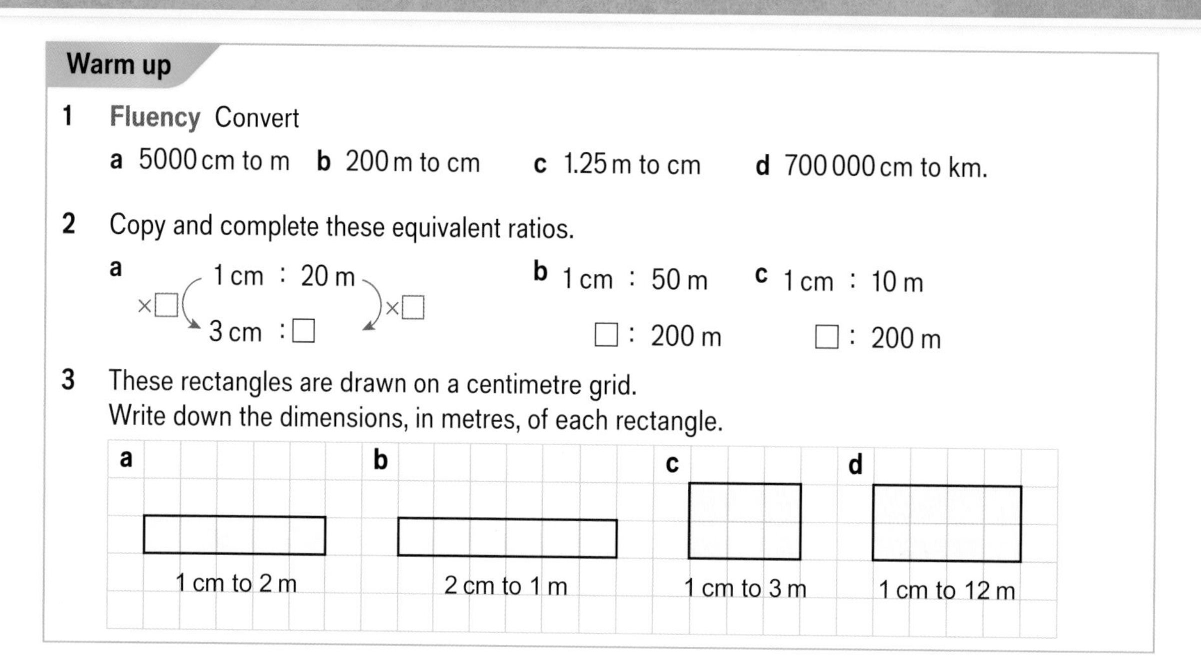

Warm up

1 **Fluency** Convert

a 5000 cm to m **b** 200 m to cm **c** 1.25 m to cm **d** 700 000 cm to km.

2 Copy and complete these equivalent ratios.

a 1 cm : 20 m, ×□, 3 cm : □, ×□

b 1 cm : 50 m, □ : 200 m

c 1 cm : 10 m, □ : 200 m

3 These rectangles are drawn on a centimetre grid.
Write down the dimensions, in metres, of each rectangle.

a 1 cm to 2 m **b** 2 cm to 1 m **c** 1 cm to 3 m **d** 1 cm to 12 m

4 Here is a sketch of two rooms.

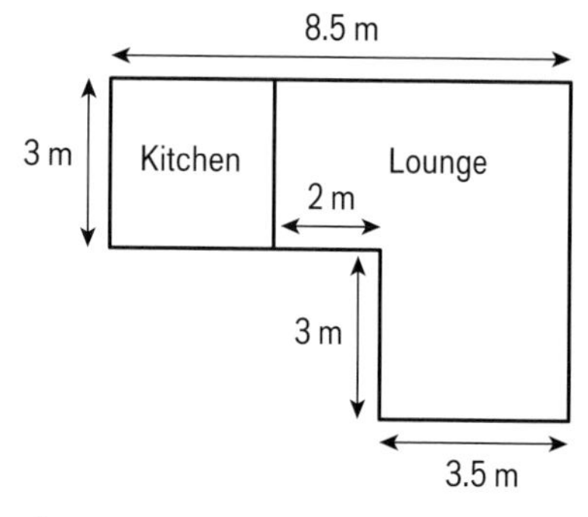

On a centimetre grid, draw a scale diagram of the rooms using the scale 2 cm represents 1 m.

5 On a scale drawing, 1 cm represents 5 m.

a On the same drawing, what length in metres do these represent?

i 4 cm **ii** 0.5 cm
iii 3.5 cm **iv** 8.2 cm

b What line length in cm on the drawing represents

i 15 m **ii** 7.5 m
iii 4 m **iv** 9 m?

6 The scale drawing shows a tree next to a building.

a Measure the tree and the building.

b How many times taller than the tree is the building?

The tree has height 5 m.

c **Reasoning** Work out an estimate for the height of the building in metres.

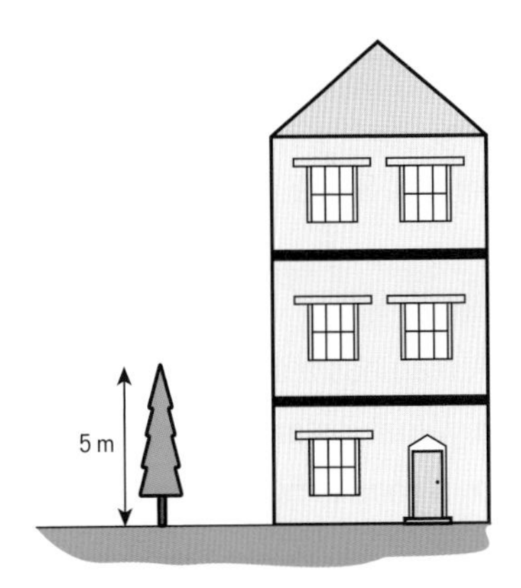

Exam-style question

7 The picture shows a bus next to a house.
The bus has a length of 10 m.

The bus and the house are drawn to the same scale.
Work out an estimate for the height, in metres, of the house. **(2 marks)**

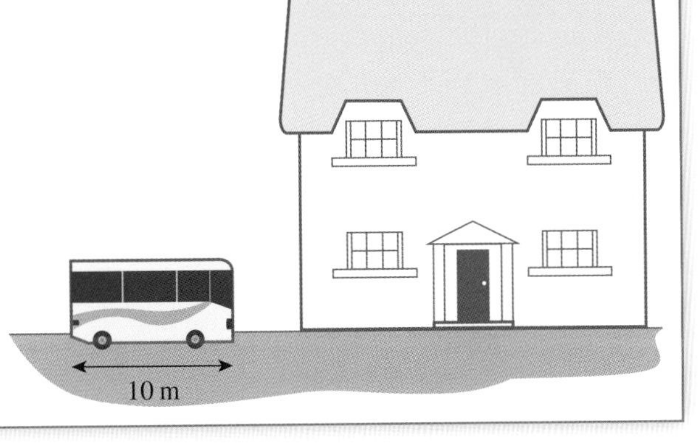

8 The height of a man is 1.9 m.
The scale diagram shows a table, a man and a door.

Estimate the height in metres of

a the table

b the door

Exam-style question

9 A map has a scale of 1 cm to 12 km.
On the map, the distance between Bristol and Swansea is 10.7 cm.
What is the real distance, to the nearest km, between Bristol and Swansea? **(2 marks)**

10 **Reasoning** A planner makes a map of four new villages, using a scale of 1 cm to 2 km.
Bradfield is 15 km north of Amphill. Carsbridge is 24 km west of Amphill.

a Draw a scale diagram showing these three villages.

b Dreyton is 9.6 km south-east of Amphill. Draw this on your diagram.

c Use your diagram to find the distance in km from

i Bradfield to Dreyton

ii Dreyton to Carsbridge

Key point

A **scale** is a ratio that shows the relationship between a length on a map or drawing and the actual length. Scale 1 : 25 000 means 1 cm on the map represents 25 000 cm in real life.

Example

A map has a scale of 1 : 20 000.

a What distance in metres does 4 cm on the map represent?

b A road is 500 m long. How long will the road be on the map?

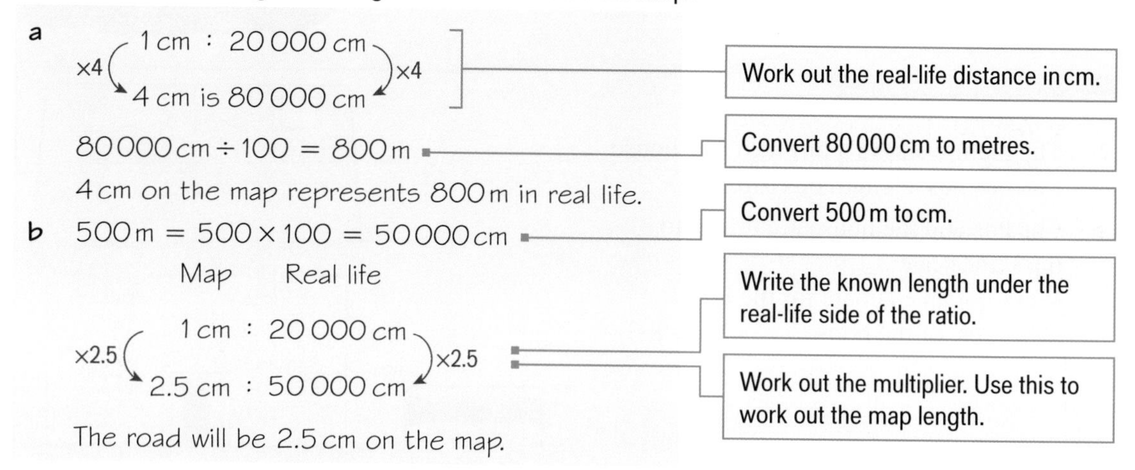

11 A scale drawing of a room has a scale of 1 : 25.

a The length of the room in the drawing is 24 cm. How long is the actual room?

b The width of the real room is 4 m. What is the width of the room on the drawing?

Exam-style question

12 The diagram shows a scale drawing of a garden.

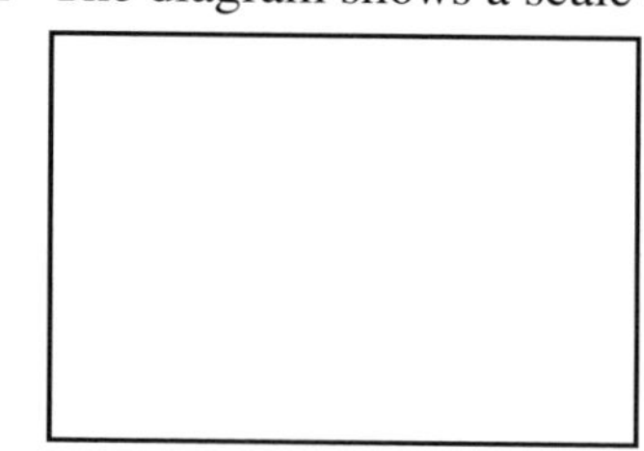

The scale of the drawing is 1 : 200.
Work out the perimeter of the real garden.
Give your answer in metres. **(5 marks)**

13 An archaeologist makes a scale drawing of her dig site using a scale of 1 : 10.
She digs a trench in the site 1 m wide by 3.5 m long.

a What are the dimensions of the trench on the scale drawing?

b An object on the drawing is 83 mm long. How long is this object in real life?

14 This diagram shows a scale plan of Jackie's bedroom, drawn on a centimetre grid. Her bed is 2 m by 1.5 m.

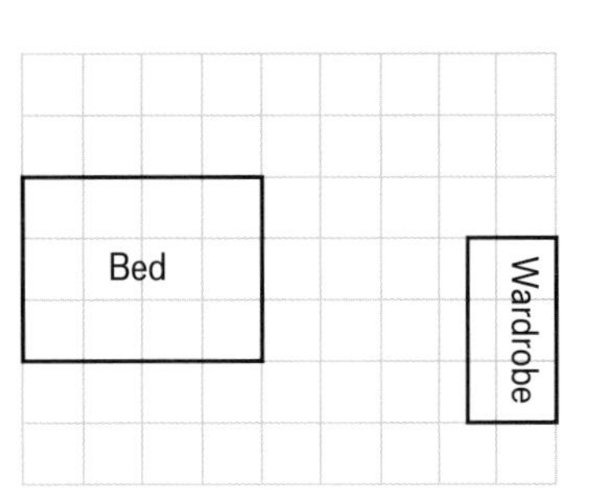

a Copy and complete this statement:
1 cm on the diagram represents □ m = □ cm.

b Write the scale used as a ratio $1 : x$.

c Use the scale to work out

i the dimensions of the bedroom

ii the dimensions of the wardrobe

iii the area of carpet which covers the whole floor

15 On a scale diagram, 5 cm represents a real-life length of 1.25 m.

a Write the two lengths as a ratio, scale diagram : real-life.

b Write as a ratio in the form $1 : m$.

16 A plane makes cargo deliveries in Ireland.
The map of the area has a scale of 200 km : 5 cm.

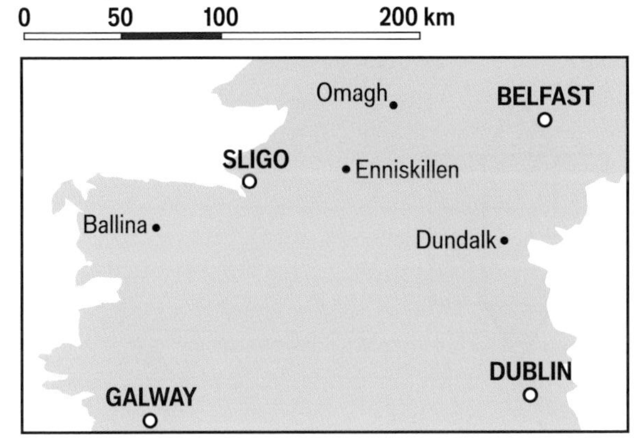

Q16 hint

Measure the distance on the map in centimetres. Then use the map scale to work out the real distance.

a Using the map, work out how far the plane flies from

i Dublin to Belfast

ii Belfast to Galway

iii Galway back to Dublin

b **Reflect** Explain why these distances may not be exact.

17 The scale factor on a map is 1 : 50 000.

a What distance in metres does 5 cm represent?

b What distance in kilometres does 7 cm represent?

c How many centimetres on the map is a real distance of 12.5 km?

Q17 hint

1 km = 1000 m
and 1 m = 100 cm

18 The scale on a map is 1 : 25 000.

a What length on the map is a 12 km walk?

b How many km is a walk represented by a length of 32 cm?

15.5 Accurate drawings 2

- Accurately draw angles and 2D shapes using a ruler, protractor and compasses.
- Construct a polygon inside a circle.
- Draw accurate nets.

Warm up

1 **Fluency** Here is a triangle.

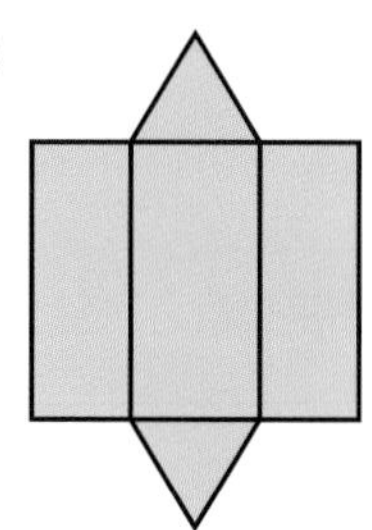

a What is the length of
  i AB ii BC?

b Write down the size of
  i angle ACB ii angle BAC

2 What 3D shape will each of these nets make?

a

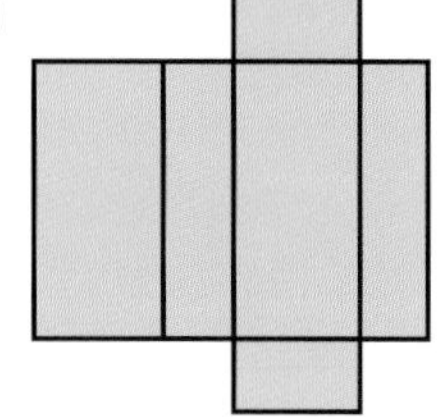

b

Key point

A **net** is a 2D shape that folds to make a 3D shape.

3 **Reasoning** This is one net of a cube.
Sketch two more nets for a cube.

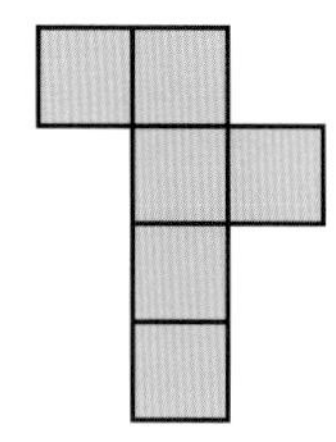

4 **Reasoning** Match each 3D shape to its net.

a

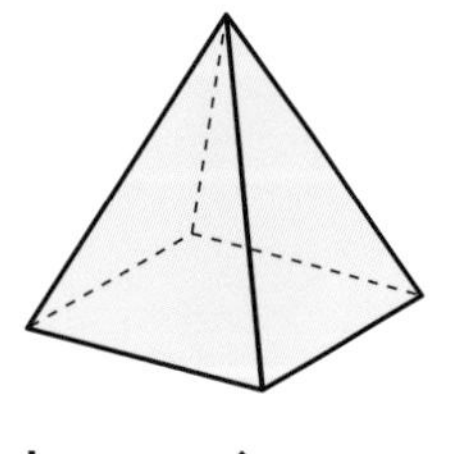

b

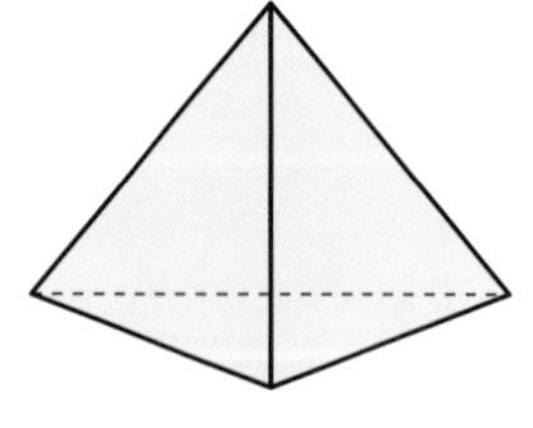

c

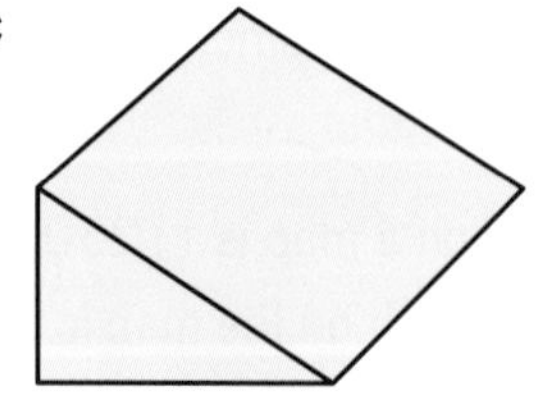

i

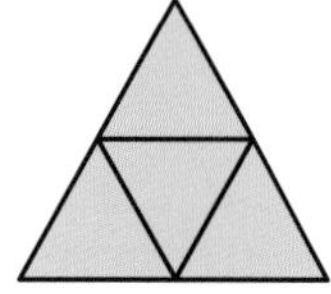

ii

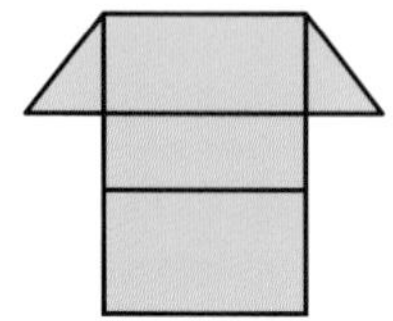

iii

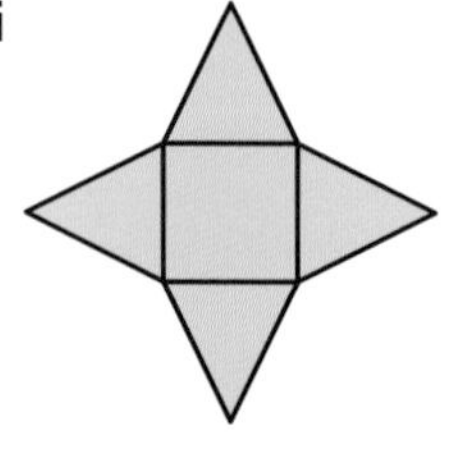

5 **a** Sketch the net for this square-based pyramid.
Label the lengths on your sketch.
b Draw the square face on a centimetre grid.
c Construct the triangular faces accurately using a rule and compasses.

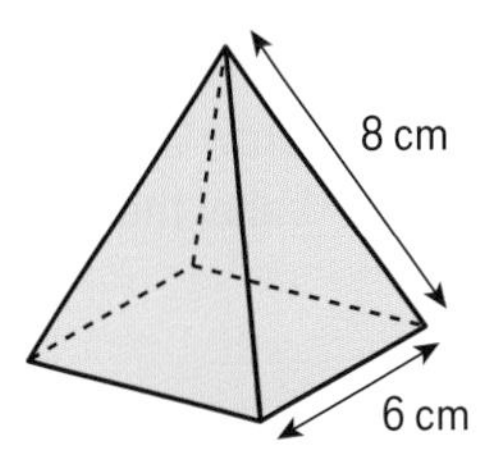

Exam-style question

6 Here is a box for chocolates.

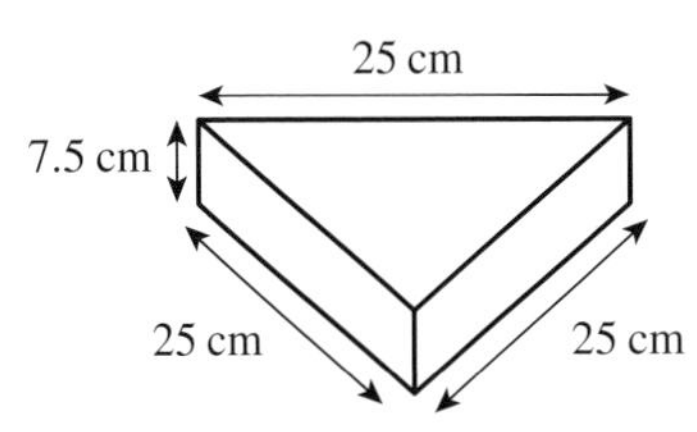

Use a ruler and compasses to draw a scale diagram of an accurate net for the box.
Use a scale of 1 cm to represent 5 cm. **(3 marks)**

Exam tip

Use a sharp pencil. Do not erase construction arcs.

7 In triangle FGH, $FG = 10.3$ cm, $GH = 9.4$ cm and angle $FGH = 62°$.
a Draw a sketch of triangle FGH showing all three given measurements.
b Now make an accurate drawing of triangle FGH.
c Measure the length of FH.

8 Here is a quadrilateral, $ABCD$.

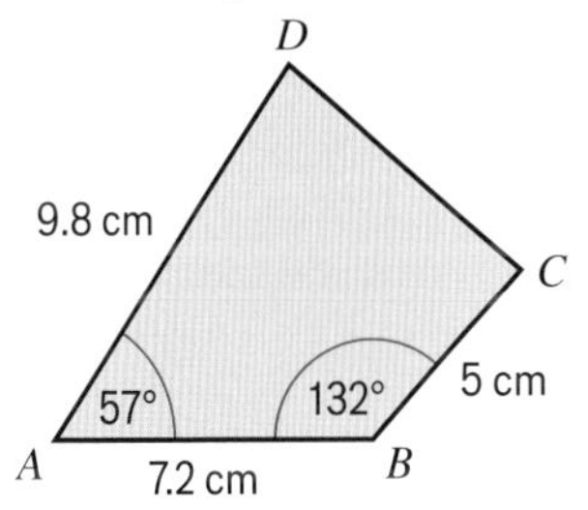

a Make an accurate drawing of this quadrilateral.
b On your drawing, measure
i the length of CD **ii** the size of angle BCD

Exam-style question

9 The diagram shows a prism with a cross-section in the shape of an isosceles trapezium.

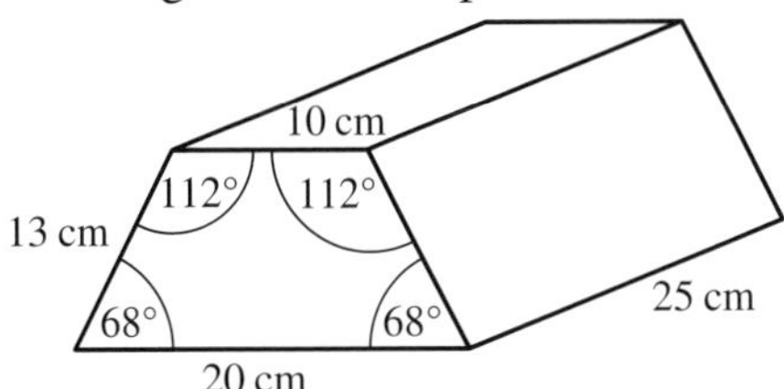

On a centimetre grid, draw an accurate front elevation and side elevation of this prism.
Use a scale of 1 cm to 2 cm. **(4 marks)**

10 Follow these instructions to construct a regular hexagon inside a circle.

a Draw a circle with radius 7 cm.
Label the centre O.

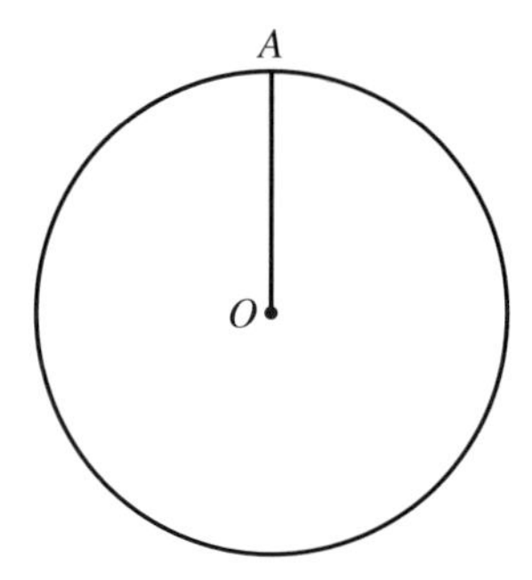

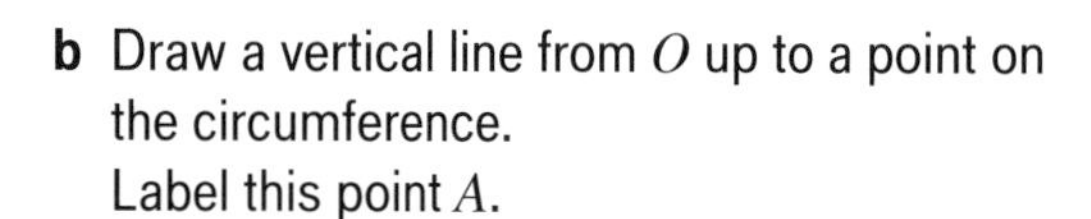

b Draw a vertical line from O up to a point on the circumference.
Label this point A.

c To work out the angle you need to measure in the middle, divide 360° by the number of sides on a hexagon.
Call this angle x.

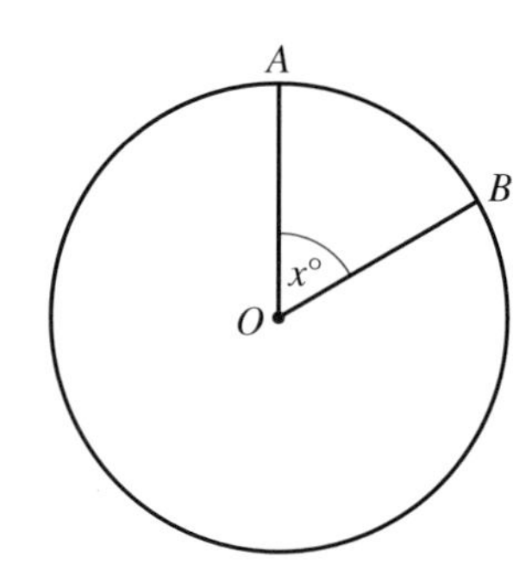

d Starting at line OA, measure angle x at the centre of the circle.
Draw a line from O to the circumference through your measured angle.
Label the new point on the circumference B and the angle $x°$.

e Join points A and B with a straight line.
Line AB is the first side of your hexagon.

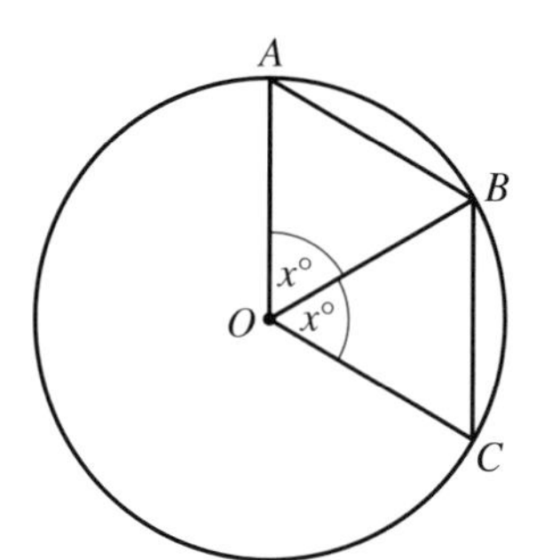

f Repeat parts **d** and **e** until you have drawn the whole hexagon.

11 Construct a regular nonagon inside a circle.

12 **Problem-solving** $JKLM$ is a cyclic quadrilateral.
All its vertices are on the circumference of a circle, centre O.
$OJ = OK = OL = OM = 6$ cm
$JOK = 72°$ and $KOL = 48°$.
Line KOM is a diameter of the circle.

a Using a ruler, compasses and a protractor, draw $JKLM$.

b On your drawing, measure angle JOM and angle KLM.

15.6 Constructions

- Bisect angles and lines using rulers and compasses.
- Find the shortest distance from a point to a line.

ActiveLearn Homework

Warm up

1 **Fluency** Which pair of lines are perpendicular to each other?

a

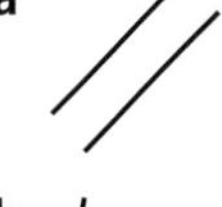

b

c

2 Use a ruler and compasses to construct an equilateral triangle with sides 6 cm.

Key point

Constructions are accurate drawings made using a ruler and a pair of compasses.
To **bisect a line** means to cut a line exactly in half.
A **perpendicular bisector** cuts a line in half at right angles.

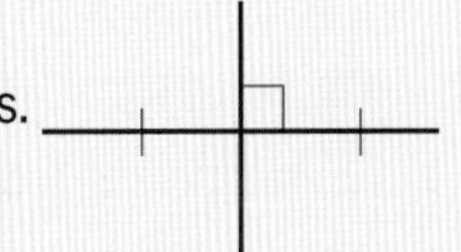

Example

Draw a line AB that is 6 cm long. Then construct its perpendicular bisector.

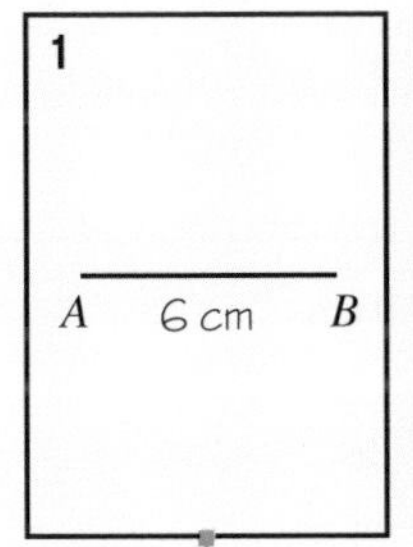

Use a ruler to draw the line.

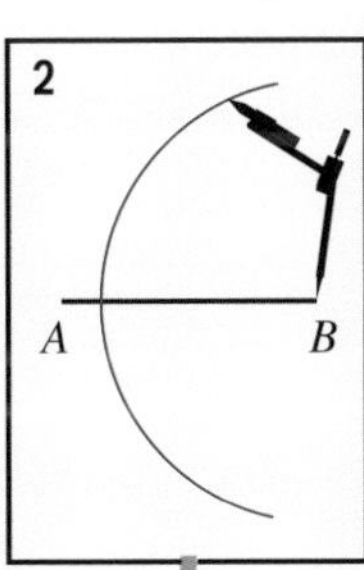

Open your compasses to more than half the length of the line. Place the point at B and draw an arc above and below the line.

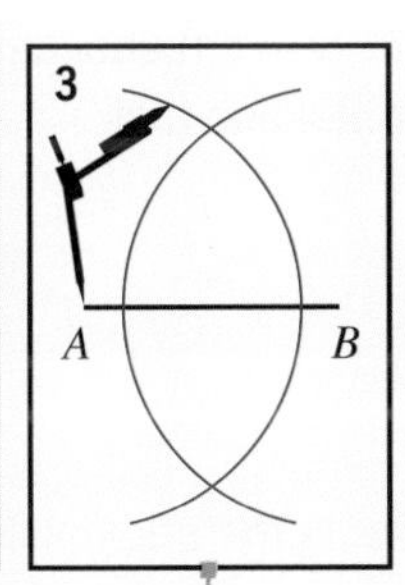

Keeping the compasses open to the same distance, move the point of the compasses to A and draw another arc.

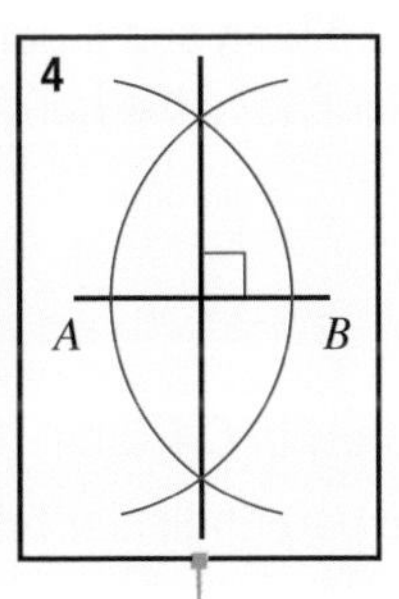

Join the points where the arcs intersect. Do not rub out your construction arcs.

3 **a** Draw a line AB 7 cm long. Construct the perpendicular bisector of AB.

b Use a ruler and protractor to check that it bisects AB at right angles.

Key point

You can also use a ruler and compasses to **construct** a perpendicular from a point to a line.

Example

Draw a horizontal line 8 cm long and a point P that is 3 cm above the line. Then construct a line that is perpendicular to the horizontal line that passes through point P.

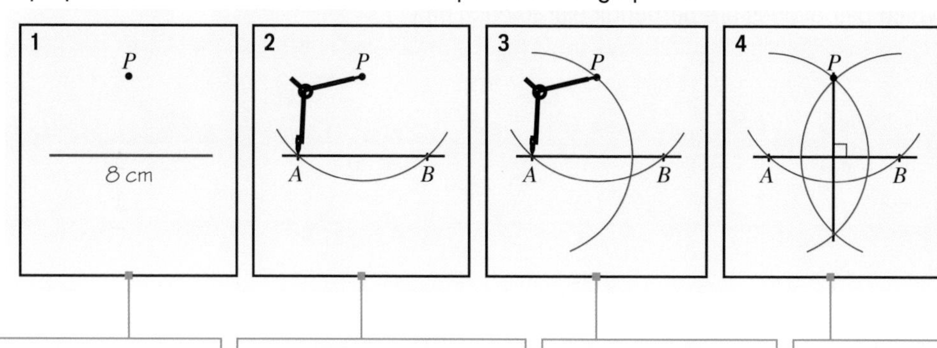

1	2	3	4
Use a ruler to draw an 8 cm line and then mark point P, 3 cm above this line.	Open your compasses to a radius larger than the distance from P to the line (the larger the distance, the more accurate your diagram). Then, with your compasses on point P, draw an arc that cuts the line twice. Label the two intersection points A and B.	Put the compasses on point A and draw an arc above and below the horizontal line. Repeat, with compasses open the same distance and with the compass point at B.	Join the intersection points of these two arcs. This line will go through P and will be perpendicular to the horizontal line.

4 Draw a horizontal line 9 cm long and a point Q, 4 cm above this line. Construct a perpendicular line to the horizontal line that passes through point Q.

Q

9 cm

5 **a** On the diagram you drew in **Q4**, label

i the point where the perpendicular bisects the original line as point X

ii the two intersection points as A and B

b Add in point C somewhere between points A and X.

c Add in point D somewhere between point B and the other end of the line.

d Measure the distances QC, QD and QX.

e **Reasoning** What is the shortest distance from Q to the horizontal line?

Q, C, A, X, B, D

Key point

The **shortest** path from a point to a line is **perpendicular** to the line.

Exam-style question

6 A town planner wants to build a new road from the sports centre to the main road. The new road needs to be the shortest distance possible.

Exam tip

Remember, 'construct' means 'use compasses and ruler'.

main road

sports centre

Scale: 1 cm represents 500 m.

a Copy or trace the diagram. Construct a line to show the position of the new road. **(2 marks)**

b Find the length of the new road, in kilometres. **(2 marks)**

7 **a** Copy or trace this diagram.

P

b Put your compasses on point P and draw arcs so that P is the midpoint of the line between the arcs.

P

c Construct the perpendicular bisector for the part of the line between the two arcs.

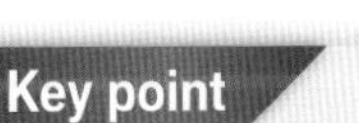

An **angle bisector** cuts an angle exactly in half.

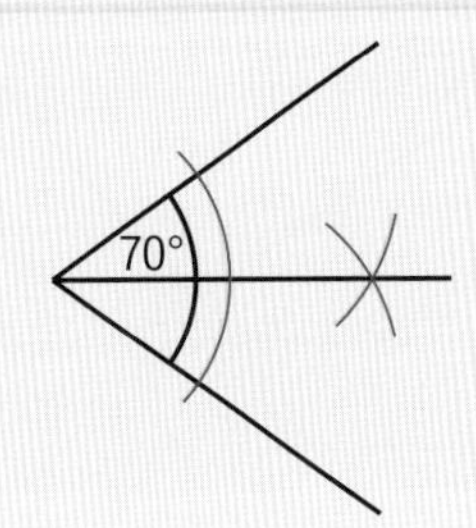

Example

Draw an angle of 80°. Construct the angle bisector.

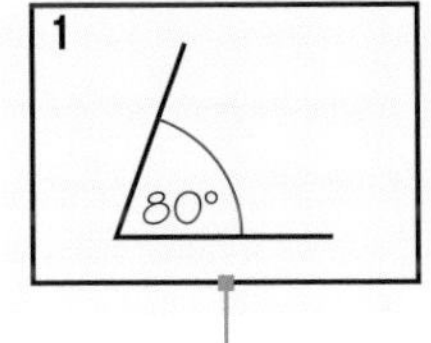

Draw the angle using a protractor.

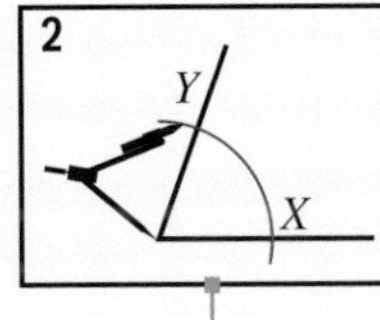

Open your compasses and place the point of the compasses at the vertex of the angle. Draw an arc that cuts both arms of the angle. Label these intersection points X and Y.

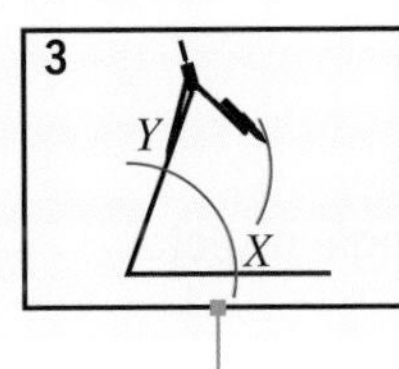

Keep the compasses open at the same length. Put the point of the compasses on Y and draw an arc in the middle of the angle.

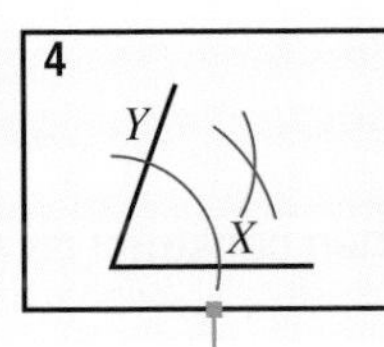

Repeat step 3 with the point of the compasses on X.

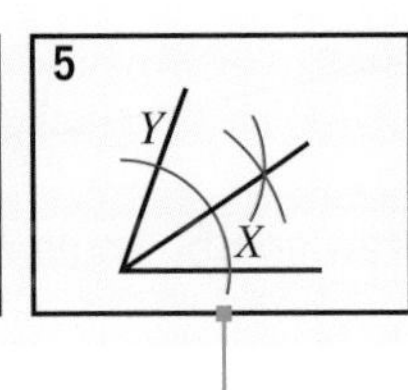

Join the intersection of these two arcs with the vertex of the angle. This straight line is the angle bisector.

8 **a** Draw an angle of 70° using a protractor. Construct the angle bisector.

b Check, by measuring with a protractor, that your angle bisector cuts the angle exactly in half.

9 Trace this angle and then use a ruler and compasses to construct the bisector of the angle. You must show all your construction lines.

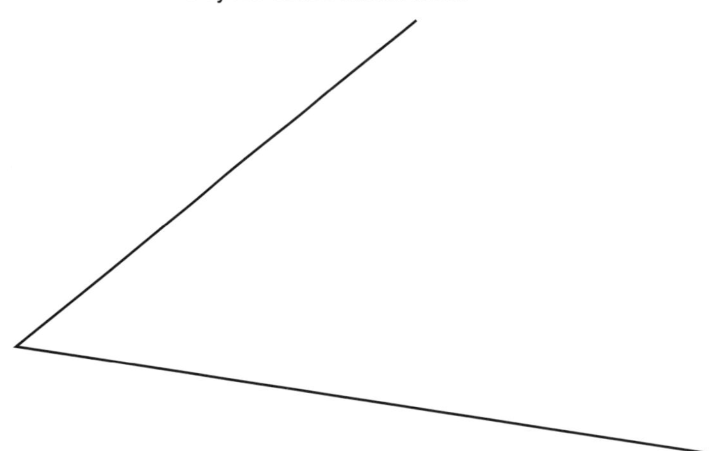

10 Trace each triangle.
Then construct the bisector of angle Q in each triangle.

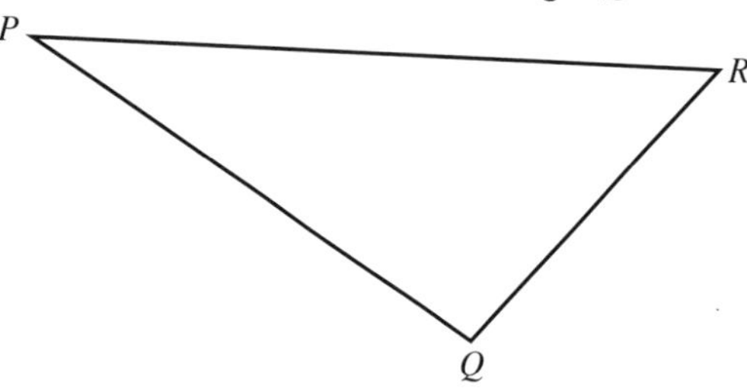

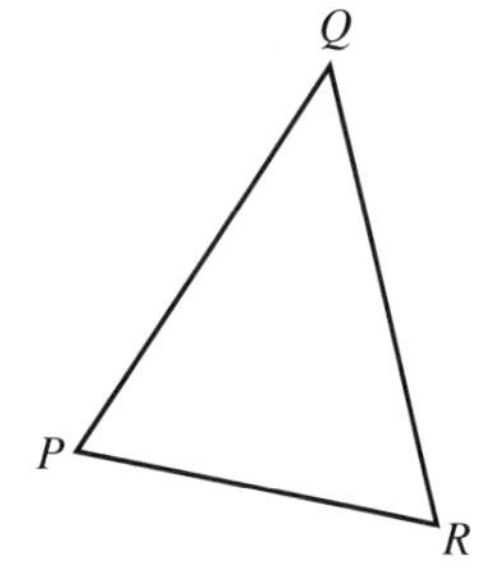

11 Draw an 85° angle. Then construct its angle bisector.

12 **Reasoning** Use your knowledge of constructing perpendicular bisectors and angle bisectors to construct a 90° and a 45° angle.

13 **a** Draw a 60° angle by constructing an equilateral triangle.

b Bisect your 60° angle to construct a 30° angle.

c **Reflect** Explain how you could construct a 120° angle.

15.7 Loci and regions

- Draw loci for the path of points that follow a given rule.
- Identify regions bounded by loci to solve practical problems.

Warm up

1 **Fluency** The scale on a map is 1 cm represents 5 km.

a How many kilometres does a distance of 4 cm on the map represent?

b How many centimetres on the map is a real distance of 7.5 km?

2 Draw a line 76 mm long. Construct its perpendicular bisector.

Key point

A **locus** is a set of all points that obey a given rule. This produces a path followed by the points. The plural of locus is **loci**.

3 a Draw a point P. Draw more points that are all 5 cm from P. When you join these points up, what shape do you make?

b What is the locus of points a fixed distance of 5 cm from a point P?

Key point

The locus of all points a given distance (d) from a fixed point (P) is a circle, with centre P and radius d.

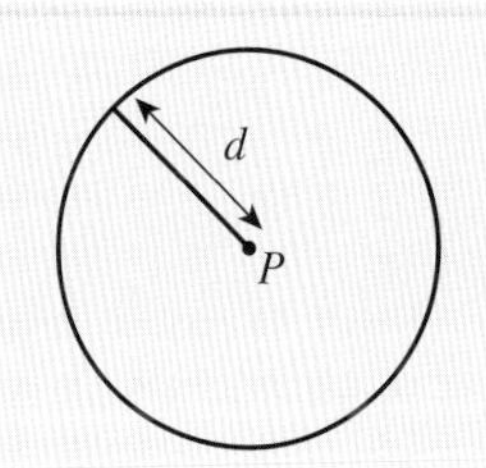

4 Draw a line AB 8 cm long. Draw points that are all 3 cm from the line.

Key point

The locus of all points 3 cm from a line, AB, is two parallel lines 3 cm from the line and a semicircle at each end with radius 3 cm.

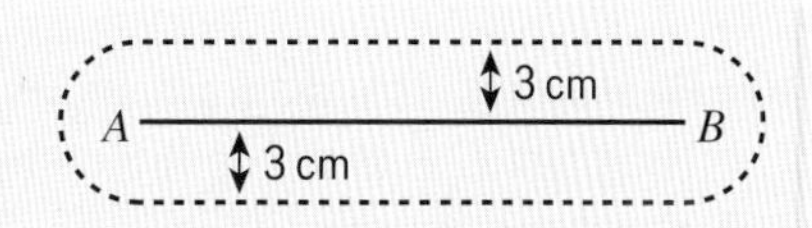

5 Draw a line 6 cm long. A point moves so that it is always exactly 2 cm away from the line. Construct the locus.

Key point

A locus that is equidistant (the same distance) from two given points, X and Y, is the perpendicular bisector of the line between the two points.

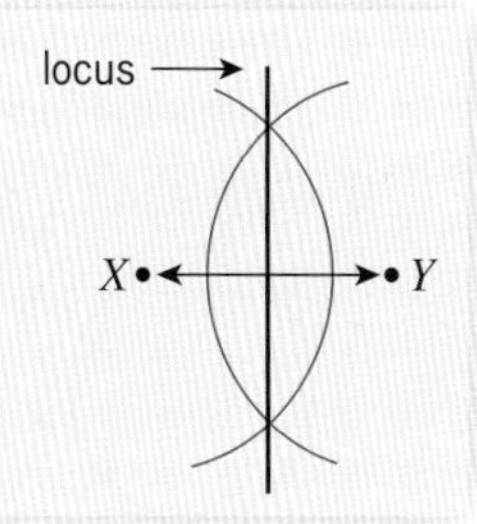

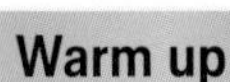

6 **a** Draw two points X and Y.

b Construct the perpendicular bisector of XY.

c Choose any point on the perpendicular.
Use a ruler to check that this point is equidistant from X and Y.

7 A road is **equidistant** from two towns A and B.

a Draw dots to represent towns A and B, 8 cm apart.

b Construct the path of the road.

Key point

A locus of all points that are equidistant from two intersecting lines, OX and OY, is the angle bisector of the two lines.

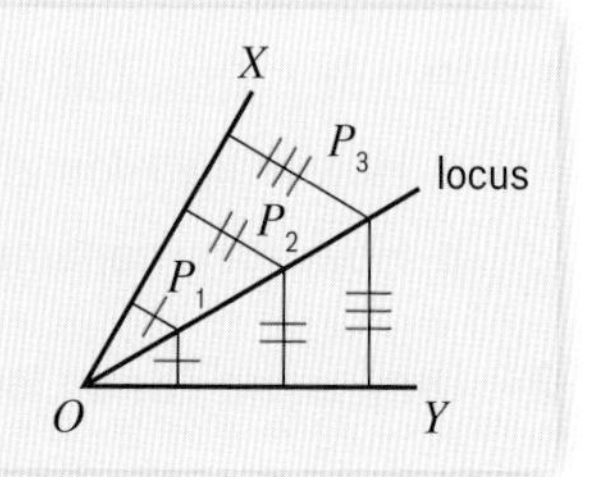

8 Draw an 80° angle, ABC. Make each arm of the angle at least 6 cm long.
Construct the locus of points equidistant from AB and BC.

Exam-style question

9 A, B and C are three points on a map.

× A

C ×

× B

Scale: 1 cm represents 100 m.
Point P is 350 m from point C.
Point P is equidistant from points A and B.
On a trace of the map, show one of the possible positions for point P. **(3 marks)**

Exam tip

Use the scale. Show your constructions clearly.

10 **Reasoning** This diagram shows a wheel.
The point C is at the centre of the wheel and the point P lies on its edge.

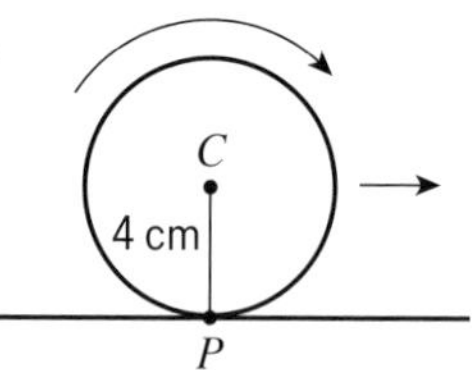

Q10a hint

Think about the distance from the centre of the wheel to the surface as the wheel moves.

a The wheel rolls along a flat surface.
Sketch the locus of point C as the wheel rolls along the surface.

b Copy and complete this sketch of the position of P as the wheel rolls.

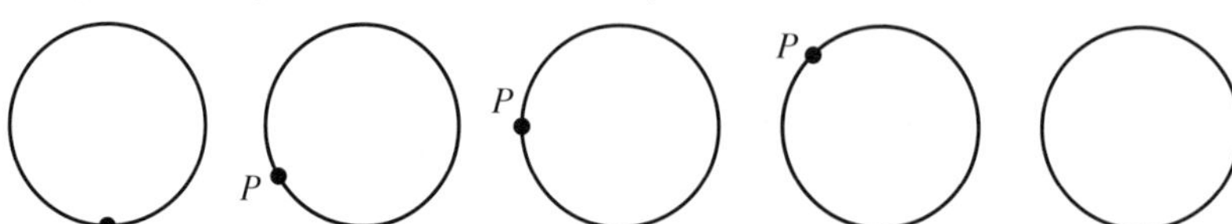

c Sketch the locus of point P as the wheel rolls along the surface.

11 A square piece of card is placed on a horizontal surface.

a The card is first rotated 90° clockwise about vertex D.
Copy or trace the diagram in colour 1. Draw the new position of the square in colour 1.
Draw the locus of the vertex A in colour 2.

b The card is then rotated 90° clockwise about vertex C and then 90° clockwise about vertex B.
Draw the new positions of the card in colour 1 and the remainder of the locus of vertex A in colour 2.

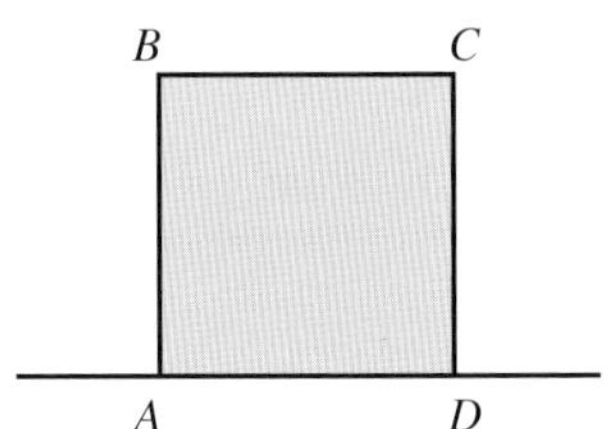

Key point

A **region** is an area bounded by loci.

12 Reasoning Draw points A and B, 7.5 cm apart. Now draw a circle centre A radius 4 cm and a circle centre B radius 5 cm. Shade the region where the circles intersect.
Which of these statements is true?

A The shaded region represents the area which is more than 4 cm from point A and more than 5 cm from point B.

B The shaded region represents the area which is less than 4 cm from point A and less than 5 cm from point B.

C The shaded region represents the area which is less than 5 cm from point A and less than 4 cm from point B.

Exam-style question

13 The diagram shows an accurate scale drawing of two towns, Exe and Wye.

× Wye

Exe ×

Scale: 1 cm to 2 km

A new cinema is to be built. The cinema will be

- less than 6 km from Exe and
- less than 9 km from Wye

Copy the diagram and shade the region where the cinema can be built. **(3 marks)**

Exam tip

Draw your circles to scale. Show all your working.

14 Problem-solving A man in his boat at point B reports seeing a swimmer in distress.
A helicopter sets out to search for the swimmer in the region of sea less than 500 metres from point B, and more than 50 m from the land.
Copy the diagram and shade in the area where the helicopter will search for the swimmer.

Scale: 1 : 10 000

B •

sea

land

15 Problem-solving A new mobile phone tower is to be built equidistant from two towns that are 6 km apart.

a Draw the position of the two towns 6 km apart. Use a scale of 1 cm : 1 km.

Phones less than 5 km from the tower receive a signal.

b Draw a scale diagram showing all possible locations of the tower so that both towns receive a signal.

16 a The diagram shows the perpendicular bisector of line AB.
Copy and complete:
The points in the shaded region are closer to point ________ than to point ________.

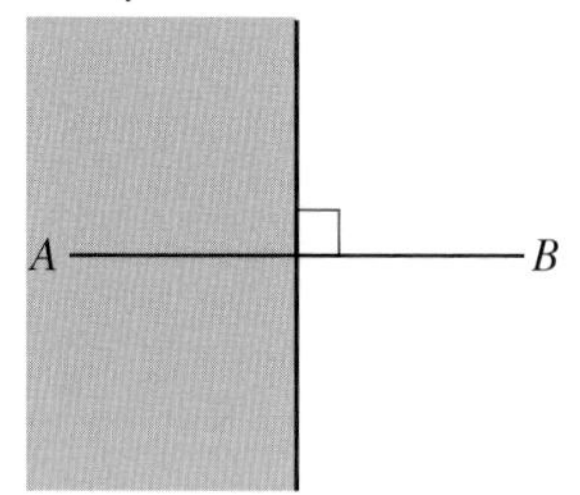

b The diagram shows the angle bisector of angle PQR.
Copy and complete:
The points in the shaded region are closer to line _______ than to line ________.

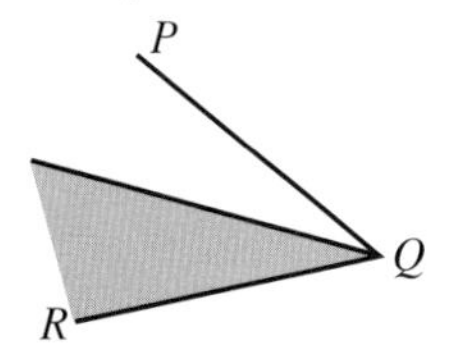

Exam-style question

17 The diagram represents a triangular garden ABC.
The scale of the diagram is 1 cm represents 1 m.
A tree is to be planted in the garden so that it is

- nearer to AB than to AC
- within 2 m of point A

Copy the diagram and shade the region where the tree may be planted. **(3 marks)**

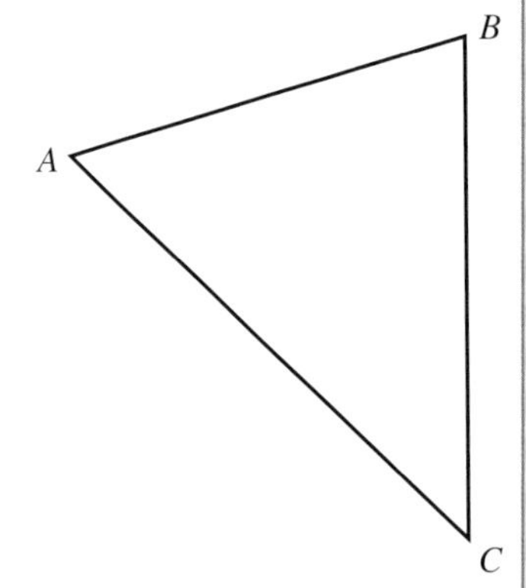

Exam tip

Read the question carefully to ensure that you draw the correct loci. Shade the correct region on your diagram.

15.8 Bearings

- Find and use three-figure bearings.
- Use angles on parallel lines to work out bearings.
- Solve problems involving bearings and scale diagrams.

Warm up

1 Fluency Work out the size of angles a, b, c and d.

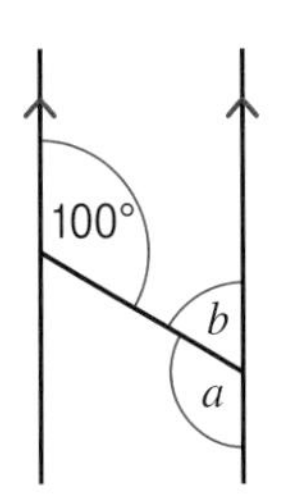

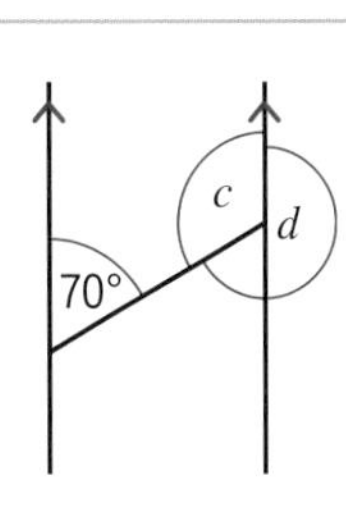

Key point

A **bearing** is an angle measured in degrees clockwise from north.
A bearing is always written using three digits.
This bearing is 025°.

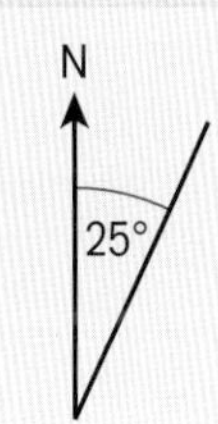

2 Find the bearing of B from A in these diagrams.
Always measure the angle of a bearing at the 'from' point.

a

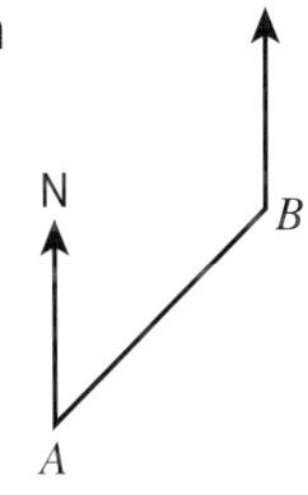

b

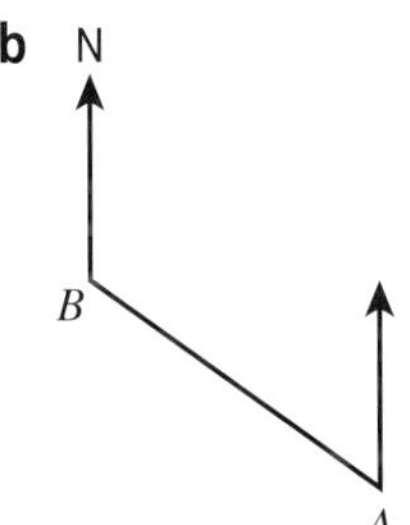

c

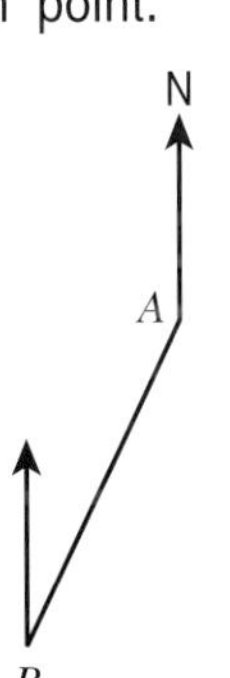

d

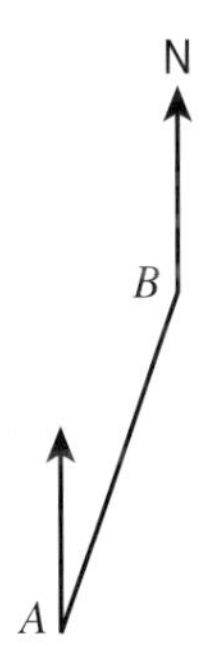

Exam-style question

3 Here is a map.
A straight road joins the two villages, Downley and Wakeford.

a Work out the real distance between the two villages. **(2 marks)**

b Find the bearing of Wakeford from Downley. **(1 mark)**

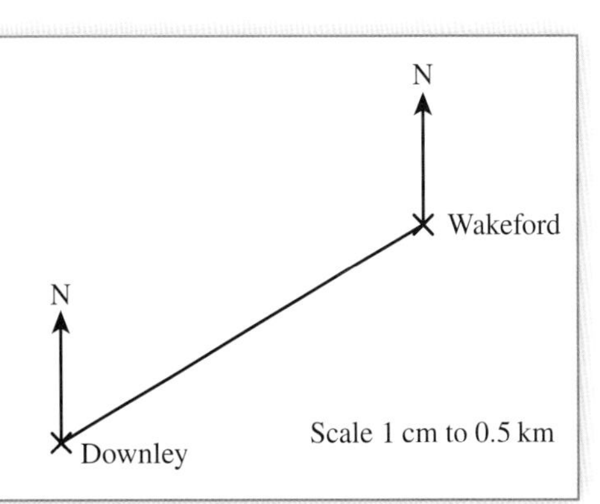

4 The diagram shows the position of two ships A and B and the harbour, H.

a What is the bearing of ship B from the harbour?

b What is the bearing of ship A from ship B?

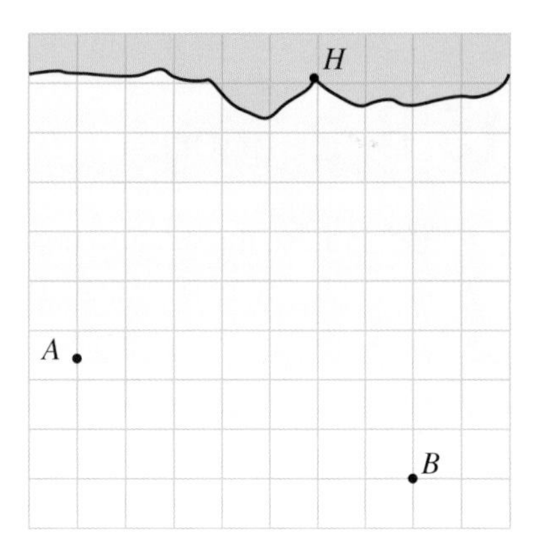

Example

Town C is 40 km from town A on a bearing of 100° from A.
Draw this accurately using a scale of 1 cm to 10 km.

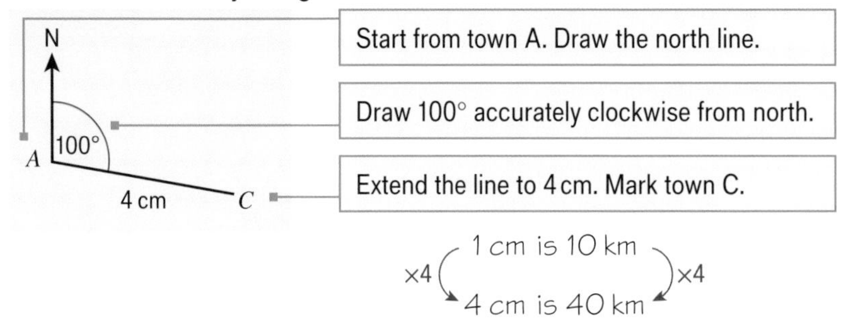

×4 (1 cm is 10 km → 4 cm is 40 km) ×4

5 Draw these bearings accurately. Use the scale 2 cm to 100 km.

a Venice Marco Polo airport is 420 km away from Rome Ciampino airport on a bearing of 356°.

b Naples airport is 175 km away from Rome Ciampino airport on a bearing of 128°.

6 Town P is on a bearing of 280° from a lookout post (L) 5 km away.
Town Q is on a bearing of 180° from L, 3 km away.

a Sketch P, L and Q. Label angles and distances.

b Draw an accurate diagram showing the towns and the lookout post.
Use the scale 1 cm to 1 km.

Q6b hint
Draw a straight line between the two towns.

c Find the distance between towns P and Q.

d Draw north lines at P and Q. Find

i the bearing of Q from P

ii the bearing of P from Q

e **Reflect** Did the sketch help you draw an accurate diagram?
Write a sentence explaining how.

7 **Problem-solving** The bearing of a yacht from a lighthouse is 162°.
The bearing of the yacht from the harbour is 225°.
The lighthouse is 81 km due west of the harbour.

a Draw an accurate diagram of the lighthouse, the yacht and the harbour.
Use a scale of 1 cm : 10 km.

b Find the bearing of the harbour from the yacht.

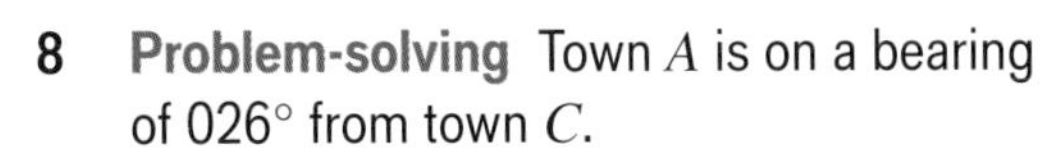

8 **Problem-solving** Town A is on a bearing of 026° from town C.

The distance between town A and C is 9.2 miles and the distance between towns B and C is 14.6 miles.

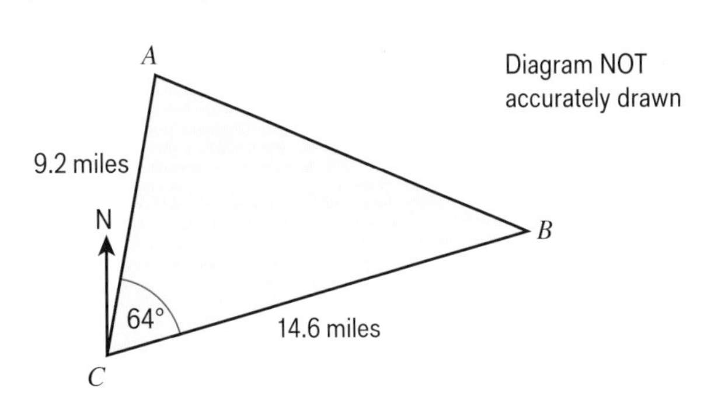

a Draw an accurate scale diagram of the position of towns A, B and C. Use a scale of 1 cm to 1 mile.

Use your diagram to find

b the bearing of town A from B

c the distance of town A from B

9 **a** Use angles in parallel lines to work out the bearing of Q from P.

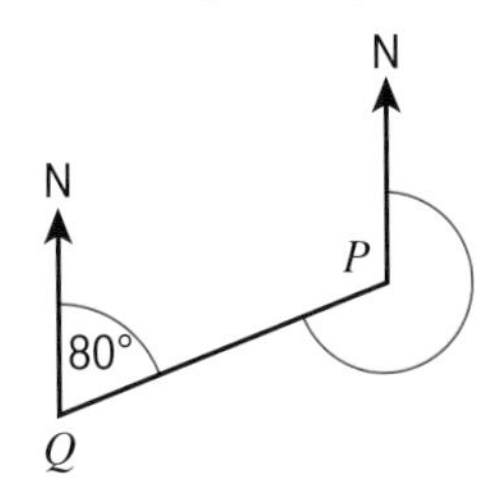

b Work out the bearing of

i A from B

ii B from A

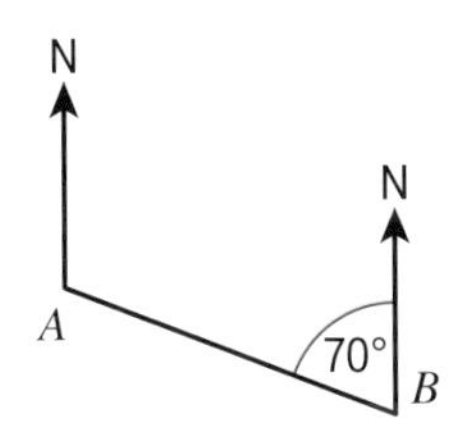

10 Work out the bearing of X from Y.

a

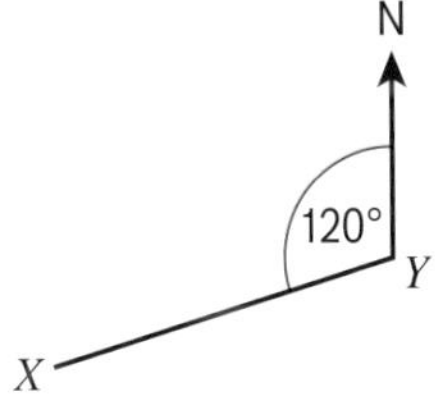

b

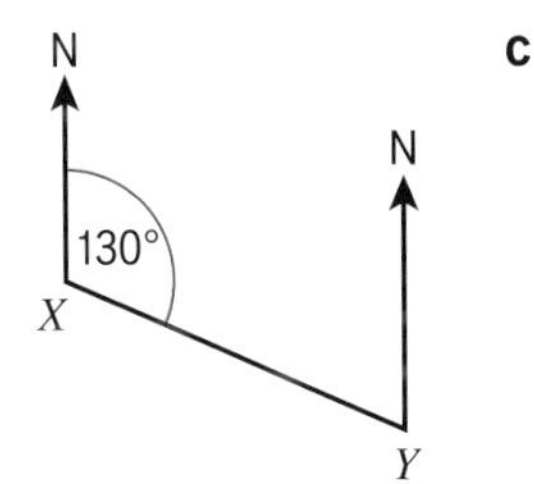

c

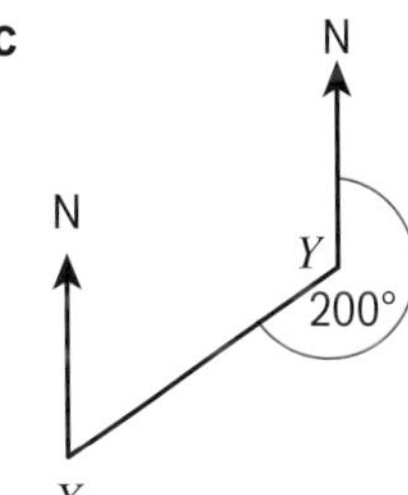

d

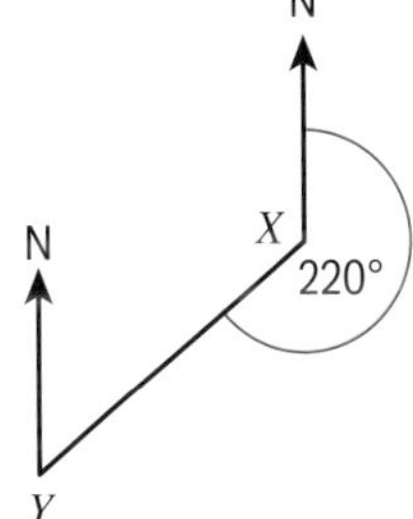

11 Work out the bearing of

a A from B

b B from C

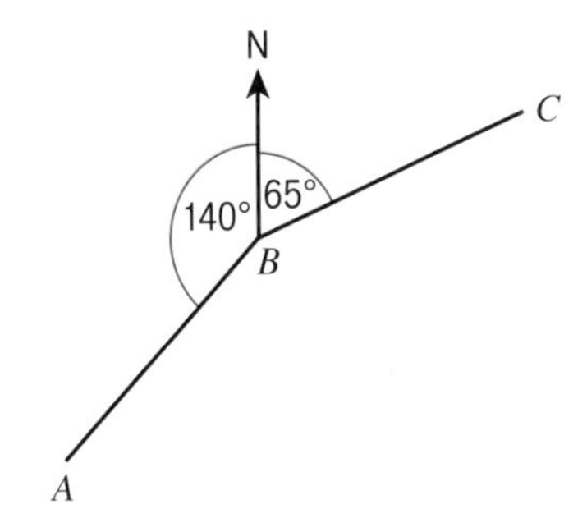

15 Check up

ActiveLearn Homework

3D solids

1 Here is a triangular prism.

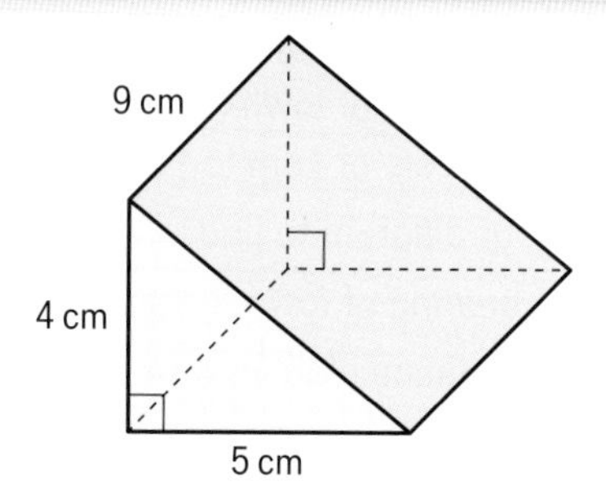

a How many vertices does this triangular prism have?

b How many edges does it have?

c How many faces does it have?

2 Here are the plan, front elevation and side elevation views of a solid shape drawn on squared paper.

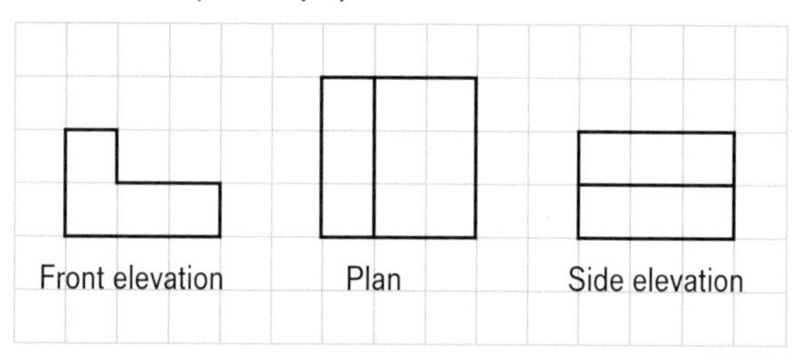

Sketch the shape. Label its lengths.

3 Sketch the prism in **Q1**. Draw in any planes of symmetry.

Constructions

4 Construct this triangle using a ruler and compasses.

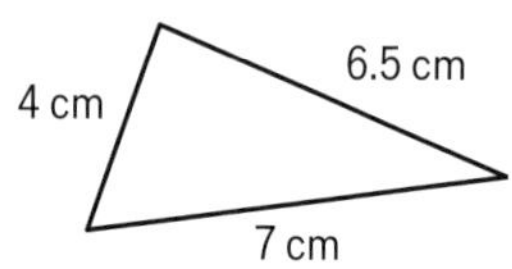

5 All edges of this triangle-based pyramid have length 5 cm.
Draw an accurate net for this pyramid.

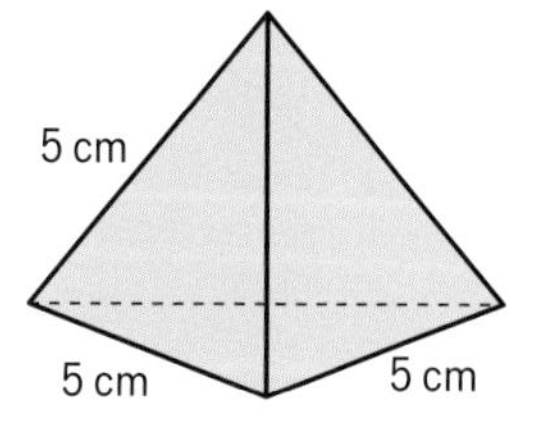

6 Draw a line 10 cm long. Use a ruler and compasses to construct its perpendicular bisector.

7 Draw a 110° angle using a ruler and protractor. Use a ruler and compasses to bisect the angle.

Loci and regions

8 **a** Draw the locus of points 4 cm away from a fixed point.

b Shade the region less than 4 cm from the point.

9 Draw two points that are exactly 10 cm apart.
Construct the locus of points that are equidistant from these two points.

10 Trace this diagram.

a Construct the locus of points which are equidistant from lines AB and BC.

b Shade in the region that is closer to line AB than line BC.

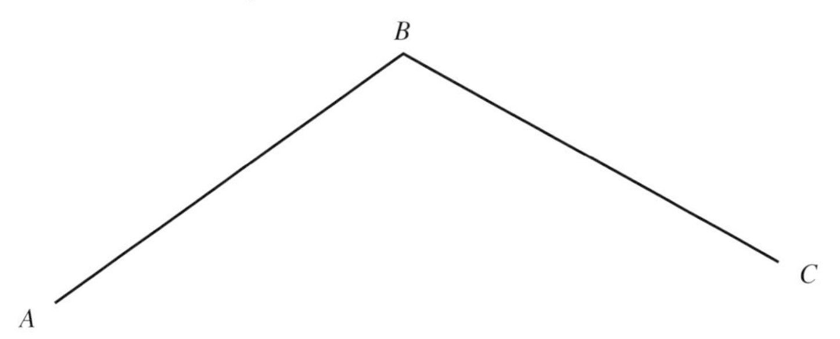

Scale drawings and bearings

11 This sketch shows a plan for a new youth club.

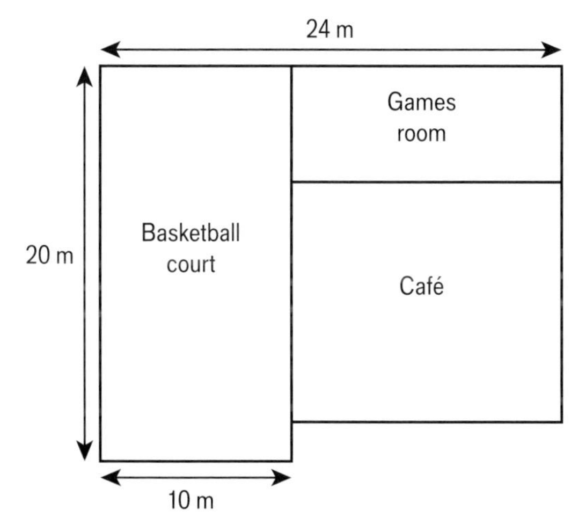

a Jim is making an accurate scale drawing using a scale of 1 cm to 2 m.
What are the dimensions of the basketball court on the scale drawing?

b The side of the café that connects to the basketball court is 6 cm wide on the scale drawing.
How wide is it in real life?

12 A map has scale 1 : 50 000.

a What is the actual length of a road 6 cm long on the map?

b A footpath is 8 km long. How long is the footpath on the map?

13 **a** Draw point P in the middle of a new page.

b Draw point Q on a bearing of 245° from P.

c Find the bearing of P from Q.

14 **a** Write down the bearing of Q from P.

b Find the bearing of P from Q.

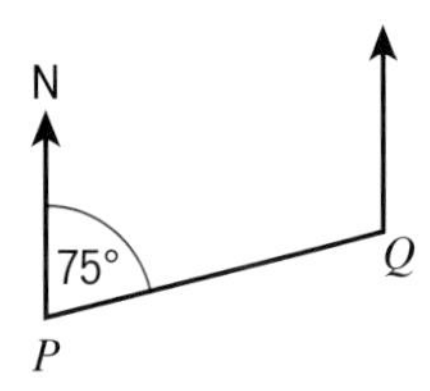

15 **Reflect** How sure are you of your answers? Were you mostly

Just guessing  Feeling doubtful  Confident 

What next? Use your results to decide whether to strengthen or extend your learning.

Challenge

16 Matthew has 1 cm cubes.

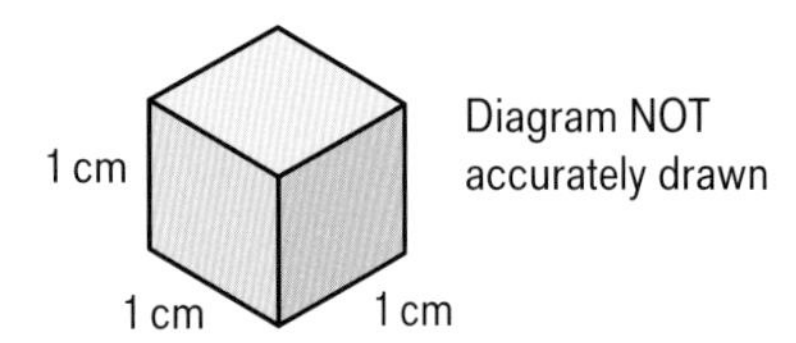

He uses some of the cubes to make this solid shape.

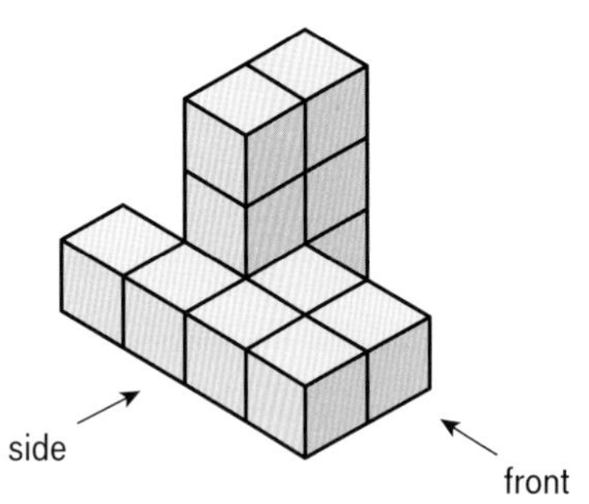

a He rearranges the cubes to make a cuboid with height 1 cm.
What are the dimensions of Matthew's cuboid?

b How many different possible solutions are there?

Cardel rearranges the same blocks to make a different cuboid.
None of the dimensions are 1 cm.

c What are the dimensions of Cardel's cuboid?

d On squared paper, draw the plan, side elevation and front elevation views of the original 3D object.

15 Strengthen

ActiveLearn Homework

3D solids

1 Match each letter to one of the words in the cloud.

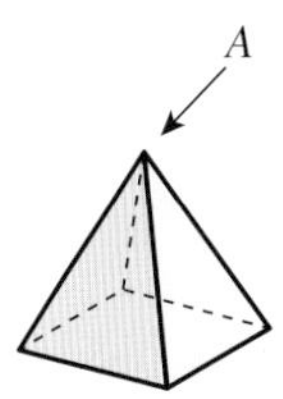

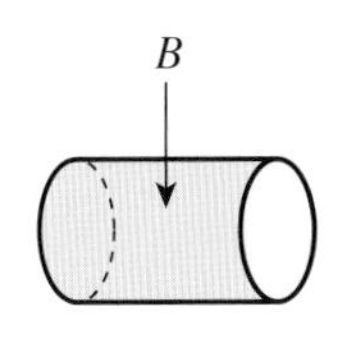

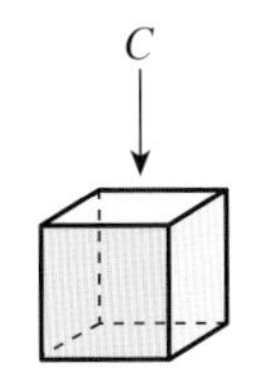

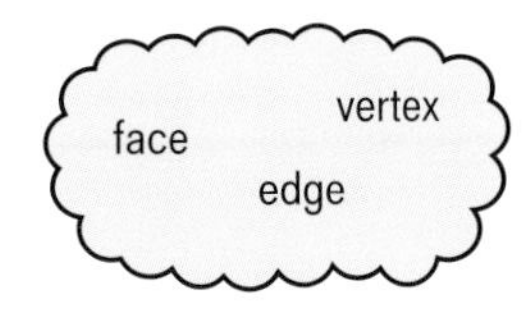

2 Here are the front elevation, side elevation and plan views of a 3D solid.

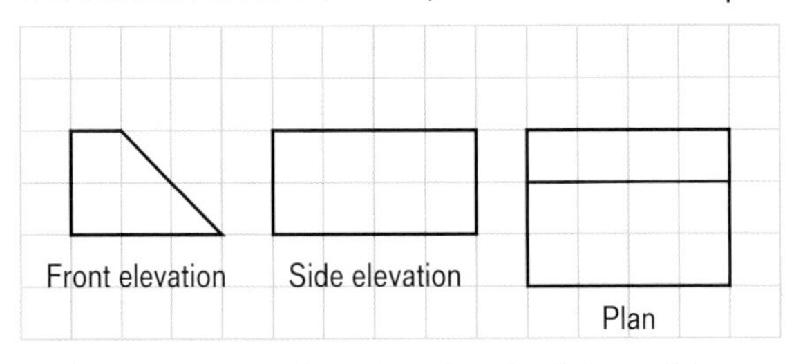

a Copy and complete this sketch of the solid.

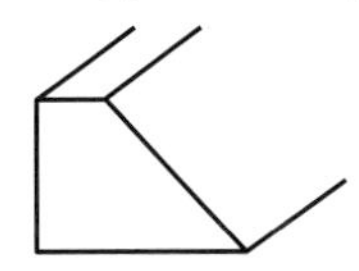

b Use the lengths from the 2D views to work out the dimensions of the 3D shape. Label them on your sketch. In the sketch, make all horizontal lines going in the same direction parallel.

c Draw the cross-section halfway along your solid. Label it 'plane of symmetry'.

Constructions

1 Follow these instructions to construct accurately a triangle with sides 5 cm, 6 cm and 7 cm. Make the longest side the base.

a Use a ruler to draw accurately the longest side.

b Open your compasses to exactly 6 cm and draw an arc from one end of the line.

c Open your compasses to exactly 5 cm and draw an arc from the other end.

d Use the point where the arcs cross to create the triangle.

7 cm

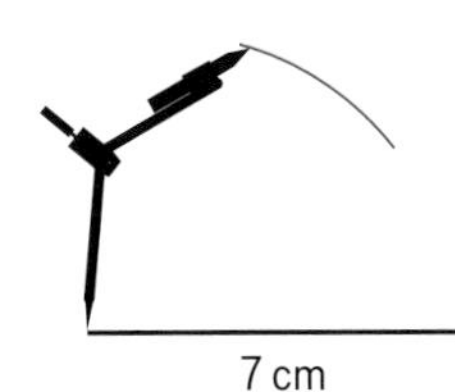

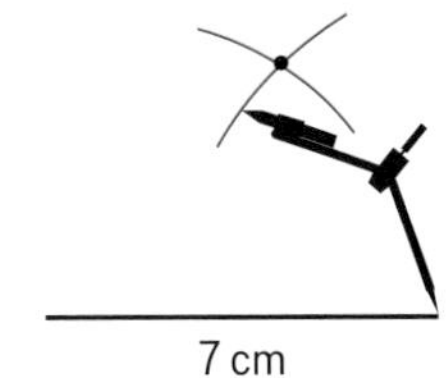

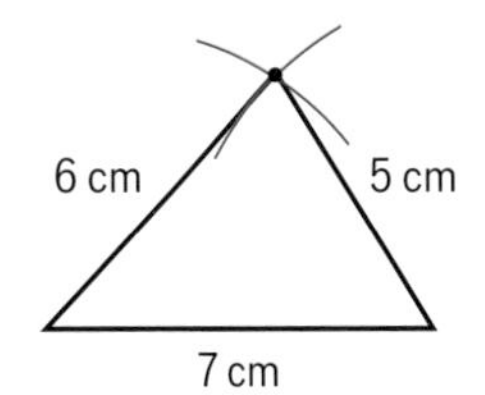

2 Draw a line 10 cm long.
Follow these instructions to construct the perpendicular bisector of your line.

a Draw the line. Open your compasses to more than half the length of the line.

b Draw the first arc.

c Draw the second arc. Join the crossed arcs with a straight line.

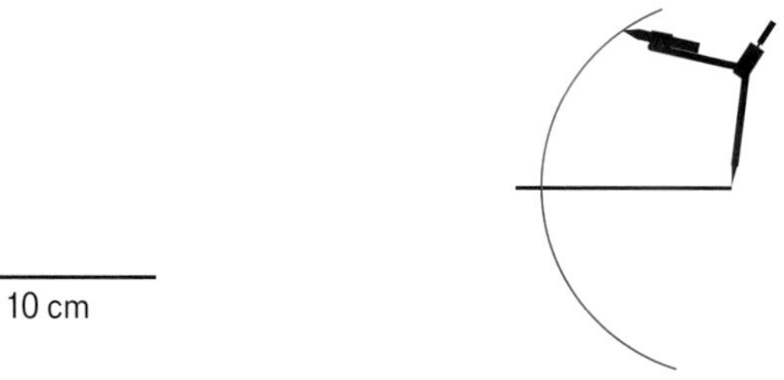

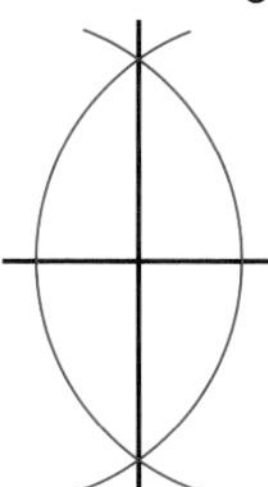

d Check your answer by measuring to make sure that the angle is 90° and that the line has been cut into two equal pieces.

3 Draw a 50° angle. Follow these instructions to construct the bisector of the angle.

a Draw an arc.

b Draw the first arc between the two sides of the angle.

c Draw the second arc.

d Draw the angle bisector.

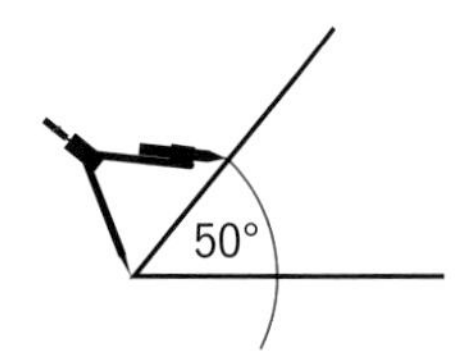

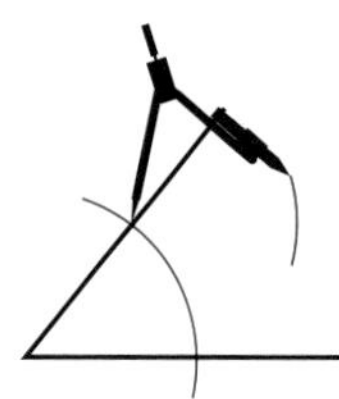

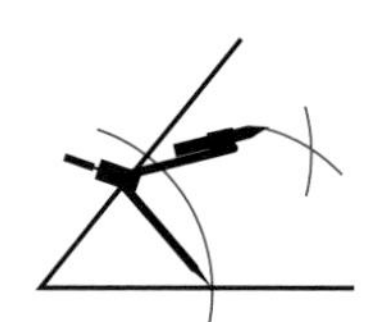

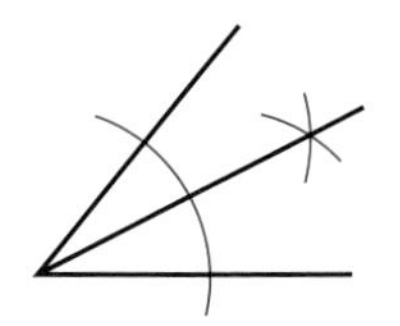

Loci and regions

1 Use compasses to draw a circle, centre C, radius 5 cm, on a centimetre grid.
Copy and complete:
The points inside the circle are _________ than 5 cm from point C.

2 Copy this diagram. Construct the angle bisector.
Shade the region on the side of the angle bisector that is closer to FG than to GH.

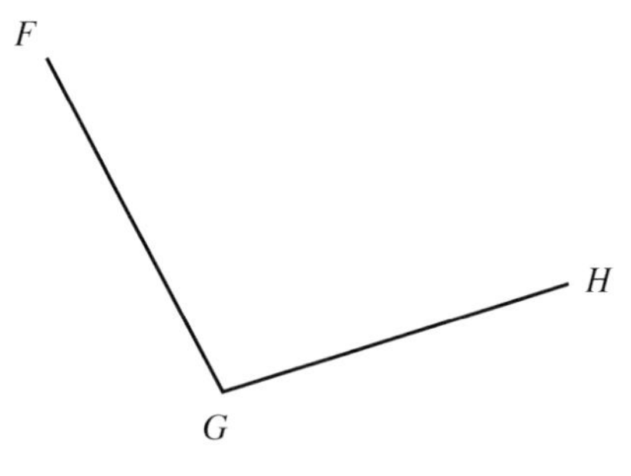

Scale drawings and bearings

1 This sketch shows a plan for a room.
Elsie is making an accurate scale drawing of the room.
She uses a scale of 1 cm to represent 2 m.
On the scale drawing, what is

i the length of the room

ii the width of the room?

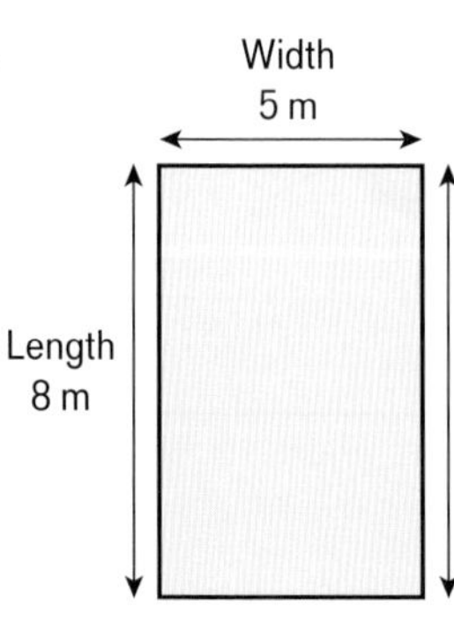

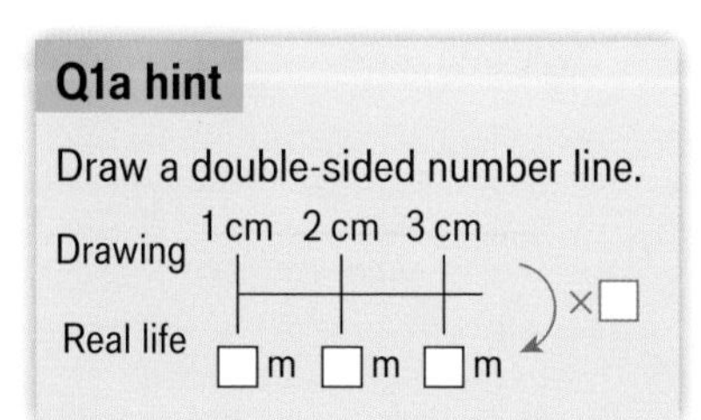

2 The length of a hike on a map is 20 cm.
The map scale is 1 : 50 000.

a Copy and complete. 1 cm represents

50 000 cm
÷100 ÷100
= □ m
÷1000 ÷1000
= □ km

b Copy and complete this double-sided number line.
Show the multiplier you use to move from one side of the number line to the other.

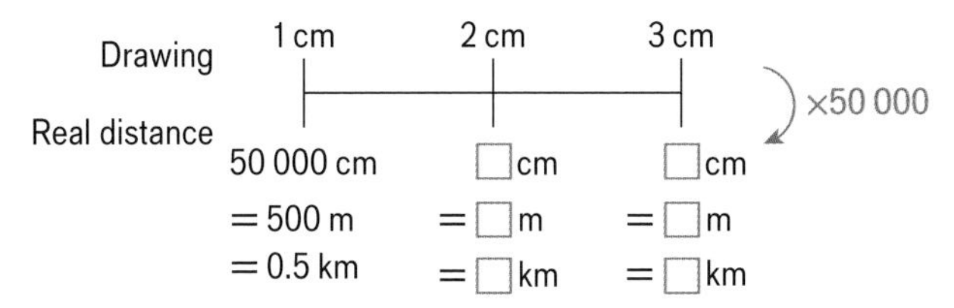

c How many kilometres long is the actual hike?

d The actual distance of another hike is 4.5 km. How long is this on the map?

3 Write the bearing of B from A in each of these diagrams.

a

N
B
27°
A

b

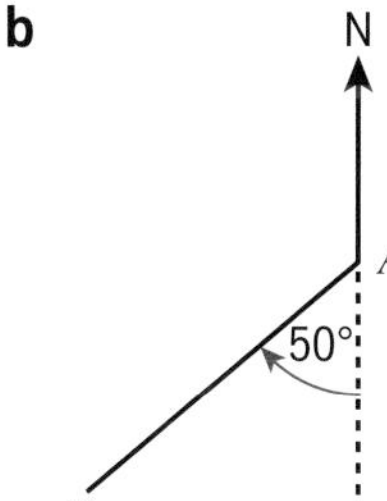

c

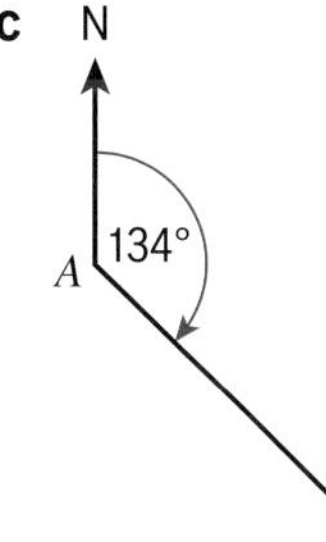

d

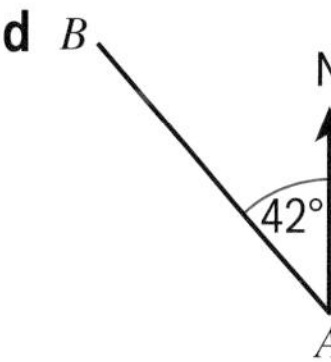

Q3 hint

Always measure bearings clockwise from north and write them as three-figure numbers. Bearing angles that are less than 100° need to use zeros to fill in the missing digits.

4 **a** Draw a point P in the middle of a page.

b Draw a north line at P.

c Measure and draw an angle of 200° clockwise from the north line.

d Mark any point Q on your line, on a bearing of 200° from P.

e Draw a north line at Q.

f Measure the bearing of P from Q.

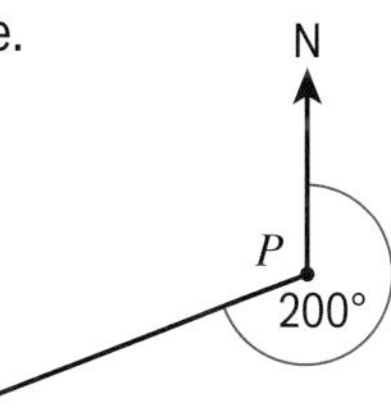

5 The bearing of B from A = 040°.

a Use alternate angles to find angle x.

b What is the bearing of A from B?

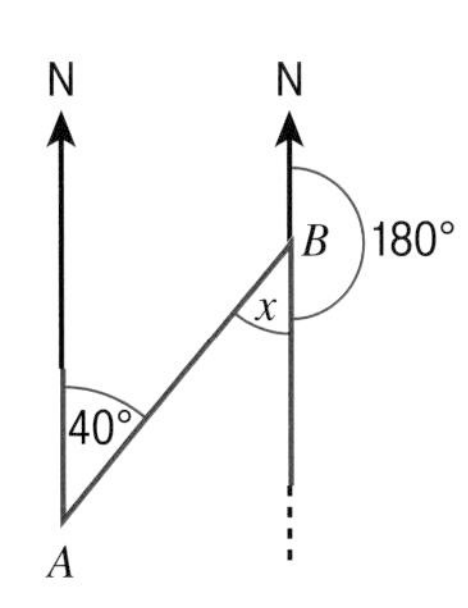

15 Extend

1 **Reasoning** Simon is designing a new kitchen and wants to draw a scale diagram of it on a sheet of A4 paper. The dimensions of the kitchen are 5 m by 8.5 m.
Which of these scales should Simon use to draw the scale diagram? Explain why.

A 1 : 200 **B** 1 : 40 **C** 1 : 25 **D** 1 : 125 **E** 1 : 5

Exam-style question

2 Sara plans to drive from Bristol to Cambridge.
The map below shows the roads between Bristol and Cambridge.
Using her ruler, Sara marks the map with the approximate route she will drive.

Exam tip

Write what you are doing with each calculation. E.g. Total distance = ☐ cm = ☐ km.
This distance will use ☐ litres.

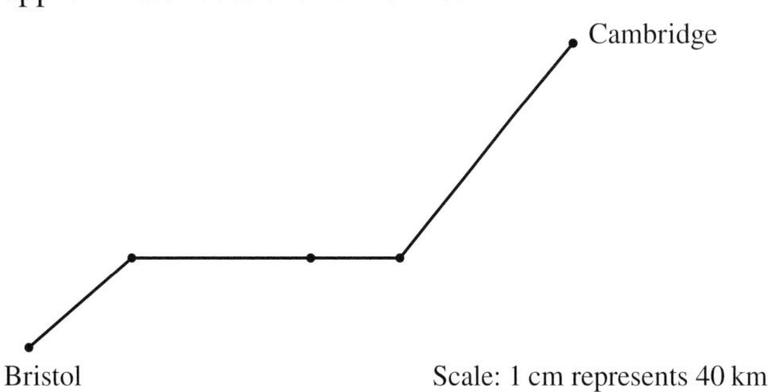

The car does 12 km per litre of fuel. A full tank holds 45 litres of fuel.
She starts with half a tank of fuel.
Does the car have enough fuel in the tank for the journey from Bristol to Cambridge?

(5 marks)

3 **Reasoning** This is a sketch of a disabled ramp 2.5 metres wide.

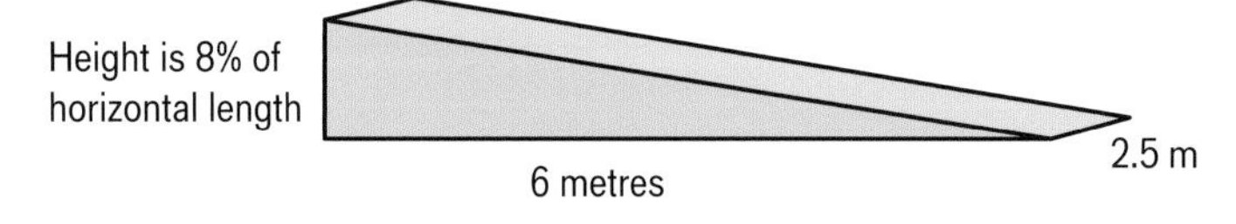

a Draw an accurate scale diagram of the plan, front elevation and side elevation views of the ramp.

b Find the sloping length of the ramp to the nearest centimetre.

4 **Reasoning**
The towns of Finchfield and Greenborough each have a television transmitter. The transmitter at Greenborough (G) has a range of 35 miles and the one at Finchfield (F) has a range of 28 miles.

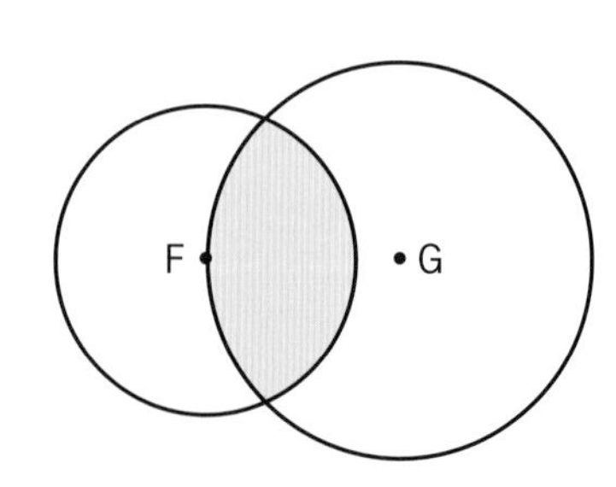

a What do the circles represent?

b Explain what the shaded region represents.

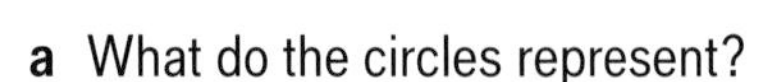

c Work out the distance from Greenborough to Finchfield.

Exam-style question

5 Here is a map.
The map shows two towns, Lowford and Upton.
A company is going to build a lorry park.
The lorry park will be less than 40 km from Lowford and less than 60 km from Upton.
On a copy of the map, shade the region where the company can build the lorry park.

• Lowford

• Upton

Scale: 1 cm represents 20 km

Exam tip

Draw accurately, with a sharp pencil.

(3 marks)

6 a Reasoning Write down the bearing of

i due north **ii** due west

b A boat sails due south.
What bearing does it sail on?

c A plane flies on a bearing of 090°.
What direction does it fly?

7 Problem-solving The diagram shows a sketch of the position of towns D, E and F.

Town D is 86 km north of town E. Town F is 37 km due east of town E.
Construct an accurate scale diagram showing positions D, E and F using a protractor and a ruler.

a What is the bearing of town D from town F?

b What is the distance in km from town D to town F?

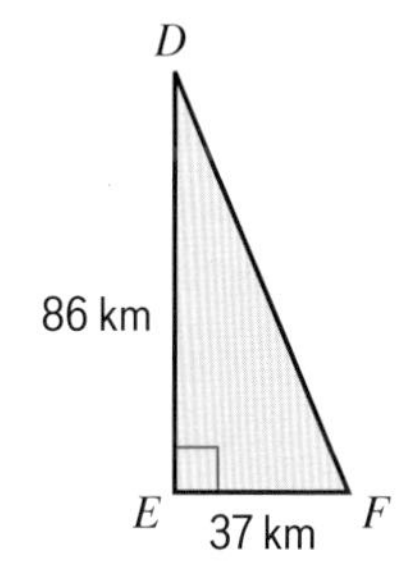

8 Reasoning On a map, the distance between Edinburgh and Carlisle is 10.2 cm.
The real distance, in km, between Edinburgh and Carlisle is 153 km.
Work out the scale of the map.

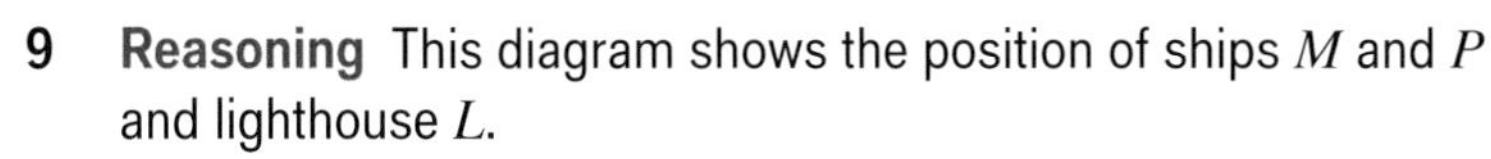

9 Reasoning This diagram shows the position of ships M and P and lighthouse L.

a Construct a scale diagram of the position of L, M and P using a ruler and pair of compasses. Use a scale of 1 cm to represent 1 km.

b Use your diagram to find the bearing of ship M and ship P from the lighthouse.

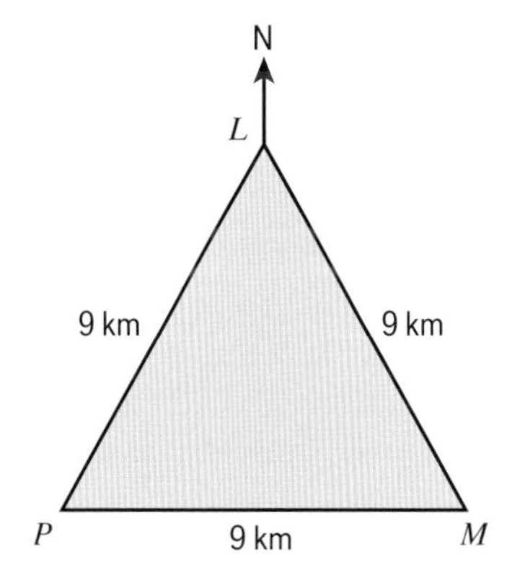

10 A ship starts at A and sails 50 km south then 30 km east.
Then it sails back to its starting point.

a Copy the diagram. Draw a north line at C.

b Use $\tan\theta = \frac{\text{opp}}{\text{adj}}$ to calculate the angle ACB.

c Work out the bearing the ship sails on to A from C.

d Use Pythagoras' theorem to calculate the distance AC.

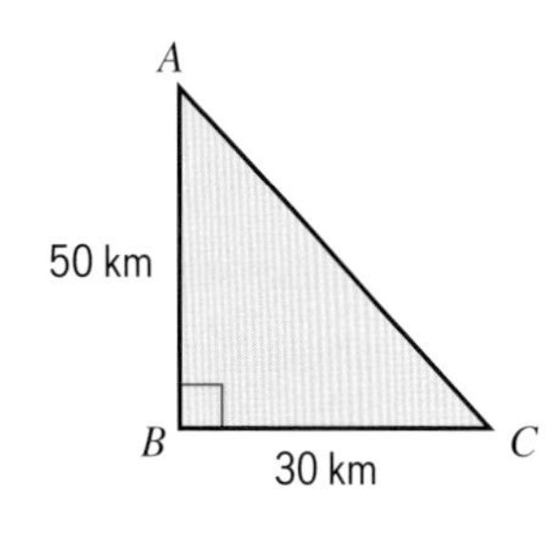

15 Test ready

Summary of key points

To revise for the test:

- Read each key point, find a question on it in the mastery lesson, and check you can work out the answer.
- If you cannot, try some other questions from the mastery lesson or ask for help.

Key points

1. The flat surfaces of 3D shapes are called **faces**, the lines where two faces meet are called edges and the corners at which the **edges** meet are called **vertices**.
   The singular of vertices is **vertex**. → **15.1**
2. A **dimension** is the size of something in a particular direction.
   Length, width, height and diameter are all dimensions of shapes or solid objects. → **15.1**
3. **Pyramids** have a base that can be any shape and sloping triangular sides that meet at a point.
   A **right prism** has two parallel faces. All the other faces are perpendicular to these. → **15.1**
4. A **plane** is a flat (2D) surface. A solid shape has a **plane of symmetry** when a plane cuts the shape in half so that the part of the shape on one side of the plane is an identical reflection of the part on the other side of the plane. → **15.2**
5. The **plan** is the view from above an object.
   The **front elevation** is the view of the front of an object.
   The **side elevation** is the view of the side of an object. → **15.2**
6. You can draw an accurate diagram of a triangle with a ruler and protractor if you know three measurements. → **15.3**
   - An **ASA** triangle has a given **A**ngle, a **S**ide length and another **A**ngle.
   - An **SAS** triangle is one where you are given two **S**ide lengths and the **A**ngle in between.
   - In an **SSS** triangle, you are given all three **S**ide lengths but none of the angles.
   - In a **RHS** triangle, you are given the **R**ight angle, the **H**ypotenuse length and another **S**ide length. The **hypotenuse** is the longest side of a right-angled triangle. → **15.3**
7. To **construct** means to draw accurately using a ruler and compasses. → **15.3**
8. A **scale** is a ratio that shows the relationship between a length on a map or drawing and the actual length.
   Scale 1 : 25 000 means 1 cm on the map represents 25 000 cm in real life. → **15.4**
9. A **net** is a 2D shape that folds up to make a 3D shape. → **15.5**
10. **Constructions** are accurate drawings made using a ruler and pair of compasses.
    To **bisect a line** means to cut a line exactly in half.
    A **perpendicular bisector** cuts a line in half at right angles. → **15.6**
11. You can also use a ruler and compasses to **construct** a perpendicular from a point to a line. → **15.6**
12. The **shortest** path from a point to a line is **perpendicular** to the line. → **15.6**
13. An **angle bisector** cuts an angle exactly in half. → **15.6**

Key points

14 A **locus** is a set of all points that obey a given rule.
This produces a path followed by the points. The plural of locus is **loci**. → 15.7

15 The locus of all points a given distance (d) from a fixed point (P) is a circle, with centre P and radius d. → 15.7

16 The locus of all points 3 cm from a line, AB, is two parallel lines 3 cm from the line and a semicircle at each end with radius 3 cm. → 15.7

17 A locus that is equidistant (the same distance) from two given points, X and Y, is the perpendicular bisector of the line between the two points. → 15.7

18 A locus of all points that are equidistant from two intersecting lines, OX and OY, is the angle bisector of the two lines. → 15.7

19 A **region** is an area bounded by loci. → 15.7

20 A **bearing** is an angle measured in degrees clockwise from north.
A bearing is always written using three digits. → 15.8

Sample student answer

Exam-style question

Construct the locus of points that are equidistant from AB and BC. **(1 mark)**

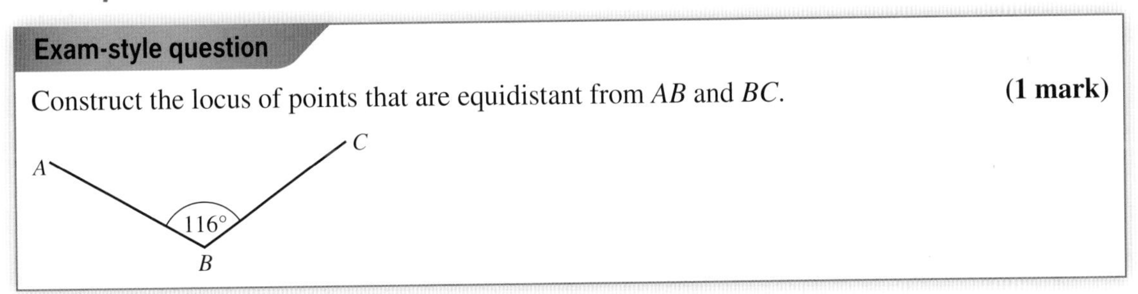

a What does 'construct' mean?

b How can you tell that the student has used a protractor to measure the angle, instead of constructing the angle bisector?

15 Unit test

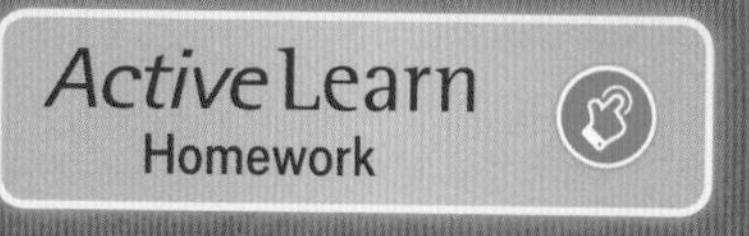

1 Here is a sketch of a shed.

a How many vertices does the shed have? **(1 mark)**

b Trace the sketch.
Draw in any planes of symmetry. **(1 mark)**

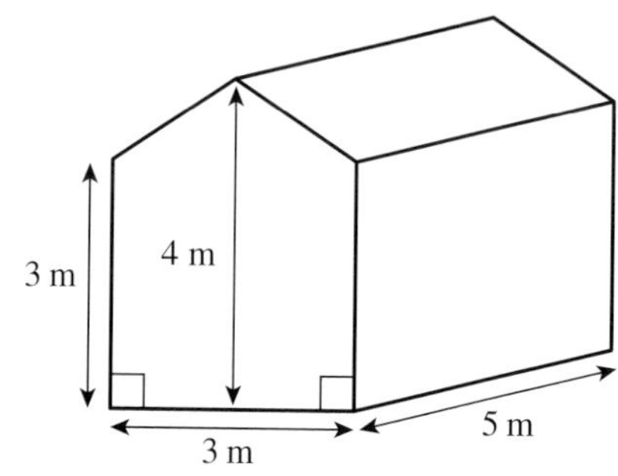

2 The diagram shows a tree and a man.
The man is of average height.
The tree and the man are drawn to the same scale.

a Write down an estimate for the real height, in metres, of the man. **(1 mark)**

b Find an estimate for the real height, in metres, of the tree. **(2 marks)**

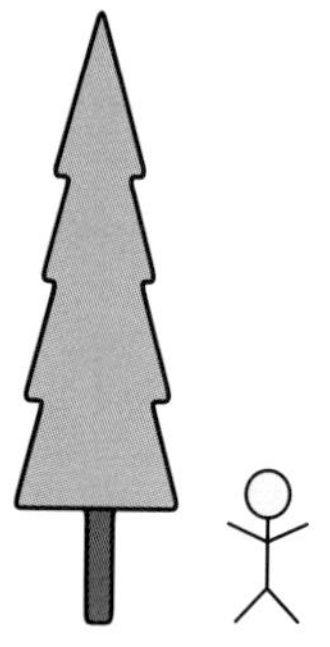

3 Draw a line 8 cm long.
Construct the locus of points which are 2 cm away from the line. **(2 marks)**

4 Draw two points, G and H, 7 cm apart.
Construct the locus of all points equidistant from G and H on your diagram. **(2 marks)**

5 Construct the triangle STU with sides $ST = 5.3$ cm, $TU = 6.7$ cm and $SU = 8.4$ cm. **(2 marks)**

6 Point P is 6 cm along an 11 cm line.
Copy the diagram and construct the perpendicular through the point P. **(2 marks)**

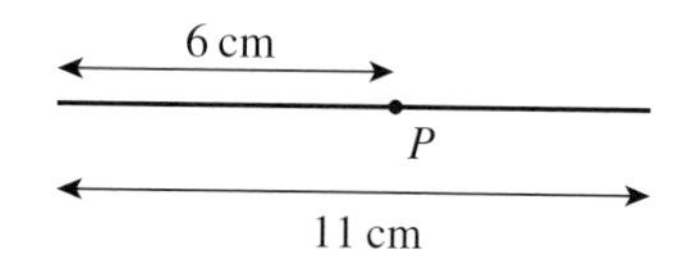

7 Here is a plan of a room drawn to a scale of 1 : 60.
Jim is going to put a lamp in the room.
The lamp has to be more than 1.5 m from A, and closer to DC than to BC.
Copy the diagram. Show, by shading the diagram, the region where Jim can put the lamp. **(3 marks)**

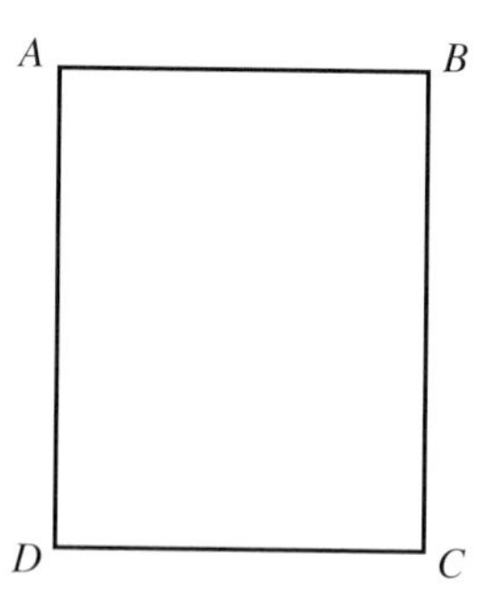

8 **a** Write the bearing of A from B. **(1 mark)**

b Work out the bearing of B from A. **(2 marks)**

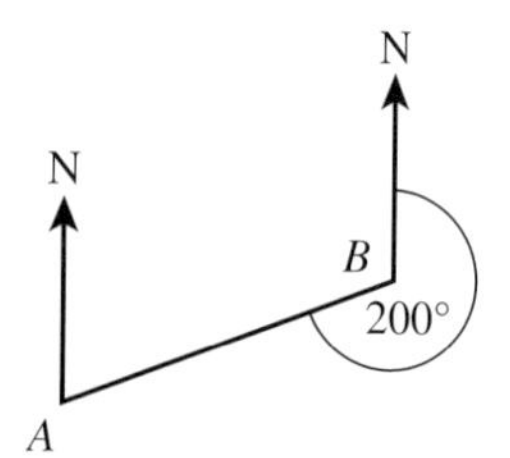

9 The diagram shows a line AB and a point C.

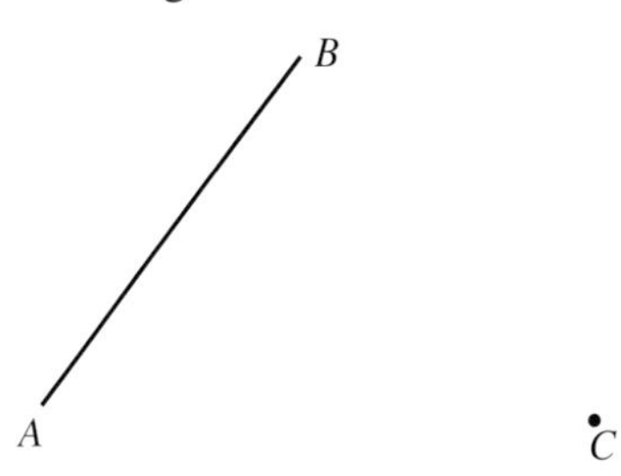

Find the shortest distance from the point C to the line AB. **(2 marks)**

10 Here is a solid square-based pyramid, $VABCD$.

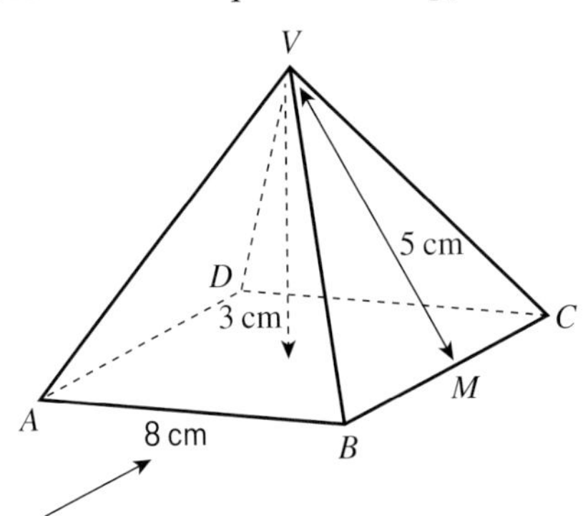

The base of the pyramid is a square of side 8 cm.
The height of the pyramid is 3 cm.
M is the midpoint of BC and $VM = 5$ cm.

a On a centimetre grid, draw an accurate front elevation of the pyramid from the direction of the arrow. **(2 marks)**

b Work out the area of one of the triangular faces of the pyramid. **(2 marks)**

11 Aircraft B and C are flying into airport A.
Both planes are flying directly towards the airport.
B is 216 km from the airport on a bearing of 242° from A.
C is 300 km from the airport on a bearing of 345° from B.

a Draw a diagram showing A, B and C. Use a scale of 1 cm = 20 km. **(3 marks)**

b Aircraft X has recently taken off from the airport.
X is 86 km from A on a bearing of 038° from B.
Draw in the position of aircraft X on your diagram. **(1 mark)**

c Aircraft Y flies on a bearing of 300° from A.
Explain why this is not a good course for aircraft Y to take. **(1 mark)**

(TOTAL: 30 marks)

12 Challenge Draw an accurate net for a tetrahedron with side length 4 cm.

13 Reflect In this unit you have done a lot of drawing.
Write down at least three things to remember when doing drawings in mathematics.
Compare your list with a classmate. What else can you add to your list?

16 Quadratic equations and graphs

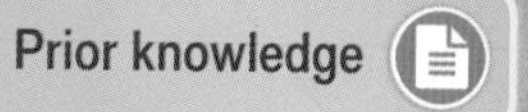

16.1 Expanding double brackets

- Multiply double brackets.
- Recognise quadratic expressions.
- Square single brackets.

ActiveLearn
Homework

Warm up

1 Fluency Simplify

a $-9 \times a$ **b** $3b \times -4$ **c** $z \times -3z$ **d** $2x \times 3x$

e $-3y + 11y$ **f** $7t - 9t - t$ **g** $m^2 + 2m + 7m + 4$ **h** $y^2 - 9y - 11y - 3$

2 Expand

a $2(a+2)$ **b** $3(x+4)$ **c** $m(m+8)$ **d** $b(2b-5)$

3 Reasoning

i Write an expression for the area of each small rectangle.

ii Write an expression for the area of the large rectangle. Collect like terms and simplify.

iii Complete the expressions for the area of each rectangle.

a

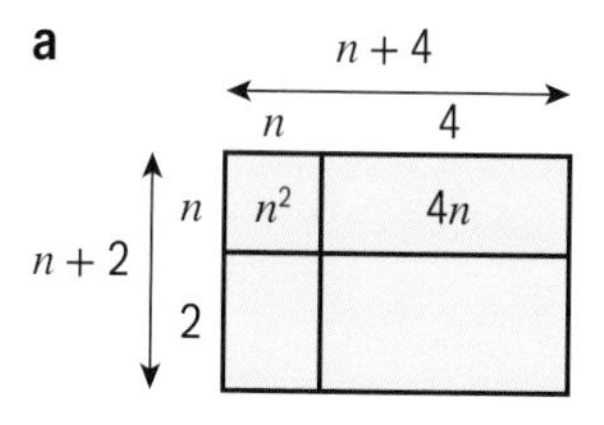

$(n+2)(n+4) = n^2 + \square + \square$

b

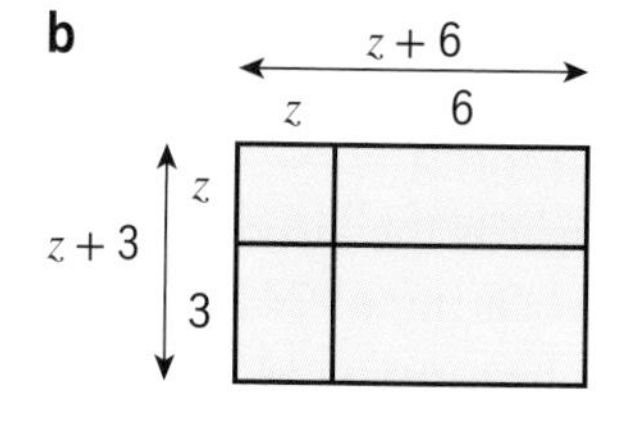

$(z+3)(z+6) = \square + \square + \square$

c

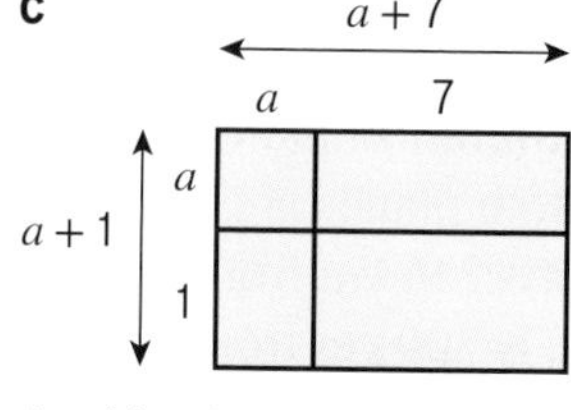

$(\quad)(\quad) = \square + \square + \square$

Key point

To **expand** or multiply double brackets, multiply each term in one bracket by each term in the other bracket.

Example

Expand and simplify $(x+2)(x+6)$.

Grid method

$\times$	x	$+2$
x	x^2	$+2x$
$+6$	$+6x$	$+12$

$= x^2 + 2x + 6x + 12$

$= x^2 + 8x + 12$

FOIL method

FOIL: Firsts, Outers, Inners, Lasts

$(x+2)(x+6) = x^2 + 6x + 2x + 12$

$= x^2 + 8x + 12$

4 Expand and simplify

a $(x+1)(x+2)$ b $(t+3)(t+4)$ c $(q+6)(q+9)$ d $(z+12)(z+1)$

e $(m+8)(m+11)$ f $(y+7)(y+10)$ g $(j+5)(j+1)$ h $(r+9)(r+6)$

i **Reflect** Which method did you use: the grid method or FOIL? Explain why.

5 Expand and simplify

a $(z+1)(z-2)$ b $(m-5)(m+6)$ c $(a+4)(a-9)$ d $(n-10)(n+7)$

e $(x-2)(x-3)$ f $(y-6)(y-1)$ g $(b-9)(b+2)$ h $(k-2)(k-8)$

6 **Reasoning** What is the missing term?

$(x+3)(x+\square) = x^2+8x+15$

7 **Reasoning** Isabella says that expanding $(x+3)(x-7)$ gives the same quadratic expression as expanding $(x-7)(x+3)$.

Is she correct? Give reasons for your answer.

8 **Reasoning** Rex expands and simplifies $(a-4)(a-7)$.

He says that the answer is $a^2+11a-28$.

Is Rex correct? Give reasons for your answer.

9 Expand

a $3(2+t)$ b $3x(2x+5)$ c $(m+3)(m+10)$

Key point

Expanding double brackets often gives a quadratic expression.

A quadratic expression always has a squared term (with a power of 2). It cannot have a power higher than 2.

It may have a term with a power of 1 that is the same letter as the squared term.

It may also have a constant (number) term.

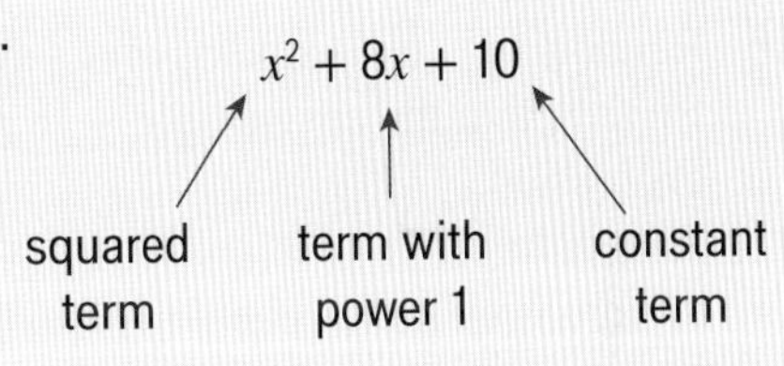

10 **Reasoning** Which of these are quadratic expressions? Give reasons for your answers.

a x^2+3x-2 b x^3+x^2-3 c y^2+9y d $5-2x$

e $16+2z-z^2$ f $x+2xy+y$ g $2z^2+z$ h m^2

Key point

To square a single bracket, multiply it by itself, then expand and simplify.

$(x+1)^2 = (x+1)(x+1)$

11 Expand and simplify

a $(x+2)^2$ b $(a+5)^2$ c $(y-9)^2$ d $(m-4)^2$ e $(t-3)^2$

12 Square these expressions. Simplify your answers.

a $x+6$ b $n+12$ c $q-4$ d $t-10$ e $y+3$

13 **Problem-solving** Match the cards that have equivalent expressions.

A $(x + 1)^2$	**B** $(x + 7)^2$	**C** $(x - 5)^2$	**D** $(x - 1)^2$
E $x^2 - 10x + 25$	**F** $x^2 - 2x + 1$	**G** $49 + 14x + x^2$	**H** $x^2 + 2x + 1$

Exam-style question

14 The length of a side of a square paving slab is $x + 4$.

$x + 4$

$x + 4$

a Work out an expression for the area of the paving slab.
Simplify your answer. **(2 marks)**

b 10 paving slabs are used to make a patio.
Write an expression for the total area of the patio.
Simplify your answer. **(1 mark)**

Exam tip

Expand and simplify all expressions. Show your working clearly.

15 **Problem-solving** The length of a rectangle is $n + 6$ and its width is $n + 3$.
Show that the area of the rectangle is $n^2 + 9n + 18$.

16 **Reasoning** Luke expands and simplifies $(x + 4)(x - 4)$.
He says that the answer is $x^2 - 16$ and this is a quadratic expression.
Is he correct?
What do you notice about the two terms in the answer?

Q16 hint

'Show that ...' means 'Show your working.'

17 Write and simplify an expression for the area of this rectangle.
$(3x + 2)(x + 1) =$

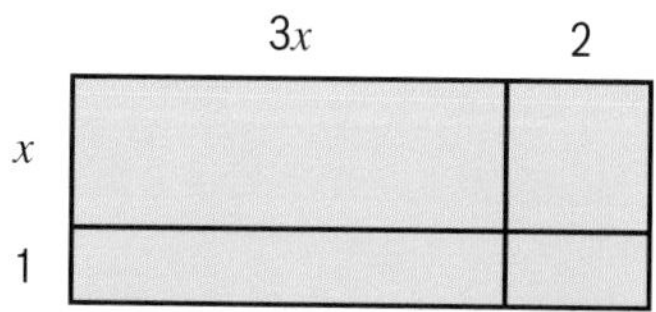

18 Use the grid method or FOIL to expand and simplify

a $(x + 2)(2x + 1)$ **b** $(x - 2)(2x + 1)$ **c** $(x + 1)(2x + 3)$ **d** $(x + 1)(2x - 3)$

e $(2x + 1)(3x + 2)$ **f** $(2x + 1)(3x - 2)$ **g** $(2x - 1)(4x + 3)$ **h** $(2x - 1)(5x - 2)$

Exam-style question

19 Expand and simplify $(5x + 2)(3x - 1)$. **(2 marks)**

16.2 Plotting quadratic graphs

- Plot graphs of quadratic functions.
- Recognise a quadratic function.
- Use quadratic graphs to solve problems.

Warm up

1 **Fluency** $x = -2$, $m = 4$ and $t = 5$.
Evaluate

a x^2 **b** $x^2 + x$ **c** $t^2 + 2t$ **d** $m^2 + m - t$

2 What is the equation of the mirror line?

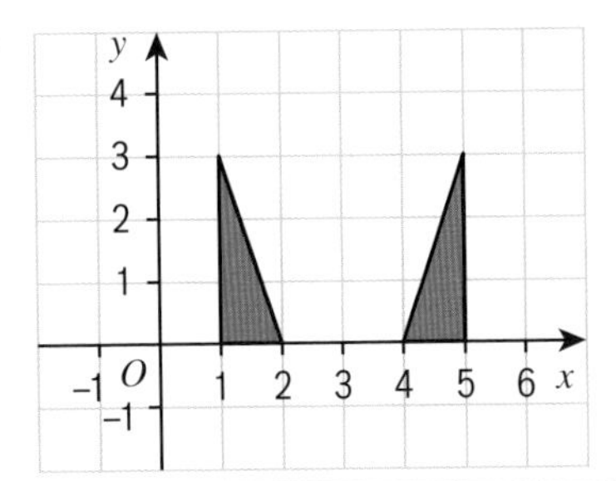

Key point

A **function** describes the relationship between variables.
For each **input** value there is an **output** value.

3 **a** Copy the table of values for the function $y = x^2$.

x	−4	−3	−2	−1	0	1	2	3	4
y									

b Work out each y value by substituting each x value into the equation, e.g. when $x = -3$, then $y = x^2 = (\square)^2 = \square$

c Plot the graph of the function $y = x^2$.
Join the points with a smooth curve.
Label your graph with the equation $y = x^2$.

4 **a** Copy and complete the table of values for $y = x^2 - 3$.

x	−4	−3	−2	−1	0	1	2
x^2	16				0		
−3	−3				−3		
y	13				−3		

b Plot the graph of $y = x^2 - 3$. Join the points with a smooth curve. Label your graph.

c **Reflect** Sketch what do you think the graph of $y = x^2 + 5$ would look like.
Make a table of values and plot the graph of $y = x^2 + 5$ to see if you are correct.

Exam-style question

5 **a** Complete the table of values for $y = x^2 - x - 2$.

x	−3	−2	−1	0	1	2	3
y	10			−2	−2		

(2 marks)

b Draw the graph of $y = x^2 - x - 2$ for values of x from −3 to 3. **(2 marks)**

6 **Reasoning** **a** Copy and complete the table of values for $y = x^2 + 2x - 4$.

x	−4	−3	−2	−1	0	1	2
x^2		9					
$+2x$		−6					+4
−4		−4			−4		
y		−1				−1	

b On graph paper, draw the graph of the function $y = x^2 + 2x - 4$ for values from −4 to 2.

Key point

The **turning point** of the curve is where it turns in the opposite direction.

Example

For the graph of $y = x^2 - 6x + 5$ write down

a the coordinates of the y-intercept

b the equation of the line of symmetry

c the turning point

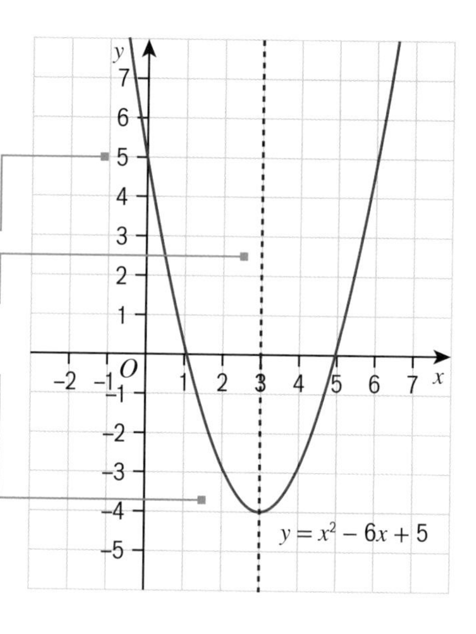

The y-intercept is the point where the graph crosses the y-axis.

Sketch in the **line of symmetry**. Write its equation.

Write down the coordinates of the point where the curve turns.

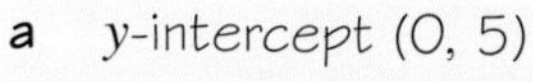

a y-intercept (0, 5)

b line of symmetry $x = 3$

c coordinates of the turning point (3, −4)

7 For each graph, write down

i the y-intercept **ii** the equation of the line of symmetry **iii** the turning point

a

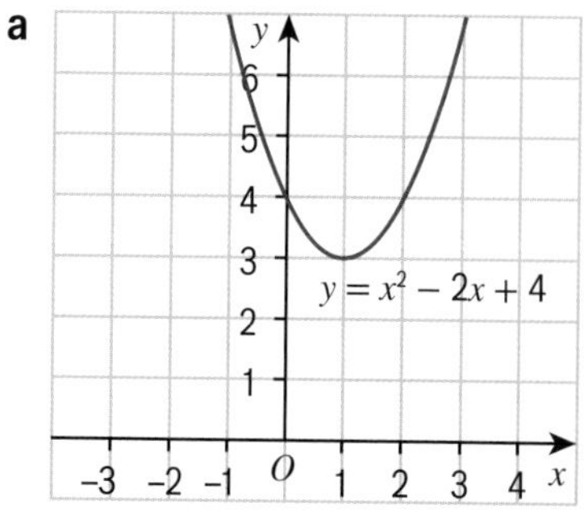

b

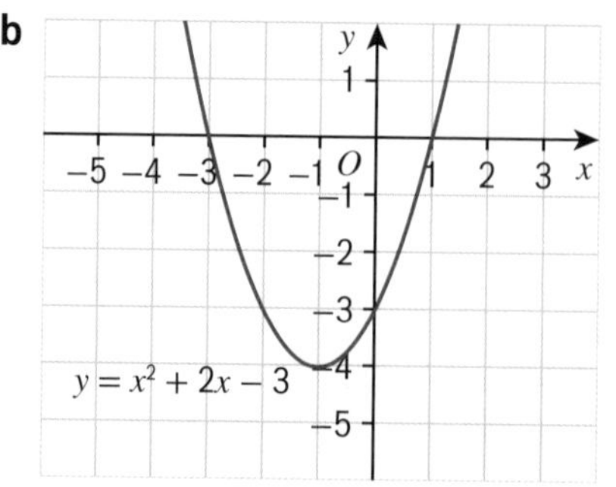

8 a Copy and complete the table. Draw the graph of the function $y = -x^2$.

x	−3	−2	−1	0	1	2	3
y	−9				−1		

b Write down i the y-intercept ii the equation of the line of symmetry

c **Reflect** Explain what is the same about the graphs of $y = x^2$ and $y = -x^2$. What is different?

Key point

A **quadratic function** has a symmetrical ∪-shaped curve called a **parabola**.
A quadratic function with a $-x^2$ term has a symmetrical ∩-shaped curve.
The curve always has a minimum or maximum turning point.

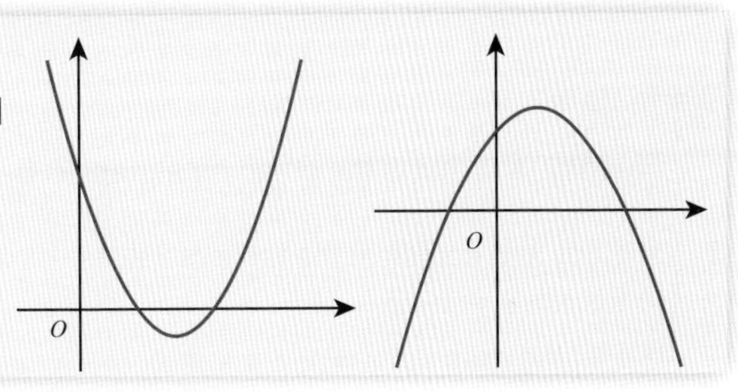

9 **Reasoning** Which of these are graphs of quadratic functions?

a
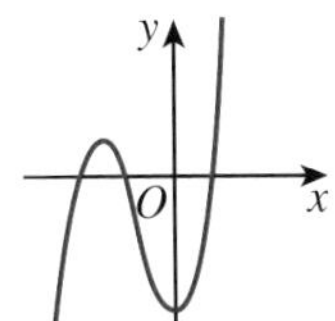

b
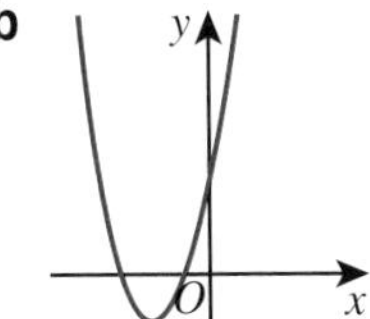

c
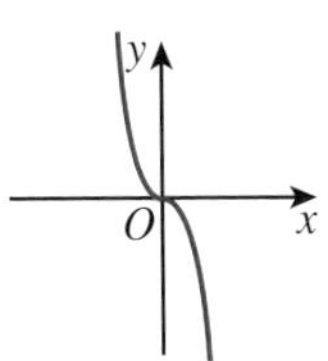

d
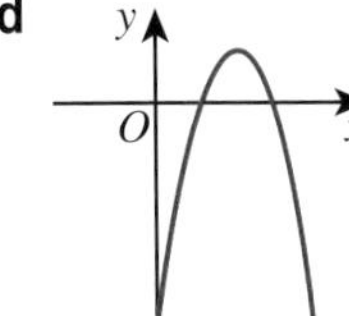

10 **Problem-solving / Reasoning**
The graph shows the height of a baseball against time.

a i What is the maximum height that the baseball reaches?

ii At how many seconds does it reach its maximum height?

b At what times is the ball 12.5 m above the ground?

c When does the ball hit the ground?

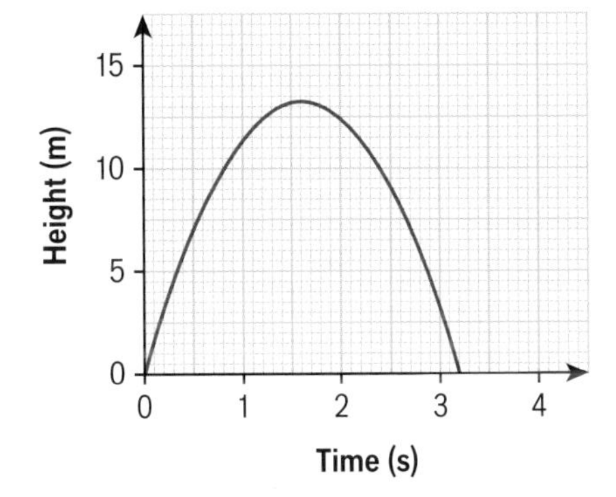

11 **Reasoning** The graph $y = x^2$ can be used to work out the area of a type of tile, where the y-axis represents area in cm^2, and the x-axis represents side length in cm.

a Use your graph of $y = x^2$ from **Q3** to work out area of a tile with side length

i 1 cm ii 4 cm iii 2.5 cm

b What shape is the tile? Explain how you know.

12 **Problem-solving** A toy rocket is fired from the ground. The table shows the height, h, of the rocket and the time, t, after it is fired.

Time, t (seconds)	0	1	2	3	4
Height, h (metres)	0	38	66	84	92

a Plot a graph of the height of the rocket against time. Put t on the horizontal axis and h on the vertical axis. Draw a smooth curve through the points.

b From your graph, estimate how long the rocket takes to reach a height of 20 metres.

c From your graph, estimate the height of the rocket at 2.5 seconds.

16.3 Using quadratic graphs

- Solve quadratic equations $ax^2+bx+c=0$ using a graph.
- Solve quadratic equations $ax^2+bx+c=k$ using a graph.

Warm up

1 **Fluency** **a** What is the origin on a graph? **b** What is another name for the line $y=0$?

2 **a** Copy and complete the table of values for $y=x^2-x-2$.

b Plot the graph of $y=x^2-x-2$.

x	−3	−2	−1	0	1	2
y						

Key point

To solve the equation $ax^2+bx+c=0$, read the x-coordinates where the graph $y=ax^2+bx+c$ crosses the x-axis. The values of x that satisfy the equation are called **roots**.

3 Look at your graph from **Q2**.

a Write down the coordinates of the points where the graph crosses the x-axis.

b The roots of $x^2-x-2=0$ are the x-coordinates where the graph cross the x-axis. Write down the roots of $x^2-x-2=0$.

4 Use the graphs to solve the equations

a $x^2-x-6=0$

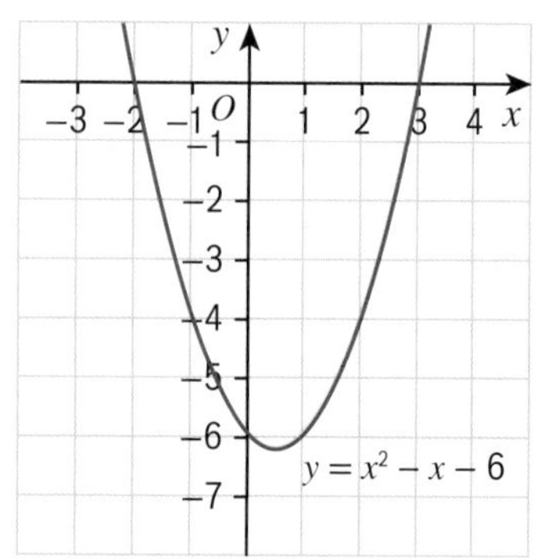

b $x^2+3x-10=0$

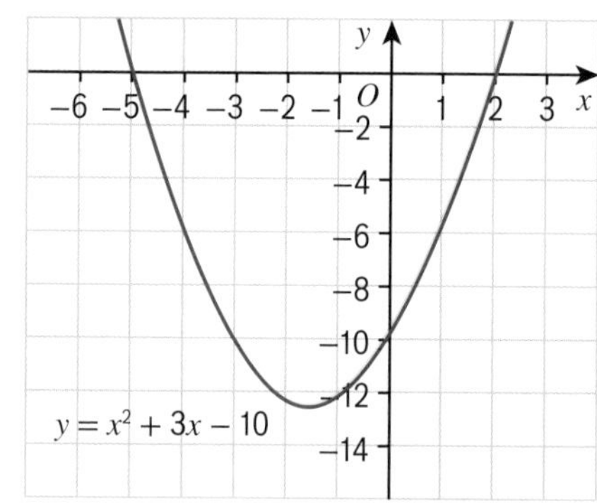

c $x^2-4x+4=0$

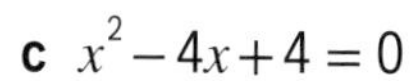
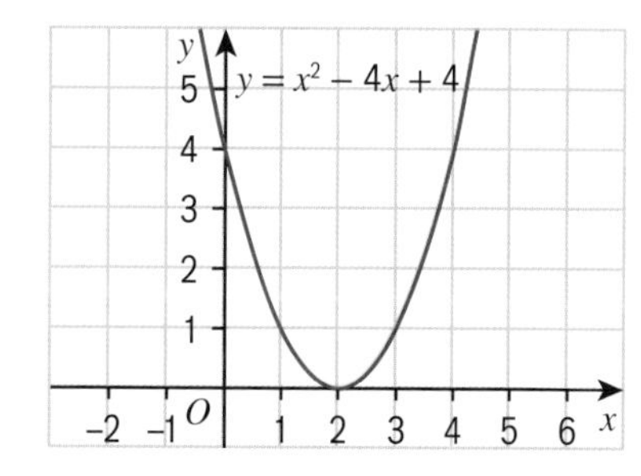

d $4x^2+4x-3=0$

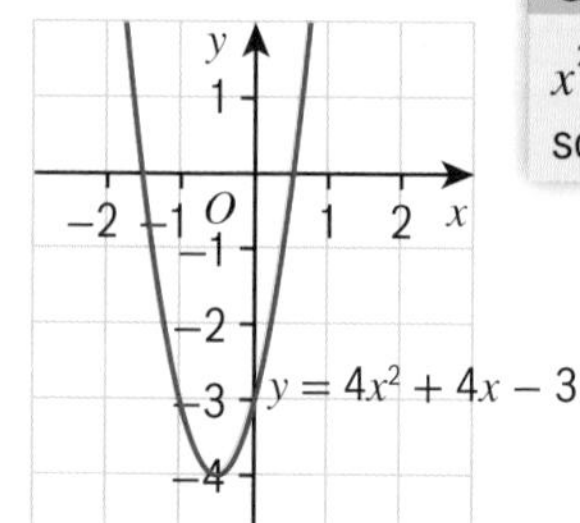

Q4c hint

$x^2-4x+4=0$ only has one solution.

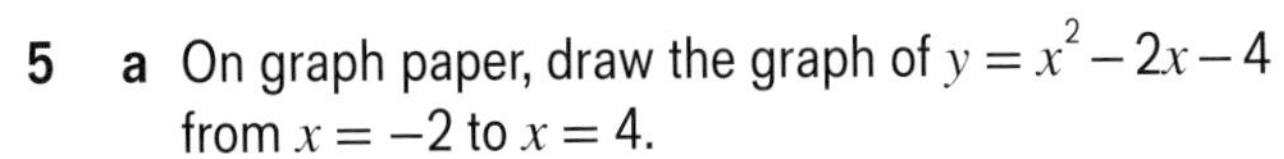

5 **a** On graph paper, draw the graph of $y = x^2 - 2x - 4$ from $x = -2$ to $x = 4$.

b From your graph, estimate the solutions to $x^2 - 2x - 4 = 0$.

Q5a hint

Draw a table of values.

Exam-style question

6 **a** Complete this table of values for $y = x^2 - 2x - 1$.

x	−3	−2	−1	0	1	2	3
y		7	2		−2		

(2 marks)

b On a copy of the grid, draw the graph of $y = x^2 - 2x - 1$ for values of x from −3 to 3.

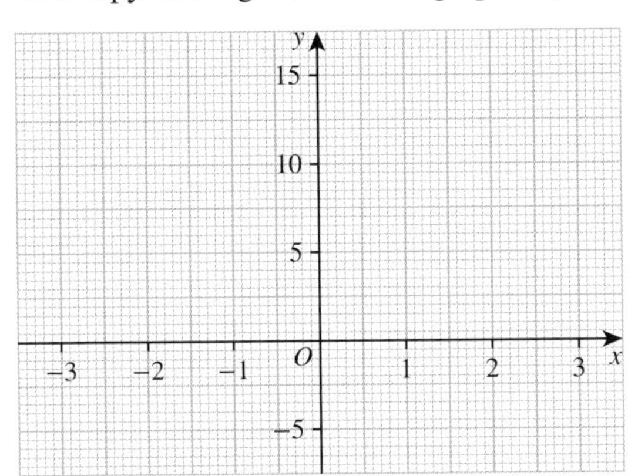

(2 marks)

c Use the graph to estimate a solution to $x^2 - 2x - 1 = 0$. **(1 mark)**

Example

Draw the graph of $y = x^2 - 2x - 2$.
Use the graph to solve the equation $x^2 - 2x - 2 = 6$.

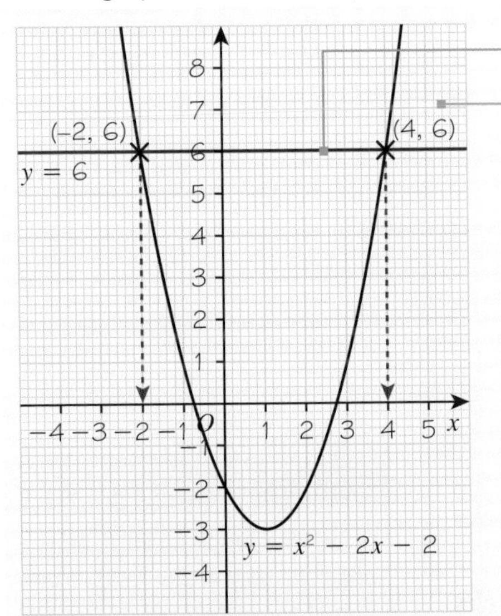

Plot and draw the graph of $y = x^2 - 2x - 2$. Draw the line $y = 6$ on the graph.

This means $y = 6$.

The solutions to the equation are $x = -2$ and $x = 4$.

Write down the x-**coordinates** where the line and curve cross.

7 **a** Use these graphs to solve the equation $x^2 + 3 = 4$.

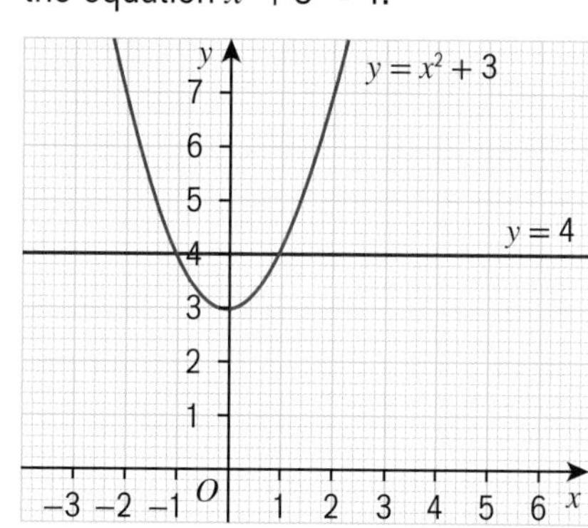

b Use these graphs to solve the equation $x^2 - 4x - 1 = -4$.

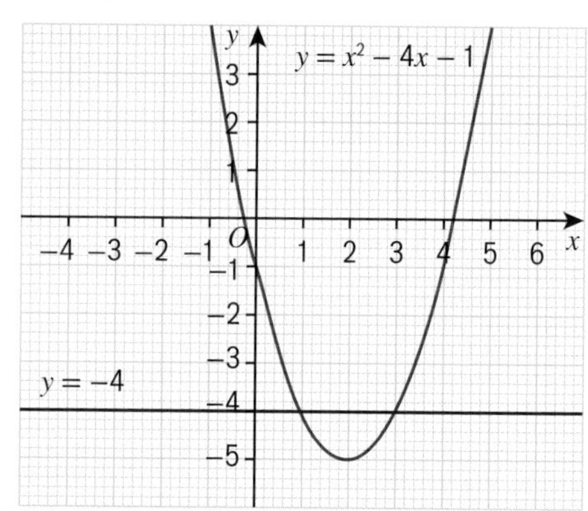

8 **a** On graph paper draw the graph of $y = x^2 - 2x - 5$ for values of x between -2 and 4.

b Use the graph to estimate the solutions to $x^2 - 2x - 5 = 0$.

c Use the graph to estimate the solutions to $x^2 - 2x - 5 = 3$.

9 **a** Use this graph to solve the equation $-x^2 + 3x = 0$.

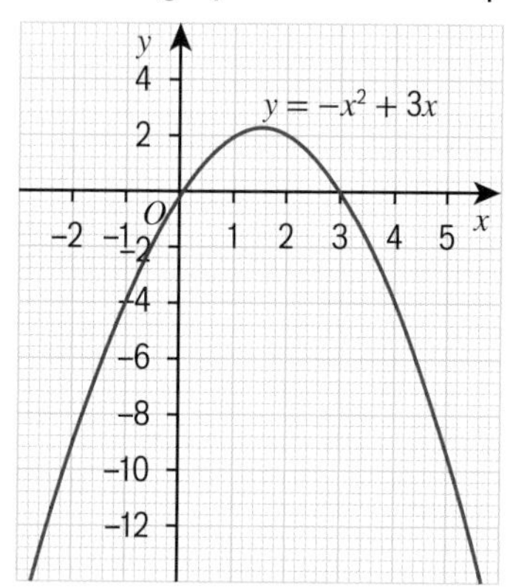

b Find estimates of the solutions to $-x^2 + 3x = -2$.

Exam-style question

10 **a** Copy and complete the table of values for $y = x^2 + 3x + 1$.

x	-4	-3	-2	-1	0	1	2
y		1			1	5	

(2 marks)

b Draw the graph of $y = x^2 + 3x + 1$ for values of x from $x = -4$ to 2. **(2 marks)**

c Use your graph to find estimates of the solutions of $x^2 + 3x + 1 = 2$. **(2 marks)**

Exam tip

Draw your graph as large as possible, so you can read values from it accurately.

11 **a** Use your graph of $y = x^2 - 2x - 1$ from **Q6** to solve the equation $x^2 - 2x - 1 = 2$.

b Why are there no solutions to the equation $x^2 - 2x - 1 = -3$?

16.4 Factorising quadratic expressions

- Factorise quadratic expressions.

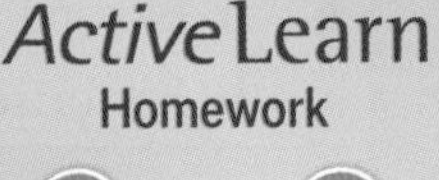

Warm up

1 Fluency Expand

a $(x+6)(x+1)$ **b** $(x-5)(x+8)$ **c** $(x-2)(x-1)$ **d** $(x+5)^2$

2 Fluency Write the negative factor pairs for

a 2 **b** 4 **c** 6 **d** 5 **e** 10

3 'Product' means multiply. The product of 2 and 3 is 6. Write down a pair of numbers

a whose product is 8 and whose sum is 6
b whose product is 12 and whose sum is 7
c whose product is –6 and whose sum is 1
d whose product is –10 and whose sum is –9

4 Copy and complete

a $(\square+2)(x+4)=x^2+6x+8$
b $(x+1)(x+\square)=x^2+8x+7$
c $(x-\square)(x+3)=x^2-2x-15$
d $(\square-7)(x-\square)=x^2-11x+28$

Q4 hint

Use the grid method and write in the terms you are given to help you work out what is missing.

		+2
x	x^2	
+4		8

Key point

$$x^2+6x+5=(x+1)(x+5)$$

factorise (left to right); expand (right to left)

Example

Factorise x^2+5x+6.

x^2+5x+6

$(x \quad)(x \quad)$ — Write a pair of brackets with x in each one. This gives the x^2 term when multiplied.

$1\times6 \quad 2\times3$ — Work out all the factor pairs of 6, the number term.

$1+6=7 \quad 2+3=5$ — Work out which factor pair will **add** to give 5, the number in the x term.

$(x+2)(x+3)$ — Then write one of the numbers in each of the brackets with x.

Check: $(x+2)(x+3)=x^2+5x+6$ ✓ — The expression is now factorised. Expand the brackets to check it is correct.

5 Factorise

a $x^2+7x+12$ **b** x^2+6x+8 **c** x^2+4x+3 **d** $x^2+7x+10$

Exam-style question

6 a Expand and simplify $(3x-1)(4x+3)$. **(2 marks)**

b Factorise x^2+3x+2. **(2 marks)**

Exam tip

Expand your answer to check it is correct.

7 a Write down all the factor pairs of –6.

b Which factor pair of –6 adds to give 1?

c Factorise x^2+x-6.

8 Factorise

a $x^2+3x-10$ b $x^2-3x-10$ c x^2-2x-3 d x^2+2x-3

e x^2-x-20 f x^2+x-20 g $x^2+8x-20$ h $x^2-8x-20$

i $x^2+11x+30$ j $x^2+5x-14$ k x^2-3x-4 l x^2-6x+8

9 **Problem-solving** An expression for the area of a square mat in cm^2 is $y^2+18y+81$.

a Find an expression for the length of one side of the mat.

b Work out the area when $y=3$.

10 Factorise

a $x^2+8x+16$ b x^2+6x+9 c x^2-2x+1 d x^2-6x+9

11 **Problem-solving** An expression for the area of a rectangle is $x^2+10x-24$.
Work out expressions for the length and width of the rectangle.

12 Expand and simplify

a $(x+2)(x-2)$ b $(x+3)(x-3)$ c $(x+4)(x-4)$ d $(x+10)(x-10)$

e **Reflect** Write down what you notice about the x term in these quadratics.

Key point

The **difference of two squares** is a quadratic expression with two squared terms, where one term is subtracted from the other. For example

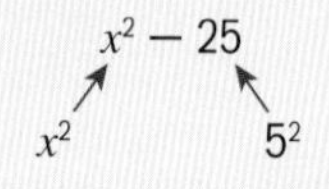

13 **Reasoning** Write three different expressions that are the difference of two squares using the terms in the box.

x^2	x	–4	+10	–30	–36	+9	$-y^2$

14 **Problem-solving** Josie cuts a small square of card from a larger square.

a Write an expression for the remaining area of the large square.

b What type of expression is your answer to **a**?

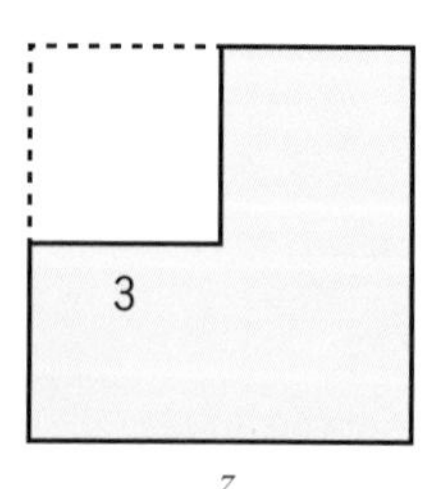

15 Factorise

a x^2-25 b x^2-36 c x^2-64 d x^2-y^2

Exam-style question

16 a Expand and simplify $(n+3)(n-4)$. **(2 marks)**

b Factorise p^2-81. **(1 mark)**

16.5 Solving quadratic equations algebraically

- Solve quadratic equations algebraically.

ActiveLearn Homework

Warm up

1 **Fluency** Find the positive and negative square root of

a 9 **b** 49 **c** 100 **d** 144

2 Factorise

a $x^2+7x+10$ **b** x^2+x-12 **c** x^2-9 **d** $x^2+12x+36$

3 Solve

a $d-8=10$ **b** $x+1=0$ **c** $p-4=0$ **d** $t-11=0$

Key point

Solutions to quadratic equations can be found algebraically as well as from a graph.

4 Square root both sides of each equation, to find one positive and one negative solution.

a $x^2=36$ **b** $y^2=81$ **c** $z^2=100$ **d** $t^2=16$

5 Copy and complete to solve $x^2-9=0$.

$x^2-9=0$ (+9 both sides)

$x^2=\square$

$x=\square$ or $x=-\square$

6 Solve

a $x^2-25=0$ **b** $y^2-1=0$ **c** $n^2-49=0$ **d** $m^2-121=0$

7 Rearrange each equation so you only have x^2 on the left-hand side. Then solve it.

a $x^2-6=30$ **b** $x^2+3=52$ **c** $x^2-20=61$ **d** $x^2+6=70$

8 Copy and complete to solve each equation.

a

$2x^2=32$ (÷2 both sides)

$x^2=\square$

$x=\square$ or $x=-\square$

b

$3x^2=12$ (÷3 both sides)

$x^2=\square$

$x=\square$ or $x=-\square$

Example

Solve $x^2+2x-15=0$.

$x^2+2x-15=0$

$(x+5)(x-3)=0$ — Factorise the quadratic expression.

$x+5=0$ or $x-3=0$ — As the two expressions multiply to make 0, at least one of them must equal zero.

$x=-5$ or $x=3$ — Solve both equations.

9 Solve

a $2x^2 = 50$ b $3x^2 = 75$ c $4x^2 - 64 = 0$ d $5x^2 - 5 = 0$

e $2x^2 + 7 = 25$ f $2x^2 - 5 = 67$ g $3x^2 + 10 = 13$ h $5x^2 - 3 = 17$

10 Solve by factorising

a $x^2 + 8x - 20 = 0$ b $x^2 + 2x - 15 = 0$ c $x^2 - 6x + 5 = 0$ d $x^2 + 4x - 21 = 0$

11 Solve

a $x^2 + 5x + 4 = 0$ b $x^2 - 5x + 4 = 0$ c $x^2 + 3x - 4 = 0$ d $x^2 - 3x - 4 = 0$

e $x^2 + 9x + 18 = 0$ f $x^2 + x - 6 = 0$ g $x^2 - 6x + 5 = 0$ h $x^2 - 10x + 16 = 0$

12 **Reasoning** Match each equation to the correct solution.

A $x = -5$ and $x = -7$	**B** $x = -7$	**C** $x = 1$ and $x = -1$	**D** $x = -6$ and $x = 2$

E $x^2 + 12x + 35 = 0$	**F** $x^2 - 1 = 0$	**G** $x^2 + 14x + 49 = 0$	**H** $x^2 + 4x - 12 = 0$

Exam-style question

13 Solve $x^2 - x - 30 = 0$. **(3 marks)**

14 Copy and complete to solve each equation.

a $x^2 - 2x = 0$
$x(x - \square) = 0$
$x = 0$ or $x - \square = 0$
$x = 0$ or $x = \square$

b $y^2 + 4y = 0$
$y(y + \square) = 0$
$y = 0$ or $y + \square = 0$
$y = 0$ or $y = \square$

15 Solve

a $n^2 + 7n = 0$ b $t^2 - 6t = 0$ c $x^2 - 9x = 0$ d $y^2 + 2y = 0$

16 **Problem-solving** Write a quadratic equation with solution $x = 3$ and $x = 4$.

17 a Plot the graph of $y = x^2 - 4$ for values of x from -3 to $+3$.

b Use your graph to solve the equation $x^2 - 4 = 0$.

c Solve $x^2 - 4 = 0$ by factorising. Show that the solutions match the solutions from your graph.

18 a Write an expression for the area of rectangle $ABCD$.

b The area of rectangle $ABCD$ is $24 cm^2$.
Write an equation for the area of $ABCD$.

c Rearrange your equation so the right-hand side is 0.
Solve the equation to find two possible values of x.

d **Reasoning** Which of the values of x you found in part **c** gives negative side lengths for $ABCD$?

e Which is the sensible solution for x?

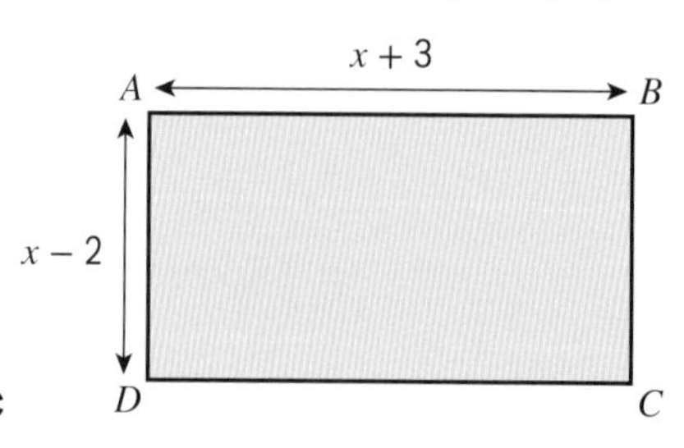

16 Check up

ActiveLearn Homework 

Quadratic graphs

1 Which of these are graphs of quadratic functions?

a

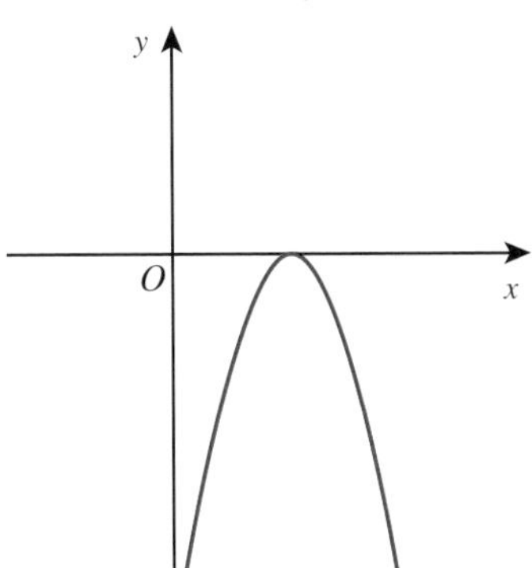

b

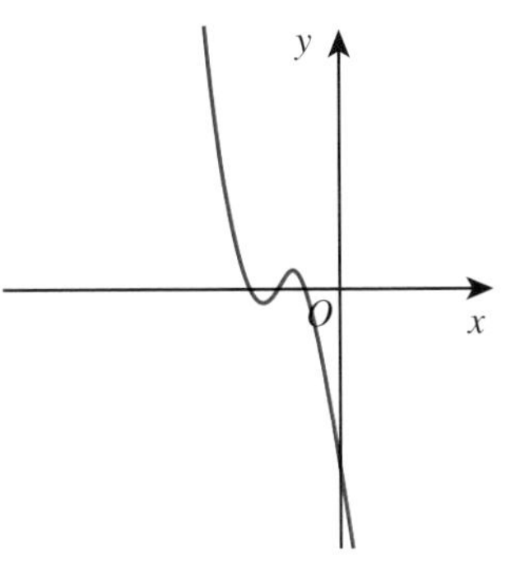

c

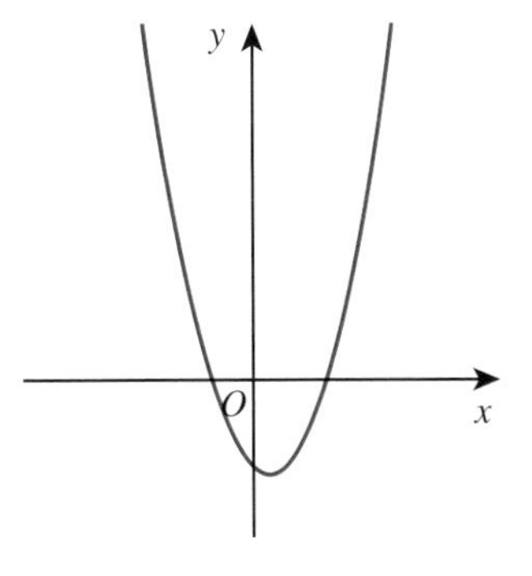

2 **a** Copy and complete the table of values for $y = x^2 - 6x$.

x	−1	0	1	2	3	4	5	6	7
y									

b Draw the graph of $y = x^2 - 6x$ for values of x from −1 to +7.

c Write the equation of the line of symmetry for your graph.

d Write the coordinates of the turning point.

e Write down the coordinates of the y-intercept.

f Use your graph to find the roots of $x^2 - 6x = 0$.

3 Here are the graphs of $y = x^2 + 5x + 3$ and $y = 7$.

a Use the graphs to estimate the solutions to $x^2 + 5x + 3 = 0$.

b Use the graphs to estimate the solutions to $x^2 + 5x + 3 = 7$.

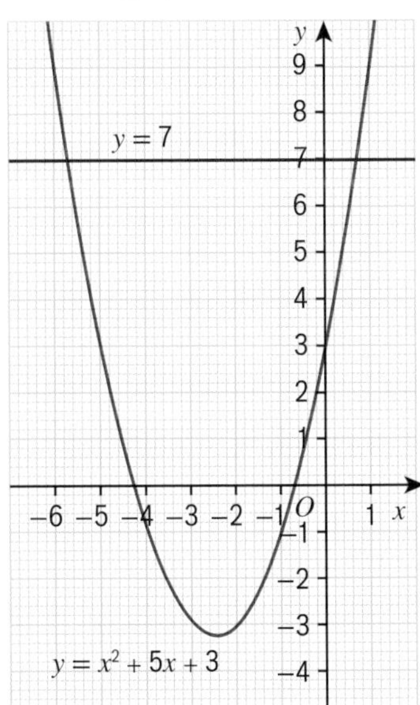

4 The graph shows the height, h metres, of a stone thrown up in the air, at time t seconds.

Use the graph to answer these.

a Estimate the height of the stone at 0.5 seconds.

b At how many seconds does the stone reach its maximum height?

c At what times is the stone at a height of 1 m?

d Estimate the number of seconds from $t = 0$ until the stone hits the ground.

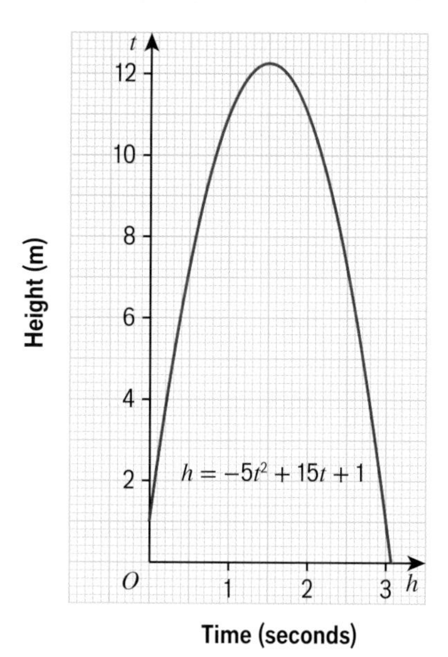

Quadratic equations

5 Expand and simplify

a $(t+6)(t+4)$

b $(f-5)(f+7)$

c $(n-7)^2$

d $(x+3)(4x-1)$

6 Factorise $x^2 - 49$.

7 Solve

a $x^2 - 9 = 0$

b $2x^2 - 72 = 0$

c $y^2 - 3y = 0$

8 Factorise

a $x^2 + 9x + 20$

b $x^2 + 2x - 8$

9 Solve

a $x^2 + 13x + 30 = 0$

b $x^2 - 7x + 6 = 0$

10 **Reflect** How sure are you of your answers? Were you mostly

What next? Use your results to decide whether to strengthen or extend your learning.

Challenge

11 Write ten different quadratic expressions, each with two of these solutions.
Solutions: $x = 3$ $x = -3$ $x = 2$ $x = -2$

16 Strengthen

Active Learn Homework

Quadratic graphs

1 **a** Copy and complete the table of values for $y = x^2 + 1$.

x	−3	−2	−1	0	1	2	3
x^2	9	4		0			
+1	+1	+1	+1	+1	+1	+1	+1
y	10		2		2		10

b Draw a set of axes that includes the lowest and highest values from all of your x- and y-coordinates.

c Plot the graph of the function. Draw a smooth curve through the points. Label your graph: $y = x^2 + 1$

2 **a** Copy and complete the table of values for $y = x^2 + x$.

x	−3	−2	−1	0	1	2
x^2	9		1	0		
$+x$	−3	−2		0		2
y	6	2			2	

b Plot the graph of the function.
This table of values doesn't show the lowest coordinate (turning point) on the curve.
Draw the turning point by estimating where it will lie between the two lowest coordinates in your table.

c Label your graph.

3 Name the labelled points on the graph of $y = ax^2 + bx + c$.
Choose from:

y-intercept | line of symmetry | turning point | roots

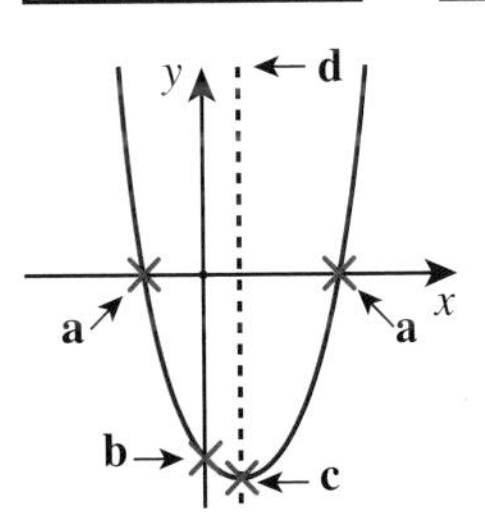

Q3 hint

a This is where the curve crosses the x-axis.
b This is where the curve crosses the y-axis.
c This is the minimum (or maximum) point of the curve.
d This line divides the curve into two identical halves. It has the equation x = 'a number'.

4 On this graph find

a the y-intercept

b the coordinates of the turning point

c the equation of the line of symmetry

d the roots of $x^2 - 6x + 8 = 0$

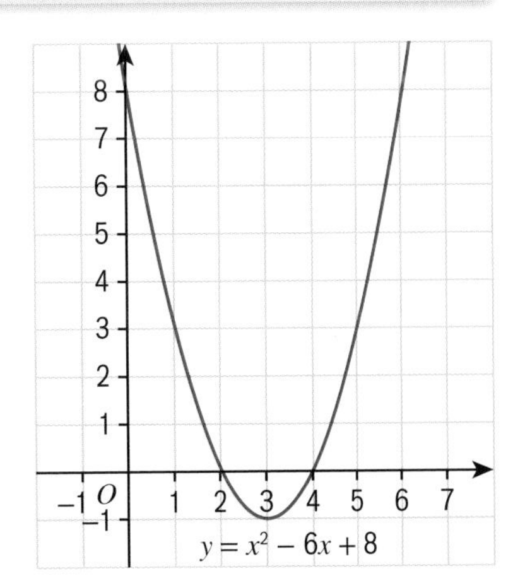

5 **a** Copy and complete the table of values for $y = x^2 + 5x + 5$.

x	−5	−4	−3	−2	−1	0
x^2						
$+5x$						
$+5$						
y						

b Plot the graph of the function. Label your graph.

c **i** Draw the line $y = 1$ on your graph.

ii The solutions to $x^2 + 5x + 5 = 1$ are the x-coordinates at the points where the curve $y = x^2 + 5x + 5$ and the line $y = 1$ cross.
Find the solutions to $x^2 + 5x + 5 = 1$.

d Estimate the solutions of the equation of the equation $x^2 + 5x + 5 = 0$.

6 Quadratic graphs are ∪ or ∩ shaped and symmetrical.
Which of these are graphs of quadratic functions?

a
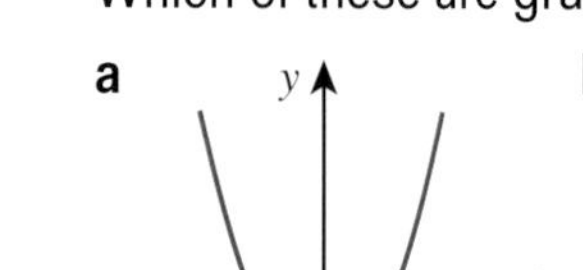

b
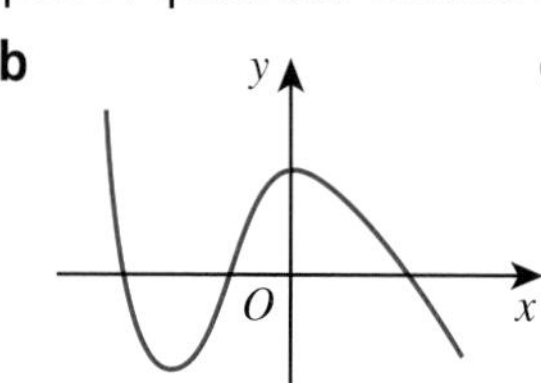

c
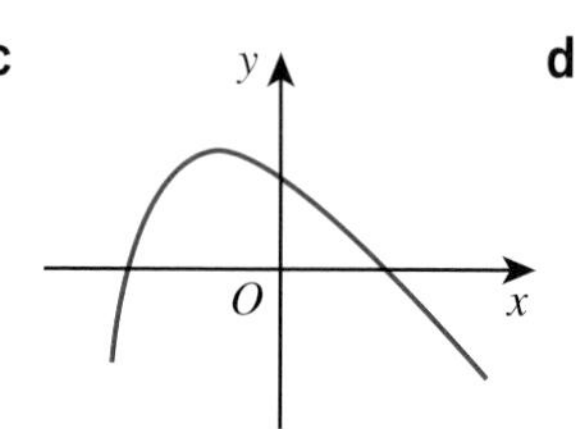

d
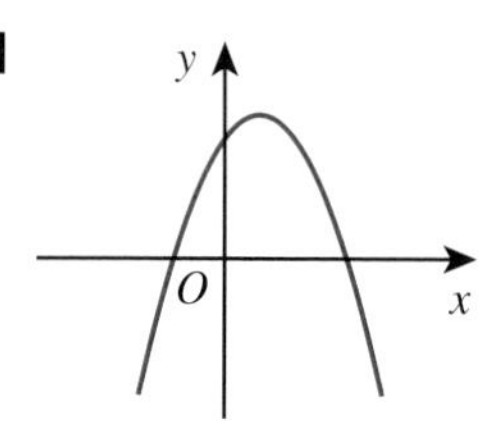

Quadratic equations

1 Emma uses this method to expand double brackets.
Draw a smiley face to help you to expand the brackets:
Copy and complete her calculation.

eyebrows
$(a + 2)$ $(a + 3)$
nose
smile

Eyebrow 1: $a \times a = \square$
Eyebrow 2: $2 \times 3 = \square$
Nose: $2 \times a = \square$
Smile: $a \times 3 = \square$

Eyebrow 1 + eyebrow 2 + nose + smile = $\square + \square + \square + \square = \square + \square + \square$

2 Expand and simplify

a $(a + 2)(a + 3)$ **b** $(t + 1)(t + 4)$
c $(x + 1)(x + 7)$ **d** $(y + 5)(y + 10)$
e $(z + 1)(z - 2)$ **f** $(f - 7)(f + 9)$
g $(x + 1)(x - 1)$ **h** $(x - 4)(x + 4)$

3 Expand and simplify

a $(m + 4)^2$ **b** $(x + 8)^2$ **c** $(g - 10)^2$ **d** $(y - 2)^2$

4 Use the method from **Q1** to expand and simplify

a $(x + 2)(3x + 1)$ **b** $(x - 2)(3x - 1)$
c $(2x + 1)(x - 5)$ **d** $(2x + 1)(x + 4)$
e $(2x + 1)(3x + 1)$ **f** $(2x - 1)(3x + 1)$
g $(3x + 4)(4x + 3)$ **h** $(3x + 2)(4x - 3)$

5 Copy and complete

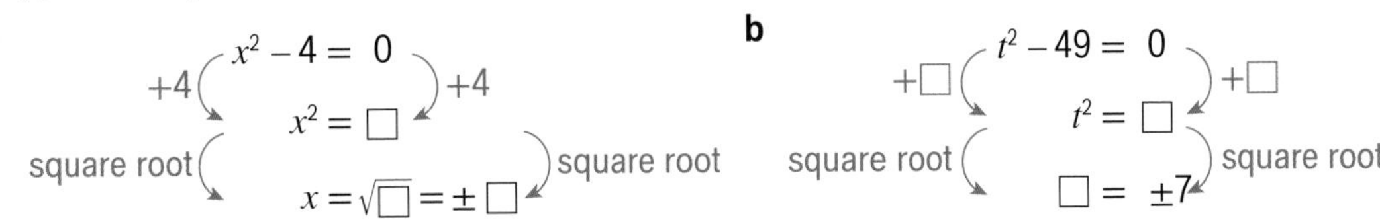

6 Match the expressions to their factorisations. Expand the brackets to check.

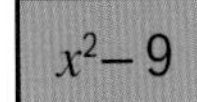

$x^2 - 9$	$(x - 2)(x + 2)$	$(x + 4)(x - 4)$	$x^2 - 49$
$(x + 7)(x - 7)$	$x^2 - 16$	$(x - 3)(x + 3)$	$x^2 - 4$

7 Factorise

a $x^2 - 16$ **b** $p^2 - 1$ **c** $y^2 - 81$ **d** $k^2 - 100$

8 **a** Write the factor pairs of 12.

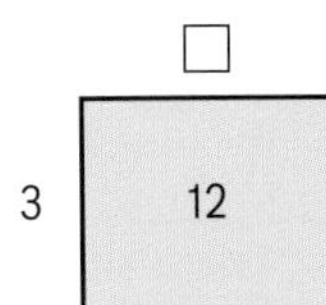

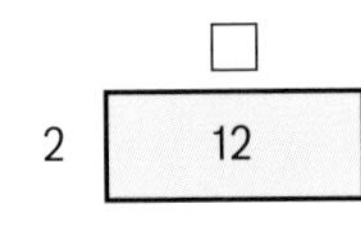

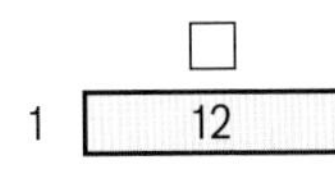

b Write the sum of each factor pair.

c Use your factor pairs to factorise

i $x^2 + 7x + 12$ **ii** $x^2 + 8x + 12$ **iii** $x^2 + 13x + 12$

9 **a** Copy and complete these factor pairs of –15.

–1 and $+\square$ 1 and $-\square$ –3 and $+\square$ +3 and $-\square$

b Write the sum of each factor pair.

c Use your factor pairs to factorise

i $x^2 - 14x - 15$ **ii** $x^2 - 2x - 15$

iii $x^2 + 2x - 15$ **iv** $x^2 + 14x - 15$

10 Work out

a 3×0 **b** 0×5 **c** 0×0 **d** $a \times 0$ **e** $0 \times b$

11 Copy and complete

a $x^2 + 6x + 8 = 0$

$(x + 2)(x + 4) = 0$

$x + 2 = 0$ or $x + 4 = 0$

↓ ↓

$x = \square$ or $x = \square$

b $x^2 + 3x - 18 = 0$

$(x + \square)(x - \square) = 0$

$x + \square = 0$ or $x - \square = 0$

↓ ↓

$x = \square$ or $x = \square$

Q11 hint

When two expressions multiply to give zero, one or both of the expressions is equal to zero.

16 Extend

1 **a** Copy and complete the table and plot the graph of the function $y = -x^2 + 3x + 4$ on a graph paper grid.

x	−2	−1	0	1	2	3	4	5
$-x^2$	−4							
$+3x$	−6							
$+4$	+4							
y	−6							

b Use your graph.

i Write down the y-intercept of the graph.

ii Write down the equation of the line of symmetry.

iii Estimate the turning point of the graph.

iv Estimate the solutions to $-x^2 + 3x + 4 = 0$.

Exam-style question

2 **a** Factorise $8x + 12$. **(1 mark)**

b Factorise $y^2 - 16$. **(1 mark)**

3 **Problem-solving** A hanging cable can be modelled by the equation $y = x^2 - 5x + 10$, where x is the horizontal distance in metres and y is the height in metres.
The initial height of the cable is equal to its final height.

a Copy and complete the table of values and plot the graph of the function $y = x^2 - 5x + 10$ on graph paper.

x	0	1	2	3	4	5
y						

b i What is the maximum height of the cable?

ii Estimate the minimum height of the cable.

iii Work out the horizontal distance that the cable spans.

4 A skateboard ramp can be modelled by the equation $y = \frac{1}{4}x^2 - x + 1$, where x is the horizontal distance in metres and y is the height in metres. The initial height of the skateboard ramp is equal to its final height.

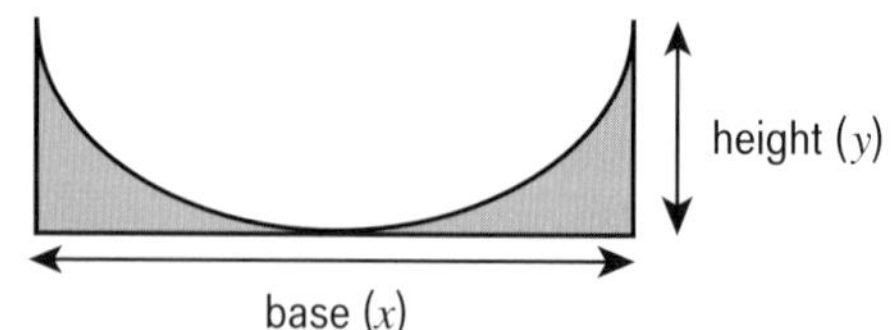

a Copy and complete the table of values and plot the graph of the function $y = \frac{1}{4}x^2 - x + 1$.

x	0	1	2	3	4
y					

b From the graph:

i What is the maximum height of the ramp?

ii How long is the base of the ramp?

iii What is the minimum height of the ramp from the ground?

5 **Problem-solving** A ball is thrown into the air. After s, seconds, its height, h, in metres above ground, is given by the formula $h = -(4s + 1)(s - 2)$.

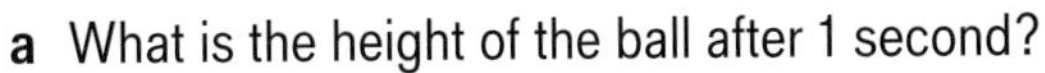

a What is the height of the ball after 1 second?

b What is the height of the ball after 2 seconds?

c Use your answer to part **b** to say what has happened to the ball after 2 seconds.

Q5a hint

After 1 second means $s = 1$. Substitute into the equation to get h.

Exam-style question

6

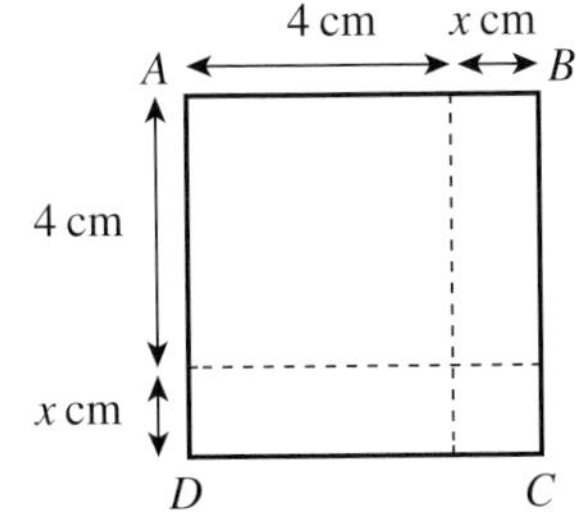

The area of square $ABCD$ is $18\,\text{cm}^2$.
Show that $x^2 + 8x = 2$. **(3 marks)**

7 **Problem-solving** The diagram shows a photo in a frame.
The photo and frame are both square.

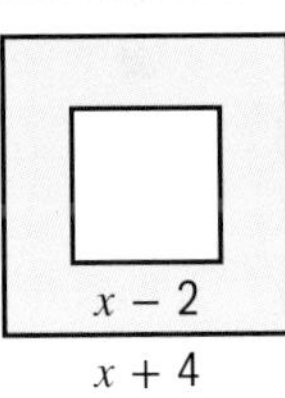

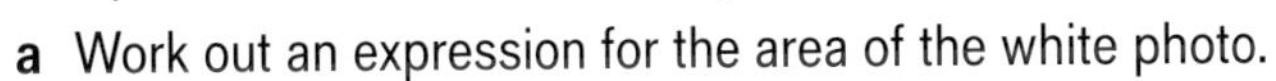

a Work out an expression for the area of the white photo.

b Using your answer to part **a**, work out and simplify an expression for the area of the blue frame surrounding the photo.

8 **Problem-solving** The area of the square and the area of the triangle are equal.

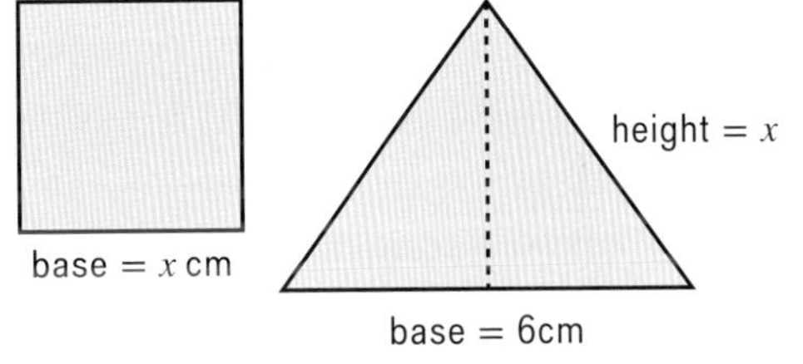

a Write and simplify an equation in x to represent the areas of the shapes.

b Solve your equation.

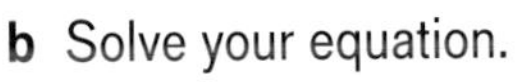

c Work out the areas of the shapes.

9 Here is the graph of $y = 2x^2 - 3x - 7$.
Use the graph to estimate the solutions to

a $2x^2 - 3x - 7 = 0$

b $2x^2 - 3x - 7 = -4$

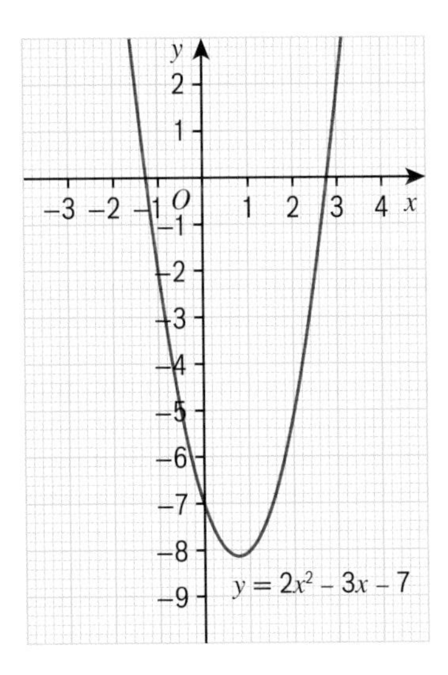

16 Test ready

Summary of key points

To revise for the test:

- Read each key point, find a question on it in the mastery lesson, and check you can work out the answer.
- If you cannot, try some other questions from the mastery lesson or ask for help.

Key points

1. To expand or multiply double brackets, multiply each term in one bracket by each term in the other bracket. → **16.1**
2. A quadratic expression always has a squared term (with a power of 2). It cannot have a power higher than 2. It may have a term with a power of 1 that is the same letter as the squared term. It may also have a constant (number) term. ax^2+bx+c has a squared term, ax^2, a term with power 1, bx, and a constant term, c. → **16.1**
3. To square a single bracket, multiply it by itself, then expand and simplify. $(x+1)^2=(x+1)(x+1)$ → **16.1**
4. A **function** describes the relationship between variables. For each **input** value there is an **output** value. → **16.2**
5. The turning point of the curve is where it turns in the opposite direction. → **16.2**
6. A quadratic function has a symmetrical ∪-shaped curve called a **parabola**. A quadratic function with a $-x^2$ term has a symmetrical ∩-shaped curve. → **16.2**
7. A quadratic curve always has a minimum or maximum turning point. → **16.2**
8. To solve the equation $ax^2+bx+c=0$ using a graph, read the x-coordinates where the graph crosses the x-axis. These are called **roots**. → **16.3**
9. To solve the equation $ax^2+bx+c=$ 'a number' using a graph, read the x-coordinates where the graph $y=ax^2+bx+c$ crosses the line $y=$ 'a number'. → **16.3**
10. factorise (→) $x^2+6x+5=(x+1)(x+5)$ (←) expand → **16.4**
11. The **difference of two squares** is a quadratic expression with two squared terms, where one term is subtracted from the other. For example x^2-25. → **16.4**
12. Solutions to quadratic equations can be found algebraically by factorising as well as from a graph. → **16.5**

Sample student answers

Exam-style question

1 **a** Complete the table of values for $y = x^2 - 3x$.

x	−2	−1	0	1	2	3	4
y	10		0				

(2 marks)

b Draw the graph of $y = x^2 - 3x$ for values of x from $x = -2$ to $x = 4$. **(2 marks)**

Student A

x	−2	−1	0	1	2	3	4
y	10	−2	0	−2	−2	0	4

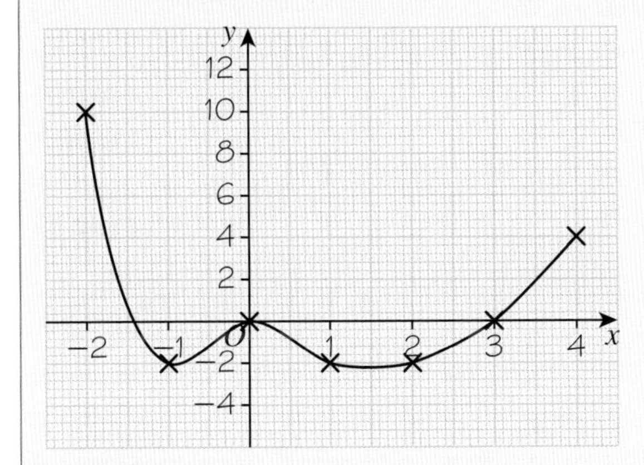

Student B

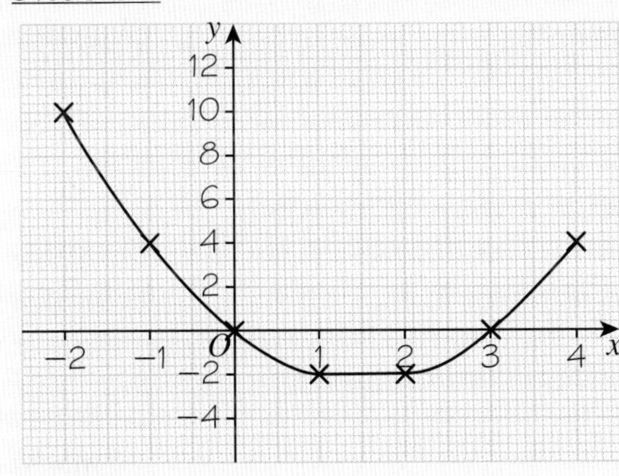

a How can you tell from the graph that one of the y-values in Student A's table must be wrong?

b Which point is incorrect?

c How could Student B improve their graph?

Exam-style question

2 Factorise $x^2 + 7x + 12$. **(2 marks)**

$x(x + 7) + 12$

Explain why this is not the correct factorisation.

16 Unit test

1 Which of these are graphs of quadratic functions?

a i

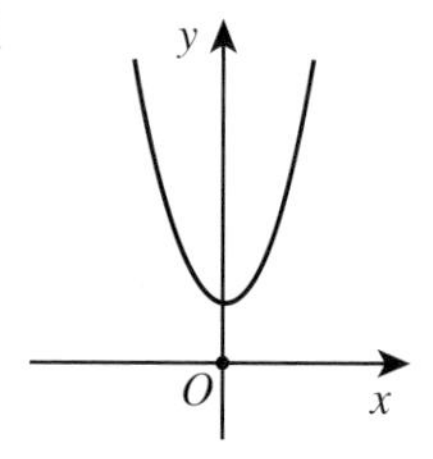

ii

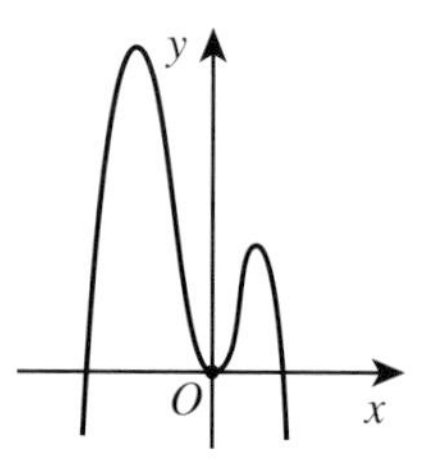

iii

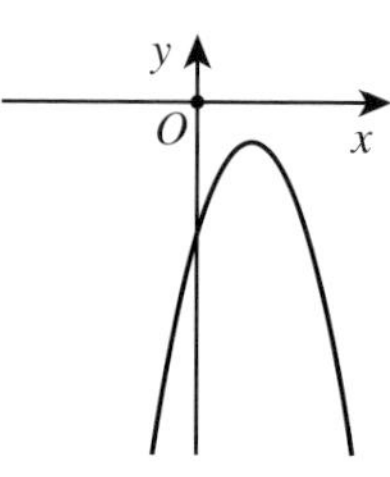

(3 marks)

b i $(a+1)(a+3)$ **ii** $7(y-2)$ **iii** $x(9x^2+3x)$ **(3 marks)**

2 Here is the graph of $y = x^2 - 5x$.

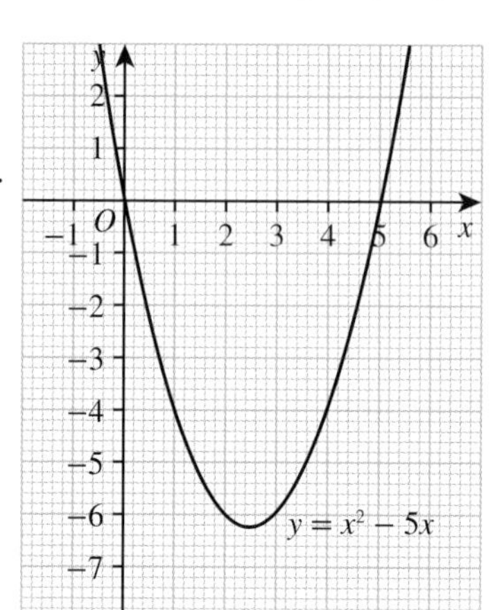

a Write down the y-intercept of the graph. **(1 mark)**

b Write the equation of the line of symmetry. **(1 mark)**

c Estimate the coordinates of the turning point of the graph. **(1 mark)**

d Use the graph to find the roots of the equation $x^2 - 5x = 0$. **(2 marks)**

3 a Copy and complete the table of values for $y = x^2 + 7x + 6$. **(2 marks)**

x	−7	−6	−5	−4	−3	−2	−1	0
y	6		−4				0	

b On graph paper, draw a graph of the function $y = x^2 + 7x + 6$ for values of x from −7 to 0. **(2 marks)**

c Use your graph to estimate the solutions to $x^2 + 7x + 6 = 3$. **(2 marks)**

4 Expand and simplify

a $(y+10)(y+1)$ **(1 mark)**

b $(z+7)(z-8)$ **(1 mark)**

c $(x+1)(x-1)$ **(1 mark)**

d $(n-9)^2$ **(1 mark)**

5 Solve $a^2 - 25 = 0$. **(2 marks)**

6 a Solve $3x^2 = 48$. **(2 marks)**

b Expand and simplify $(2x-3)(3x+1)$. **(2 marks)**

c Factorise $x^2 + 4x + 4$. **(1 mark)**

7 Solve $x^2 + 6x - 40 = 0$. **(3 marks)**

8 An expression for the area of a rectangular tile is $x^2 - 10x + 24\,\text{cm}^2$.
Write expressions for the length and width of the tile. **(3 marks)**

9 Solve $2y^2 - 98 = 0$. **(2 marks)**

10 The graph shows the height, h metres, of a practice missile fired from a ship, at time, t seconds.

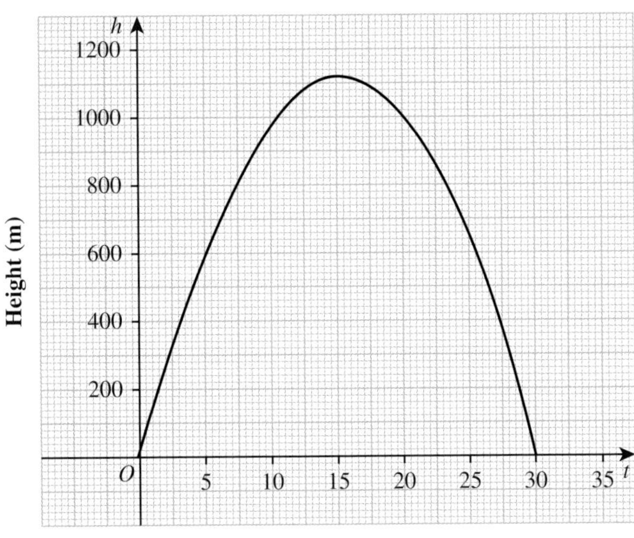

From the graph:

a Estimate the height of the missile at 10 seconds. **(1 mark)**

b What is the maximum height the missile reaches? **(1 mark)**

c Estimate at what times the missile is at a height of 600 m. **(1 mark)**

d How long does it take for the missile to fall from its maximum height into the sea? **(1 mark)**

(TOTAL: 40 marks)

11 **Challenge** A length of 40 m of fencing is used to make a rectangle.
The area of the rectangle can be modelled by the function $A = 20w - w^2$, where A is the area enclosed and w is the width of the rectangle in metres.

a Draw a graph of the function $A = 20w - w^2$.
Put w on the x-axis and A on the y-axis.

b Use your graph to find the width of the enclosure when the area is a maximum.

c Find the length and width that give the rectangle with maximum area.

12 **Reflect** For this unit, copy and complete these sentences.
I showed I am good at _____.
I found ___ hard.
I got better at _____ by ___.
I was surprised by ______.
I was happy that _____.
I still need help with _____.

17 Perimeter, area and volume 2

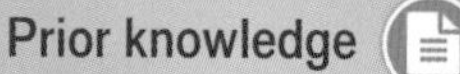

Prior knowledge

17.1 Circumference of a circle 1

- Calculate the circumference of a circle.
- Solve problems involving the circumference of a circle.

ActiveLearn
Homework

Warm up

1 **a** Round 3.7816 to 1 decimal place (1 d.p.) and to 2 d.p.

b Round 3487 to 3 significant figures (3 s.f.)

2 $m = 3pq$
Work out the value of m when $p = 3$ and $q = 5$.

3 Work out $\frac{58.6}{14}$. Round your answer to 2 d.p.

Key point

The **circumference** is the **perimeter** of a circle.

4 **a** The radius of the London Eye is 60 metres. Work out its diameter.

b A Frisbee has a diameter of 10 inches. Work out its radius.

c The diameter of a circle is d. Write a formula to give the radius r.

d The radius of a circle is r. Write a formula to give the diameter d.

5 **Problem-solving** The length of a tennis court is 78 feet. 1 foot = 12 inches.
360 tennis balls fit touching each other along the length of the court.
What is the diameter of a tennis ball in inches?

6 **Reasoning** The table shows the diameter and circumference of some circular objects.

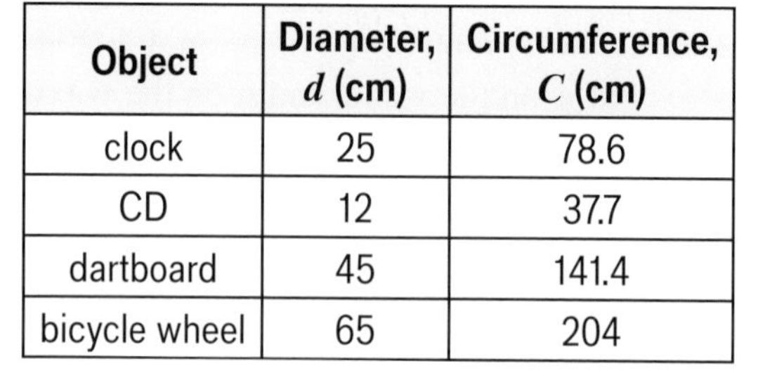

Object	Diameter, d (cm)	Circumference, C (cm)
clock	25	78.6
CD	12	37.7
dartboard	45	141.4
bicycle wheel	65	204

a Work out the value of $\frac{C}{d}$ for each object.
Round your answers to 2 d.p.

b **Reflect** Write down what do you notice about the values of $\frac{C}{d}$.

c Copy and complete the formula $\frac{C}{d} = \square$

$C = \square \times d$

d The diameter of the centre circle on a football pitch is 18.3 metres.
Use your formula to work out the circumference of the centre circle.
Round your answer to 1 d.p.

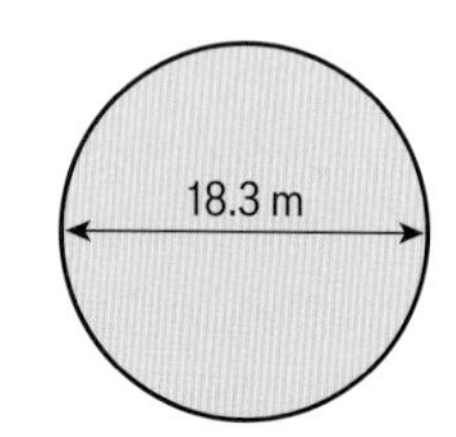

Key point

The Greek letter π (pronounced pi) is the ratio of the circumference of a circle to the diameter. Its decimal value never ends, but starts as 3.141 592 653 589 7...
The formula for the circumference of a circle is $C = \pi d$.
If you know the radius you can use $C = 2\pi r$.

7 Work out the circumference of each circle. Use the π button on your calculator.
Round your answers to 1 d.p.

a

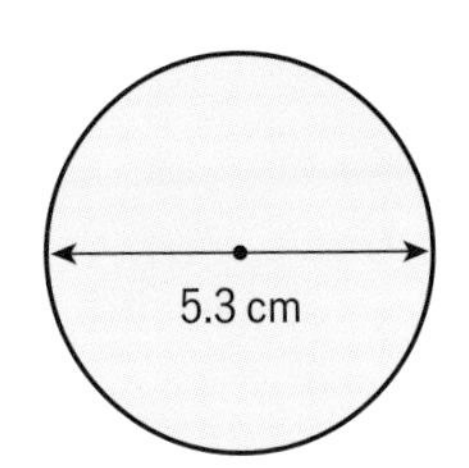

b

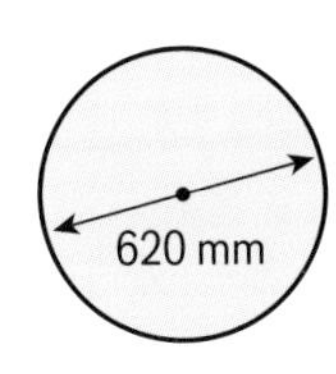

c

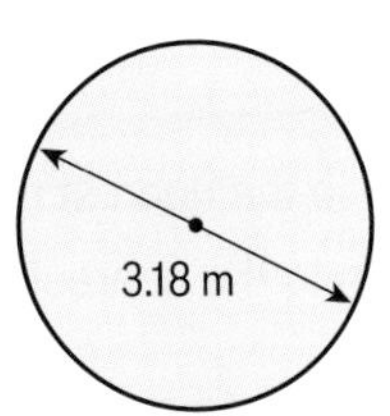

d

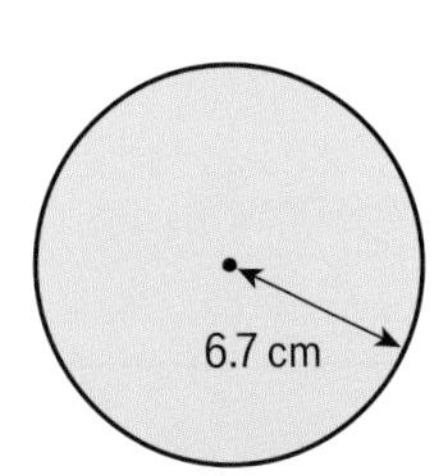

e

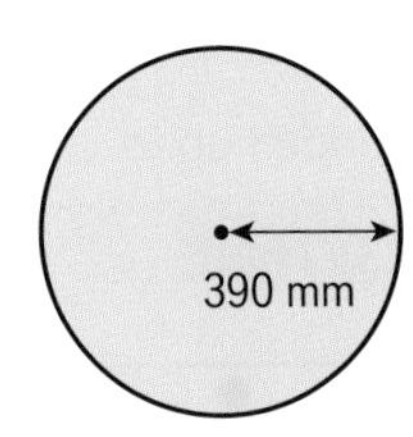

f

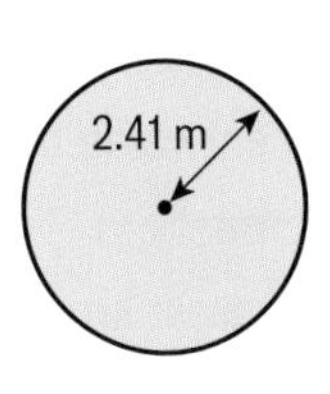

Exam-style question

8 The diameter of this circle is 11 cm.
Work out the circumference of this circle.

11 cm

Diagram NOT accurately drawn

Give your answer correct to 3 significant figures. **(2 marks)**

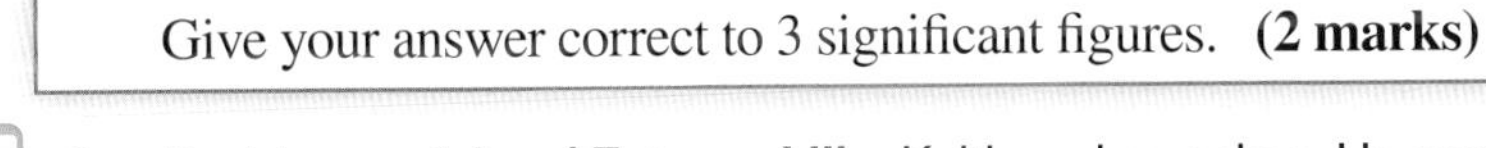

Exam tip

Write down all the numbers on your calculator display before you round.

9 **Problem-solving / Future skills** Keith makes cakes. He uses a cake tin with a 20 cm diameter. Keith puts ribbon around each cake. He allows an extra 2 cm of ribbon for joining.

a How much ribbon does he need for each cake?

b He buys the ribbon in rolls of 25 metres.
How many cakes can he decorate from one roll?

Q9b hint

25 m = □ cm

10 **Problem-solving / Future skills** Petra puts edging round a circular flower bed with diameter 460 cm. The edging comes in 245 cm lengths. How many lengths does she need?

11 The radius of a racing bike wheel is 340 mm.

a How far does the wheel travel in one revolution?

b How many revolutions will the wheel make in a 500 m sprint race?

12 A crop circle has radius 90 metres. Write the circumference in terms of π.

13 Madhu says the circumference of this circle is 14π cm.
Paul says the circumference of this circle is 7π cm.
Reasoning Who is correct? Explain your answer.

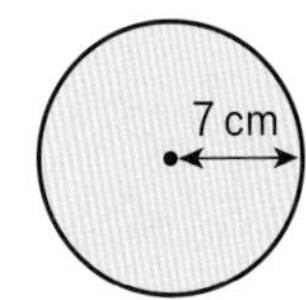

17.2 Circumference of a circle 2

- Calculate the circumference and radius of a circle.
- Write error intervals for rounded and truncated values.

Warm up

1 Fluency

a Round 2.5 m to the nearest metre.

b Round 3.49 m to the nearest metre.

2 a Rearrange $A = lw$ to make l the subject of the formula.

b Work out the value of l correct to 2 significant figures (2 s.f.) when $A = 30.5$ and $w = 4.2$.

3 a What is the smallest value that can be rounded up to

i 8 cm to the nearest cm

ii 20 km to the nearest km

iii 38 mm to the nearest mm?

b Write a value that rounds down to

i 8 cm to the nearest cm

ii 20 km to the nearest km

iii 38 mm to the nearest mm

Q3a i hint

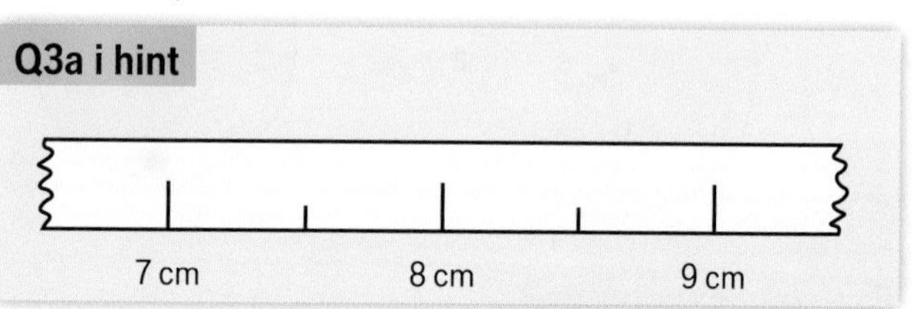

Key point

Measurements given to the nearest whole unit may be inaccurate by up to one half of a unit below and one half of a unit above.
For example, the range of possible values for a length given as 3 cm to the nearest cm is 2.5 cm $\leqslant$ length $<$ 3.5 cm.

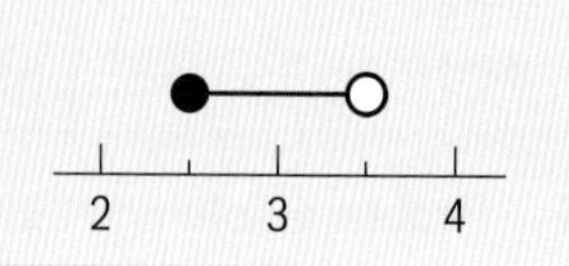

4 The length of a pen is 15 cm to the nearest cm.
Write down the minimum possible length of the pen.

5 Bob's height is 1.63 m correct to the nearest cm.

a What is his minimum possible height in metres?

b What is the minimum possible height that would round to 1.64 m?

c Write an inequality $\square \leqslant h < \square$ to show Bob's possible heights.

6 There are 25 300 people at a concert, rounded to the nearest 100.
Write an inequality to show the possible numbers of people.

Q6 hint

Rounded to the nearest 100, so $\frac{1}{2}$ of 100 above and below.

7 Using inequalities, write an error interval to show the possible values for

a 4.36 m (rounded to 2 d.p.)

b 720 (rounded to the nearest 10)

c 15.7 (rounded to 3 s.f.)

d 450 (rounded to 2 s.f.)

Key point

To **truncate** a number to 1 digit, you remove the other digits without rounding.
5.694 truncated to 1 digit is 5.

8 Truncate each number to **i** 1 digit **ii** 2 digits

a 4.03	**b** 4.562	**c** 4.86	**d** 4.91
e 4.997	**f** 5.01	**g** 5.09	**h** 5.12

9 A number n truncates to 4.

a What is the lowest value n could have?

b What is the lowest number that truncates to 5?

c Use your answer to **a** and **b** to complete this error interval for n.

$\square \leqslant n < \square$

Exam-style question

10 Maisie used her calculator to work out the value of a number, N.
The answer on her calculator display began
7.2
Complete the error interval for N.
$\_\_\_\_\_\_ \leqslant N < \_\_\_\_\_\_$ **(2 marks)**

Exam tip

Write the values clearly on the dotted lines.

Exam-style question

11 The height of a book is 26 cm correct to the nearest cm.
The height of the shelf in the bookcase is 260 mm correct to the nearest mm.
Explain why the book may not fit upright on the shelf. **(2 marks)**

Exam tip

Show all of your working and explain what your working shows at each stage of your calculation.

12 a Write down the first 6 digits of π. Use your calculator.

b Round your answer from part **a** to 3 d.p.

c Use your value from part **b** to work out the circumference of this circle.

d Use the π button on your calculator to work out the circumference of the same circle.

e **Reflect** Which value for the circumference is more accurate?

5 cm

13 Calculate an estimate for the circumference of this circle, using π rounded to the nearest whole number.

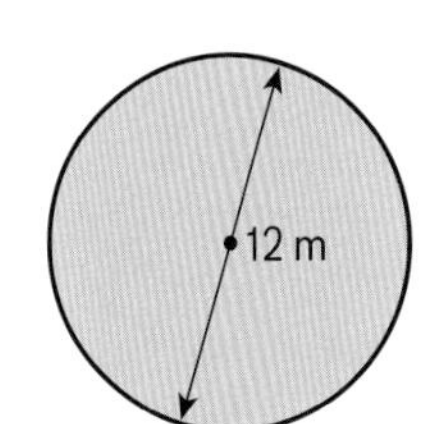

Exam-style question

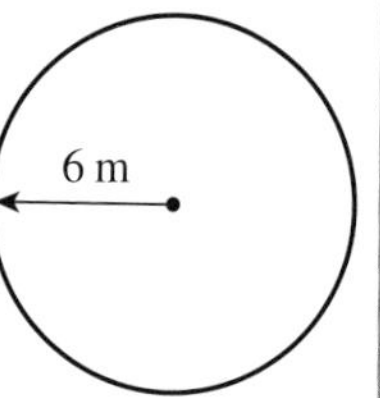

14 Lucas has a pond in the shape of a circle of radius 6 m.
He is going to put a wire fence round the pond.
Wire fence comes in 4 metre rolls.

a Work out an estimate for the number of rolls of wire fence Lucas needs.
You must show your working. **(4 marks)**

b Is your estimate for part **a** an underestimate or an overestimate?
Give a reason for your answer. **(1 mark)**

Key point

To give an answer to an appropriate degree of accuracy you usually state it to the same degree of accuracy as the least accurate measurement in the question.

15 Work out the circumference of each of these circular objects.
Give your answers to an appropriate degree of accuracy.

a A pond with diameter 4.8 m.

b A battery with diameter 6 mm.

c The world's largest pancake with radius 750.5 cm.

Q15a hint

4.8 m is given to 1 d.p. so your answer should be to 1 d.p.

Example

The circumference of a circle is 60.8 cm. Work out the radius of the circle.

$C = 2\pi r$

$60.8 = 2\pi r$ — Substitute the values that you know.

$\frac{60.8}{2\pi} = r$ — Rearrange to make r the subject.

$r = 9.67662054$ — Enter $\frac{60.8}{2\pi}$ as a fraction on your calculator.

radius = 9.7 cm (to 1 d.p.) — Write the answer to the same degree of accuracy as the measurement given. Remember to include the units.

16 Work out, to 1 d.p., the diameter of a circle with circumference

a 350 cm **b** 2.8 m **c** 24 cm

Q16 hint

Start by substituting $C = \pi d$.

17 Work out, to 3 s.f., the radius of a circle with circumference

a 670 m **b** 3 km **c** 52 cm

18 **Reasoning** The circumference of a circle is 28 cm correct to the nearest cm.

a Write the error interval for the circumference.

b Work out the smallest possible value for the radius, correct to 1 d.p.

c Write the error interval for the radius.

17.3 Area of a circle

- Work out the area of a circle.
- Work out the radius or diameter of a circle.
- Solve problems involving the area of a circle.
- Give answers in terms of π.

*Active*Learn
Homework

Warm up

1 **Fluency** Estimate these by rounding each value to 1 significant figure (1 s.f.)

a 3.14×4^2 **b** 2.3×5.9 **c** $9.2 \div 1.1$ **d** $18.6 \div 4.5$

2 Substitute into each formula to work out the unknown quantity.

a $p = mh^2$ when $m = 10$ and $h = 2$

b $t = as^2$ when $a = 3$ and $s = 6$

3 Work out r when

a $r^2 = 10.4$ **b** $r^2 = \frac{16}{9}$ **c** $3r = 24$ **d** $\pi r = 52$

Key point

The formula for the area A of a circle with radius r is $A = \pi r^2$.

Example

A circle has radius 6.4 cm.
Work out the area of the circle. Give your answer correct to 3 s.f.

$A = \pi r^2$

$= \pi \times 6.4^2$ — Write the substitution. Input it into your calculator.

$= 128.6796351$ — Write down all the figures on the calculator display.

$= 129\text{ cm}^2$ (to 3 s.f.) — Round the answer to the required accuracy. Remember the units.

4 Work out the area of each circle. Round your answers to an appropriate degree of accuracy.

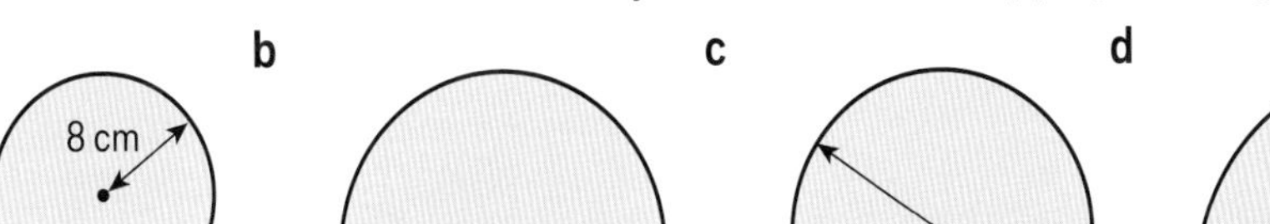

a 8 cm

b 6.5 m

c 420 cm

d 546.8 m

e **Reflect** Leo calculates the answer to part **b** is 417.0 m^2.
What mistake has he made?

Q4c hint
Work out the radius first.

5 **a** Estimate the area of each of these circular objects.

i The head of a nail with a radius of 3 mm.

ii A trampoline with a diameter of 14 feet.

iii The Caldera Crater at Yellowstone Lake with a diameter of 5 km.

b Calculate each area in part **a** leaving your answers in terms of π.

c Calculate each area in part **a** correct to 2 s.f.

Q5a hint

$\pi \times 3^2 = 9\pi \approx 9 \times \square$

6 **Reasoning** Rita says the area of the circle is $25\pi\,\text{cm}^2$.
Jack says the area of the circle is $100\pi\,\text{cm}^2$.
Who is correct? Explain your answer.

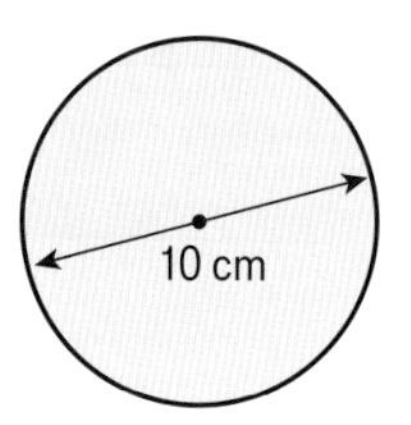

7 **Future skills** A circular space has a radius of 4.5 m.

a Work out the area of the space.

b Gavin sows 50 g of grass seed per m^2 in this space.
How much grass seed does Gavin need to cover this space?

8 **Problem-solving** A scientist wants to estimate the number of slugs in a field with an area of $4050\,\text{m}^2$.
She marks out a circle of radius 1.5 m.

a Work out the area of the circle.

b She finds 35 slugs inside the circle. Estimate the number of slugs in the field.

Example

The area of a circle is $98.5\,\text{cm}^2$. Work out the radius of the circle.
Give your answer correct to 1 d.p.

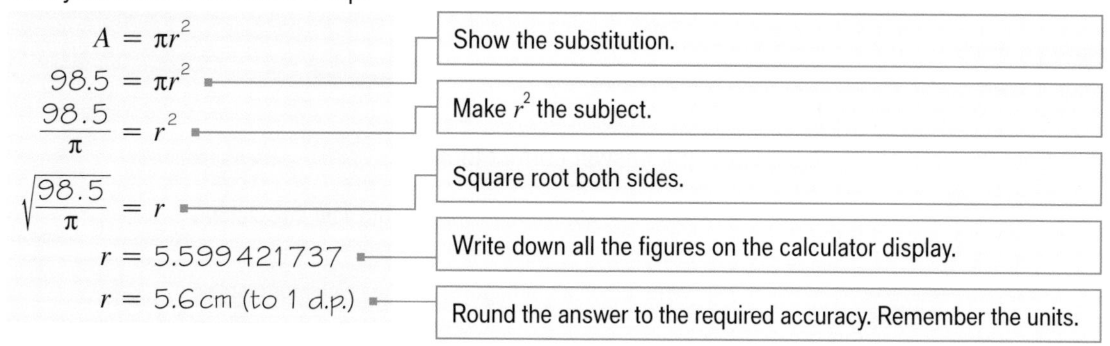

$A = \pi r^2$

$98.5 = \pi r^2$ — Show the substitution.

$\frac{98.5}{\pi} = r^2$ — Make r^2 the subject.

$\sqrt{\frac{98.5}{\pi}} = r$ — Square root both sides.

$r = 5.599421737$ — Write down all the figures on the calculator display.

$r = 5.6$ cm (to 1 d.p.) — Round the answer to the required accuracy. Remember the units.

9 For each circle work out

i the radius **ii** the diameter

a

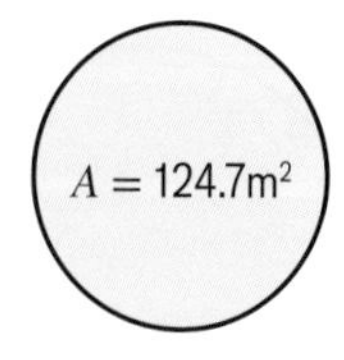

b

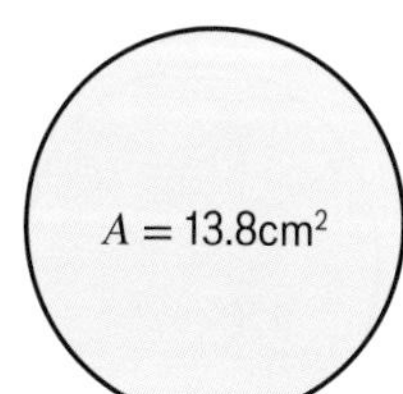

c

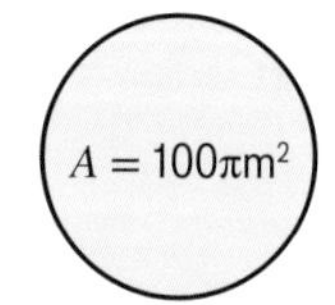

d

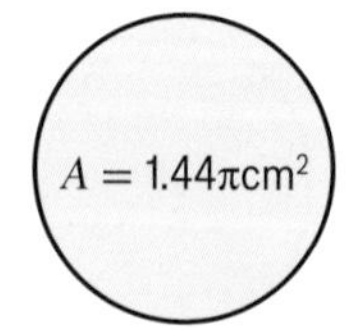

10 **Problem-solving** A sprinkler covers a circular area of $380\,\text{m}^2$.
The sprinkler is in the centre of a circle.
How far from the sprinkler does the water reach?
Give your answer to an appropriate degree of accuracy.

11 **Problem-solving** A circular rug has an area of $1.13\,m^2$.
Find the diameter of the rug in cm.

Exam-style question

12 There is a circular pond of radius 3.1 m in a rectangular garden.
The garden is 10 m by 15 m.
Max is going to plant daffodil bulbs in all of the area not covered by the pond.
Daffodil bulbs are sold in packs of 80.
Each pack of bulbs will cover $1.5\,m^2$ of garden.

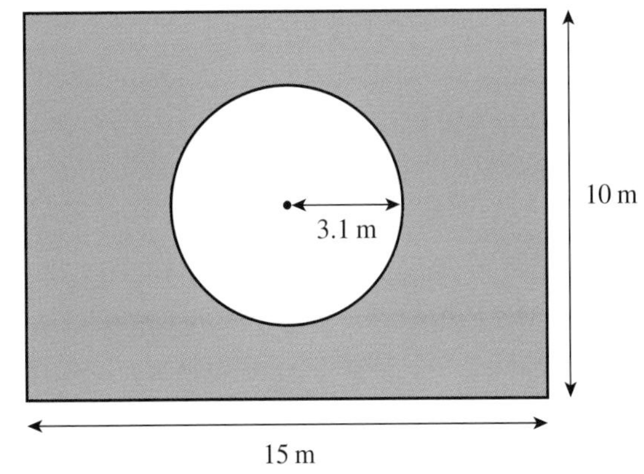

a Work out an estimate for the number of packs of bulbs Max needs.
You must show your working. **(4 marks)**

b Is your estimate for part **a** an underestimate or an overestimate?
Give a reason for your answer. **(1 mark)**

13 **Problem-solving** This circular mirror has a circular metal frame.
Work out the area of the metal frame correct to 3 s.f.

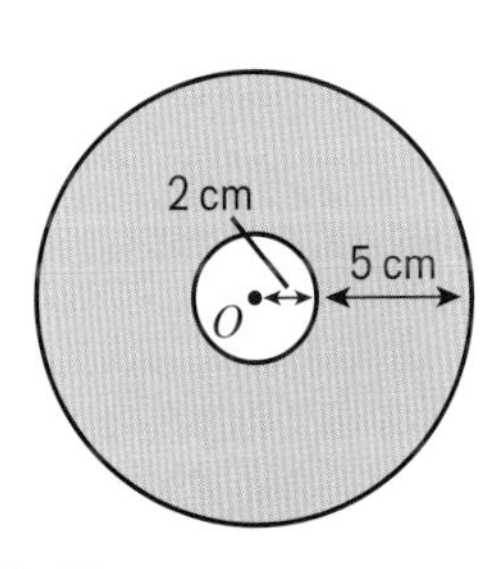

Q13 hint

Required area = area of large circle − area of small circle

14 The diagram shows a pattern made from two circles.
Each circle has a centre O.
Work out, in terms of π

a the area of the whole pattern

b the area of the inner circle

c the shaded area

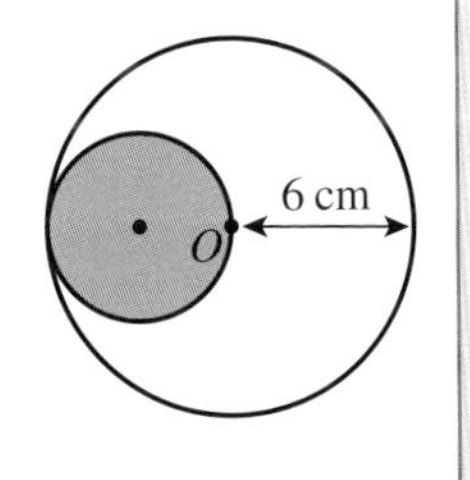

Exam-style question

15 The diagram shows a logo made from two circles.
The large circle has centre O.
Malia says that exactly $\frac{1}{4}$ of the logo is shaded.
Is Malia correct?
You must show your working. **(4 marks)**

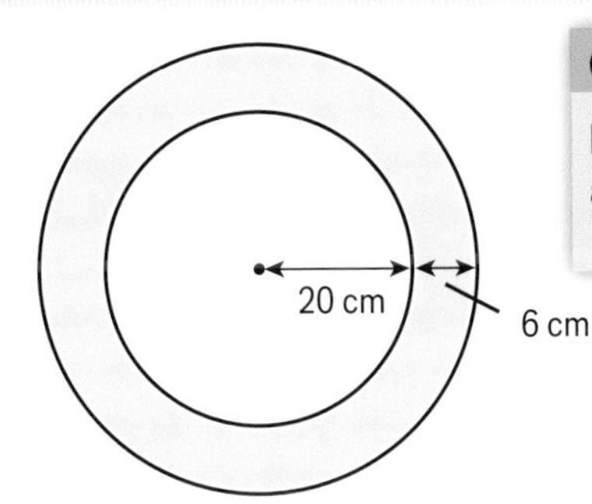

Exam tip

Learn the formulae for the area and the circumference of a circle.

16 **Reflect** You have used three formulae for circles:

$C = 2\pi r \qquad A = \pi r^2 \qquad C = \pi d$

Write yourself some tips to help you remember them.

17.4 Semicircles and sectors

- Understand and use maths language for circles and perimeters.
- Work out areas and perimeters of sectors of circles.

Warm up

1 **Fluency** What is the angle at the centre of

a a circle **b** a semicircle **c** a quarter circle?

2 Write each fraction in its simplest form.

a $\frac{180}{360}$ **b** $\frac{90}{360}$ **c** $\frac{15}{60}$ **d** $\frac{270}{360}$

3 Work out

a $\frac{1}{4}$ of 360 **b** $10 \times \square = 360$ **c** $\frac{360}{5}$ **d** $360 \div \square = 90$

Key point

A **chord** is a line that touches the circumference at each end.
An **arc** is a part of the circumference of a circle.
A **segment** is a part of a circle between an arc and a chord.
A **sector** is a slice of a circle between an arc and two radii.
A **tangent** is a line outside a circle that touches the circle at only one point.

4 Copy these diagrams. Write chord, arc, segment, sector and tangent in the correct places.

a **b** 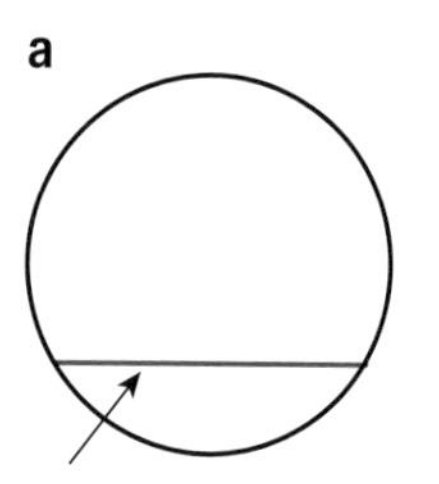

c 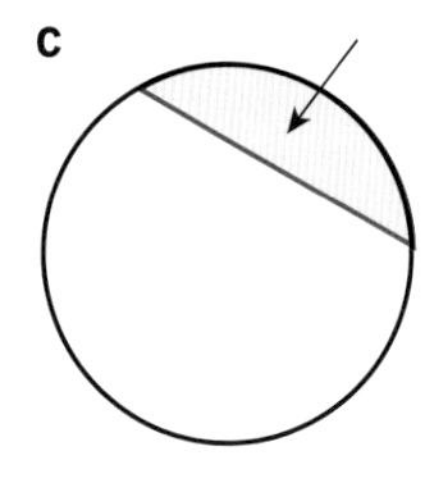

d **e** 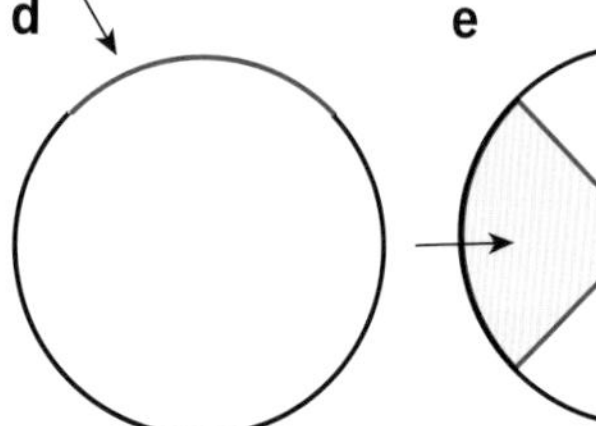

5 Work out the area of each semicircle and quarter circle.
Give your answers

i in terms of π

ii correct to 1 decimal place (1 d.p.)

Q5 hint
Find the area of the whole circle.

a 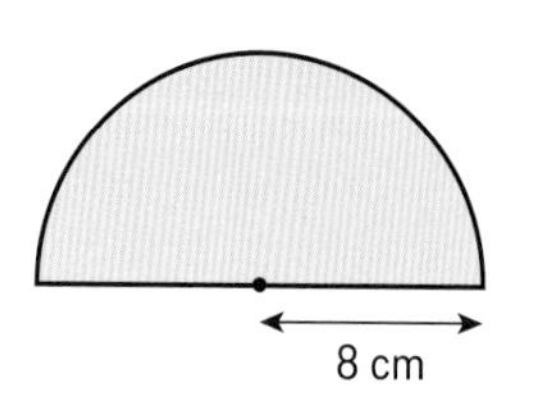

b 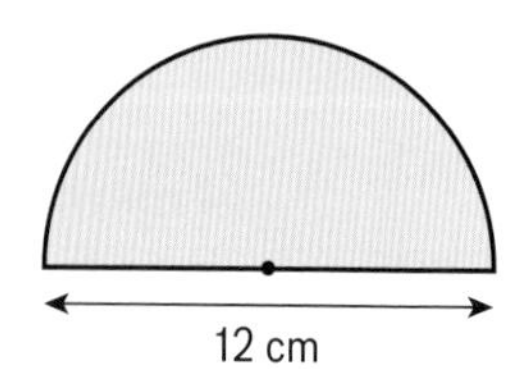

c 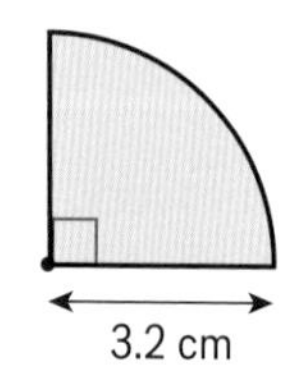

d 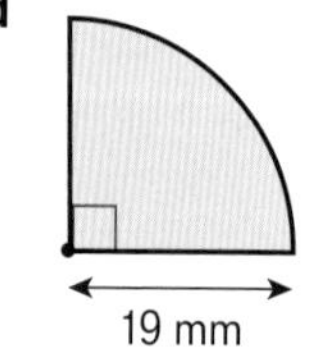

6 The diagram shows a semicircle.

a Work out what the circumference would be if it were a whole circle.

b Halve your answer to part **a** to find the length of the arc.

c Work out the perimeter of the semicircle by adding the arc length and the diameter.

6 cm

Q6a hint

arc

diameter

7 Work out the perimeter of

a this semicircle in terms of π

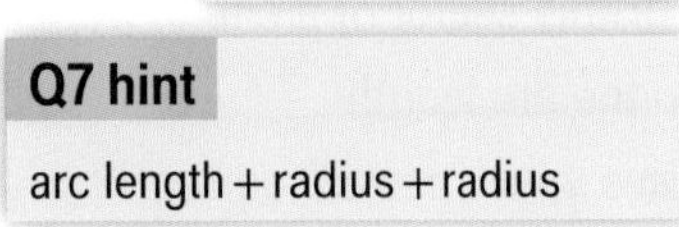
Q7 hint

arc length + radius + radius

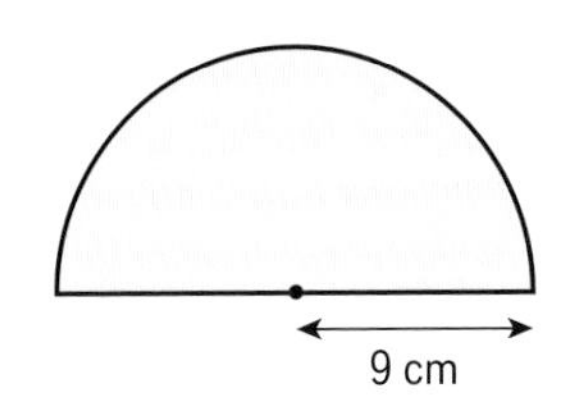

b this quarter circle to 1 d.p.

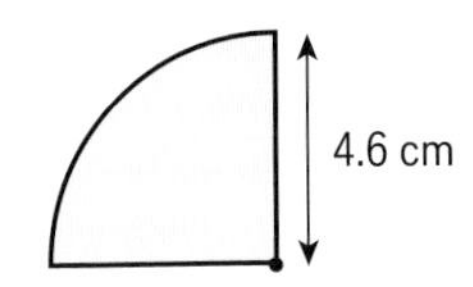

8 **Problem-solving** The minute hand on a clock is 7 cm long.
Work out, to the nearest whole cm, the distance the point end moves in

a 1 hour **b** 4 hours **c** 30 minutes **d** 15 minutes **e** 105 minutes

9 **Reasoning** The diagram shows a pizza with a radius of 12 cm and a slice cut out.

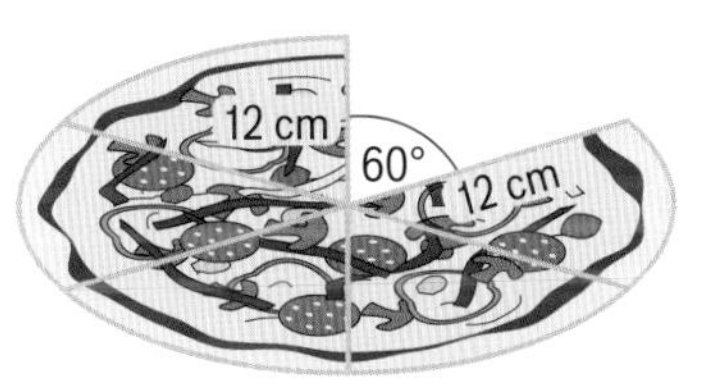

a Work out the area of the whole pizza in terms of π.

b Sam cuts slices of pizza with an angle size of 60°.
How many slices can he cut?

c What fraction of the pizza is each slice?

d Repeat parts **b** and **c** for these angle sizes.

i 30° **ii** 36° **iii** 120° **iv** 45°

Q9b hint

$\frac{360}{60} = \square$

Exam-style question

10 A farmer has a field in the shape of a semicircle of diameter 40 m.

40 m

The farmer asks Pip to build a fence around the edge of the field.
Pip tells the farmer how much it will cost.

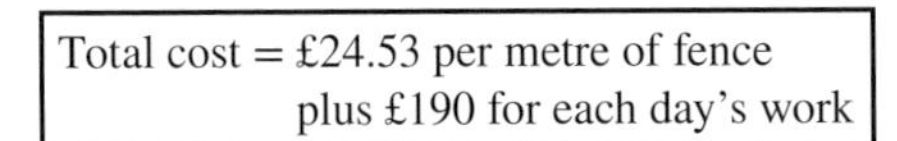
Total cost = £24.53 per metre of fence
plus £190 for each day's work

Pip takes three days to build the fence.
Work out the total cost. **(5 marks)**

Exam tip

When you have finished your working out, check that you have answered the question that is being asked.

Key point

For a sector of a circle with an angle of $x°$ and radius r:

Area of sector $= \frac{x}{360} \times \pi r^2$

Arc length $= \frac{x}{360} \times 2\pi r$

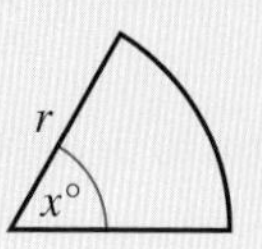

Example

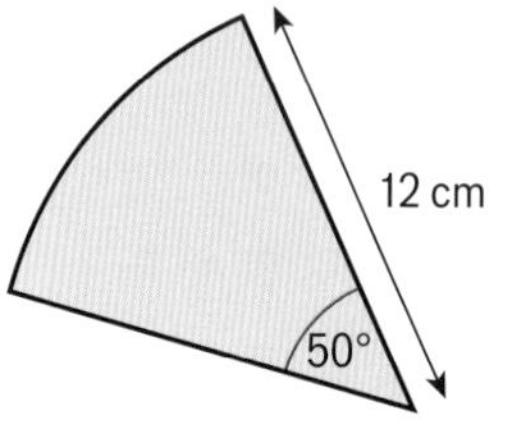

For this sector of a circle, work out

a the arc length **b** the perimeter **c** the area

Give your answers correct to 3 s.f.

a Arc length $= \frac{x}{360} \times 2\pi r$

$= \frac{50}{360} \times 2 \times \pi \times 12$ — Substitute angle size 50 and radius 12.

$= 10.47197551$ — Write down all the figures on your calculator display.

$= 10.5$ cm (to 3 s.f.) — Give the answer correct to 3 s.f. and include the units.

b Perimeter $= 10.47197551 + 12 + 12$ — Arc length + radius + radius. Use the unrounded value for the arc length.

$= 34.47197551$

$= 34.5$ cm (to 3 s.f.) — Give the answer correct to 3 s.f. and include the units.

c Area $= \frac{x}{360} \times \pi r^2$

$= \frac{50}{360} \times \pi \times 12^2$

$= 62.83185307$

$= 62.8\text{ cm}^2$ (to 3 s.f.)

11 For each sector of a circle, work out

i the arc length **ii** the perimeter **iii** the area

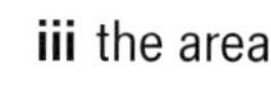

a

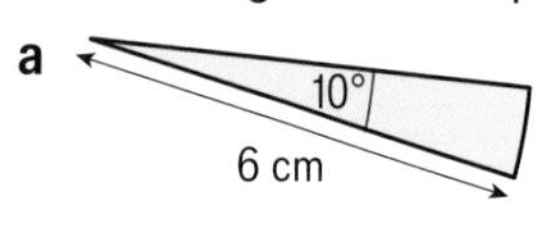

b

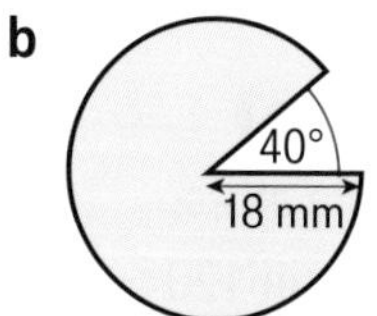

Q11b ii hint

Work out the angle of the sector.

c

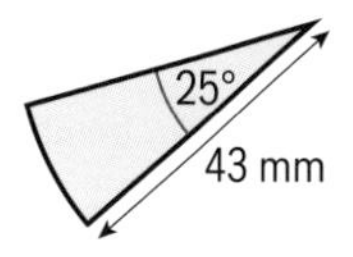

d

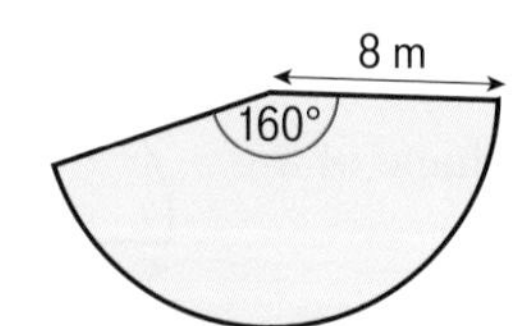

e

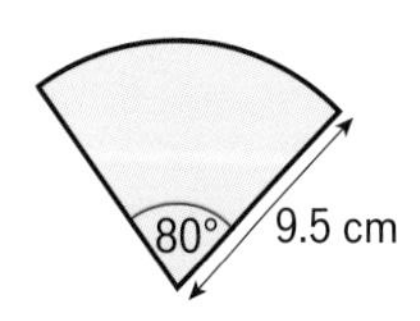

f

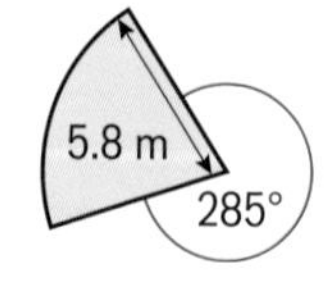

Give your answers correct to 3 s.f.

17.5 Composite 2D shapes and cylinders

- Solve problems involving areas and perimeters of 2D shapes.
- Work out the volume and surface area of cylinders.

Warm up

1 **Fluency** Sketch a net of this cylinder. What shapes are its faces?

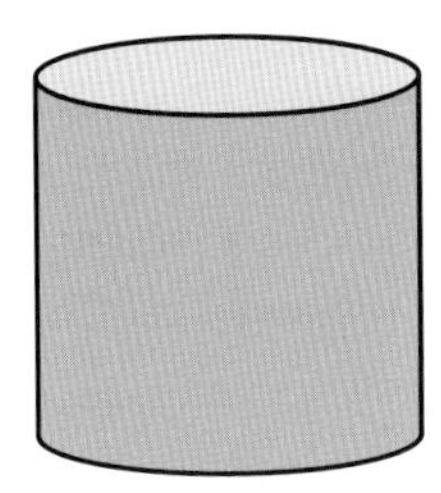

2 Work out the volume of this prism.

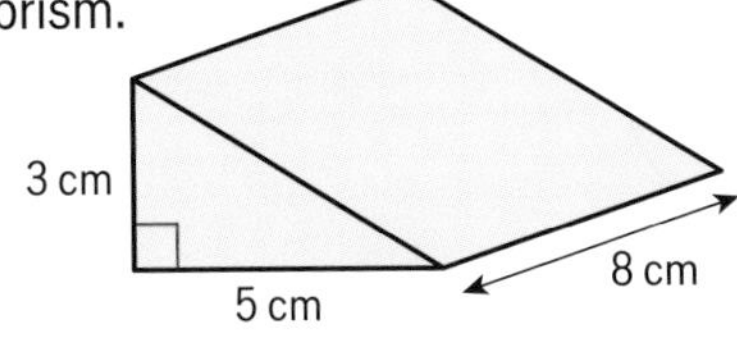

3 Find the area of each shape. Leave your answers in terms of π where appropriate.

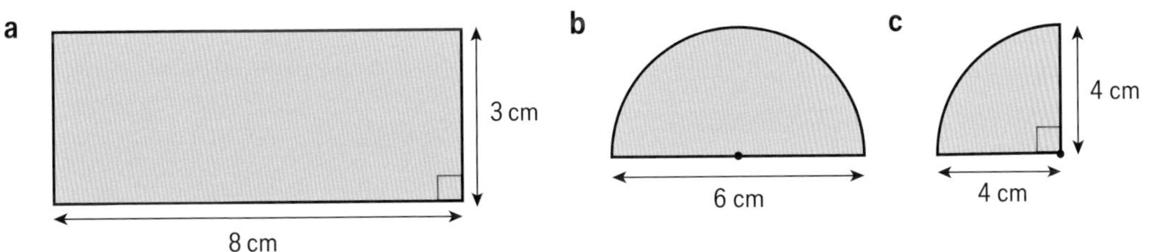

4 Find the perimeter of each shape in **Q3**.
Leave your answers in terms of π where appropriate.

Example

For this shape work out, correct to 1 decimal place (1 d.p.),

a the perimeter

b the area

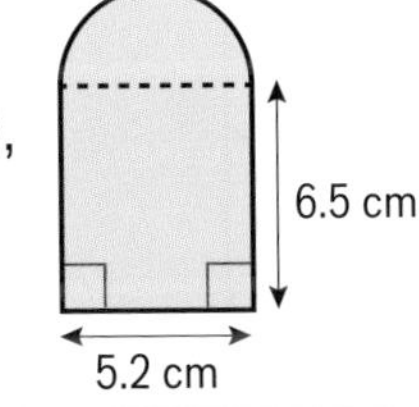

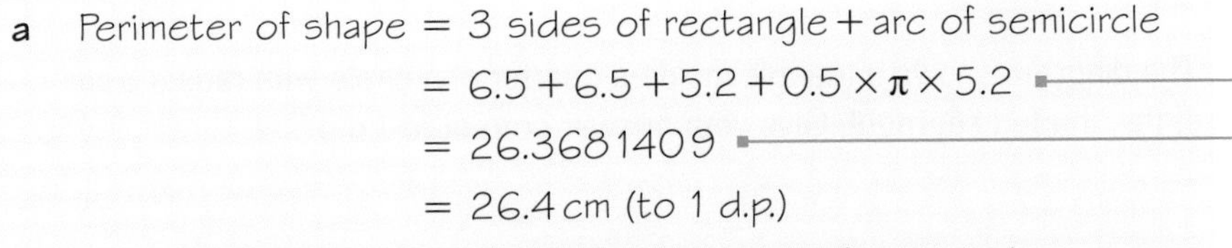

a Perimeter of shape = 3 sides of rectangle + arc of semicircle

$= 6.5 + 6.5 + 5.2 + 0.5 \times \pi \times 5.2$

$= 26.3681409$

$= 26.4$ cm (to 1 d.p.)

b Area of shape = area of rectangle + area of semicircle

$= 5.2 \times 6.5 + 0.5 \times \pi \times 2.6^2$

$= 44.418\,583\,17$

$= 44.4\text{ cm}^2$ (to 1 d.p.)

Write down all the figures on your calculator display.

Round the answer to the required accuracy. Remember the units.

5 Work out the perimeter and area of each of these shapes.
Give your answers to an appropriate degree of accuracy.
Show all your working.

a

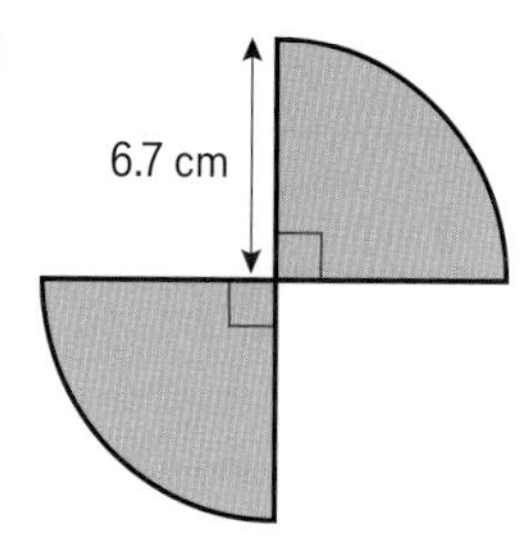

b

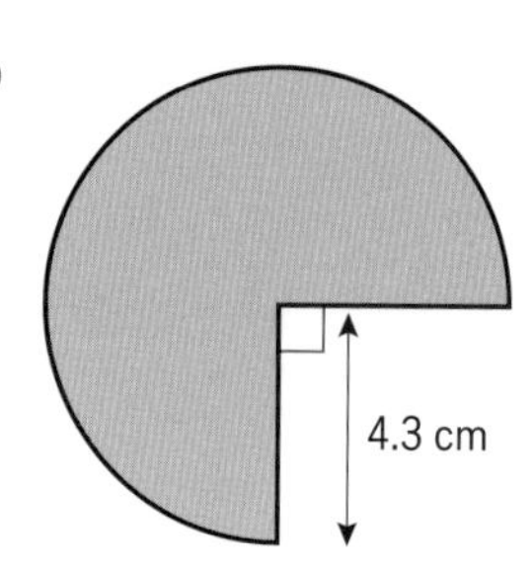

6 The diagram shows a school sports field.
The field is a rectangle with semicircular ends.
The rectangle is 80 m by 60 m.

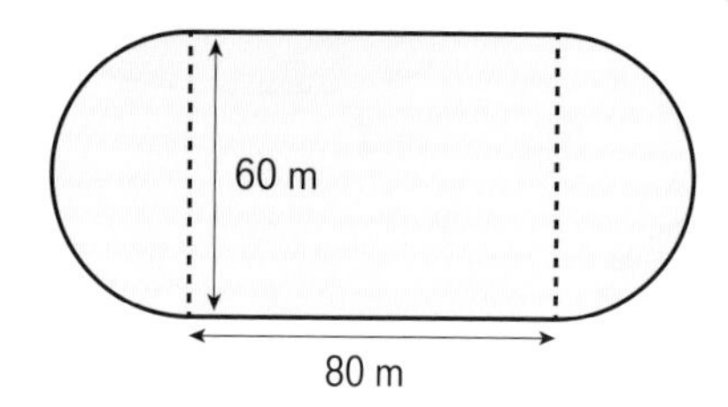

a What is the diameter of the semicircle at each end?

b Work out the area of the field. Give your answer to the nearest square metre.

c Work out the perimeter of the field. Give your answer to the nearest metre.

d How many times must Bob run round the field to run at least 1 km?

7 **Problem-solving** The diagram shows the logo for a new company letterhead.
The logo is symmetrical.
Each section is a quarter circle.
Calculate the area that will be shaded. Give your answer correct to 2 d.p.

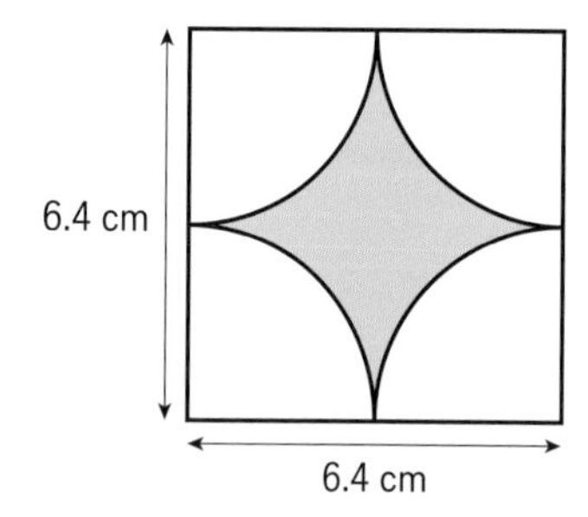

8 **Problem-solving** The diagram shows a triangle inside a quarter of a circle with radius 3 cm.
Work out the area of the shaded segment. Give your answer correct to 1 d.p.

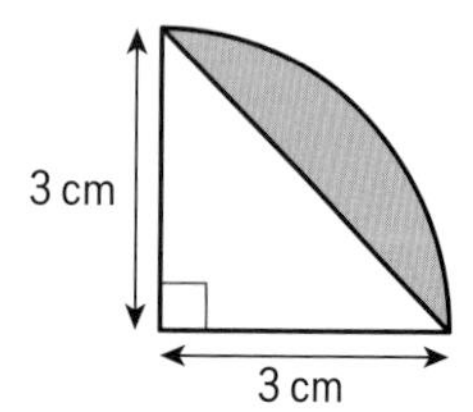

9 **a** The diagram shows a cylinder. What shape is its cross-section?

b Write a formula for the area of its cross-section.

c Write a formula for the volume of a cylinder.

$V = \square \times h$

Key point

The formula for the volume of a cylinder is $V = \pi r^2 h$.

10 Work out the volume of each cylinder. Give your answers to 3 significant figures (3 s.f.).

a

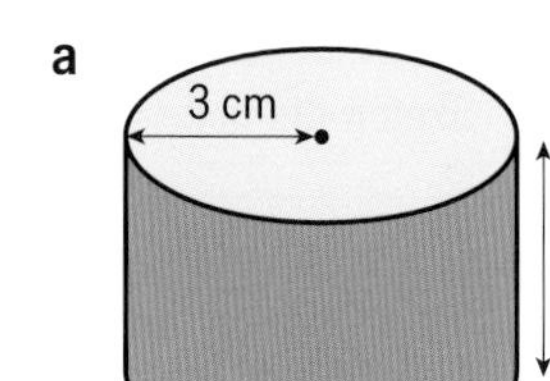

b

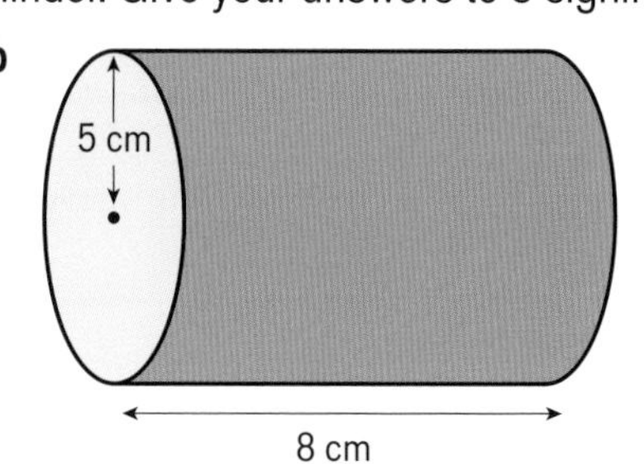

c

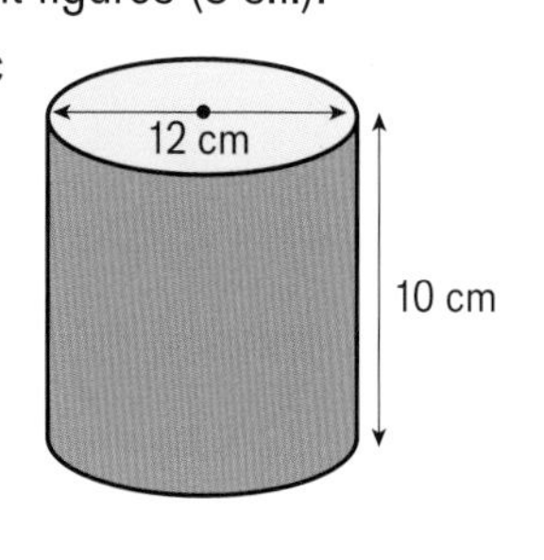

11 A cylindrical tank has a radius of 1.5 m and a height of 3 m.

a What is the volume of the tank?

b The tank is full of water. How many times can a bucket that holds 20 litres be filled from the tank?

Q11b hint

$1\,m^3 = 1000$ litres

12 **Reasoning** Which of these two cylinders has the greater volume?

A

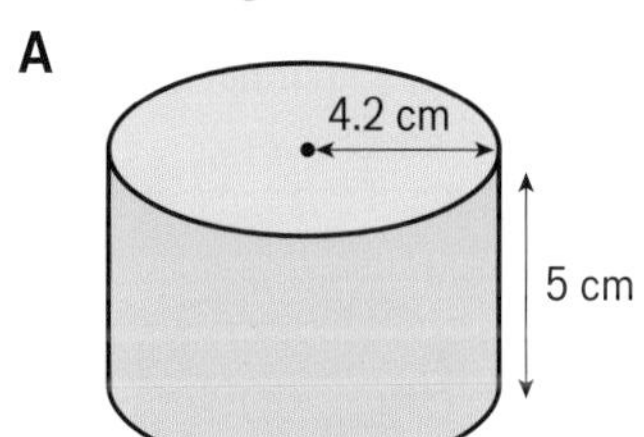

B

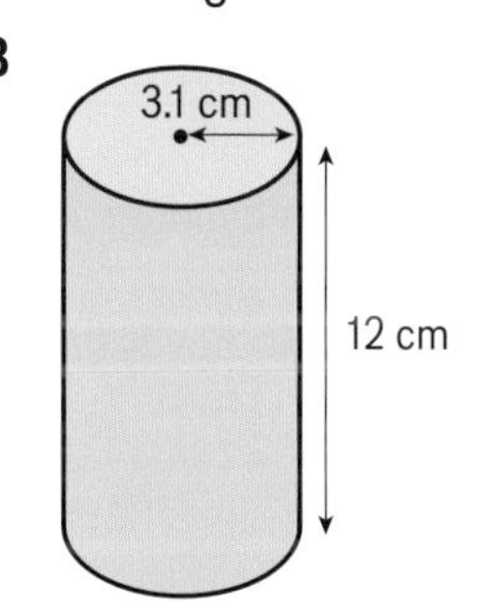

Exam-style question

13 A water butt is in the shape of a cylinder with a diameter of 510 mm and a height of 660 mm.
How many litres of water does the water butt contain when it is completely full?
1 litre = $1000\,cm^3$ **(4 marks)**

Exam tip

You will need to know the formula for the volume of a cylinder and understand how to use it.

14 **Problem-solving** A cylindrical oil drum has a radius of 50 cm and a height of 160 cm.

a Work out the volume of the oil drum.

The drum is completely filled with oil.
Oil has a density of $4.3\,g/cm^3$.

b Work out the mass of the oil in the tank in kilograms.

Q14b hint

Density = $\frac{\text{mass}}{\text{volume}}$

Key point

The **surface area** of a prism is the total area of all its faces.

Example

Work out the surface area of this cylinder.

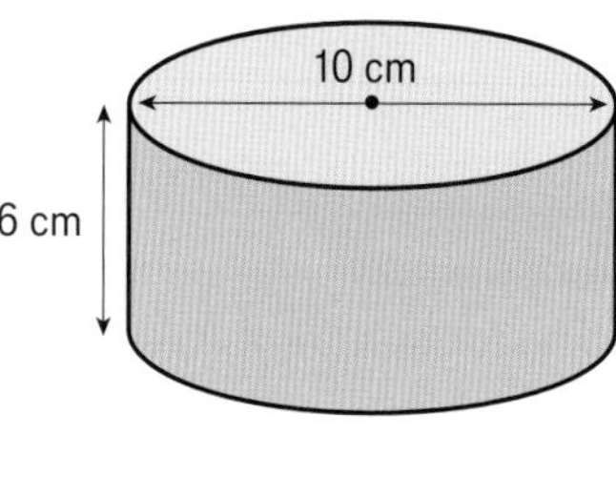

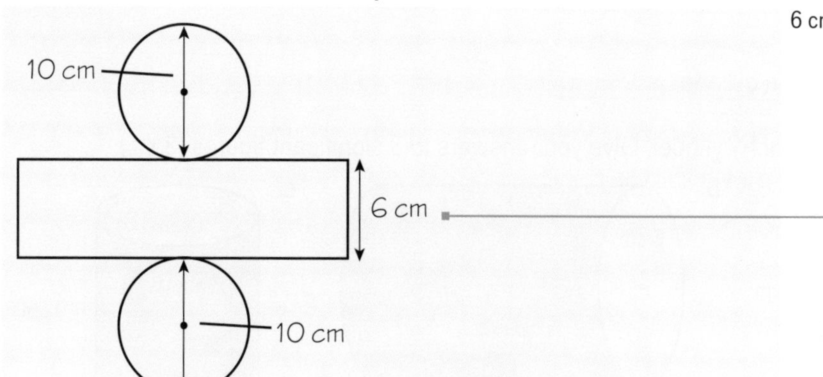

Sketch the net.

The length of the rectangle must be the same as the circumference of the circle.

Circle area $= \pi \times 5^2 = 78.53981634\,\text{cm}^2$

Circumference of circle $= \pi \times 10 = 31.41592654\,\text{cm}$

Rectangle area $= 6 \times 31.41592654 = 188.4955592\,\text{cm}$

6 × circumference of circle

Total surface area of cylinder $= 2 \times$ circle area + rectangle area

$= 2 \times 78.53981634 + 188.4955592$

Use all the digits in the calculation.

$= 345.5751919$

$= 345.6\,\text{cm}^2$ (to 1 d.p.)

Round the final answer to a suitable level of accuracy.

15 Work out the total surface area of each cylinder **A** and **B** in **Q12**.

16 Work out the area of the label needed to cover the curved surface of this tin can.

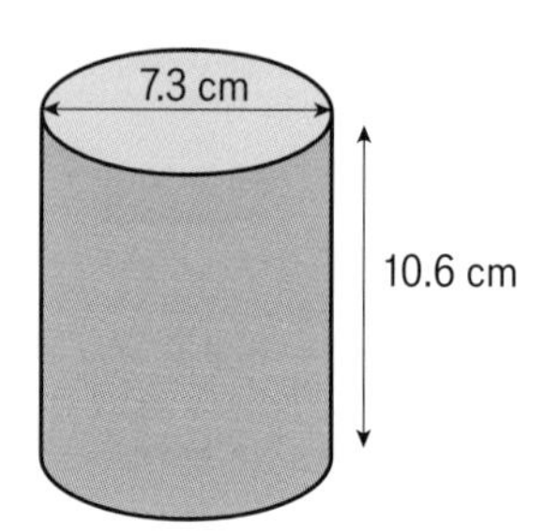

17 **Problem-solving** A polytunnel is used for growing vegetables. Its shape can be approximated to a half-cylinder. Its cross-section has a diameter of 10 feet and its length is 60 feet.

Work out in terms of π

a its volume in cubic feet

b its surface area in square feet

Q17b hint

The base of the half-cylinder is not part of the surface area of the polytunnel.

18 **Reflect** Write some tips to help you remember the formulae for the volume and surface area of a cylinder.

Q18 hint

Diagrams may help.

17.6 Pyramids and cones

- Work out the volume of a pyramid.
- Work out the surface area of a pyramid.
- Work out the volume of a cone.
- Work out the surface area of a cone.

Warm up

1 **Fluency** How many faces does a square-based pyramid have? What shapes are they?

2 Work out the area of each shape.

a

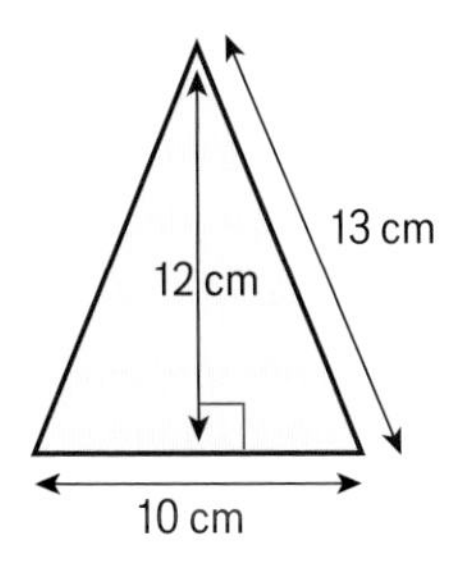

b

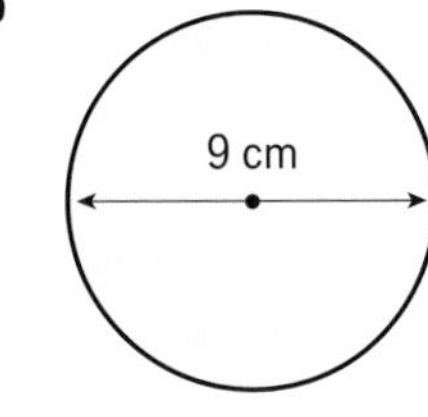

3 Write the answers in terms of π.

a $\frac{1}{3}$ of 18π b $5\pi \times 9$ c $\frac{1}{3} \times 6^2 \times \pi$ d $\frac{1}{3} \times 4^2 \times \pi \times 15$

Key point

The volume of a pyramid $= \frac{1}{3} \times$ area of base $\times$ vertical height

Example

A pyramid has a square base of side 7 cm and a vertical height of 12 cm. Work out the volume of the pyramid.

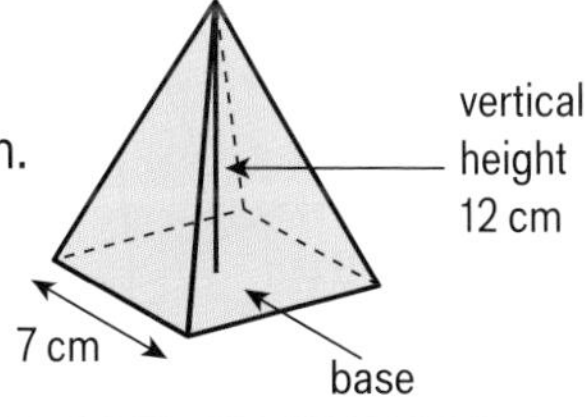

$$\text{Volume of pyramid} = \frac{1}{3} \times \text{area of base} \times \text{vertical height}$$
$$= \frac{1}{3} \times 7 \times 7 \times 12$$
$$= 196\text{ cm}^3$$

The base is a 7 cm square, so area $= 7 \times 7\text{ cm}^2$.
Height is 12 cm.

4 Work out the volume of each pyramid.

a

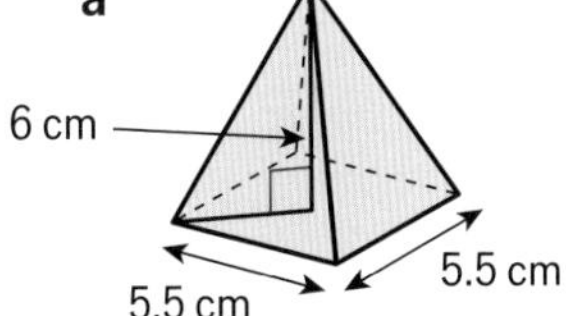

b

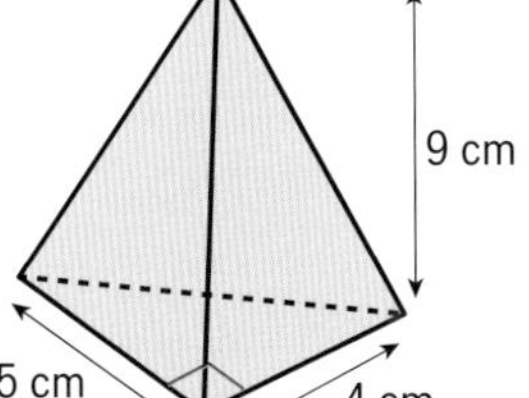

c

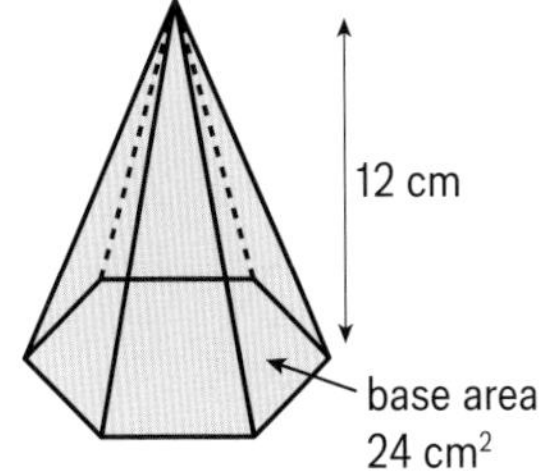

Q4b hint

The base is a triangle.

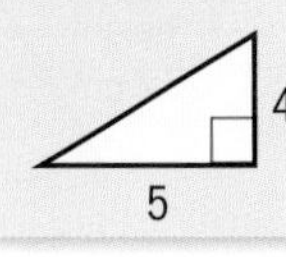

5 **Problem-solving** The Great Pyramid at Giza has a square base 475 yards by 475 yards.
Its height is 286 yards.
Work out the volume of the Great Pyramid. Give your answer in cubic yards.

Example

Work out the surface area of this square-based pyramid.

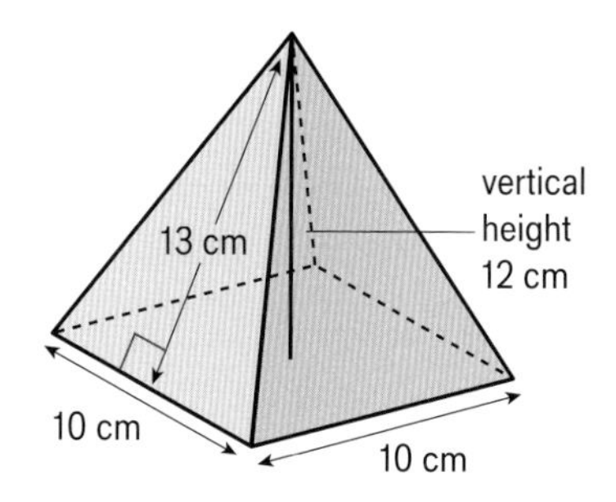

Surface area is the total area of all the faces.
The net has 1 square and 4 triangles.

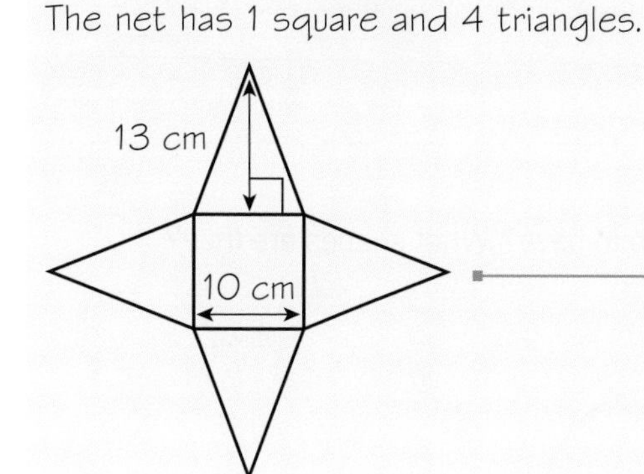

Sketch the net if it helps.

Area of a triangular face $= \frac{1}{2} \times 10 \times 13$
$= 65\text{ cm}^2$

Area of triangle $= \frac{1}{2}bh$

Area of 4 triangles $= 4 \times 65$
$= 260\text{ cm}^2$

Work out the area of all the triangular faces.

Area of base $= 10 \times 10$
$= 100\text{ cm}^2$

Work out the area of the base.

Total area $= 260 + 100 = 360\text{ cm}^2$

Work out the total surface area.

6 Work out the surface area of each of these square-based pyramids.

a

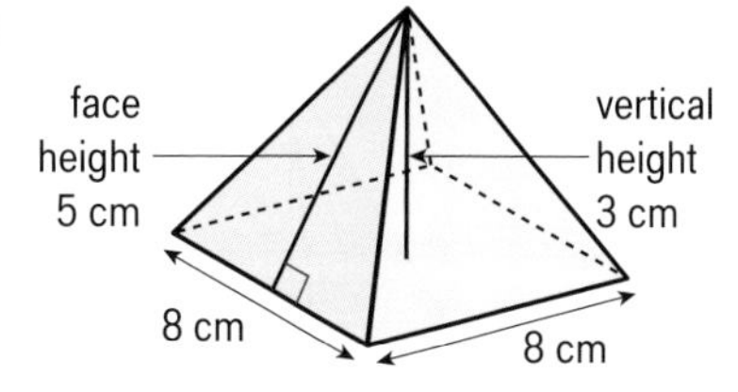

b

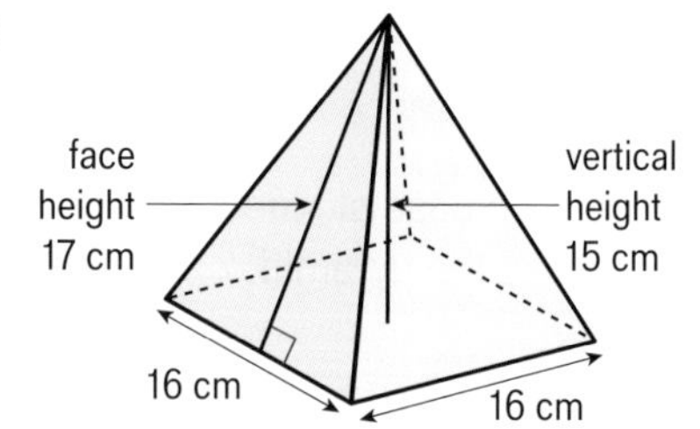

7 Work out the volume of each pyramid in **Q6**.

8 By rounding the dimensions to 1 significant figure (1 s.f.) work out an estimate of

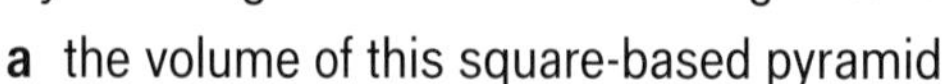

a the volume of this square-based pyramid

b the surface area of the pyramid

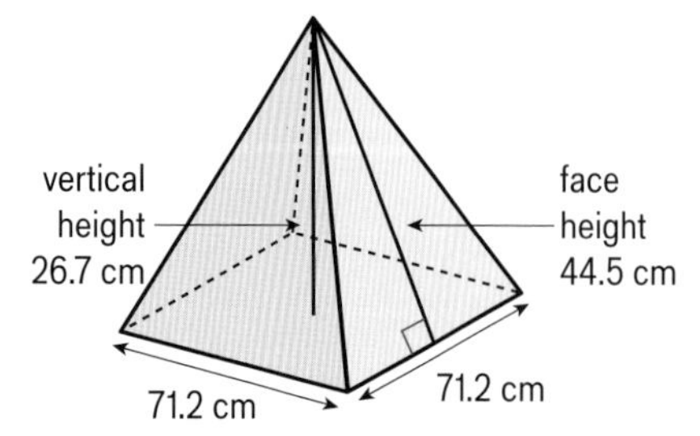

Key point

A cone is a pyramid with a circular base.
The volume of a cone = $\frac{1}{3}$ × area of base × vertical height

9 **a** What shape is the base of a cone?

b What is the formula for the area of a circle of radius r?

c **Reasoning** Write the formula for the volume, V, of a cone with base radius r and height h.

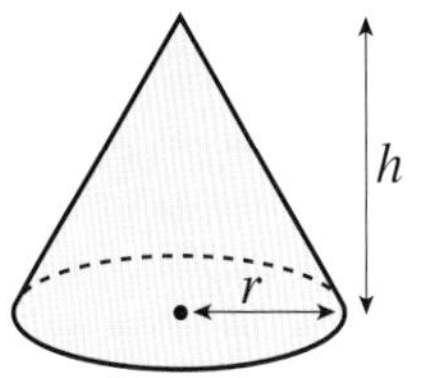

10 Work out the volume of each cone.

a

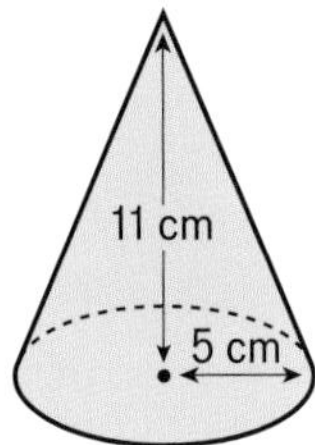

b

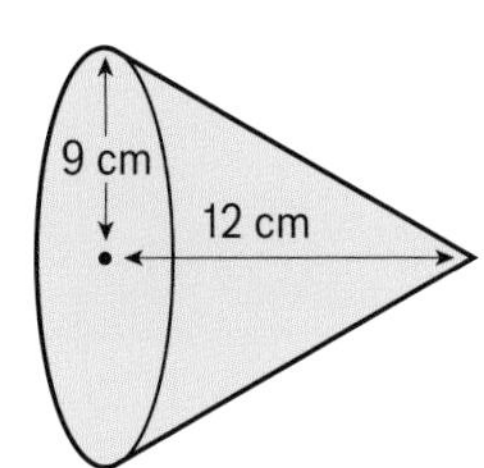

c

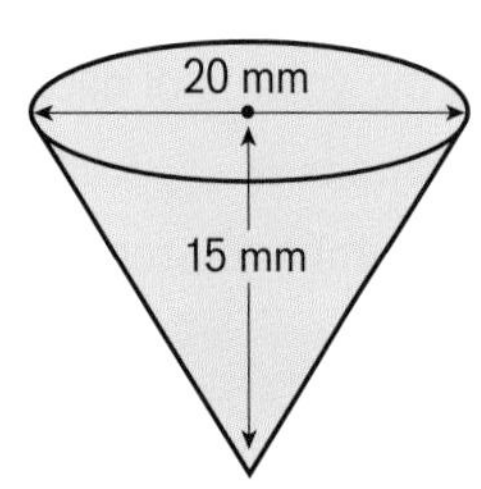

11 **Problem-solving** A paper cone cup has a diameter of 8 cm and a height of 10 cm.

a What is the radius of the base of the cone?

b What is the volume of the cone? Give your answer correct to 3 s.f.

c How many of these cups can be filled from a 5 litre container?

Q11c hint

$1\text{ cm}^3 = 1\text{ ml}$
1 litre = 1000 ml

Key point

The area of the curved surface of a cone = π × base radius × slant height = πrl
The **slant height** of a cone is the length of the sloping side.

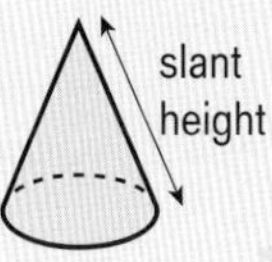

Example

Work out the total surface area of this cone in terms of π.

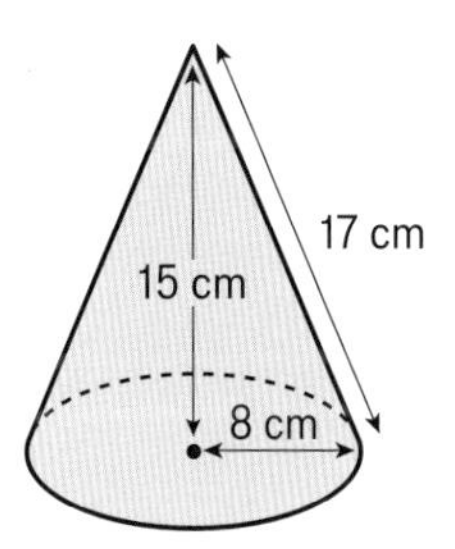

Area of curved surface = π × base radius × slant height
$= \pi \times r \times l$
$= \pi \times 8 \times 17$ — Use the slant height, not the vertical height.
$= 136\pi$

Area of base $= \pi \times 8^2$
$= 64\pi$

Total area $= 136\pi + 64\pi$ — Total surface area = area of curved surface + area of base
$= 200\pi\text{ cm}^2$

12 Work out the total surface area of each cone

i in terms of π

ii as a number correct to 3 s.f.

a

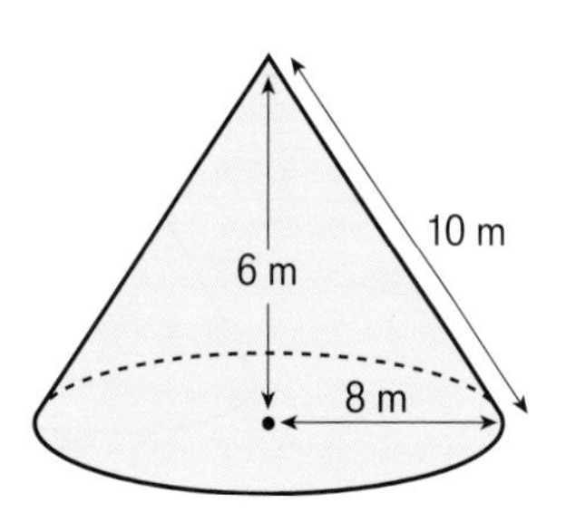

b

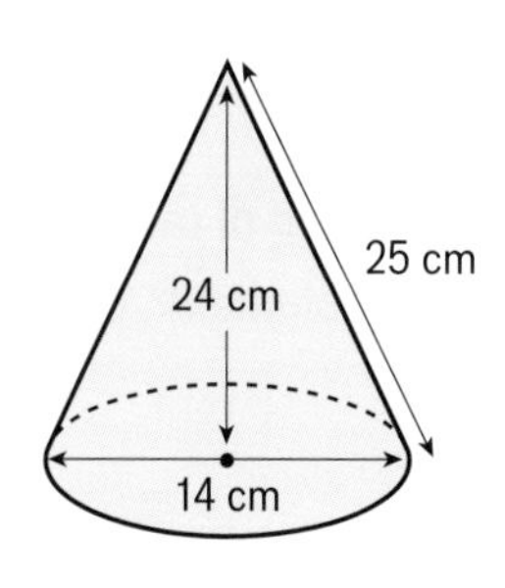

c

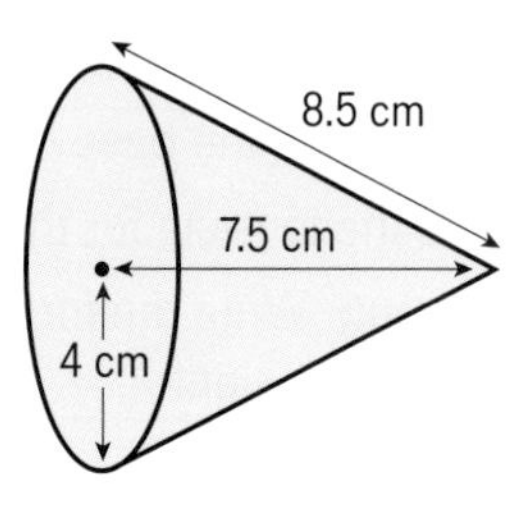

13 Work out the volume of each cone in **Q12**

i in terms of π

ii as a number correct to 3 s.f.

Exam-style question

14 The diagram shows the cone-shaped roof of a building.

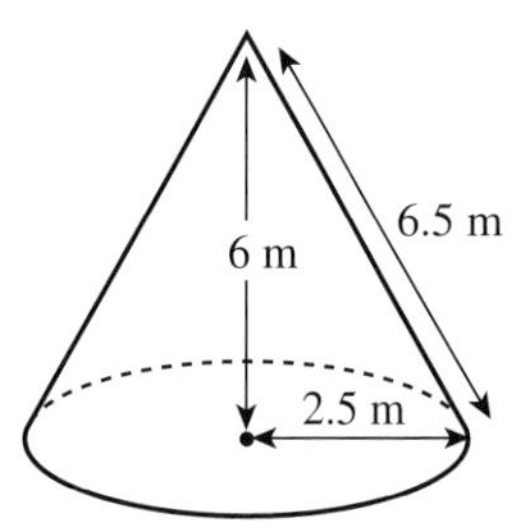

A litre of paint covers $15\,\text{m}^2$.
How many litres of paint does Uzma need to paint the curved surface of the roof? **(4 marks)**

Curved surface area of cone $= \pi rl$

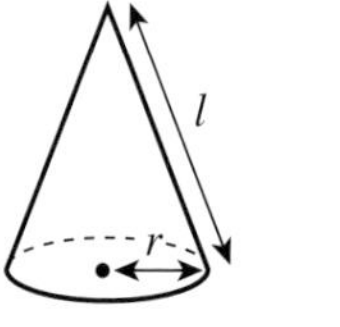

Exam tip

Always read the question carefully. Here you need to find the area of the curved part of the cone, not the total surface area.

15 a Use Pythagoras' theorem to work out the slant height of the cone.
Copy and complete the working.
$3^2 + 4^2 = \square$
Slant height, $l = \sqrt{\square} = \square$

b Work out the area of the curved surface in terms of π.

c Work out the area of the base in terms of π.

d Work out the total surface area of the cone in terms of π.

e Work out the total surface area of the cone as a decimal correct to 1 d.p.

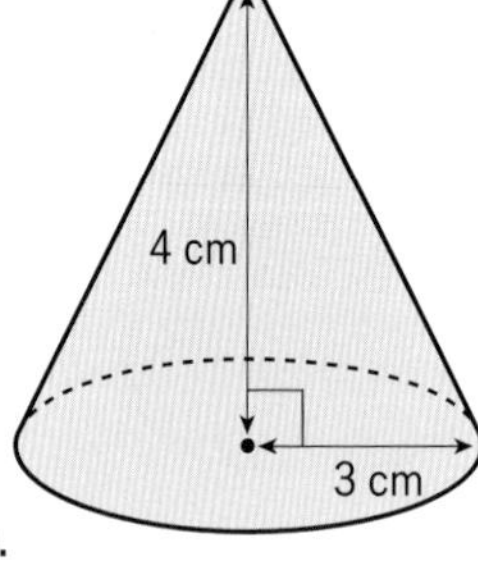

16 Work out the total surface area of the cone correct to 3 s.f.

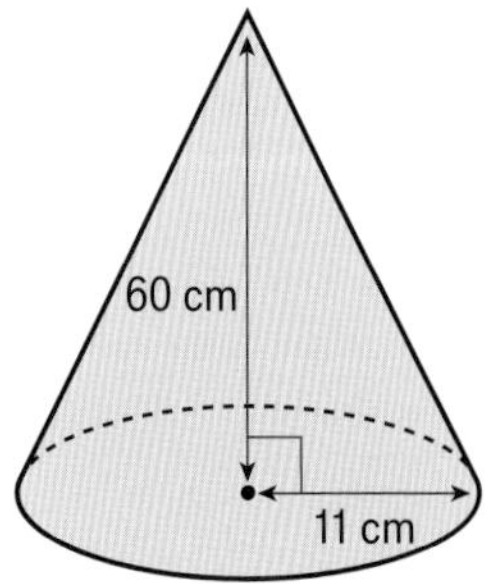

17.7 Spheres and composite solids

- Work out the volume and surface area of a sphere.
- Work out the volume and surface area of composite solids.

ActiveLearn Homework

Warm up

1 **Fluency** Write down the formulae for

a the volumes of a cuboid, pyramid, cylinder and cone

b the area of a circle and the curved surface area of a cone

2 Work out

a $\frac{1}{3}$ of 24 b $\frac{4}{3}$ of 24 c 2^3 d 4×6^2

3 Use a calculator to work out

a $\frac{4}{3} \times 6$ b $\frac{4}{3} \times 2^3$ c $\frac{4}{3} \times \pi$ d $\pi \times 5.2^2$

Key point

A **sphere** is a solid where all points on the surface are the same distance from the centre.

The volume of a sphere $= \frac{4}{3}\pi r^3$

Spherical means in the shape of a sphere.

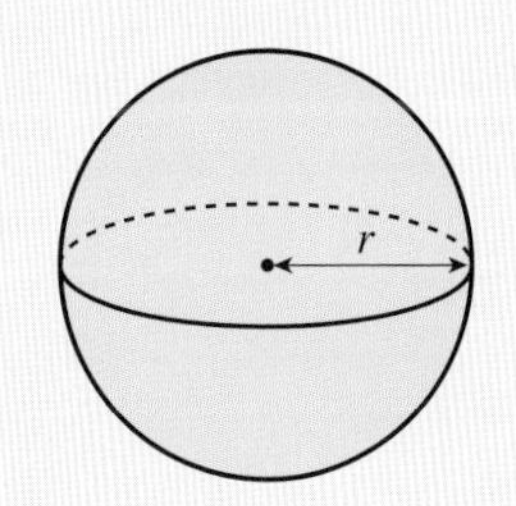

Example

The radius of a spherical ball is 11 cm.
Work out the volume of the ball to the nearest cm^3.

Volume of ball $= \frac{4}{3}\pi r^3$

$= \frac{4}{3} \times \pi \times 11^3$ — Put this in your calculator.

$= 5575.279763$ — Give the answer in full first.

$= 5575\text{ cm}^3$ — Round to the nearest cm^3. Include the units.

4 A spherical marble has a radius of 1.2 cm.
Work out the volume of the marble

a in terms of π

b to the nearest cm^3

5 **Problem-solving** A spherical snooker ball has a diameter of 52.5 mm.
Work out the volume of the snooker ball to the nearest mm^3.

Q5 hint
Work out the radius first.

6 **Reasoning** A hemisphere has a radius of 6 m.
Moira says the volume is 144π m^3.
Jill says the volume is 36π m^3.
Who is correct? What mistake has the other person made?

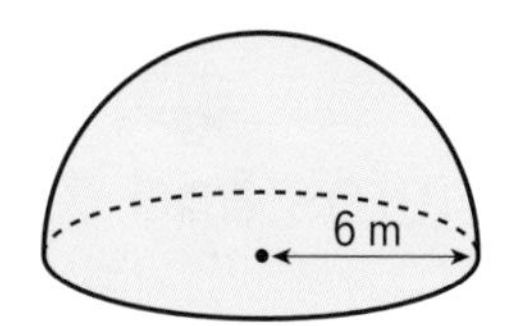

Key point

The surface area of a sphere $= 4\pi r^2$

7 Work out the surface area of the marble in **Q4**.
Give your answer to the nearest cm^3.

8 **Problem-solving** The planet Venus can be modelled as a sphere with a diameter of 12 104 km.
By rounding all values to 1 significant figure (1 s.f.), work out an estimate of the surface area of the planet Venus.

Example

Work out the total surface area of a hemisphere with radius 15 cm.
Give your answer in terms of π.

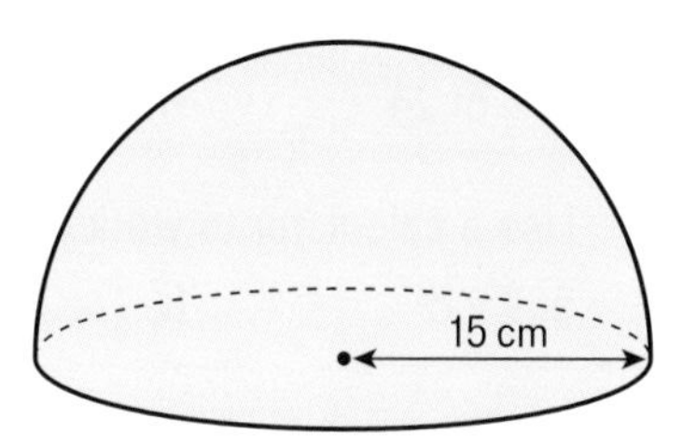

Surface area of sphere $= 4 \times \pi \times 15^2$
$= 900\pi$

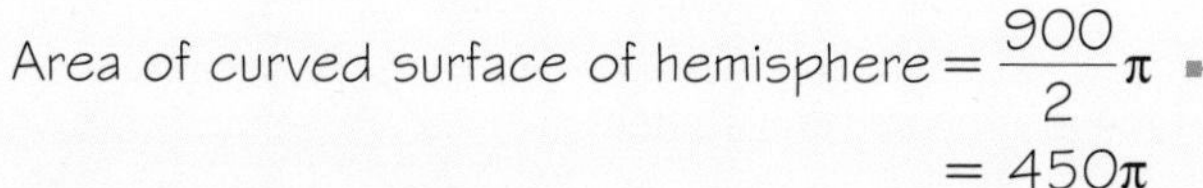

Area of curved surface of hemisphere $= \frac{900}{2}\pi$
$= 450\pi$

Halve the answer as a hemisphere is half a sphere.

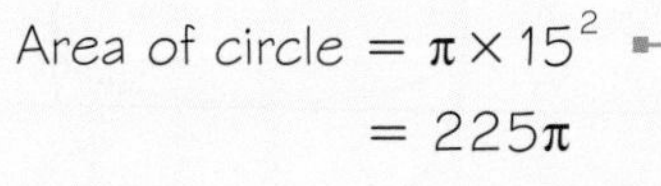

Area of circle $= \pi \times 15^2$
$= 225\pi$

The base of the hemisphere is a circle with radius 15 cm.

Total surface area $= 450\pi + 225\pi$
$= 675\pi$ cm^2

The total area is the curved surface area + the circle area.

9 Work out the total surface area of the hemisphere in **Q6**. Give your answer

a in terms of π

b to 2 decimal places (2 d.p.)

10 **Problem-solving** Work out the volume of this solid.

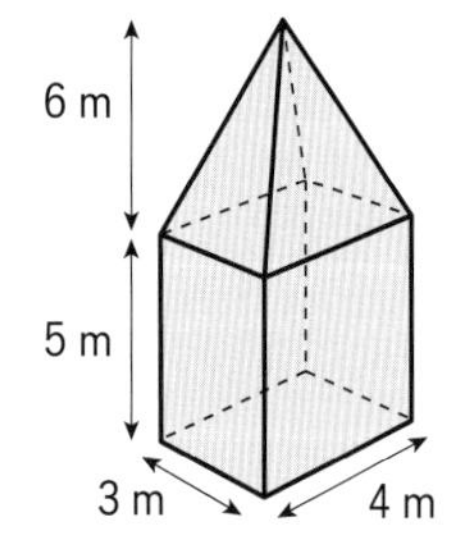

Q10 hint

Work out the volumes of the pyramid and the cuboid separately.

Exam-style question

11 The diagram shows a storage tank.

The storage tank consists of a hemisphere on top of a cylinder.
The height of the cylinder is 20 metres.
The radius of the cylinder is 2 metres.
The radius of the hemisphere is 2 metres.
Calculate the total volume of the storage tank.
Give your answer correct to 3 significant figures.

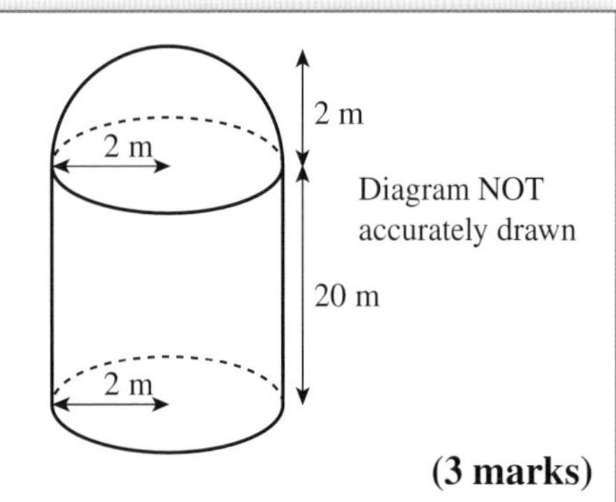

(3 marks)

Exam tip

Do not round the answers to any calculations until you get your final answer.

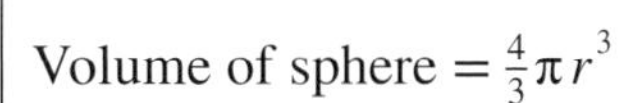

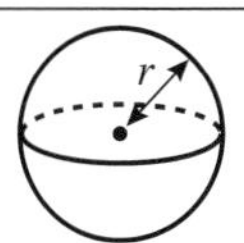

12 Work out the volume of each solid. Give your answers in terms of π.

a

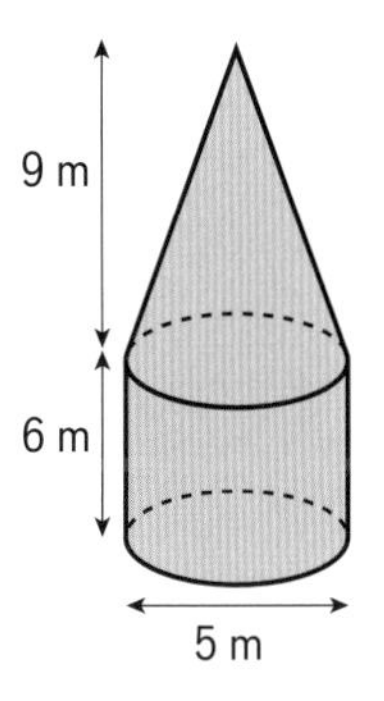

b

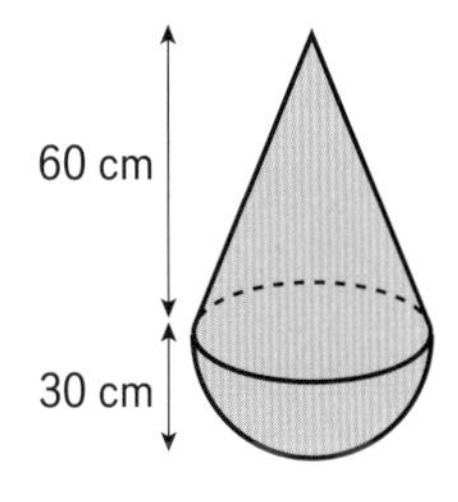

c

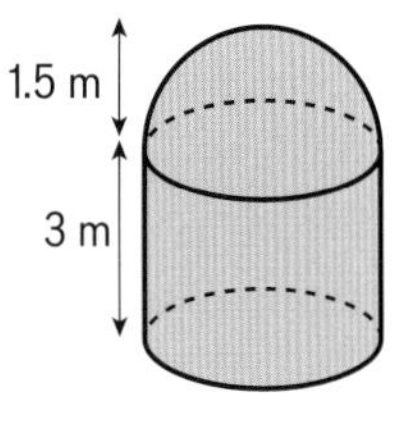

13 Problem-solving Work out the total surface area of this solid

a in terms of π

b as a decimal correct to 1 d.p.

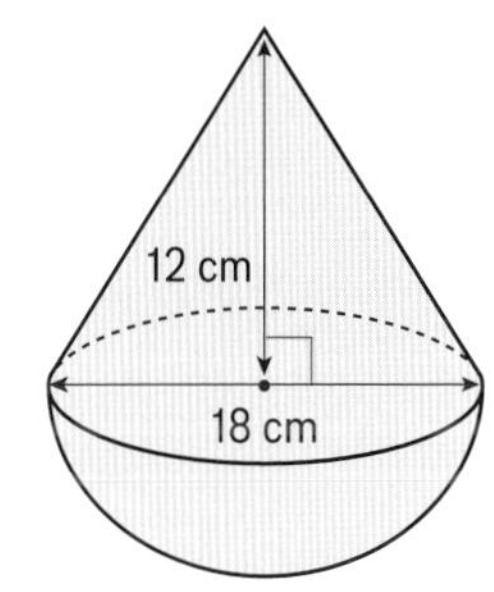

Q13b hint

The circle where the cone and the hemisphere join is not part of the surface area.

14 Work out the total surface area of each solid

i in terms of π

ii correct to 3 s.f.

a

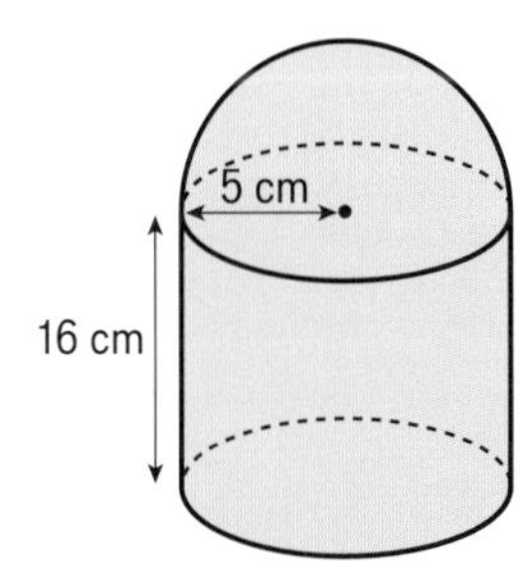

b

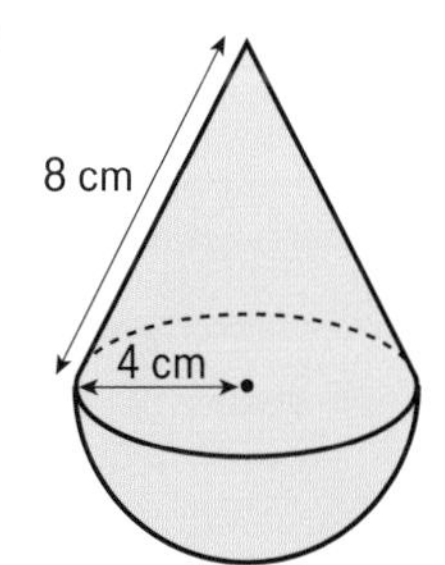

c

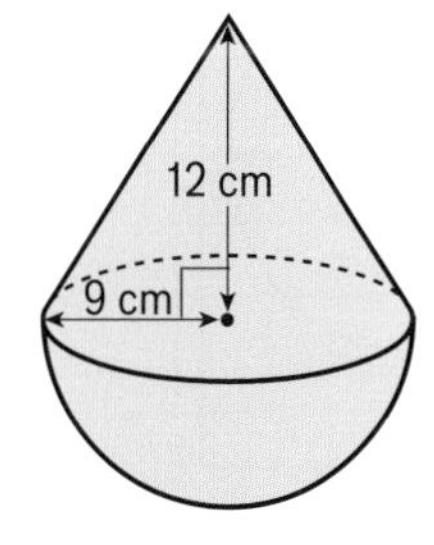

15 Reflect What did you find easiest in this lesson? What was most difficult? What made them easy/difficult?

17 Check up

ActiveLearn Homework

Accuracy

1 **a** A measurement, m, is rounded to the nearest cm.
The result is 11 cm. Using inequalities, write down the error interval for m.

b Write an inequality for the error interval of 5.32 m rounded to 2 decimal places (2 d.p.).

2 The length of a pen is 14 cm correct to the nearest cm.
The length of a box is 142 mm correct to the nearest mm.
Explain why the pen may not fit in the box.

Circles and sectors

3 Work out the circumference of a circle with a radius of 6 cm

a in terms of π **b** as a number correct to 1 d.p.

4 The circumference of a circle is 44 cm.
Work out the diameter of the circle. Give your answer to a sensible degree of accuracy.

5 Work out the area of a circle with radius 5 cm.
Give your answer correct to 3 significant figures (3 s.f.).

6 A circle has area 49π m^2. Work out its radius.

7 The diagram shows a semicircle with a radius of 16 cm.
Work out correct to 3 s.f.

a its perimeter **b** its area

16 cm

8 Work out the shaded area correct to the nearest mm^2.

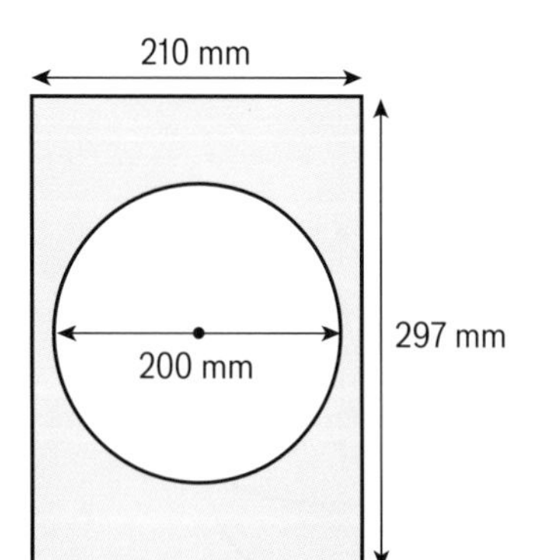

9 The diagram shows a sector with a radius of 7 cm and an angle of 45°.
Work out to a suitable accuracy

a the area of the sector

b the length of the arc AB

c the perimeter of the sector

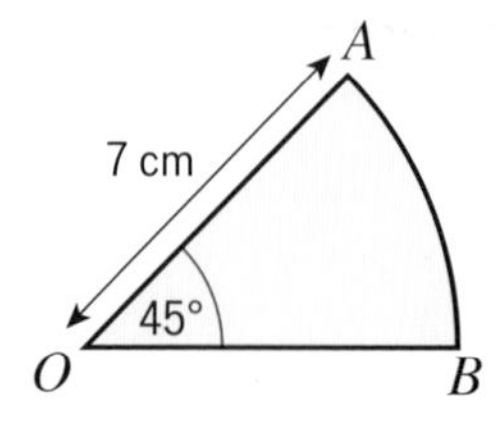

Volumes and surface areas

10 Work out the volume of each of these solids. Give your answers

i in terms of π **ii** correct to 3 s.f.

a

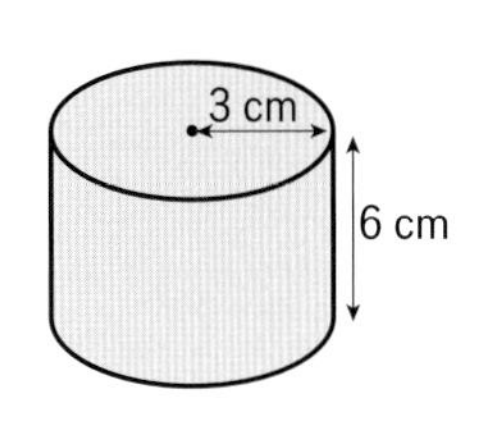

b

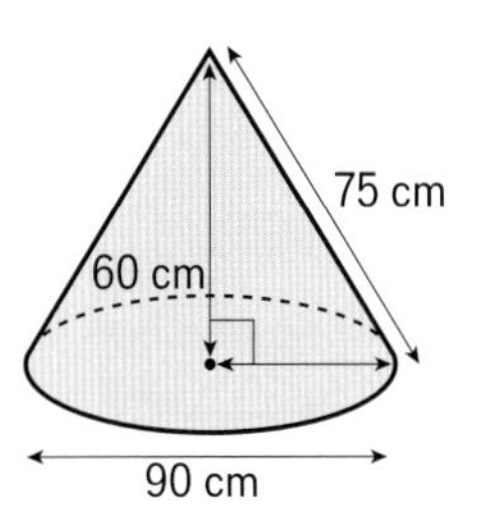

Volume of sphere $= \frac{4}{3}\pi r^3$

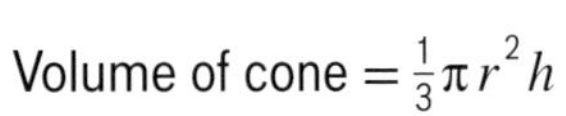

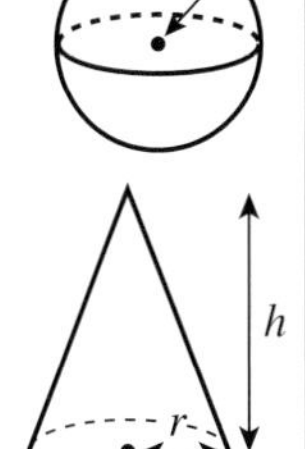

c

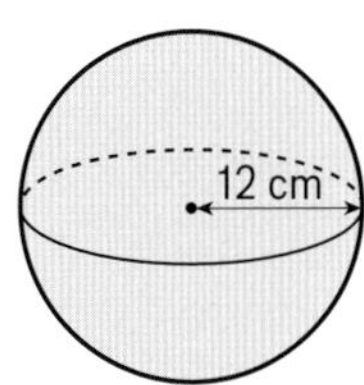

11 Work out the total surface area of each solid in **Q10**.
Give your answers correct to 3 s.f.

12 Work out the volume and surface area of this rectangular-based pyramid.
Give your answers correct to the nearest cm^3 or cm^2.

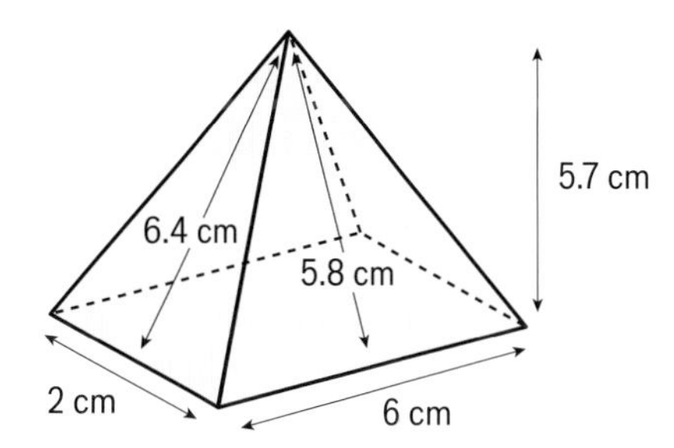

13 **Reflect** How sure are you of your answers? Were you mostly

Just guessing ☹ Feeling doubtful 😐 Confident ☺

What next? Use your results to decide whether to strengthen or extend your learning.

Challenge

14 **a** Work out the volume of this cone and this cylinder, in terms of π.

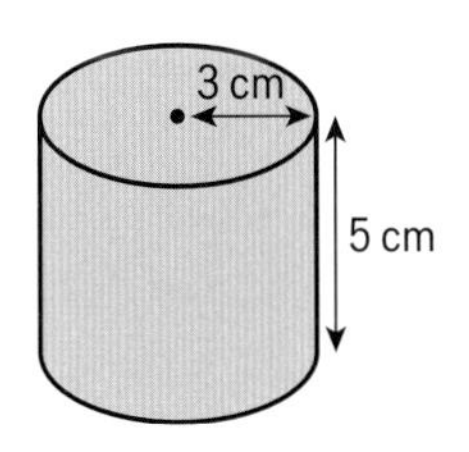

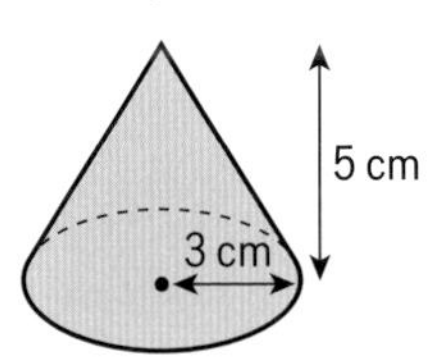

b What is the relationship between the volume of a cone and a cylinder of equal base radius and height?

c Explain how the formulae for the volume of a cone and the volume of a cylinder show this relationship.

15 A circle has a diameter of $(x + 9)$ cm and an area of $36\pi\,cm^2$.
Work out the value of x.

17 Strengthen

*Active*Learn Homework

Accuracy

1 a What is the smallest possible value for 4 cm, rounded to the nearest cm?

b Write an inequality $\square \leqslant n < \square$ for the error intervals for

i $n = 4$ cm, rounded to the nearest cm

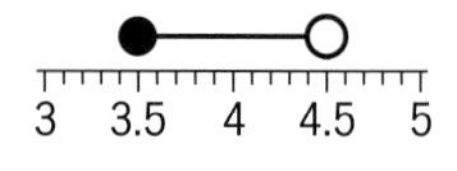

ii $n = 2.5$ cm, rounded to 1 decimal place (1 d.p.)

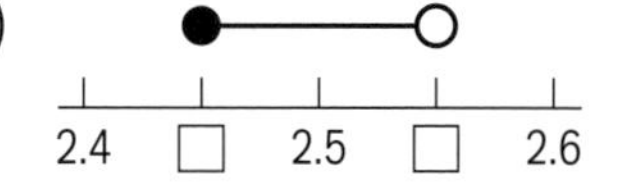

iii $n = 5.46$ cm, rounded to 2 d.p.

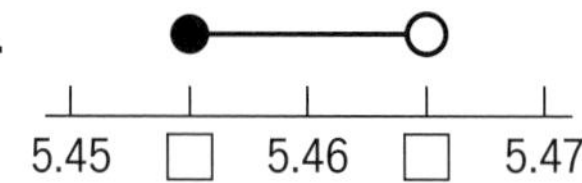

Circles and sectors

1 Use the formula with diameter d in it to work out the circumference of a circle with diameter 12 cm

a in terms of π, $C = \square\pi$

b correct to 1 d.p.

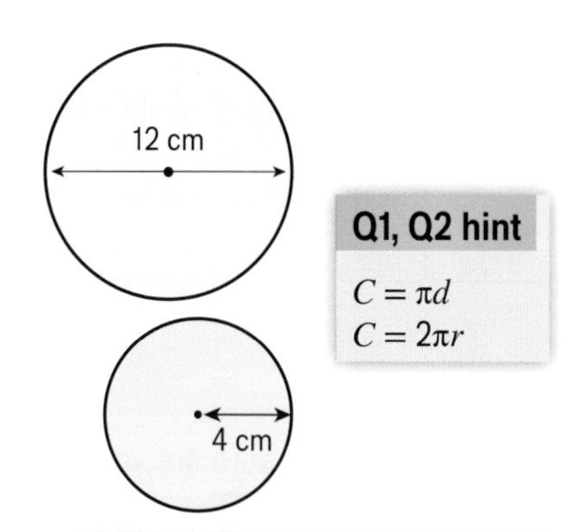

Q1, Q2 hint

$C = \pi d$

$C = 2\pi r$

2 Use the formula with radius r in it to work out the circumference of a circle with radius 4 cm

a in terms of π

b as a number correct to 1 d.p.

3 The circumference of a circle is 66 cm.

a Write down the formula connecting the circumference C and the diameter d.

b Work out the diameter of the circle. Give your answer to the nearest cm.

Q3b hint

Use an inverse function machine.

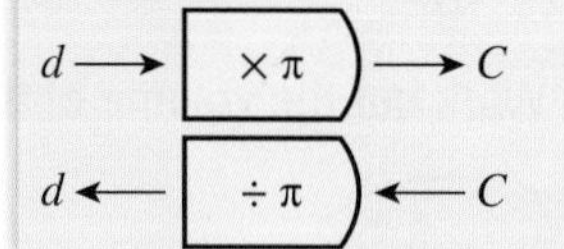

4 A circle has a diameter of 20 cm.

a Work out the radius.

b Copy and complete: Area $= \pi \times \square^2$

c Work out the area of the circle correct to 1 d.p.

Q4b hint

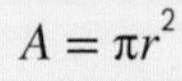

$A = \pi r^2$

5 A circle has area 25π m^2. Use an inverse function machine to work out

a its radius

b its diameter

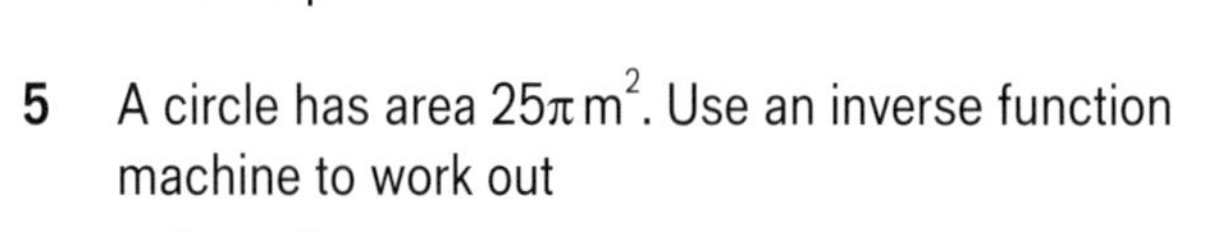

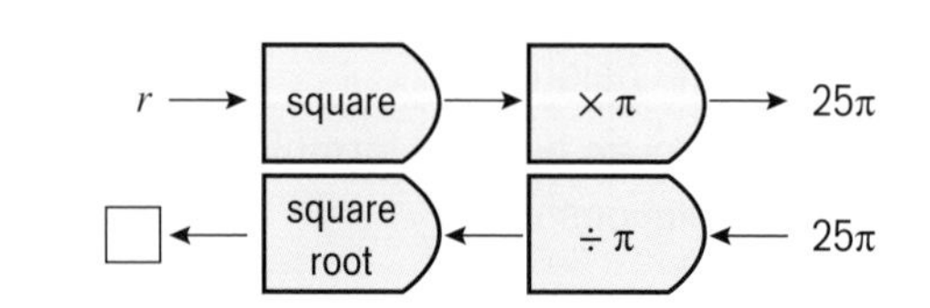

6 What fraction of each circle is shaded?

a

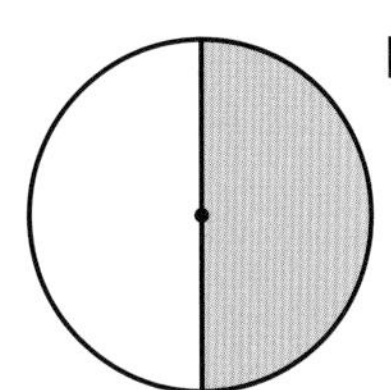

b

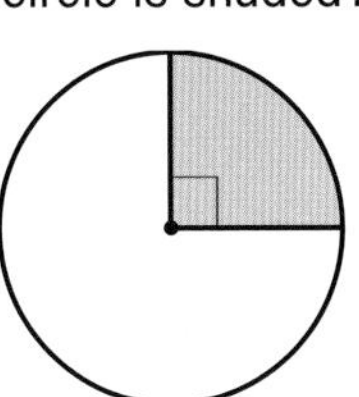

c

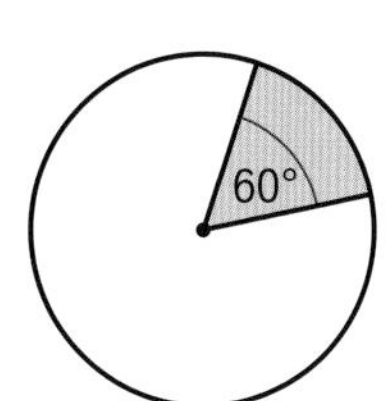

d

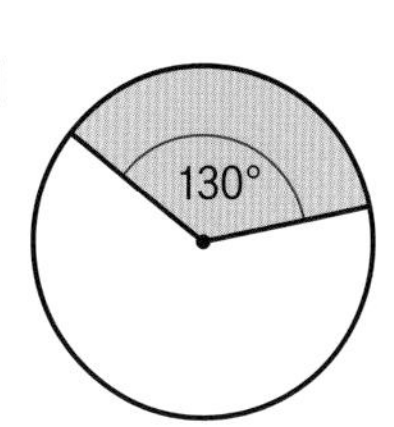

Q6c hint

$\frac{\square}{360} = \frac{\square}{\square}$

7 The diagram shows a semicircle with a radius of 10 cm.

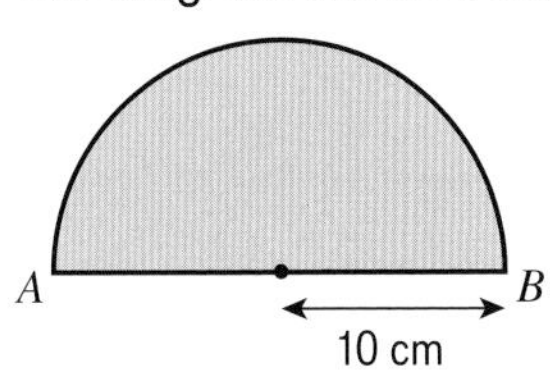

a Work out the area and circumference of a whole circle with radius 10 cm.

b Work out in terms of π

i the area of the semicircle

ii the arc length AB

iii the diameter AB

iv the perimeter of the semicircle

Q7b i and ii hint

What fraction of the circle is it?

Q7b iv hint

Perimeter =
arc length + diameter

8 Work out, to the nearest cm^2, the area

a of the rectangle

b of the circle

c of the shaded part

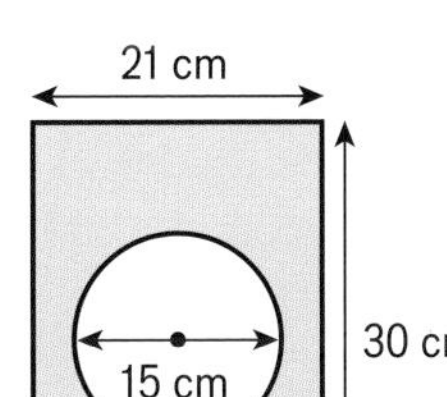
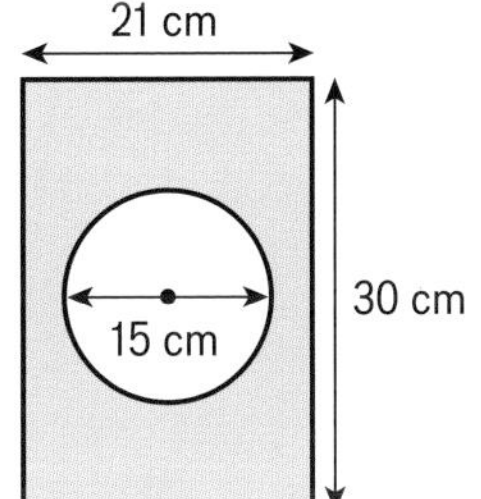

Q8c hint

Shaded area =
area of rectangle – area of circle

9 **a** Work out the circumference of a circle with radius 3 cm.

b What fraction of the whole circle is this sector?

c Work out the length of the arc AB.

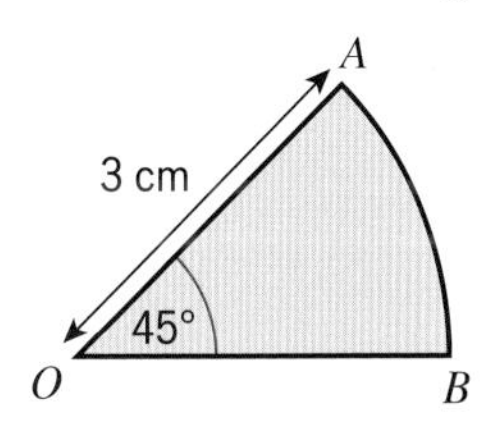

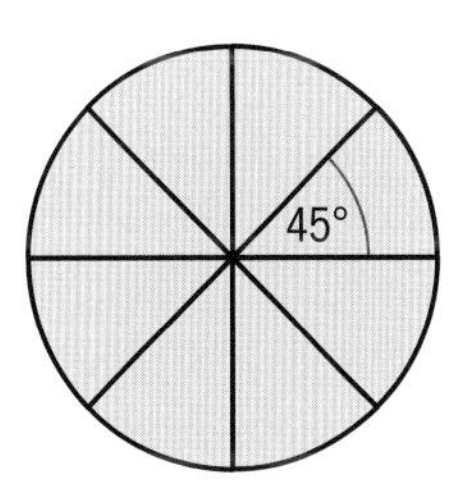

10 Work out in terms of π

a the area of a whole circle with radius 12 cm

b the area of the sector

c the circumference of a whole circle with radius 12 cm

d the arc length of the sector

e the perimeter of the sector

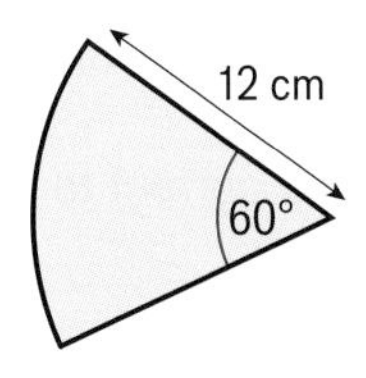

Q10a hint

$A = \square^2 \times \pi$

Volumes and surface areas

1 For this cylinder

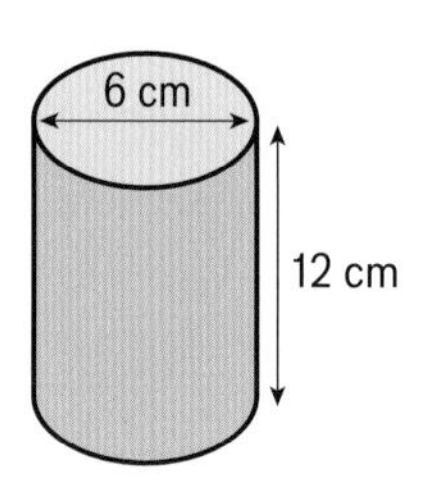

a sketch the cross-section and label the radius

b work out the area of the cross-section

c work out the volume (area of cross-section × height) to the nearest whole number

2 **a** Sketch the faces of this cylinder.

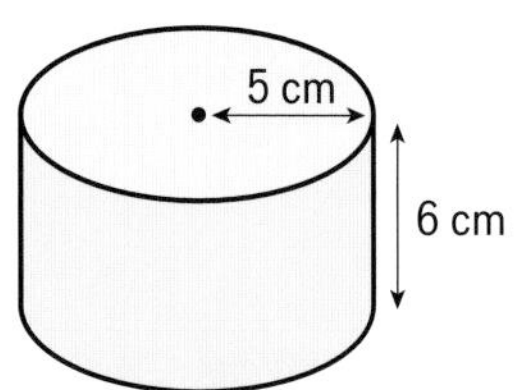

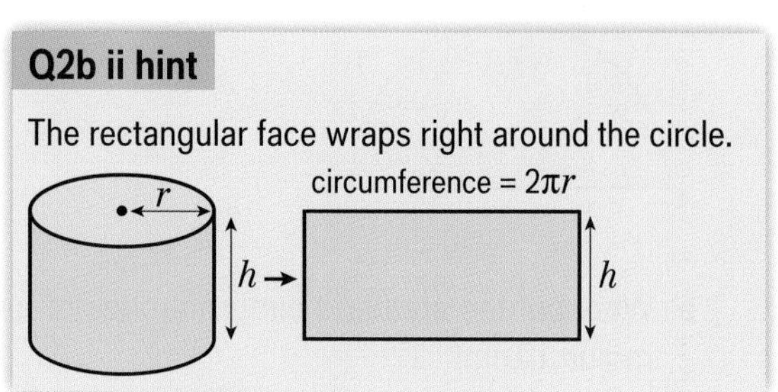

b Work out

i the area of the circle

ii the length of the rectangle

iii the area of the rectangle

iv the total surface area

3 For this rectangular-based pyramid

a work out the area of the base

b write down the vertical height

c work out the volume

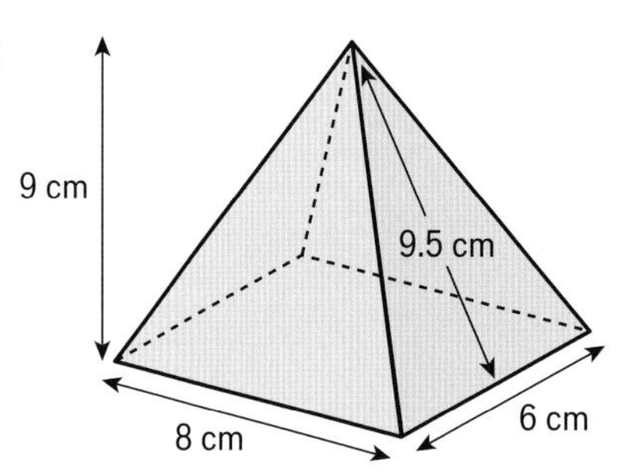

Q3c hint

Volume = $\frac{1}{3}$ × area of base × vertical height

4 For this cone

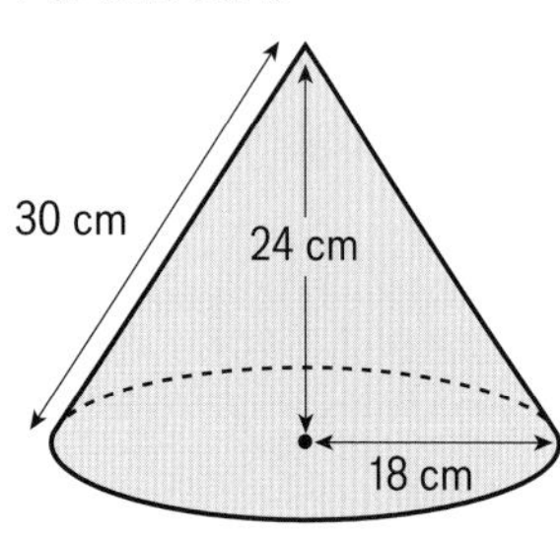

Volume of cone $= \frac{1}{3}\pi r^2 h$

Curved surface area of cone $= \pi rl$

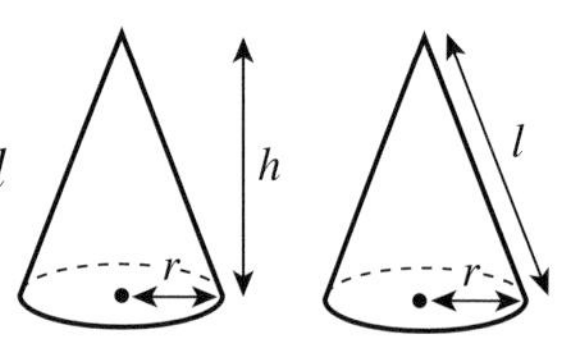

a work out the area of the base in terms of π

b decide which value is the vertical height

c work out the volume correct to 3 s.f.

d work out the curved surface area in terms of π

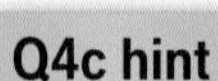

Volume = $\frac{1}{3}$ × area of base × vertical height

5 **a** Work out the volume of this sphere correct to 1 d.p.

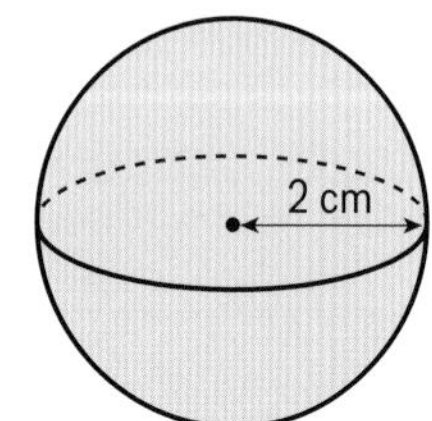

Volume of sphere $= \frac{4}{3}\pi r^3$

Surface area of sphere $= 4\pi r^2$

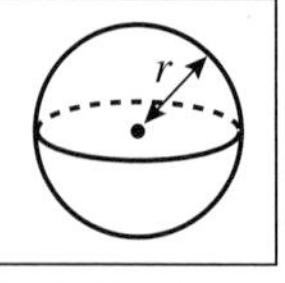

Q5 hint

Volume is in cm^3: $\frac{4}{3}\pi r^3$
Area is in cm^2: $4\pi r^2$

b Work out the surface area of the sphere correct to 1 d.p.

17 Extend

1 A 170 g tube of sweets contains about 100 sweets.
There is an error of ±10% in the mass, m, of the tube of sweets. There is an error of ±5% in the number of sweets.

a Work out the maximum and minimum possible masses of the tube of sweets.

b Work out the maximum and minimum numbers of sweets.

Q1a hint

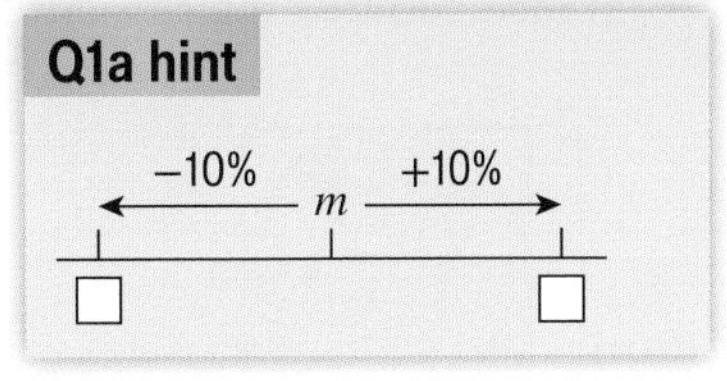

2 Mike's pond is a circle of diameter 5.7 m.
Pond edging is sold in 1.2 m strips.
Each strip costs £6.98.
Work out the total cost of the edging for the pond.

Q2 hint

Start by working out the length of edging needed to go around the pond.

3 **Reasoning** The line joining (−2, 5) and (6, 5) is the diameter of a circle.
Work out the area of the circle in square units correct to 1 decimal place (1 d.p.).

Q3 hint

Work out the length of the line first.

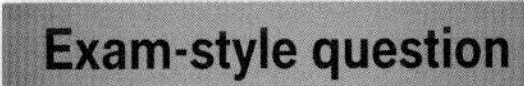

Exam-style question

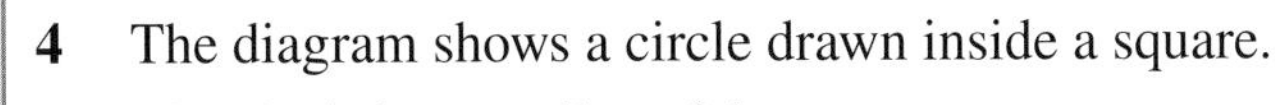

4 The diagram shows a circle drawn inside a square.
The circle has a radius of 4 cm.
The square has a side of length 8 cm.
Work out the shaded area.
Give your answer to 3 significant figures (3 s.f.). **(3 marks)**

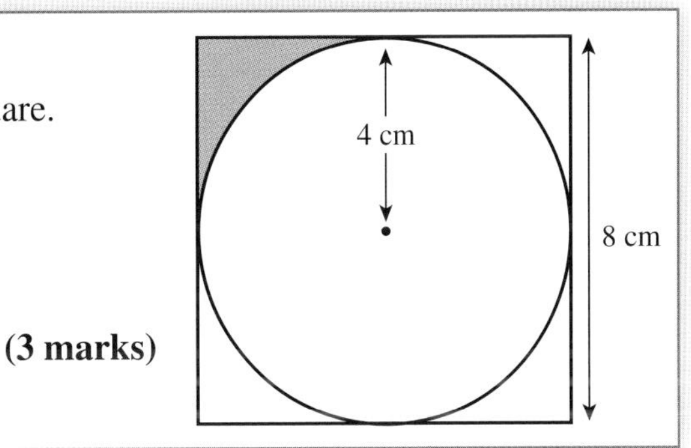

5 Georgia walks once round a circle with diameter 50 metres.
There are 5 points equally spaced on the circumference of the circle.

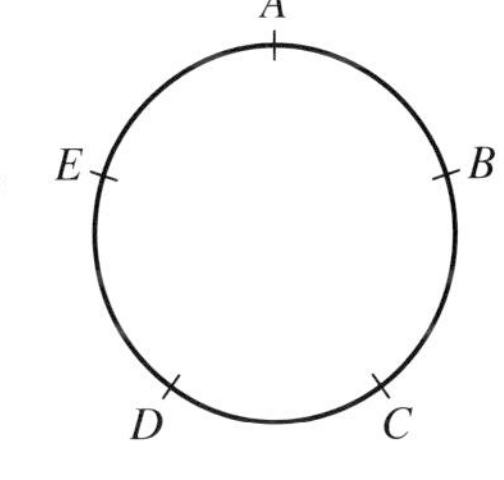

a Find the distance Georgia walks between one point and the next point.

Two of the points are moved, as shown in the diagram below.

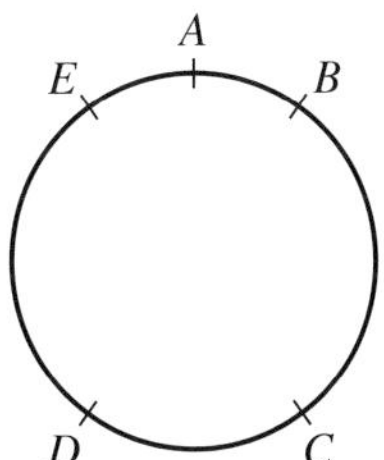

Georgia walks once round the circle again.

b Has the mean distance that Georgia walks between one point and the next point changed?
You must give a reason for your answer.

Exam-style question

6 The diagram shows a rectangle *CDEF* with sides of length 16 cm and 8 cm.
The diagram shows a semicircle with diameter *AB*.
It also shows a quarter circle with centre *B*.

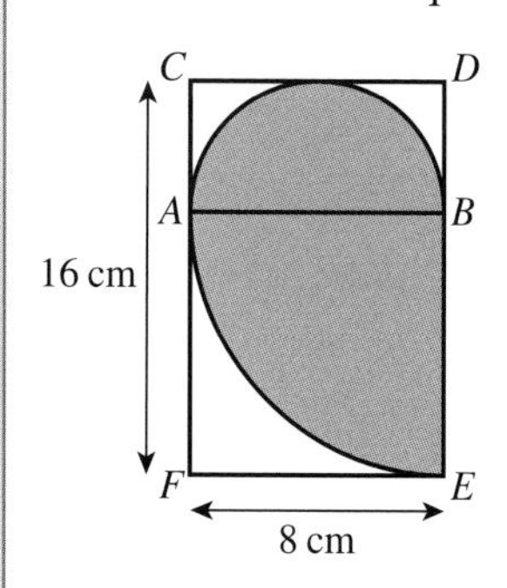

Show that $\frac{\text{shaded area}}{\text{area of rectangle}} = \frac{3\pi}{16}$ **(4 marks)**

Exam tip

To give an answer in terms of π, calculate areas in terms of π.

7 **Reasoning** The diagram shows a vase in the shape of a cylinder.

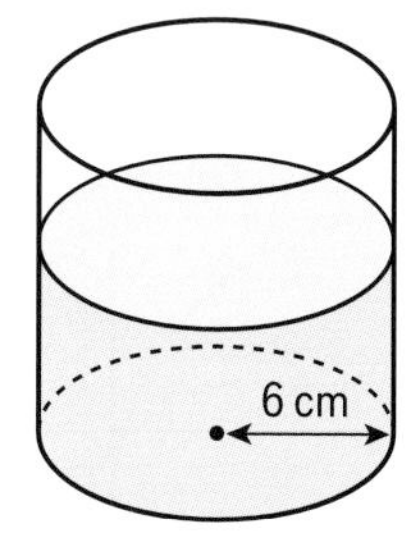

Q7 hint

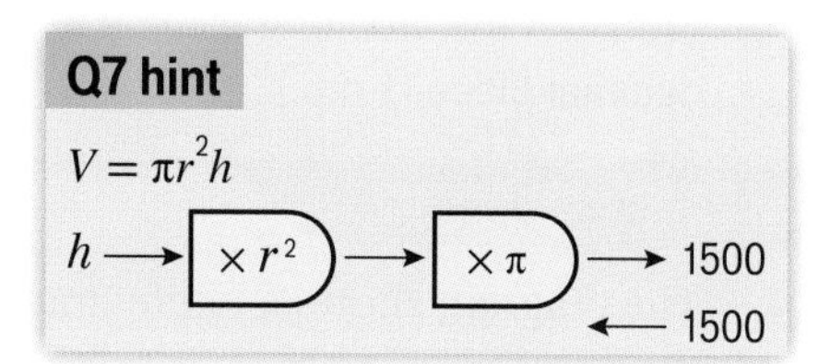

The vase has a radius of 6 cm. There is 1500 cm^3 of water in the vase.
Work out the depth of the water in the vase. Give your answer correct to 1 d.p.

8 **Problem-solving** A water tank is in the shape of a cylinder with radius 35 cm and depth 140 cm.
It is filled at the rate of 0.4 litres per second.
Does it take longer than 1 hour to fill the tank? Show your working.

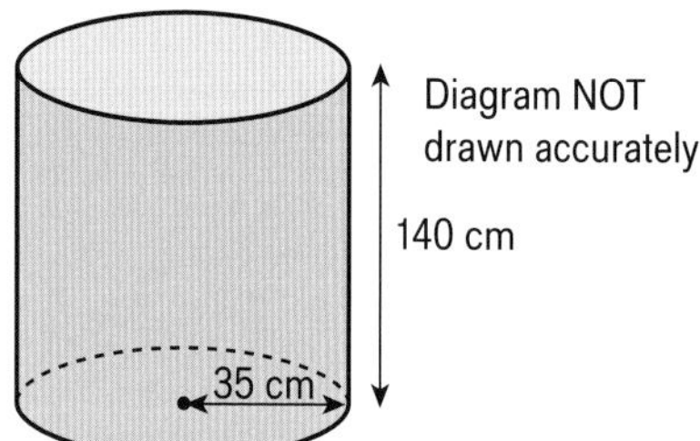

9 The diagram shows a solid made from two cones joined together.
Work out in terms of π

a the total surface area of the solid

b the volume of the shape

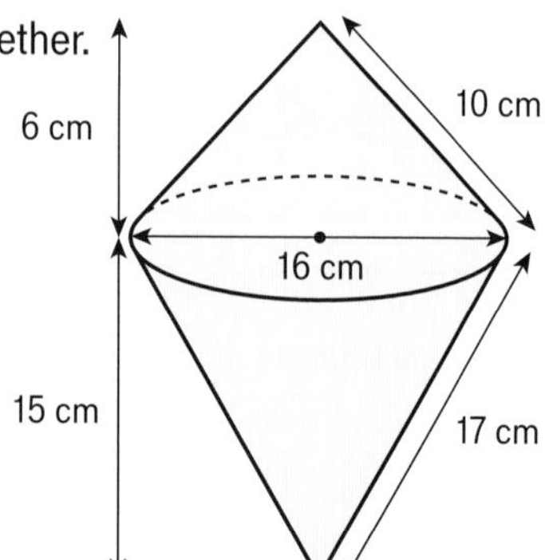

Exam-style question

10 The diagram shows a cone and a cylinder.

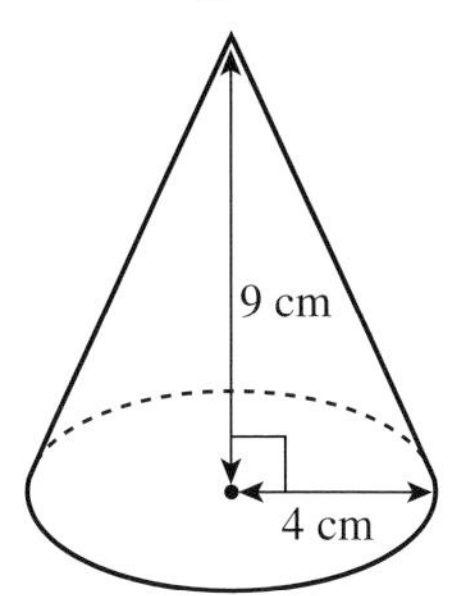

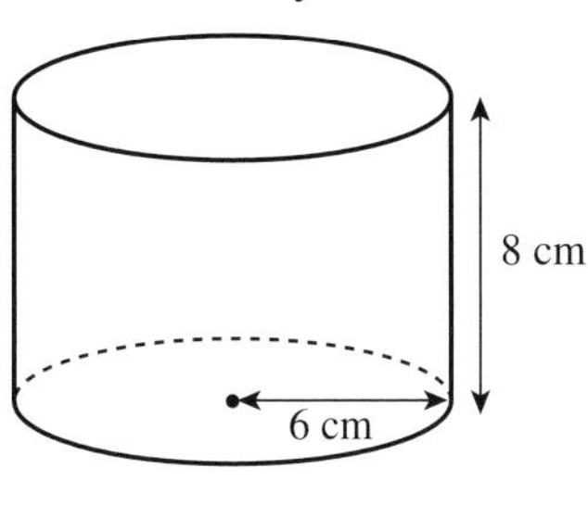

Volume of cone $= \frac{1}{3}\pi r^2 h$

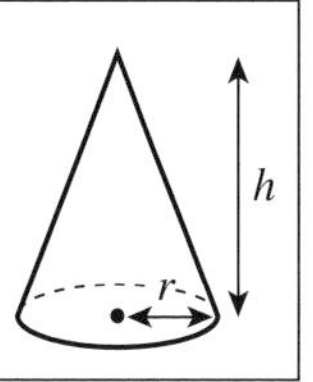

Find the ratio of the volume of the cone to the volume of the cylinder in its simplest form. **(4 marks)**

11 **Problem-solving** The diagram shows a solid sphere made of ebony.
The radius of the sphere is 2.5 cm.
The mass of the sphere is 63 g.
An object will float in water if its density is less than 1 g/cm^3.
Will this ebony sphere float in water?

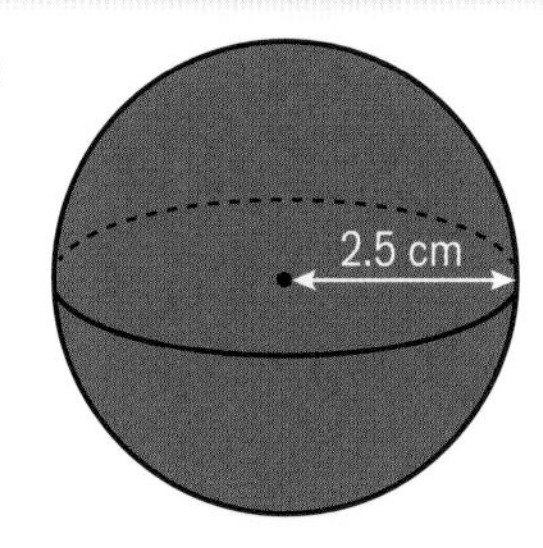

Q11 hint

Density $= \frac{\text{mass}}{\text{volume}}$

12 **Problem-solving** Brenna has some containers in the shape of hemispheres with diameter 32 cm.

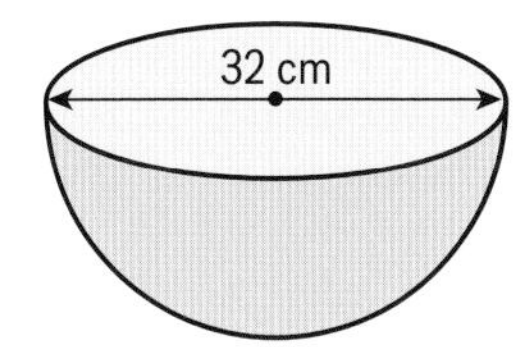

She is going to fill the containers completely with compost.
She has 90 litres of compost.
Work out how many containers Brenna can completely fill with compost.
1 litre = 1000 cm^3

13 **Problem-solving** The diagram shows a rectangular-based pyramid and a cube.

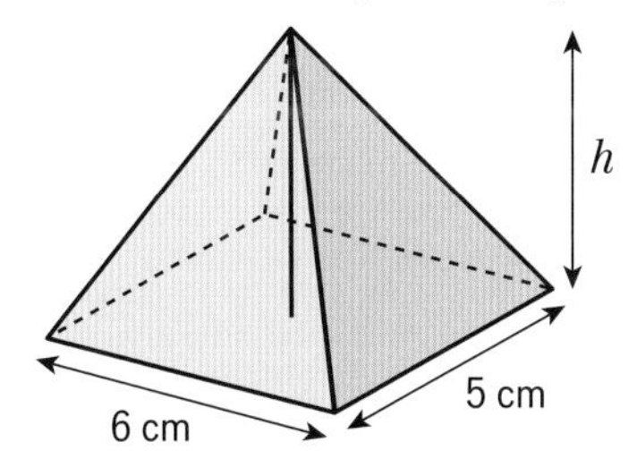

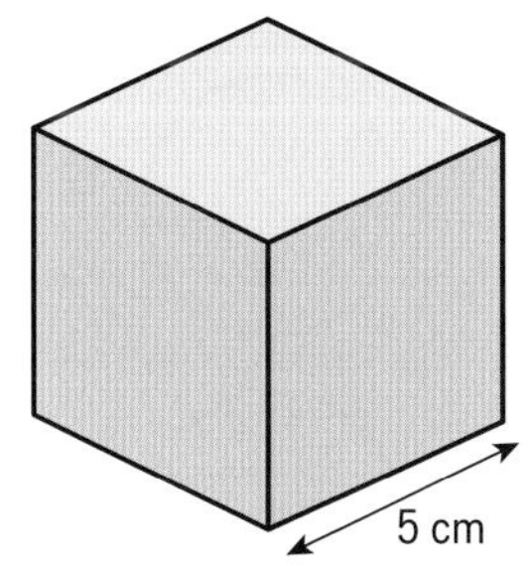

The ratio of the volume of the pyramid to the volume of the cube is 2 : 1.
Work out the height of the pyramid.

17 Test ready

Summary of key points

To revise for the test:

- Read each key point, find a question on it in the mastery lesson, and check you can work out the answer.
- If you cannot, try some other questions from the mastery lesson or ask for help.

Key points

1. Measurements given to the nearest whole unit may be inaccurate by up to one half of a unit below and one half of a unit above.
For example, the range of possible values for a length given as 3 cm to the nearest cm is $2.5\text{ cm} \leqslant \text{length} < 3.5\text{ cm}$. → 17.2
2. To **truncate** a number to 1 digit, you remove the other digits without rounding.
For example, 5.964 truncated to 1 digit is 5. → 17.2
3. An answer given to an appropriate degree of accuracy has the same number of decimal places as the most accurate measurement in the question. → 17.2
4. The **circumference** is the **perimeter** of a circle. → 17.1
5. The Greek letter π (pronounced pi) is the ratio of the circumference to the diameter of a circle. Its decimal value never ends but starts as 3.141 592 653 589 7... → 17.1
6. To find the circumference of a circle (C) when given the radius (r) or the diameter (d), use $C = \pi d$ or $C = 2\pi r$. → 17.1
7. To find the diameter of a circle when given the circumference, use $d = \frac{C}{\pi}$. → 17.2
8. To find the area (A) of a circle when given the radius, use $A = \pi r^2$. → 17.3
9. To find the radius of a circle when given the area, use $r = \sqrt{\frac{A}{\pi}}$. → 17.3
10. A **chord** is a line through a circle that touches the circumference at each end. → 17.4

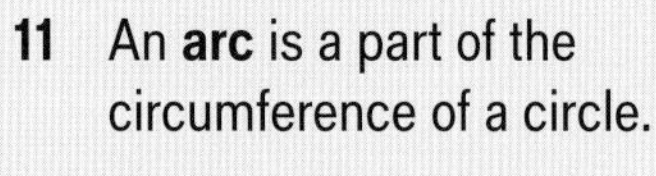

11. An **arc** is a part of the circumference of a circle. → 17.4

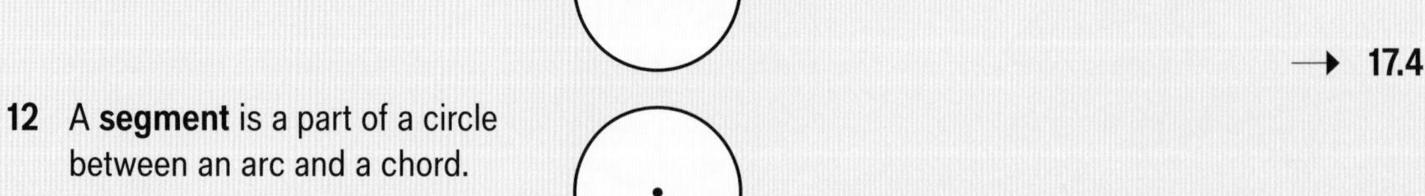

12. A **segment** is a part of a circle between an arc and a chord. → 17.4
13. A **tangent** is a line outside a circle that touches the circle at only one point. → 17.4

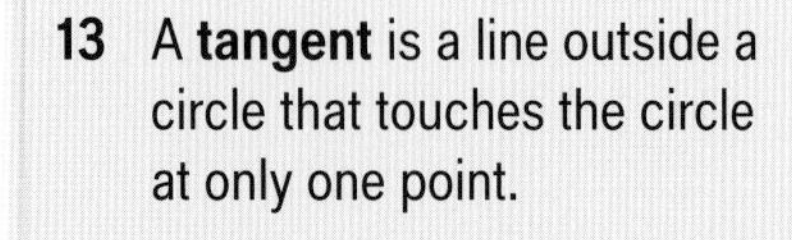

Key points

14 A **sector** is a slice of a circle between an arc and two radii. → 17.4

15 For a sector of a circle with an angle of $x°$ and radius r:

Area of sector $= \frac{x}{360} \times \pi r^2$

Arc length $= \frac{x}{360} \times 2\pi r$

Perimeter of sector = arc length + $2r$ → 17.4

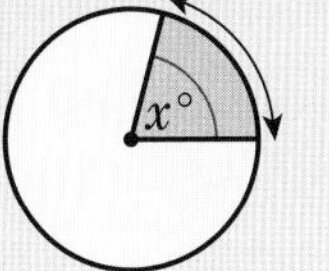

16 The formula for the volume of a **cylinder** is $V = \pi r^2 h$. → 17.5

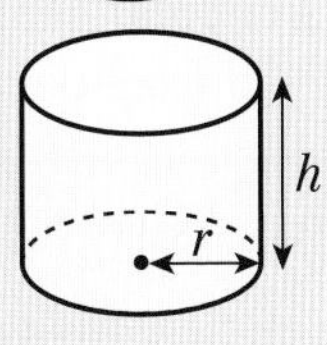

17 The surface area of a prism is the total area of all its faces. → 17.5

18 The volume of a pyramid $= \frac{1}{3} \times$ area of base $\times$ vertical height → 17.6

19 The volume of a cone $= \frac{1}{3} \times$ area of base $\times$ vertical height → 17.6

20 The area of the curved surface of a cone

$= \pi \times$ base radius $\times$ slant height $= \pi rl$

The **slant height** of a cone is the length of the sloping side. → 17.6

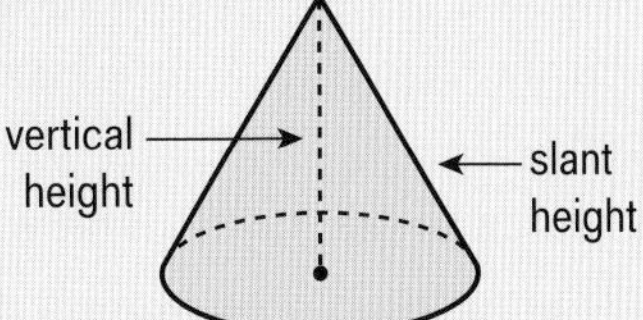

21 The volume of a **sphere** $= \frac{4}{3}\pi r^3$ → 17.7

22 The surface area of a sphere $= 4\pi r^2$ → 17.7

23 A **hemisphere** is half of a sphere. → 17.7

24 Total surface area of a hemisphere

= curved surface area + area of circular base

$= 2\pi r^2 + \pi r^2$ → 17.7

Sample student answers

Exam-style question

The diagram shows a square $ABCD$ with side of length 12 cm.
It also shows the arc of a circle, centre A.

Work out the shaded area. Give your answer to 3 s.f.

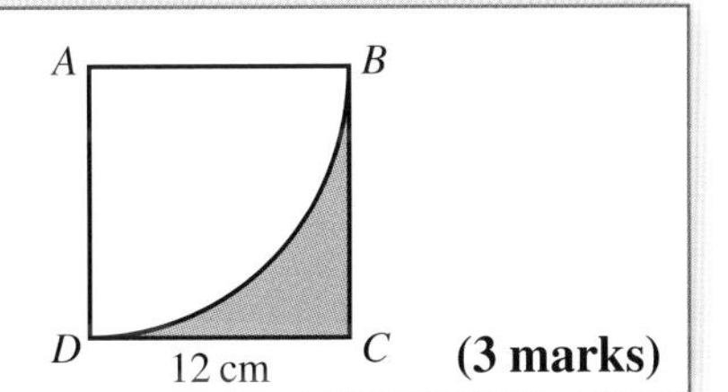

(3 marks)

Student A

Area of square $= 12 \times 12 = 144\text{ cm}^2$

Area of circle $= \pi \times 12^2 = 144\pi$

Area of quarter circle $= 36\pi$

Shaded area $= 144 - 36\pi = 30.9\text{ cm}^2$

Student B

Area of square $= 12 \times 12 = 144\text{ cm}^2$

Area of circle $= 2 \times \pi \times 12 = 24\pi$

Area of quarter circle $= 6\pi$

Shaded area $= 144 - 6\pi = 125\text{ cm}^2$

a Which student has the correct answer?

b Explain the mistake the other student has made.

17 Unit test

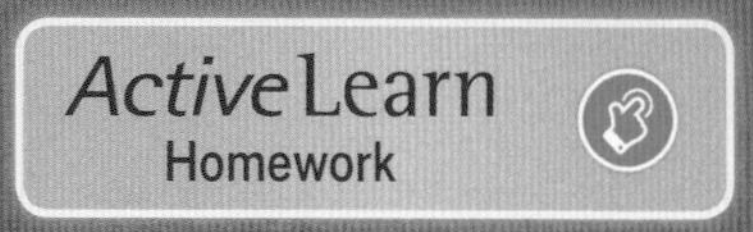

1 A field is 80 m long, to the nearest 10 m.
Write an inequality to show the error interval for l, the length of the field. **(2 marks)**

2 The circumference of a circular garden is 88 m.
Fred wants to make a path from one point on the circumference to another through the centre of the garden.
How long will the path be? Give your answer to the nearest metre. **(2 marks)**

3 Work out the area of a circle with a radius of 9 cm.
Give your answer correct to 1 decimal place (1 d.p.). **(2 marks)**

4 A circle has an area of $121\pi\,m^2$. Work out its radius. **(1 marks)**

5 The diagram shows a design for a window.
The window has a rectangular bottom and a semicircular top.

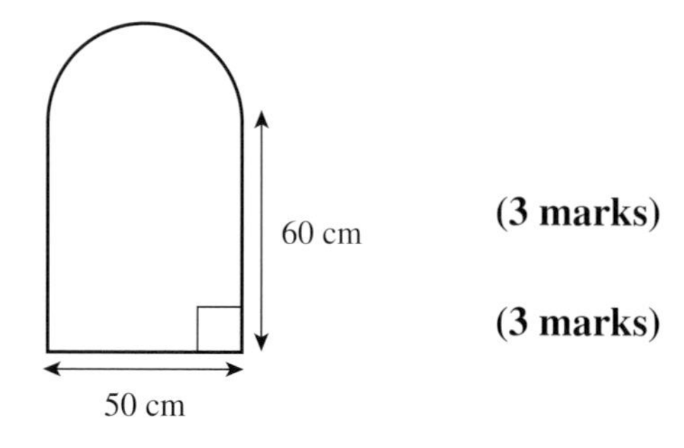

a Work out the area of the glass in the window. **(3 marks)**

b A frame of lead beading goes around the glass.
How much lead beading is needed? **(3 marks)**

Give your answers to 3 significant figures (3 s.f.).

6 *OAB* is the sector of a circle with centre *O*.

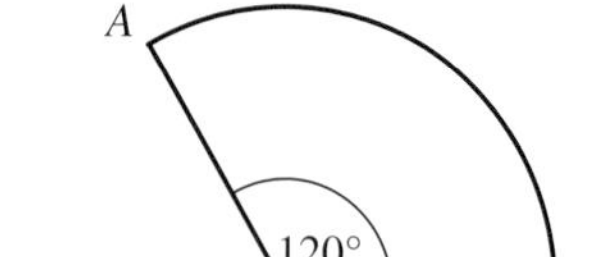
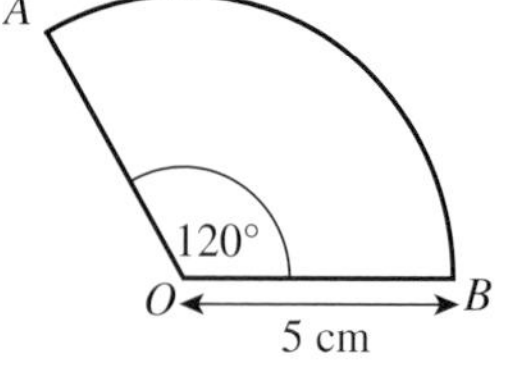

Work out in terms of π

a the perimeter of sector *OAB* **(2 marks)**

b the area of sector *OAB* **(2 marks)**

7 Tom has to cover 2 tanks completely with paint.
Each tank is in the shape of a cylinder with a top and a bottom.
The tank has a diameter of 2.4 m and a height of 1.1 m.

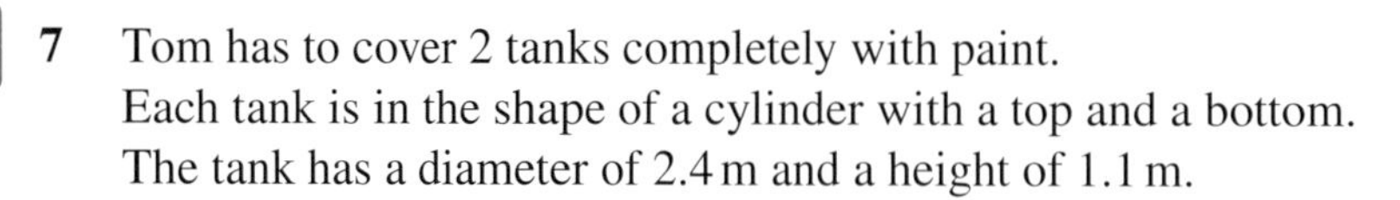
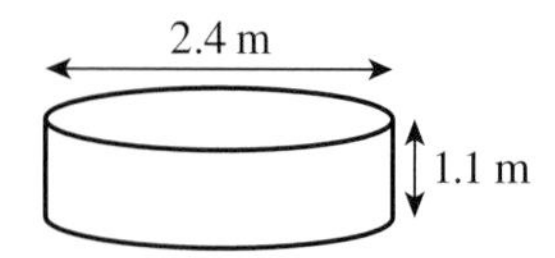

Tom has 8 tins of paint.
Each tin of paint covers $4\,m^2$.
Has Tom got enough paint to completely cover the 2 tanks?
You must show how you get your answer. **(5 marks)**

8 The diagram shows a container used to store rainwater.
The container is in the shape of a cylinder of radius 50 cm.
The height of the rainwater in the container is 70 cm.
A gardener uses 80 litres of the rainwater.
1 litre = 1000 cm^3
Work out the new volume of the rainwater in the container.
Give your answer correct to 3 s.f.

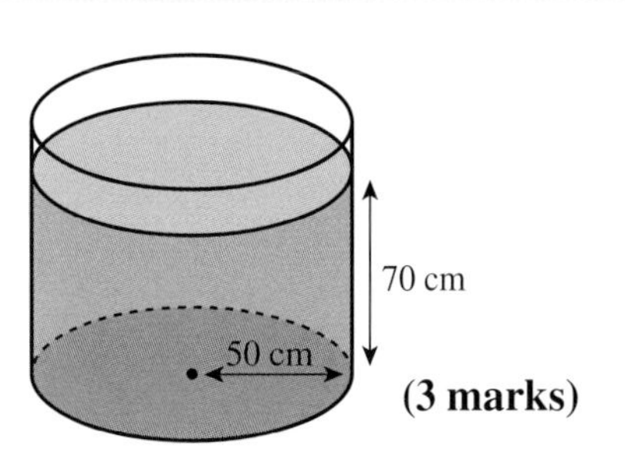

(3 marks)

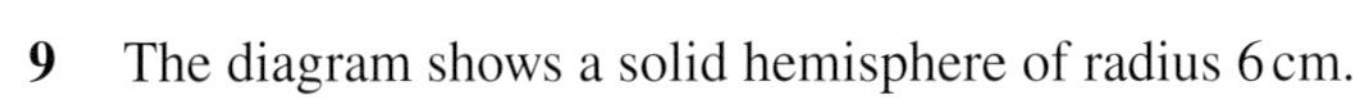

9 The diagram shows a solid hemisphere of radius 6 cm.

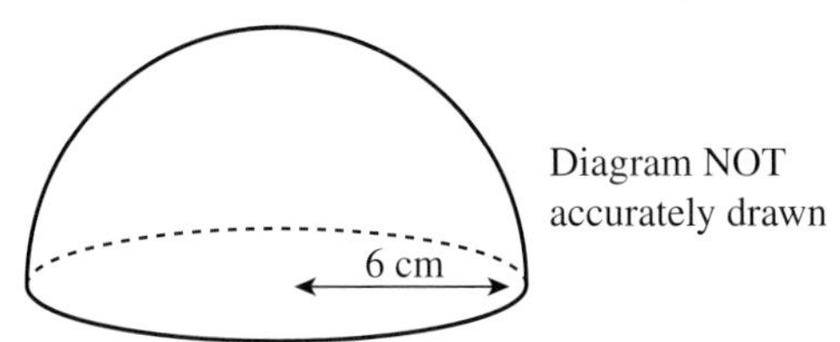

Surface area of sphere = $4\pi r^2$

Find the total surface area of the solid hemisphere.
Give your answer in terms of π. **(3 marks)**

10 The diagram shows a square-based pyramid.

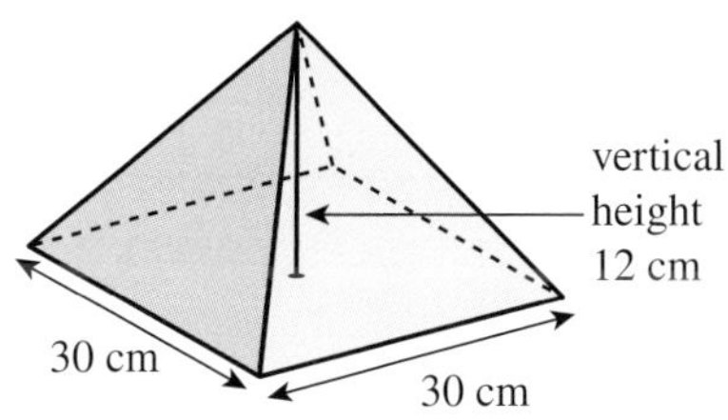

Work out the volume of the pyramid. **(2 marks)**

11 The diagram shows a cone.

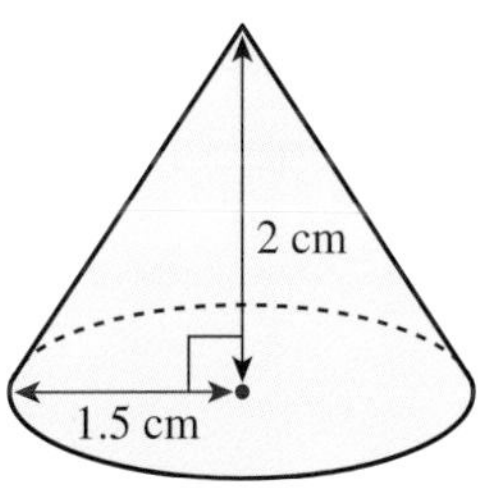

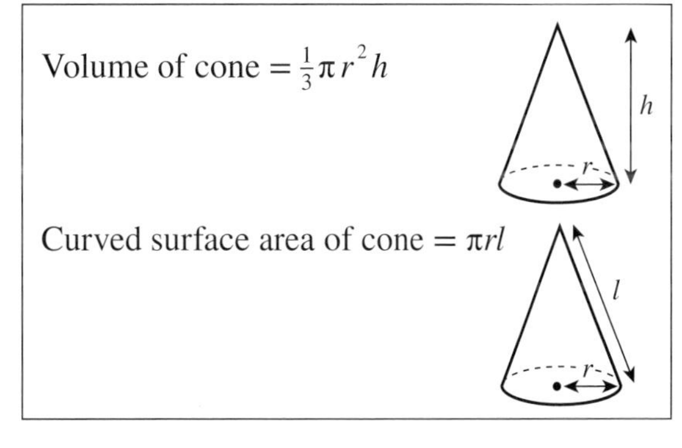

Work out

a its volume **(2 marks)**

b its total surface area **(3 marks)**

(TOTAL: 35 marks)

12 **Challenge** A sheet of A1 paper is 594 mm by 841 mm.
The sheet of paper can be rolled in two different ways to make two different cylinders.
Which cylinder has the greater volume, *A* or *B*?
Show working to explain.

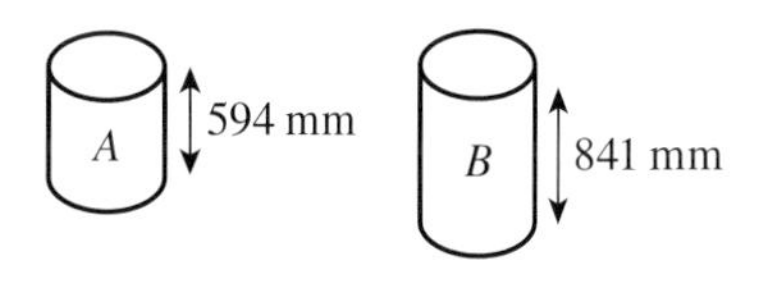

13 **Reflect** Look back at this unit.
Which lesson made you think the hardest?
Write a sentence to explain why.

Q13 hint

Begin your sentence with:
Lesson ____ made me think the hardest because ____

Mixed exercise 5

ActiveLearn Homework

1 Problem-solving Here are the names of some solids.

A	Triangular prism
B	Square-based pyramid
C	Cuboid
D	Tetrahedron
E	Hexagonal prism
F	Cone

Copy and complete the Venn diagram where

$\mathscr{E}$ is the universal set

X is the set of solids with at least one triangular face

Y is the set of solids with at least 8 edges

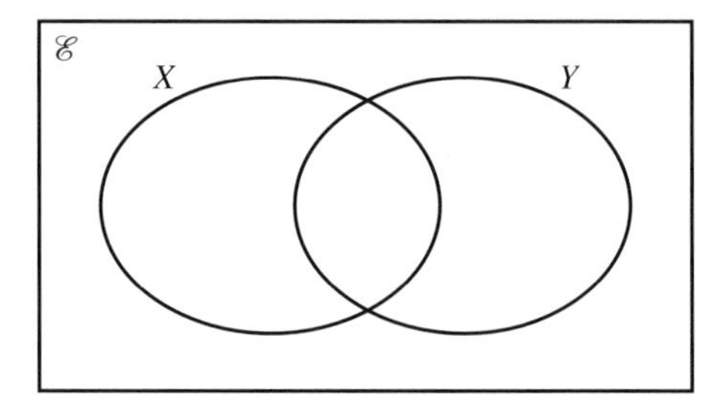

Exam-style question

2 The length of a bus is 13.75 metres.
Zach buys a scale model of the bus.
The scale of the model is 1 : 50.
Work out the length of the scale model of the bus.
Give your answer in centimetres. **(3 marks)**

3 Problem-solving The plan, front elevation and side elevation of a solid 3D shape are all squares of the same size.
Name the solid.

4 Problem-solving Draw a rectangle that measures 7 cm by 5 cm.
Using a pair of compasses and a ruler, construct one line of symmetry.

Exam-style question

5 The diagram shows a prism with a cross-section in the shape of a trapezium.
Draw the front elevation and the side elevation of the prism.
Use a scale of 2 cm to 1 m. **(4 marks)**

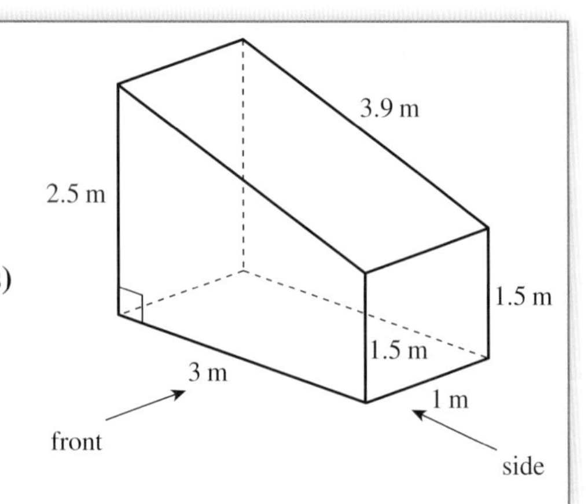

Exam-style question

6 The circumference of a circle is 135 mm to the nearest millimetre.
Copy and complete the error interval for the circumference of the circle.

□ mm $\leqslant$ circumference < □ mm **(2 marks)**

7 **Problem-solving** Here is a pyramid with a square base.
The sloping faces are congruent isosceles triangles.

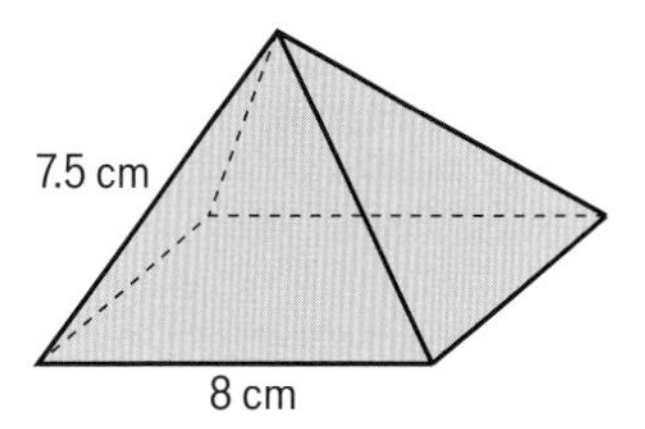

a Draw a full-size accurate plan of the pyramid.

b Using compasses and a ruler, construct an accurate drawing of one of the triangular sloping faces of the pyramid.

8 **Problem-solving** Match each pair of equivalent expressions.

a $(x+5)(x-6)$ **i** $x^2+7x-30$

b $(x+6)(x+5)$ **ii** $x^2+11x+30$

c $(x-5)(x-6)$ **iii** $x^2-7x-30$

d $(x+10)(x-3)$ **iv** x^2-x-30

e $(x+2)(x-15)$ **v** $x^2-11x+30$

f $(x+3)(x-10)$ **vi** $x^2-13x-30$

9 **Reasoning** Sophie is asked to construct the line from point P that is perpendicular to line AB.

She places a protractor on line AB and positions it so that point P is at 90° to the line AB, as shown.

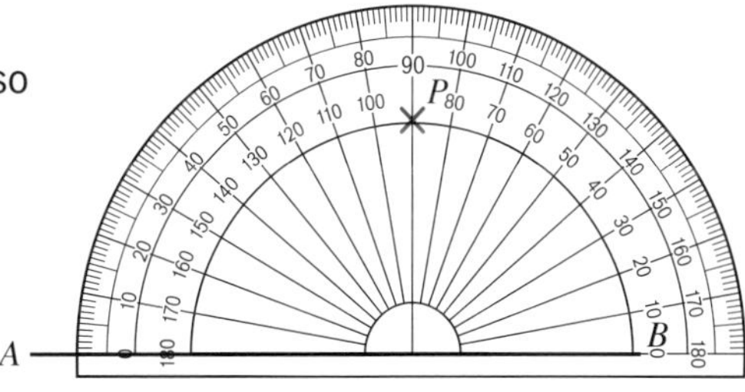

a What has Sophie done wrong?

b Copy the original diagram and construct the line from point P that is perpendicular to line AB.

10 **Reasoning** The diagram shows two shapes, A and B, made of rectangles.

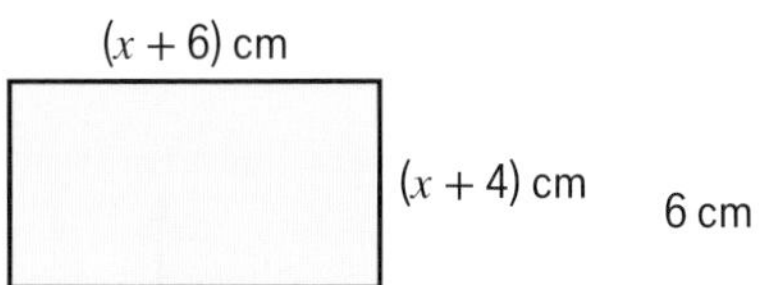

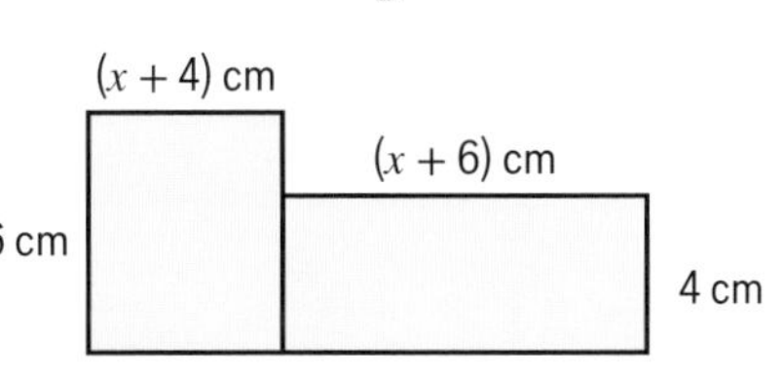

Here are two expressions for the areas of A and B.

i $6(x+4)+4(x+6)$ **ii** $(x+6)(x+4)$

a Match each expression to the correct diagram.

b Expand and simplify the expression for the area of shape A.

c Expand and simplify the expression for the area of shape B.

11 Reasoning John is asked to factorise $x^2 - x - 6$.
John writes:

$x^2 - x - 6 = (x + 3)(x - 2)$

Explain why John is incorrect.

12 Problem-solving Here is part of a map.

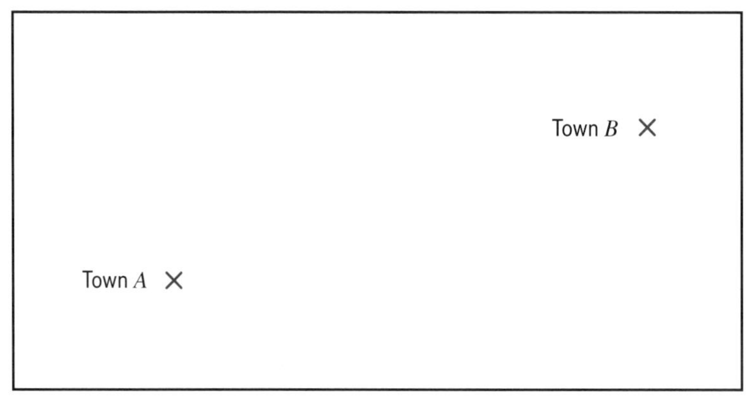

Scale: 1 cm represents 4 km

The map shows two towns, A and B. A company is going to build a supermarket.
The supermarket will be more than 20 km from Town A and less than 14 km from Town B.
Copy the diagram. Find and shade the region on the map where the company can build the supermarket.

Exam-style question

13 This accurate scale drawing shows the positions of two airports, A and B.

Sayed flies a plane from airport A on a bearing of 075°.
He flies for $1\frac{1}{2}$ hours at an average speed of 700 km/h to airport C.
Find the distance, in km, of airport C from airport B and the bearing of airport C from airport B.

(5 marks)

Scale: 1 cm represents 200 km

14 Problem-solving A circular stained-glass window is edged in lead.
It has 8 lead lines from its centre, as shown in the diagram.
The length of each lead line from the centre is 28 cm.
Work out the total length of the lead in the window.
Give your answer correct to 3 significant figures.

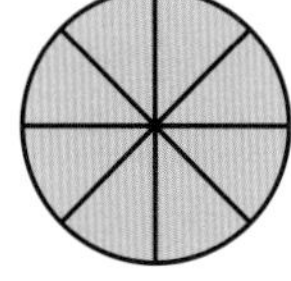

15 **a** Copy and complete the table of values for $y = x^2$.

x	−3	−2	−1	0	1	2	3
y							

b Draw the graph of $y = x^2$ for values of x from −3 to 3.

c **Reasoning** Use your graph to estimate the value of $\sqrt{5}$.

16 **Reasoning** Show that the solutions to the quadratic equation $x^2 + 3x - 28 = 0$ are $x = -7$ or $x = 4$.

Exam-style question

17 P is a quarter circle of radius 11 cm.
Q is a circle.
The area of P is 4 times the area of Q.
Show that the radius of Q is 2.75 cm. **(3 marks)**

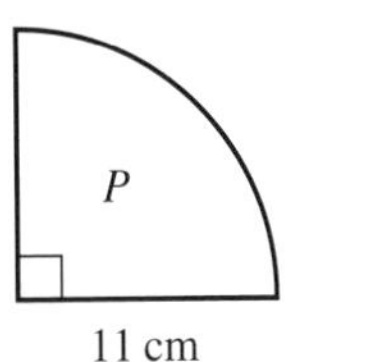

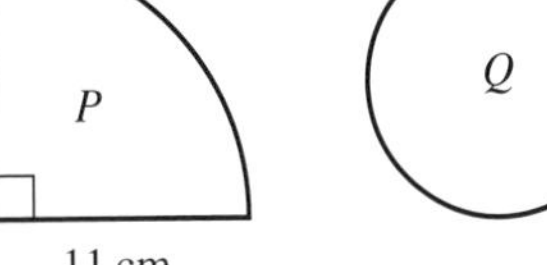

18 **Problem-solving** The diagram shows a circle touching a square.
The area of the square is $100\,\text{cm}^2$.
Work out the shaded area.
Give your answer correct to 3 significant figures.

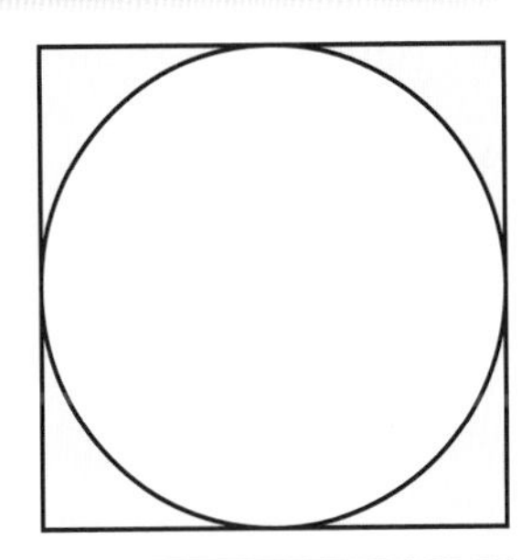

19 **Reasoning** The diagram shows cone A and cone B.

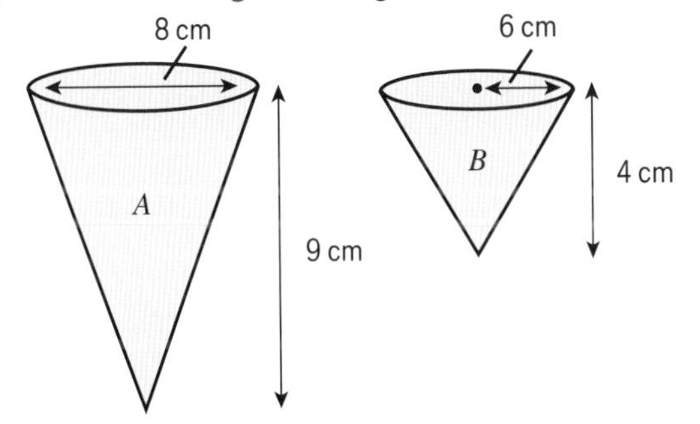

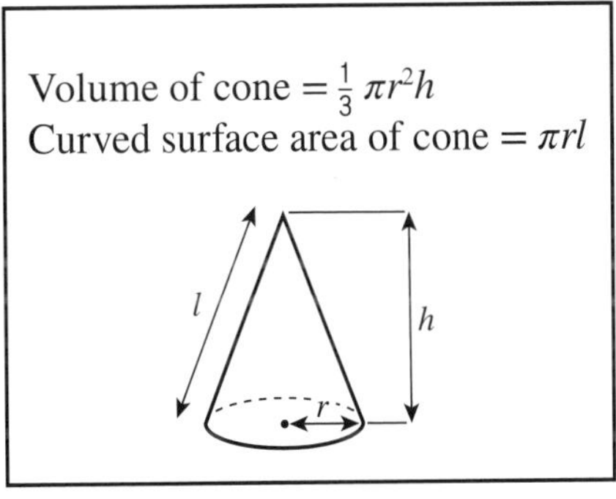

a Show that the volume of cone A is the same as the volume of cone B.

b Use Pythagoras' theorem to work out the slant height of each cone.

c Do the cones have the same surface area?
You must show your working.

20 **Reasoning** The inside of a bowl is a hemisphere with radius 20 cm.

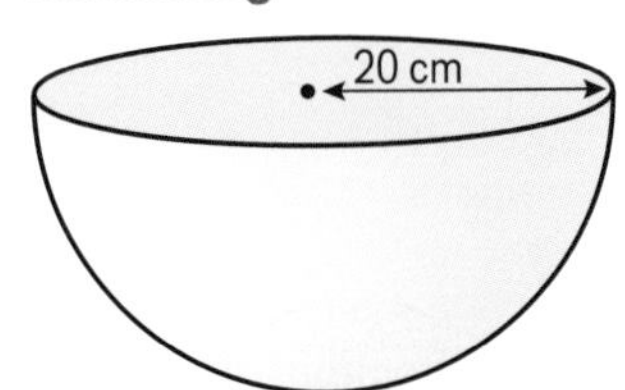

Volume of sphere = $\frac{4}{3}\pi r^3$

Water fills the bowl at a rate of 2 litres per minute.
Karis says, 'It takes less than a quarter of an hour to fill the bowl with water.'
Is Karis correct? You must show working to explain your answer.

18 Fractions, indices and standard form

Prior knowledge 

18.1 Multiplying and dividing fractions

- Multiply and divide mixed numbers and fractions.

ActiveLearn Homework  

Warm up

1 **Fluency** What is $\frac{25}{20}$ as a mixed number in its simplest form?

2 Write $3\frac{4}{5}$ as an improper fraction.

3 Work these out. Parts **a** and **d** are started for you.

a $3\times2\frac{1}{2}=3\times\frac{5}{2}=\frac{3\times\square}{2}=\frac{\square}{2}=\square\frac{1}{2}$ **b** $2\frac{2}{3}\times6$ **c** $\frac{3}{7}\times\frac{4}{5}$

d $\frac{1}{4}\div\frac{3}{5}=\frac{1}{4}\times\frac{5}{3}=\frac{\square}{\square}$ **e** $10\div\frac{3}{8}$ **f** $5\div1\frac{3}{7}$

4 Match the equivalent fractions and decimals.

a $\frac{9}{10}$ **b** $\frac{3}{5}$ **c** $\frac{3}{100}$ **d** $\frac{9}{5}$

A 1.8 **B** 0.9 **C** 0.6 **D** 0.03

5 Write each mixed number as an improper fraction. Then work out its reciprocal.

a $1\frac{3}{7}$ **b** $3\frac{1}{7}$ **c** $3\frac{2}{7}$ **d** $7\frac{2}{3}$

F H

6 Write the reciprocal of each decimal.

a 0.3 **b** 1.3

c 0.33 **d** 2.33

Q6 hint

First write the decimal as a fraction or mixed number.

Exam-style question

7 Find the value of the reciprocal of 1.25.
Give your answer as a decimal. **(1 mark)**

Exam tip

Clearly show every step of your working.

8 Work these out.

a $\left(\frac{1}{2}\right)^2$ **b** $\left(\frac{1}{3}\right)^2$ **c** $\left(\frac{2}{3}\right)^2$ **d** $\left(\frac{2}{3}\right)^3$

Q8a hint

$\left(\frac{1}{2}\right)^2=\frac{1}{2}\times\frac{1}{2}$

Key point

To multiply or divide mixed numbers, change the mixed numbers to improper fractions first.

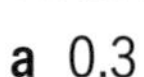

9 Work these out. Give your answer in its simplified form.

a $1\frac{1}{4}\times\frac{2}{5}$ **b** $1\frac{1}{4}\times\frac{3}{5}$ **c** $2\frac{1}{4}\times\frac{2}{5}$

Q9 hint

First change the mixed number to an improper fraction.

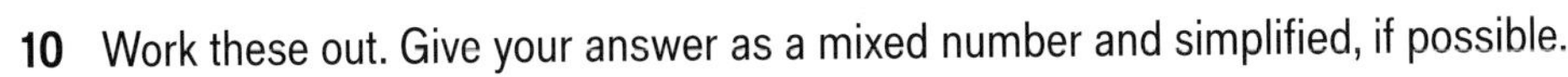

10 Work these out. Give your answer as a mixed number and simplified, if possible.

a $2\frac{1}{4}\times\frac{3}{5}$ **b** $\frac{3}{5}\times2\frac{3}{4}$ **c** $3\frac{3}{4}\times\frac{2}{5}$ **d** $\frac{4}{5}\times3\frac{3}{4}$

11 Work these out. Where possible, simplify before multiplying.

a $2\frac{1}{7}\times\frac{1}{10}$ **b** $\frac{10}{11}\times1\frac{2}{5}$ **c** $1\frac{1}{8}\times\frac{4}{9}$

d $\frac{4}{5}\times3\frac{2}{3}$ **e** $\frac{2}{7}\times2\frac{5}{11}$

Q11a hint

$\frac{15}{7}\times\frac{1}{10}=\frac{{}^{3}\cancel{15}\times1}{{}_{2}\cancel{10}\times7}$

12 Work these out.

a $1\frac{3}{5}\times2\frac{2}{5}$ **b** $2\frac{2}{3}\times1\frac{2}{3}$ **c** $1\frac{1}{9}\times1\frac{3}{5}$ **d** $2\frac{2}{5}\times7\frac{1}{2}$ **e** $(1\frac{1}{5})^2$

Example

Work out $3\frac{1}{5}\div\frac{7}{10}$.

$$3\frac{1}{5}\div\frac{7}{10}=\frac{16}{5}\div\frac{7}{10}$$

$$=\frac{16}{5}\times\frac{10}{7}$$ To divide by a fraction, multiply by its reciprocal.

$$=\frac{16\times\cancel{10}^{2}}{{}_{1}\cancel{5}\times7}$$ Simplify if possible.

$$=\frac{32}{7}=4\frac{4}{7}$$ Write the answer as a mixed number.

13 Work these out.

a $1\frac{3}{5}\div\frac{1}{2}$ **b** $1\frac{1}{8}\div\frac{2}{3}$ **c** $2\frac{3}{5}\div\frac{7}{10}$

d $1\frac{4}{5}\div\frac{1}{10}$ **e** $\frac{4}{5}\div1\frac{4}{10}$ **f** $1\frac{2}{3}\div2\frac{3}{5}$

g $2\frac{1}{2}\div2\frac{1}{4}$ **h** $3\frac{2}{3}\div2\frac{1}{5}$ **i** $2\frac{1}{4}\div5\frac{1}{2}$

Q13 hint

First change mixed numbers to improper fractions.

14 Use a calculator to work out

a $4\frac{3}{5}\times\frac{6}{7}$ **b** $18\frac{1}{2}\times14\frac{4}{9}$ **c** $19\div2\frac{3}{5}$ **d** $18\frac{4}{5}\div\frac{1}{3}$

15 **Problem-solving** A roll of ribbon is cut into pieces measuring $1\frac{1}{4}$ inches.
How many whole pieces can be cut from a $32\frac{1}{2}$ inch roll?

16 **Problem-solving** A tile measures $1\frac{1}{2}$ inches by $2\frac{1}{4}$ inches. What is the area of the tile?

17 **Reasoning** Each of these calculations is missing a single digit.
Use the inverse operation to work out the missing digits.

a $\frac{2}{5}\times1\frac{\square}{3}=\frac{8}{15}$ **b** $\square\frac{1}{5}\times2\frac{1}{2}=8$ **c** $\frac{2}{\square}\div1\frac{1}{6}=\frac{4}{7}$ **d** $1\frac{1}{3}\div2\frac{\square}{5}=\frac{5}{9}$

Exam-style question

18 A necklace is made of beads which are $1\frac{1}{4}$ inches long.
Each bead costs 30p and the thread costs £2.30.
The necklace is $21\frac{1}{4}$ inches long.
What is the cost of making the necklace? **(3 marks)**

Exam tip

Break the question into steps. Show your working and clearly state your final answer.

18.2 The laws of indices

- Know and use the laws of indices.

Warm up

1 **Fluency** Work out
a $-2+3$ b $-2-3$ c $2--3$ d $2-3$

2 Work out
a 3^2 b 3^3 c 9^2
d 10^4 e 1^5 f 5^3

3 Write as a single power
a $2^2\times2^4$ b $10^5\times10$ c $7^6\div7^2$ d $\frac{4^6}{4^2}$ e $(6^2)^3$

4 Evaluate $\frac{11^5\times11}{11^4}$.

5 Work out
a $\left(\frac{1}{6}\right)^2$ b $\left(\frac{1}{2}\right)^4$ c $\left(\frac{2}{5}\right)^3$ d $\left(\frac{1}{10}\right)^4$

Key point

To raise the power of a number to another power, multiply the indices.

6 The answer to each question is one of the following.

3^2 3^4 3^5 3^6 3^8 3^{12}

Choose the correct answer for each expression.
a $(3^2)^3$ b $(3^2)^2\times3$ c $(3^4\times3^2)^2$ d $(3^8\div3^6)^3$
e $(3^{12}\div3^{10})\times3^3$ f $(3^4\div3^2)^2$ g $(3\times3)^2\times(3^2)^4$ h $\frac{3^5\times3^7}{(3^2)^3}$

7 Write as a power of 3. The first one is started for you.
a $9^4=(3^{\square})^4=3^{\square}$
b 9^5
c 27^2
d 27^6

Q7c hint

$27=3^{\square}$

8 **Reasoning** Edward writes $9^6=(3^2)^6=3^8$.
Write a sentence explaining what Edward has done wrong.

9 **Reasoning** Copy and complete the table.

Power of 2	2^3	2^2	2^1	2^0	2^{-1}	2^{-2}	2^{-3}
Value	8				$\frac{1}{2^1} = \square$	$\frac{1}{2^2} = \square$	$\frac{1}{2^3} = \square$

10 **a** Use your calculator to work out
i 4^0 **ii** 100^0 **iii** 57^0

b **Reflect** Write a sentence stating what you notice about numbers raised to the power 0.

Q10 hint

Use the x^y key.

Key point

Any number (or term) raised to the power 0 is equal to 1.

11 **a** Use your calculator to work out **i** 4^{-1} **ii** 5^{-1} **iii** 10^{-1}

b Write each of your answers in part **a** as a fraction.

c **Reflect** Write a sentence stating what you notice about numbers raised to the power of −1.

Key point

Any number (or term) raised to the power −1 is the reciprocal of the number.
For example, $3^{-1} = \frac{1}{3}$.

12 Write as a fraction **a** 8^{-1} **b** 7^{-1} **c** 6^{-1} **d** x^{-1}

13 **a** Use your calculator to work out **i** 10^{-2} **ii** 10^{-3} **iii** 10^{-4}

b Write each of your answers in part **a** as a fraction.

c **Reflect** Write a sentence explaining what you notice about numbers raised to a negative power.

Key point

3^{-2} is the same as $(3^{-1})^2$.
To find a negative power, find the reciprocal and then raise the value to the positive power.

Example

Work out the value of

a 3^{-2} **b** $\left(\frac{2}{3}\right)^{-3}$

a $3^{-2} = \left(\frac{1}{3}\right)^2$ — The reciprocal of 3 is $\frac{1}{3}$.

$= \frac{1}{3} \times \frac{1}{3}$ — Square the fraction.

$= \frac{1 \times 1}{3 \times 3}$

$= \frac{1}{9}$

b $\left(\frac{2}{3}\right)^{-3} = \left(\frac{3}{2}\right)^3$ — The reciprocal of $\frac{2}{3}$ is $\frac{3}{2}$.

$= \frac{3}{2} \times \frac{3}{2} \times \frac{3}{2}$ — Work out $\left(\frac{3}{2}\right)^3$.

$= \frac{3 \times 3 \times 3}{2 \times 2 \times 2}$

$= \frac{27}{8}$

$= 3\frac{3}{8}$ — Write the answer as a mixed number.

14 Work out the value of **a** 2^{-3} **b** 7^{-2} **c** 8^{-2} **d** 5^{-3}

15 Work out the value of

a $\left(\frac{1}{2}\right)^{-3}$ **b** $\left(\frac{1}{5}\right)^{-2}$ **c** $\left(\frac{2}{9}\right)^{-2}$ **d** $\left(\frac{5}{3}\right)^{-3}$

Q16 hint

Write the mixed number as an improper fraction.

Exam-style question

16 Work out the value of $\left(1\frac{1}{4}\right)^{-2}$. **(2 marks)**

Exam tip

When asked for a value, give an answer that is an integer, fraction or decimal with no power.

17 Write these expressions as a single power.
The first one is started for you.

a $2^5 \times 2^{-2} = 2^{5-2} = 2^{\square}$ **b** $(2^3)^{-1}$

c $8^{-3} \times 8^3$ **d** $4^7 \div 4^{-3}$

e $\frac{12^5}{12^{-2}}$ **f** $(11^8 \div 11^{-5}) \times 11$ **g** $(3^{-2})^2 \times 3^{-4}$

Q17a hint

The laws of indices for negative numbers are the same as for positive numbers.

18 Work out the value of

a $\frac{9^3 \times 9}{9^2}$ **b** $\frac{4^5 \times 4^{-2}}{4^3}$ **c** $\frac{20^8}{(20^3)^2}$

19 Write as a single power

a $a^{-2} \times a^5$ **b** $b^{-2} \times b^{-1}$ **c** $\frac{m^{-3}}{m^2}$

d $\frac{n^2}{n^{-3}}$ **e** $\frac{x^3 \times x^{-2}}{x^4}$ **f** $\frac{c^{-3} \times c^5}{c^2}$

Q19 hint

Write answers with negative powers as $\frac{1}{x^{\square}}$.

20 Problem-solving $x = 4$. Work out the value of

a $x^{-3} \times x^2$ **b** $\frac{x^5}{x^3}$ **c** x^0

21 Problem-solving Write these numbers from smallest to largest.

10^3 100^2 10^{-3} 100^{-2} 1^4 1^{-5}

Exam-style question

22 Write these numbers in order of size.
Start with the smallest number.

5^{-4} 0.5 5^{-1} 0.5^2 **(2 marks)**

Exam tip

When writing numbers in order, always pay attention to whether you are asked to start with the smallest or the largest.

23 Work out

a $6^{-1} \times 12$ **b** 300×10^{-1} **c** 1800×10^{-4} **d** 18×23^0

24 Use your calculator to work out these calculations.
Where necessary, give your answers to 3 significant figures.

a $0.5^{-2} \times 10^{-3}$ **b** $18^7 \div 20^6$ **c** $4.5^{-2} + 9.8$

d $\frac{3^{-8} \times 5^7}{8^{-2}}$ **e** $2.5^7 \times 3.5^2 + 8.4^{-1}$ **f** $1.4^3 - 0.9^2$

18.3 Writing large numbers in standard form

- Write large numbers in standard form.
- Convert numbers from standard form into ordinary numbers.

ActiveLearn Homework

Warm up

1 **Fluency** Work out the value of

a 10^2 b 10^3 c 10^4 d 10^5

2 Work out

a 5×100 b 3.5×1000 c 8.75×10 d $9.03\times100\,000$

3 Write in figures

a 1 million b 1 billion

Q3b hint

1 billion = 1000 million

4 Copy and complete.

a $4.7\times\square=47$ b $8.19\times\square=81.9$ c $3.07\times\square=307$

Key point

Standard form is a way of writing very large or very small numbers.

A number in standard form looks like this.

8.4×10^5

This part is a number between 1 and 10. There is a single non-zero digit before the decimal point. (8.4)

This part is a power of 10. (10^5)

5 Write each number as an ordinary number.
The first one has been started for you.

a $3\times10^2=3\times100=\square$ b 5×10^6 c 7×10^4 d 9×10^{11}

Example

Write 4000 in standard form.

$4000 = 4\times1000$ — Write the number as a number between 1 and 10 multiplied by a power of 10.

$= 4\times10^3$ — Write the power of 10 using indices.

6 Write each number in standard form.

a 500 b 300 000 c 9 000 000 000 d 7 million

7 Write each number as an ordinary number.

a $4.7\times10^4=4.7\times10\,000=$ b 9.21×10^6

c 8.3×10^{11} d 9.23×10^5 e 6.3×10^{11}

f 9.05×10^6 g 6.702×10^9 h 4.07×10^8

Q7f hint

The 0 between the 9 and the 5 is essential: $9.05\neq9.5$

Example

Write 45 600 in standard form.

$45\,600 = 4.56 \times 10^4$

4.56 lies between 1 and 10.
Multiply by the power of 10 needed to give the original number.
4 5 6 0 0

8 **Problem-solving** Match the ordinary numbers to those in standard form.

a 79 000	**d** 7 910 000 000	**A** 7.91×10^3	**D** 7.91×10^7
b 7910	**e** 79 100 000	**B** 7.9×10^{10}	**E** 7.91×10^9
c 790 000 000	**f** 79 000 000 000	**C** 7.9×10^4	**F** 7.9×10^8

9 Write each number using standard form.

a 450 000 **b** 32 000 **c** 150 000 **d** 7 250 000
e 6 291 000 **f** 1 500 000 **g** 703 000 000 **h** 76 billion

10 Neptune is 4 503 000 000 km (to the nearest million km) from the Sun.
Write this distance in standard form.

11 **Reasoning**

a Which of these numbers is written in standard form? Explain.

A 42×10^2 **B** 33.4×10^3 **C** 6.9×10^4 **D** 0.4×10^3

b Rewrite any numbers not written in standard form, so that they are in standard form.

Q11b hint

$42 \times 10^2 = 4.2 \times 10 \times 10^2$
$= 4.2 \times ...$

Exam-style question

12 Write these numbers in order of size.
Start with the largest number.

3.2×10^5 320 3.2×10^8 3.2×10^4 3.2×10^9

(2 marks)

Exam tip

When comparing numbers, write them all in the same way. For example, here convert all to standard form or all to ordinary numbers.

Exam-style question

13 **a** Write 705 000 000 in standard form. **(1 mark)**

b Write 3.45×10^7 as an ordinary number. **(1 mark)**

Exam tip

You can use a calculator to check your answer.

14 **Future skills** The table shows the meaning of prefixes for large numbers.
Use the table to write these measurements in standard form.

a 8 terabytes in bytes **b** 4.6 Mm in metres
c 17.7 Gl in litres **d** 0.95 kilograms in grams.

Prefix	Letter	Number
tera-	T	1 000 000 000 000
giga-	G	1 000 000 000
mega-	M	1 000 000
kilo-	k	1000

15 **Reasoning** Write < or > in the boxes.

a $3.2 \times 10^4 \square 5.4 \times 10^8$ **b** $5.02 \times 10^7 \square 5.2 \times 10^7$ **c** $6.35 \times 10^3 \square 7.35 \times 10^2$

18.4 Writing small numbers in standard form

- Write small numbers in standard form.
- Convert numbers from standard form with negative powers into ordinary numbers.

*Active*Learn
Homework

Warm up

1 **Fluency** Work out

a $3000 \div 10$ **b** $920 \div 100$ **c** $891 \div 1000$ **d** $32.45 \div 1000$

2 Write these numbers as decimals.

a $\frac{1}{1000}$ **b** $\frac{1}{10^6}$ **c** 10^{-1} **d** 10^{-2}

Q2c and d hint
Write as fractions first.

3 Which of these are the same as

a $\div 10$ **b** $\div 100$?

 $\times \frac{1}{10}$
 $\times 10^{-2}$
 $\times 0.1$
 $\times \frac{1}{100}$
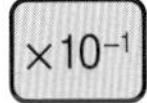 $\times 10^{-1}$
$\times 0.01$

4 Write each number as an ordinary number. The first one has been started for you.

a $3 \times 10^{-5} = 3 \times 0.00001 = \square$ **b** 8×10^{-2} **c** 4×10^{-1} **d** 7×10^{-11}

Example

Write 0.000 05 in standard form.

$0.00005 = 5 \times 0.00001$
$= 5 \times 10^{-5}$

Write the number as a number between 1 and 10 multiplied by a power of 10.

5 Write each number in standard form.

a 0.03 **b** 0.005 **c** 0.0001 **d** 0.000 000 3

6 Write each number as an ordinary number.

a $4.5 \times 10^{-5} = 4.5 \times 0.00001 = \square$ **b** 3.8×10^{-6} **c** 8.34×10^{-9} **d** 1.401×10^{-1}

Key point

To write a small number in standard form:
- Place the decimal point after the first non-zero digit.
- How many places has this moved the digit? This is the negative power of 10.

Example

Write 0.003 52 in standard form.

$0.00352 = 3.52 \times 10^{-3}$

3.52 lies between 1 and 10.
Multiply by the power of 10 needed to give the original number.
0 . 0 0 3 5 2

7 Write each number in standard form.

a 0.052 b 0.000 71 c 0.000 569 d 0.002 41

e 0.000 014 f 0.001 09 g 0.000 030 4 h 0.6102

8 Match these numbers to their equivalent written in standard form.

a 0.003 03	e 0.303	A 3.3×10^{-4}	E 3.03×10^{3}
b 0.000 33	f 0.33	B 3.3×10^{-9}	F 3.03×10^{-6}
c 0.000 003 03	g 3030	C 3.3×10^{-1}	G 3.3×10^{4}
d 0.000 000 003 3	h 33 000	D 3.03×10^{-3}	H 3.03×10^{-1}

9 **Reasoning** A question in an exam reads

Write 0.000 907 in standard form.

Here are some of the incorrect answers given.

Lucy: 0.907×10^{-3} Ali: 9.7×10^{-4} Sam: 9.07×10^{4} Tom: 9.07^{-4}

a For each answer explain what the student did wrong.

b What is the correct answer?

10 **Problem-solving** The table gives the radius of one atom of different elements.

a Write the radii as numbers in standard form.

b Write the atoms in order of size, starting with the smallest.

Element	Radius (m)
lithium	0.000 000 000 145
sodium	0.000 000 000 18
phosphorus	0.000 000 000 1
nitrogen	0.000 000 000 065
tin	0.000 000 000 145

Exam-style question

11 **a** Write 0.000 000 705 in standard form. **(1 mark)**

b Write 3.2×10^{-4} as an ordinary number. **(1 mark)**

12 **Future skills** The table shows the meaning of prefixes for small numbers.
Write the following measurements in standard form.

a 7 picograms in grams

b 1.4 μs in seconds

c 593 dm in metres

d 10.5 nV in volts

e 0.38 milliamps in amps

f 9.9 centilitres in litres

Prefix	Letter	Number
deci-	d	0.1
centi-	c	0.01
milli-	m	0.001
micro-	μ	0.000 001
nano-	n	0.000 000 001
pico-	p	0.000 000 000 001

13 **Reasoning** The diameter of a hydrogen atom is 0.000 000 000 05 m.
Enter this number into your calculator and press the = key.
How does the calculator answer compare with the number written in standard form?

18.5 Calculating with standard form

- Multiply and divide numbers in standard form.
- Add and subtract numbers in standard form.

Warm up

1 **Fluency** Work out

a $\frac{8-4}{5}$ **b** $3\times4-2$ **c** $\frac{7}{5+6}$ **d** $\frac{3\times8}{2}$

2 Rewrite each number in standard form.

a 32×10^4 **b** 180×10^7 **c** 0.9×10^4 **d** 59.6×10^{-6}

3 Write as a single power of 10

a $10^4\times10^5$ **b** $10^3\times10^{-2}$ **c** $10^8\div10^4$ **d** $\frac{10^{-3}}{10^{-4}}$

4 Work out

a i 13×15 **ii** 1.3×1.5

b i $56\div8$ **ii** $5.6\div8$

c $39\,000+120\,000$

d $0.0018-0.00006$

5 Work out the answers to these calculations.
Give your answers in standard form.

a $3\times2\times10^5=\square\times10^5$ **b** $5\times1.5\times10^{-7}$

c $12\times5\times10^3$ **d** $9\times2\times10^{-6}$

e $8\times10^{-3}\div2=\square\times10^{-3}$ **f** $6\times10^3\div3$

g $4\times10^8\div8$ **h** $1\times10^{-11}\div5$

Q5 hint

Check that your answer is in standard form.

Key point

To multiply and divide numbers in standard form, use the laws of indices to simplify the power of 10.

Example

Giving your answer in standard form, work out $12\times10^3\times3\times10^2$.

$12\times10^3\times3\times10^2=12\times3\times10^3\times10^2$ — Rewrite the multiplication, grouping the numbers together and the powers of 10 together.

$=36\times10^5$ — Multiply the numbers together. Use the laws of indices to combine the powers of 10.

$=3.6\times10^6$ — If the number part is not between 1 and 10, rewrite the number in standard form.

6 Giving your answers in standard form, work out

a $3\times10^7\times7\times10^3$ **b** $5\times10^2\times4\times10^{11}$

c $1.5\times10^7\times2\times10^{-1}$ **d** $(4\times10^{-2})\times(3\times10^6)$

e $(3\times10^9)\times(3.2\times10^{-3})$ **f** $(7\times10^{-6})\times(1.11\times10^{-5})$

Q6d hint

$(4\times10^{-2})\times(3\times10^6)$
$=4\times10^{-2}\times3\times10^6$

Example

Giving your answer in standard form, work out

$$\frac{9\times10^4}{3\times10^2}$$

$$\frac{9\times10^4}{3\times10^2}=\frac{9}{3}\times\frac{10^4}{10^2}$$

Write the calculation as two fractions, grouping the numbers together and the powers of 10 together.

$$=3\times10^2$$

Divide the numbers and use the laws of indices for the powers of 10. Check that the answer is in standard form.

7 Giving your answers in standard form, work out

a $\frac{8\times10^{-7}}{2\times10^3}$ **b** $\frac{6\times10^{-9}}{2\times10^3}$ **c** $\frac{2\times10^{-10}}{4\times10^{-5}}$

d $(8.8\times10^8)\div(4\times10^3)$ **e** $(9.01\times10^7)\div(1\times10^4)$ **f** $(2.4\times10^4)\div(3\times10^{-2})$

Q7d hint

$\frac{8.8\times10^8}{4\times10^3}$

8 Giving your answer as an ordinary number, work out

a $3.4\times10^7\times2\times10^{-4}$ **b** $\frac{4.5\times10^3}{1.5\times10^{-1}}$ **c** $\frac{4\times10^8\times1.2\times10^{-5}}{6\times10^{-2}}$

Exam-style question

9 Work out the value of $\frac{2.44\times10^3\times1.3\times10^5}{5.2\times10^{-4}}$.

Give your answer in standard form. **(2 marks)**

Exam tip

You can use your calculator to work out the answer. Then don't forget to write it in standard form.

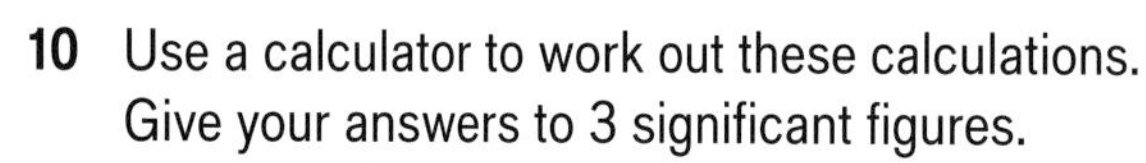

10 Use a calculator to work out these calculations.
Give your answers to 3 significant figures.

a $5.3\times10^7\times7.2\times10^3$ **b** $4.2\times10^7\times3.56\times10^{-2}$

c $9.1\times10^{-4}\times3.8\times10^{-6}$ **d** $(5.6\times10^3)\div(3.4\times10^8)$

e $(3.2\times10^4)\div(5.02\times10^{-5})$ **f** $\frac{5.4\times10^{13}}{3.82\times10^{-5}}$

Q10 hint

To round a standard form number to 3 significant figures, only round the number part that is between 1 and 10.

11 **Problem-solving** A snail crawls 1×10^{-3} kilometres in one hour.
How far will it travel in 3 days? Give your answer as an ordinary number in metres.

12 **Problem-solving** A grain of rice has a mass of 2×10^{-5} kg.
How many grains of rice are there in 1 kg?

Key point

To add and subtract numbers in standard form, write both numbers as ordinary numbers, add or subtract, and then convert back to standard form.

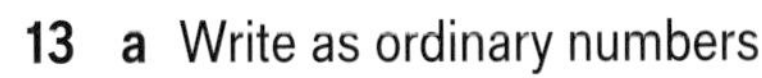

13 **a** Write as ordinary numbers

  i 3.2×10^7 **ii** 1.9×10^8

b Use your answers to part **a** to work out
$3.2 \times 10^7 + 1.9 \times 10^8$

c Write your answer to part **b** in standard form.

Q13b hint

Write as a column addition.

14 **a** Use the method in **Q13** to work these out, giving your answers in standard form.

  i $9 \times 10^{10} + 4 \times 10^{11}$ **ii** $5 \times 10^8 + 7 \times 10^6$

  iii $6.13 \times 10^7 + 7.2 \times 10^3$ **iv** $9.2 \times 10^{-3} + 3.2 \times 10^{-2}$

  v $5.4 \times 10^{-7} + 7.6 \times 10^{-5}$

b Check your answers to part **a** using a calculator.

Q14a iv and v hint

Write both numbers as decimals.

15 **a** Write as decimals

  i 1.9×10^{-4} **ii** 3.4×10^{-5}

b Use your answer to part **a** to work out
$1.9 \times 10^{-4} - 3.4 \times 10^{-5}$

c Write your answer to part **b** in standard form.

Q15 hint

Write as a column subtraction.

16 **a** Use the method in **Q15** to work these out, giving your answer in standard form.

  i $3 \times 10^9 - 7 \times 10^8$ **ii** $6 \times 10^{-7} - 8 \times 10^{-9}$

  iii $5.3 \times 10^5 - 8.1 \times 10^3$ **iv** $3.2 \times 10^{-4} - 4.7 \times 10^{-6}$

  v $4.7 \times 10^{-10} - 5.09 \times 10^{-12}$ **vi** $9.99 \times 10^{-15} - 4.4 \times 10^{-18}$

b Check your answers to part **a** using a calculator.

17 Giving your answer in standard form, use a calculator to work out

  a $5.2 \times 10^{-7} - 4.11 \times 10^{-12}$

  b $3.2 \times 10^{11} + 5.004 \times 10^{14} - 4.19 \times 10^{13}$

18 Use a calculator to work out these calculations.
Give your answers to 3 significant figures.

  a $\dfrac{3.2 \times 10^4 - 7.6 \times 10^3}{4.1 \times 10^{-2}}$ **b** $\dfrac{5.4 \times 10^{-3}}{2.5 \times 10^6 + 3.5 \times 10^8}$

Q18 hint

Enter the calculation as a fraction.

Exam-style question

19 The distance from Mercury to the Sun is approximately 58 000 000 000 m.
Convert this distance to km.
Give your answer in standard form. **(2 marks)**

20 The distance between Earth and Mars is estimated to be 57.6 million km.

  a Write this distance in metres.

  b Write your distance from part **a** in standard form.

  c The speed of light is approximately 3×10^8 m/s.

  Use the formula time $= \dfrac{\text{distance}}{\text{speed}}$ to calculate an estimate for the time taken for light to travel from Mars to Earth.

18 Check up

ActiveLearn Homework 

Reciprocals and fractions

1 Find the reciprocal of 1.4. Give your answer in its simplest form.

2 Work out **a** $\frac{4}{5}\times1\frac{5}{7}$ **b** $4\frac{1}{4}\times\frac{2}{3}$ **c** $1\frac{1}{2}\times2\frac{2}{3}$

3 Work out **a** $4\frac{1}{2}\div\frac{3}{4}$ **b** $3\frac{1}{3}\div1\frac{5}{6}$

Indices

4 Write as a single power **a** $(5^2)^3\times5$ **b** $(2^{-3})^4$ **c** $x^{-2}\times x^4$ **d** $\frac{3^{-8}}{3^5}$

5 Work out the value of

a 4^{-1} **b** 2^{-3} **c** y^0 **d** $\left(\frac{3}{5}\right)^{-2}$ **e** $\left(\frac{3}{4}\right)^0$

Standard form

6 Write each number as an ordinary number.

a 4×10^8 **b** 5.26×10^3 **c** 3.5×10^{-1} **d** 8.099×10^{-7}

7 Write each number in standard form.

a 190 000 **b** 1 050 000 000 **c** 0.000 007 **d** 0.000 045 2

8 Giving your answers in standard form, work out

a $4\times10^4\times2\times10^5$ **b** $2.2\times10^{-2}\times3\times10^7$ **c** $5\times10^3\times3\times10^4$

9 Work out

a $(9\times10^8)\div(3\times10^2)$ **b** $(4\times10^4)\div(8\times10^{-3})$ **c** $(7.7\times10^{-8})\div(2.2\times10^{-2})$

10 Work out **a** $3\times10^{12}+2.4\times10^{11}$ **b** $9.1\times10^{-5}-3.5\times10^{-7}$

11 Use a calculator to work out **a** $4.5\times10^{-3}\times9.2\times10^{-6}$ **b** $\frac{4.5\times10^7}{2.3\times10^{-2}+5.7}$

Give your answers in standard form to 3 significant figures.

12 **Reflect** How sure are you of your answers? Were you mostly

Just guessing ☹ Feeling doubtful 😐 Confident 

What next? Use your results to decide whether to strengthen or extend your learning.

Challenge

13 Use each of the numbers 1, 2, 3, 4, 5 and 6 once in the following calculation to give the largest solution possible.

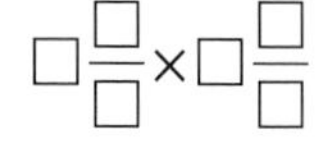

Q13 hint

For example, $1\frac{2}{3}\times4\frac{5}{6}$

18 Strengthen

ActiveLearn Homework 

Reciprocals and fractions

1 Copy and complete to work out the reciprocal of these numbers.

a $0.2 = \frac{\square}{10}$ reciprocal $= \frac{10}{\square} = \square$ **b** $0.5 = \frac{\square}{\square}$ reciprocal $= \frac{\square}{\square} = \square$

c $0.25 = \frac{\square}{100}$ reciprocal $= \frac{100}{\square} = \square$ **d** $0.05 = \frac{\square}{\square}$ reciprocal $= \frac{\square}{\square} = \square$

e $1.3 = 1\frac{\square}{10} = \frac{13}{10}$ reciprocal $= \frac{\square}{\square}$ **f** $2.1 = 2\frac{\square}{\square} = \frac{\square}{\square}$ reciprocal $= \frac{\square}{\square}$

2 **a** Write $3\frac{2}{5}$ as an improper fraction.

b Copy and complete the calculation.

$$3\frac{2}{5} \times \frac{2}{3} = \frac{\square}{5} \times \frac{\square}{\square}$$
$$= \frac{\square}{\square} \times \frac{\square}{\square}$$
$$= \frac{\square}{\square} = \square\frac{\square}{\square}$$

3 Use the method in **Q2** to work out

a $2\frac{1}{2} \times \frac{1}{3}$ **b** $3\frac{1}{5} \times \frac{2}{5}$ **c** $\frac{3}{7} \times 4\frac{1}{2}$

Q3c hint

$\frac{3}{7} \times 4\frac{1}{2} = 4\frac{1}{2} \times \frac{3}{7}$

4 Copy and complete the calculation.

$$3\frac{1}{3} \times 1\frac{3}{4} = \frac{10}{\square} \times \frac{\square}{4}$$
$$\frac{\overset{\square}{\cancel{10}}}{\square} \times \frac{\square}{\underset{\square}{\cancel{4}}}$$
$$= \frac{\square}{\square} \times \frac{\square}{\square}$$
$$= \frac{\square}{\square} = \square\frac{\square}{\square}$$

Q4 hint

Divide the 10 and the 4 by 2.

5 Use the method in **Q4** to work out

a $1\frac{1}{2} \times \frac{1}{3}$ **b** $2\frac{4}{5} \times \frac{5}{7}$ **c** $\frac{2}{3} \times 5\frac{1}{4}$

6 Copy and complete the calculation.

$$1\frac{1}{4} \times 2\frac{1}{3} = \frac{\square}{4} \times \frac{\square}{3}$$
$$= \frac{\square \times \square}{\square \times \square}$$
$$= \frac{\square}{\square} = \square\frac{\square}{\square}$$

7 Use the method in **Q6** to work out

a $1\frac{3}{5}\times1\frac{1}{3}$ **b** $2\frac{1}{3}\times4\frac{1}{2}$ **c** $1\frac{1}{4}\times1\frac{1}{3}$

8 Janice starts a problem.

$$2\frac{1}{3}\div\frac{2}{3}=\frac{7}{3}\div\frac{2}{3}$$
$$=\frac{7}{3}\times\ldots$$

Copy and complete the calculation.

9 Use the method in **Q8** to work out

a $3\frac{1}{5}\div\frac{1}{4}=\frac{\square}{5}\div\frac{1}{4}=$ **b** $\frac{4}{5}\div2\frac{1}{2}$ **c** $1\frac{1}{2}\div\frac{2}{3}$

Indices

1 **a** **i** Copy and complete $8^2\div8^2=\frac{8^2}{8^2}=\frac{\square}{\square}=\square$

ii Copy and complete $8^2\div8^2=8^{\square-\square}=8^{\square}$

iii Use your answers to parts **i** and **ii** to find the value of 8^0.

b Repeat part **a** for $10^2\div10^2$ to find the value of $10^{\square}$.

c Copy and complete the rule:
When you write a number to the power 0, the answer is ____.

2 **a** **i** Copy and complete $7^2\div7^5=7^{2-5}=7^{\square}$

ii Copy and complete

$$7^2\div7^5=\frac{7^2}{7^5}=\frac{\not{7}\times\not{7}}{\not{7}\times\not{7}\times7\times7\times7}=\frac{1}{7^{\square}}$$

iii Use your answers to parts **i** and **ii** to copy and complete $7^{-3}=\frac{1}{7^{\square}}$

Q2a hint

To divide powers of the same number, subtract the indices.

b **i** Copy and complete $4^3\div4^5=4^{3-5}=4^{\square}$

ii Copy and complete

$$4^3\div4^5=\frac{4^3}{4^5}=\frac{\not{4}\times\not{4}\times\not{4}}{\not{4}\times\not{4}\times\not{4}\times4\times4}=\frac{1}{4^{\square}}$$

iii Use your answers to parts **i** and **ii** to copy and complete $4^{-2}=\frac{1}{4^{\square}}$

3 Work out

a 8^{-1} **b** 5^{-1} **c** 9^0 **d** $\left(\frac{3}{4}\right)^0$ **e** $\left(\frac{1}{2}\right)^{-1}$

4 Copy and complete to write as a single power

a $2^3\times2^{-4}=2^{3-4}=2^{\square}$ **b** $4^{-2}\times4^5=4^{-2+5}=4^{\square}$

c $8^{-9}\times8^2=8^{-9+2}=8^{\square}$ **d** $19^{-7}\times19^{-3}=19^{-7-3}=19^{\square}$

5 Copy and complete to write as a single power

a $7^{-2} \div 7^{3} = 7^{-2-3} = 7^{\square}$ **b** $12^{8} \div 12^{-2} = 12^{8--2} = 12^{\square}$

c $6^{2} \div 6^{-5} = 6^{2--5} = 6^{\square}$ **d** $2^{-7} \div 2^{-4} = 2^{-7--4} = 2^{\square}$

6 Write as a single power

a $(3^2)^4 = 3^{(2\times4)} = 3^{\square}$ **b** $(5^2)^7$

c $(19^3)^{10}$ **d** $(6^2)^3$

e $(5^{-2})^3$ **f** $(5^3)^{-4}$

g $(9^{-2})^{-3}$ **h** $(6^{-4})^{-2}$

Q6e hint

The rule is the same if one or both powers are negative.

7 Copy and complete

a $3^{-1} = \frac{1}{3}$

$3^{-2} = \left(\frac{1}{3}\right)^2 = \frac{1}{\square} \times \frac{1}{\square} = \frac{1}{\square}$

$3^{-3} = \left(\frac{1}{3}\right)^3 = \frac{1}{\square} \times \frac{1}{\square} \times \frac{1}{\square} = \frac{1}{\square}$

b $\left(\frac{2}{5}\right)^{-1} = \frac{5}{2}$

$\left(\frac{2}{5}\right)^{-2} = \left(\frac{5}{2}\right)^2 = \frac{5}{2} \times \frac{5}{2} = \frac{\square}{\square}$

$\left(\frac{2}{5}\right)^{-3} = \left(\frac{5}{2}\right)^3 = \frac{5}{2} \times \frac{5}{2} \times \frac{5}{2} = \frac{\square}{\square}$

8 Use the method in **Q7** to work out the value of

a $\left(\frac{2}{3}\right)^{-2}$ **b** $\left(\frac{1}{4}\right)^{-3}$ **c** $\left(\frac{4}{7}\right)^{-2}$ **d** $\left(\frac{3}{10}\right)^{-3}$

Standard form

1 Copy and complete this table of powers of 10.

10^{-6}	10^{-5}	10^{-4}	10^{-3}	10^{-2}	10^{-1}	10^{0}	10^{1}	10^{2}	10^{3}	10^{4}	10^{5}	10^{6}
		0.0001			0.1		10		1000	10 000		

2 Use the table you completed for **Q1** to write each number as an ordinary number.

a $3\times10^{-4} = 3\times0.0001 = \square$ **b** $5\times10^{2} = 5\times\square = \square$

c 9×10^{-4} **d** 1.2×10^{3}

e 5.7×10^{-5} **f** 1.12×10^{2}

g 9.03×10^{-4} **h** 1.01×10^{6}

3 A number written in standard form looks like this.

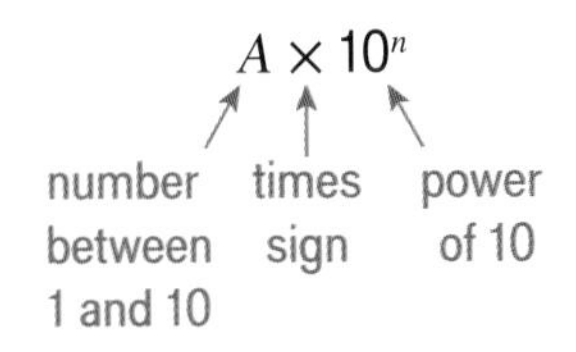

Q3a hint

×10 ×10 ×10 ×10

1 . 8

1 8 0 0 0

Copy and complete to write each number in standard form.

a $18\,000 = 1.8\times10^{\square}$ **b** $960\,000 = 9.6\times10^{\square}$

c $40\,000 = 4\times10^{\square}$ **d** $9\,000\,000 = \square\times10^{\square}$

e $751\,000 = \square\times10^{\square}$ **f** $1\,080\,000 = \square\times10^{\square}$

Q3e hint

First decide which of these lies between 1 and 10: 751, 75.1, 7.51, 0.751

4 Write each number in standard form.

a $0.06 = 6 \times 10^{\square}$

b $0.000\,04 = \square \times 10^{\square}$

c $0.0036 = 3.6 \times 10^{\square}$

d $0.000\,12 = \square \times 10^{\square}$

e $0.234 = \square \times 10^{\square}$

f $0.000\,005\,08 = \square \times 10^{\square}$

Q4a hint

Divide by how many 10s to get 0.06?

÷10 ÷10

6

0 . 0 6

Q4e hint

First decide which of these lies between 1 and 10: 234, 23.4, 2.34

5 Copy and complete

$3 \times 10^7 \times 2 \times 10^3 = 3 \times 2 \times 10^7 \times 10^3$

$= \square \times 10^{7+3}$

$= \square \times 10^{\square}$

6 Use the method in **Q5** to work out

a $5 \times 10^2 \times 1 \times 10^9$

b $2 \times 10^{-8} \times 2 \times 10^2$

c $3.4 \times 10^{-4} \times 2 \times 10^{-5}$

Q6b hint

10^{-8+2}

7 Rewrite these numbers in standard form.
The first and third are started for you.

a $15 \times 10^5 = 1.5 \times \square \times 10^5 = 1.5 \times 10^{\square+\square} = 1.5 \times 10^{\square}$

b 29×10^3

c $36.1 \times 10^4 = 3.61 \times \square \times 10^4 = 3.61 \times 10^{\square+\square} = 3.61 \times 10^{\square}$

d 42.8×10^6

e 17.2×10^{-3}

8 Giving your answers in standard form, work out these calculations.

a $4 \times 10^5 \times 3 \times 10^{-2}$

b $9 \times 10^{-2} \times 8 \times 10^4$

c $7 \times 10^8 \times 5 \times 10^5$

d $2.1 \times 10^{11} \times 6 \times 10^{-15}$

Q8 hint

Make sure your final answer is a number between 1 and $10 \times 10^{\square}$

9 Copy and complete

$$\frac{9 \times 10^7}{3 \times 10^3} = \frac{9}{3} \times \frac{10^7}{10^3} = 3 \times 10^{7-3} = 3 \times 10^{\square}$$

10 Use the method in **Q9** to work these out (give your answers in standard form).

a $\dfrac{8 \times 10^8}{2 \times 10^3}$

b $\dfrac{6.9 \times 10^5}{3 \times 10^{-2}}$

c $(8 \times 10^{-4}) \div (2 \times 10^3)$

d $\dfrac{1 \times 10^5}{2 \times 10^{-2}}$

Q10c hint

Write the division as $\dfrac{8 \times 10^{-4}}{2 \times 10^3}$

Q10d hint

Make sure your final answer is in standard form.

11 **a** Write as ordinary numbers

i 5.2×10^4

ii 3.5×10^3

b Use your answer to part **a** to copy and complete these calculations.

i $5.2 \times 10^4 + 3.5 \times 10^3$

```
   □ □ □ □ □
+    □ □ □ □
____________

____________
```

ii $5.2 \times 10^4 - 3.5 \times 10^3$

```
   □ □ □ □ □
-    □ □ □ □
____________

____________
```

c Write your answers to **b i** and **b ii** in standard form.

12 **a** Write as ordinary numbers (decimals)

i 1.5×10^{-3}

ii 2.23×10^{-5}

b Use your answer to part **a** to copy and complete these calculations.

i $1.5 \times 10^{-3} + 2.23 \times 10^{-5}$

```
  0. □ □ □ □
+ 0. □ □ □ □ □ □ □
__________________

__________________
```

ii $1.5 \times 10^{-3} - 2.23 \times 10^{-5}$

```
  0. □ □ □ □ 0 0 0
- 0. □ □ □ □ □ □ □
__________________

__________________
```

c Write your answers to **b i** and **b ii** in standard form.

13 Use a calculator to work out

i 6.871×10^{-5}

ii $6.871 \times 10^{-5} \times 1.55 \times 10^2$

iii $\dfrac{6.871 \times 10^{-5}}{1.55 \times 10^2}$

For each calculation

a write all the digits on your calculator display

b write your answer in standard form, rounding the number between 1 and 10 to 3 significant figures

14 Use a calculator to work out

a $7.904 \times 10^{11} - 3.9 \times 10^8$

b $(7.8 \times 10^5) \div (1.86 \times 10^{-5} + 9.79 \times 10^{-3})$

c $\dfrac{8.3 \times 10^{-8} + 1.33 \times 10^{-5}}{4.55 \times 10^5}$

Give your answers in standard form to 3 significant figures.

Q14b and c hint

Use brackets on your calculator.

18 Extend

1 Write these numbers in order, smallest first.

$\frac{7}{9}$ $\frac{2}{3} \div 3\frac{4}{5}$ 0.98 57% $3\frac{1}{3} \div 1\frac{2}{3}$

2 There are $1\frac{3}{4}$ pints in a litre.
There are 8 pints in a gallon.
How many litres are there in a gallon?

3 **a** $2n \times 3 = 24$
What is the value of n?

b $a^{-1} \times 12 = 2$
What is the value of a?

4 $a = -2$, $b = -3$ and $c = 4$. Work out the value of

a $(10^a)^b$ **b** the reciprocal of a **c** c^b **d** $c^b \times c^c$

e $c^a \div c^b$ **f** $12b^0$ **g** $\frac{1}{a} \times a$ **h** $(bc)^a$

5 **Reasoning**

a Write as a power of 2

i 4 **ii** 4^3 **iii** $(4^3)^8$

b Write as a power of 5

i 125 **ii** 125^{-3} **iii** $(125^{-3})^9$

c Write $(81^4)^{-5}$ as a power of 3.

Q5a ii hint
$4^3 = (2^{\square})^3$

6 Simplify

a $2x^{-1}$ **b** $(2x)^{-1}$ **c** $2x^{-2}$ **d** $(2x)^{-2}$

7 Use a calculator to evaluate $(0.002)^{-3}$. Write your answer in

a words **b** standard form

8 Use a calculator to work out

$$\frac{0.3^{-7} + 8.2^{-4}}{3.02^{15} - 5.2^{-3}}$$

Give your answer in standard form to 3 significant figures.

9 **Reasoning** A cuboid has sides of length 0.01 m, 0.03 m and 0.03 m.
Work out

a the volume **b** the surface area

Give your answers in standard form.

Q9 hint
Sketch the cuboid and label the dimensions.

10 **Problem-solving** How many

a kB in 1 GB

b μg in 1 mg

c μV in 1 GV?

Q10a hint
1 kilobyte (kB) = 10^3 bytes,
1 gigabyte (GB) = 10^9 bytes

Q10b hint
1 microgram (μg) = 10^{-6} g,
1 milligram (mg) = 10^{-3} g

Exam-style question

11 Work out the reciprocal of 2.5×10^{-6}. **(3 marks)**

Exam tip

Show your working so it is clear that you understand the terminology (like reciprocal) and notation (like numbers written in standard form).

12 Problem-solving
A litre of water contains approximately 3.35×10^{25} molecules of water.
What is the volume of one molecule of water?

Q12 hint

How many cm^3 in 1 litre?

13 Reasoning The distance from Earth to a distant comet (in kilometres) is recorded at five equally spaced intervals over a year.

3.2×10^9 $\quad 3.91 \times 10^9$ $\quad 4.201 \times 10^9$ $\quad 3.99 \times 10^9$ $\quad 4.29 \times 10^9$

Work out the mean distance between Earth and the comet.

14 Reasoning A satellite travels 4×10^4 km in 1.5 hours.
Work out the speed of the satellite. Give your answer in standard form.

Exam-style question

15 Write these numbers in order, largest first.

$2\frac{1}{3} \times 3\frac{3}{5}$ $\quad 6.5 \times 10^2$ $\quad (3^2)^2$ $\quad 5.6 \times 10^4 - 3.2 \times 10^3$ **(5 marks)**

16 The diameter of Earth is approximately 1.3×10^4 km.
Modelling Earth as a sphere, estimate

a the surface area of Earth

b the volume of Earth

Give your answers in standard form.

c About 70% of Earth's surface is water.
What is its land area?

Q16 hint

For a sphere of radius r, the surface area $= 4\pi r^2$ and the volume $= \frac{4}{3}\pi r^3$. Two significant figures is accurate enough for an estimate.

17 $a = 2 \times 10^{-3}$, $b = 3 \times 10^{-2}$ and $c = 5 \times 10^{-4}$. Work out the value of

a $a + b$ **b** abc **c** $\frac{ab}{c}$ **d** c^2

18 Reasoning A bacterium cell has a surface area of 2×10^{-6} cm^2.
In a Petri dish there are ten cells.
The number of cells doubles each day.
What is the total surface area of the bacteria after

a 1 day **b** 2 days **c** 10 days?

Give your answers in standard form.

Q18 hint

Assume that the cells do not join together.

Exam-style question

19 $p^2 = \frac{q - r}{r}$

$q = 1.6 \times 10^9$

$r = 4 \times 10^4$

Find the value of p. Give your answer in standard form correct to 2 significant figures. **(3 marks)**

Exam tip

Check that your answer is the value of p and not the value of p^2.

18 Test ready

Summary of key points

To revise for the test:

- Read each key point, find a question on it in the mastery lesson, and check you can work out the answer.
- If you cannot, try some other questions from the mastery lesson or ask for help.

Key points

1	When multiplying or dividing mixed numbers, change the mixed number to an improper fraction first.	→ 18.1
2	To raise the power of a number to another power, multiply the indices. For example, $(5^3)^4 = 5^{3\times4} = 5^{12}$.	→ 18.2
3	Any number raised to the power 0 is equal to 1. For example, $12^0 = 1$.	→ 18.2
4	Any number (or term) raised to the power −1 is the reciprocal of the number. For example, $4^{-1} = \frac{1}{4}$, $\left(\frac{2}{3}\right)^{-1} = \frac{3}{2}$.	→ 18.2
5	To find a negative power, find the reciprocal and then raise the value to the positive power. For example, $5^{-2} = \left(\frac{1}{5}\right)^2$.	→ 18.2
6	**Standard form** is used to write very large or very small numbers.	→ 18.3
7	A number written in standard form is a value between 1 and 10 multiplied by a power of 10.	→ 18.3, 18.4
8	Large numbers written in standard form have a positive power of 10. For example, $3.5 \times 10^9 = 3\,500\,000\,000$.	→ 18.3
9	Small numbers written in standard form have a negative power of 10.	→ 18.4
10	To multiply or divide numbers in standard form, use the laws of indices to simplify the power of 10.	→ 18.5
11	To add or subtract numbers in standard form, write both numbers as ordinary numbers, add or subtract, and then convert back to standard form.	→ 18.5

Sample student answers

Exam-style question

1 One sheet of paper is 5×10^{-3} cm thick.
Malik wants to put 600 sheets of paper into the paper tray in his printer.
The paper tray is 2.5 cm deep.
Is the paper tray deep enough for 600 sheets of paper?
You must explain your answer. **(3 marks)**

Student A

$$600 \times 5 \times 10^{-3} = 6 \times 100 \times 5 \times 10^{-3}$$
$$= 6 \times 10^2 \times 5 \times 10^{-3}$$
$$= 6 \times 5 \times 10^2 \times 10^{-3}$$
$$= 30 \times 10^{-1}$$
$$= 3\text{ cm}$$

No, the paper tray is not deep enough.

Student B

$$2.5 \div (5 \times 10^{-3}) = \frac{2.5}{5 \times 10^{-3}}$$
$$= \frac{1}{2 \times 10^{-3}}$$
$$= \frac{1}{0.002}$$
$$= 5000$$

Yes, the paper tray is deep enough.

Which student gives the better answer and why?

Exam-style question

2 Work out $\frac{0.06 \times 0.0004}{0.003}$.
Give your answer in standard form. **(3 marks)**

Student A

$$\frac{6 \times 10^{-2} \times 4 \times 10^{-4}}{3 \times 10^{-3}} = \frac{6 \times 4 \times 10^{-2} \times 10^{-4}}{3 \times 10^{-3}}$$
$$= \frac{24 \times 10^{-6}}{3 \times 10^{-3}}$$
$$= 8 \times 10^{3}$$

Student B

$$0.06 \times 0.0004 = 0.000024$$

$$\frac{0.000024}{0.003} = \frac{0.000024}{0.003} \xrightarrow{\times 1\,000\,000} \frac{24}{3000}$$

(numerator and denominator each $\times 1\,000\,000$)

$$= \frac{8}{1000}$$

a Student A and Student B used different methods to answer the question.
Which method do you prefer? Explain.

b Both students' final answers are incorrect. Explain why.

18 Unit test

ActiveLearn Homework

1 Work out

a $3\frac{1}{5}\times\frac{3}{4}$ **(2 marks)**

b $4\frac{2}{3}\times2\frac{1}{7}$ **(2 marks)**

2 Work out

a $1\frac{1}{2}\div\frac{2}{5}$ **(2 marks)**

b $3\frac{1}{3}\div1\frac{3}{5}$ **(2 marks)**

3 Write as a single power

a $3^{-7}\times3^{2}$ **(1 mark)**

b $\frac{10^{8}}{10^{-3}}$ **(1 mark)**

c $(11^{-2})^{5}$ **(1 mark)**

4 Write each number as an ordinary number.

a 3.04×10^{7} **(1 mark)**

b 2.1×10^{-3} **(1 mark)**

5 Write each number in standard form.

a 9 070 000 000 **(1 mark)**

b 0.000 031 4 **(1 mark)**

6 $a = 2$. Work out the value of

a a^{0} **(1 mark)**

b a^{1} **(1 mark)**

c $(a^{2})^{3}$ **(1 mark)**

d a^{-2} **(1 mark)**

7 Work out

a $\left(\frac{1}{8}\right)^{2}$ **(2 marks)**

b $\left(\frac{3}{4}\right)^{-3}$ **(2 marks)**

8 A spider web is 0.000 006 m thick.
Write this number in standard form. **(1 mark)**

9 Use a calculator to work out $7234\times69\,147$.

a Write your answer to 3 significant figures. **(1 mark)**

b Write your answer to part **a** in standard form. **(1 mark)**

10 Work out the reciprocal of 0.001. Give your answer in standard form. **(2 marks)**

11 Write these numbers in order of size, smallest first. You must show your working.

4.5×10^{2} $\quad (0.05)^{-2}$ $\quad \frac{(5^{3})^{2}}{5^{2}}$ **(3 marks)**

12 Work out $5.2^2 \times 8.5^{-8}$.
Write your answer in standard form correct to 2 significant figures. **(1 mark)**

13 Work out

a $3 \times 10^5 \times 2 \times 10^2$ **(2 marks)**

b $\dfrac{4 \times 10^{-9}}{2 \times 10^4}$ **(2 marks)**

c $8 \times 10^8 \times 7 \times 10^{-3}$ **(2 marks)**

d $2 \times 10^3 \times 3.1 \times 10^{-7}$ **(2 marks)**

Give your answers

i in standard form

ii as ordinary numbers

14 Giving your answers in standard form, work out

a $8.1 \times 10^7 + 2 \times 10^5$ **(2 marks)**

b $6.07 \times 10^{-8} - 4.2 \times 10^{-9}$ **(2 marks)**

15 The mass of an atom of gold is 3.18×10^{-22} g.
How many atoms of gold are there in 1 gram?
Give your answer in standard form to 3 significant figures. **(2 marks)**

16 The table shows the results of weighing 100 feathers.

Mass (g)	Frequency
1.2×10^{-2}	22
1.3×10^{-2}	37
1.4×10^{-2}	35
1.5×10^{-2}	6

Work out the mean mass of a feather.
Give your answer as an ordinary number. **(4 marks)**

(TOTAL: 50 marks)

17 Challenge The answer to a question is 4^{-5}.
Write as many questions as possible that give this answer.
Try to use

- laws of indices
- standard form
- reciprocals.

18 Reflect Look back at the topics in this unit.

a Which one are you most confident that you have mastered?
What makes you feel confident?

b Which one are you least confident that you have mastered?
What makes you least confident?

c Discuss a question you feel least confident about with a classmate.
How does discussing it make you feel?

19 Congruence, similarity and vectors

Prior knowledge 

19.1 Similarity and enlargement

- Understand similarity.
- Use similarity to solve angle problems.

ActiveLearn Homework 

Warm up

1 **Fluency** When an enlargement makes

a a shape bigger, the scale factor is ______ than 1

b a shape smaller, the scale factor is ______ than 1

2 Which of these fractions are equivalent to $\frac{3}{4}$?

$\frac{4}{5}$ $\frac{6}{8}$ $\frac{5}{6}$ $\frac{15}{20}$

3 Find the scale factor of the enlargement that maps

i shape A onto shape B

ii shape B onto shape A

a

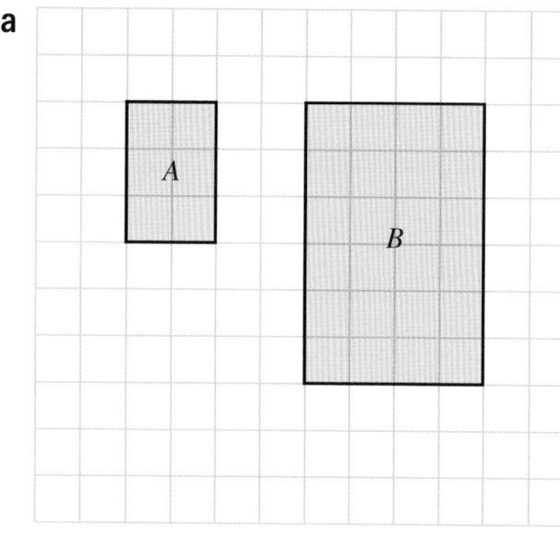

b

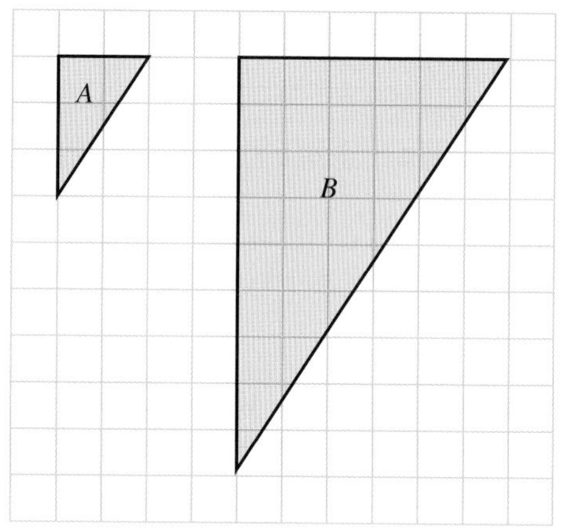

c

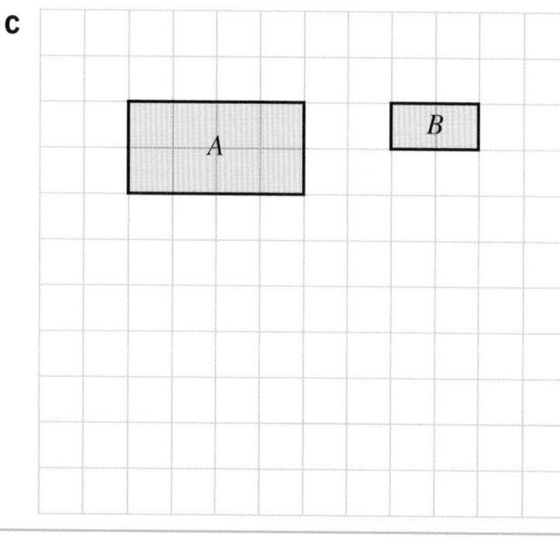

Key point

When one shape is an enlargement of another, the shapes are **similar**.

4 **Reasoning** Which of these rectangles are similar? Explain how you know.

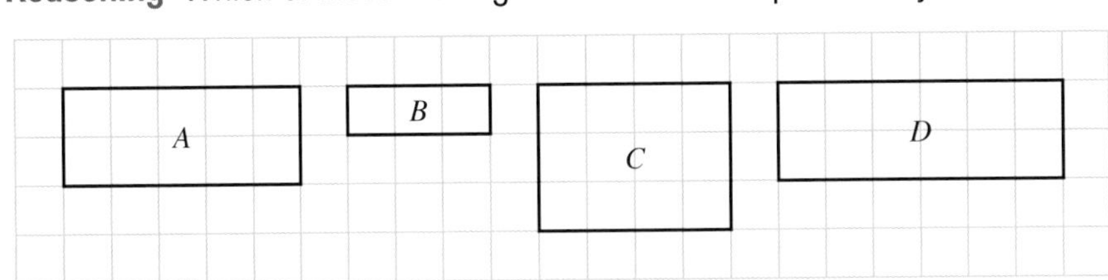

Q4 hint

Which two rectangles are enlargements of each other?

Exam-style question

5 A small photograph has a length of 4 cm and a width of 3 cm.
Preeti enlarges the small photograph to make a large photograph.
The large photograph has a length of 10 cm.

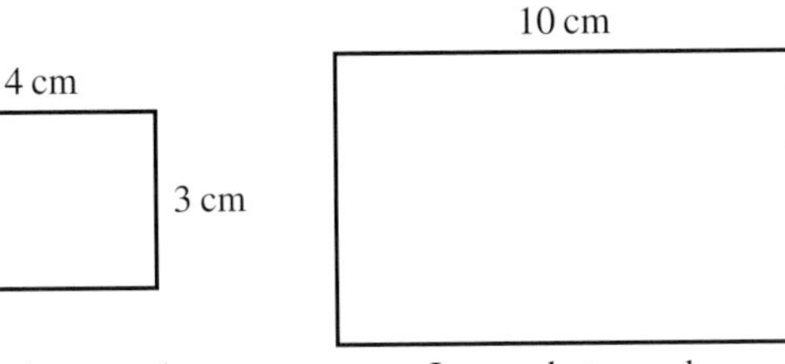

Diagram NOT accurately drawn

Exam tip

Make sure you show your working.

The two photographs are similar rectangles.
Work out the width of the large photograph. **(3 marks)**

6 **Reasoning** Here are two rectangles.
Jamal says the rectangles are similar because
$1 + 2 = 3$ and $5 + 2 = 7$.
Is Jamal correct? Explain.

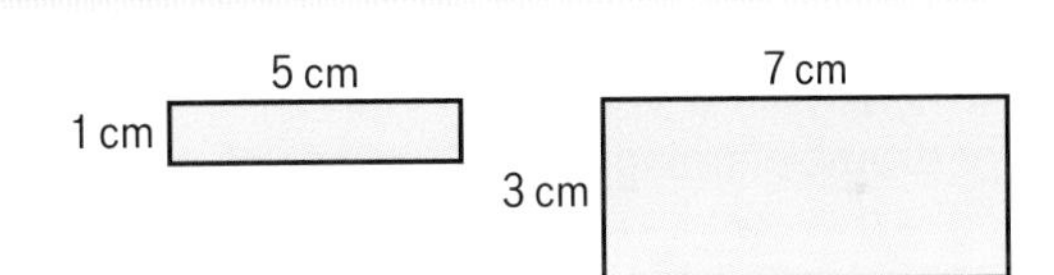

7 **Reasoning** Identify the two similar triangles.
Give a reason for your answer.

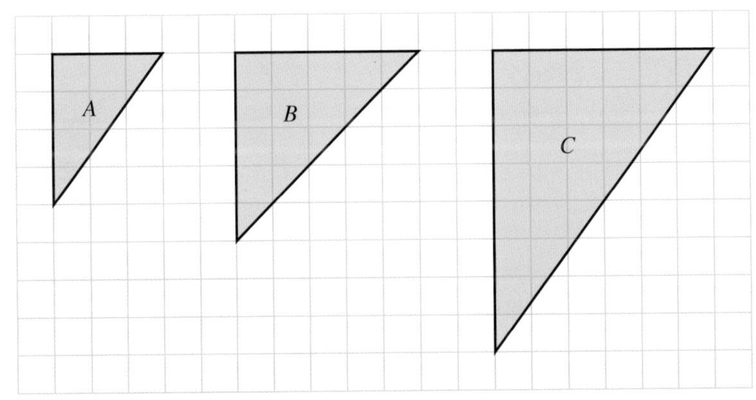

8 a **Reflect** Measure the angles in the two similar triangles in **Q7**.
What do you notice?
b Measure the angles in the other triangle.
c **Reflect** Explain how you can use angles to decide if two shapes are similar.

Key point

These triangles are similar.
Corresponding sides are shown in the same colour.
Corresponding angles are shown with the same angle markers.

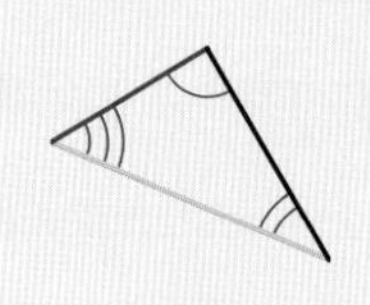

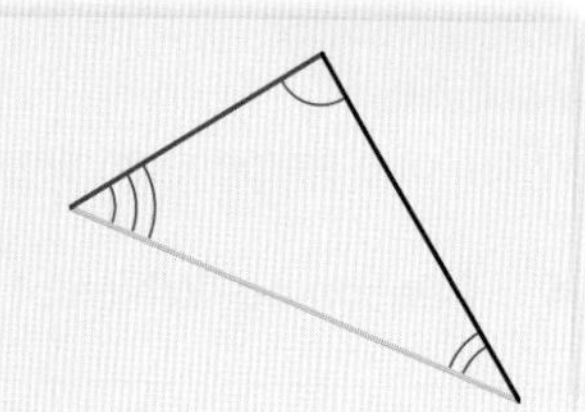

9 Triangle XYZ is similar to triangle ABC.

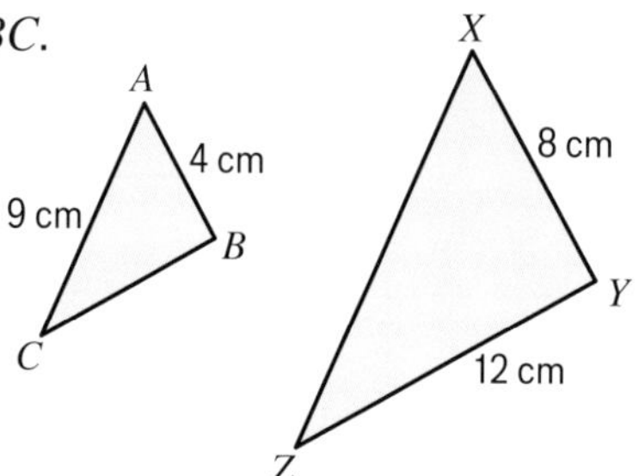

a Which side corresponds to
  i AB **ii** XZ
  iii BC?

b What is the scale factor of the enlargement that maps
  i triangle ABC to triangle XYZ
  ii triangle XYZ to triangle ABC?

c Work out the length of
  i XZ **ii** BC

d Which angle is the same as
  i angle CAB **ii** angle XZY **iii** angle ABC?

e Write $\frac{AB}{AC}$ as a fraction. Then write $\frac{XY}{XZ}$ as a fraction. What do you notice?

Q9b hint
Compare the lengths of corresponding sides

Q9e hint
Simplify the fraction $\frac{XY}{XZ}$

Key point

For similar shapes:
Corresponding sides are all in the same ratio.
Corresponding angles are equal.
For example, triangles ABC and DEF are similar, so

$$\frac{AB}{DE} = \frac{AC}{DF} = \frac{BC}{EF}$$

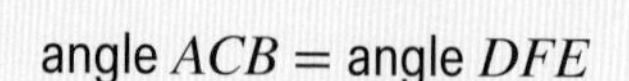

angle BAC = angle EDF angle ABC = angle DEF angle ACB = angle DFE

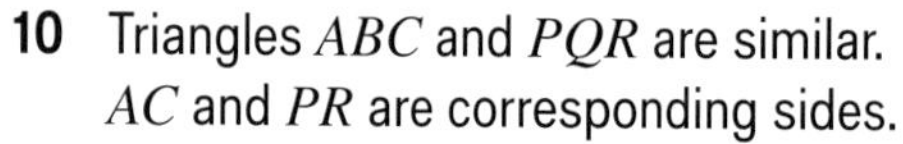

10 Triangles ABC and PQR are similar.
AC and PR are corresponding sides.

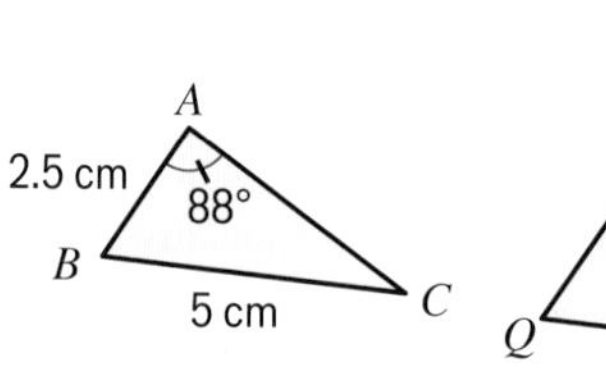

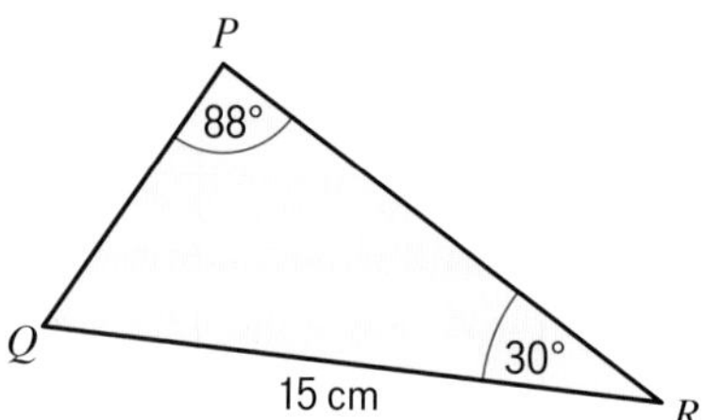

a Which side corresponds to BC?

b Work out the length of PQ.

c Work out the size of
  i angle ACB
  ii angle ABC

11 **Reasoning** **a** Are triangles PQR and UVW similar? Explain your answer.
Sketch triangle UVW in the same orientation (same way round) as triangle PQR.
Mark it with all the information you know.

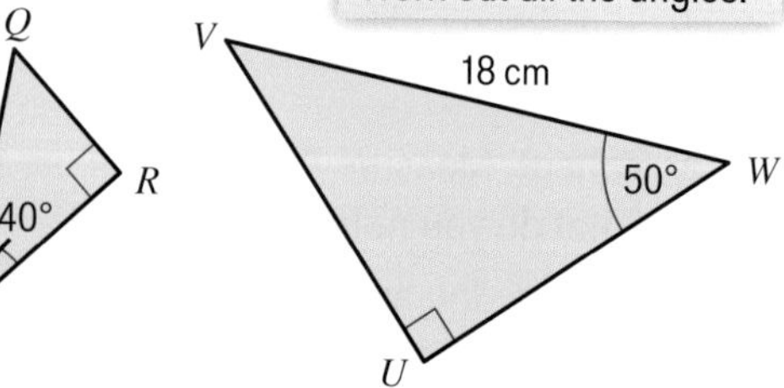

b Which side corresponds to RQ?

c Copy and complete.
  i $VW = \square \times PQ$ **ii** $UW = 3 \times \square$ **iii** $QR = \square \times \square$

Q11a hint
Work out all the angles.

19.2 More similarity

- Find the scale factor of an enlargement.
- Use similarity to solve problems.

Warm up

1 Fluency

Which angle in triangle PQR is equal to the angles in triangle ABC?

a A **b** B **c** C

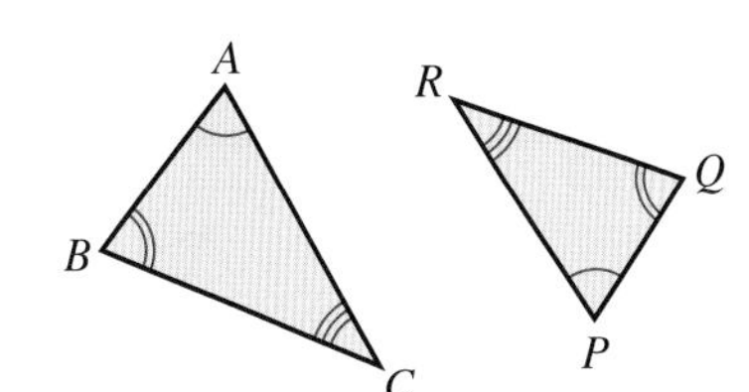

2 On squared paper draw two shapes that are similar to shape A.

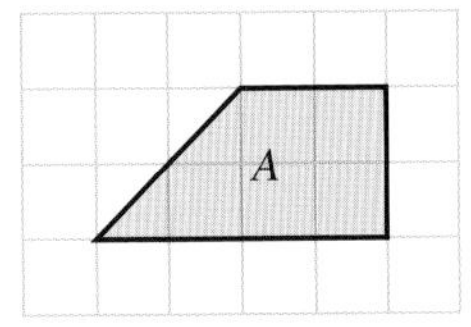

3 Solve $\frac{x}{16} = \frac{3}{12}$.

4 Reasoning Copy the diagram.
Label all of the angles of 53°.
Choose a reason from the box for each.

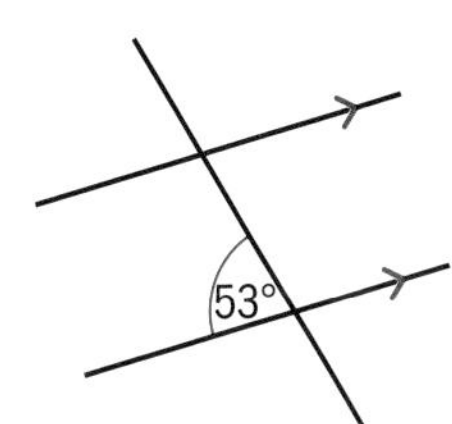

corresponding angle
vertically opposite angle
alternate angle

Key point

Angle facts can be used to find missing angles.
Corresponding sides can be used to work out the scale factor of an enlargement and calculate missing lengths.

5 **a Reasoning** Explain how you know that these triangles are similar.

b Write the scale factor of the enlargement that maps triangle A to triangle B.

c Use the scale factor to find the value of x.

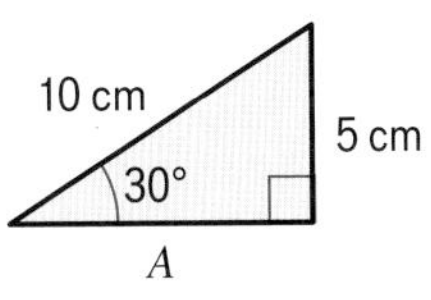

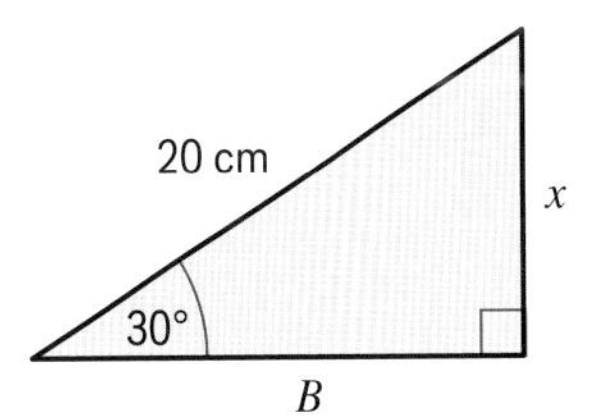

6 **a** Use the diagram in **Q5** to copy and complete

$$\frac{x}{20} = \frac{5}{\square}$$

b Solve the equation in part **a** to find the value of x.

Q6b hint

You should get the same answer as you got in **Q5**.

7 **a** **Reasoning** Explain how you know that this triangle is similar to the triangles in **Q5**.

b Write the value of the ratio $\frac{\text{opp}}{\text{hyp}}$.

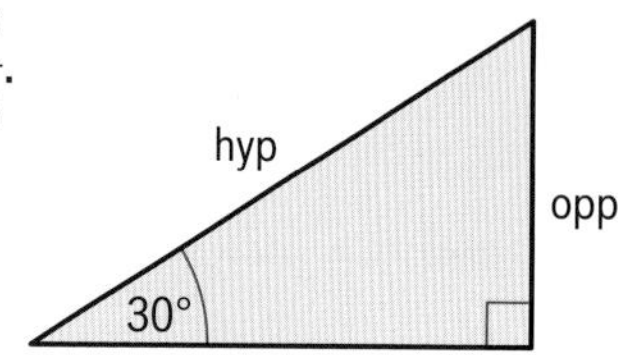

Q7b hint

The value of $\frac{\text{opp}}{\text{hyp}}$ is always the same when the angle is 30° because the triangles are similar.

c The ratio $\frac{\text{opp}}{\text{hyp}}$ is called the sine of the angle.

Check that $\sin 30° = \frac{1}{2}$

8 **Reasoning**

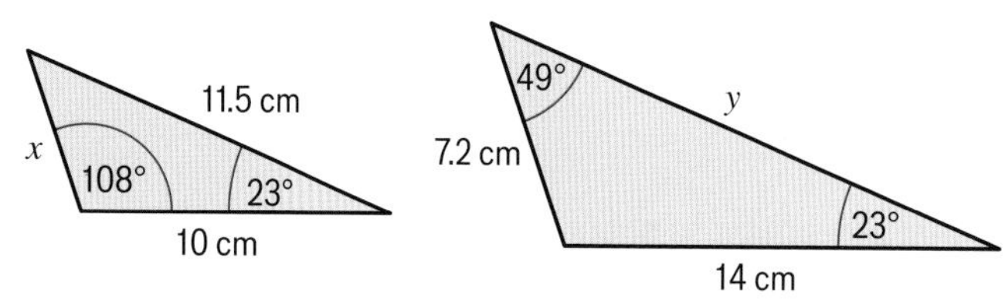

a Explain how you know that these triangles are similar.

b Write the scale factor of the enlargement.

c Find the values of x and y.

9 **Reasoning** These trapezia are mathematically similar.

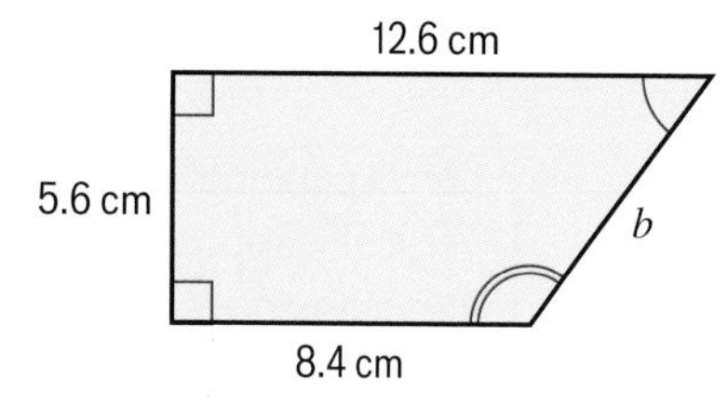

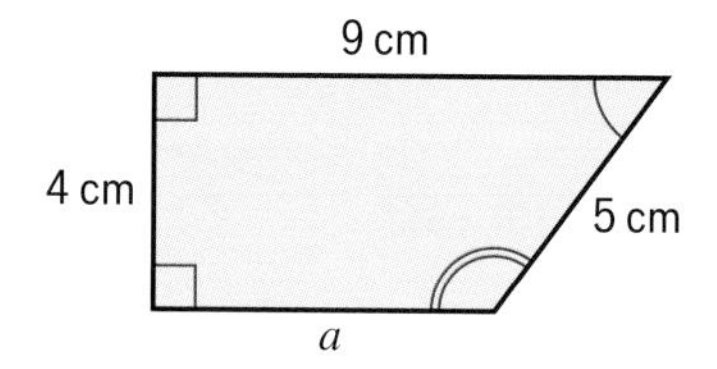

Find the values of a and b.

Exam-style question

10 Triangle ABC and triangle DEF are mathematically similar.

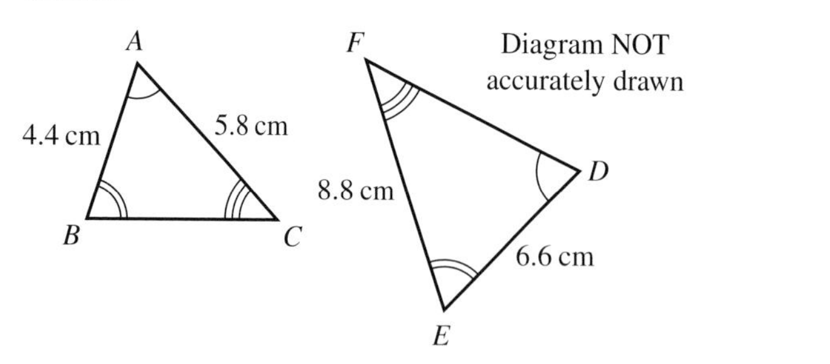

a Work out the length DF. **(2 marks)**

b Work out the length BC. **(2 marks)**

Q10 hint

AB and EF are not corresponding sides.

Exam tip

When comparing similar shapes, it can help to sketch them in the same orientation.

Example

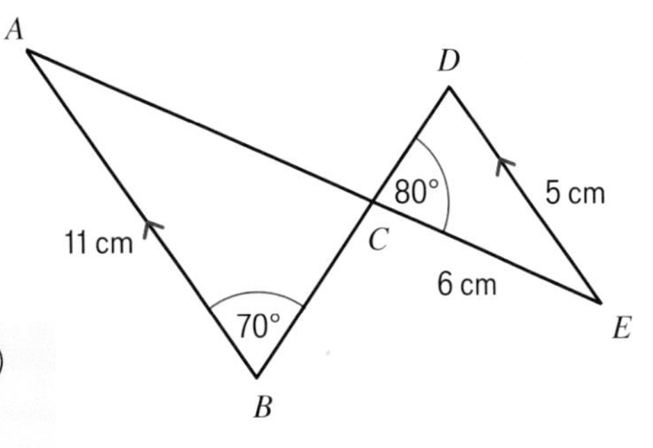

The lines AB and DE are parallel.

a Show that triangles ABC and EDC are similar.

b Write the scale factor of the enlargement that maps triangle ABC to triangle EDC.

c Work out the length of AC.

a Angle $ACB = 80°$ (vertically opposite angles)
Angle $CDE = 70°$ (alternate angles)
Triangles ABC and EDC are similar because they have the same angles.

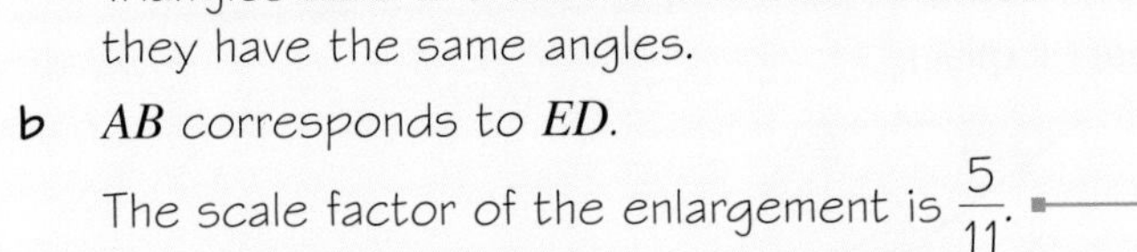

b AB corresponds to ED.
The scale factor of the enlargement is $\frac{5}{11}$.

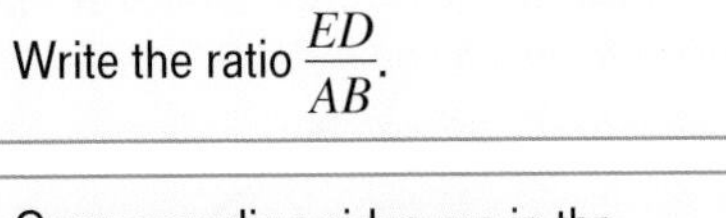

Write the ratio $\frac{ED}{AB}$.

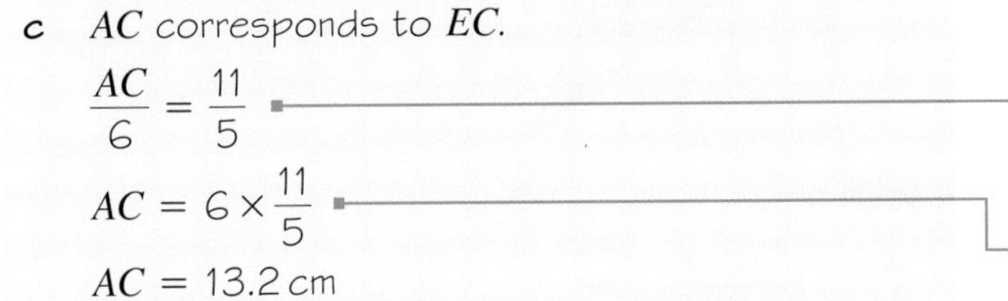

c AC corresponds to EC.

$\frac{AC}{6} = \frac{11}{5}$

$AC = 6 \times \frac{11}{5}$

$AC = 13.2$ cm

Corresponding sides are in the same ratio.
Write the unknown length, AC, on the top of the fraction.

Solve the equation.
Multiply both sides by 6.

11 **Reasoning** The lines PQ and ST are parallel.
Triangles PQR and TSR are similar.

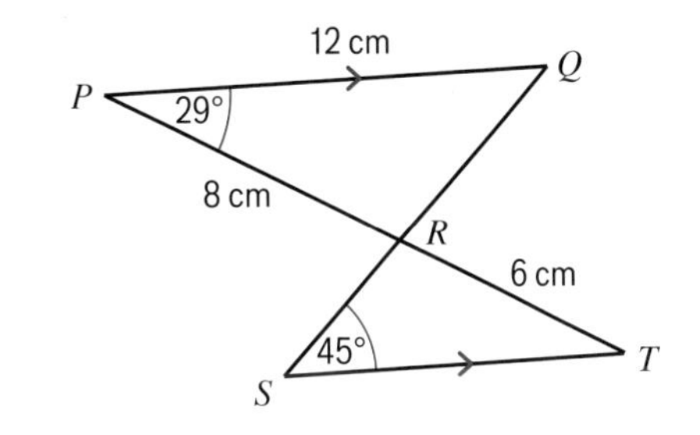

a Which angle corresponds to angle PQR?

b Find the size of angle PRQ.

c Which side corresponds to PR?

d Write the scale factor of the enlargement that maps triangle PQR to triangle TSR.

e Work out the length of ST.

12 **Reasoning** The lines PQ and ST are parallel.

a Copy and complete to show that triangles PQR and RST have the same angles, so they are similar.
Angle PQR = Angle □
Angle QPR = Angle □
Angle PRQ = Angle □

b $QS = 8$ cm
Work out the length of
i QR **ii** RS

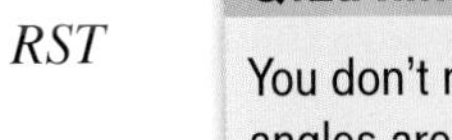

Q12a hint

You don't need to know what the angles are.

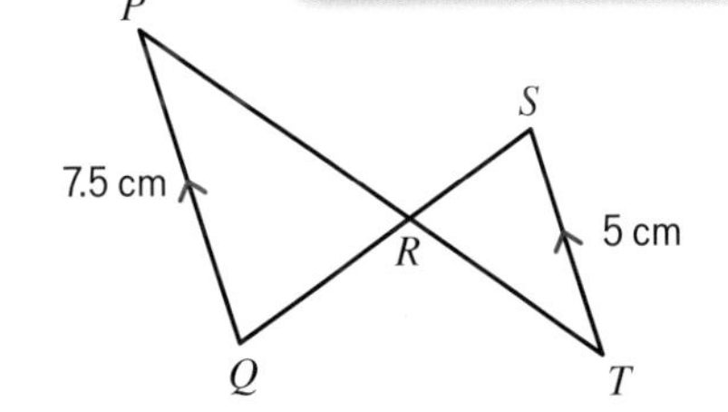

Exam-style question

13 The lines DE and BC are parallel.
Triangles ABC and ADE are similar.

a Work out the size of angle BAC. **(2 marks)**

b Work out the length of AB. **(2 marks)**

c Work out the length of BD. **(2 marks)**

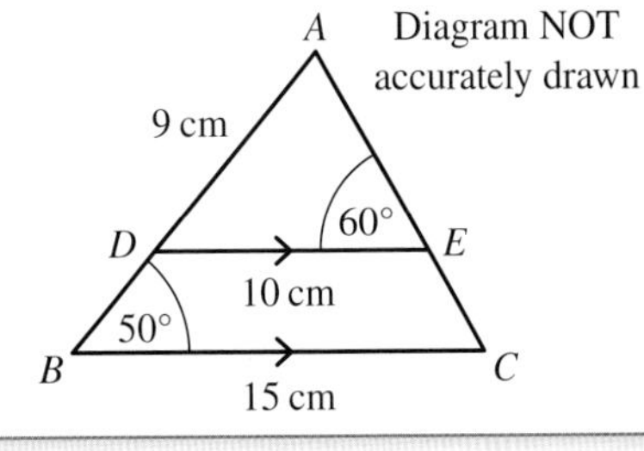

19.3 Using similarity

- Determine when two shapes are definitely not (or may not be) similar.
- Understand the similarity of regular polygons.
- Calculate perimeters of similar shapes.

Warm up

1 **Fluency** Are these two triangles similar? Explain.

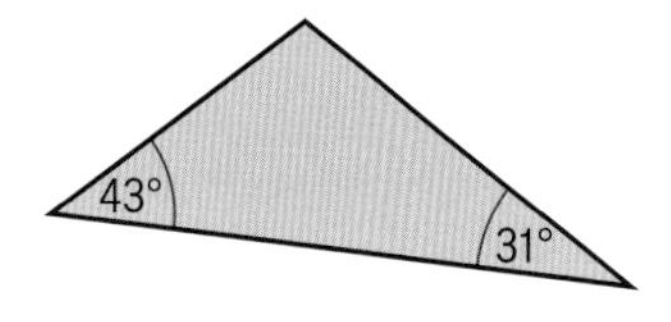

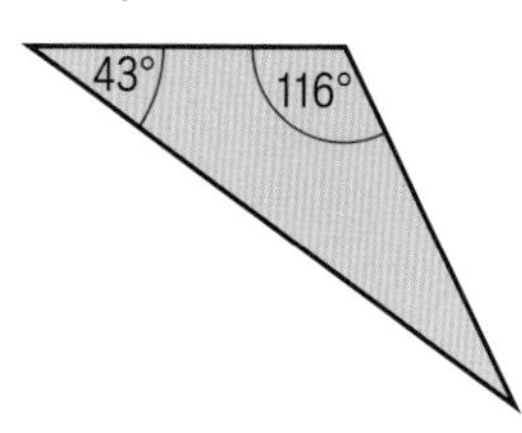

2 These rectangles are similar. Find the value of x.

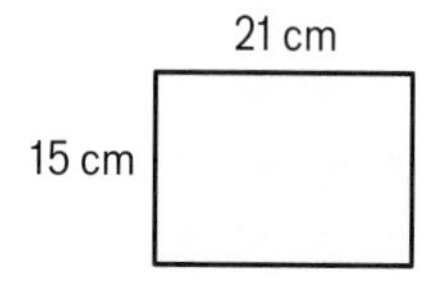

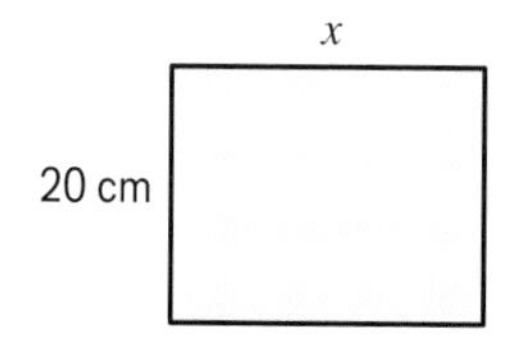

3 Use Pythagoras' theorem to work out the values of x and y.

a

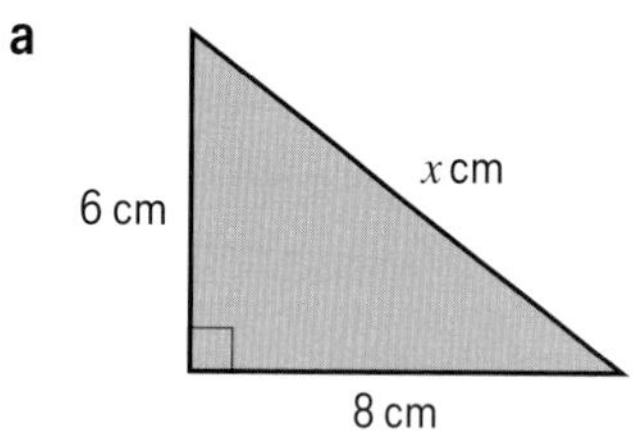

b

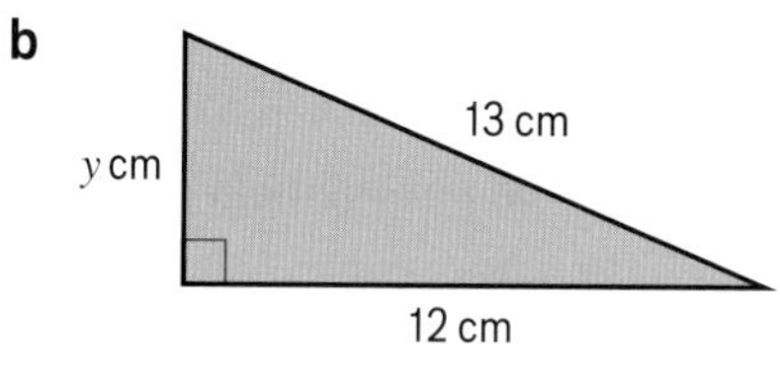

4 **Reasoning** Sam says triangles ABC and DEF are similar, because

$BC \times 2 = DE$ and $AC \times 2 = EF$

$3 \times 2 = 6$ $\quad$ $5 \times 2 = 10$

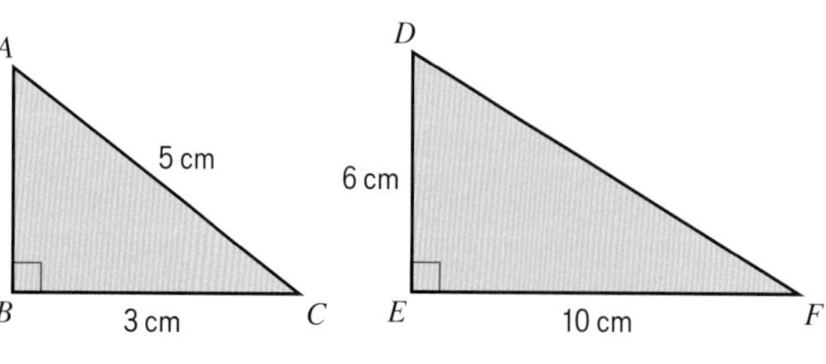

a Work out lengths AB and DF.

b Does $AB \times 2 = DF$?

c Are triangles ABC and DEF similar? Explain.

d What mistakes has Sam made?

Q4 hint

Which are the corresponding sides in triangles ABC and DEF?

5 **Reasoning** For each question, write 'Yes' or 'Need more information'.

a Two triangles share one identical angle. Are they similar?

b Two triangles share two identical angles. Are they similar?

c One triangle has one side that is double the length of one of the sides of another triangle. Are they similar?

d Two triangles have two pairs of sides in the same ratio. Are they similar?

6 **Reasoning** These pentagons have the same angles.
Are they similar?
Explain your answer.

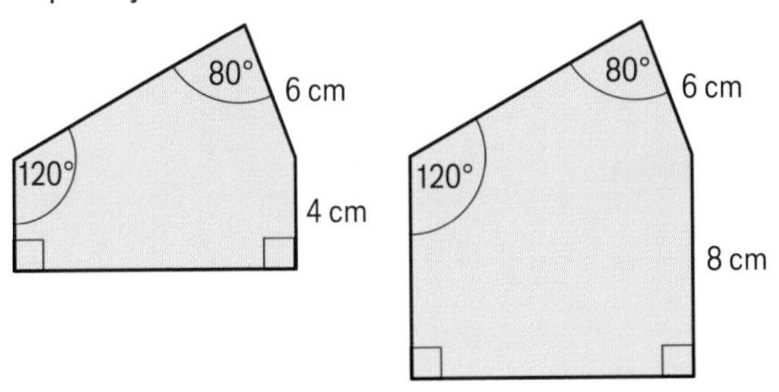

7 **Reasoning** These hexagons have sides in the same ratio.
Are they similar?
Explain your answer.

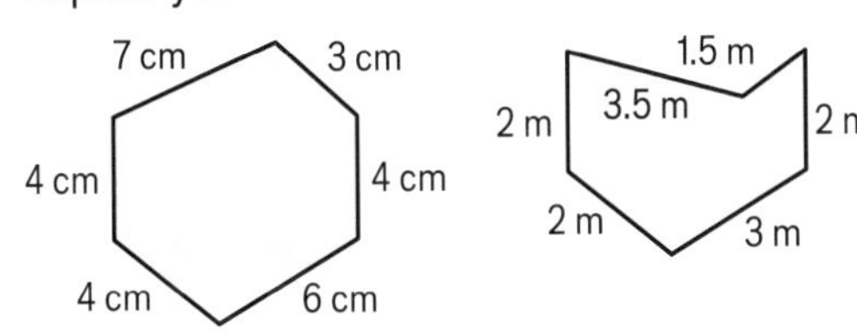

8 **Reasoning** For each question, write 'Yes' or 'Need more information'.

a Two polygons share two identical angles. Are they similar?

b Two polygons have all sides in the same ratio. Are they similar?

c Two polygons share the same angles and have sides in the same ratio. Are they similar?

9 **Reasoning** Here are two regular pentagons.
Are they similar?
Explain your answer.

Q9 hint

A regular polygon has angles that are all the same size and sides that are all the same length.

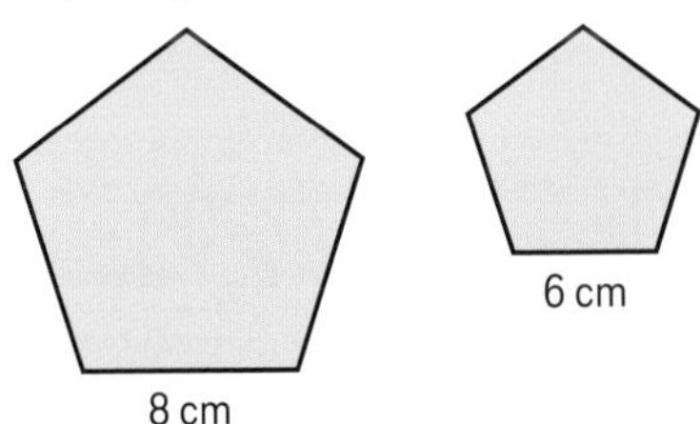

10 **Reasoning**

a Are all regular pentagons similar? Explain your answer.

b Are all regular hexagons similar? Explain your answer.

11 Problem-solving In the diagram, shape A is mapped to shape B by an enlargement.

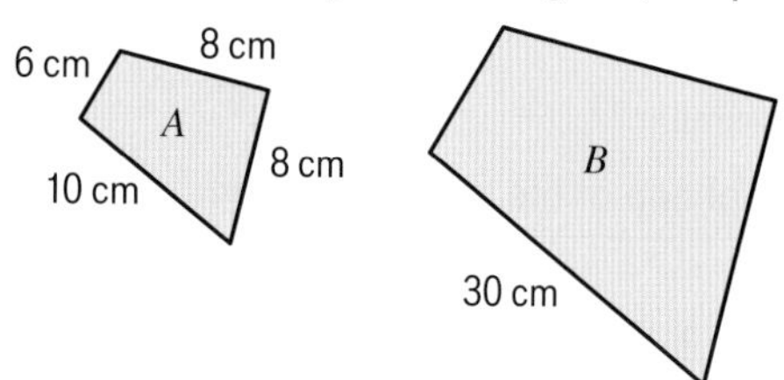

a Find the scale factor of the enlargement.

b Work out the lengths of the unknown sides of shape B.

c Work out the perimeter of shape A.

d Work out the perimeter of shape B.

e Copy and complete.
Perimeter of shape $B = \square \times$ perimeter of shape A

f Look at your answers to parts **a** and **e**.
What do you notice?

Key point

When a shape is enlarged, the perimeter of the shape is enlarged by the same scale factor.

12 Problem-solving
These two windows are mathematically similar.
The perimeter of window A is 8 m.
Work out the perimeter of window B.

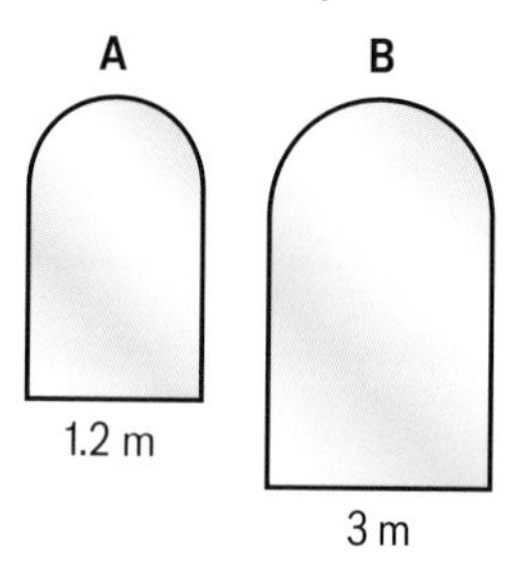

Exam-style question

13 The diagram shows two flower beds made in the shape of similar kites.

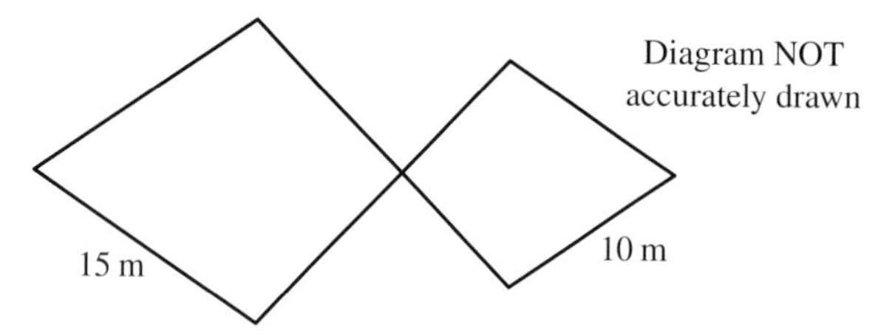

The perimeter of the larger flower bed is 48 m.
Work out the perimeter of the smaller flower bed.

(2 marks)

Exam tip

In questions on similarity, use the words 'scale factor' in your answer.

19.4 Congruence 1

- Recognise congruent shapes.
- Use congruence to work out unknown angles.

ActiveLearn Homework

Warm up

1 **Fluency** One of these triangles is congruent to triangle ABC. Which one?

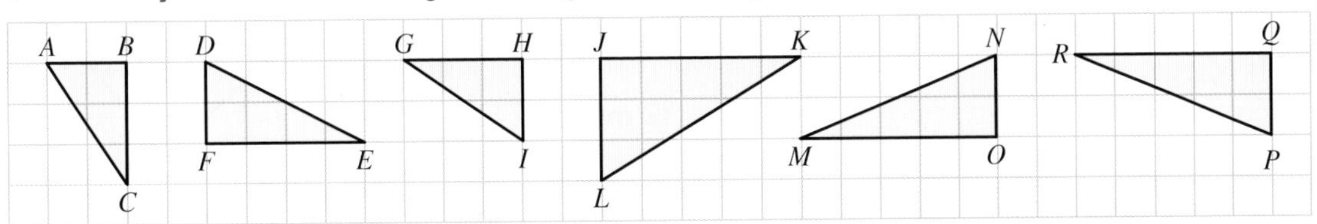

2 In **Q1**, triangles MNO and PQR are congruent.
Complete the sentences.
a Side NO and side ___ are corresponding sides.
b Angle MNO and angle ___ are corresponding angles.

Q2 hint
Sketch triangles MNO and PQR in the same orientation.

3 **a** Construct these triangles accurately using a ruler and compasses.
b Trace one of the triangles.
Does it fit exactly on the other?
c Are your two triangles congruent?

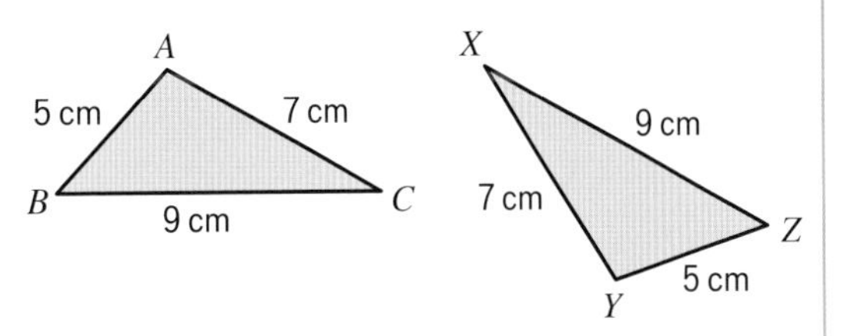

4 **Reasoning** Here are two pairs of congruent triangles. Work out the sizes of angles x and y.

a **b**

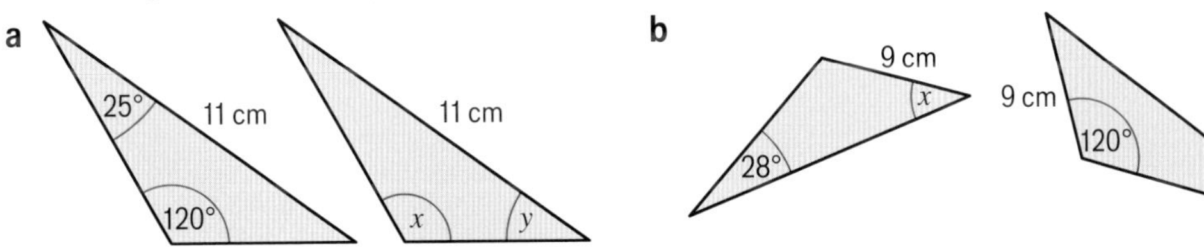

Q4a hint
The angle opposite the side that is 11 cm will be the same in each triangle.
To find angle y, use the fact that angles in a triangle add to 180°.

Key point
When two shapes are congruent, one can be rotated or reflected to fit exactly on the other.
Triangles are **congruent** if they have equivalent
- SSS (all three sides)
- SAS (two sides and the included angle)
- ASA (two angles and the included side)
- RHS (right angle, hypotenuse, side)

An included side is between two angles. An included angle is between two sides.

5 Which of these is the reason for congruency for triangles ABC and XYZ in **Q3**?
SSS SAS ASA RHS

6 **Reasoning** Each pair of triangles is congruent. Give a reason for congruency.

a

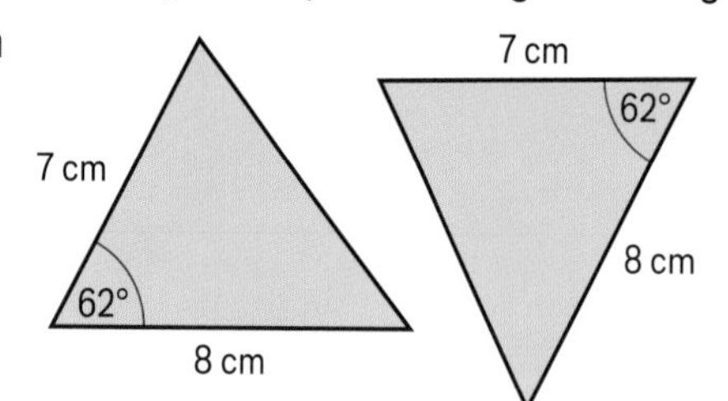

b

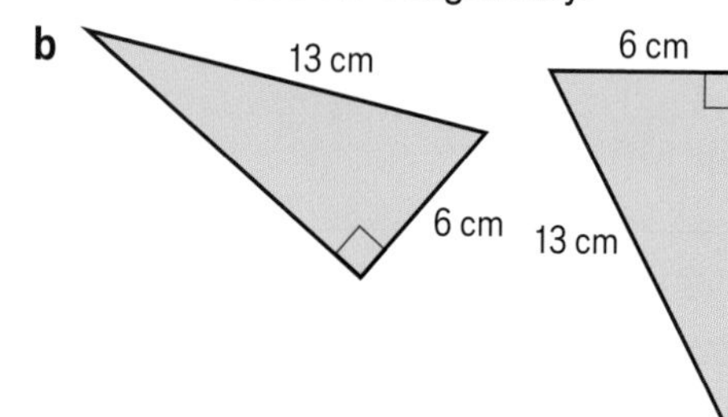

Exam-style question

7 Miguel says these two triangles are congruent.
Is Miguel correct?
Explain. **(2 marks)**

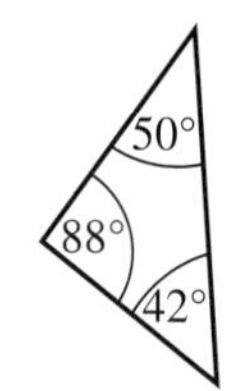

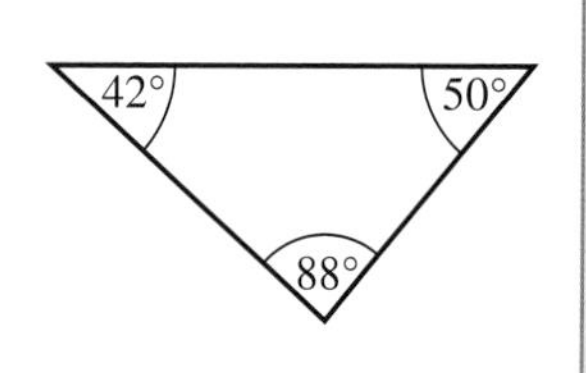

Key point

Triangles where all angles are the same (AAA) may not be congruent.
However, they are always similar.

8 **Reasoning** Construct these triangles accurately using a ruler and compass.

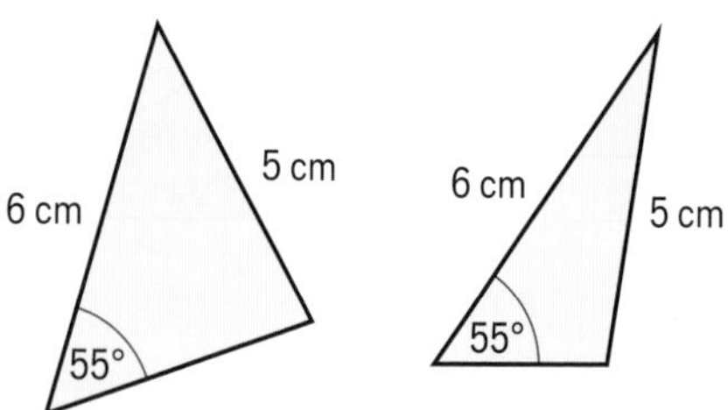

Are your two triangles congruent? Explain.

Key point

Triangles with two equivalent sides and an equivalent non-included angle (SSA) may not be congruent.

9 **Reasoning** Which of these triangles are congruent? Give a reason for congruency.

A

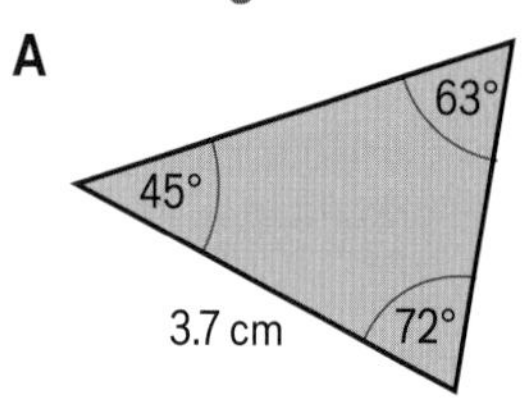

B

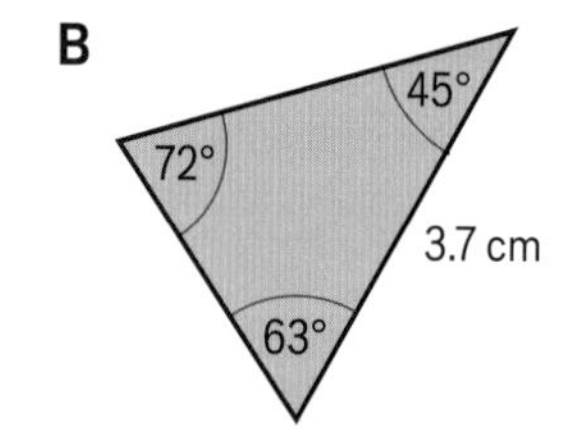

C

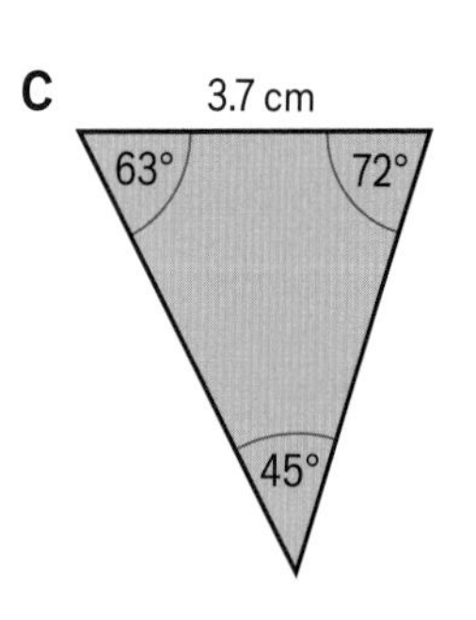

D

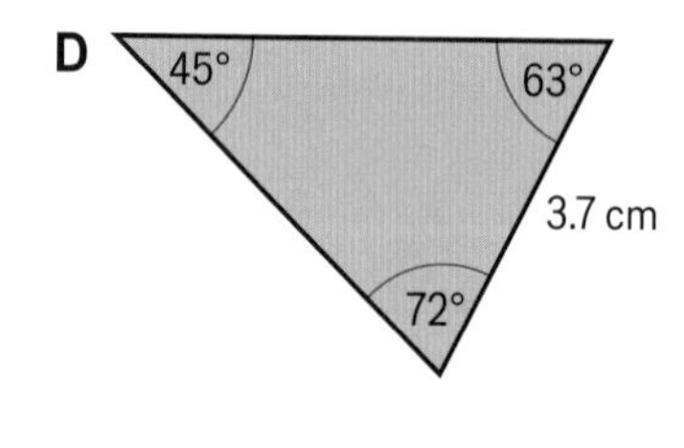

19.5 Congruence 2

- Use congruence to work out unknown sides and angles in triangles and shapes made of triangles.

ActiveLearn Homework

Warm up

1 Fluency Which sides and which angles are equal in these shapes?

a isosceles triangle **b** rectangle **c** rhombus **d** parallelogram

A, *B*, *C*   *D*, *E*, *F*, *G*   *H*, *I*, *J*, *K*

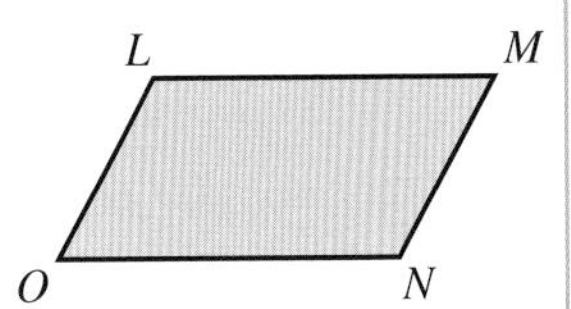

2 Find the values of x, y and z in each diagram. Choose your reasons from the box.

a

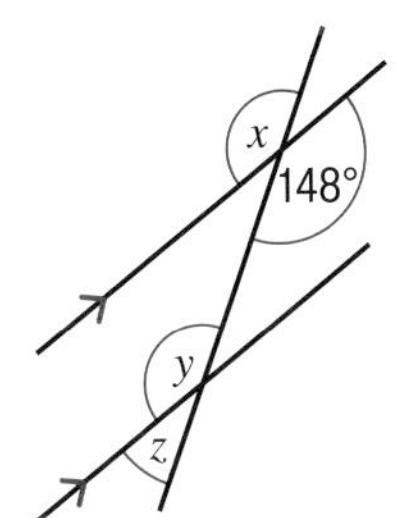

b

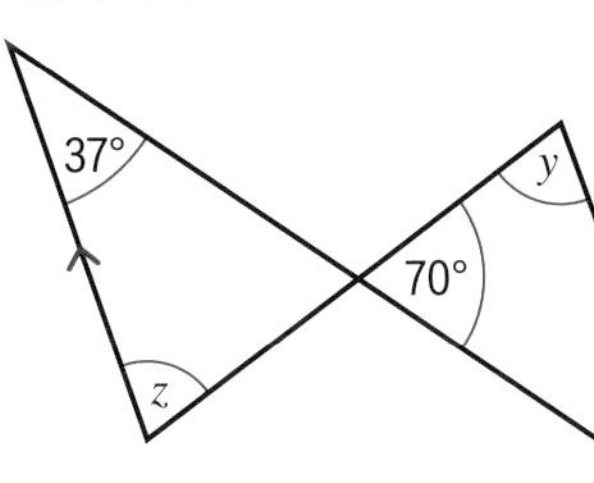

Vertically opposite angles are equal.
Alternate angles are equal.
Corresponding angles are equal.
Angles on a straight line add up to 180°.
Angles in a triangle add up to 180°.

3 M is the midpoint of AB.
M is also the midpoint of CD.

a Which length is the same as AM?

b Which length is the same as MD?

c Which angle is the same as angle AMC?

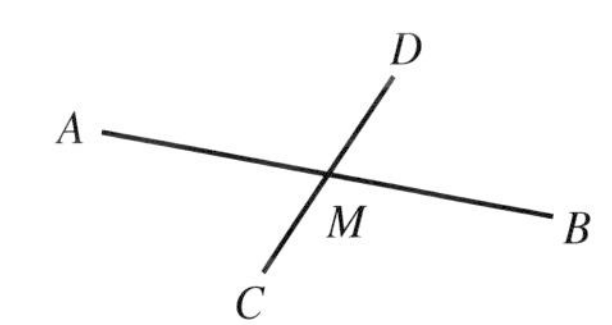

4 Which triangle is congruent to triangle A?

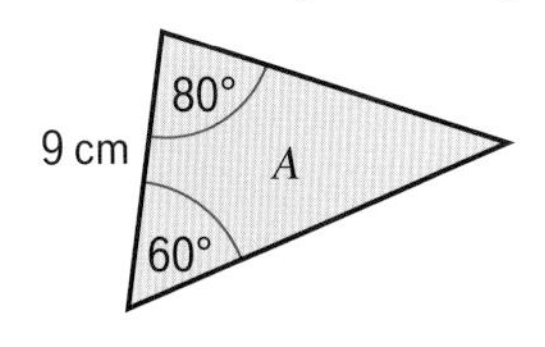

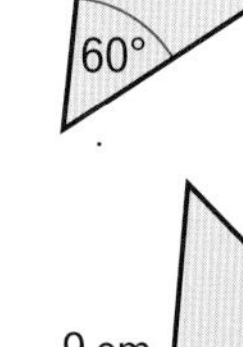

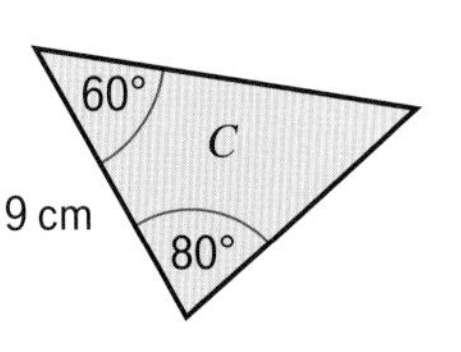

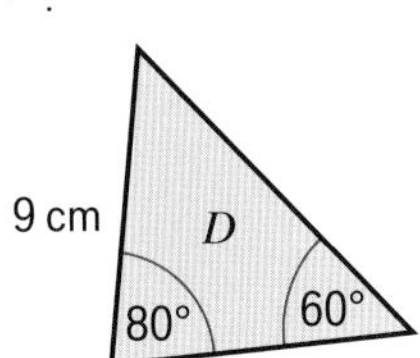

Q4 hint

Do they rotate or reflect to fit exactly on top of each other?

Example

XYZ is an isosceles triangle. M is the midpoint of YZ.

a What can you say about triangles XMY and XMZ?

b Find the size of angle MXZ.

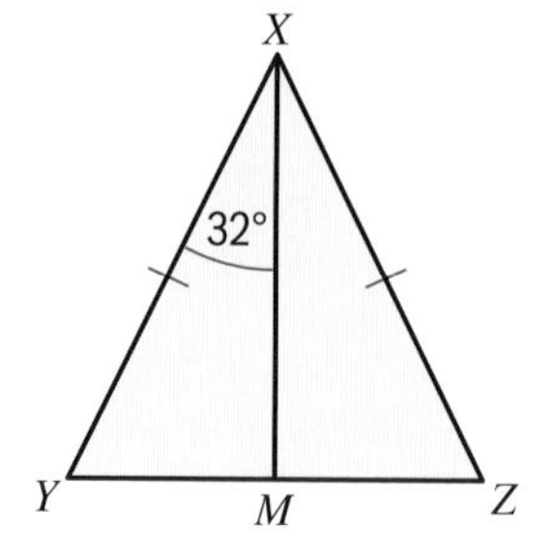

a Triangle XMY is congruent to triangle XMZ, because XM cuts triangle XYZ in half.

b Angle MXZ = angle MXY (corresponding positions in congruent triangles)
Angle $MXZ = 32°$

5 **Problem-solving** Triangles X and Y are congruent.

a What is length l?

b Find the size of

i angle a

ii angle b

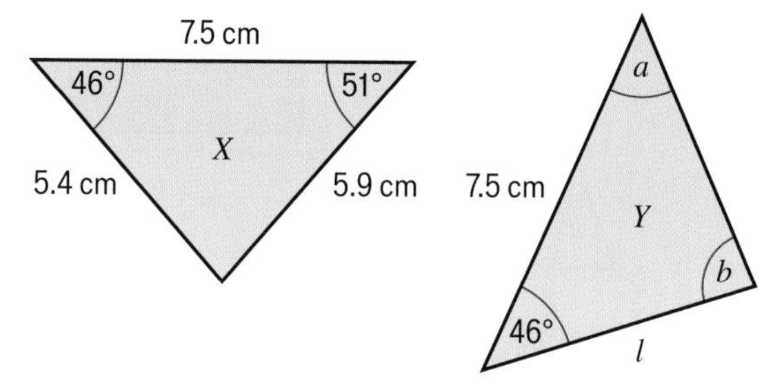

6 **Problem-solving** $ABCD$ is a rectangle.

a Which angle is the same as angle DBA?
Give a reason for your answer.

b What can you say about triangles DAB and BCD?

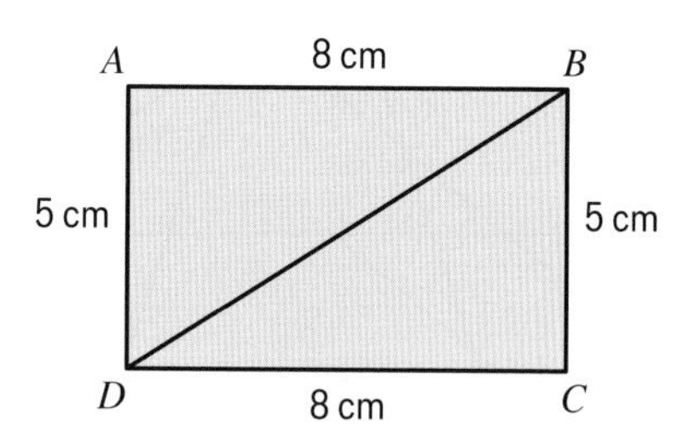

7 **Problem-solving** The diagram shows a rhombus $WXYZ$.

a Find the size of angle XYZ.
Give a reason for your answer.

b Are triangles XWZ and XYZ congruent?
If so, give a reason for congruency.

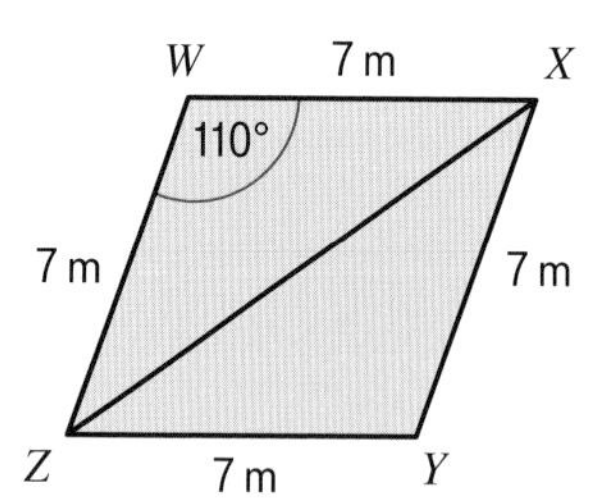

Q7a hint

What do you know about opposite angles in a rhombus?

8 **Problem-solving** $ABCD$ is an arrowhead.

a Which side do triangles ABC and ACD share?

b Are triangles ABC and ACD congruent?
If so give a reason for congruency.

c Find the size of angle

i ADC

ii CAD

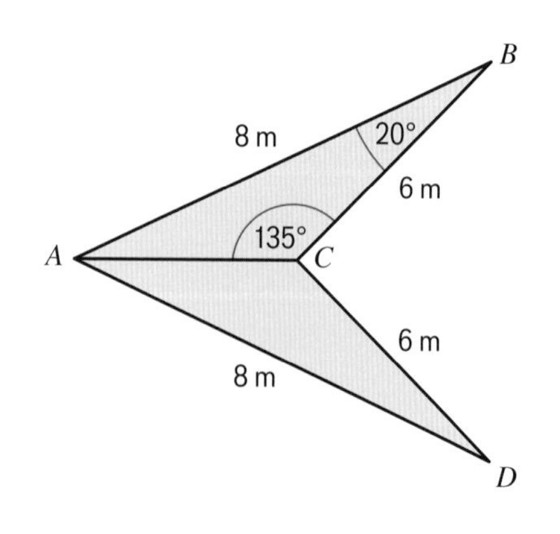

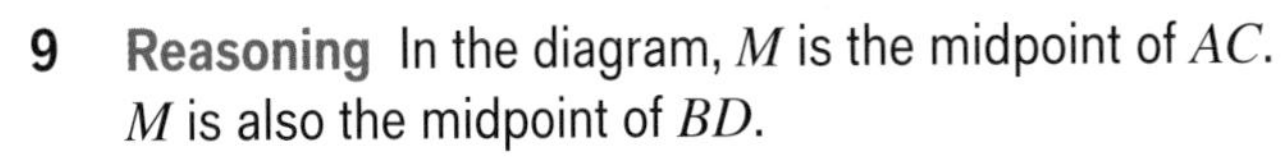

9 **Reasoning** In the diagram, M is the midpoint of AC. M is also the midpoint of BD.

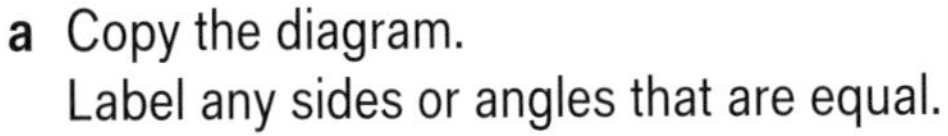

a Copy the diagram. Label any sides or angles that are equal.

b Are triangles ABM and CDM congruent? Give a reason for your answer.

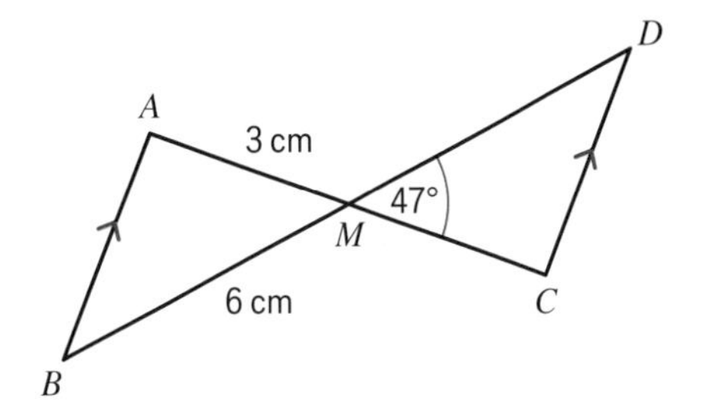

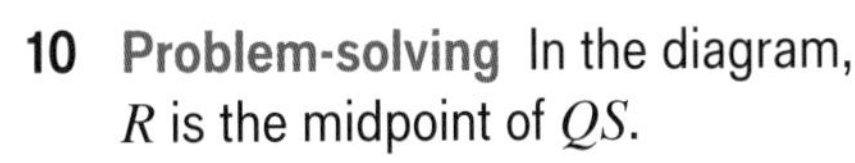

10 **Problem-solving** In the diagram, R is the midpoint of QS.

a Are triangles PQR and TSR congruent?

b Find the length of RT. Give reasons for your answer.

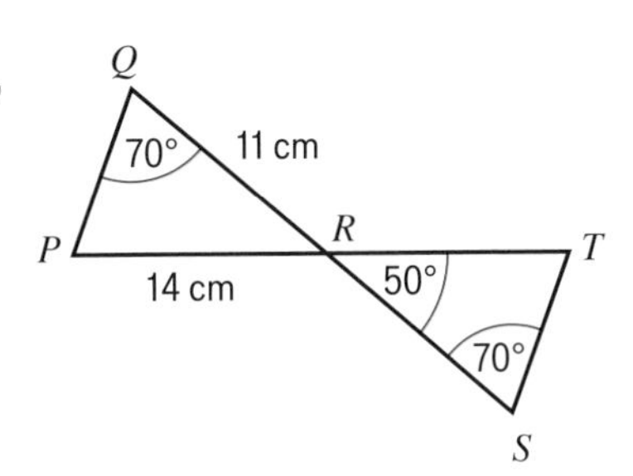

Q10 hint

Copy the diagram. Label any sides or angles that are equal.

Exam-style question

11 In this diagram, C is the centre of the circle, radius 6 cm.

a Write down the size of angle RCS. **(2 marks)**

b Write the lengths of CQ, CR and CS. **(2 marks)**

c Are triangles PCQ and CRS congruent? **(2 marks)**

Give reasons for your answers.

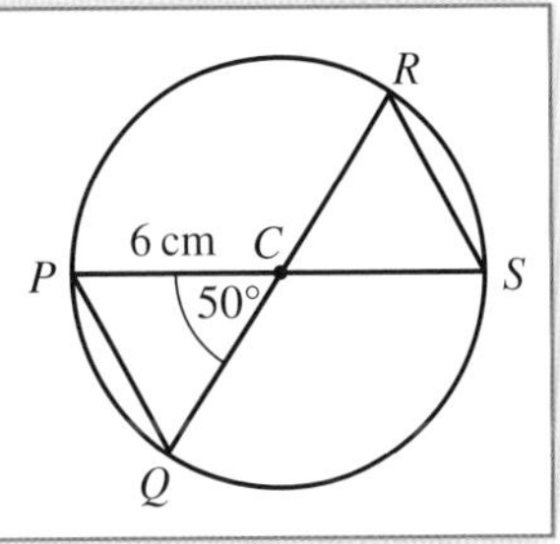

12 **Reasoning** In the diagram, $ABCD$ is a parallelogram.

a Find the size of angle DBC.

b Find the size of angle BDC.

c Is triangle ABD congruent to triangle CDB?

d Find the length of CD.

Give reasons for your answers.

13 **Reasoning**

a Which angle in triangle NOL is 64°?

b Are triangles LMN and NOL congruent?

Give reasons for your answers.

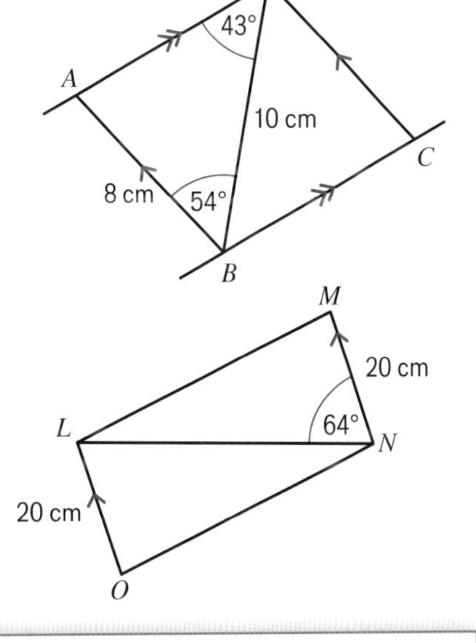

Exam-style question

14 a Work out the size of angle RSQ. **(2 marks)**

b Are triangles PQS and RQS congruent? **(2 marks)**

c $PQ = 19$ cm
Which other side has length 19 cm? **(2 marks)**

Give reasons for your answers.

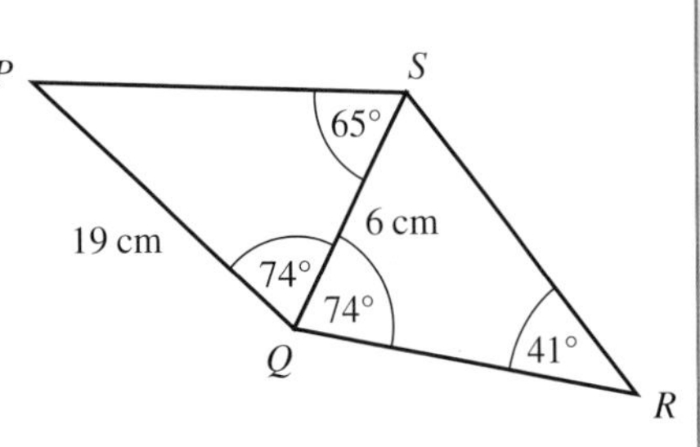

19.6 Vectors 1

- Add vectors.
- Find the resultant of two vectors.

ActiveLearn Homework

Warm up

1 **Fluency** Work out **a** $-2+4$ **b** $3-5$ **c** $6--4$ **d** $-5-2$

2 Copy the grid and shape A.

a Translate shape A by $\begin{pmatrix}3\\4\end{pmatrix}$. Label the image B.

a Translate shape B by $\begin{pmatrix}2\\-3\end{pmatrix}$. Label the image C.

c Describe fully the single translation that maps shape A onto shape C.

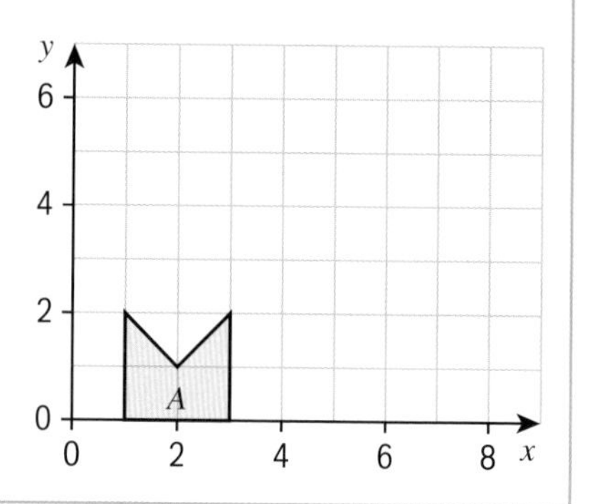

3 **Reflect** Describe how you can use the column vectors for the two translations (A to B, and B to C) in **Q2a** and **Q2b** to get the column vector for the translation (A to C) in your answer to **Q2c**.

Key point

To add two column vectors, add the top numbers and add the bottom numbers.

4 Add these column vectors. The first one is started for you.

Give your answers in the form of a single vector $\begin{pmatrix}\square\\\square\end{pmatrix}$.

a $\begin{pmatrix}2\\5\end{pmatrix}+\begin{pmatrix}3\\1\end{pmatrix}=\begin{pmatrix}2+\square\\5+\square\end{pmatrix}=\begin{pmatrix}\square\\\square\end{pmatrix}$ **b** $\begin{pmatrix}4\\1\end{pmatrix}+\begin{pmatrix}3\\2\end{pmatrix}$ **c** $\begin{pmatrix}5\\4\end{pmatrix}+\begin{pmatrix}2\\0\end{pmatrix}$

d $\begin{pmatrix}8\\-3\end{pmatrix}+\begin{pmatrix}-1\\2\end{pmatrix}$ **e** $\begin{pmatrix}-5\\4\end{pmatrix}+\begin{pmatrix}5\\-4\end{pmatrix}$

Key point

Two translations can be combined into a single translation by adding the column vectors.

5 Shape A is translated to shape B by the column vector $\begin{pmatrix}3\\1\end{pmatrix}$.

Shape B is translated to shape C by the column vector $\begin{pmatrix}-1\\4\end{pmatrix}$.

Find the column vector of the single translation from shape A to shape C.

Key point

A vector has a magnitude (size) and direction.
The diagram shows a vector that starts at A and finishes at B.

You can write this vector as $\overrightarrow{AB}=\begin{pmatrix}3\\2\end{pmatrix}$.

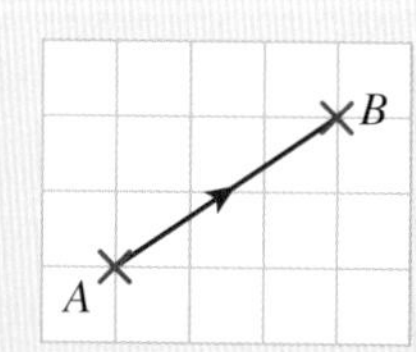

6 Write as column vectors

a $\overrightarrow{PQ}$

b $\overrightarrow{XY}$

c $\overrightarrow{UV}$

7 **a** Write as column vectors

i $\overrightarrow{AB}$ **ii** $\overrightarrow{BC}$ **iii** $\overrightarrow{AC}$

b **Reasoning**

Show that $\overrightarrow{AB} + \overrightarrow{BC} = \overrightarrow{AC}$.

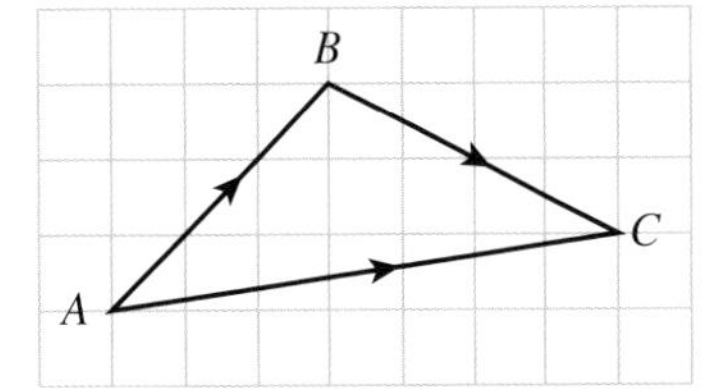

Key point

A to B followed by B to C is equivalent to A to C.

$\overrightarrow{AB} + \overrightarrow{BC} = \overrightarrow{AC}$

The sum of two vectors is called the resultant.

$\overrightarrow{AC}$ is the **resultant** of $\overrightarrow{AB}$ and $\overrightarrow{BC}$.

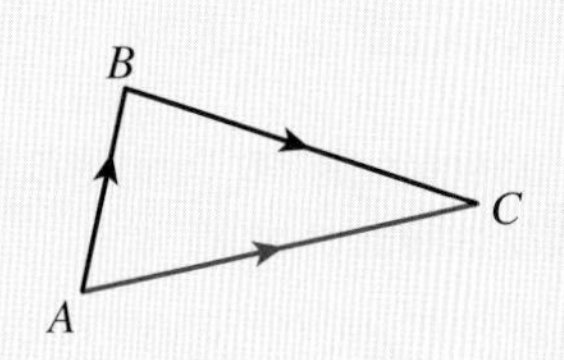

8 Copy and complete.

a $\overrightarrow{NQ} + \overrightarrow{QT} = \square$ **b** $\overrightarrow{MP} + \square = \overrightarrow{MR}$ **c** $\square + \overrightarrow{TZ} = \overrightarrow{VZ}$

Key point

You can show vectors using different notation. A single letter may be used to represent a vector. The letter is shown in **bold type**: $\mathbf{a} = \begin{pmatrix} -2 \\ 4 \end{pmatrix}$

You can't write bold in handwriting, so you should underline a letter that represents a vector. For **a**, write $\underline{a}$.

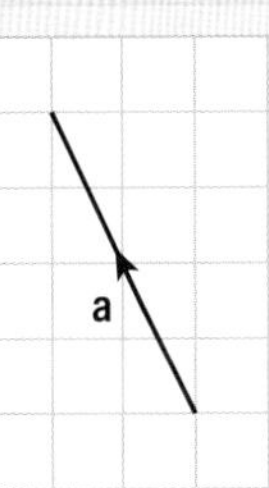

9 Here are two vectors, **p** and **q**.

a Copy vector **p**.

b Draw vector **q** at one end of vector **p** so that the lines and arrows follow on.

c Draw in the third side of the triangle and label it **p** + **q**.

d Repeat the steps in parts **a**–**c**, but draw **q** at the other end of **p**, so that the lines and arrows still follow on.

e **Reflect** Does it matter which end of **p** you draw **q**? Is **p** + **q** the same vector? Explain.

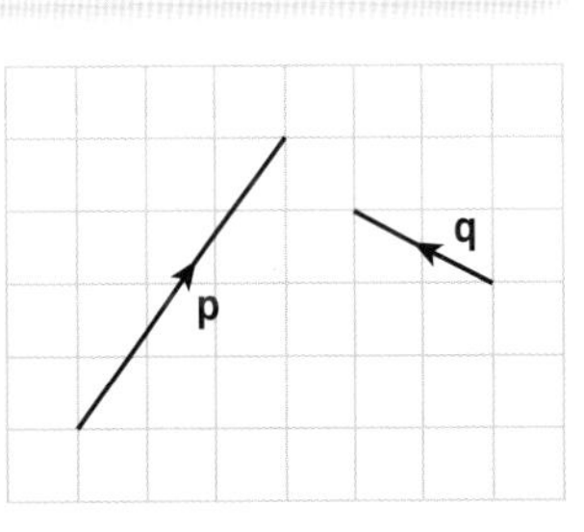

Exam-style question

10 **a** $\mathbf{a} = \begin{pmatrix} -3 \\ -8 \end{pmatrix}$ $\mathbf{b} = \begin{pmatrix} 2 \\ -6 \end{pmatrix}$

Work out **a** + **b** as a column vector. **(2 marks)**

b The resultant of $\overrightarrow{AB}$ and $\overrightarrow{BC}$ is $\begin{pmatrix} -1 \\ 4 \end{pmatrix}$.

$\overrightarrow{BC} = \begin{pmatrix} 3 \\ 0 \end{pmatrix}$

Work out $\overrightarrow{AB}$. **(2 marks)**

Exam tip

Make sure you write column vectors like this $\begin{pmatrix} \square \\ \square \end{pmatrix}$, not $\left(\frac{\square}{\square}\right)$ or $(\square, \square)$.

19.7 Vectors 2

- Subtract vectors.
- Find multiples of a vector.
- Identify two column vectors that are parallel.
- Solve problems using vectors.

Warm up

1 **Fluency** Work out **a** $-2-4$ **b** $-3--7$ **c** -2×5 **d** -3×-4

2 $ABCD$ is a rectangle. AC and BD are diagonals of rectangle $ABCD$. AC and BD intersect at X.

a Which side of the rectangle is parallel and equal in length to

i AB **ii** BC?

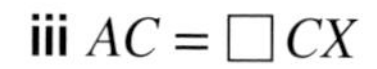

b Copy and complete with the missing number

i $AC = \square AX$ **ii** $BD = \square BX$ **iii** $AC = \square CX$ **iv** $BD = \square DX$

3 Copy the diagram and label all of the vectors.

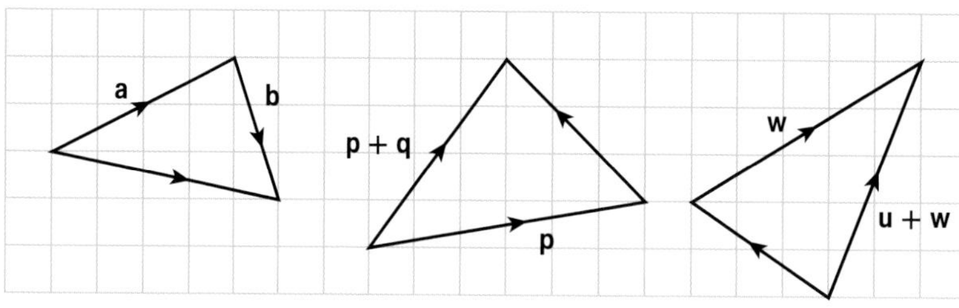

4 Find the resultant of $\begin{pmatrix}4\\-1\end{pmatrix}$ and $\begin{pmatrix}-3\\5\end{pmatrix}$.

Key point

$-\mathbf{a}$ is the negative of $\mathbf{a}$ and points in the opposite direction.

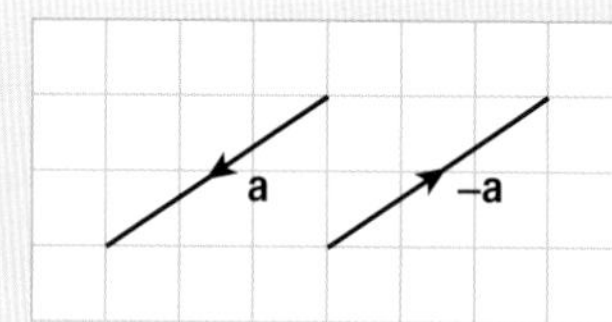

5 $\mathbf{p} = \begin{pmatrix}4\\2\end{pmatrix}$ $\mathbf{q} = \begin{pmatrix}1\\3\end{pmatrix}$ $\mathbf{r} = \begin{pmatrix}-3\\-1\end{pmatrix}$

Copy the diagram and label the vectors using $\mathbf{p}$, $\mathbf{q}$, $\mathbf{r}$, $-\mathbf{p}$, $-\mathbf{q}$, $-\mathbf{r}$, $\mathbf{p}+\mathbf{q}$ and $\mathbf{q}+\mathbf{r}$.

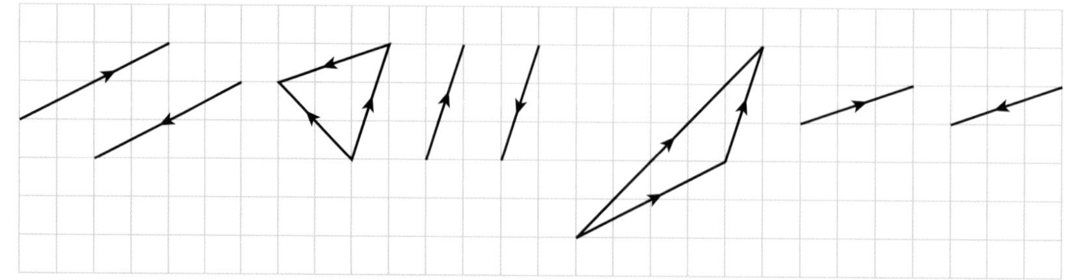

Key point

a − **b** is the same as **a** + (−**b**).

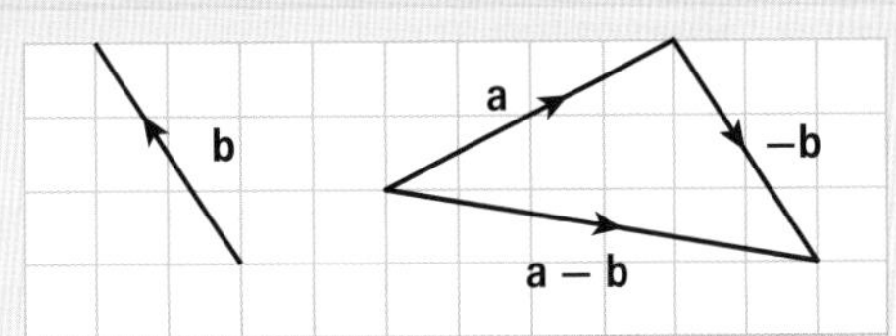

6 **Reasoning** Copy the diagram and label all of the vectors.

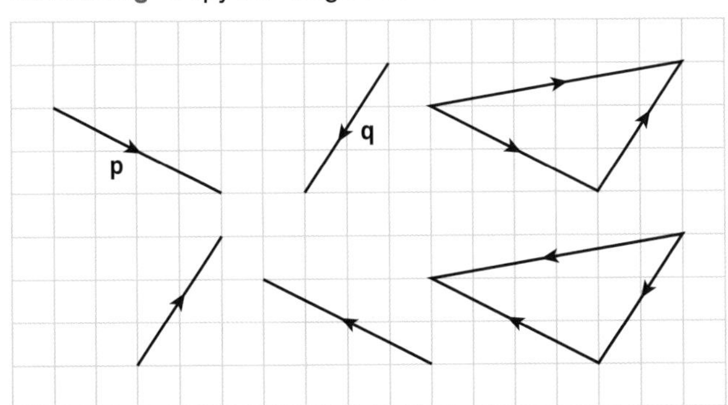

Key point

To subtract column vectors, subtract the top numbers and subtract the bottom numbers.

7 Subtract these column vectors. The first one is started for you.

a $\begin{pmatrix}8\\5\end{pmatrix} - \begin{pmatrix}3\\4\end{pmatrix} = \begin{pmatrix}8-3\\5-4\end{pmatrix} = \begin{pmatrix}\square\\\square\end{pmatrix}$ b $\begin{pmatrix}-2\\3\end{pmatrix} - \begin{pmatrix}1\\3\end{pmatrix}$ c $\begin{pmatrix}4\\-3\end{pmatrix} - \begin{pmatrix}-1\\2\end{pmatrix}$

8 $\mathbf{p} = \begin{pmatrix}3\\2\end{pmatrix}$ $\mathbf{q} = \begin{pmatrix}-2\\5\end{pmatrix}$ $\mathbf{r} = \mathbf{p} - \mathbf{q}$

a Write **r** as a column vector. b Show **p**, −**q** and **r** as a triangle on a grid.

9 $\mathbf{x} = \begin{pmatrix}7\\-4\end{pmatrix}$ $\mathbf{y} = \begin{pmatrix}3\\-6\end{pmatrix}$ $\mathbf{z} = \mathbf{x} - \mathbf{y}$

a Write **z** as a column vector. b Show **x**, −**y** and **z** as a triangle on a grid.

Key point

You can multiply a vector by a number. For example, if $\mathbf{a} = \begin{pmatrix}4\\3\end{pmatrix}$ then $2\mathbf{a} = \begin{pmatrix}8\\6\end{pmatrix}$ and $-3\mathbf{a} = \begin{pmatrix}-12\\-9\end{pmatrix}$.

10 $\mathbf{p} = \begin{pmatrix}2\\5\end{pmatrix}$ $\mathbf{q} = \begin{pmatrix}3\\-2\end{pmatrix}$ Write as column vectors

a 2**p** b 3**q** c 2**p** + 3**q** d −4**p**
e −5**q** f −4**p** − 5**q** g **p** − (−**q**) h 3**p** − 2**q**

Exam-style question

11 $\mathbf{r} = \begin{pmatrix}3\\0\end{pmatrix}$ $\mathbf{t} = \begin{pmatrix}-2\\3\end{pmatrix}$

Work out **r** − 2**t** as a column vector. **(2 marks)**

Exam tip

When working with negative numbers, always double-check your arithmetic.

12 $\mathbf{x} = \begin{pmatrix}3\\2\end{pmatrix}$

Show on a grid a **x** b 2**x** c −**x** d −2**x**

13 **Problem-solving / Reasoning** $\mathbf{q} = \begin{pmatrix} -2 \\ 4 \end{pmatrix}$

Write as column vectors

a a vector in the same direction as **q** but twice as long

b a vector the same length as **q** but in the opposite direction

c a vector 3 times as long as **q** but in the opposite direction

d a vector parallel to **q** but half as long

e **Reflect** There are two possible answers to part **d**. What is the other answer?

Q13e hint

A **parallel vector** may have the same direction or the opposite direction.

14 **Problem-solving** In quadrilateral $ABCD$, $\overrightarrow{AB} = \mathbf{a}$, $\overrightarrow{BC} = \mathbf{b}$

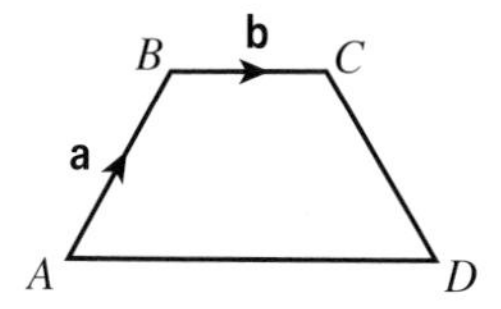

a Find in terms of **a** and **b**, the vector $\overrightarrow{AC}$.

b Find in terms of **a**, the vector $\overrightarrow{BA}$.

c Find in terms of **b**, the vector $\overrightarrow{CB}$.

d Find in terms of **a** and **b**, the vector $\overrightarrow{CA}$.

e $\overrightarrow{AD}$ is parallel to $\overrightarrow{BC}$, and it is twice as long. Find in terms of **b**, the vector $\overrightarrow{AD}$.

Q14a hint

Start at A. Move along $\overrightarrow{AB}$. Move along $\overrightarrow{BC}$. Finish at C.

Key point

Equal vectors have the same magnitude (size), for example the same length. They are also parallel and have the same direction.

15 **Reasoning** Which of these vectors are equal?

16 **Problem-solving** $PQRS$ is a rectangle.
QS is one diagonal of the rectangle.
$\overrightarrow{PQ} = \mathbf{p}$ and $\overrightarrow{QR} = \mathbf{q}$

P p Q q S R

a Draw the rectangle. Mark the vector (with an arrow and letter) on side
i PS **ii** SR

b Express in terms of **p** and **q**, the vector $\overrightarrow{SQ}$.

c Draw the other diagonal PR. Mark the point where PR and SQ intersect X.

d **Reasoning** Ali says $\overrightarrow{SX}$ is $\frac{1}{2}\mathbf{p} - \frac{1}{2}\mathbf{q}$. Is Ali correct? Explain.

Exam-style question

17 $CDEF$ is a parallelogram.
The diagonals of the parallelogram intersect at O.
$\overrightarrow{OC} = \mathbf{c}$ and $\overrightarrow{OD} = \mathbf{d}$

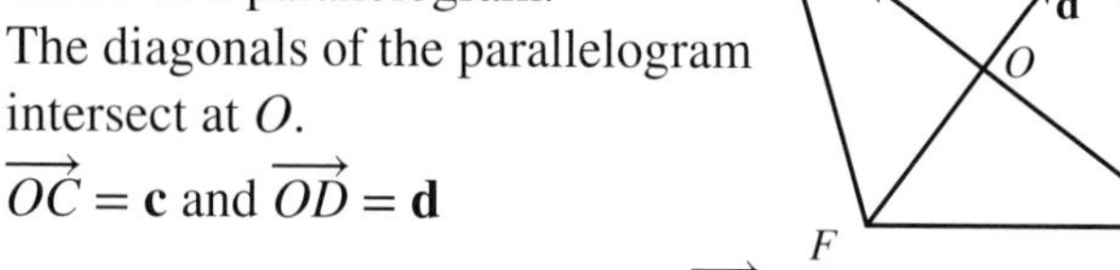

a Find in terms of **c**, the vector $\overrightarrow{EC}$. **(1 mark)**

b Find in terms of **c** and **d**, the vector $\overrightarrow{CD}$. **(1 mark)**

c Find in terms of **c** and **d**, the vector $\overrightarrow{DE}$. **(1 mark)**

Exam tip

Make sure you show that **c** and **d** are vectors by underlining c and d.

19 Check up

ActiveLearn Homework

Similarity and enlargement

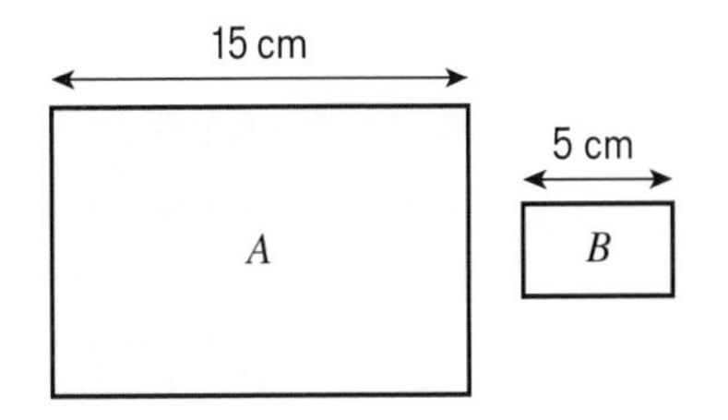

1 Rectangle A is similar to rectangle B.

a Work out the scale factor of the enlargement that maps A to B.

b The perimeter of A is 48 cm.
Work out the perimeter of B.

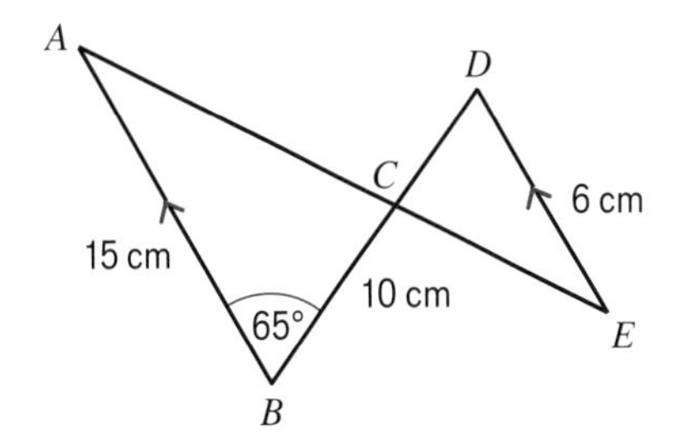

2 Triangle ABC is similar to triangle EDC.

a Find the scale factor of the enlargement that maps triangle EDC to triangle ABC.

b Find the length of CD.

3 Triangles ABC and PQR are similar. BC and QR are corresponding sides.

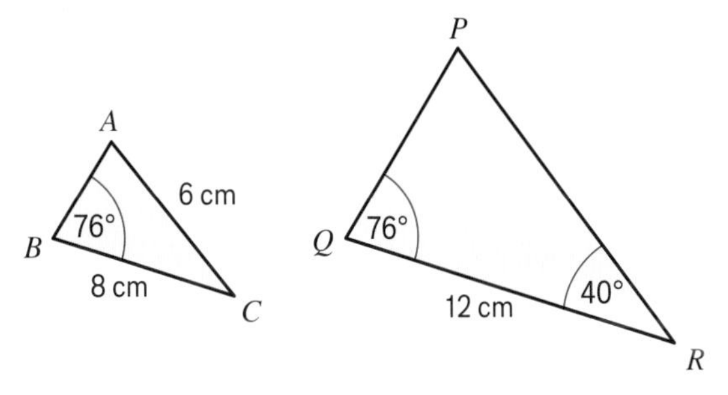

a Which side corresponds to AB?

b Find the length of PR.

c Find the size of

i angle ACB

ii angle BAC

Congruence

4 Decide whether each pair of triangles is congruent.
Give a reason for each answer.

a

b

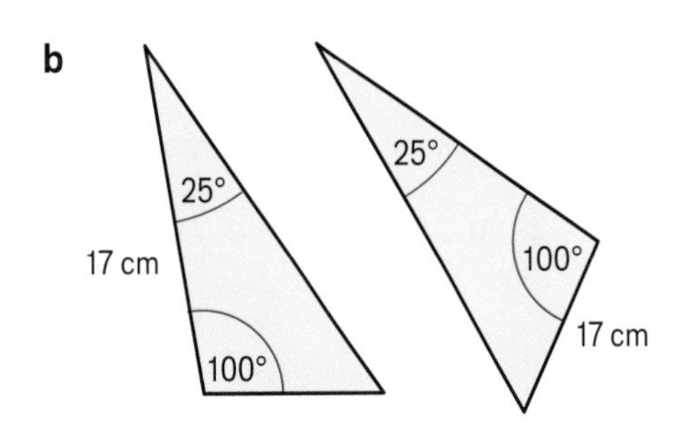

5 In the diagram, AB is parallel to CD.

a Work out angles a and b.
Give reasons for your answers.

b Is triangle ABD congruent to triangle BCD?
Give a reason for your answer.

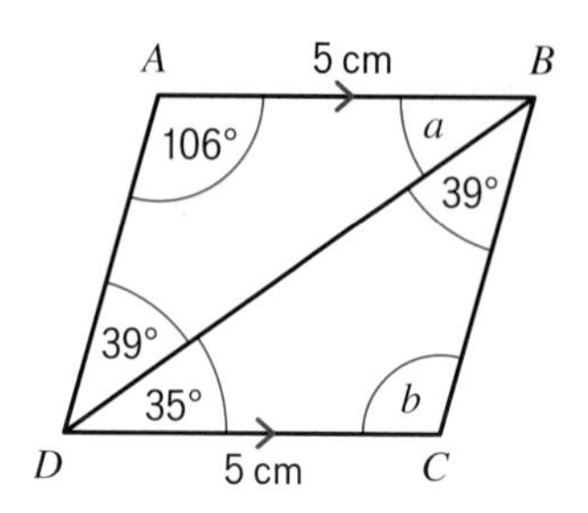

Vectors and translations

6 Shape A is translated to shape B by the column vector $\begin{pmatrix} 2 \\ 5 \end{pmatrix}$.

Shape B is translated to shape C by the column vector $\begin{pmatrix} -1 \\ 3 \end{pmatrix}$.

Find the column vector of the single translation from shape A to shape C.

7 Find the resultant of these vectors.

a $\overrightarrow{UV}$ and $\overrightarrow{VW}$

b $\begin{pmatrix} 4 \\ -3 \end{pmatrix}$ and $\begin{pmatrix} 2 \\ 3 \end{pmatrix}$

8 $\mathbf{a} = \begin{pmatrix} 3 \\ -1 \end{pmatrix}$ $\quad \mathbf{b} = \begin{pmatrix} -2 \\ 4 \end{pmatrix}$

Write as column vectors

a $3\mathbf{a}$

b $\mathbf{a} - 2\mathbf{b}$

9 Here are two vectors, **c** and **d**.
On squared paper, draw and label the vectors

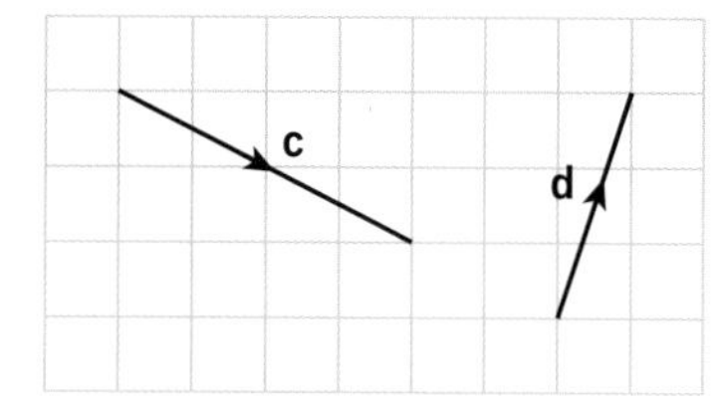

a  $-\mathbf{c}$

b $2\mathbf{d}$

c $\mathbf{c} + \mathbf{d}$

d $\mathbf{c} - \mathbf{d}$

10 **Reflect** How sure are you of your answers? Were you mostly

Just guessing ☹ Feeling doubtful 😐 Confident ☺

What next? Use your results to decide whether to strengthen or extend your learning.

Challenge

11 $ABCDEF$ is a regular hexagon.

$\overrightarrow{AB} = \mathbf{a}$ $\quad \overrightarrow{BC} = \mathbf{b}$ $\quad \overrightarrow{CD} = \mathbf{c}$

Find

a $\overrightarrow{ED}$ **b** $\overrightarrow{EF}$ **c** $\overrightarrow{FA}$

d $\overrightarrow{AC}$ **e** $\overrightarrow{DA}$

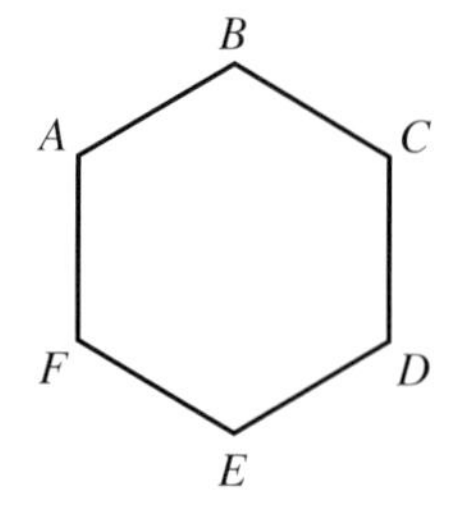

19 Strengthen

ActiveLearn Homework 

Similarity and enlargement

1 **a** On squared paper, draw a rectangle 2 by 3. Label it A.

b Enlarge the rectangle by scale factor 2. Label it B.

c Work out **i** the perimeter of A **ii** the perimeter of B

d Copy and complete. A to B, scale factor 2.
Perimeter of A to perimeter of B, scale factor □.

e Enlarge rectangle A by scale factor 3. Label it C.

f Predict the perimeter of C. Work out the perimeter to check your prediction.

2 Each pair of shapes is similar.

i

ii

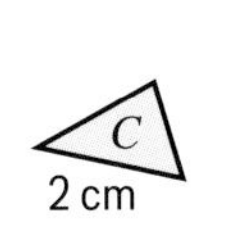

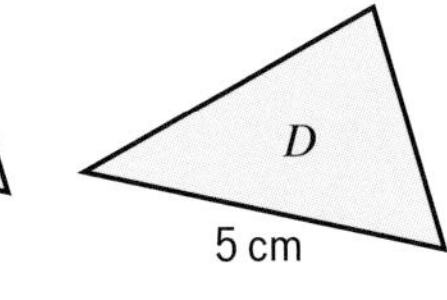

iii

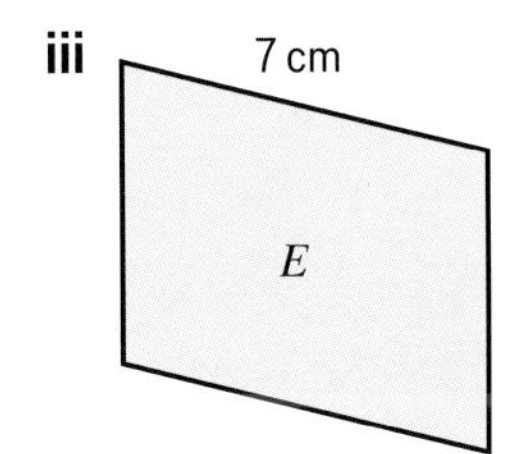

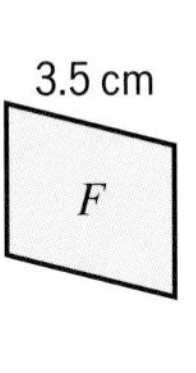

a Copy and complete to work out the scale factor of the enlargement that maps A to B.

length of A = 5 cm
corresponding length of B = 20 cm) ×□

b Use the method in part **a** to work out the scale factor that maps

i C to D

ii E to F

Q2b ii hint

×□
↑
This is a fraction

3 The perimeter of triangle A in **Q2** is 12 cm.
What is the perimeter of triangle B?
The perimeter of triangle C is 12 cm.
What is the perimeter of triangle D?

Q3 hint

To find lengths on triangle B, multiply lengths on triangle A by □.

4 **Reasoning**
The diagram shows two triangles.

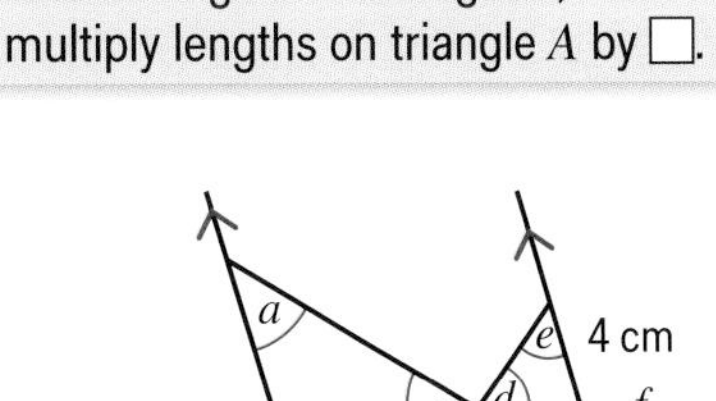
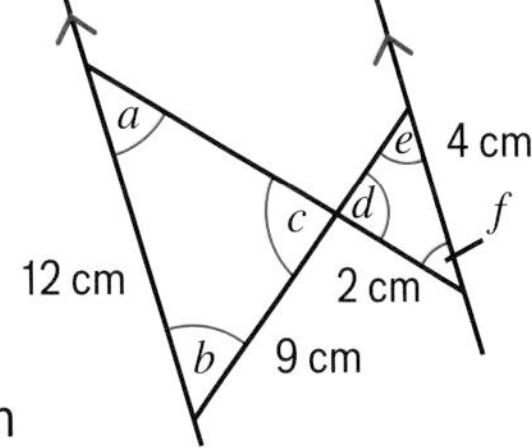

a Explain why

i $a = f$ **ii** $b = e$ **iii** $c = d$

Triangles with equal angles are similar.

b **i** Trace the triangles and then sketch them in the same orientation (the same way up).

ii Label the measurements you know.

iii Use a pair of corresponding sides to work out the scale factor.

iv Find the missing lengths.

5 Triangles ABC and XYZ are similar. BC and YZ are corresponding sides.

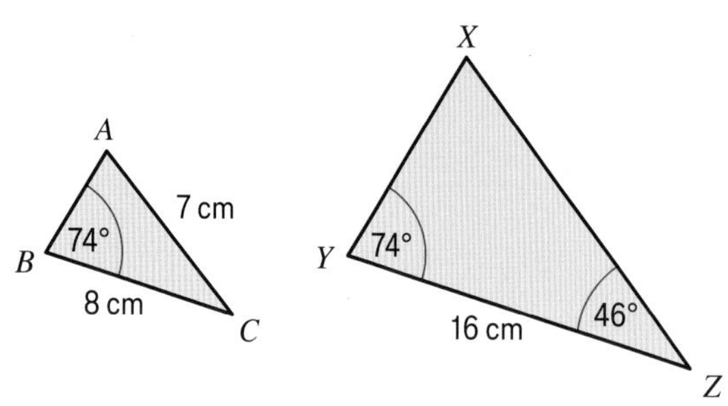

Q5 hint

Find the matching angle in the larger triangle.

a Which side corresponds to AB?

b Work out the scale factor.

c Use the scale factor you worked out in part **b** to find the length of XZ.

d Find the size of **i** angle ACB **ii** angle BAC

Congruence

1 **Reasoning** Draw diagrams rotating or reflecting the triangles to show they are identical.

a

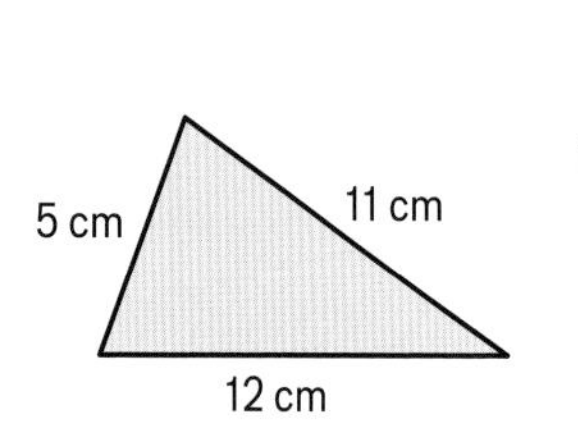

11 cm 12 cm 5 cm

Q1a hint

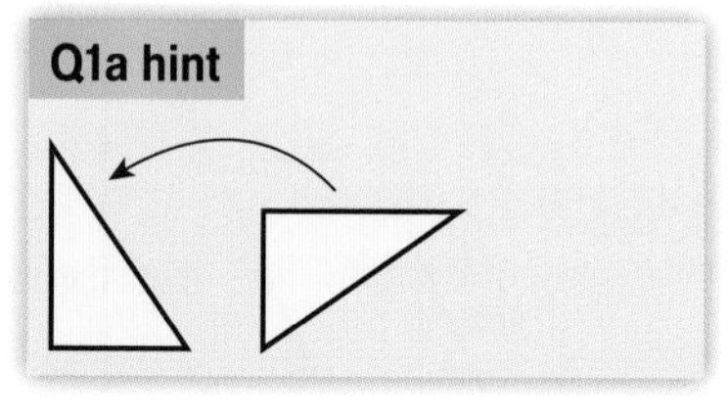

b

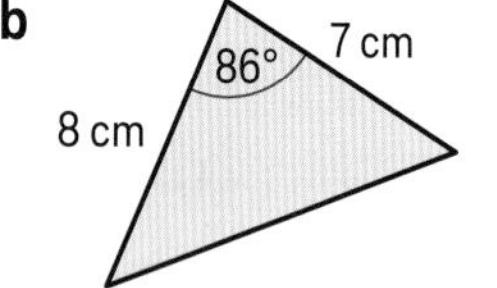

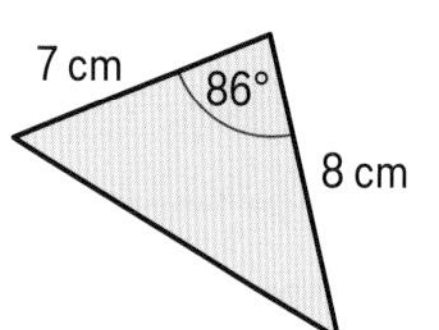

Q1b hint

Draw a mirror line.

2 **a** Copy the diagram and label the missing angles.

b Draw each triangle accurately using a ruler and protractor. Are they congruent? Give a reason for your answer.

c Which side in triangle SVU has length 12 cm?

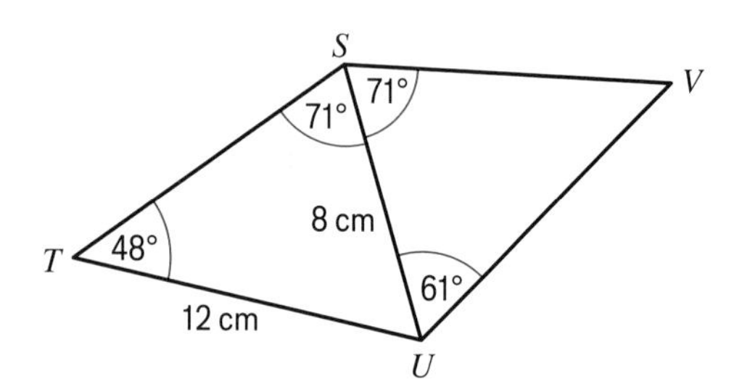

Vectors and translations

1 Copy and complete to write as a single column vector.

a $\begin{pmatrix}1\\2\end{pmatrix}+\begin{pmatrix}5\\3\end{pmatrix}=\begin{pmatrix}1+5\\2+3\end{pmatrix}=\begin{pmatrix}\square\\\square\end{pmatrix}$

b $\begin{pmatrix}-2\\-1\end{pmatrix}+\begin{pmatrix}1\\3\end{pmatrix}=\begin{pmatrix}-2+1\\-1+3\end{pmatrix}=\begin{pmatrix}\square\\\square\end{pmatrix}$

c $\begin{pmatrix}4\\-7\end{pmatrix}+\begin{pmatrix}-6\\-1\end{pmatrix}=\begin{pmatrix}4+-6\\-7+-1\end{pmatrix}=\begin{pmatrix}\square\\\square\end{pmatrix}$

d $\begin{pmatrix}5\\1\end{pmatrix}-\begin{pmatrix}2\\3\end{pmatrix}=\begin{pmatrix}5-2\\1-3\end{pmatrix}=\begin{pmatrix}\square\\\square\end{pmatrix}$

e $\begin{pmatrix}6\\-2\end{pmatrix}-\begin{pmatrix}-3\\-4\end{pmatrix}=\begin{pmatrix}6--3\\-2--4\end{pmatrix}=\begin{pmatrix}\square\\\square\end{pmatrix}$

Q1c hint

$4+-6$ is the same as $4-6$

2 **Reasoning** Point X is translated to point Y by the vector $\begin{pmatrix} 4 \\ -1 \end{pmatrix}$.

Point Y is translated to point Z by the vector $\begin{pmatrix} -3 \\ 5 \end{pmatrix}$.

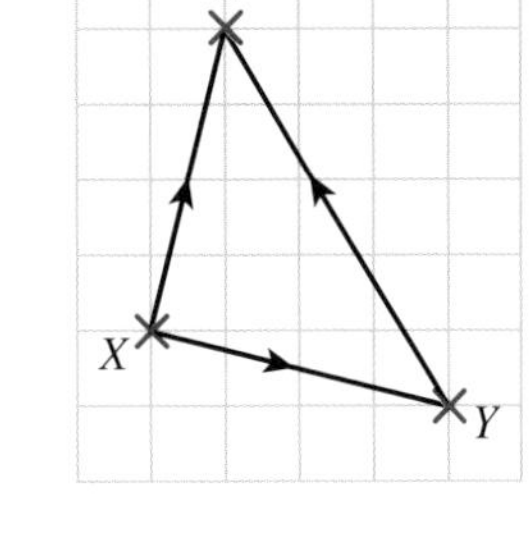

a Find the vector of the translation X to Z.

b Work out $\begin{pmatrix} 4 \\ -1 \end{pmatrix} + \begin{pmatrix} -3 \\ 5 \end{pmatrix}$.

What do you notice?

Q2a hint

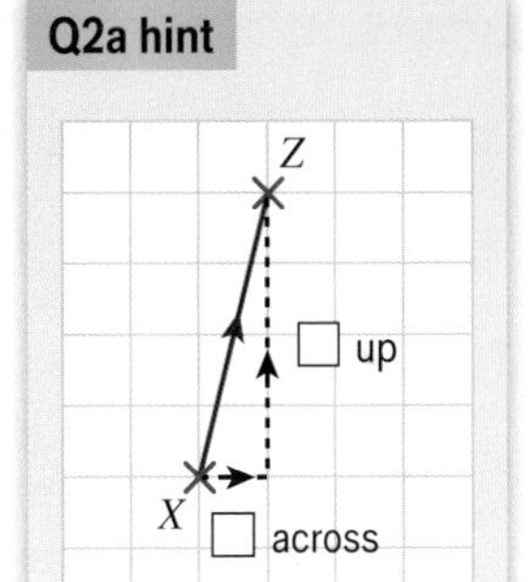

3 Point A is translated to point B by the vector $\begin{pmatrix} 2 \\ 5 \end{pmatrix}$.

Point B is translated to point C by the vector $\begin{pmatrix} -3 \\ 2 \end{pmatrix}$.

Find the vector of the single translation from point A to point C.

4 Find the resultant of these vectors.
$\overrightarrow{PR}$ and $\overrightarrow{RX}$

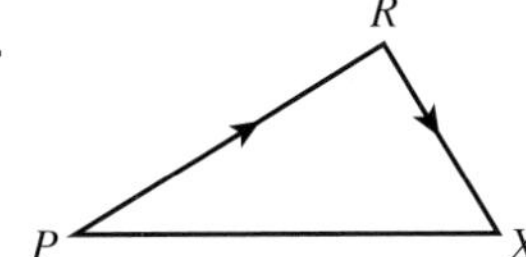

Q4 hint

Which vector starts at P and ends at X?

5 $\mathbf{p} = \begin{pmatrix} 4 \\ 1 \end{pmatrix}$ $\mathbf{q} = \begin{pmatrix} 2 \\ -6 \end{pmatrix}$

Copy and complete to write as column vectors

a $3\mathbf{p} = \begin{pmatrix} 3\times4 \\ 3\times1 \end{pmatrix} = \begin{pmatrix} \square \\ \square \end{pmatrix}$

b $-2\mathbf{q} = \begin{pmatrix} -2\times2 \\ -2\times-6 \end{pmatrix} = \begin{pmatrix} \square \\ \square \end{pmatrix}$

c $3\mathbf{p} - 2\mathbf{q}$

d $-3\mathbf{p} - 2\mathbf{q}$

Q5c and d hint

Use your answers to **a** and **b**.

6 Here are vector **a** and vector **b**.

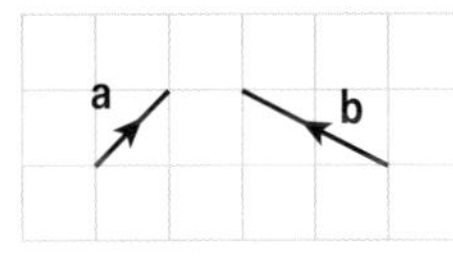

a Move your finger along vector **a** in the direction of the arrow.

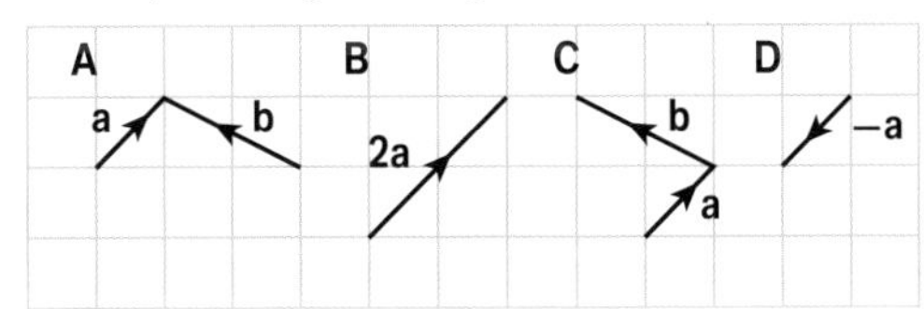

b Choose the diagram from **A–D** that shows a vector where

i you move along the same line but in the opposite direction to **a**

ii you move along the same line but twice the distance of **a**.

c On squared paper draw and label the vectors **i** $-\mathbf{b}$ **ii** $2\mathbf{b}$

d Choose the diagram from **A–D** that shows a vector where you move along **a** and then $-\mathbf{b}$.
This is vector $\mathbf{a} - \mathbf{b}$.

e There is one diagram left from **A–D** that you have not chosen.
What vector does it show?

7 $\mathbf{r} = \begin{pmatrix} -2 \\ 5 \end{pmatrix}$ $\mathbf{s} = \begin{pmatrix} 3 \\ 4 \end{pmatrix}$ Write as column vectors

a $\mathbf{r} + \mathbf{s}$ **b** $\mathbf{r} - \mathbf{s}$ **c** $\mathbf{s} - \mathbf{r}$ **d** $2\mathbf{s} + \mathbf{r}$

19 Extend

1 **Reasoning** X has coordinates (20, 15).
Y has coordinates (40, 10).
A straight line is drawn through these points.
Find the coordinates of the point where this line crosses

a the y-axis **b** the x-axis

> **Q1 hint**
> Draw a diagram and look for similar triangles.

2 **Reasoning** Triangle ABC is similar to triangle DEC.

a Find the scale factor of the enlargement that maps triangle DEC to triangle ABC.

b Find the length of

i BC **ii** BE

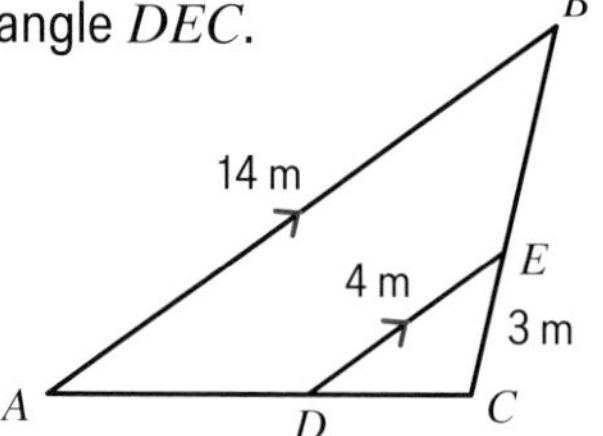

> **Q2b ii hint**
> $BE = BC - EC$

3 **a** Show that triangle ABC and triangle ADE have the same angles, so they are similar.

b Work out the scale factor of the enlargement that maps triangle ADE to triangle ABC.

c Work out the length of BD.

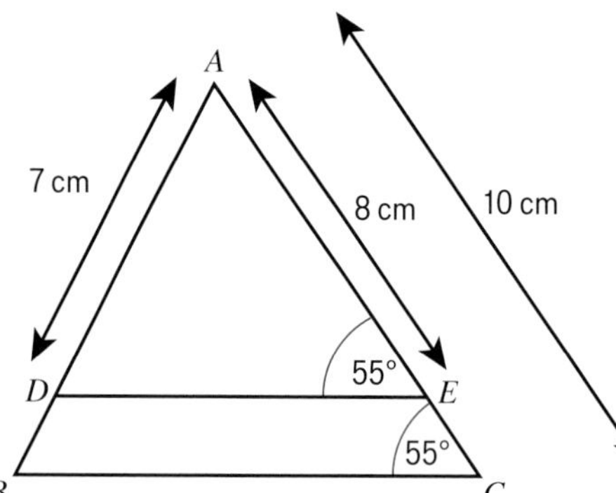

> **Q3a hint**
> How do you know that DE and BC are parallel?

> **Q3 hint**
> Sketch the two separate triangles.

Exam-style question

4 ABC and EDC are straight lines.
AE is parallel to BD.
$CD = 6.5$ cm
$CE = 11.7$ cm
$BD = 3$ cm

a Work out the length of AE. **(2 marks)**

$AC = 8.1$ cm

b Work out the length of AB. **(2 marks)**

> **Exam tip**
> Show working by writing down all your calculations, even when you use a calculator to work out the answer.

5 Work out the missing lengths in these right-angled triangles.

i **ii** **iii** **iv** **v**

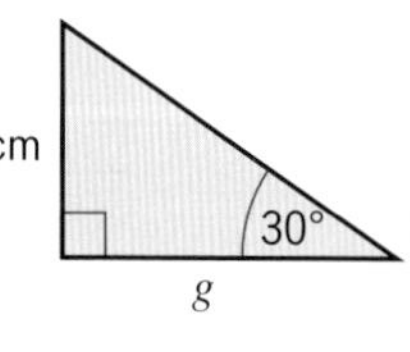

a Which are congruent? **b** Which are similar?

Exam-style question

6 a Find the length of AC. **(2 marks)**

b Find the length of FD. **(1 mark)**

c Are triangles ABC and DEF congruent? **(3 marks)**

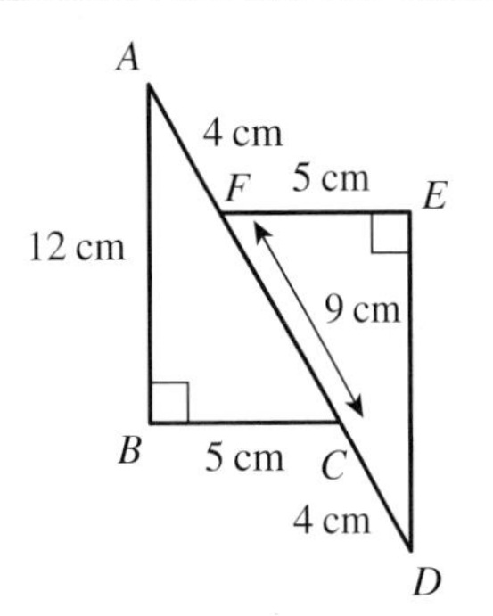

7 Simplify $\overrightarrow{FJ}+\overrightarrow{JT}+\overrightarrow{TV}$.

Q7 hint

Simplify by combining the vectors into a single vector.

8 Write in terms of **a** and/or **b**

a **v**

b **w**

c **x**

d **y**

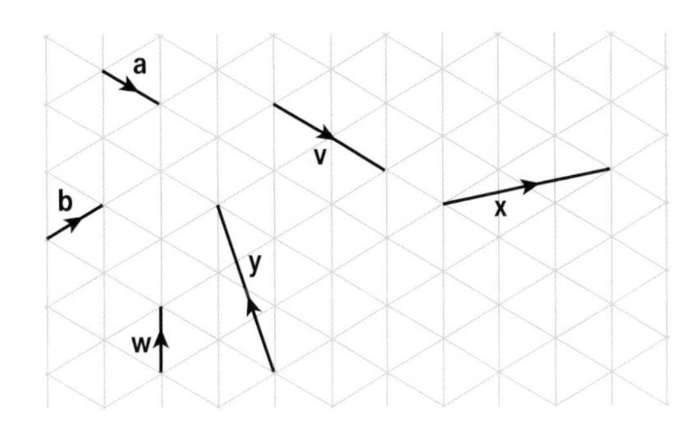

9

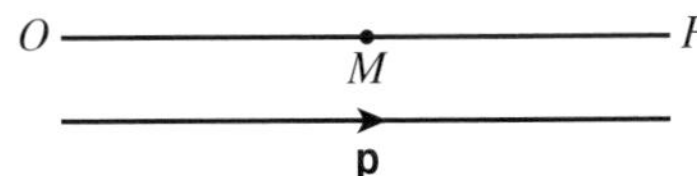

$\overrightarrow{OP}=\mathbf{p}$

M is the midpoint of OP.

a Write down $\overrightarrow{OM}$ in terms of **p**.

Q9a hint

$\overrightarrow{OP}=\mathbf{p}$

$\overrightarrow{OM}=\square\mathbf{p}$

$\overrightarrow{OP}=\begin{pmatrix}8\\0\end{pmatrix}$

b Express $\overrightarrow{OM}$ as a column vector.

10 ABC is a triangle.

$\overrightarrow{BA}=\mathbf{a}$ and $\overrightarrow{BM}=\mathbf{b}$

M is the midpoint of $\overrightarrow{BC}$.

Write down in terms of **a** and/or **b**

a $\overrightarrow{AM}$

b $\overrightarrow{BC}$

c $\overrightarrow{AC}$

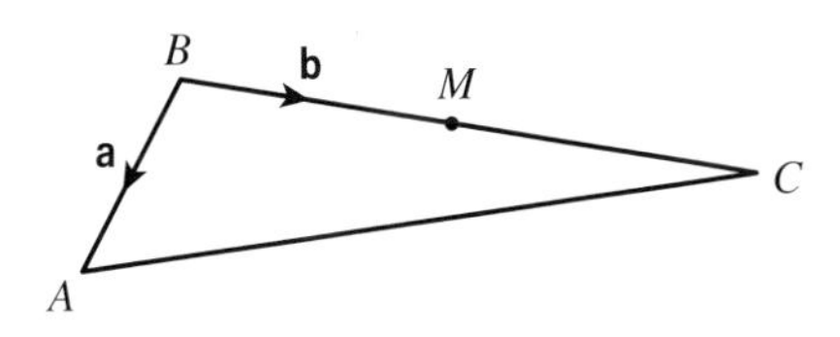

19 Test ready

Summary of key points

To revise for the test:

- Read each key point, find a question on it in the mastery lesson, and check you can work out the answer.
- If you cannot, try some other questions from the mastery lesson or ask for help.

Key points

1 When one shape is an enlargement of another, the shapes are **similar**. → 19.1

2 These triangles are similar.
Corresponding sides are shown in the same colour.
Corresponding angles are shown with the same angle markers. → 19.1

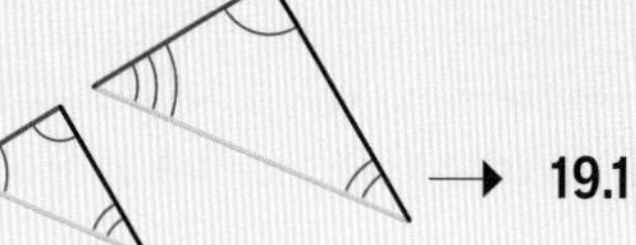

3 For similar shapes:
Corresponding sides are all in the same ratio. **Corresponding angles** are equal.
For example, triangles ABC and DEF are similar, so

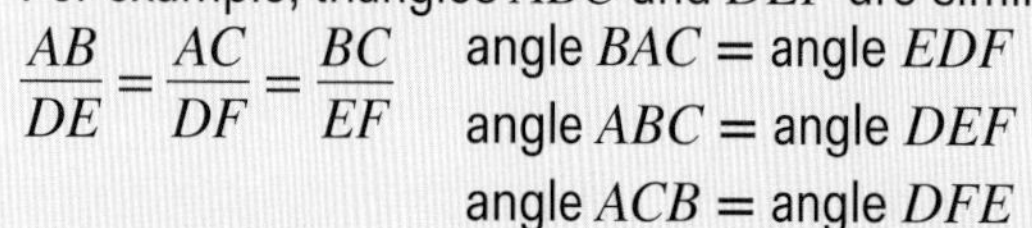

$\frac{AB}{DE} = \frac{AC}{DF} = \frac{BC}{EF}$ angle BAC = angle EDF
angle ABC = angle DEF
angle ACB = angle DFE

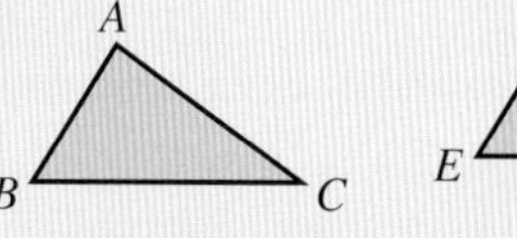

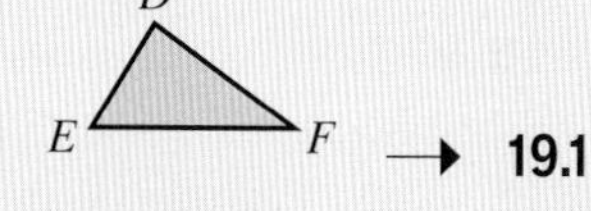

→ 19.1

4 Angle facts can be used to find missing angles. Corresponding sides can be used to work out the scale factor of an enlargement and calculate missing lengths. → 19.2

5 When a shape is enlarged, the perimeter of the shape is enlarged by the same scale factor. → 19.3

6 When two shapes are congruent, one can be rotated or reflected to fit exactly on the other.
Triangles are **congruent** if they have equivalent

- SSS (all three sides)
- SAS (two sides and the included angle)
- ASA (two angles and the included side)
- RHS (right angle, hypotenuse, side)

An included side is between two angles. An included angle is between two sides. → 19.4

7 Triangles where all angles are the same (AAA) may not be congruent.
However, they are always similar. → 19.4

8 Triangles with two equivalent sides and an equivalent non-included angle (SSA) may not be congruent. → 19.4

9 To add column vectors, add the top numbers and add the bottom numbers. → 19.6

10 Two translations can be combined into a single translation by adding the column vectors. → 19.6

11 A vector has a magnitude (size) and direction.
The diagram shows a vector that starts at A and finishes at B.
You can write this vector as $\overrightarrow{AB} = \begin{pmatrix} 3 \\ 2 \end{pmatrix}$. → 19.6

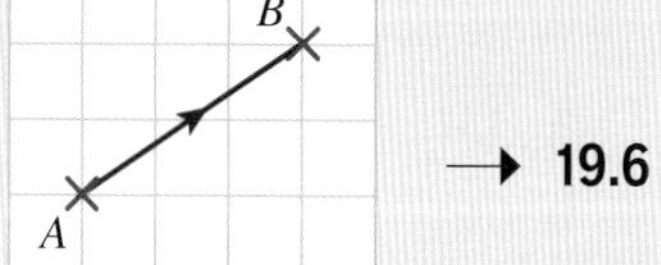

12 A to B followed by B to C is equivalent to A to C.
$\overrightarrow{AB} + \overrightarrow{BC} = \overrightarrow{AC}$
The sum of two vectors is called the resultant.
$\overrightarrow{AC}$ is the **resultant** of $\overrightarrow{AB}$ and $\overrightarrow{BC}$. → 19.6

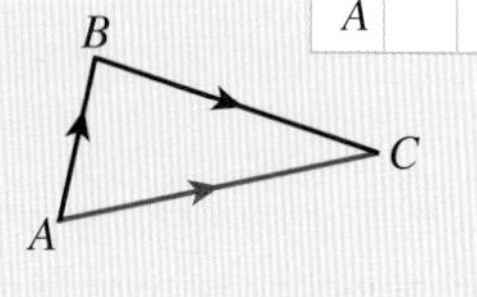

Key points

13 You can show vectors using different notation. A single letter may be used to represent a vector. The letter is shown in **bold type**: $\mathbf{a} = \begin{pmatrix} -2 \\ 5 \end{pmatrix}$. You can't write bold in handwriting, so you should underline a letter that represents a vector. For **a**, write $\underline{a}$. → **19.6**

14 $-\mathbf{a}$ is the negative of **a** and points in the opposite direction. → **19.7**

15 Equal vectors have the same magnitude (size), for example the same length. They are also parallel and have the same direction. → **19.7**

16 $\mathbf{a} - \mathbf{b}$ is the same as $\mathbf{a} + (-\mathbf{b})$. → **19.7**

17 To subtract column vectors, subtract the top numbers and subtract the bottom numbers. → **19.7**

18 You can multiply a vector by a number. For example, if $\mathbf{a} = \begin{pmatrix} 4 \\ 3 \end{pmatrix}$, then $2\mathbf{a} = \begin{pmatrix} 8 \\ 6 \end{pmatrix}$ and $-3\mathbf{a} = \begin{pmatrix} -12 \\ -9 \end{pmatrix}$. → **19.7**

19 Equal vectors have the same magnitude (size), that is, the same length. They are also parallel and have the same direction. → **19.7**

Sample student answers

Exam-style question

1 A is the point (1, 4). B is the point (4, 6). Find the vector $\overrightarrow{AB}$.

Give your answer as a column vector $\begin{pmatrix} x \\ y \end{pmatrix}$. **(2 marks)**

$\begin{pmatrix} 3 \\ 2 \end{pmatrix}$

Describe two things the student has done to help them find the vector accurately.

Exam-style question

2

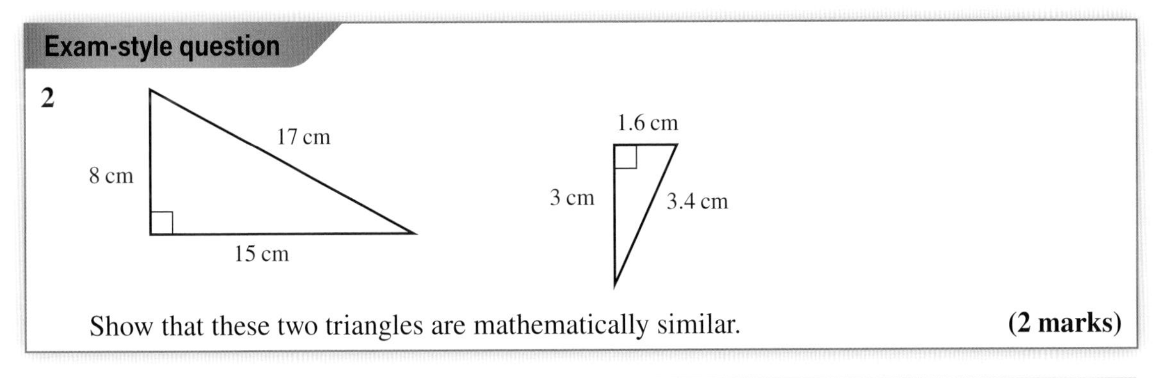

Show that these two triangles are mathematically similar. **(2 marks)**

$15 \div 3 = 5$ Scale factor 5

The student is unlikely to receive full marks for their answer.
Explain why and write an answer that would get full marks.

19 Unit test

1 Triangle ADE is similar to triangle ABC.

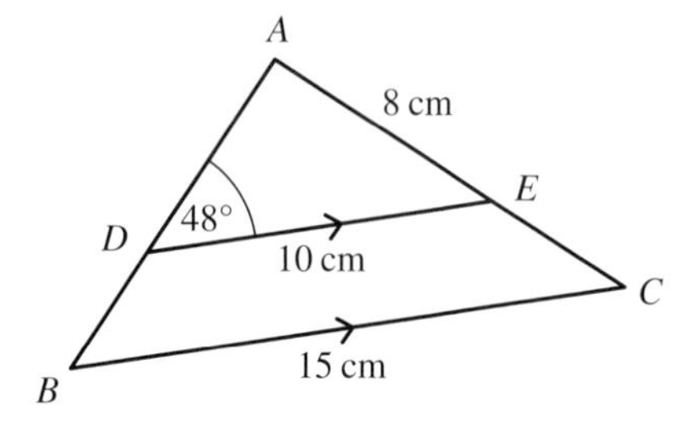

a Write the scale factor of the enlargement that maps triangle ADE to triangle ABC. **(2 marks)**

b Find the size of angle ABC.
Give a reason for your answer. **(2 marks)**

c Work out the length of EC. **(2 marks)**

2 Shapes A and B are regular hexagons.
The height of B is 3 times the height of A.
The perimeter of A is 18 cm. Find the perimeter of B. **(2 marks)**

3 These two triangles A and B are mathematically similar.

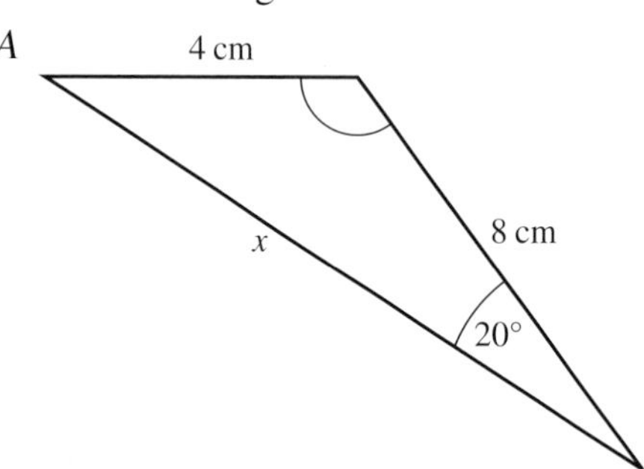

Find x and y. **(4 marks)**

4 ABC and PQR are triangles.

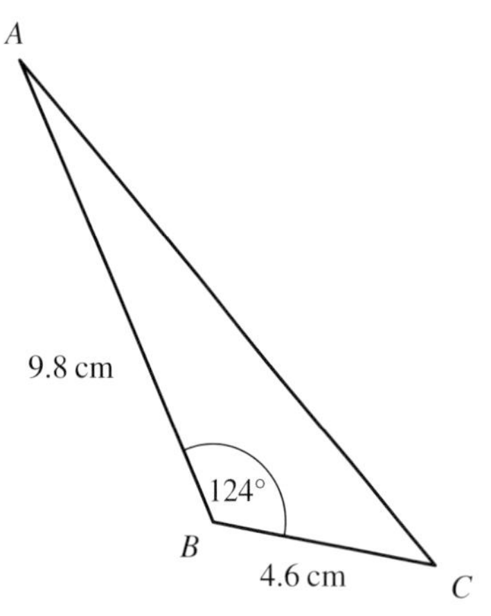

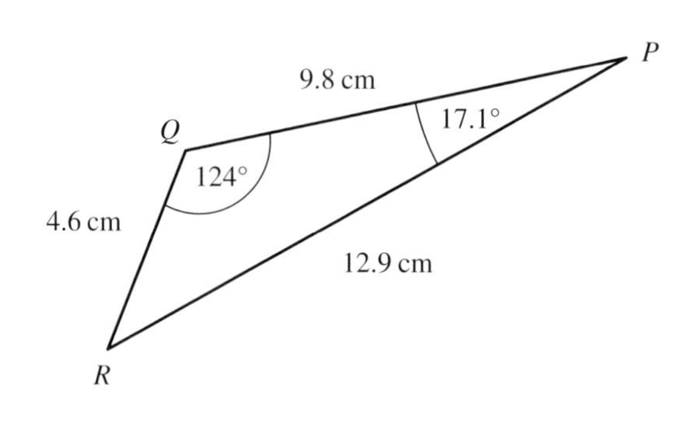

a Are triangles ABC and PQR congruent?
Give a reason for your answer. **(2 marks)**

b Write the length of AC. **(1 mark)**

c Work out the size of angle ACB. **(1 mark)**

5 Write as a single column vector

$\begin{pmatrix} 3 \\ -2 \end{pmatrix} - \begin{pmatrix} -2 \\ 0 \end{pmatrix}$ **(2 marks)**

6 *ABCD* is a parallelogram.

a Is triangle *ABC* congruent to triangle *CDA*? **(2 marks)**

b Which angle is the same as angle *ABC*? **(2 marks)**

Give reasons for your answers.

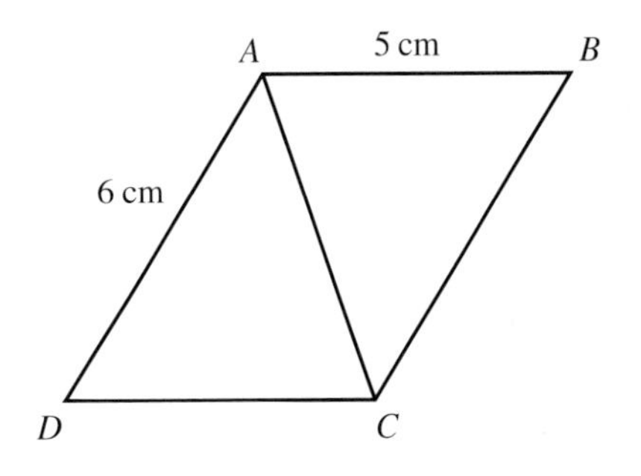

7 Find the resultant of

$\begin{pmatrix} 7 \\ -2 \end{pmatrix}$ and $\begin{pmatrix} -3 \\ -5 \end{pmatrix}$ **(2 marks)**

8 Find a vector that is

a in the same direction as $\begin{pmatrix} 3 \\ -4 \end{pmatrix}$ but twice as long **(2 marks)**

b the same length as $\begin{pmatrix} -6 \\ 5 \end{pmatrix}$ but in the opposite direction **(2 marks)**

c parallel to $\begin{pmatrix} 4 \\ -8 \end{pmatrix}$ but half as long **(2 marks)**

9 Work out $2\mathbf{a} + \mathbf{b}$.

$\mathbf{a} = \begin{pmatrix} -2 \\ 6 \end{pmatrix}$ $\mathbf{b} = \begin{pmatrix} 5 \\ -7 \end{pmatrix}$ **(2 marks)**

10 *ABCD* is a rhombus.
BD is a diagonal of rhombus *ABCD*.
$\overrightarrow{AB} = \mathbf{a}$ and $\overrightarrow{BC} = \mathbf{b}$

a Find in terms of **a** the vector $\overrightarrow{CD}$. **(1 mark)**

b Find in terms of **a** and **b** the vector $\overrightarrow{BD}$. **(1 mark)**

c Show that $\overrightarrow{AC} \neq \overrightarrow{BD}$. **(1 mark)**

(TOTAL: 35 marks)

11 Challenge Which of these are **i** similar **ii** congruent **iii** neither? Explain.

a A pair of right-angled triangles, each with one angle of 40°.

b A pair of right-angled triangles, each with hypotenuse that is 5 cm, and one side that is 3 cm.

c A pair of isosceles triangles, one with an angle of 100°, and the other with only one angle of 40°.

d A pair of equilateral triangles, each with side length 3 cm.

12 Reflect 'Notation' means symbols. Mathematics uses a lot of notation.

For example: = means 'is equal to' ° means degrees ∟ means a right angle

Look back at this unit. Write a list of all the maths notation used.
Why do you think this notation is important?

20 More algebra

Prior knowledge

ActiveLearn Homework

20.1 Graphs of cubic and reciprocal functions

- Draw and interpret graphs of cubic functions.
- Draw and interpret the graph of $y = \frac{1}{x}$.

Warm up

1 Fluency Work out the reciprocal of

a 4 **b** $\frac{1}{3}$ **c** 0.5 **d** −5 **e** $-\frac{1}{4}$

2 Here is the graph of $y = x^2 + 2x - 3$.
Use the graph to find the roots of the equation $x^2 + 2x - 3 = 0$.

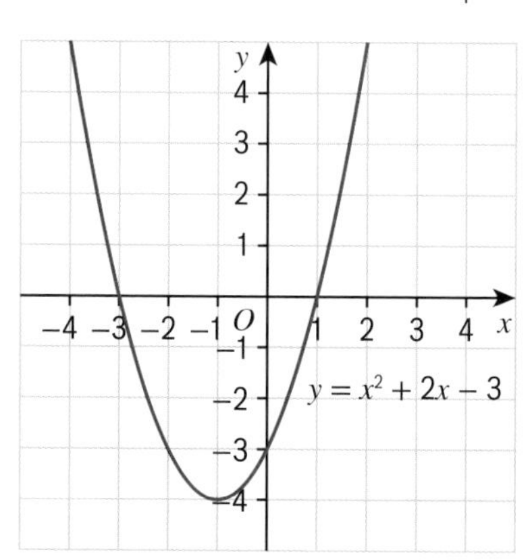

Key point

A **cubic function** contains a term in x^3 but no higher power of x.
It can also have terms in x^2 and x and number terms.

3 Copy and complete this table of values for $y = x^3$.

x	−3	−2	−1	0	1	2	3
y	−27					8	

4 a On graph paper, draw a coordinate grid with the x-axis from −3 to +3 and the y-axis from −30 to +30.

b Now plot the points from your table of values for $y = x^3$ in **Q3**.

c Join the points with a smooth curve. Label your graph with its equation.

5 a Use your graph from **Q4** to estimate

i 2.3^3 **ii** $\sqrt[3]{15}$

b Use a calculator to work out

i 2.3^3 **ii** $\sqrt[3]{15}$

c Which are more accurate: part **a** answers or part **b** answers? Explain.

6 **a** Make a table of values for $y = -x^3$ for values of x from -3 to $+3$.

b Plot the graph of $y = -x^3$.

c **Reflect**

i What is the same and what is different about the graphs of $y = x^3$ and $y = -x^3$?

ii What type of symmetry do they have?

Q6a hint

Draw a table like the one in **Q3**. Draw the axes from **Q4** again.

7 **a** Copy and complete this table of values for $y = x^3 + 2$.

x	-3	-2	-1	0	1	2	3
x^3		-8					27
+2	+2	+2	+2	+2	+2	+2	+2
y		-6					29

b Draw a pair of axes on graph paper. Make sure all the x- and y-values in the table are on your axes. Plot the graph of $y = x^3 + 2$.

c Make a table of values for $y = x^3 - 1$.

d Plot the graph of $y = x^3 - 1$ on the same axes as your graph of $y = x^3 + 2$.

e Write down what is the same about your two graphs and what is different.

f **Reflect** What would the graph of $y = x^3 + 5$ look like? What about $y = x^3 - \frac{1}{2}$?

Key point

A cubic function can have one, two or three roots.

8 Use your graph of $y = x^3 + 2$ from **Q7** to find the solution of $x^3 + 2 = 0$.

Q8 hint

Read off the x-value where the graph of $y = x^3 + 2$ crosses the x-axis.

Exam-style question

9 **a** Complete the table of values for $y = x^3 - 5x$.

x	-3	-2	-1	0	1	2	3
y			4	0			12

(2 marks)

Exam tip

You might find it easier to draw your own table with a row for each part of the function
Plot the points using crosses.
Join them with a smooth curve.
Always use a sharp pencil.

x	
x^3	
$-5x$	
y	

b On a copy of the grid, draw the graph of $y = x^3 - 5x$ from $x = -3$ to $x = 3$.

(2 marks)

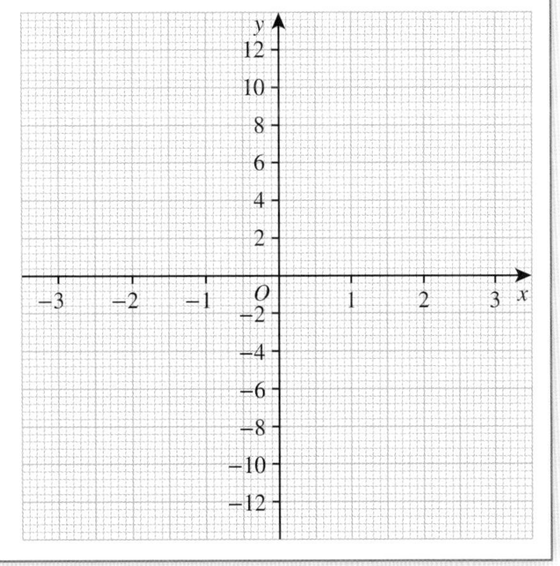

10 **a** Copy and complete the table of values for $y = \frac{1}{x}$.

x	−4	−3	−2	−1	$-\frac{1}{2}$	$-\frac{1}{4}$	$\frac{1}{4}$	$\frac{1}{2}$	1	2	3	4
y	$-\frac{1}{4}$				−2			2	1			$\frac{1}{4}$

b Plot the points. Join the points in each part to form smooth curves.

c Label your graph $y = \frac{1}{x}$.

d Use your graph to estimate the value of x when $y = 3.5$.

Key point

The x- and y-axes are **asymptotes** to the curve $y = \frac{1}{x}$.

An asymptote is a line that the graph gets closer and closer to, but never actually touches.

11 Explain why you cannot read the value of y when $x = 0$ from your graph of $y = \frac{1}{x}$.

Exam-style question

12 **a** Complete the table of values for $y = \frac{4}{x}$.

x	0.5	1	1.5	2	2.5	4	5
y		4		2			0.8

b Copy the grid.

Draw the graph of $y = \frac{4}{x}$ for values of x from 0.5 to 5. **(2 marks)**

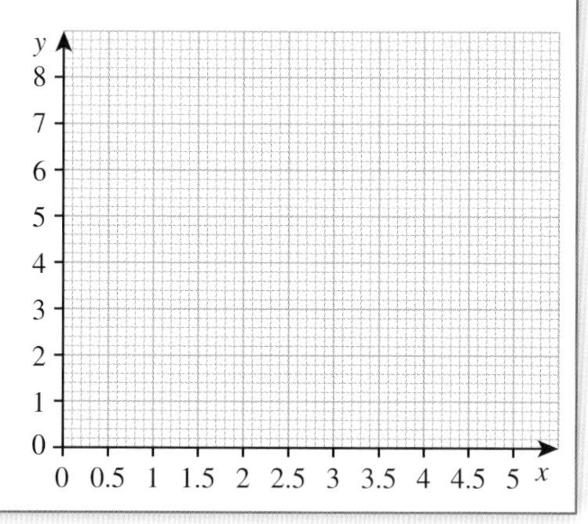

13 **Reflect** Look back at the graphs you drew in **Q4**, **Q6** and **Q10**.

a Sketch the graphs of $y = x^3$ and $y = \frac{1}{x}$. How can you remember which is which?

b How can you find the graph of $y = -x^3$ from your sketch in part **a**?

14 **Reasoning** Match each equation to a graph.

a $y = x^2$ **b** $y = x^3$ **c** $y = 4x$

d $y = \frac{1}{x}$ **e** $y = -x^3$ **f** $y = -2x$

i

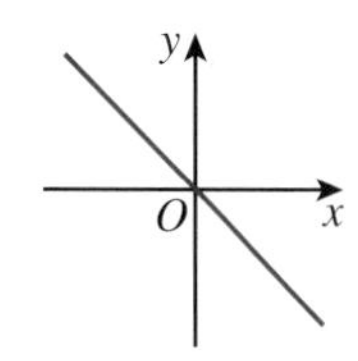

ii

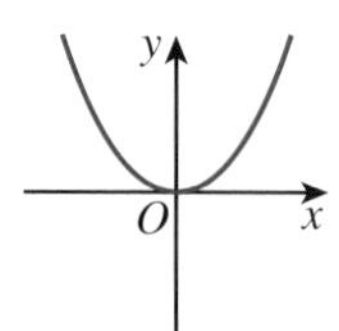

iii

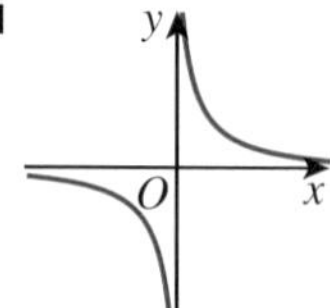

iv

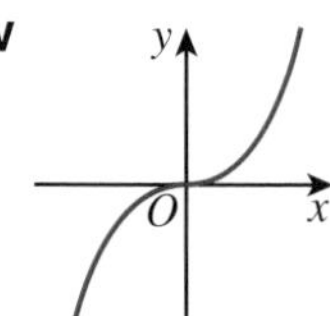

v

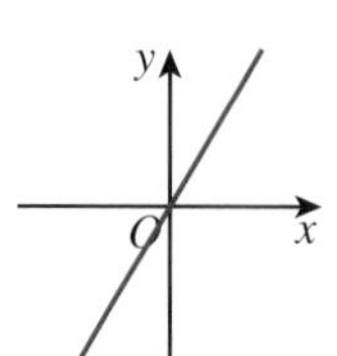

vi

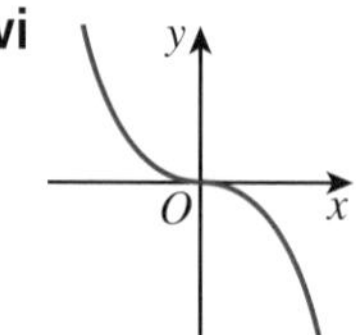

20.2 Non-linear graphs

- Draw and interpret non-linear graphs to solve problems.

Warm up

1 **Fluency** Match each graph to its equation: $y = x$ and $y = \frac{1}{x}$.

a

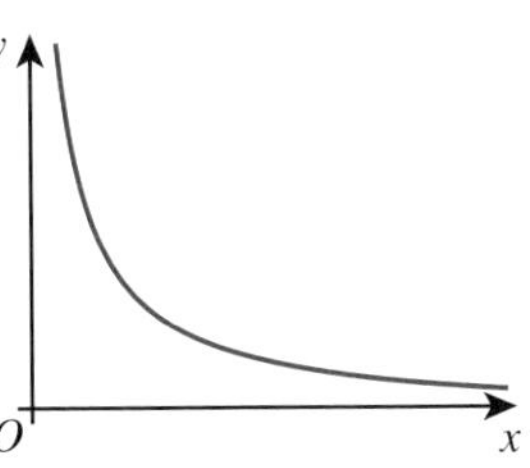

b

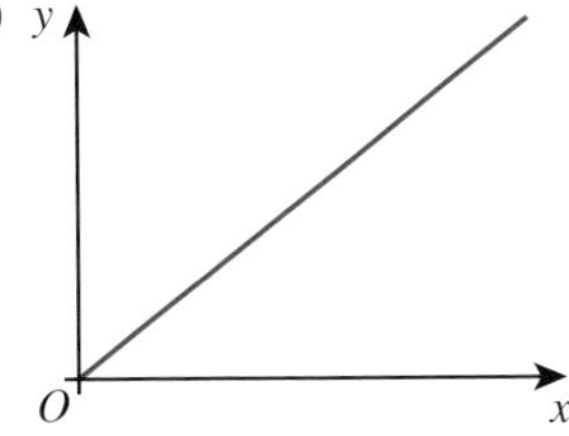

2 What is the volume of a cube with side

a 2 cm **b** 5 cm **c** x cm?

3 Write 1.3 hours in hours and minutes.

Key point

A non-linear graph is any graph that is not a straight line. So a curved graph is a non-linear graph. You can estimate values from a non-linear graph.

4 **Reasoning** This solid is made from two cubes.

a Write an equation for the total volume, y, of the solid.

b Make a table of values for your equation in **a**, for values of x from 0 to 4.

c Draw the graph.

d From your graph, estimate the volume of the solid when $x = 2.7$ cm.

e A solid like this has volume 40 cm^3.
Estimate the value of x.

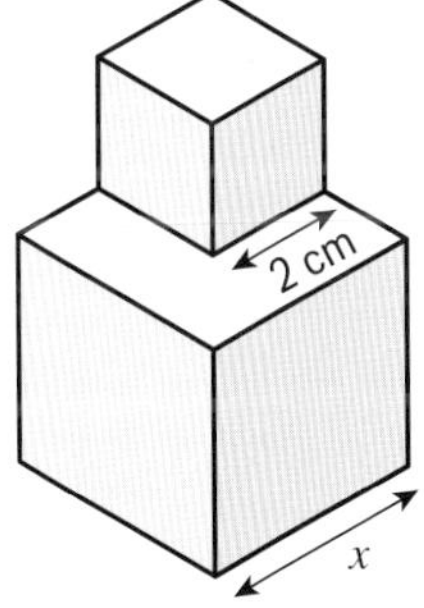

5 **Reasoning** The graph shows the count rate against time for a radioactive material, phosphorus-32. The count rate measures the number of radioactive emissions per second.

a What is the count rate after 4 weeks?

b How many weeks does it take for the count rate to reduce to 10?

c The half-life of a radioactive material is the time it takes for the count rate to halve.
What is the half-life of phosphorus-32?

d Does the count rate ever reach zero?
Explain.

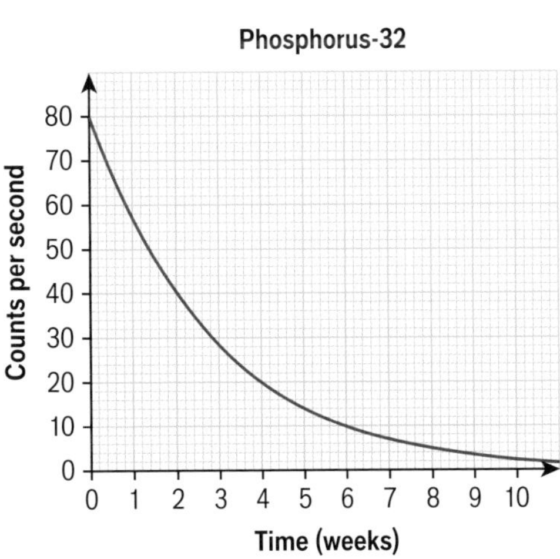

Key point

When x and y are in direct proportion, as x increases y increases.
$y = kx$

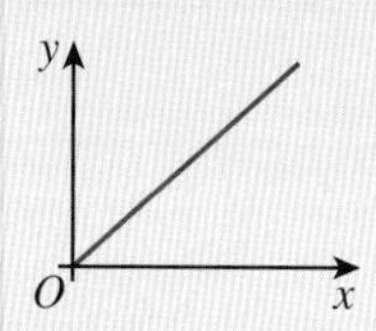

When x and y are inversely proportional, as x increases y decreases.
$y = \frac{k}{x}$

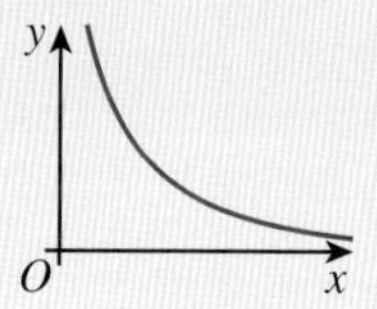

Example

This graph shows the number of days it takes different numbers of workers to build a swimming pool.

a Describe the relationship between the number of workers and the number of days.

b Estimate how long it takes 8 workers to build a swimming pool.

c How many workers are needed to build a swimming pool in 4 days?

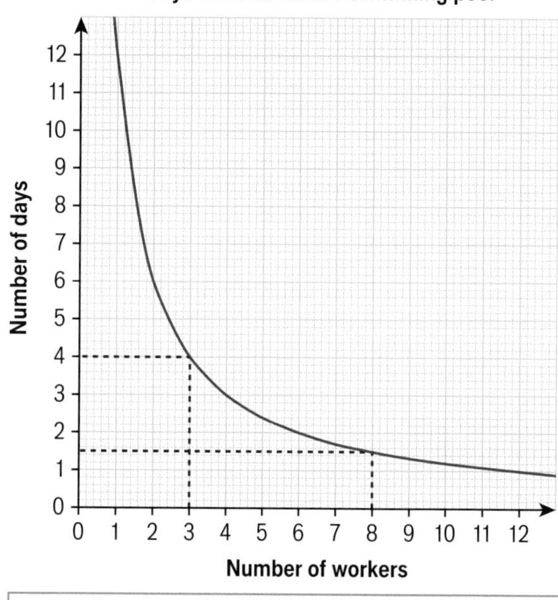

a Number of workers and number of days are in inverse proportion.

b 8 workers will take 1.5 days.

c 4 days will need 3 workers.

Does the graph show direct or inverse proportion?

Use the graph to find the number of days for 8 workers.

Use the graph to find the number of workers for 4 days. If the answer is a decimal then round up.

6 **Problem-solving** This graph shows how long it takes different numbers of cleaners to clean the bedrooms in a hotel.

a Describe the relationship between the number of cleaners and the number of hours.

b Estimate how long it takes 12 cleaners to clean the bedrooms.

c The bedrooms need to be cleaned between 10 am and 4 pm.
How many cleaners are needed?

Q6c hint

10 am–4 pm is ☐ hours.

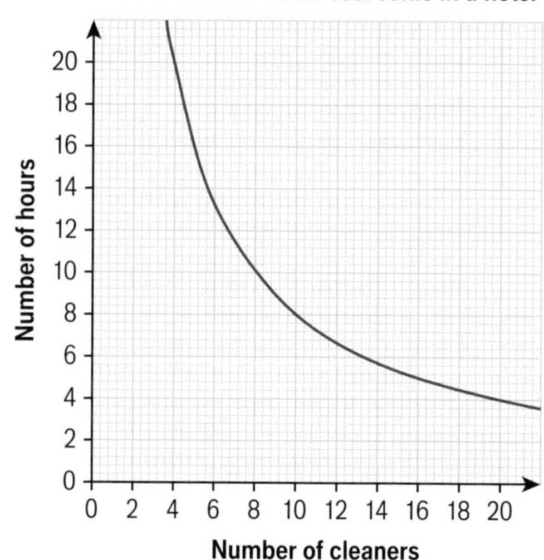

7 **Reasoning** The graph shows the amount of water in a tank as it empties.

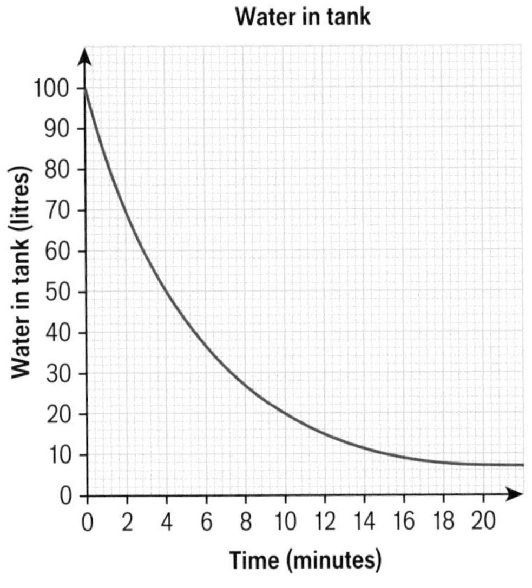

a How much water was there in the tank to start with?

b How much water was in the tank after 10 minutes?

c Estimate how many minutes it took for

i half the water in the tank to empty out

ii 75 litres to empty out

d Compare the change in volume between 0 and 2 minutes with the change between 8 and 10 minutes.
Show that the tank empties less quickly as the volume of water in it decreases.

Exam-style question

8 The table shows the times taken to drive 120 km at different speeds:

Speed (km/h)	20	30	40	60	80
Time (hours)	6	4	3	2	1.5

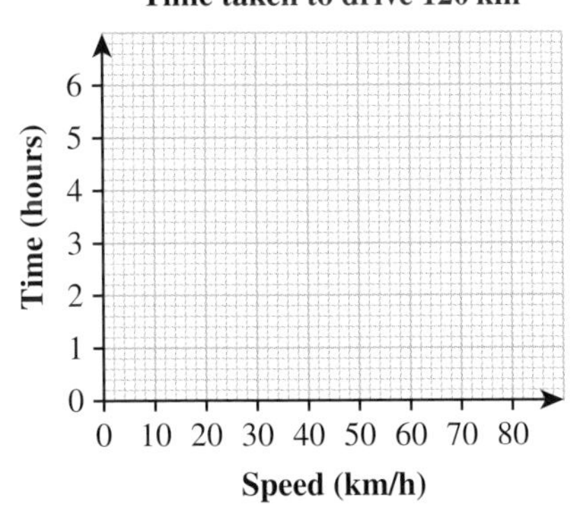

a Describe the relationship between speed and time. **(2 marks)**

b Estimate how long it takes to drive 120 km at 35 km/h. **(1 mark)**

c Liberty took $2\frac{1}{2}$ hours to drive 120 km.
Estimate her average speed for the journey. **(1 mark)**

Exam tip

If the exam question gives you a grid, use it to draw a graph even if the question does not ask you to.

9 **Problem-solving** The graph shows how the number of cells in a yeast sample grows over a 4-hour period.

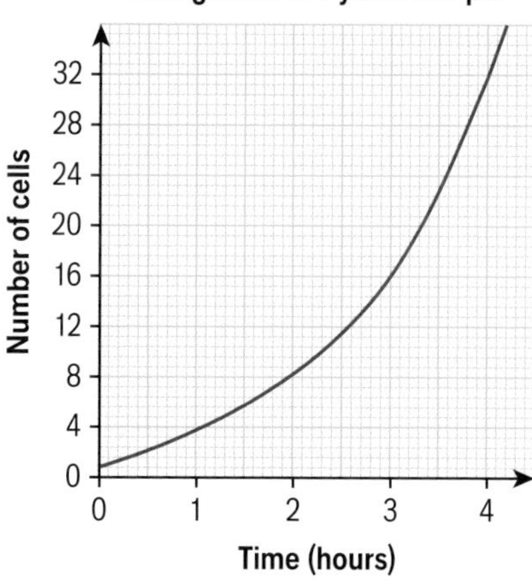

a Copy and complete the table to show the numbers of cells.

Time (hours)	0	1	2	3	4
Number of cells	2				

b Describe the sequence of the number of cells.
Give the first term and the term-to-term rule.

c Use the graph to estimate the number of cells after $3\frac{1}{4}$ hours.

d Estimate when the number of yeast cells reaches 28.

10 The graph shows the value of an investment over a 5-year period.

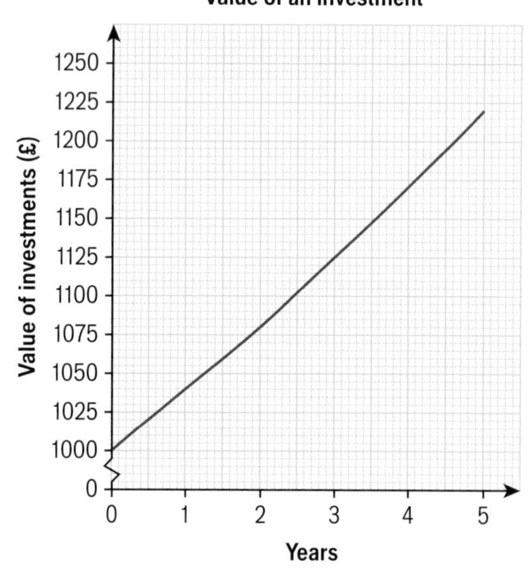

a What was the initial value of the investment?

b Estimate the value of the investment after 5 years.

c By how much did the value increase in the first year?

d The rate of interest remained the same for the five years.

Use the formula percentage change $= \frac{\text{actual change}}{\text{original amount}} \times 100$ to work out the percentage interest rate.

20.3 Solving simultaneous equations graphically

- Solve simultaneous equations by drawing a graph.
- Write and solve simultaneous equations.

Warm up

1 **a** One coffee costs £3 and a lemonade £2. Write an expression for the price of
  i x coffees **ii** y lemonades **iii** x coffees and y lemonades
 b x coffees and y lemonades cost £9. Write an equation to show this.

2 **a** Draw the graphs of $y = 3$ and $2x + y = 7$ on the same pair of x- and y-axes from 0 to 10.
 b Write the coordinates of the point where the lines cross.

Key point

Simultaneous equations are equations that are both true for a pair of variables (letters).
To find the solution to simultaneous equations graphically:
1 Draw the graphs on the same pair of axes.
2 Find the point where the lines cross (the point of **intersection**).

3 **a** Draw the graph of $2x + y = 9$.
 b Draw the graph of $y = 1$ on the same axes.
 c The solution to the simultaneous equations $2x + y = 9$ and $y = 1$ is at the intersection of the graphs.
 Write the solution $x = \square$ and $y = \square$ for the point of intersection.

4 Draw two graphs on the same axes to solve these simultaneous equations.
 $3x + 2y = 9$
 $x - 2y = -1$

5 Solve these simultaneous equations graphically.
 $4x + y = -5$
 $2x + 3y = 10$

6 Use the graphs to solve the pairs of simultaneous equations.
 a $x - y = 2$
  $x + y = 8$
 b $x - y = 2$
  $x + 4y = 11$
 c $x + y = 8$
  $x + 4y = 11$
 d **Reflect** Are there any values of x and y that satisfy all three equations simultaneously? Explain.

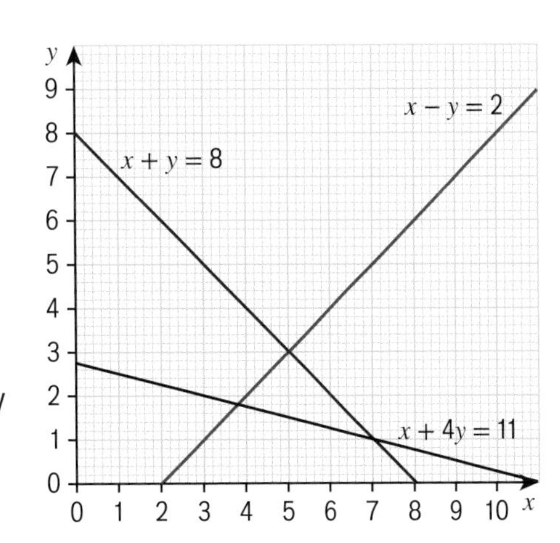

Exam-style question

7 The graph of $2y+x=4$ is shown on this grid.

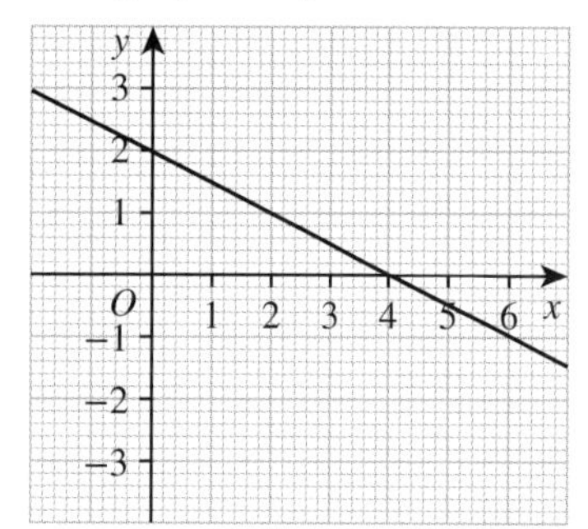

Exam tip

Plot points accurately with crosses. Join the points with a sharp pencil and ruler.
Write your solutions clearly:
$x = \square$, $y = \square$

a Copy the graph grid and the line $2y+x=4$.
Draw the graph of $y=\frac{1}{2}x-3$ on the grid. **(2 marks)**

b Solve these simultaneous equations graphically.

$2y+x=4$
$y=\frac{1}{2}x-3$ **(1 mark)**

8 **Reasoning** By drawing their graphs, show that the simultaneous equations $x+2y=4$ and $x+2y=-7$ have no solution.

Q8 hint

Write a sentence beginning:
They do not have a solution because ...

Example

The sum of two numbers is 4 and their difference is 2.
Find the two numbers.

$x+y=4$
$x-y=2$

Use x and y for the two numbers.
Write two equations, one for the sum and one for the difference.

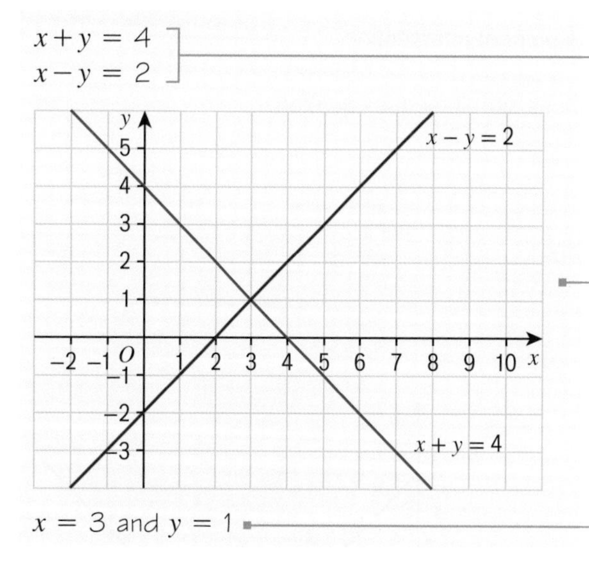

Draw the graphs of $x+y=4$ and $x-y=2$.
Read the solution from the point of intersection.

$x=3$ and $y=1$

Check: $3+1=4$, $3-1=2$ ✓

9 The sum of two numbers is 23 and their difference is 13.
Draw graphs to find the numbers.

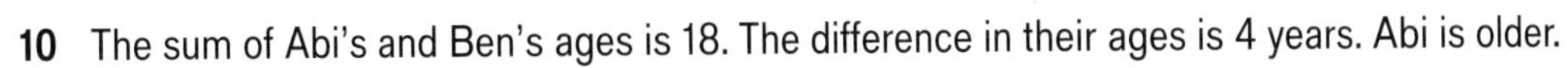

10 The sum of Abi's and Ben's ages is 18. The difference in their ages is 4 years. Abi is older.

a Write an equation for the sum of their ages.
Use x for Abi's age and y for Ben's age.

b Write an equation for the difference in their ages.

c Draw the graph of each equation.

d Find Abi's and Ben's ages from your graph.

11 **a** Cinema tickets for 1 adult and 1 child cost £16.
Write an equation to represent this.
Use x for the price of an adult's ticket and y for the price of a child's ticket.

b At the same cinema, tickets for 1 adult and 4 children cost £34.
Write an equation to represent this.

c Copy these axes.
Continue labelling the axes as needed.
Draw the graph of each equation.

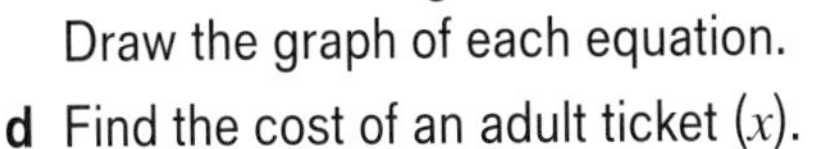

d Find the cost of an adult ticket (x).

e Find the cost of a child ticket (y).

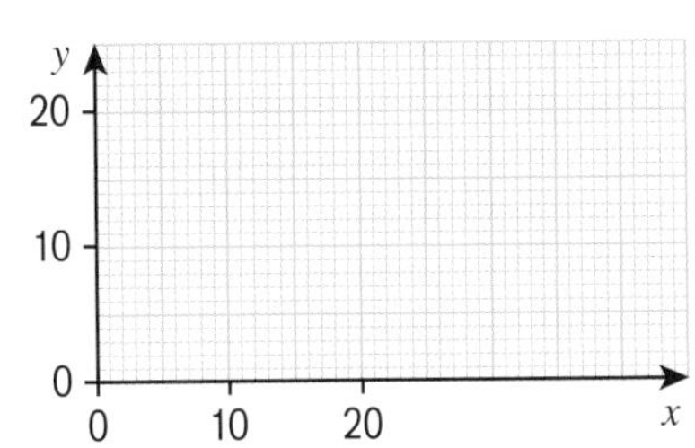

12 **Problem-solving** A Scout leader plans a trip to a theme park.
Scout group: 2 adult tickets, 7 child tickets, total cost £76.
Cub group: 4 adult tickets, 9 child tickets, total cost £112.

a What do x and y represent in this equation for the Scout group: $2x + 7y = 76$?

b Write an equation using x and y for the Cub group.

c Draw graphs of the equations from parts **a** and **b**.
Find the cost of an adult ticket and a child ticket.
Write your answer in £s.

13 **Problem-solving** A school trip for Year 10 uses 2 minibuses and 1 coach for 60 people.
A trip for Year 11 uses 1 minibus and 2 coaches for 84 people.

a Write equations for Years 10 and 11.
Use x for the number of people in a minibus and y for the number in a coach.

b Solve your simultaneous equations graphically to work out the number of people in a minibus and the number of people in a coach.

14 **Problem-solving** In a coffee shop:
2 lattes and 1 Americano cost £6.80
3 lattes and 2 Americanos cost £11.10

Write a pair of simultaneous equations and solve them graphically, to find the cost of

a a latte **b** an Americano

15 **Reflect** Look back at your answers in this lesson.
How easy is it to read accurate answers from a graph?
Which set of simultaneous equations would be easier to solve accurately from a graph:

A $x + y = 7$
$3x - y = 1$

or

B $x + y = 115$
$2x - 3y = 5$

Explain your answer.

20.4 Solving simultaneous equations algebraically

- Solve simultaneous equations algebraically.

Warm up

1 **Fluency** Simplify

a $2x - 2x$ **b** $2x - -2x$ **c** $-2x + 2x$ **d** $-2x - -2x$

2 Copy and write in the missing + or −.

a $3y \square -3y = 0$ **b** $5x \square 5x = 0$ **c** $-6x \square 6x = 0$ **d** $-4y \square -4y = 0$

3 Solve these equations.

a $3y = 18$ **b** $-4x = 6$ **c** $2y = -8$ **d** $5x - 3 = 7$

4 Here are two simultaneous equations

$2x + y = 9$
$y = 1$

a Substitute $y = 1$ into the equation $2x + y = 9$.

b Solve your equation in x from part **a**.

c Write the solutions to the simultaneous equations, $x = \square$, $y = \square$.

d In Lesson 20.3 **Q3**, you drew graphs to solve these equations. Are your solutions the same?

Key point

To solve simultaneous equations by the elimination method, add or subtract the equations to eliminate either the x or the y terms.

Example

Solve these simultaneous equations algebraically: $3x + 2y = 9$ and $x - 2y = -1$

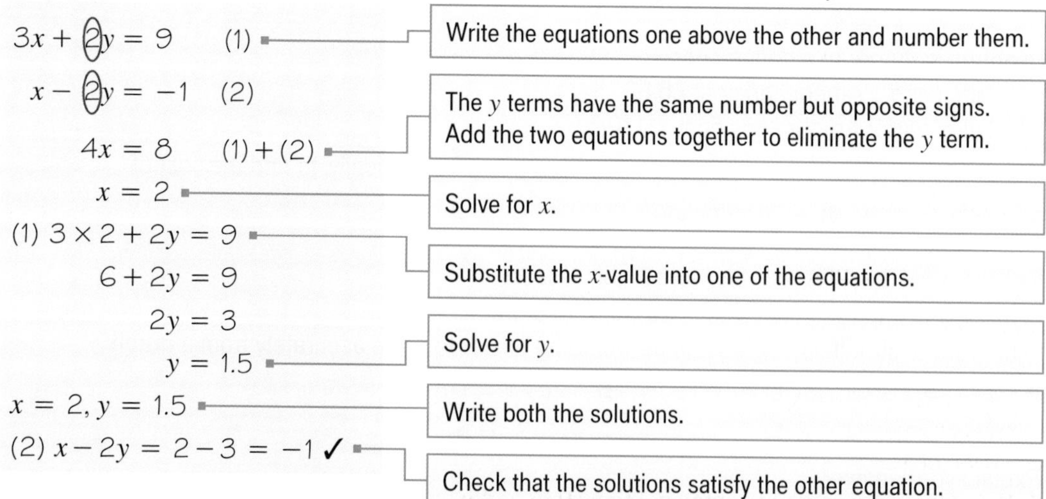

5 **a** Solve these simultaneous equations algebraically.

$x - y = 2$
$x + y = 8$

b In Lesson 20.3 **Q6a**, you solved these equations using graphs. Are your solutions the same?

6 Solve these simultaneous equations algebraically. Start by adding the two equations to eliminate terms.

a $5x + y = 8$
$3x - y = 0$

b $-2x + 3y = 19$
$2x + y = 1$

c $4x - y = 8$
$-4x + 5y = 4$

d $6x - 2y = -6$
$3x + 2y = 0$

7 Solve each pair of simultaneous equations. Add or subtract the equations to eliminate the x terms, or the y terms.

a $2x - y = 12$
$-2x - 3y = -4$

b $4x + y = 2$
$8x + y = 3$

c $-x + 2y = 10$
$-x - 3y = -13$

d **Reflect** Explain how you decide whether to add or subtract the equations to eliminate terms.

Exam-style question

8 Solve the simultaneous equations

$3x - y = 2$
$3x + 5y = -7$ **(2 marks)**

Exam tip

Number the equations.
Show clearly how you eliminate one of the letter terms.
Write your answer $x = \square$, $y = \square$

9 **Problem-solving** The sum of two numbers is 24 and their difference is 14. Write and solve a pair of simultaneous equations to find the two numbers.

10 Copy and complete to solve the simultaneous equations.

a

$5x - 2y = 4$ (1)

$2x - y = -1$ (2)

×2 ↓ ×2

$\square x - 2y = -\square$ (3)

b Subtract equation (3) from equation (1).

$$\begin{array}{rcl} 5x - 2y &=& 4 \\ -\ \square x - 2y &=& -\square \\ \hline x + 0 &=& \square \\ \hline \end{array}$$

c Solve for x.

d Substitute your value of x into equation (1) to find y.

11 Solve the simultaneous equations.
Give your answers as decimals where necessary.

a $3x+y=11$
$2x-3y=-11$

b $3x-y=13$
$x-5y=9$

c $3x+y=3$
$x-y=-1$

d $x+7y=8$
$3x-4y=4$

12 Solve these simultaneous equations:

$2x+3y=11$ (1)
$3x+4y=15$ (2)

a First multiply equation (1) by 3.
Label this equation (3).

b Now multiply equation (2) by 2.
Label this equation (4).

c Solve the simultaneous equations (3) and (4).

13 Repeat **Q12**, but this time:

a Multiply equation (1) by 4. Label this equation (3).

b Multiply equation (2) by 3. Label this equation (4).

c Solve the simultaneous equations (3) and (4).

d **Reflect** Did you find **Q12** or **Q13** easier to solve these equations?
Explain why.

14 Charlie is paid £55 for 6 hours' work plus 1 hour's overtime.
For 10 hours' work plus 2 hours' overtime he is paid £95.
Write two simultaneous equations and solve them to work out how much he is paid for

a 1 hour's work

b 1 hour's overtime

15 Follow the steps to find the equation of the straight line that passes through the points $A(2, 4)$ and $B(1, 10)$.

a Copy and complete these two equations for the line.
Use the x and y values from each coordinate pair in
$y=mx+c$
At point A: $4=\square m+c$
At point B: $\square=m+c$

Q15a hint

The points lie on the line, so their coordinates 'fit' the equation for the line.

b Solve the two simultaneous equations to find the values of m and c.

c Substitute the values of m and c into $y=mx+c$ to write the equation of the line.

16 **Problem-solving** Find the equation of the straight line that passes through

a $C(2, 5)$ and $D(4, 7)$

b $E(2, 8)$ and $F(-4, -1)$

20.5 Rearranging formulae

- Change the subject of a formula.

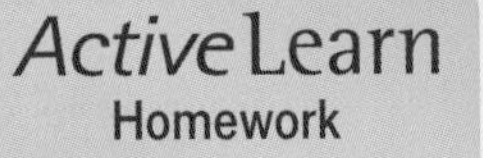

Warm up

1 **Fluency** What is the inverse operation of

a $+x$ **b** $-z$ **c** $\times y$ **d** $\times -s$ **e** $\div m$ **f** $\square^2$ **g** $\sqrt{\ }$

2 Solve these equations.

a $5x - 4 = 11$ **b** $\frac{3x}{2} = 6$ **c** $x + 2 = 5x - 6$

3 **a** Which of these lines are parallel?

i $y = \frac{1}{2}x + 4$ **ii** $y = 4x + \frac{1}{2}$ **iii** $y = 2x + \frac{1}{2}$ **iv** $y = \frac{1}{2}x - \frac{1}{2}$

b Which of the lines have the same y-intercept?

Key point

The subject of a formula is the letter on its own on one side of the equals sign.

4 What is the subject of each formula?

a $C = 3d$ **b** $d = \frac{m}{v}$ **c** $V = IR$ **d** $A = lw$

Key point

You can solve an equation or change the subject of a formula using inverse operations.

5 Make the letter in brackets the subject of each formula.
The first two are started for you.

a $A = 5w$ [w]

$\frac{A}{\square} = \square$

b $M = \frac{V}{2}$ [V]

$\square M = \square$

c $C = 3d$ [d] **d** $V = IR$ [R]

e $s = \frac{d}{t}$ [d] **f** $p = \frac{r}{q}$ [r] **g** $d = \frac{m}{v}$ [m] **h** $P = \frac{F}{A}$ [F]

6 Make the letter in brackets the subject of each formula.
The first one is started for you.

a $p = \frac{4}{q}$ [q]

$pq = \square$

$q = \frac{\square}{\square}$

b $P = \frac{F}{A}$ [A] **c** $d = \frac{m}{v}$ [v] **d** $R = \frac{V}{I}$ [I]

7 This is a distance, speed, time triangle used to help remember formulae.

The triangle shows that $S = \frac{D}{T}$ $\quad D = ST \quad$ $T = \frac{D}{S}$

Rearrange $S = \frac{D}{T}$ to make

a D the subject of the formula

b T the subject of the formula

c **Reflect** Do you get the same formulae?

Which is easier to remember – the triangle or how to change the subject of $S = \frac{D}{T}$?

d The formula for density is density $= \frac{\text{mass}}{\text{volume}}$.

Draw a density (D), mass (M), volume (V) triangle. Rearrange $D = \frac{M}{V}$ to make

i M the subject **ii** V the subject

Example

a Make a the subject of $v = u + at$. **b** Make z the subject of $x = y - az$.

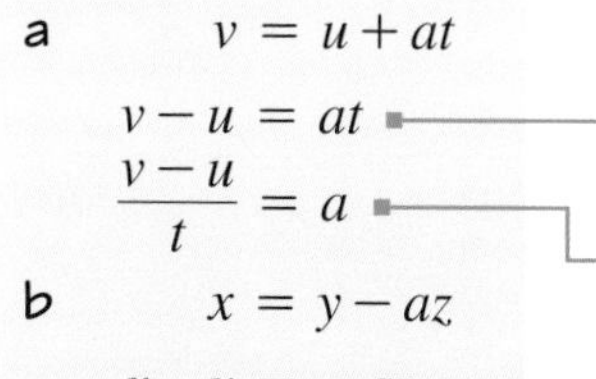

a
$$v = u + at$$
$$v - u = at$$
Rearrange so the term including a is on its own on one side of the equals sign.
$$\frac{v-u}{t} = a$$
Use the inverse operation to make a the subject.

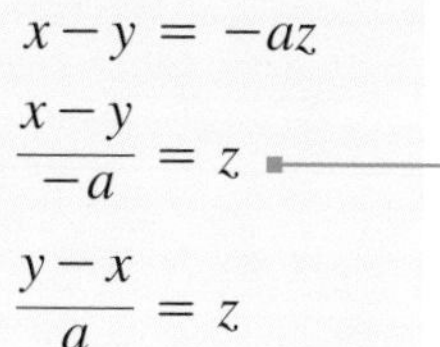

b
$$x = y - az$$
$$x - y = -az$$
$$\frac{x-y}{-a} = z$$
Multiply top and bottom by −1 to remove the negative sign from the denominator: $\frac{-1\times(x-y)}{-1\times -a} = \frac{-x+y}{a} = \frac{y-x}{a}$
$$\frac{y-x}{a} = z$$

8 Change the subject of each formula to the letter in brackets.

a $v = u + at$ [t] **b** $x = -y + mz$ [z] **c** $l = q - rp$ [p] **d** $f = e - gh$ [h]

9 **a** Rearrange the equation of the line $6x + 2y - 5 = 0$ into the form $y = mx + c$.

b What is the gradient and y-intercept of the line?

10 **Problem-solving** Which of these lines pass through the point (0, 5)?

a $y = 2x + 5$ **b** $2y - 5x = 5$ **c** $5y + 3x = 5$ **d** $5y + 2x = 25$

Exam-style question

11 The equation of the line L_1 is $y = 4x - 1$.
The equation of the line L_2 is $3y - 12x + 9 = 0$.
Show that these two lines are parallel. **(2 marks)**

Exam tip

Show clearly how you rearrange the equation.

12 Make the letter in brackets the subject of each formula.

a $C = 2\pi r$ [r] **b** $V = lwh$ [l] **c** $V = lwh$ [w] **d** $A = 2\pi rh$ [h]

13 **Problem-solving** Jane wants to draw three circles with circumferences 15 cm, 20 cm and 25 cm.
How far should she open her compasses for each circle?
Round your answers to a sensible degree of accuracy.

14 Make the letter in brackets the subject of each formula.

a $I = \frac{PRT}{100}$ [T] **b** $\frac{PV}{T} = k$ [T] **c** $A = \frac{bh}{2}$ [b] **d** $P = 2(l + w)$ [w]

e $X = m(m + n)$ [n] **f** $M = \frac{n+2}{5}$ [n] **g** $P = \frac{3(Q-t)}{2}$ [Q] **h** $A = 2\pi r(r + h)$ [h]

15 Make x the subject of each formula. The first one is started for you.

a $y = \frac{x+1}{a} + 2$ **b** $z = \frac{x-1}{b} + 5$ **c** $m = \frac{x+3}{n} + p$

$y - \square = \frac{x+1}{a}$
$\square(y - \square) = x + 1$
$\square(y - \square) - \square = x$

16 Rearrange each formula so that any y terms are one side of the = sign. Then make y the subject. The first one is started for you.

a $4y - g = y + 6$ **b** $2y + 7 = 5y - q$ **c** $2(y - t) = 4y + 3$

$4y - y = \square + 6$
$\square y = \square + 6$

Q16c hint
Expand the brackets first.

Example

a Make x the subject of $y = ax^2$.

a $y = ax^2$
$\frac{y}{a} = x^2$
$\sqrt{\frac{y}{a}} = x$ — Take the square root of both sides.

b Make x the subject of $y = \sqrt{4x}$.

b $y = \sqrt{4x}$
$y^2 = 4x$ — Square both sides.
$\frac{y^2}{4} = x$

17 Make the letter in brackets the subject.

a $y = x^2$ [x] **b** $y = 5z^2$ [z] **c** $y = \frac{1}{2}x^2$ [x] **d** $A = \pi r^2$ [r]

e $y = \sqrt{x}$ [x] **f** $t = \sqrt{3s}$ [s] **g** $P = \sqrt{t + r}$ [t] **h** $m = \sqrt{\frac{t}{2}}$ [t]

i $A = 4\pi r^2$ [r] **j** $V = \pi r^2 h$ [r] **k** $V = \frac{1}{3}\pi r^2 h$ [h] **l** $V = \frac{4}{3}\pi r^3$ [r]

18 Reasoning These formulae connect an object's final velocity (v) with its initial velocity (u), acceleration (a), distance travelled (s) and time (t).

a Make u the subject of $v^2 = u^2 + 2as$. **b** Make a the subject of $v^2 = u^2 + 2as$.

c Make u the subject of $s = ut + \frac{1}{2}at^2$. **d** Make a the subject of $s = ut + \frac{1}{2}at^2$.

19 Pythagoras' theorem states that $c^2 = a^2 + b^2$.
Make b the subject of $c^2 = a^2 + b^2$.

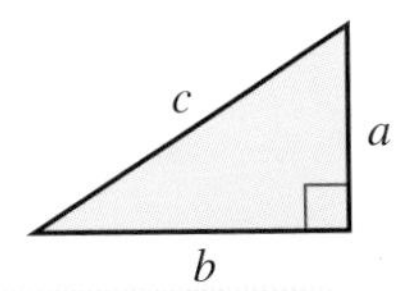

Exam-style question

20 Make x the subject of the formula $y = \sqrt{\frac{x+2}{5}}$. **(3 marks)**

Exam tip
Get rid of square root signs first.

20.6 Proof

- Identify expressions, equations, formulae and identities.
- Prove results using algebra.

Warm up

1 **Fluency** Are these expressions, formulae or equations?

a $2x+3$ b $5x=20$ c $C=2\pi r$

d $2v-3u+6$ e $38=\pi r^2$ f $x^2+3x+2=0$

2 Expand a $x(x-2)$ b $(x+1)(x-4)$ c $(x+2)^2$

3 Factorise a $3x+3$ b x^2-3x c $4x-2$

4 Work out the area of this shape.

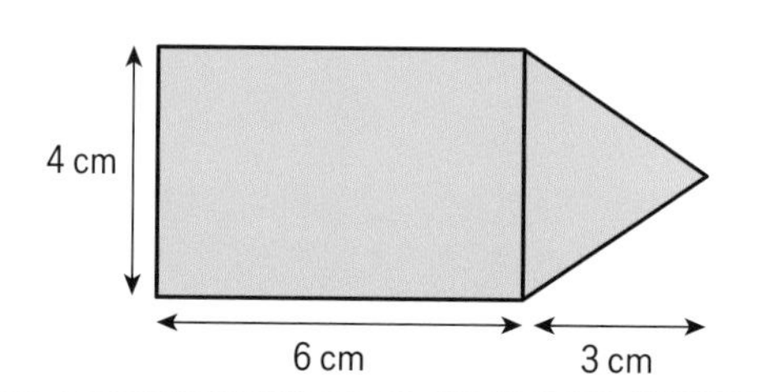

Key point

An **equation** has an equals sign. You can solve it to find the value of the letter.
An **identity** is similar, but is true for *all* values of x and uses the symbol '$\equiv$'.

5 **Reasoning**

a Solve $3x+5=26$.

b Is $3x+5=26$ true for all values of x?

c Is $3x+5=26$ an equation or an identity?

Q5c hint

Remember an identity is true for all values of x. An equation is only true for some values of x.

6 **Reasoning** Decide whether these are expressions, formulae, equations or identities.

a $2x+11=19$ b $v=u+at$ c $3x(x-1)=0$ d x^2-2x+5

Key point

To show a statement is an identity, expand and simplify the expressions on one or both sides of the identity sign, until the two expressions are the same.

Example

Show that $(x-3)^2-2 \equiv x^2-6x+7$.

$\equiv$ means 'is identically equal to'.

LHS $\equiv (x-3)^2-2$

$\equiv (x-3)(x-3)-2$

$\equiv x^2-6x+9-2$

$\equiv x^2-6x+7 \equiv$ RHS

Expand the brackets on the left-hand side (LHS).

Collect like terms to get the expression on the right-hand side (RHS).

7 **Reasoning** Show that

a $x(x^2 - 4) \equiv x^3 - 4x$

b $(x - 2)^2 + 4x \equiv x^2 + 4$

c $x^2 + 6x + 25 \equiv (x + 5)^2 - 4x$

d $2x^3 - 5x^2 \equiv x^2(2x - 5)$

e $(x + 4)(x - 4) \equiv x^2 - 16$

f $x^2 - a^2 \equiv (x + a)(x - a)$

8 **Reasoning** Follow the steps to show that $(x + 2)(x + 3) + x + 3 \equiv (x + 3)^2$.

a Multiply out the brackets on the LHS.

b Simplify the LHS by collecting like terms.

c Multiply out the brackets on the RHS.

d Show that the LHS and RHS are the same.

9 **Reasoning** A rectangular card with length $x + 3$ and width $x + 1$ has a smaller rectangle cut out of the middle.
The smaller rectangle has length $x - 1$ and width x.

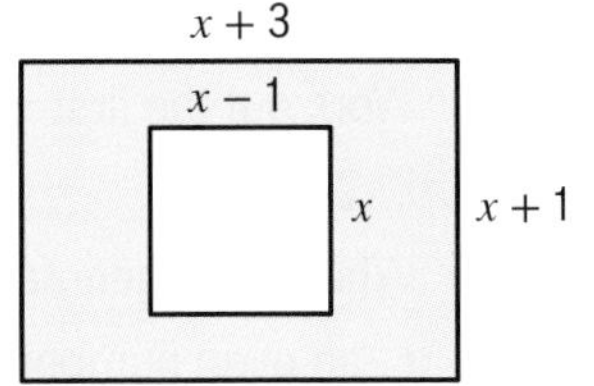

a Write an expression for the total area of the rectangle before the middle was cut out.

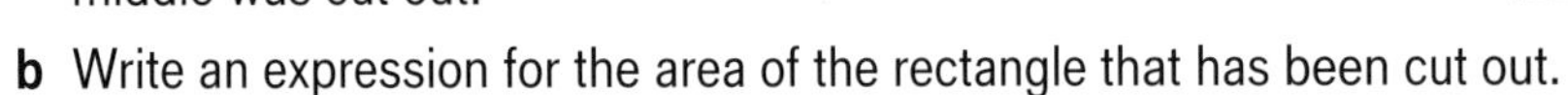

b Write an expression for the area of the rectangle that has been cut out.

c Subtract your expression from part **b** from your expression from part **a** to show that the area of the remaining card is $5x + 3$.

Exam-style question

10 Show that the area of this pentagon can be written as $2x^2 + x - 1$. **(4 marks)**

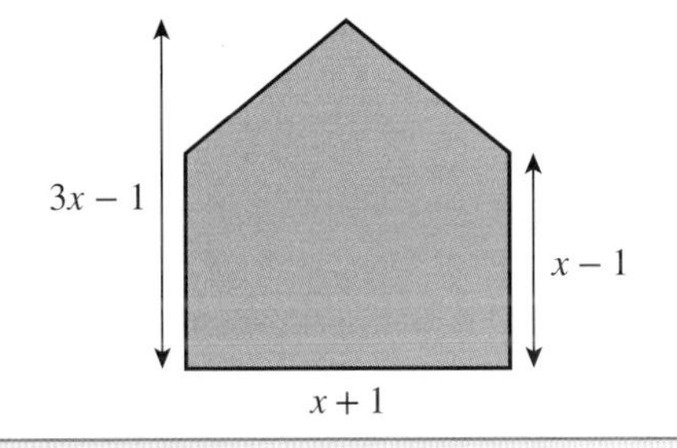

Exam tip

Write any other lengths you work out on the diagram.
Simplify algebraic expressions by collecting like terms.
End with the statement $= 2x^2 + x - 1$.

Key point

Consecutive integers follow on one after the other, like 1, 2, 3 or –5, –4, –3.

11 **Reasoning**

a Write any three consecutive integers. Make the first one equal to n.
Write expressions for the other two numbers: n, $n + \square$, $n + \square$.

b Write a different set of three consecutive numbers.
Make the second one equal to n and write expressions for the other two numbers.

12 **Reasoning** The median of a set of five consecutive integers is n.

a Write expressions for the five integers.

b Show that the range of the five integers is 4.

c Show that the mean of the five integers is n.

13 Reasoning Follow these steps to show that the sum of any three consecutive integers is a multiple of 3.

a Write expressions for the three integers as n, ___ , ___ .

b Work out the sum of these integers.

c Simplify the sum as much as possible, then factorise it.

d Copy and complete: The sum of three consecutive numbers is ________ , which is ____ $\times 3$, which is a multiple of 3.

14 Reasoning Show that the sum of any four consecutive numbers is a multiple of 2.

Q14 hint

Follow the steps in **Q13**.

15 Reasoning

a Work out the first five terms of the sequence with general (nth) term

i $2n$ ii $2n-1$

b What is the general term for

i an even number ii an odd number?

c What type of number is $2n+1$?

Q15b hint

Look at your sequences in part **a**.

Key point

An even number is a multiple of 2.
$2m$ and $2n$ are both general terms for even numbers where m and n are integers.

16 Reasoning Copy and complete this proof to show that the sum of two even numbers is even.

Two even numbers are $2m$ and $2n$.

$2m+2n = \square(\square+\square)$

This is a multiple of $\square$, so it is an even number.

17 Reasoning Show that the sum of two odd numbers is even.

Q17 hint

Use $2m+1$ and $2n+1$ as the odd numbers.

18 Reasoning

a Show that the product of two even numbers is even.

b Hence show that the square of an even number is an even number.

19 Reasoning Given that $2(x-a) = x+4$, where a is an integer, show that x must be an even number.

Q19 hint

Expand the brackets. Rearrange to make x the subject.
$2 \times$ an integer = an even number.

Exam-style question

20 x and y are even numbers.

a Give an example to show that the value of $2(x+y)$ is a multiple of 4. **(2 marks)**

b Show that, when x and y are even numbers, the value of $2(x+y)$ will always be a multiple of 4. **(2 marks)**

Exam tip

Here, 'Give an example' means choose two even numbers for x and y and work out $2(x+y)$.

20 Check up

ActiveLearn Homework

Graphs

1 **a** Copy and complete this table of values for $y = x^3 - 2$.

x	−3	−2	−1	0	1	2	3
y							

b On graph paper, draw a coordinate grid with the x-axis from −3 to +3 and the y-axis from −30 to +30.

c Draw the graph of $y = x^3 - 2$ from $x = -3$ to $x = +3$.

2 **a** Draw the graph of $y = \frac{1}{x}$ from $x = -3$ to $x = +3$.

x	−3	−2	−1	−0.5	0.5	1	2	3
y								

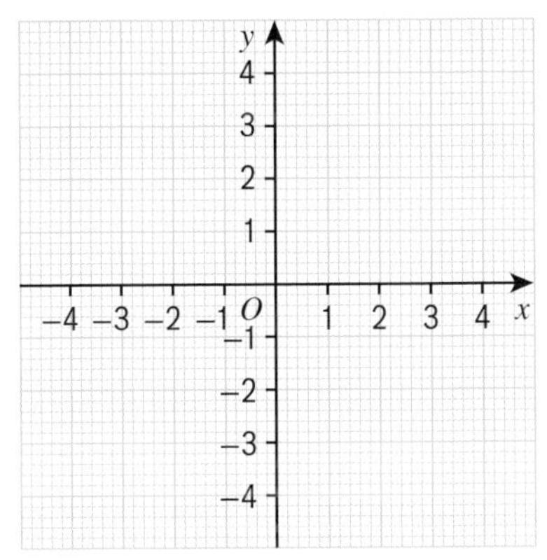

b Use your graph to estimate

i the value of y when $x = -1.5$

ii the value of x when $y = 4$

3 The graph shows how the volume of a fixed amount of a gas varies as the pressure changes.
Estimate

a the volume when the pressure is 4 kilopascals

b the pressure when the volume is 10 litres

c the change in volume when the pressure is increased from 2 kilopascals to 6 kilopascals

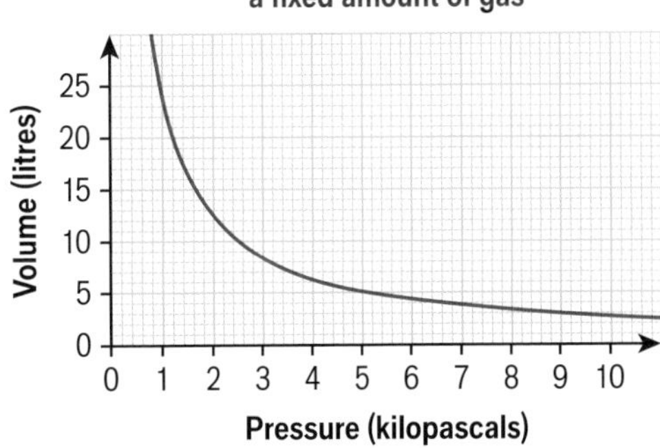

Simultaneous equations

4 A cup of coffee costs £y and a glass of juice costs £x. Three coffees and two juices cost £9. Write an equation to show this information.

5 Use the graph to find the solutions to the simultaneous equations $2x + y = 7$ and $x + 2y = 8$.

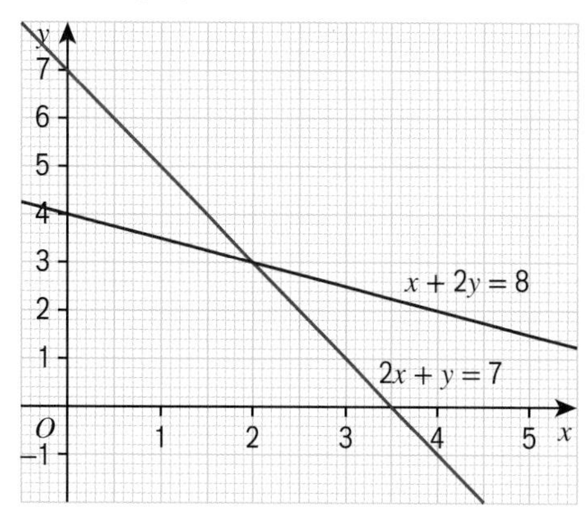

6 The sum of two numbers is 24. Their difference is 8.

a Write two equations for the two numbers. Use x and y.

b Solve your equations to find the two numbers.

7 Solve the simultaneous equations $2x + 3y = 14$ and $5x + y = 22$ algebraically.

Using algebra

8 Make the letter in brackets the subject of each formula.

a $m = pr + t$ $[t]$ **b** $z = ac + y$ $[c]$ **c** $p = 3(m + n)$ $[m]$ **d** $v = \frac{pq}{r}$ $[r]$

9 Make the letter in brackets the subject.

a $R = t^2$ $[t]$ **b** $n = \frac{pt}{x}$ $[p]$ **c** $y = 4 - \frac{x}{2} + z$ $[x]$ **d** $6x + 4 = y + x$ $[x]$

10 Show that $xy(x - y) \equiv x^2y - xy^2$ is an identity.

11 **Reflect** How sure are you of your answers? Were you mostly

Just guessing ☹ Feeling doubtful 😐 Confident ☺

What next? Use your results to decide whether to strengthen or extend your learning.

Challenge

12 In this puzzle, the rows and columns add to the totals given. Work out the value of ★, Δ and □.

Δ	★	□	1	10
2	Δ	3	□	11
★	□	Δ	★	12
□	2	Δ	Δ	13
11	11	14	10	

Q12 hint

Look for rows or columns with only two shapes. Write and solve a pair of simultaneous equations.

20 Strengthen

ActiveLearn Homework

Graphs

1

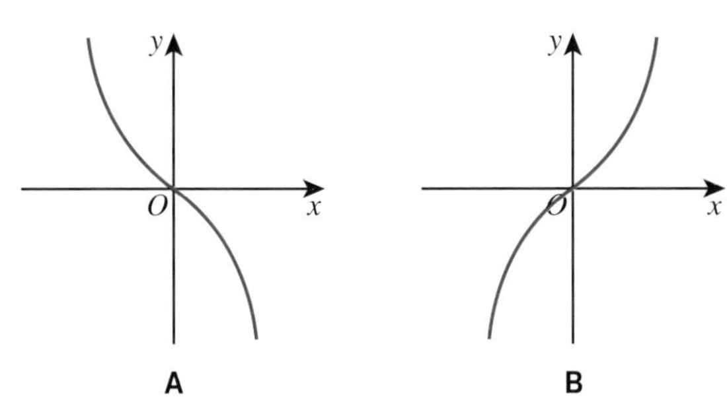

a For Graph A, when $x = 1$, is the y-value positive or negative?

b For Graph B, when $x = 1$, is the y-value positive or negative?

c Which graph is $y = x^3$ and which is $y = -x^3$?

d Copy the graphs and label them with their equations.

2 Shona makes a table of values and plots the graph of $y = x^3 - 1$.

a On her graph, which points do not fit the shape of an x^3 graph?

b Find the incorrect points in Shona's table of values.
Work out the correct values.

x	−3	−2	−1	0	1	2	3
y	−28	7	−2	−1	0	7	−26

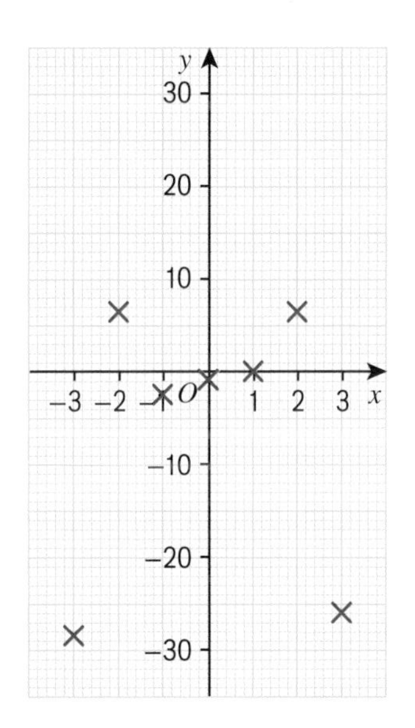

3 **a** Copy and complete this table of values for the graph of $y = x^3 + 1$.

x	−3	−2	−1	0	1	2	3
y	−26					9	

b Copy the coordinate grid from **Q2** and then draw the graph of $x^3 + 1$ on the grid.

Q3b hint

First sketch the shape you expect the graph to be – is it like $y = x^3$ or like $y = -x^3$?
Check that your points fit the shape before you join them.

4 Match the graphs to their equations.

$y = x$ $\qquad$ $y = x^2$ $\qquad$ $y = \frac{1}{x}$

a

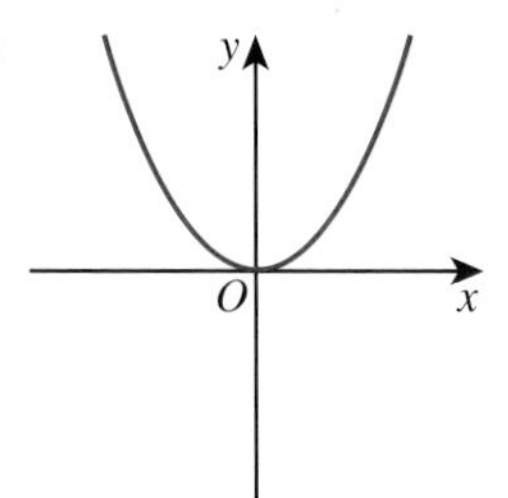

b

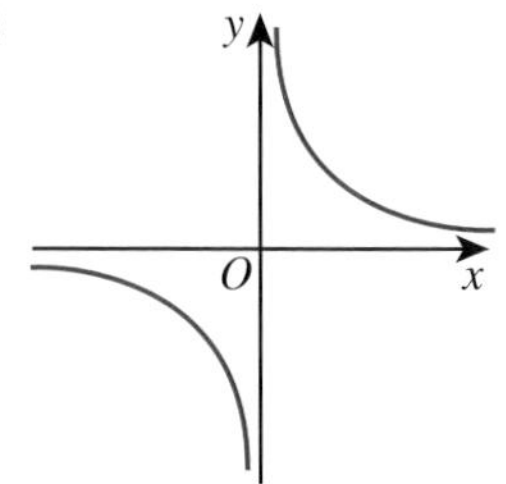

c

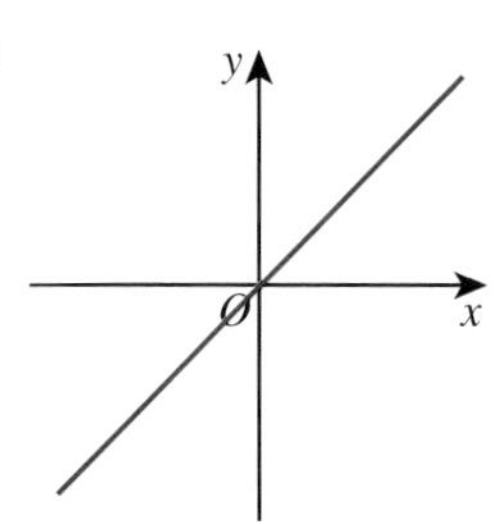

5 Dan makes a table of values and plots the graph of $y = \frac{1}{x}$.

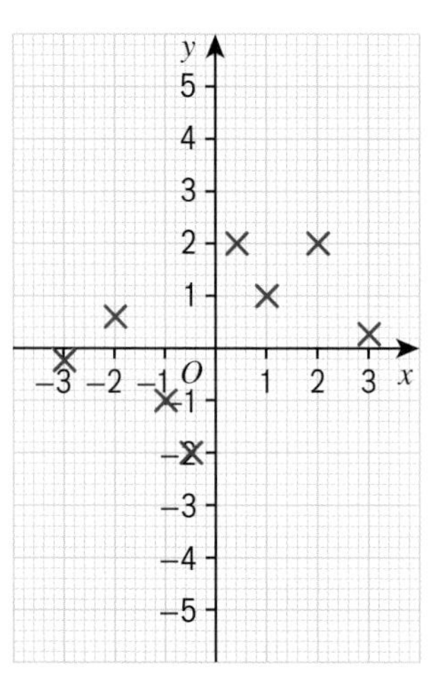

a From the graph, which points do you think are incorrect?

b Find the incorrect points in Dan's table of values.
Work out the correct values.

x	−3	−2	−1	0.5	0.5	1	2	3
y	−0.3	0.5	−1	−2	2	1	2	0.3

c Copy the coordinate grid and plot the correct points.
Join them with smooth curves to draw the graph of $y = \frac{1}{x}$.

6 The graph shows the numbers of rats recorded in a colony.

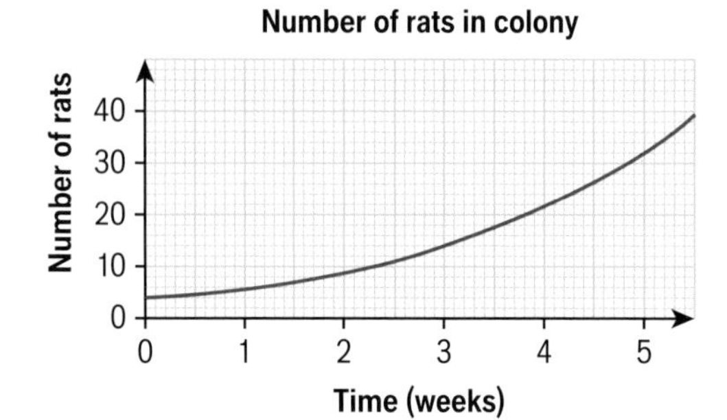

a What does the horizontal axis show?

b What does the vertical axis show?

c The number of rats must be a whole number.
Estimate the number of rats at

i 3 weeks

ii 5 weeks

d Describe the change in the number of rats from week 3 to week 5.

Q6d hint

Is it an increase or decrease? By how much? Write a sentence beginning: The number of rats ...

Simultaneous equations

1 Write an equation for each statement.

a 2 pens and 3 notebooks cost £14. Call the cost of a pen x and the cost of a notebook y. Use a bar model.

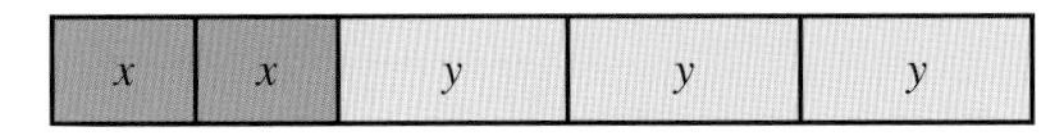

$2x + 3y = \square$

b 3 bananas and 5 apples cost £10.

c 6 adult tickets and 10 child tickets cost £70.

d 2 large bags and 4 small bags weigh 18 kg.

e 5 washers and a bolt measure 65 mm.

2 **a** Find the intersection of the two graph lines with the equations $x + y = 8$ and $x + 5y = 16$. Write down the x-value and y-value at the intersection, $x = \square$ and $y = \square$.

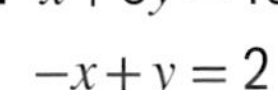

b Use the method in part **a** to find the solutions to the simultaneous equations.

i $-x + y = 2$
$x + y = 8$

ii $x + 5y = 16$
$-x + y = 2$

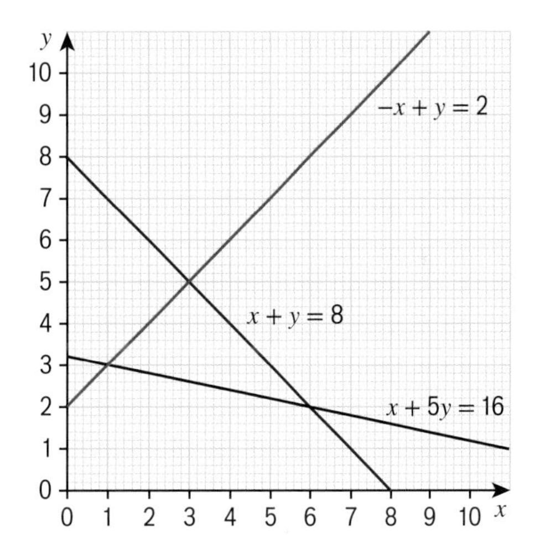

3 For each pair of these simultaneous equations, try (1) adding them and (2) subtracting them. One method will eliminate x or y. Use this method to find both x and y.

a $2x + y = 22$
$x + y = 12$

b $x + y = 20$
$x - y = 8$

c $2x - y = 11$
$x - y = 4$

d $-x + y = 3$
$x + 3y = 29$

Q3a hint

Adding – does this help here?

$$\begin{array}{rrcrcr} & 2x & + & y & = & 22 \\ + & x & + & y & = & 12 \\ \hline & 3x & + & 2y & = & 34 \end{array}$$

4 **a** Copy this pair of simultaneous equations.

$2x + 3y = 9$ (1)
$x + 2y = 5$ (2)

b Circle the coefficients of x. Is one a multiple of the other?

c Multiply equation (2) to give $2x$, as in equation (1).

$2x + 3y = 9$ (1)
$x + 2y = 5$ (2)
$\times\square$ $\times\square$
$2x + \square = \square$ (3)

d Solve the simultaneous equations (1) and (3).

Using algebra

1 Rearrange each formula to make a the subject.

a $s = at + 3$

b $p = qy - a$

c $v = 4x + a$

d $y = 4a + b$

Q3a hint

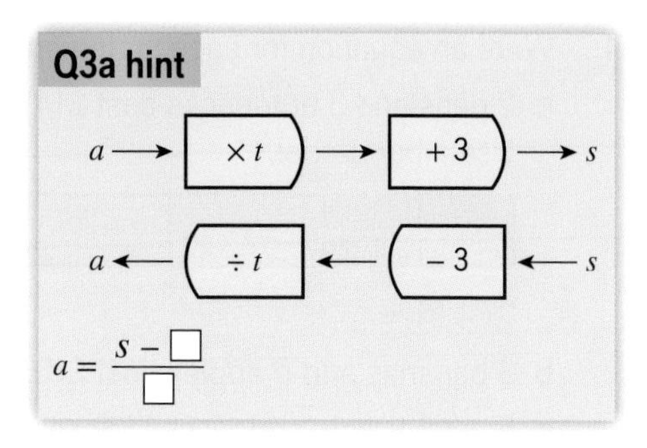

2 Rearrange each formula to make d the subject.
Expand any brackets first.

a $x = 2(d + 3)$

b $s = 2(d - t)$

c $y = \frac{dt}{2}$

d $t = \frac{kr}{d}$

3 Rearrange each formula to make n the subject.

a $n^2 = 81$ **b** $n^2 = c^2$ **c** $n^2 = r$

d $T = n^2$ **e** $6y = n^2$

Q3a hint

The inverse operation for 'square' is 'square root'.

4 **a** Solve the equation $\frac{3x}{9} = 6$ using the balancing method.

b Use balancing to rearrange each formula to make m the subject.

i $n = \frac{3m}{p}$ **ii** $k = \frac{mt}{2}$ **iii** $r = \frac{mp}{q}$ **iv** $v = \frac{bm}{z}$

5 **a** Solve $5x - 4 = 2 + 2x$ using the balancing method.

b Use the balancing method to make x the subject of each formula.

i $2x - y = x + 6$ **ii** $5x - y = x + 7y$ **iii** $3x - 7 = 2x - y$

6 **a** Solve $5 = 3 + \frac{x}{2}$ using the balancing method.

b Use the balancing method to make x the subject of each formula.

i $y = 5 + \frac{x}{3}$ **ii** $z = 2 + \frac{x}{a}$ **iii** $t = \frac{x}{b} + 4 + y$

7 Expand and/or simplify to show that LHS = RHS in each identity.

a $2a + 3b + 6a - 2b \equiv 8a + b$

b $x^2 + 3x - 2x + 4x^2 - 5 \equiv 5x^2 + x - 5$

c $2(x + 4) \equiv 2x + 8$

d $x^2(y + 2) - 4y \equiv y(x^2 - 4) + 2x^2$

Q7 hint

LHS = left-hand side
RHS = right-hand side

20 Extend

1 **a** **Reasoning** Plot these readings from a science experiment on a graph.

x	1	2	5	10	20	30
y	4.0	3.5	3.2	3.1	3.05	3.03

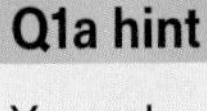
Q1a hint

You only need the y-axis to go from 3 to 4. Use a large scale so you can plot the points accurately.

b Are the two quantities in direct or inverse proportion or neither? Explain.

2 **Reasoning** The graph shows the population of an ant colony.

a Estimate the number of ants in week 3.

b Estimate the increase in their number from week 3 to week 5.

c Between which two weeks was the number of ants increasing fastest?

d When do you think there will be more than 400 ants in the colony?

Number of ants in a colony

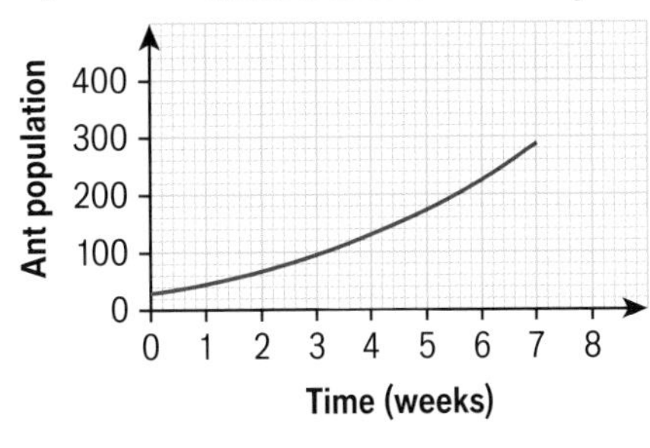

3 **a** Draw the graph of $y = x^3 - 2x + 5$ for $-3 \leqslant x \leqslant 3$.

b Use your graph to find the solutions to the equation $x^3 - 2x + 5 = 0$ that lie between -3 and 3, by reading the x-value where the graph line crosses the x-axis.

4 The table shows the masses of some copper samples.
The density of copper $= 8.92\text{ g/cm}^3$.
Use the formula
density $= \frac{\text{mass}}{\text{volume}}$, $d = \frac{m}{v}$,
to calculate the volume of each sample.
Give your answers to 1 d.p.

Sample	Mass, m (g)	Volume, v (cm^3)
a	25	
b	60	
c	90	
d	125	
e	240	
f	350	

Q4 hint

First rearrange the formula to make v the subject.

5 Luke invests £500 for 4 years in a savings account.
By the end of 4 years he has received a total of £80 simple interest.
Work out the annual rate of simple interest.

Q5 hint

$I = \frac{PRT}{100}$

6 Make x the subject of each formula.

a $m + n = p - rx$

b $a(x - b) = c$

c $d(e + x) = fx$

d $g(h + x) = mx - n$

e $p(x - q) = r(x + s)$

f $a(a + x) = cd$

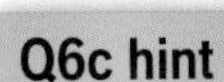
Q6c hint

$d(e + x) = fx$
$de + \square x = fx$
$de = fx - \square x$
$de = x(f - \square)$
$\frac{de}{f - \square} = \square$

7 **Reasoning** In this grid, a 2 by 2 square is highlighted.

1	2	3	4	5
6	7	8	9	10
11	12	13	14	15
16	17	18	19	20

a Show that adding the diagonal numbers $8+14$ and $9+13$ gives the same total.

b Choose another 2 by 2 square on the grid.
Add the diagonal numbers.
Do they give the same total?

c Write expressions for the numbers in the grid.

d Use algebra to show that, for any 2×2 square on this grid, the sum of both diagonals will always be equal.

8 n	9 $n + \square$
13 $n + \square$	14 $n + \square$

8 **Reasoning** The diagram shows a rectangle and a square.
All the measurements are in centimetres.
The total area of the rectangle and the square is $35\,\text{cm}^2$.

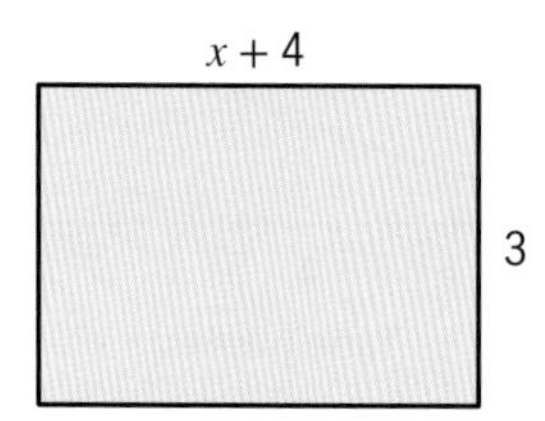

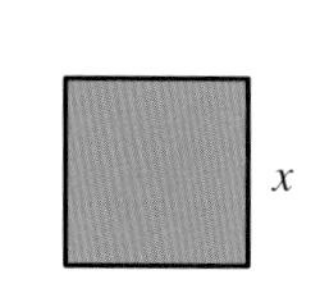

a Show that $x^2 + 3x = 23$.

b Show that the value of x is between 3 and 4.

Q8a hint

Write an expression for the combined area and put it equal to 35.

Q8b hint

Try $x = 3$ and $x = 4$ in the equation in part **a**.

Exam-style question

9 The diagram shows a cube and a cuboid.

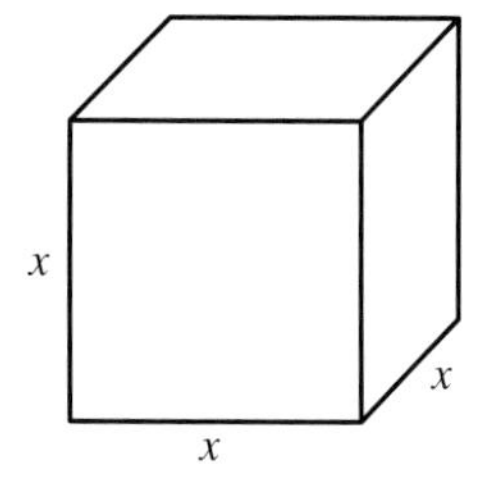

Diagram NOT accurately drawn

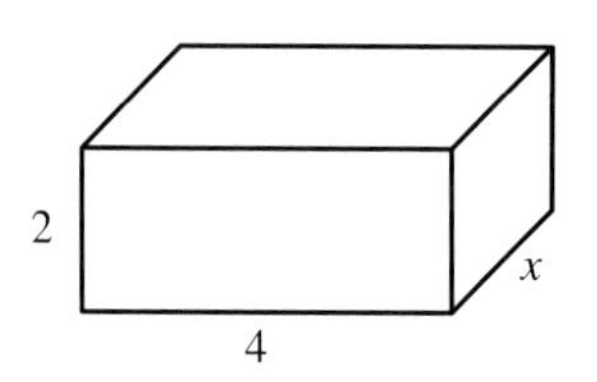

All the measurements are in cm.
The volume of the cube is $100\,\text{cm}^3$ more than the volume of the cuboid.
Show that $x^3 - 8x = 100$. **(2 marks)**

Exam tip

When a question gives you a diagram, you need to use it. Write an equation connecting the volumes of the two cuboids.

Exam-style question

10 a Give an example to show that the square of an odd number is an odd number. **(2 marks)**

b Show that the square of an odd number will always be an odd number. **(2 marks)**

Exam tip

'Show that ... will always be ...' means 'prove algebraically'.

20 Test ready

Summary of key points

To revise for the test:

- Read each key point, find a question on it in the mastery lesson, and check you can work out the answer.
- If you cannot, try some other questions from the mastery lesson or ask for help.

Key points

1 A **cubic function** contains a term in x^3 but no higher power of x.
It can also have terms in x^2 and x and number terms.
When a cubic function is equal to zero, it can have one, two or three solutions – the x-values where the graph crosses the x-axis.

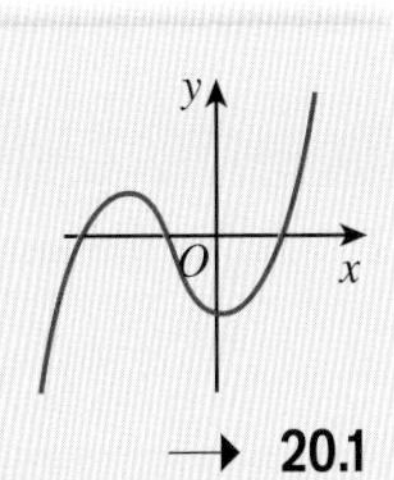

→ 20.1

2 For the reciprocal function $y = \frac{1}{x}$, the x- and y-axes are **asymptotes** to the curve.
An asymptote is a line that the graph gets closer and closer to, but never actually touches.

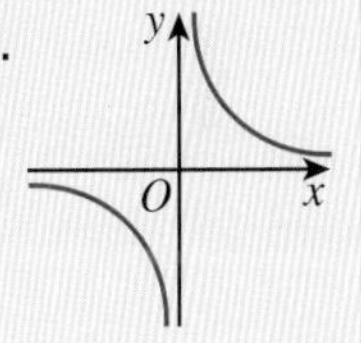

→ 20.1

3 A non-linear graph is any graph that is not a straight line.
So a curved graph is a non-linear graph.
You can estimate values from a non-linear graph. → 20.2

4 When two quantities are in **direct proportion**, they are linked by an equation of the form $y = kx$ and their graph is a straight line through the origin. → 20.2

5 When two quantities are in **inverse proportion**, they are linked by an equation of the form $y = \frac{k}{x}$ and their graph is a reciprocal graph. → 20.2

6 The point where two (or more) lines cross is called the **point of intersection**. → 20.3

7 **Simultaneous equations** are equations that are both true for a pair of variables (letters). → 20.3

8 To find the solution to a pair of simultaneous equations graphically:
- Draw the graphs on the same pair of axes.
- Find the point of intersection. → 20.3

9 To solve simultaneous equations by the elimination method, add or subtract the equations to eliminate either the x or the y terms.
You may need to multiply one or both equations first, to get equal coefficients of x or y. → 20.4

10 To find the equation of a line between two points, substitute the values of x and y from each pair of coordinates into $y = mx + c$ to get two simultaneous equations.
Solve to find m and c. → 20.4

11 The **subject** of a formula is the letter on its own on one side of the equals sign. → 20.5

12 You can solve an equation or change the subject of a formula using inverse operations. → 20.5

13 An **equation** has an equals sign. You can solve it to find the value of the letter. → 20.6

Key points

14 An **identity** looks similar to an equation, but is true for *all* values of x and uses the symbol '$\equiv$'. → **20.6**

15 To show that a statement is an identity, expand and simplify the expressions on one or both sides until the two expressions are the same. → **20.6**

16 **Consecutive integers** follow on one after the other, like 1, 2, 3 or –5, –4, –3 or n, $n+1$, $n+2$. → **20.6**

17 As an even number is a multiple of 2, $2m$ and $2n$ are both **general terms** for even numbers. → **20.6**

Sample student answers

Exam-style question

1 Solve the simultaneous equations

$5x+2y=7$

$3x-y=16$

(3 marks)

$5x+2y=7$ (1)

$3x-y=16$ (2)

(1) + (2) $8x+y=23$

What mistake has the student made?

Exam-style question

2 Make a the subject of $v^2=u^2+2as$. **(2 marks)**

$v^2=u^2+2as$

$v^2-u^2=2as$

$v^2-u^2-2s=a$

What mistake has the student made?

Exam-style question

3 Find a.

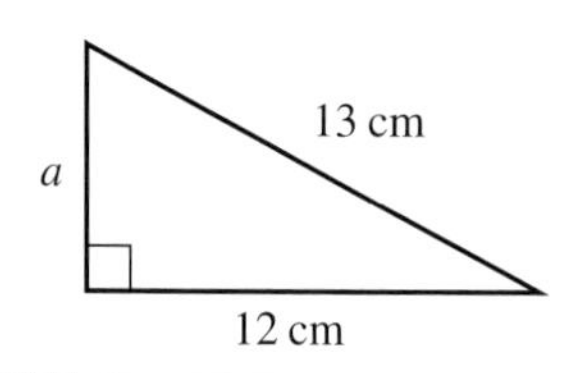

(3 marks)

$c^2=a^2+b^2$

$c^2-b^2=a^2$

$a=c-b$

$=13-12=1\text{cm}$

What mistake has the student made? Write the correct answer.

20 Unit test

1 Make x the subject of each formula.

a $z = 4(x + y)$ **(1 mark)**

b $t = \frac{rs}{x}$ **(2 marks)**

2 Jim is 10 years older than Elise. The sum of their ages is 58.
Write and solve a pair of simultaneous equations to work out

a Jim's age **(2 marks)**

b Elise's age **(1 mark)**

3 The graph shows how atmospheric pressure varies with the height above sea level.

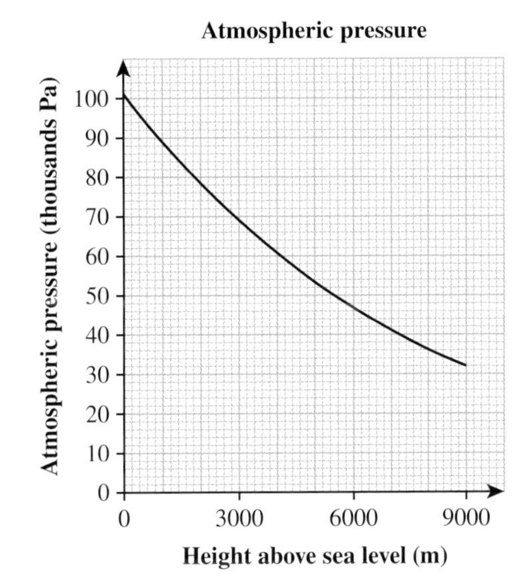

a Estimate

i the atmospheric pressure at the summit of Everest, 8848 m above sea level.

ii the fall in atmospheric pressure on the climb from Base Camp at 5380 m to the summit. **(2 marks)**

b Climbers carry oxygen to use at atmospheric pressures below 44 000 Pa.
At what height above sea level should a climber start to use oxygen? **(1 mark)**

c Work out the percentage fall in atmospheric pressure when the height above sea level increases from 0 m to 1000 m. **(2 marks)**

4 Make r the subject of each formula.

a $N = \sqrt{r - s}$ **(1 mark)**

b $F = \frac{2 + r}{x + 3}$ **(2 marks)**

5 Max plots this graph of his results from a science experiment.
Describe the relationship between the two variables. **(1 mark)**

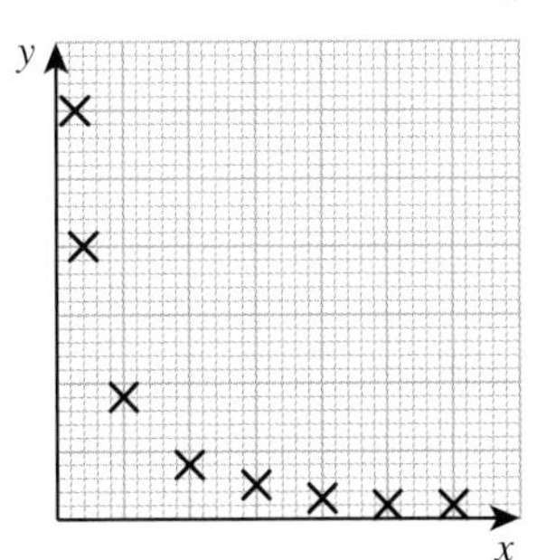

6 **a** Use the graph to solve the simultaneous equations

$3x+2y=10$

$x-2y=2$ **(2 marks)**

b The lines $x+5y=9$ and $x-2y=2$ meet at point M.
Use algebra to find the coordinates of M accurately. **(3 marks)**

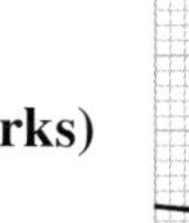

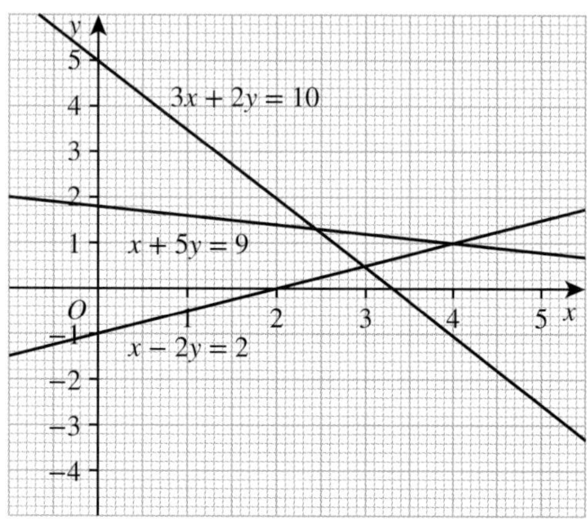

7 **a** Draw the graph of $y=x^3+3x-7$ for $-2 \leqslant x \leqslant 3$. **(3 marks)**

x						
y						

b Use your graph to find the solution to the equation $x^3+3x-7=0$ that lies between -2 and 3. **(1 mark)**

8 Solve the simultaneous equations

$5x+y=35$

$x-3y=15$ **(3 marks)**

9 Show that $(n+2)^2 \equiv (n+1)^2+2n+3$. **(3 marks)**

10 A taxi firm charges a fixed cost and an amount per km.
A 6 km journey costs £8.80 and a 10 km journey costs £12.
Work out the cost of a 15 km journey. **(4 marks)**

11 Make x the subject of

a $T=\sqrt{\frac{2s}{x}}$ **(3 marks)**

b $n=3\sqrt{5-3x}$ **(3 marks)**

(TOTAL: 40 marks)

12 **Challenge**

a Choose any 2×2 square in a 100 square.

b Multiply the two pairs of diagonal numbers.
Work out the difference between your two products.

c Show that your result in part **b** is true for any 2×2 square in the 100 square.

Q12 hint

A 100 square has the numbers 1–100 in rows of 10.

13 **Reflect** Write down a word that describes how you feel

a before a maths test

b during a maths test

c after a maths test

Hint

Here are some possible words: OK, worried, excited, happy, focused, panicked, calm.
Beside each word, draw a face, ☺ or ☹, to show if it is a good or a bad feeling.
Discuss with a classmate what you could do to change ☹ feelings to ☺ feelings.

Mixed exercise 6

1 Work out $\left(1\frac{1}{3}+2\frac{3}{4}\right)\div 2\frac{1}{2}$.
Give your answer as a mixed number in its simplest form.

2 **Reasoning** In the box are some equations, expressions, formulae and identities.

$14 = 3x + 2$	$3(x + 2)$	$S = \frac{D}{T}$	$3(x + 2) = 21$
$3x + 6 = 3(x + 2)$	$r^2 = 4$	$3x + 2$	$A = \pi r^2$

List each of these under the four categories: equations, expressions, formulae and identities.

Exam-style question

3 Here are six graphs.

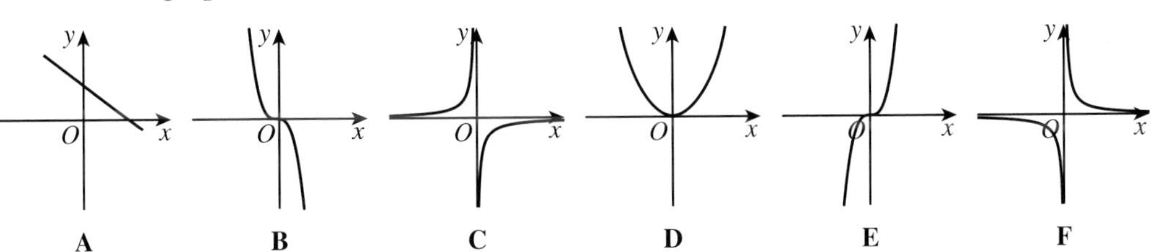

Write down the letter of the graph that could have the equation

a $y = x^3$ **(1 mark)**

b $y = \frac{1}{x}$ **(1 mark)**

4 **Problem-solving** Sarah is a baker and uses milk in some of her products.
She works for $5\frac{1}{2}$ days each week.
She uses $2\frac{3}{4}$ litres of milk per day.
How much milk does Sarah use each week?
You must show your working.

5 **Problem-solving** The table shows the populations of 4 cities, A, B, C and D.
Write the cities in descending order based on their population.

City	Population
A	6 830 000
B	9.4×10^6
C	8.2 million
D	5.87×10^6

6 **Reasoning** Prove that $5(2n+1)+3(2n+5)$ is always a multiple of 4 when n is a positive integer.

7 **Reasoning** You can change temperatures from degrees Fahrenheit (F) to degrees Celsius (C) using the formula $C = \frac{5(F-32)}{9}$.
Make F the subject of the formula.

8 **Problem-solving** ACE is a straight line.
Triangles ABC and CDE are similar.
Triangle CDE is an enlargement of triangle ABC with scale factor 1.5.

$CD = DE$
$AB = 4$ cm, $CE = 3.6$ cm

a Work out the length of CD.
b Work out the length of AE.
c Work out the size of angle x.

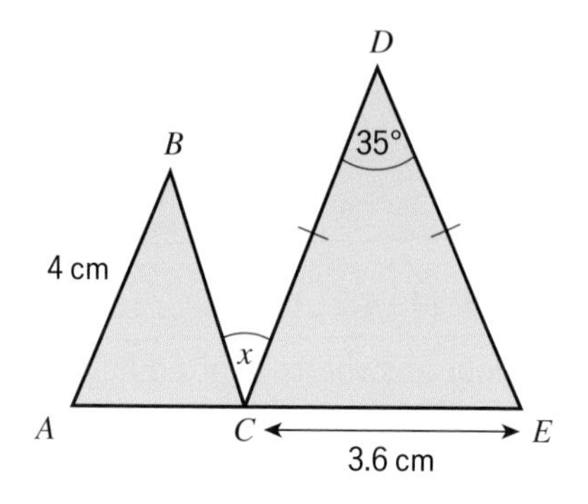

9 **Problem-solving** Write these numbers in order from smallest to largest.

6×10^{-5} $\quad 7 \times 10^{-6}$ $\quad 0.002$ $\quad 9 \times 10^{-3}$

10 **Reasoning** The diagram shows a rectangular piece of card.

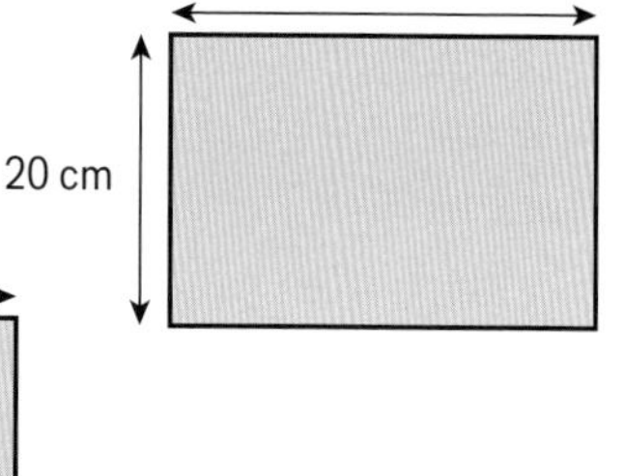

Zach cuts a smaller rectangle from the card, as shown by the dotted line below.

Zach says that the smaller rectangle is mathematically similar to the card he started with.
Is Zach correct?
You must give a reason for your answer.

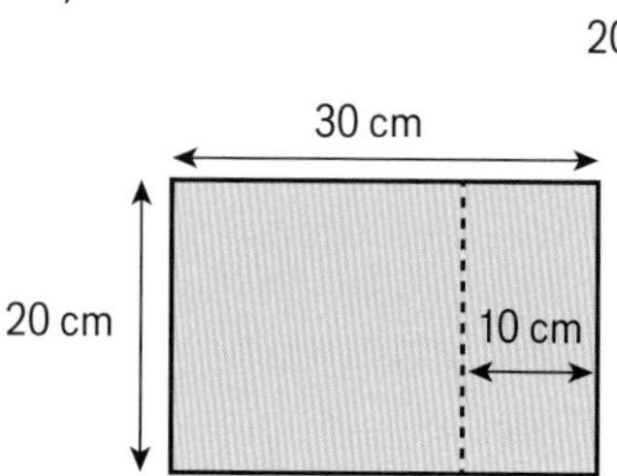

11 The diagram shows a trapezium.
All measurements are in centimetres.

Show that the area of the trapezium is $x^2 + 5x + 4$ cm^2.

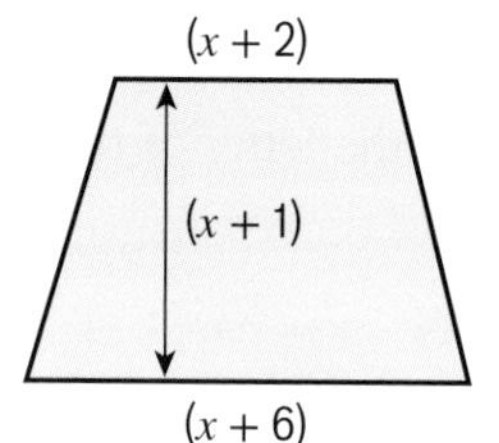

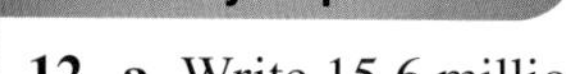

12 **a** Write 15.6 million in standard form. **(1 mark)**

b Write 3.75×10^{-4} as an ordinary number. **(1 mark)**

Amira was asked to compare the following two numbers.

$P = 5.345 \times 10^6$ and $Q = 4.102 \times 10^7$

She says, '5.345 is bigger than 4.102 so P is bigger than Q.'

c Is Amira correct? You must give a reason for your answer. **(1 mark)**

13 **Reasoning** $ABCDE$ and $EFGHI$ are both regular pentagons.

a Explain why pentagons $ABCDE$ and $EFGHI$ are similar.
b Work out the size of angle x.

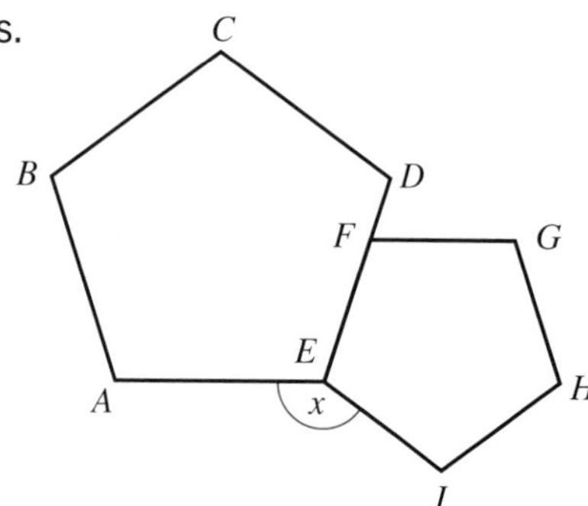

14 Reasoning Khaled is investigating some prisms with the same volume.
The graph shows the lengths and areas of cross-section of his prisms.

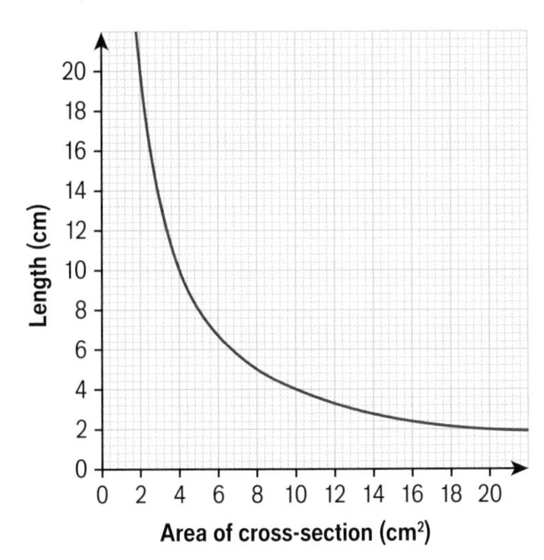

a Describe the relationship between the area of the cross-section and the length of the prisms.

b Use the graph to estimate the length of a prism with cross-sectional area $8\,\text{cm}^2$.

c What is the volume of all of Khaled's prisms?

15 Reasoning The graph of $x+2y=6$ is shown on the grid.

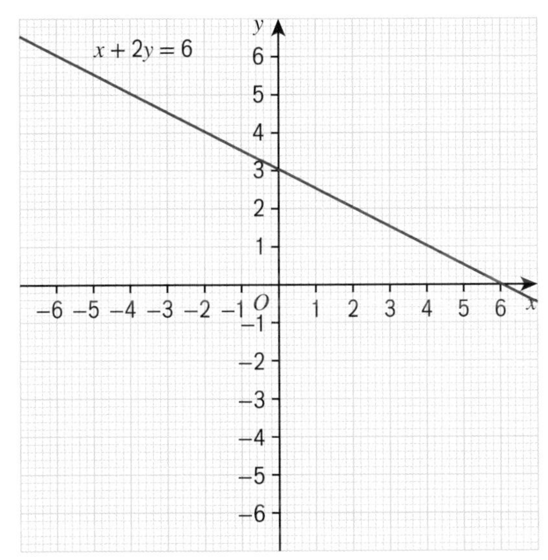

Copy the graph and use it to solve the simultaneous equations $x+2y=6$ and $y=x-3$.

Exam-style question

16 Work out $(3.762\times10^{-8})\div(2\times10^{-6})$.
Give your answer in standard form. **(2 marks)**

17 Reasoning $\mathbf{a}=\begin{pmatrix}-2\\5\end{pmatrix}$, $\mathbf{b}=\begin{pmatrix}7\\-1\end{pmatrix}$, $\mathbf{c}=\begin{pmatrix}1\\3\end{pmatrix}$

a Work out $\mathbf{a}+\mathbf{b}+\mathbf{c}$.

$2\mathbf{a}+\mathbf{b}=x\mathbf{c}$, where x is an integer.

b Work out the value of x.

Exam-style question

18 PQR and XYZ are similar right-angled triangles.

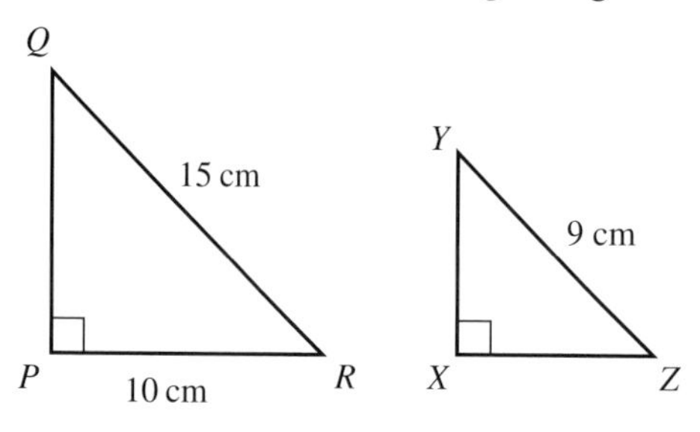

Angle PQR = angle XYZ

a Work out the length of XZ. **(2 marks)**

Triangle ABC is congruent to CDE.

$AE = 12\,\text{cm}$

$AC = 7\,\text{cm}$

b Work out the length BD. **(2 marks)**

Angle $BAC = 35.5°$

c Work out the size of angle CED. **(2 marks)**

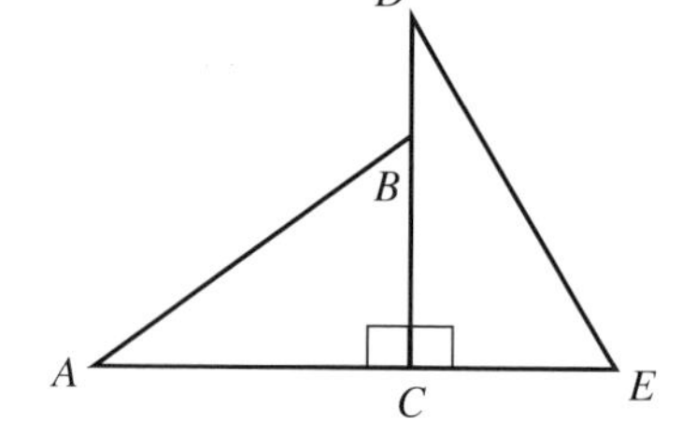

19 Problem-solving In a café, 3 teas and 1 coffee cost £8.30.
In the same café, 1 tea and 2 coffees cost £7.10.
Work out the cost of 1 tea and the cost of 1 coffee in the café.

Exam-style question

20 Here are two column vectors.

$$\mathbf{a} = \begin{pmatrix} 4 \\ 3 \end{pmatrix} \quad \mathbf{b} = \begin{pmatrix} 2 \\ -1 \end{pmatrix}$$

Draw and label the vector $\mathbf{a} - 3\mathbf{b}$ on squared paper. **(3 marks)**

21 Problem-solving Write these numbers in order from smallest to largest.

2^{-3} 0.0002 2^0 0.2 2^{-1}

22 Problem-solving The diagram shows a logo made using four congruent right-angled triangles.
Work out the perimeter of the logo.

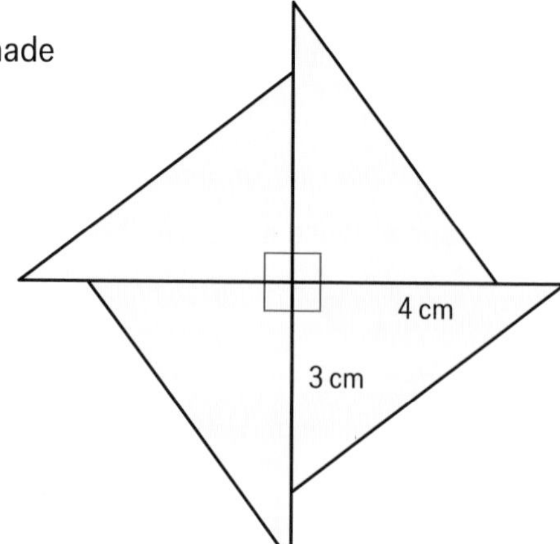

23 Reasoning L_1 and L_2 are intersecting straight lines.
Line L_1 has equation $y = 3x - 7$.
Line L_2 goes through the points (−1, 6) and (7, −2).
Work out where lines L_1 and L_2 intersect.
You must show all your working.

Answers

Prior knowledge answers

UNIT 11 Ratio and proportion

11.1 Writing ratios

1 6

2 **a** 8 **b** 7 **c** 5

3 45

4 2 : 5

5 **a**

b

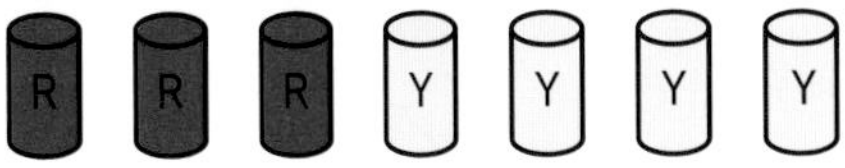

c Students' own answers, e.g. No.

6 **a** There is 1 purple bead for every 4 blue beads.
b 6
c **i** 6 **ii** 24

7 **a** 1 : 3 **b** 2 : 1 **c** 3 : 1 **d** 1 : 6
e 3 : 4 **f** 9 : 4 **g** 3 : 8 **h** 1 : 10
i 4 : 5

8 Students' own answers, e.g. He has reversed the numbers (it should be 90 : 3), and he has not simplified (it should be 30 : 1).

9 3 : 10

10 1 : 3

11 Yes, 4 : 32 is 1 : 8

12 A, B, and D are equivalent.
C and E are equivalent.

13 **a** 4 : 5 : 3 **b** 6 : 4 : 5 **c** 8 : 6 : 5 **d** 2 : 5 : 7

14 25 : 11 : 36

15 2 : 5 : 9

11.2 Using ratios 1

1 **a** 100 **b** 10 **c** 1000 **d** 0.5

2 **a** 50 **b** 30 **c** 40

3 **a** 15 **b** 37.1 **c** 937

4 **a** 6 **b** 7 **c** 16 **d** 25

5 10

6 40 g

7 No, he will use 300 g.

8 £120

9 320 m

10 **a** 9
b Students' own answers, e.g. No, because 9 staff and 36 residents is still a ratio of 1 : 4.

11 **a** 1 : 15 **b** 5 : 6 **c** 10 : 3 **d** 16 : 29

12 100

13 **a** 5 : 62 **b** 5 : 1 **c** 9 : 10 **d** 3 : 7

14 **a** Old, ratio simplifies to 4 : 3
b New, ratio simplifies to 16 : 9

15 80

11.3 Ratios and measures

1 **a** 35 cm^2 **b** 8 cm^3

2 **a** 1000 **b** 1000 **c** 1000
d 60 **e** 60 **f** 3600

3 **a** 1 : 3 **b** 1 : 3 **c** 1 : 9
d Students' own answers, e.g. Every length is in the same ratio, and the areas are in the square of that ratio.

4 **a** 3 : 5 **b** 9 : 25 **c** 27 : 125
d Students' own answers, e.g. The ratio of areas is the square of the ratio of lengths, and the ratio of volumes is the cube of the ratio of lengths.

5 1 : 6

6 **a** 6 : 25 **b** 1 : 4 **c** 2 : 5 **d** 240 : 1

7 1 : 4

8 **b** km : m = 1 : 1000 **c** l : ml = 1 : 1000
d m : mm = 1 : 1000

9 **a** 2500 **b** 34 **c** 3800 **d** 7.3 m

10 **a** 8 km **b** 10 miles **c** 41.95 km

11 No; she needs $1500 \times 36 \times 0.0254 = 1371.6$ m of wool, but she only has $6 \times 225 = 1350$ m.

12 **a** \$690 **b** £350
c $1300 ÷ 1.38 = £942.03 so £950 is worth more.

13 **a** £3935 **b** It would go down.

11.4 Using ratios 2

1 **a** 4 : 1 **b** 2 : 5

2

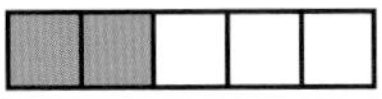

3 **a** 5 **b** 2.5 **c** 1.5 **d** 0.7

4 **a** 12.47 **b** 3.158

5 **a** 2 : 7
b Students' own answers, e.g. 2 : 8, 10 : 40.

6 **a** £12, £6 **b** £6, £36 **c** £12, £15
d 14 kg, 21 kg **e** 25 m, 35 m **f** 3 litres, 4.5 litres

7 **a** 4.875 g **b** 1.625 g

8 **a** 12.5 litres **b** 7.5 litres
c Students' own answers, e.g. The answers to **a** and **b** should add to 20.

9 **a** £16, £24, £32 **b** 20 g, 30 g, 50 g
c 90 ml, 120 ml, 150 ml

10 5 litres, 10 litres, 20 litres

11 £60

12 No, he has enough cement and sand but not enough gravel (he needs 75 kg).

13 **a** £22.86, £57.14
b 15.556 litres, 54.444 litres
c £3.13, £9.38, £12.50 (amounts total £25.01 due to rounding)
d Students' own answers, e.g. for money to the nearest penny; for litres to the nearest ml.

14 **a** 2 : 1 **b** Bob gets £120, Phil gets £60
c Students' own answers, e.g. It is fair that the prize money they each receive reflects their individual contributions.

15 Andrea gets £190, Penny gets £285.

16 6 and 18

17 £150

18 42

19 45

20 6 : 1 : 3

21 £1200

11.5 Comparing using ratios

1 a $\frac{2}{5}$ b 60%

2 a $\frac{7}{9}$ b $\frac{3}{5}$ c 8
d 25 e 6 f 15%

3 a 5 : 3 b 11 : 9 : 6 c 9 : 10 d 6 : 1

4 8

5 a 5 : 3 b $\frac{5}{8}$ c $\frac{3}{8}$
d Students' own answers, e.g. The total of the parts in the ratio in part **a** is the denominator of the fraction in parts **b** and **c**.

6 a $\frac{4}{7}$ b $\frac{3}{7}$
c Students' own answers, for example Check that when added together they equal 1.

7 a 3 : 1 b 24 c 32

8 a 2 : 5 b 15

9 No; $\frac{4}{13}$ are singers.

10 7 : 3 : 2

11 45%

12 30

13 20%

14 a $\frac{2}{5}$ b 10%
c Milk 200 g, chocolate 175 g, butter 75 g, syrup 50 g
d 1 kg using all the syrup

15 a 4 : 1 b 1 : 3.5 c $\frac{3}{4}$: 1
d 4 : 1 and $\frac{3}{4}$: 1; 1 is on the right of the colon.

16 a 0.3 : 1 b 0.4 : 1 c 1.75 : 1
d 1.78 : 1 e 0.21 : 1

17 a 1 : 2 b 0.5 : 1

18 0.2 : 1

19 Anna's squash is stronger as it has a squash-to-water ratio of 1 : 17. Jeevan's squash has a ratio of 1 : 19.

20 Josh's paint is darker as it has a red-to-white ratio of 4 : 1. Josh's has a ratio of 3.75 : 1.

21 Raj makes concrete in the ratio 7.28 : 1 and Sunil makes concrete in the ratio 5.34 : 1. Therefore Sunil's concrete has a higher proportion of cement.

11.6 Using proportion

1 a 1.8p per gram b 3.25 ml for 1p

2 a 175 b 35 c 480 d 30
e 13.5 f 1.6 g 5 h 7.2

3 a 0.22 : 1 b 2.2 : 1
c 0.4 : 1

4 a × 2, 1800 g b ÷ 2, 450 g
c 450, 1350 g
d Students' own answers, for example
i 18 people = 6 people + 6 people + 6 people, so 900 + 900 + 900 = 2700 g of mince needed
ii 15 people = 12 people + 3 people, so 1800 + 450 = 2250 g of mince needed

5 a 12 b 3 c 9 d 15

6 £62.70

7 £450

8 339.6

9 a i 0.295 45... kg per £1 ii 0.433 333... kg per £1
b 2.6 kg size

10 a £4.20 and £4.35 b 450 g
c Students' own answers, e.g. Cost per gram: 450 g pack is £0.028/g, 300 g is £0.029/g.

11 2 litres

12 Stationery for You costs 6 × 3.19 = £19.14; Paper etc. costs 5 × 2 × 1.95 = £19.50 so Stationery for you is cheaper.

11.7 Proportion and graphs

1 a $y = 5x$ b 7

2 a 4 b $y = 4x$

3 Yes, when x is zero, y is zero and when x is doubled so is y.

4 a

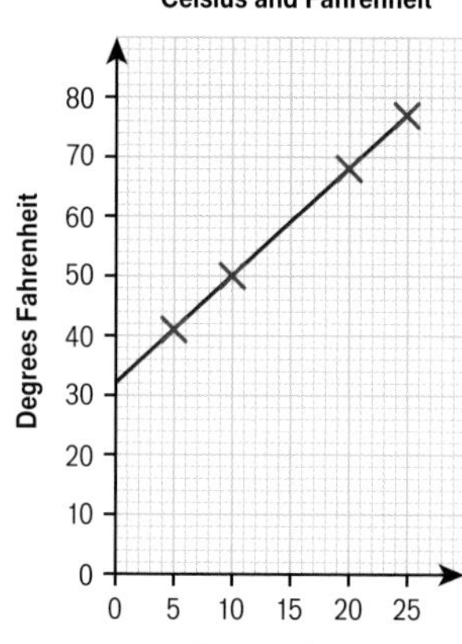

b 32 °F c No, line does not go through origin.

5 a Yes b $\frac{4}{1000}$ or 0.004 c £4
d Students' own answers, e.g. 1000 × 0.004 = 4.

6 a Yes b No c No d Yes

7 No, the charge for 2 hours is not double the charge for 1 hour.

8

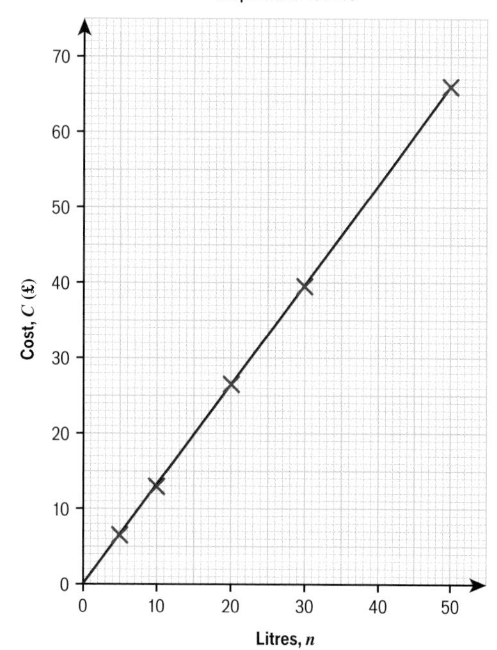

b $y = 1.32x$ c Yes
d $C = 1.32n$ e £108.24
f Students' own answers.

9 **a** e.g. 10 : 80 **b** 1 : 8 **c** 8 **d** $P = 8G$
e Students' own answers, e.g. The unit ratio, the gradient and the formula all show 8 pints to 1 gallon.

10 **a**

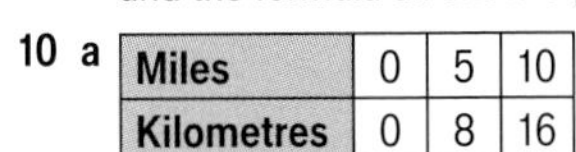

Miles	0	5	10
Kilometres	0	8	16

b

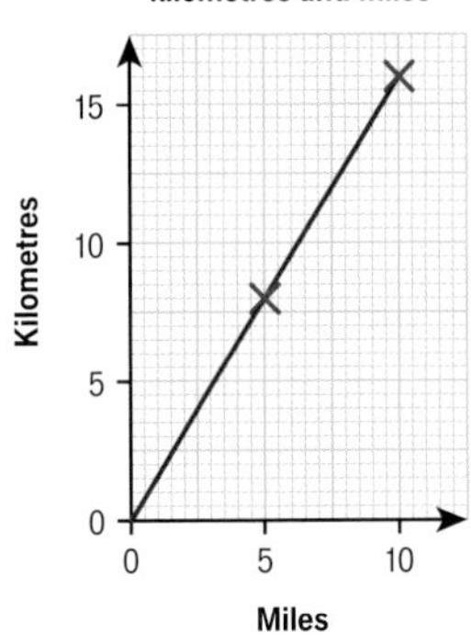

c 1 : 1.6 **d** $y = 1.6x$

11 No, when $C = 0$, $K = 273.15$

11.8 Proportion problems

1 Less
2 **a** 5 **b** 3000 **c** 7.5
3 **a** 300 **b** 3 hours 20 minutes
c 2 hours 45 minutes
4 **a** 36 hours **b** 7 hours 12 minutes
c Students' own answers, e.g. All the people work at the same rate.
5 **a** 2 hours 30 minutes **b** 3 hours 20 minutes
6 2.5 days
7 No, 7 people can transplant 56 seedlings an hour, so it will take $420 \div 56 = 7.5$ hours.
8 **a** i, iv **b** ii **c** iii, v
9 **a** 1 hour 20 minutes
b No. The time taken is not affected by the number of walkers.
10 4 kg
11 360
12 $16\frac{2}{3}$ days
13 18 hours
14 2 hours 30 minutes
15 15 people
16 2810 g

11 Check up

1 **a** 300 g **b** 75 g **c** 225 g
2 300 ml
3 **a** 6 **b** 18
4 **a** 3 : 4 **b** 3 : 2 : 6 **c** 5 : 7
5 **a** 4200 **b** 50 **c** 246 **d** 15.625
6 20 tins
7 15 m
8 **a** Naadim £24, Bal £36
b $\frac{2}{5}$ **c** $24 + 36 = 60$
9 **a** 1.3 : 1 **b** 0.25 : 1 **c** 1.5 : 1
10 **a** 18 **b** 30
11 Both the same 1 : 9.5

12 **a** Yes **b** $y = 2x$ or $B = 2U$
13 **a** 12 days **b** 2 days
14 4 kg
15 1.5 minutes
17 Students' own answers.

11 Strengthen

Simple proportion and best buys

1 **a** £6 **b** £15 **c** £24 **d** £36
2 **a** £8 **b** £16 **c** £64 **d** £80
3 **a** 100 g
b **i** £1.25 **ii** £1.20
c Jar B is better value.

Ratio and proportion

1 **a** 4 **b** 8
2 **a** ÷2, 1 : 3 **b** ÷4, 3 : 2 **c** ÷5, 2 : 3 : 7
3 **a** 3 : 4 **b** 1 : 7 **c** 3 : 10
d 6 : 5 **e** 7 : 9 **f** 1 : 25
4 **a** 380 **b** 2500 **c** 123 **d** 3
5 15 tins
6 3 oz
7 **a** Isabel £8, Freya £12 **b** £20
8 **a** 30 cm, 60 cm, 150 cm **b** $30 + 60 + 150 = 240$
9 **a** 27 **b** 45
10 **a**

q	q	q	g	g	g	g	g	g	g

b $\frac{7}{10}$ **c** $\frac{3}{10}$
11 **a**

r	r	r	m

b 3 : 1 **c** 9 **d** 36
12 **a** B **b** D **c** C **d** A
13 **a** Amar ÷20, 1 : 9; Ben ÷50, 1 : 10
b Amar has made the stronger squash. Amar 1 : 9, Ben 1 : 10

Proportion, graphs, direct and inverse proportion

1 **a** Yes **b** Yes **c** Yes
2 **a**

Kilograms	0	1	2	3
Pounds	0	2.2	4.4	6.6

b $p = 2.2k$
3 **a**

Length (m)	0	1	2	3	4	5	6	7	8	9	10
Cost (£)	0	4	8	12	16	20	24	28	32	36	40

b

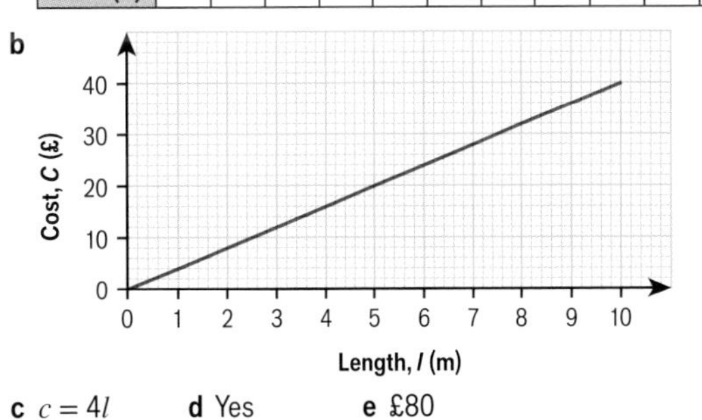

c $c = 4l$ **d** Yes **e** £80

4 a Less b More

c

Length of pipe	More or less	Mass
3 m	–	375 g
1 m	Less	125 g
2 m	More	250 g

5 a More b More

c

Number of people	More or less	Hours
3	–	4
1	More	12
2	More	6

11 Extend

1 a No, not through origin.
b Yes, straight line through origin.
c No, curved line and not through origin.
d Yes, straight line through origin.
e No, curved line.

2 Superfoods (Superfoods: 60p for 100 g Healthy4You: 68.6p for 100 g)

3 28

4 a 9 hours b £648

5 17 : 71

6 140°

7 £200

8 30

9 273 kg

10 41

11 Ollie = £51.30, Sam = £44.55, Peter = £33.75

12 3 : 4 : 8

13 Chicago (57.2p per litre)

11 Test ready

Sample student answers

1 a The student should state that the greatest amount of pink lemonade Antony can make is 250 ml.
b 120 ml of lemonade means that the value of 1 part in the ratio is 40 ml, so the greatest amount of pink lemonade Antony can make is $5 \times 40 = 200$ ml.

2 a Student 1 b Student 1

11 Unit test

1 7 : 5 (Australia : India)

2 64p or £0.64

3 36 limes

4 570 euros

5 12

6 1 tub at Mini Mart costs £2.35.
1 tub at Dave's Deli costs £2.50.
Better value at Mini Mart (providing calculation supports this).

7 6 : 1

8 25 litres

9 a $\frac{2}{9}$ b Ali £20

10 £6.40

11 a 800 g b 60 g
c No. A pie made using 1 egg would serve 1.5 people. $5 \times 1.5 = 7.5$. You would need 6 eggs to make a beef pie for 8 people.

12 a Yes, with a convincing reason e.g. straight line through the origin.
b $D = 50T$

13 45 minutes

14 1.5 hours

15 Anna saves $\frac{3}{10}$ of her salary = 30%.
Bob spends 75%, meaning he saves 25%.
$100 \div 5 = 20$. $20 \times 2 = 40$. Sally saves 40%.
They each earn the same monthly salary and Sally saves the highest percentage of her salary; therefore Sally saves the most.

16 a Triangle: 30°, 60°, 90°;
Quadrilateral: 36°, 72°, 108°, 144°;
Pentagon: 36°, 72°, 108°, 144°, 180°
b 1 : 2 : 3 : 4 : 5 : 6 : 7 : 8;
30°, 60°, 90°, 120°, 150°, 180°, 210°, 240°

17 Students' own answers.

UNIT 12 Right-angled triangles

12.1 Pythagoras' theorem 1

1 a 36 b 7 c 9 or −9

2 144 cm^2

3 a 625 b 25

4 a 8.06 b 18.6

5 a

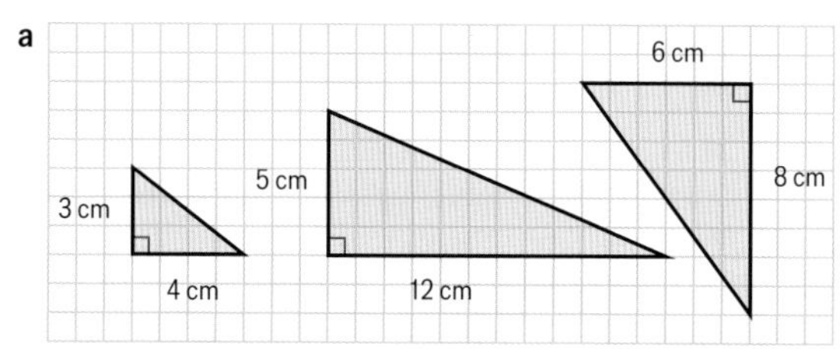

b i 5 cm, 13 cm, 10 cm ii 5 cm, 13 cm, 10 cm

6 a 6.5 cm b 34 m c 29 cm

7 a Students' own check.
b 25 cm^2 c 25 cm^2 d $5^2 = 3^2 + 4^2$
e The square on the hypotenuse has the **same** area as the sum of the areas of the **squares** on the other two sides.

8 a i, ii

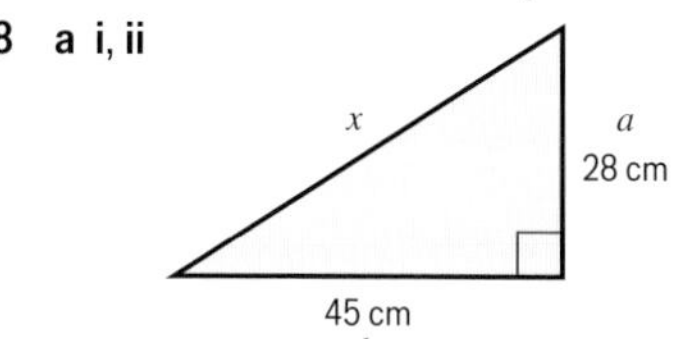

iii 53 cm

b i, ii

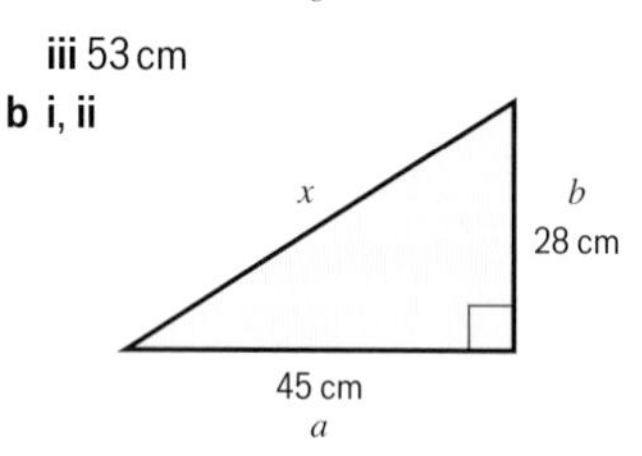

iii 53 cm

c No, because $45^2 + 28^2 = 28^2 + 45^2$

9 a 10.1 cm b 15.7 cm c 149 m
d 173 m e 16.9 km

10 a 9.35 m b 10.19 m c 23.32 m

11 8.10 m

12 **a** BC is opposite the right angle so it is the longest side, and 10.6 cm is not longer than 77 cm.

b She has forgotten to square 36 and 77.

c 85 cm

12.2 Pythagoras' theorem 2

1 AC; it is opposite the right angle.

2 11.3 m

3 **a** 12 **b** 4 **c** 6

4 **a, b**

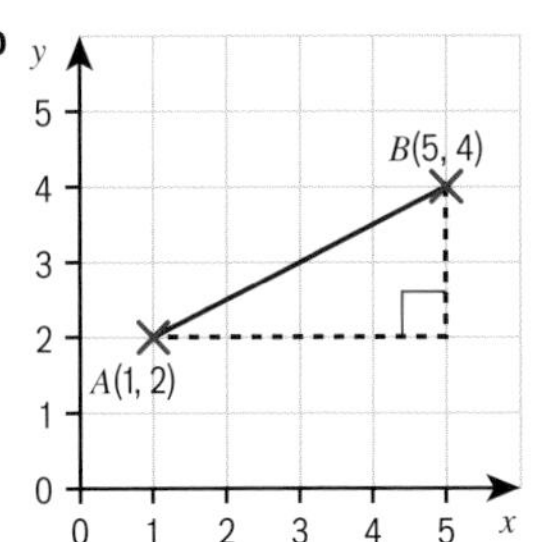

c 4.47 cm

5 **a** 5 **b** 13 **c** 17

6 **a** 13.9 cm (1 d.p.) **b** 5.8 cm (1 d.p.)

c 7.89 m (2 d.p.)

d It is sensible to give the answer to the same accuracy as the lengths that are given.

7 **a** A and D, B and C

b 8.0 cm in A and D, 5.5 cm in B and C

8 50.3 cm

9 **a** $h = 11.85$ cm **b** $x = 4.5$ cm, $y = 8.3$ cm

10 **a** $8^2 \neq 5^2 + 6^2$; not right-angled

b $10^2 = 6^2 + 8^2$; right-angled

c $6^2 \neq 4^2 + 5^2$; not right-angled

12.3 Trigonometry: the sine ratio 1

1 **a** Right-angled

b **i** $\frac{1}{2}$ **ii** 0.5 **iii** 1 : 2

2

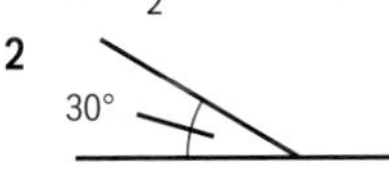

3 1.251

4 **a**

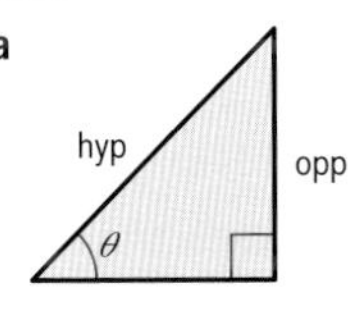

b

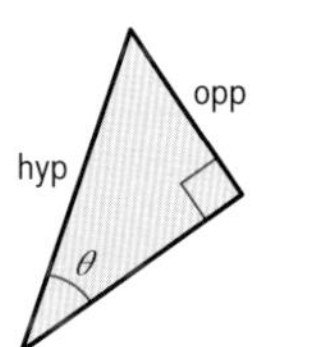

c

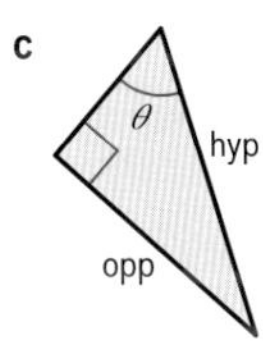

5 **a** Accurate drawings of the triangles, drawn to scale.

b

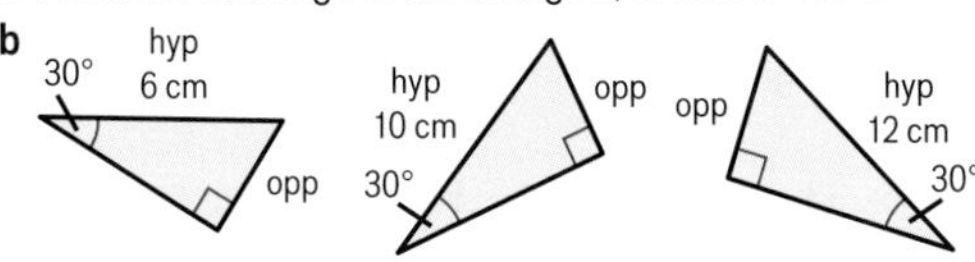

6 **a** **i** 3 cm, 5 cm, 6 cm

ii $\frac{3}{6} = 0.5$, $\frac{5}{10} = 0.5$, $\frac{6}{12} = 0.5$

b They are all the same.

7 **a** 0.766 **b** 0.616 **c** 0.729

8 0.5

9 **a** $\frac{9}{14}$ **b** $\frac{7}{10}$ **c** $\frac{3}{5}$

10 **a** 6.5 cm **b** 8.7 cm **c** 7.5 cm

d 15.1 cm **e** 18.2 cm **f** 28.0 cm

11 2.78 m

12 15.0 cm (1 d.p.)

13 No; the rope will only reach 6.9 m up the flagpole.

12.4 Trigonometry: the sine ratio 2

1 $\frac{5}{13}$

2 **a** 0.530

3 **a** 40° **b** 70° **c** 10°

4 **a** 30° **b** 65°

5 Students' own calculator check.

6 **a** 20.2° **b** 55.5° **c** 60.1°

d 66.4° **e** 38.7° **f** 38.8°

7 **a** Annie rounded after the first step and Joe rounded the final answer.

b Joe; his working uses the full calculator value, so his answer is more accurate.

8 **a** 41.8° **b** 53.8° **c** 40.8°

9 7.7°

10 50.6°

12.5 Trigonometry: the cosine ratio

1 Hypotenuse: QR, LN, ST

Adjacent: PQ, LM, RT

2 **a** **i** Right-angled **ii** Right-angled

iii Isosceles

b $BC = 0.5 \times BD$

3 **a** 9.6 cm **b** 11.8°

4 **a** Accurate drawings of the triangles, drawn to scale.

b **i**

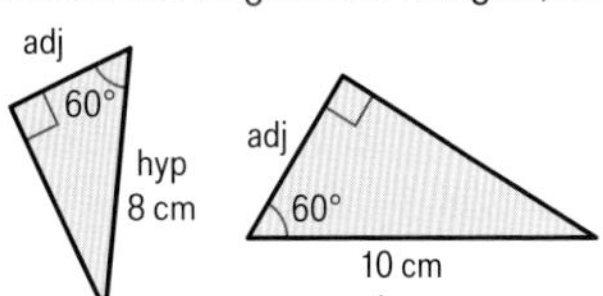

ii 4 cm, 5 cm, 4.5 cm

iii $\frac{4}{8} = 0.5$, $\frac{5}{10} = 0.5$, $\frac{4.5}{9} = 0.5$

c They are all the same.

5 **a** 0.799 **b** 0.139 **c** 0.995

6 **a** $\frac{15}{17}$ **b** $\frac{7}{25}$ **c** $\frac{9}{41}$

7 $\cos 60° = \frac{\text{adj}}{\text{hyp}} = \frac{1}{2}$; $\sin 30° = \frac{\text{opp}}{\text{hyp}} = \frac{1}{2}$; so $\cos 60° = \sin 30°$

8 **a** 13.6 cm **b** 5.6 cm **c** 10.4 cm

9 5.74 m

10 **a** 68.8° **b** 47.7°

11 **a** 55.0° **b** 51.3° **c** 16.3° **d** 51.1°

12 **b** 0.8829... then $\cos^{-1}$(0.8829...)

c $\cos^{-1}$(0.42261...) then 65° then cos 65°

13 **a**

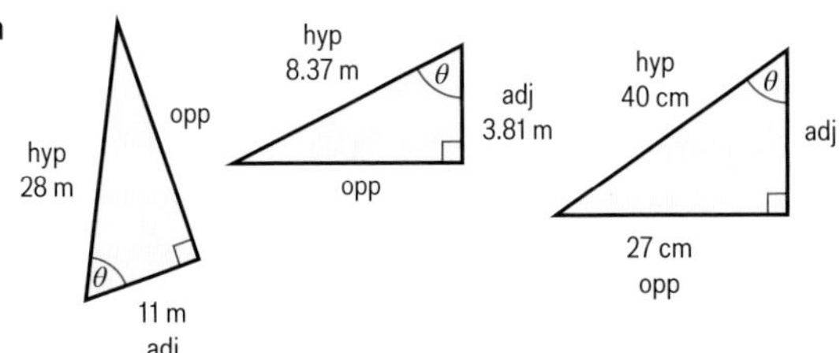

b $\cos\theta$, $\cos\theta$, $\sin\theta$

c $\cos\theta = \frac{11}{28}$, $\cos\theta = \frac{3.81}{8.37}$, $\sin\theta = \frac{27}{40}$

d 66.9°, 62.9°, 42.5°

14 40.9°

12.6 Trigonometry: the tangent ratio

1 Opposite: PR, MN, RS
Adjacent: PQ, LM, RT

2 **a** $\frac{35}{12}$ **b** $\frac{36}{77}$

3 **a** 1.732 **b** 0.247 **c** 4.041

4 **a** Accurate drawings of the triangles, drawn to scale.

b i

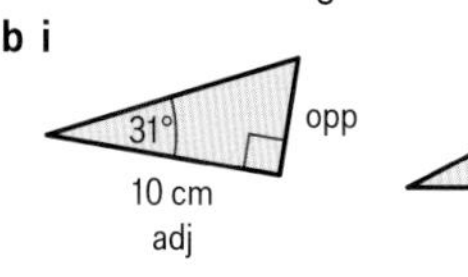

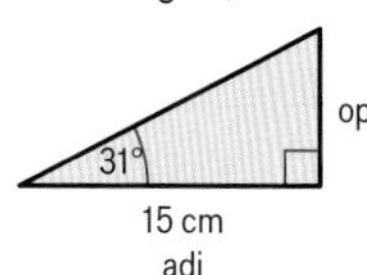

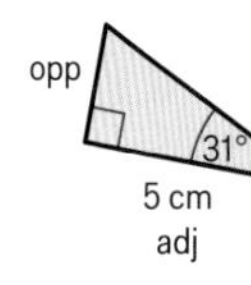

ii 6 cm, 9 cm, 3 cm

iii $\frac{6}{10} = 0.6$, $\frac{9}{15} = 0.6$, $\frac{3}{5} = 0.6$

c They are all the same.

5 **a** 0.466 **b** 1.192 **c** 3.732

6 **a** 0.6 **b** 1.667 (3 d.p.)

7 **a** $\frac{8}{15}$ **b** $\frac{24}{7}$ **c** $\frac{40}{9}$

8 **a** 11.4 **b** 35.2 **c** 29.6

9 4.77 m

10 **a** Elevation **b** Depression **c** Elevation

11 **a**

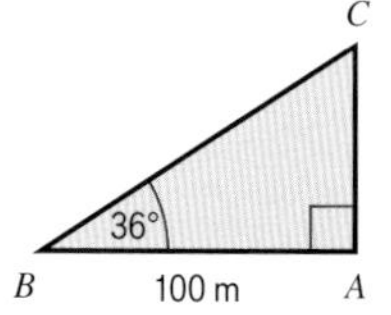

b 73 m

12 71 m

13 **a** 40.5° **b** 59.9° **c** 41.2° **d** 60.5°

14 **b** 2.355... then $\tan^{-1}$(2.355...)
c $\tan^{-1}$(1.920...) then 62.5° then tan 62.5°

15 **a** 51.3° **b** 48.4° **c** 21.9°

16 66.7° (1 d.p.)

12.7 Finding lengths and angles using trigonometry

1 **A** $\tan\theta$ **B** $\sin\theta$ **C** $\cos\theta$

2 22.6

3 $c^2 = 1^2 + 1^2$; $c^2 = 2$; $c = \sqrt{2}$ cm

4 **a** 17.9 m **b** 32.1 m **c** 50.9 m **d** 14.0 m

5 **a** 66.0° **b** 39.6° **c** 54.1°

6 Students' own answers.

7 **a** 41.4° **b** Stay the same; $\frac{12}{16} = \frac{3}{4} = \frac{15}{20}$

8 40.0 m (1 d.p.)

9 17.4 cm (1 d.p.)

10 **a i** 1 **ii** $\frac{1}{\sqrt{2}}$
b $\frac{1}{\sqrt{2}}$ **c** They are the same.

11 **a** $PR^2 = PQ^2 + QR^2$; $2^2 = PQ^2 + 1^2$; $PQ^2 = 4 - 1 = 3$; $PQ = \sqrt{3}$
b i $\frac{1}{2}$ **ii** $\frac{\sqrt{3}}{2}$ **iii** $\frac{\sqrt{3}}{2}$ **iv** $\frac{1}{2}$
c i They are the same. **ii** They are the same.

12 **a** 4 **b** 5 **c** 2

13 $a = b = 30°$, $c = d = 60°$

12 Check up

1 **a** 65 **b** 18

2 10.6 cm

3 Yes; $12.5^2 = 3.5^2 + 12^2$

4 **a** 10.2 **b** 32.8 **c** 87.5

5 3.83 m

6 2.18 m

7 **a** 41.6° **b** 40.2°

8 30.8°

9 **a** 1 **b** $\frac{1}{2}$ **c** $\frac{\sqrt{3}}{2}$ **d** 0

11 $\tan\theta = \frac{5}{5} = 1$; in any square the opposite and adjacent sides are equal so the ratio is always 1.

12 Strengthen

Pythagoras' theorem

1 **a** 15 cm **b** 20 cm **c** 19.5 m

2 **a** Hypotenuse
b $6\,\text{cm} \times 6\,\text{cm} = 36\,\text{cm}^2$
c $5\,\text{cm} \times 5\,\text{cm} = 25\,\text{cm}^2$
d $25\,\text{cm}^2 + 36\,\text{cm}^2 = 61\,\text{cm}^2$

3 $c^2 = 8^2 + 15^2$; $c^2 = 289$; $c = \sqrt{289}$; $c = 17$ km

4 **a** 26 m **b** 8.5 km

5 **a** $AB = a = 6$ cm, $BC = b = 4$ cm
b 7.2 cm

6 12.5 cm

7 **a**

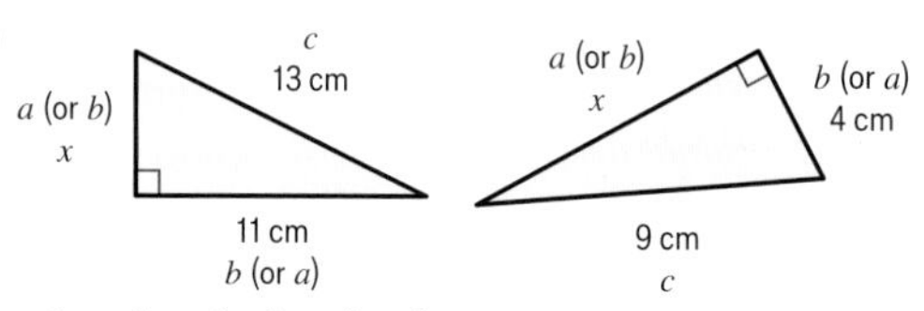

b $13^2 = x^2 + 11^2$; $9^2 = x^2 + 4^2$ **c** 6.93 cm, 8.06 cm

8 **a**

b 72.25 **c** 72.25 **d** Yes **e** Yes

Finding lengths using trigonometry

1 **a, b, c**

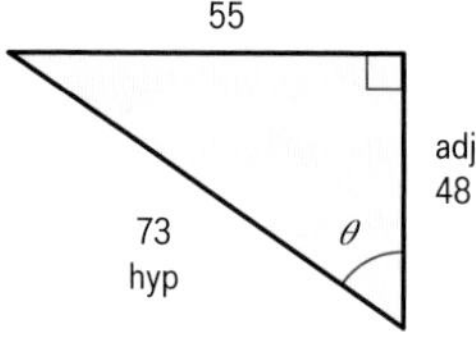

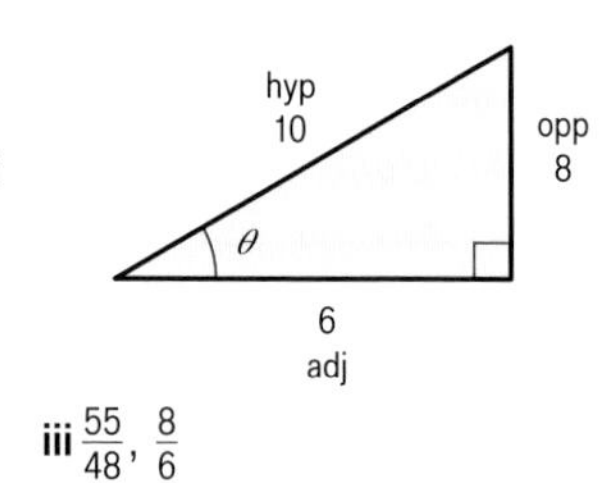

d i $\frac{55}{73}$, $\frac{8}{10}$ **ii** $\frac{48}{73}$, $\frac{6}{10}$ **iii** $\frac{55}{48}$, $\frac{8}{6}$

2 **i–vi**

a $\theta = 33°$ **b** $\theta = 58°$ **c** $\theta = 35°$

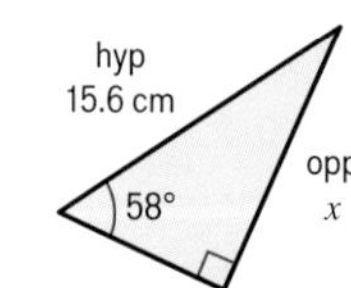

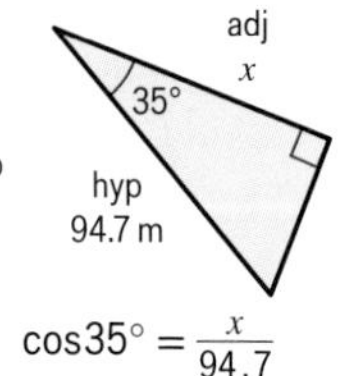

$\tan 33° = \frac{x}{42}$
$x = 42 \times \tan 33°$
$x = 27.3$ cm

$\sin 58° = \frac{x}{15.6}$
$x = 15.6 \times \sin 58°$
$x = 13.2$ cm

$\cos 35° = \frac{x}{94.7}$
$x = 94.7 \times \cos 35°$
$x = 77.6$ m

3 **a** 15.55 m **b** 4.63 m

Finding angles using trigonometry

1 a 11.5° b 78.5° c 33.7°

2 a

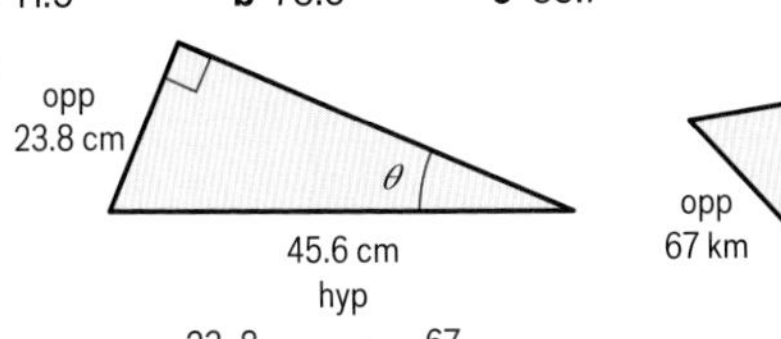

θ
opp
67 km
adj
98 km

b, c $\sin\theta = \frac{23.8}{45.6}$, $\tan\theta = \frac{67}{98}$

d 31.5°, 34.4°

3

Angle	0°	30°	45°	60°	90°
sin	0	$\frac{1}{2}$	$\frac{1}{\sqrt{2}}$	$\frac{\sqrt{3}}{2}$	1
cos	1	$\frac{\sqrt{3}}{2}$	$\frac{1}{\sqrt{2}}$	$\frac{1}{2}$	0
tan	0	$\frac{1}{\sqrt{3}}$	1	$\sqrt{3}$	

12 Extend

1 130 km

2 Yes; the angle is 69.1° which is below 75°.

3 $9.98\,\text{cm}^2$

4 11.2 cm

5 a 1.464 413 507 b 1.46

6 a Increases b Decreases

7 a $37^2 = 35^2 + 12^2$ b 71.1°

8 $648\,\text{cm}^2$

9 0.475 kg

10 54.7°

11 48.2°, 48.2°, 83.6°

12 35.9 cm

13 4 cm

12 Test ready

Sample student answers

1 a Student A

b The diagram has helped the student to match the correct sides to the lengths given, and therefore put the correct numbers into the formula. Labelling the sides a, b and c reminds them that it is the longest side which is by itself in the formula. To improve the answer further, they should have written $a = \sqrt{21.6} = 4.6475\ldots$ after $a^2 = 21.6$ to ensure all method marks in case they made a careless error with rounding.

2 a Students' own answers, e.g.
144° is an obtuse angle, and the largest angle in a right-angled triangle is 90°.

b $BC = \frac{25}{\sin 72°} = \frac{25}{0.9510\ldots} = 26.2865\ldots$

$BC = \frac{25}{\cos 18°} = \frac{25}{0.9510\ldots} = 26.2865\ldots$

c Students' own answers, e.g.
In the right-angled triangle, the side that is opposite to the 72° angle is adjacent to the 18° angle, so $\sin 72° = \cos 18°$.

12 Unit test

1 5.36 m

2 9.90 cm

3 No, because $18^2 \neq 17^2 + 6^2$

4 a 53.1° b 2.33 cm

5 a 15.17 cm b 9.38 m

6 25.6°

7 6.95 m

8 13.0°

9 7.78 cm

10 $\frac{1}{2}$

11 $43.5\,\text{cm}^2$

12 a cosine ratio b sine ratio
c Pythagoras' theorem d tangent ratio
e Either sine ratio or cosine ratio or Pythagoras' theorem

UNIT 13 Probability

13.1 Calculating probability

1 a $\frac{3}{5}$ b 0.6 c 60%

2 a 1 b $\frac{3}{4}$ c $\frac{4}{7}$ d 0.3

3 a i $\frac{3}{7}$ ii $\frac{5}{7}$ iii 0
b Green

4 a ×

0 $\frac{1}{2}$ 1

b $\frac{2}{6}$ or equivalent

5 a $\frac{1}{11}$ b $\frac{2}{11}$ c $\frac{2}{11}$

6 $\frac{5}{11}$

7 a $\frac{5}{12}$ b $\frac{4}{12}$ c $\frac{3}{12}$ d $\frac{9}{12}$
e 1

8 $\frac{3}{7}$

9 0.975

10 a Yes b No c No

11 a Yes b No c No

12 a 0.4 b 1 c 0.3

13 $\frac{1}{2}$

14 a 5 b 10 c 20

15 10

16 P(6) = 0.19, $0.4 \times 300 = 120$

17 a More blue counters because blue has a greater probability than red.
b i 12 ii 20
c 40

18 a Green 0.2, Yellow 0.2 b 15

13.2 Two events

1 a Heads or Tails b $\frac{1}{2}$

2 a (R, E) (R, C) (R, L) (R, P) (W, E) (W, C) (W, L) (W, P)
b 8

3 a $\frac{1}{8}$ b $\frac{1}{8}$ c $\frac{2}{8}$ or $\frac{1}{4}$

4 a (R, L) (B, L) (Y, L) (R, S) (B, S) (Y, S)
b 6
c i $\frac{1}{6}$ ii $\frac{3}{6}$ or $\frac{1}{2}$ iii $\frac{2}{6}$ or $\frac{1}{3}$

5 a (green, blue), (green, red), (green, yellow), (blue, red), (blue, yellow), (red, yellow)
b i $\frac{1}{6}$ ii $\frac{3}{6}$ or $\frac{1}{2}$

6 a

	Head	Tail
Head	*H, H*	*H, T*
Tail	*T, H*	*T, T*

b 4

c i $\frac{1}{4}$ **ii** $\frac{1}{2}$ **iii** $\frac{3}{4}$

d 1 – P(no heads) = P(at least one head)

7 a 10

b

	Blue	Red	Yellow	Green	Pink
Heads	*H, B*	*H, R*	*H, Y*	*G, Y*	*H, P*
Tails	*T, B*	*T, B*	*T, Y*	*T, G*	*T, P*

8 a

	Red	Green	Blue	Yellow
Red	*R, R*	*R, G*	*R, B*	*R, Y*
Green	*G, R*	*G, G*	*G, B*	*G, Y*
Blue	*B, R*	*B, G*	*B, B*	*B, Y*

b $\frac{1}{12}$ **c** $\frac{1}{2}$ **d** 0

e Mischa is wrong because you have to look at the outcomes not the individual letters. There are 12 possible outcomes and 6 of them contain at least one blue, so the probability of getting at least one blue is $\frac{6}{24}$.

9 a

	A	*B*	*C*
A	*A, A*	*A, B*	*A, C*
B	*B, A*	*B, B*	*B, C*
C	*C, A*	*C, B*	*C, C*

b $\frac{3}{9}$ or $\frac{1}{3}$ **c** 10

10 a

		Top		
		Blue	Red	Green
Skirt	**Blue**	*B, B*	*B, R*	*B, G*
	Red	*R, B*	*R, R*	*R, G*
	Green	*G, B*	*G, R*	*G, G*

b 9

c i $\frac{1}{9}$ **ii** $\frac{1}{9}$ **iii** $\frac{5}{9}$

d 18

11 a

1st throw						
6	6, 1	6, 2	6, 3	6, 4	6, 5	6, 6
5	5, 1	5, 2	5, 3	5, 4	5, 5	5, 6
4	4, 1	4, 2	4, 3	4, 4	4, 5	4, 6
3	3, 1	3, 2	3, 3	3, 4	3, 5	3, 6
2	2, 1	2, 2	2, 3	2, 4	2, 5	2, 6
1	1, 1	1, 2	1, 3	1, 4	1, 5	1, 6
	1	**2**	**3**	**4**	**5**	**6**
	2nd throw					

b i $\frac{1}{36}$ **ii** $\frac{6}{36}$ or $\frac{1}{6}$ **iii** $\frac{6}{36}$ or $\frac{1}{6}$ **iv** $\frac{11}{36}$

12 No. There are 36 possible outcomes from rolling a dice twice, and one outcome is (1, 1) so P(1, 1) = $\frac{1}{36}$.

13 a

Dice 2						
11	13	15	17	19	21	23
9	11	13	15	17	19	21
7	9	11	13	15	17	19
5	7	9	11	13	15	17
3	5	7	9	11	13	15
1	3	5	7	9	11	13
	2	**4**	**6**	**8**	**10**	**12**
	Dice 1					

b i $\frac{3}{36}$ **ii** $\frac{21}{36}$ **iii** 0 **iv** $\frac{12}{36}$

c Multiple of 3 more likely: P(multiple of 3) = $\frac{12}{36}$; P(7) = $\frac{3}{36}$

14 No, losing is more likely. There are 12 possible outcomes but only 4 are odd (1, 3, 3, 9) so the probability of losing is $\frac{8}{12}$.

15 $\frac{5}{9}$

13.3 Experimental probability

1 a $\frac{1}{6}$ **b** 10

2 a 20 **b** 70 **c** $\frac{8}{25}$

3 a

Outcome	1	2	3	4	5	6
Frequency	1	3	7	3	2	4

b 20 **c** 4 **d** $\frac{4}{20}$ or $\frac{1}{5}$ **e** 20

f Students' own answers, e.g. how often a particular outcome occurs.

4 a $\frac{25}{60}$ **b** Yes, $\frac{35}{60} = \frac{7}{12}$

5 a 18 **b** $\frac{7}{18}$

6 She could get a more accurate measure using more trials.

7 a 100 **b** $\frac{72}{100}$ or 0.72

c Belinda because she carried out more trials/repeated the experiment more times.

d $\frac{192}{260}$ or 0.74 (2 d.p.)

e 0.74 is the best estimate as it uses the most trials.

8 a $\frac{18}{60}$ or $\frac{3}{10}$ **b** $\frac{1}{4}$

c Students' own answers with suitable reasoning, e.g. No, because the frequencies are too varied, or Yes, because the range of frequencies is 0.2–0.3, which is close to the theoretical frequency of 0.25, and the sample of 60 is relatively low.

9 a $\frac{42}{120}$ **b** $\frac{1}{6}$ **c** 20

d Students' own answers, e.g. The dice are biased as 42 is more than twice as many differences of 0 as one would predict using the theoretical probability.

10 a $\frac{20}{100}$ or $\frac{1}{5}$ **b** $\frac{57}{300}$

11 a i $\frac{175}{328}$ or 0.53 (2 d.p.) **ii** $\frac{65}{328}$ or 0.20 (2 d.p.)

b $\frac{116}{175}$ or 0.66 (2 d.p.)

12 a $\frac{14}{55}$ **b** $\frac{3}{16}$

13 a i $\frac{5}{70}$ or $\frac{1}{14}$ **ii** $\frac{5}{70}$ or $\frac{1}{14}$

b $\frac{10}{20}$ **c** $\frac{17}{25}$

13.4 Venn diagrams

1 A whole number

2 **a** 2, 3, 5, 7, 11
b 3, 6, 9, 12, 15
c 1, 4, 9, 16, 25

3 **a** $A = \{1, 3, 5, 7, 9\}$ $B = \{1, 4, 9\}$
b i False **ii** True **iii** False
c 1, 9 **d** 3, 5, 7
e 1, 3, 4, 5, 7, 9 **f** 2, 6, 8

4 **a** $X = \{2, 4, 6, 8, 10, 12\}$
$Y = \{3, 6, 9, 12\}$
$\mathscr{E} = \{1, 2, 3, 4, 5, 6, 7, 8, 9, 10, 11, 12\}$
b $\mathscr{E}$ = {integers 1 to 12}
X = {multiples of 2 up to 12}
Y = {multiples of 3 up to 12}
c $\frac{6}{12}$ or $\frac{1}{2}$

5 **a** $\mathscr{E} = \{1, 2, 3, 4, 5, 6, 7, 8, 9, 10, 11, 12, 13, 14, 15\}$
$A = \{2, 4, 6, 8, 10, 12, 14\}$
$B = \{3, 6, 9, 12, 15\}$
b–d

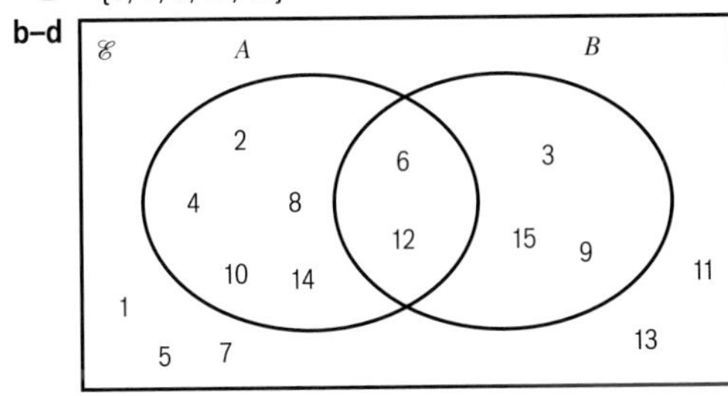

e Count them. There must be the same number as in $\mathscr{E}$. Cross them off your lists as you enter them.

6 **a** $X \cap Y = \{6, 12\}$
b $X \cup Y = \{2, 3, 4, 6, 8, 9, 10, 12\}$
c $X' = \{1, 3, 5, 7, 9, 11\}$
d $Y' = \{1, 2, 4, 5, 7, 8, 10, 11\}$
e $X' \cap Y = \{3, 9\}$

7 **a** $A \cap B = \{6, 12\}$
b $A \cup B = \{2, 3, 4, 6, 8, 9, 10, 12, 14, 15\}$
c $A' = \{1, 3, 5, 7, 9, 11, 13, 15\}$
d $B' = \{1, 2, 4, 5, 7, 8, 10, 11, 13, 14\}$
e $A' \cap B = \{3, 9, 15\}$
f $A \cap B' = \{2, 4, 8, 10, 14\}$

8 **a**

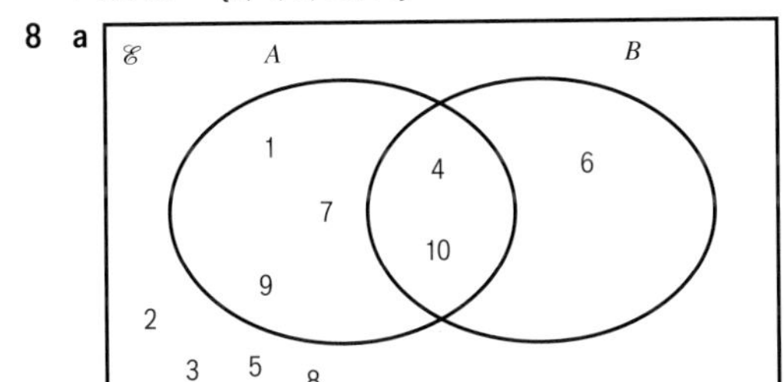

b $\frac{2}{10}$ or $\frac{1}{5}$

9 **a** 2, 4, 6, 8, 10, 12, 14, 15, 16, 18
b Even multiples of 3 between 10 and 20 **or** multiples of 6 between 10 and 20

10 **a**

$\mathscr{E}$ A B C
7 17 3 13 11 5 15 1 9 19

b 3, 5

11 **a** 12 **b** 24 **c** $\frac{8}{25}$ **d** $\frac{4}{25}$
e She has missed out the people who have mobile phones and tablets. 20 people have mobile phones.

12 **a** 44
b i $\frac{18}{44}$ **ii** $\frac{21}{44}$

13 **a**

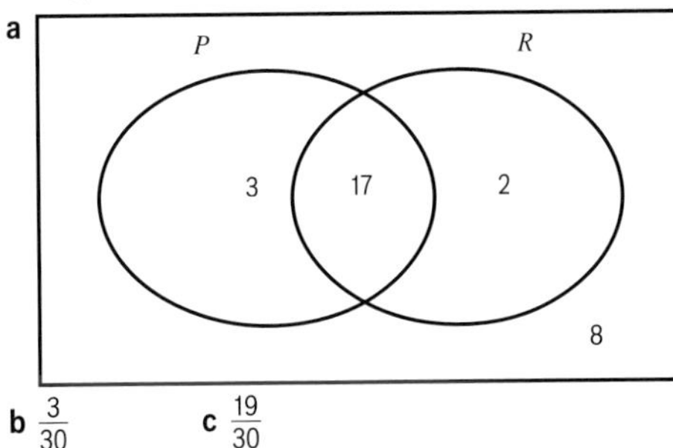

b $\frac{3}{30}$ **c** $\frac{19}{30}$

13.5 Tree diagrams

1 **a** $\frac{1}{9}$ **b** $\frac{1}{4}$ **c** $\frac{6}{9}$ or $\frac{2}{3}$

2 **a** $\frac{7}{10}$ **b** $\frac{3}{10}$

3 **a** 40
b i 18 **ii** 22 **iii** 7
iv 10 **v** 25 **vi** 15
c Numbers on each pair add to the total in the circle joining the pair.

4

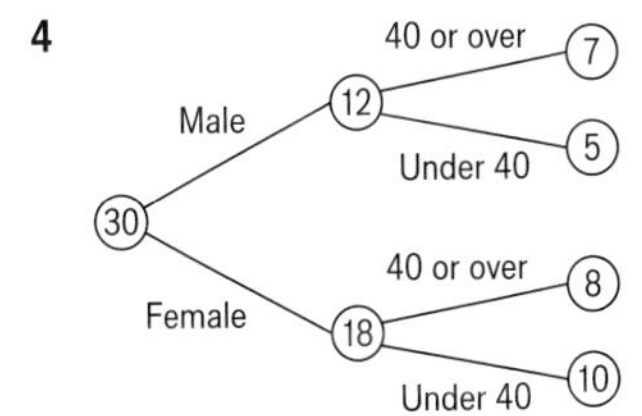

5 a

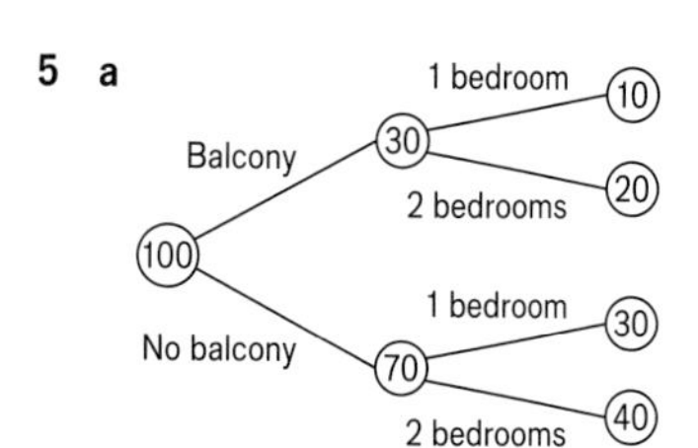

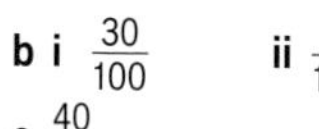

b i $\frac{30}{100}$ **ii** $\frac{40}{100}$

c $\frac{40}{70}$

d

	1 bedroom	2 bedrooms	Total
Balcony	10	20	30
No balcony	30	40	70
Total	40	60	100

Students' own answers for which they prefer and why.

6 a

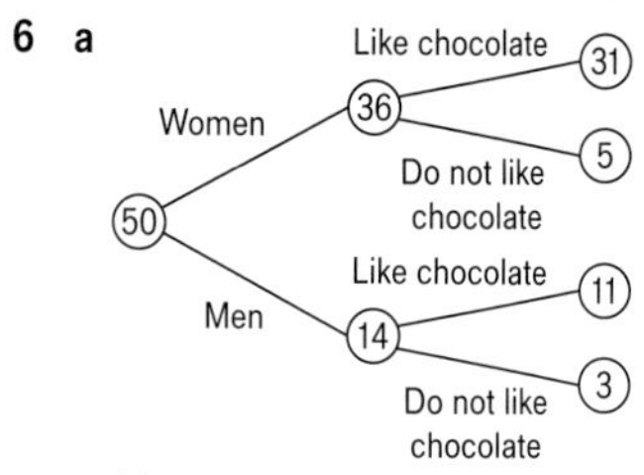

b $\frac{31}{42}$

7 a $\frac{5}{7}$ **b** 6 **c** 4 **d** $\frac{4}{6}$

e No, picking a soft centre first time affects the probability of picking a soft centre the second time.

8 a, b, c

9 a $\frac{1}{3}$ **b** $\frac{1}{9}$ **c** $\frac{4}{9}$ **d** $\frac{4}{9}$

10 a $\frac{3}{4}$

b

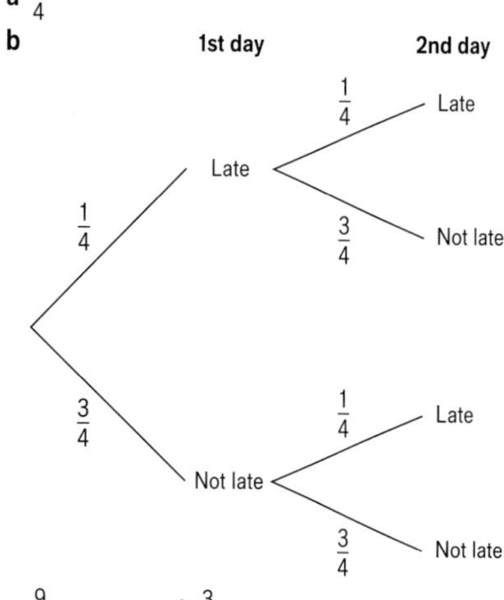

c $\frac{9}{16}$ **d** $\frac{3}{16}$

e Students' own answers, e.g. The probabilities of the four outcomes add to 1.

11 a

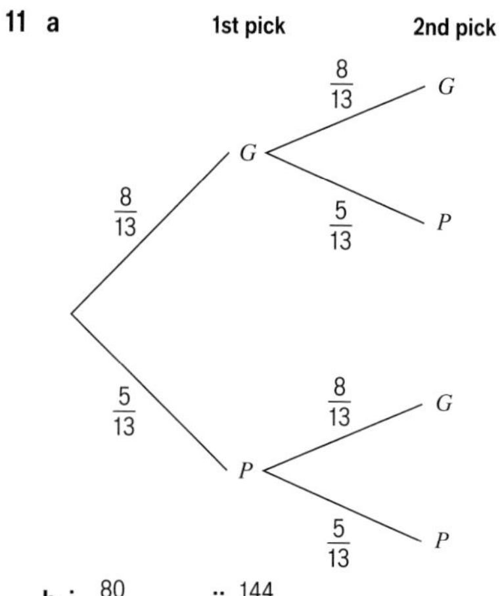

b i $\frac{80}{169}$ **ii** $\frac{144}{169}$

12 a

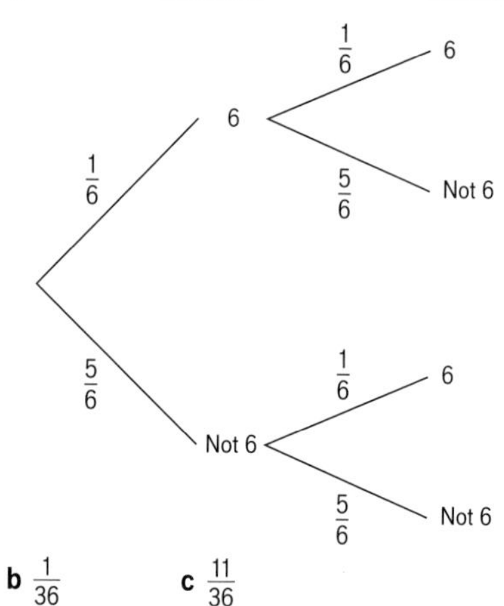

b $\frac{1}{36}$ **c** $\frac{11}{36}$

13 (1) Probabilities on pairs of branches do not add to 1.
(2) For 2nd spin, after blue first spin, the probabilities are on the wrong branches. P(red) should be 0.75.

13.6 More tree diagrams

1 a $\frac{1}{3}$ **b** $\frac{10}{42} = \frac{5}{21}$

c $\frac{26}{64} = \frac{13}{32}$ **d** $\frac{41}{64}$

2 a

1st pick 2nd pick

5/8 R: 5/8 R, 3/8 G

3/8 G: 5/8 R, 3/8 G

b $\frac{9}{64}$ **c** $1 - \frac{9}{64} = \frac{55}{64}$

3 a $\frac{4}{10}$ or $\frac{2}{5}$ **b** 9 **c** 3

d i $\frac{3}{9}$ or $\frac{1}{3}$ **ii** $\frac{6}{9}$ or $\frac{2}{3}$

4 a Dependent **b** Dependent

c Independent **d** Independent

5 a $\frac{42}{90}$ or $\frac{7}{15}$ **b** $\frac{42}{90}$ or $\frac{7}{15}$

6 a

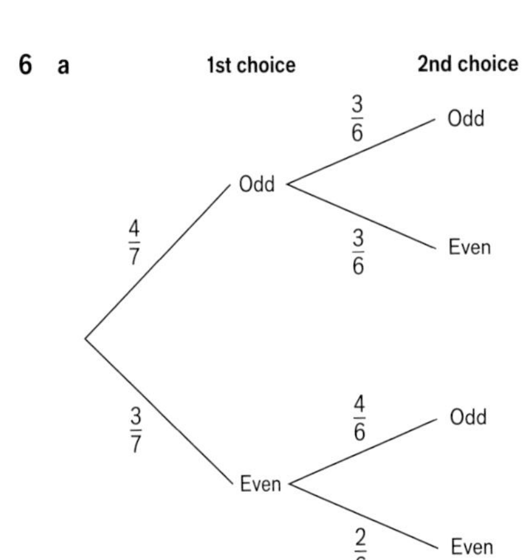

b i $\frac{6}{42}$ or $\frac{1}{7}$ **ii** $\frac{24}{42}$ or $\frac{4}{7}$

7 a

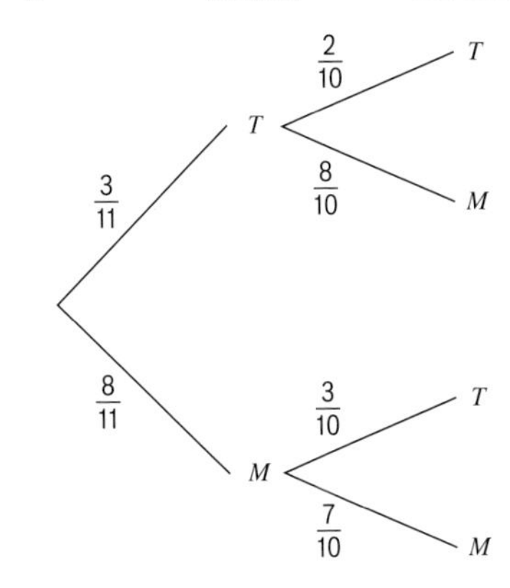

b i $\frac{6}{110}$ **ii** $\frac{104}{110}$

c Add to 1, or each is 1 – the other.

8 a

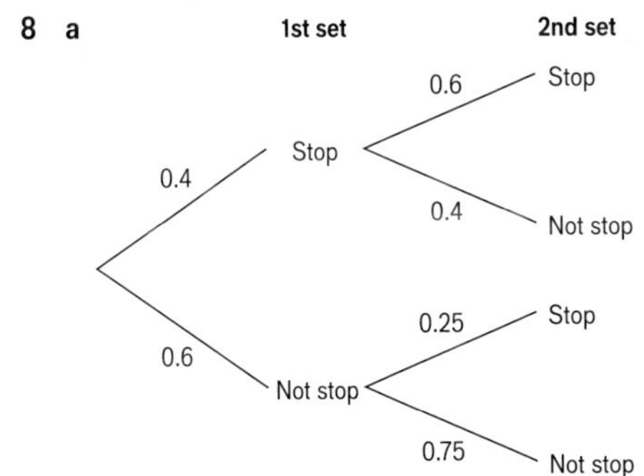

b 0.45

9 a

1st sock 2nd sock

White $\frac{5}{8}$: White $\frac{4}{7}$, Grey $\frac{3}{7}$

Grey $\frac{3}{8}$: White $\frac{5}{7}$, Grey $\frac{2}{7}$

b $\frac{26}{56}$ or $\frac{13}{28}$

13 Check up

1 $\frac{3}{4}$

2 0.5

3 $\frac{18}{38}$

4 a $\frac{25}{110}$ **b** $\frac{2}{9}$

c Bill and Fred's combined results

5 a $\frac{80}{200}$ or $\frac{2}{5}$ **b** $\frac{1}{2}$

c 100

d It may not be fair. 80 is less than the 100 heads predicted.

6 a

		Beth			
		2	**3**	**4**	**6**
Anna	**1**	3	4	5	7
	2	4	5	6	8
	4	6	7	8	10
	5	7	8	9	11

b 16

c i $\frac{3}{16}$ **ii** $\frac{11}{16}$

7 $\frac{3}{8}$

8 a 30 **b** 20 **c** $\frac{13}{30}$

9 a

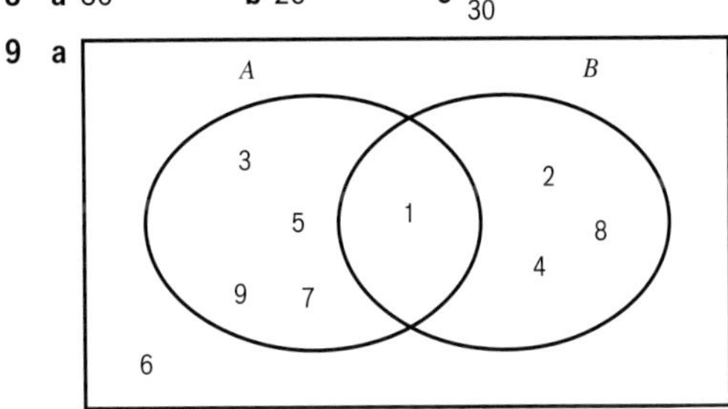

b i {1} **ii** {1, 2, 3, 4, 5, 7, 8, 9} **iii** {2, 4, 6, 8}

10 a

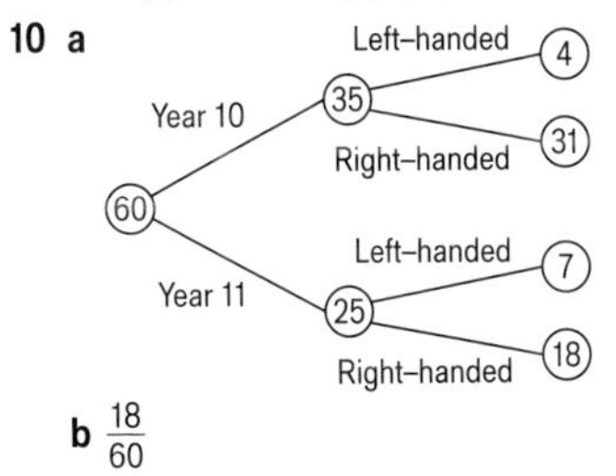

b $\frac{18}{60}$

11 a

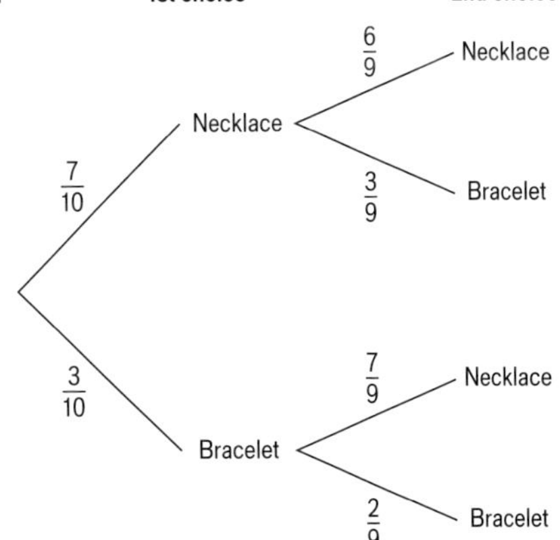

b $\frac{42}{90}$ **c** $\frac{42}{90}$

13 a No; P(win) = $\frac{5}{12}$ **b** No; P(win) = $\frac{7}{12}$

13 Strengthen

Calculating probabilities

1 $\frac{5}{6}$

2 0.3

3 **a** 18

b i $\frac{12}{30}$ **ii** $\frac{18}{30}$

Experimental probability

1 **a** $\frac{1}{6}$

b i 60 **ii** 90

c i $\frac{12}{60}$ or $\frac{1}{5}$ **ii** $\frac{30}{90}$ or $\frac{1}{3}$

d Maddie's, as its P(2) $= \frac{1}{5}$ which is closer to $\frac{1}{6}$, than Freya's P(2) $= \frac{1}{3}$.

2 **a** The experimental probability is $\frac{7}{10}$, which is a lot higher than the theoretical probability of $\frac{1}{2}$.

b The experimental probability is $\frac{102}{200}$, which is close to the theoretical probability of $\frac{1}{2}$.

c His 2nd estimate is more accurate as more trials give a more accurate result.

Probability diagrams

1 **a**

	2	4	6
3	(2, 3)	(4, 3)	(6, 3)
6	(2, 6)	(4, 6)	(6, 6)
9	(2, 9)	(4, 9)	(6, 9)

b 9

c i $\frac{1}{9}$ **ii** $\frac{4}{9}$ **iii** $\frac{3}{9}$ or $\frac{1}{3}$

d

	2	4	6
3	5	7	9
6	8	10	12
9	11	13	15

e $\frac{6}{9}$ or $\frac{2}{3}$

2 **a, b**

		Bag 1	
		10p	50p
Bag 2	10p	10, 10	(10, 50)
	20p	20, 10	20, 50
	50p	(50, 10)	50, 50

c $\frac{2}{6}$ or $\frac{1}{3}$

3 **a** 22

c i $\frac{22}{25}$ **ii** $\frac{6}{25}$ **iii** $\frac{15}{25}$ or $\frac{3}{5}$

4 **a** ii **b** iii **c** i

5 **a** 1, 4, 5, 6, 8, 9

b 1, 2, 3, 4, 10

c 1, 2, 3, 4, 5, 6, 8, 9, 10

d 1, 4

e 1, 2, 3, 4, 5, 6, 7, 8, 9, 10

6 **a–d**

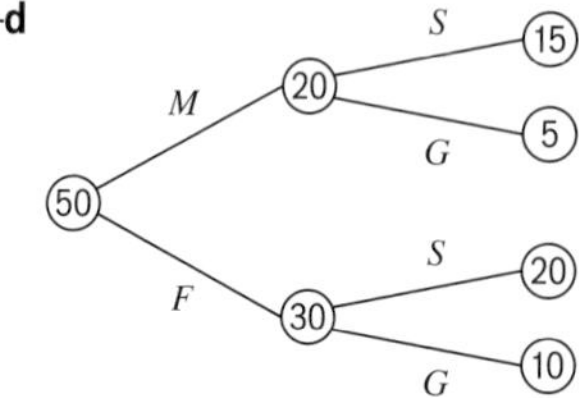

e $\frac{5}{50}$ or $\frac{1}{10}$

Dependent events

1 **a, b** There are only 9 sweets to choose from for the 2nd sweet.

a Eating a jelly sweet first means there are now $7-1=6$ left.

b Eating a fruit sweet first means there are now $3-1=2$ left.

c

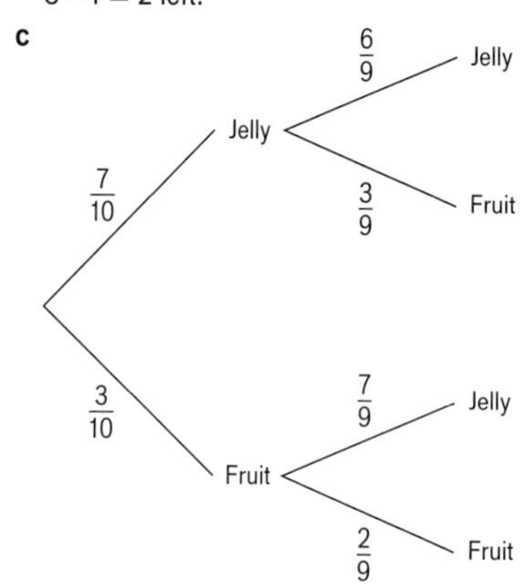

d $\frac{42}{90}$

e i (J, J) (J, F) (F, J) (F, F)

ii (J, F) and (F, J)

iii $P(J, F) = \frac{21}{90}$; $P(F, J) = \frac{21}{90}$;

iv $\frac{42}{90}$

13 Extend

1 **a** $\frac{8}{15}$ **b** 9

2 **a** 1 blue, 1.5 green, 5 red, 2.5 yellow

b P(green) $= 0.15 = \frac{3}{20}$. You could not get this with 10 counters.

c 20

3 9

	Adult ticket	Bus pass	Child	Total
Male	3	2	1	6
Female	2	4	8	14
Total	5	6	9	20

4 **a** 100 **b** $\frac{1}{100}$ **c** $\frac{1}{20}$ **d** 25

5 a

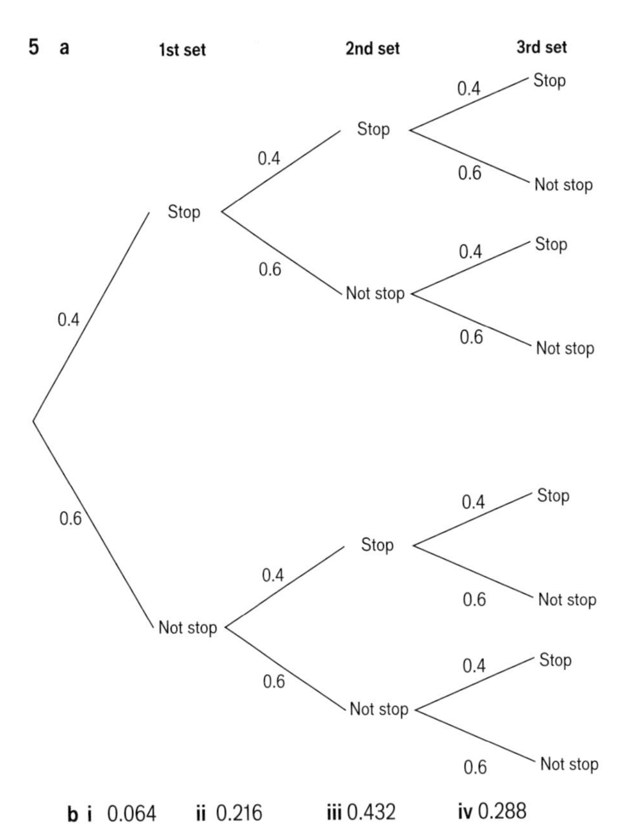

b **i** 0.064 **ii** 0.216 **iii** 0.432 **iv** 0.288

6 a 16

b E. g. Half the coins are silver, so 'half the total' must be a whole number. So the total must be even.

7 a

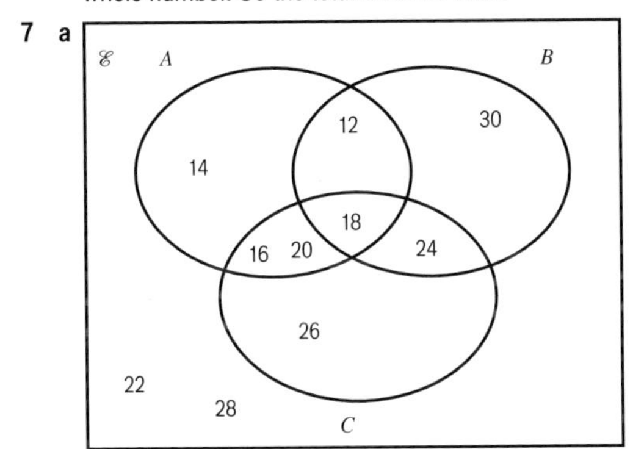

b $\frac{2}{20}$ or $\frac{1}{10}$

8 $\frac{21}{100}$

13 Test ready

Sample student answers

a Should multiply along branches, $\frac{4}{7}\times\frac{1}{6}$.

Also $\frac{4}{7}+\frac{1}{6}\neq\frac{5}{13}$, student has added numerators and denominators, $\frac{4}{7}+\frac{1}{6}=\frac{24+7}{42}=\frac{31}{42}$

b $\frac{4}{7}\times\frac{1}{6}=\frac{4}{42}=\frac{2}{21}$

13 Unit test

1 $15-2-3-6=4$ red counters

$P(R \text{ or } W)=\frac{4+2}{15}=\frac{6}{15}=\frac{2}{5}$

2 $\frac{5}{9}$

3 a

	Male	Female	Total
Long hair	2	14	16
Short hair	10	6	16
Total	12	20	32

b $\frac{10}{32}$

4 a

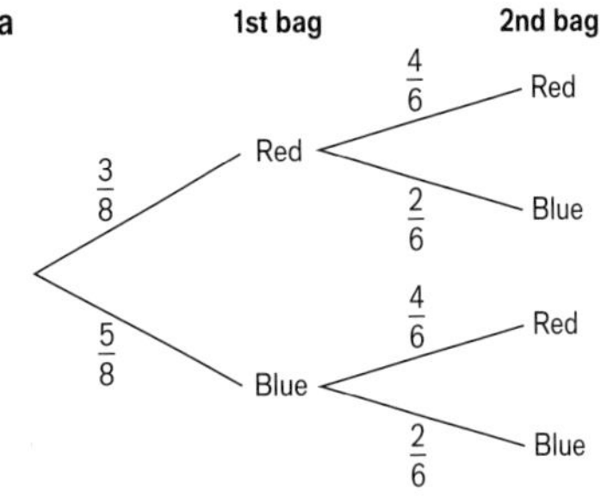

b $\frac{10}{48}$ or $\frac{5}{24}$

5 a $\frac{5}{12}$ **b** 24

c $24+12=36$

$P(P)=\frac{1}{2}$, so there are now 18 purple balls.

$18-10=8$ purple balls added.

$P(B)=\frac{1}{4}$, so there are now 9 blue balls.

$9-6=3$ blue balls added.

$12-8-3=1$ green ball added.

6 a

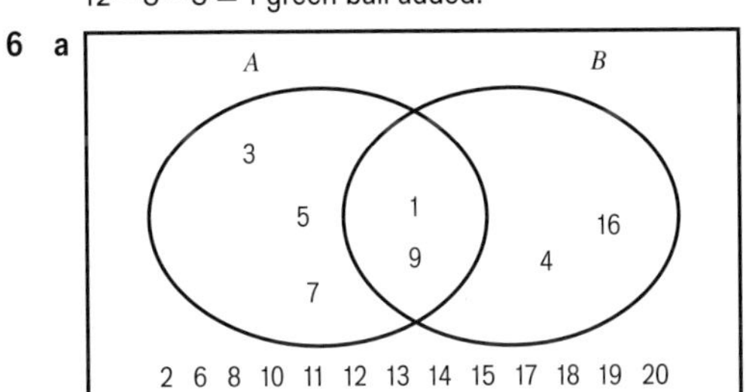

b $\frac{7}{20}$

7 a

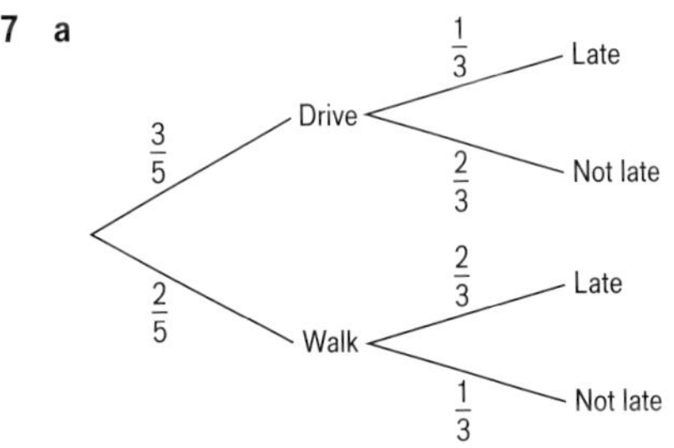

b $\frac{3}{15}$ or $\frac{1}{5}$ **c** $\frac{2}{15}+\frac{6}{15}=\frac{8}{15}$

8 a $\frac{17}{30}$ **b** $\frac{53}{90}$ **c** 2

9 Two-way table or frequency tree

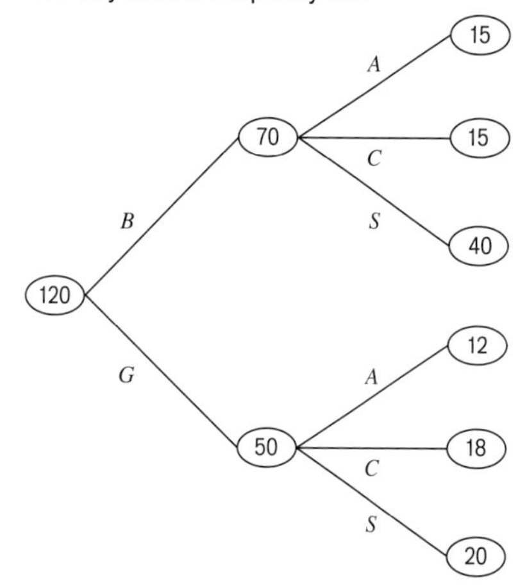

	A	C	S	Total
Boy	15	15	40	70
Girl	12	18	20	50
Total	27	33	60	120

$\frac{33}{120}$

10 Greater than 7 and less than 7 are equally likely. 15 outcomes are less than 7, 15 are greater than 7 and 6 are equal to 7.

11 Students' own answers.

UNIT 14 Multiplicative reasoning

14.1 Percentages

1 **a** 0.15 **b** 1.5 **c** 0.015

2 **a** 120% **b** 85% **c** 105% **d** 97.5%

3 **a** 1.4 **b** 0.88 **c** 1.03 **d** 0.99

4 **a** £385 **b** £336 **c** £700

5 **a** 8 **b** 80 **c** 8

6 £400

7 £9

8 £265 000

9 Matt (Matt saves £72; Paul saves £52.50)

10 **a** £40 000 **b** £7500 **c** £1000 **d** £708.33

11 Actual change = £3334.40 – £3200 = £134.40

$$\text{percentage change} = \frac{\text{actual change}}{\text{original amount}} \times 100$$

$$= \frac{134.40}{3200} \times 100 = 4.2\%$$

12 6.8%

13 3.2%

14 **a**

Item	Cost price	Selling price	Actual profit	Percentage profit
ring	£5	£8	**£3**	**60%**
bracelet	£12	£18	**£6**	**50%**
necklace	£20	£30	**£10**	**50%**
watch	£18	£25	**£7**	**38.9%**

b No

15 7.3%

16 7.46%

17 35.4%

18 **a** Students' own answers, e.g. £564 is more than double £240.
b 135%

19 120%

20 **a** 21.6% **b** Percentage decrease
c 8850 children **d** 2.6% decrease

14.2 Growth and decay

1 **a** 5^3 **b** 1.5^3 **c** 5^4 **d** 1.5^5

2 **a** **i** 1.3 **ii** 1.03 **iii** 1.035
b **i** 0.8 **ii** 0.98 0.975

3 **a** 0.7 **b** £4200 **c** 0.9
d £3780 **e** 0.63

4 **a** 1.4 **b** 1.015 **c** 1.421

5 **a** 1.05575 **b** £26 393.75

6 **a** £123 600 **b** £121 128 **c** 1.0094

7 **a** 1.11 **b** 0.93 **c** 1.0323 **d** 0.8649

8 1242

9 No; $1.2 \times 0.92 = 1.104$; this is an increase of 10.4%

10 **a** **i** 1.1×1.1 **ii** $1.1 \times 1.1 \times 1.1$ **iii** $1.1 \times 1.1 \times 1.1 \times 1.1$
b **i** 0.9×0.9 **ii** $0.9 \times 0.9 \times 0.9$
iii $0.9 \times 0.9 \times 0.9 \times 0.9$

11 **a** Students' own answers, e.g. A decrease of 10% over 3 years is $0.9 \times 0.9 \times 0.9$ or 0.9^3.
b £3790.80

12 £204

13 £183.62

14 £913.57

15 **a** Jo (Annie receives £112.94; Jo receives £171.37)
b Students' own answers, e.g. No; Jo still gets more interest.

16 16 315 bacteria to 5 s.f.

17 838.2 counts per second

18 3750 counts per minute

19 **a** 253 or 254 outlets
b Students' own answers, e.g. The nearest whole number.

14.3 Compound measures

1 **a** £10 **b** 20 km per litre

2 **a** 4 **b** 36 **c** 0.25

3 $24\,m^3$

4 **a** 1000 **b** 1 000 000

5 **a** $1.5 \times 16.50 = £24.75$ **b** £33 **c** £775.50

6 **a** **i** 1 litre **ii** 0.5 litres
b Students' own answers, e.g. Multiply by 5.
c 40 hours

7 **a** $36\,000\,cm^3$
b 7.2 minutes or 7 minutes 12 seconds

8 **a** 15 km/litre
b Students' own answers, e.g. The rate will depend on the driving conditions.

9 $\frac{10.8}{1.5} = 7.2\,g/cm^3$

10 $8\,g/cm^3$

11 $2.4\,g/cm^3$

12 **a** Gold $19.32\,g/cm^3$, platinum $21.45\,g/cm^3$
b Platinum is more dense as it has a higher density.

13 $0.6\,g/cm^3$

14 **a** $10.5 = \frac{M}{0.729}$ **b** 7.6545 g

15 **a** 5115.5 g **b** 5.1155 kg

16 a i 58.24 g ii 24.955 g
b 8.3195 g/cm^3

17 a $7.87 = \frac{5400}{V}$ b 686 cm^3

18 745 cm^3

19 17.3 N/m^2

20 90 N

21 20 N/m^2

22 Yes; new pressure 24 000 N/m^2 is two-thirds of original pressure 36 000 N/m^2

14.4 Distance, speed and time

1 a 15 km/h b 15 000 m

2 a $d = 3$ b $s = 5$ c $t = 3$

3 a 0.5 hours b 0.25 hours c 1.25 hours
d 0.1 hours

4 a 12 minutes b 1 hour 45 minutes
c 3 hours 24 minutes

5 $\frac{426}{3} = 142$ km/h

6 a 2.25 hours b 48 mph

7 a 12 km/h b 95.2 mph

8 a 0.75 hours b $46 = \frac{D}{0.75}$ c 34.5 miles

9 a 2000 miles b 17.5 km

10 a $10 = \frac{12}{T}$ b 1.2 hours
c 1 hour 12 minutes

11 a 40 seconds b 45.5 seconds

12 a 51.5 mph b 168 miles

13 1.4 m/s

14 200 seconds

15 a 45 km b less c 750 m d 12.5 m

16 a 3600 m/h b 43 200 m/h c 28 800 m/h

17 a 18 km/h b 64.8 km/h c 108 km/h

18 a 15 m/s b 20 m/s c 2.5 m/s

19 The peregrine falcon is faster. 108 m/s = 388.8 km/h, which is faster than 350 km/h.

20 a 10 miles b 6 seconds

21 a 100 km b 2.25 hours c 44.4 km/h

22 18.3 km/h

23 a i 11 ii 9
b i 2 ii 1
c i 3 ii 0.76

24

s (m)	u (m/s)	a (m/s^2)	t (s)
10	**9**	2	1
8	2	**12**	1
15	**4.5**	3	2

25 a 10 b 5

26 44.7 m/s (to 3 s.f.)

14.5 Direct and inverse proportion

1 C

2 a 1.25 b 0.8 c 0.5

3 a 1 : 4 or 0.25 : 1 b 3 : 1 or 1 : $\frac{1}{3}$
c 5.5 : 1 or 1 : $\frac{2}{11}$ d $\frac{3}{5}$: 1 or 1 : $\frac{5}{3}$

4 a

Kilograms	0	5	10
Pounds	**0**	**11**	**22**

b
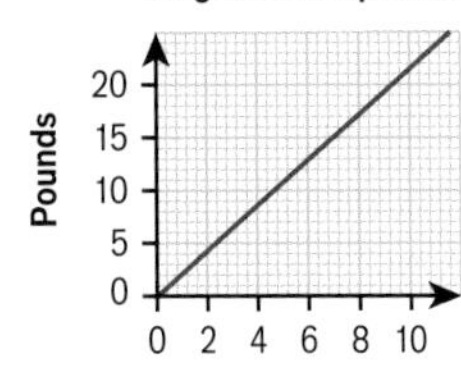

c Yes; straight-line graph through the origin
d $y = 2.2x$

5 a 4 : 1 10 : 2.5 16 : 4 20 : 5
b They are all 4 : 1. c $F = 4P$
d Students' own answers, e.g. The ratio tells you the coefficient.

6 a $x = 3y$ b $t = 6s$ c $q = 3.5p$
d $F = 2.5m$ e $b = 1.25a$ f $s = 1.75r$

7 a 8 : 10 16 : 20 24 : 30 32 : 40
40 : 50 all are equivalent to 1 : 1.25 or 0.8 : 1
b d is in direct proportion to t as the ratio $d : t$ is the same for all values.
c $\frac{d}{t} = 0.8$ or $d = 0.8t$ or $t = 1.25d$
d 20 miles

8 a i $F \propto m$ ii $V \propto x$
b i $F = km$ ii $V = kx$

9 a $x \propto t$; $x = kt$ b $k = 1.5$ c 15

10 a $r = 1.25s$ b 8.75

11 a $P = 1.25E$ b £100

12 £3795

13

X	10	20	15	**2.5**	**12.5**
Y	15	**7.5**	**10**	60	12

14 a Direct b Inverse c Neither
d Inverse e Direct f Direct

15 a i $y \propto \frac{1}{x}$ ii $P \propto \frac{1}{A}$
b i $y = \frac{k}{x}$ ii $P = \frac{k}{A}$

16 a $y \propto \frac{1}{x}, y = \frac{k}{x}$
b $k = 60$ c 4

17 $s = \frac{k}{t}$; $20 = \frac{k}{0.4}$, so $k = 20 \times 0.4 = 8$;
when $t = 0.5$,
$s = 8 \div 0.5 = 16$.

18 15 amperes

14 Check up

1 82.5%

2 16.7%

3 £2678.06

4 5643 bees

5 £405.45

6 2.8 N/cm^2

7 50 kg

8 a 46.7 km/h b 3 hours 7.5 minutes

9 6.25 m/s^2

10 Usain Bolt is faster: 12.4 m/s = 44.64 km/h, which is faster than 40 km/h.

11 59 mph

12 a

Pounds (£)	Euros (€)
150	189
400	504
320	**403.20**
280	352.80

b Yes; the ratio of $\frac{\text{pounds}}{\text{euros}}$ is constant so the number of pounds and the number of euros increase at the same rate.

c $e = 1.26p$

13 a $k = 4$ b $f = 6$

14 a $y = \frac{45}{x}$ b $y = 3$

16 Students' own answers.

14 Strengthen

Percentages

1 a 25% b 30% c 85.7%

2 a 25% b 20% c 30%

3 a £30 b 500%

4 £50

5 £300 000

6 a £60 b £88 c £160

7 a 400 g b 300 km c 400 litres

8

Year	Amount at start of year	Interest at end of year	Amount at start of year + interest = total at end of year
1	£700	700 × 0.038 = **26.6**	£726.60
2	£726.60	**726.60** × 0.038 = **27.61**	£754.21

9 a £815.89 b £65.89

10 a Increase = 1.15, decrease = 0.85

b Increase = 1.08, decrease = 0.92

c Increase = 1.026, decrease = 0.974

d Increase = 1.21, decrease = 0.79

e Increase = 1.07, decrease = 0.93

f Increase = 1.045, decrease = 0.955

g Increase = 1.35, decrease = 0.65

h Increase = 1.112, decrease = 0.888

11 a 1.2 b 44 (answer must be a whole number)

12 a 0.958 b 12.91 million square metres

Compound measures

1 a £13.20 b 6.60 × 1.25 = 8.25

c 6.60 × 1.5 = 9.90 d £303.60

2 a $V = \frac{M}{D}$ b $M = D \times V$

3

Metal	Mass (g)	Volume (cm^3)	Density (g/cm^3)
aluminium	**27**	10	2.70
copper	448	**50**	8.96
zinc	427.8	60	**7.13**

4

Force (N)	Area (cm^2)	Pressure (N/cm^2)
60	15	**4**
220	20	11
45	**5**	9

Distance, speed and time

1 a $T = \frac{D}{S}$ b $D = S \times T$

2

Distance (miles)	Time (hours)	Speed (mph)
180	4	45
145	2.5	58
150	3	**50**
120	1.25	**96**
45	**1.5**	30
154	**2.75**	56

3 a 72 miles b 1.5 hours c 48 mph

4 a Higher b Lower c Lower

5

km/h	m/h	m/min	m/s
9	**9000**	**150**	**2.5**
18	**18 000**	**300**	5
12	**12 000**	**200**	**3.3**
28.8	**28 800**	**480**	8

6 a $v = u + at$ b $v^2 = u^2 + 2as$

c $s = ut + \frac{1}{2}at^2$

7 a 2 b 4

c 25

Direct and inverse proportion

1 $W = 14$, $X = 24$, $Y = 28$, $Z = 20$

2 $W = 2$, $X = 6$, $Y = 16$, $Z = 1.6$

3 y is proportional to x; $y = kx$; $y \propto x$

y is inversely proportional to x; $y = \frac{k}{x}$; $y \propto \frac{1}{x}$

4 a $y \propto x$ b $y = kx$ c $k = 3$

d $y = 3x$ e 12

5 a $y \propto \frac{1}{x}$ b $y = \frac{k}{x}$ c 100

d $y = \frac{100}{x}$ e 25

14 Extend

1 2% then 1.5% is the better offer: 1.02 × 1.015 = 1.0353 (which is greater than 1.025); this is a 3.53% increase after two years.

2 12.2 m/s

3 2.7 g/cm^3

4 a 12 minutes b 3.1 minutes faster

5 a £281.60 b £250

6 a Rounding speed to 1 s.f. gives 20 m/s = 72 km/h. Therefore 36 km takes 0.5 hours or 30 minutes.

b An overestimate; students' own answers, e.g The answer was rounded up.

7 a 28.8 mm b 750 g

8 a 6 b 8

9 a $V = \frac{3000}{P}$

b 10 m^3 c 400 N/m^3

10 a 25 N b 2.316 kg

11 Rapid Bank: amount after 3 years = £2000 × 1.014 × 1.01 × 1.01 = £2068.76

Eco Bank: amount after 2 years = £2000 × 1.015 × 1.005 × 1.005 = £2050.35

Jed should invest in The Rapid Bank.

12 Students' own answers, e.g. The second part was shorter than the first.

14 Test ready

Sample student answers

1 **a** Student B has the correct answer.
Student A has worked out 12% of the value of the house after the increase. This is not the same as 12% of the original value and is not the correct method to calculate the original value.
Original value $\times 1.12 = £660\,800$

b They could have found 12% of their answer and added it on to see whether this came to £660 800.

2 **a i** Students' own answers, e.g. The student should have converted 1.75 minutes into 105 seconds.

ii Students' own answers, e.g. Completing an 800 m race in 0.67 seconds would be a very high speed.

b Students' own answers, e.g. No; he would have a lower speed in an 800 m than in a 2100 m race, so a 800 m race would take longer than was calculated in part **a**.

c Students' own answers, e.g. The time to complete the 800 m race should be increased.

14 Unit test

1 $18 - 15.4 = 2.6$, $\frac{2.6}{15.4} \times 100 = 16.9\%$ (1 d.p.)

2 $1785 \div 105 \times 100 = £1700$

3 $20 \times 20 = 400\text{ cm}^2$, $400 \div 10\,000 = 0.04\text{ m}^2$
Pressure $= \frac{6}{0.04} = 150\text{ N/m}^2$

4 **a** $18 \div 2.5 = 7.2$ km/h
b 7.2 km/h = 7200 m/h = 120 m/min = 2 m/s
Suvi is correct

5 $44.5 \div 15.5 = 2.87$ hours (2 d.p.)
$0.5 \div 25 = 0.02$ hours
Total distance $= 44.5 + 0.5 = 45$ km
Total time $= 2.87 + 0.02 = 2.89$ hours
Average speed $= 45 \div 2.89 = 15.6$ km/h (1 d.p.)

6 **a** $274\,500 - 250\,000 = 24\,500$
$\frac{24\,500}{250\,000} \times 100 = 9.8\%$ (1 d.p.)
b $274\,500 \times 1.018 \times 1.018 = £284\,470.94$

7 **a** $600 \times 1.017 \times 1.017 \times 1.017 = £631.12$ **b** £4.97

8 48 seconds

9 $I \times R =$ constant $9 \times 14 = 126$ $126 \div 12 = 10.5$ amperes

10 **a** W × F = constant
$1000 \times 300 = 300\,000$
$300\,000 \div 600 = 500$ m
b $300\,000 \div 842 = 356.3$ kHz (1 d.p.)

11 Volume of lemon oil $= 1.7 \div 0.85 = 2\text{ cm}^3$
Volume of honey $= 7 \div 1.4 = 5\text{ cm}^3$
Total mass $= 1.7 + 7 = 8.7$ g
Density $= 1.24\text{ g/cm}^3$

12 **a**

Year	Amount at start of year	Multiplier in index form	Total amount at end of year
1	£650	1.034	**£672.10**
2	**£672.10**	1.034^2	**£694.95**
3	**£694.95**	1.034^3	**£718.58**
4	**£718.58**	$\mathbf{1.034^4}$	**£743.01**
5	**£743.01**	$\mathbf{1.034^5}$	**£768.27**

b 1.034^{10} **c** 1.034^n
d Total amount at the end of year $n = P \times \left(1 + \frac{r}{100}\right)^n$

13 Students' own answers.

Mixed exercise 4

1 175 cm → **11.4**

2 Yes, as at 30 mph, 10 miles can be covered in 20 minutes. He leaves the supermarket at 5.35 pm so could be home by 5.55 pm. → **14.4**

3 There are 15 students who scored over 65 marks, so the probability is $\frac{15}{20} = \frac{3}{4}$ → **13.1**

4 Spinner labelled with 2 As, 4 Bs, 5 Cs and 5 Ds → **13.1**

5 Top row of table: £21, £28, £38.50
Second row of table: £10.50, £31.50, £42, £57.75 → **11.6**

6 1, 5, 5, 6 → **13.2**

7 0.2, 0.12 → **11.2, 13.1**

8 Account A interest = £1200
Account B interest = £1208 (to nearest pound)
So Sian should invest her money in account B. → **14.2**

9 No, as 18 minutes = 0.3 hours
Speed $= 20 \div 0.3 = 66.666...$ mph
which is lower than the 70 mph speed limit.
Or: time to drive 20 miles at 70 mph is
$20 \div 70 = \frac{2}{7}$ hour = 17.1 minutes (3 s.f.). → **14.4**

10 No, as $60 \div 24 = 2.5$, $275\text{ g} \times 2.5 = 687.5$ g so not enough porridge oats. → **11.6**

11 $a = \sin^{-1}\left(\frac{12.5}{15}\right) = 56.4°$ → **12.4**

12 No, using Pythagoras' theorem, the missing side is 16.3 cm (to 3 s.f.). Perimeter = 49.3 cm → **12.1**

13 **a** → **13.4**

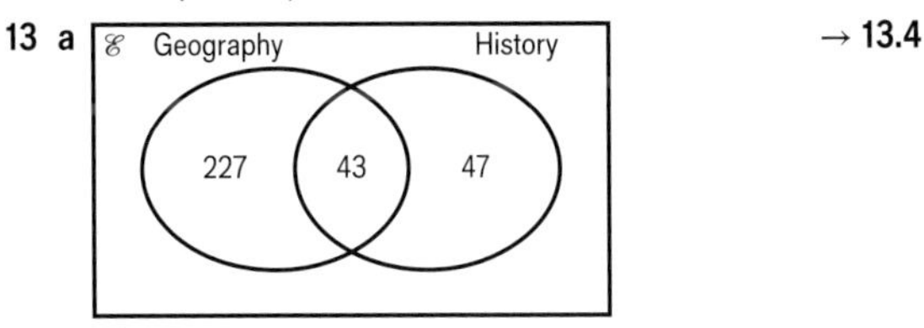

b $\frac{43}{317}$ → **13.4**

14 33 cards. Initially Angus has 75 and Khaled 105. After the exchange Angus has 108 and Khaled 72. So 33 cards have been given to Angus. → **11.4**

15 2 times or double → **14.3**

16 £54.60 → **14.1**

17 3 : 5 → **11.1**

18 **a** Because $6.5^2 + 7.2^2 = 9.7^2$
b Because $\frac{3.5}{7} = \frac{1}{2}$ and $\cos 60 = \frac{1}{2}$ → **12.2, 12.5, 12.6**

19 $x = 19.7°$, $y = 9.86°$ → **11.4, 12.5**

20 a → 13.6

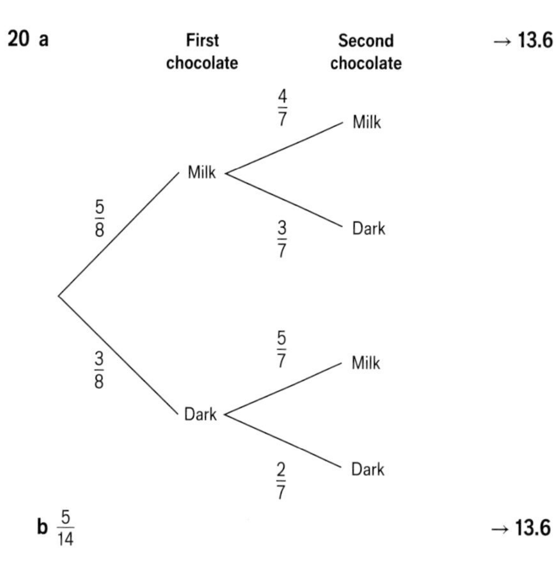

b $\frac{5}{14}$ → 13.6

UNIT 15 Constructions, loci and bearings

15.1 3D solids

1 **a** Regular pentagon **b** Regular hexagon
c Regular decagon

2 **a**

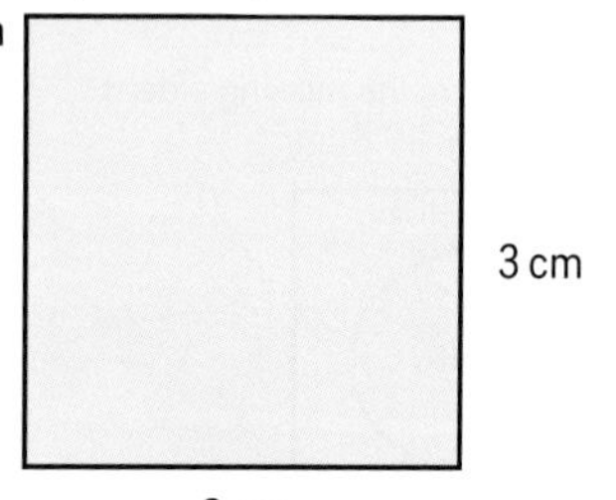

b

3 **a** 6 **b** 12

4 **a** Cuboid (rectangular prism) **b** 8

5 **a** Rectangle
b 10 cm × 4 cm, 10 cm × 3 cm, 3 cm × 4 cm

6 **a** Cube **b** Cuboid **c** Cylinder
d Sphere **e** Cone

7 **a** Equilateral triangle
b Right-angled triangles
c Rectangles
d

	Number of faces	Number of edges	Number of vertices
Cube	6	12	8
Tetrahedron	4	6	4
Square-based pyramid	5	8	5
Triangular prism	5	9	6
Hexagonal prism	8	18	12

e $F+V-E=2$

8 No, she is not correct. There are 14 faces and 36 edges.

9

10

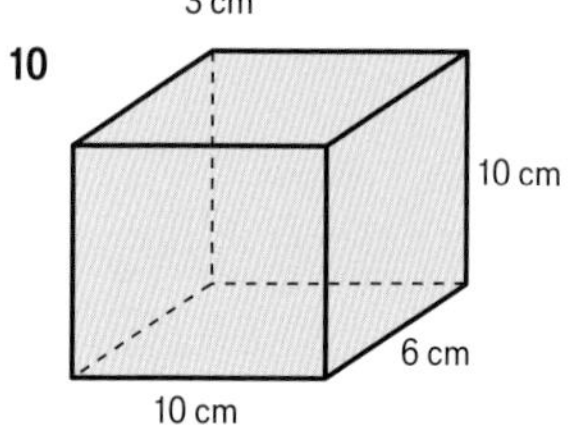

11 **a** Rectangles **b** 6

15.2 Plans and elevations

1 **a** Accurate drawing of a 7.5 cm line
b Accurate drawing of a 47 mm line
c Accurate drawing of a 6.9 cm line
d Accurate drawing of a 24 mm line

2 2, 0, 1, 6, 0

3 **a**

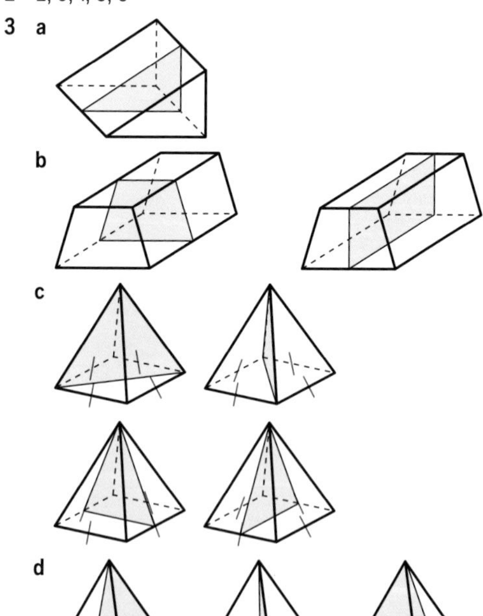

4 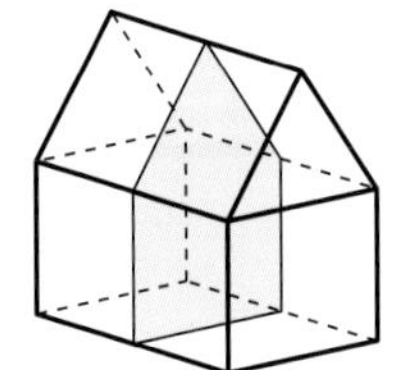 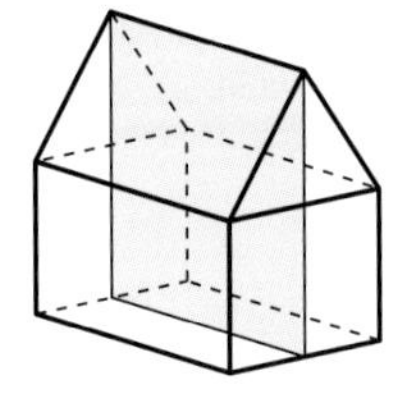

5 Infinite ways

6 a b c

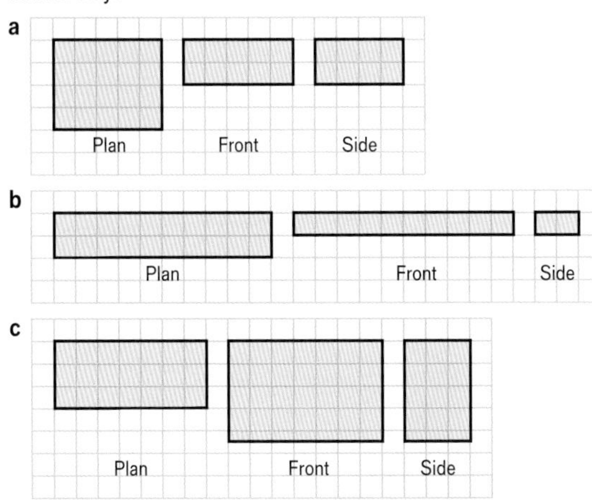

7 a i Cube or cuboid
ii Cylinder or sphere
iii Triangular prism
iv Cone

b Yes, **i** could be a cube or a cuboid and **ii** could be a cylinder or a sphere.

8

Plan

Front/side

9 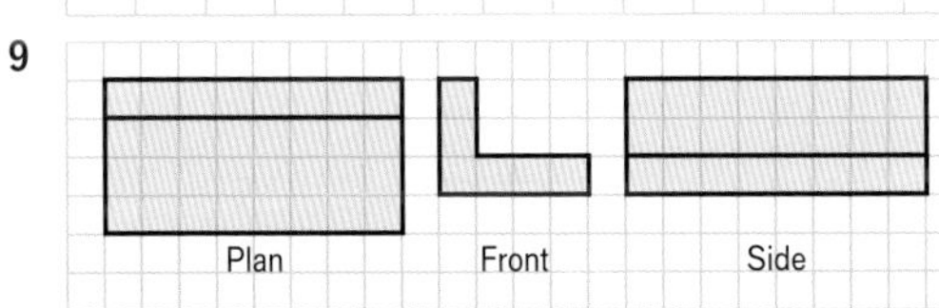

10 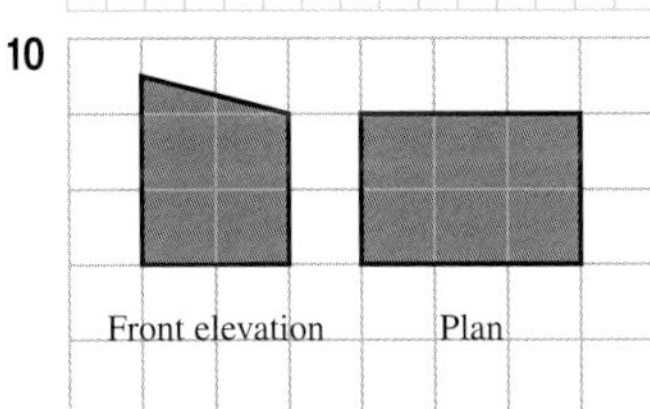

11 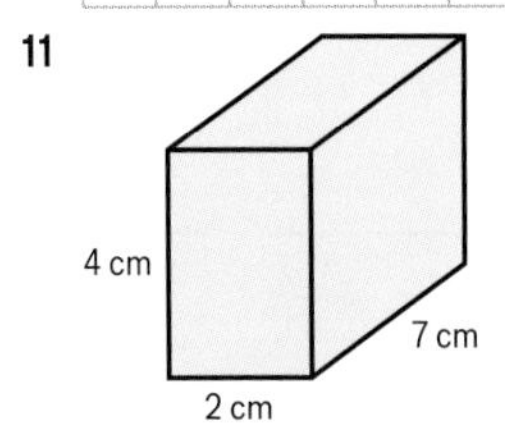

12 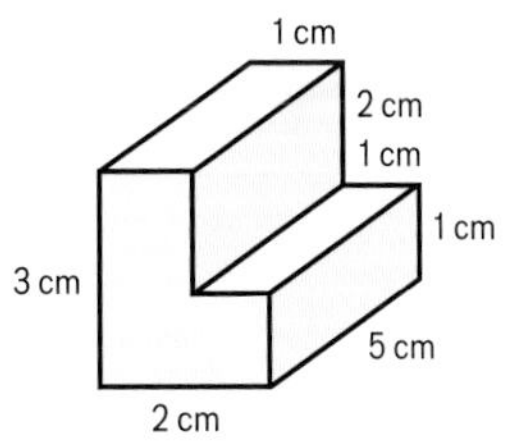

13 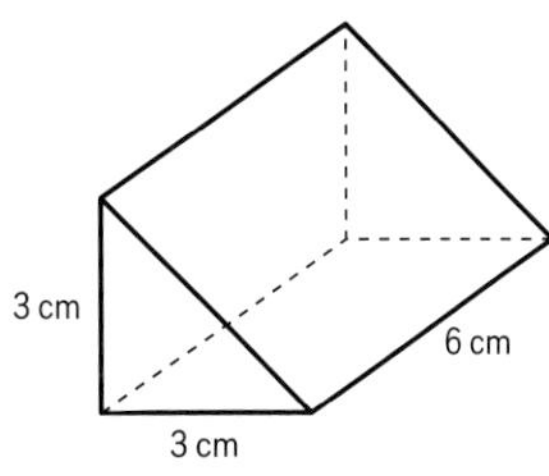

15.3 Accurate drawings 1

1 Shapes A, C and E

2 a Accurate drawing of a circle with radius 6 cm

b Accurate drawing of half a circle with radius 4.5 cm

3 Accurate drawing of the triangle

4 Accurate drawing of triangle PQR, with $PQ = 6.8$ cm, angle $RPQ = 75°$, angle $PQR = 48°$

5 Accurate drawing of triangle XYZ, with $XY = 5.6$ cm, angle $XYZ = 63°$, $YZ = 7$ cm

6 a Accurate drawing of triangle labelled ABC, with $BC = 6.3$ cm, angle $C = 42°$, $AC = 7$ cm

b 61°

7 a ASA triangles A and D, SAS triangles B, C and E

b Accurate drawings of all triangles

c A and D, C and E

d i Always true
ii Always true

8 a Accurate drawing of triangle ABC, with angles 93°, 39°, 49°

b Accurate drawing of triangle TUV, with angles 54°, 44°, 81°

9 a Accurate drawing of an equilateral triangle with side length 5 cm

b 60°

10 Accurate drawing of a triangle with sides 6 cm, 3 cm and 8 cm

11 a Accurate drawing of triangle VWX, with side lengths 5 cm, 8.5 cm and 6.9 cm

b $VW = 6.9$ cm

12 a Accurate drawings of the 5 triangles

b i Always true
ii Always true

13 a **i**,**ii**, RHS
b **i**, **iii**, ASA

15.4 Scale drawings and maps

1 a 50 m
b 20 000 cm
c 125 cm
d 7 km

2 a 60 m
b 4 cm
c 20 cm

3 a 2 m, 10 m
b 0.5 m, 3 m
c 6 m, 9 m
d 24 m, 48 m

4

8.5 m
3 m
Kitchen
Lounge
2 m
3 m
3.5 m

5 **a** **i** 20 m **ii** 2.5 m **iii** 17.5 m **iv** 41 m
b **i** 3 cm **ii** 1.5 cm **iii** 0.8 cm **iv** 1.8 cm

6 **a** 2 cm, 6 cm **b** 3 times **c** 15 m

7 Bus approximately 2 cm long, house approximately 5 cm tall; house = 2.5 × bus = 25 m (accept answers in range 24–27 m)

8 **a** 0.9–1.1 m **b** 2–2.3 m

9 128 km

10 **a**

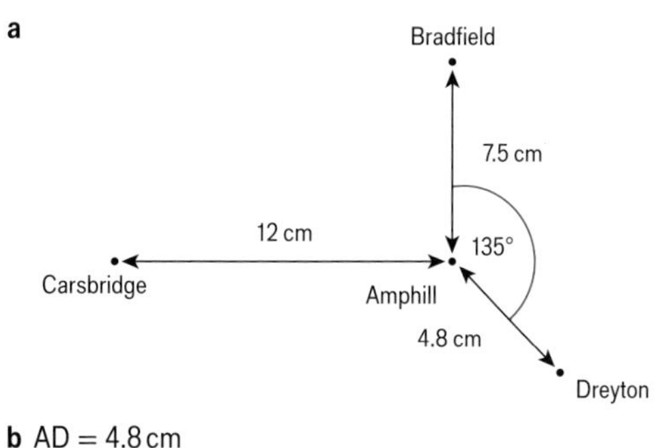

b AD = 4.8 cm
c **i** BD = 11.4 cm = 22.8 km
ii DC = 15.8 cm = 31.6 km

11 **a** 6 m **b** 16 cm

12 28 m

13 **a** 10 cm × 35 cm **b** 83 cm (= 830 mm)

14 **a** 0.5 m, 50 cm **b** 1 : 50
c **i** 4.5 m × 3.5 m **ii** 0.75 m × 1.5 m
iii 15.75 m^2

15 **a** 5 cm : 1.25 m **b** 1 : 25

16 **a** **i** 136 km **ii** 248 km **iii** 192 km
b Students' own answers, e.g. The plane is unlikely to take off from the centre of each town or city.

17 **a** 2500 m **b** 3.5 km **c** 25 cm

18 **a** 48 cm **b** 8 km

15.5 Accurate drawings 2

1 **a** **i** 6 cm **ii** 10 cm
b **i** 37° **ii** 90°

2 **a** Cuboid **b** Triangular prism

3 Any two (or equivalent) of:

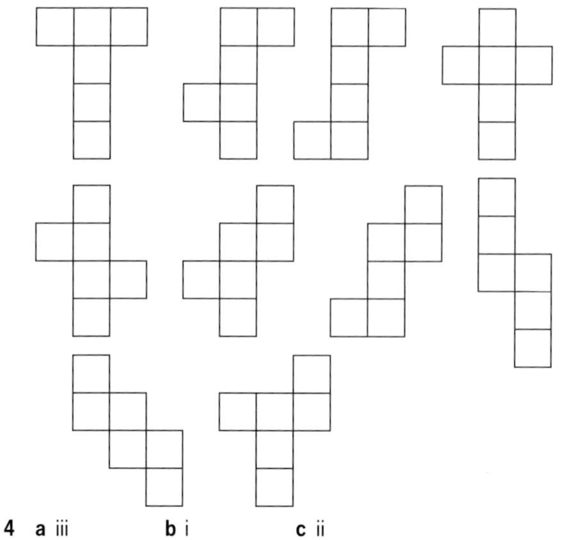

4 **a** iii **b** i **c** ii

5 **a**

6 cm
8 cm
6 cm
8 cm

b Accurate square drawn with sides 6 cm
c Accurate triangles drawn with sides 8 cm, 8 cm and 6 cm

6 Accurate scale diagram drawn with scale 1 : 5

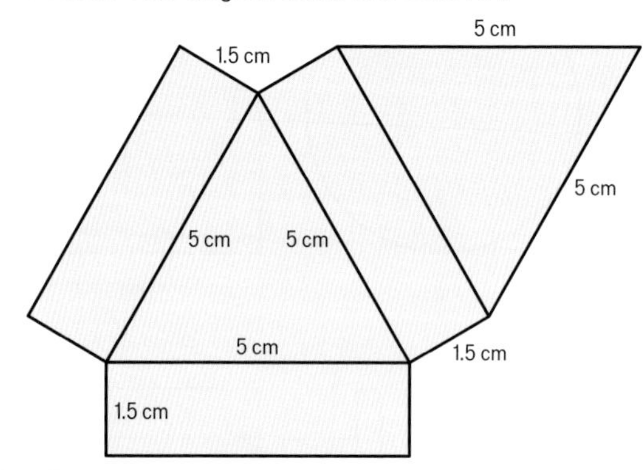

7 **a, b**

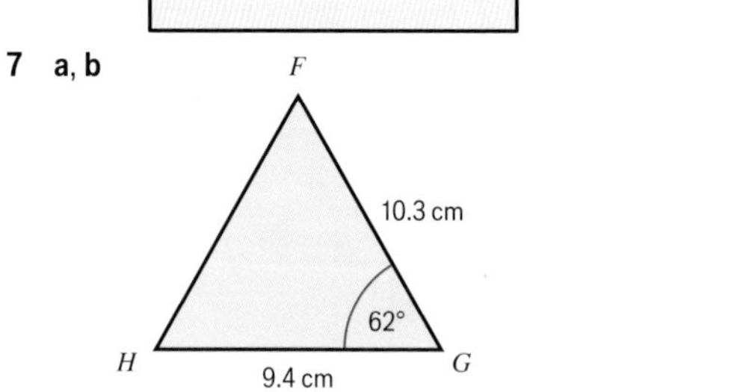

c 10.1 cm or 10.2 cm (accurately drawn, 10.18 cm to 2 d.p.)

8 **a** Accurate drawing of quadrilateral
b 6.9 cm **c** 88°

9

10 cm
112°
112°
13 cm
20 cm
12 cm
25 cm

10

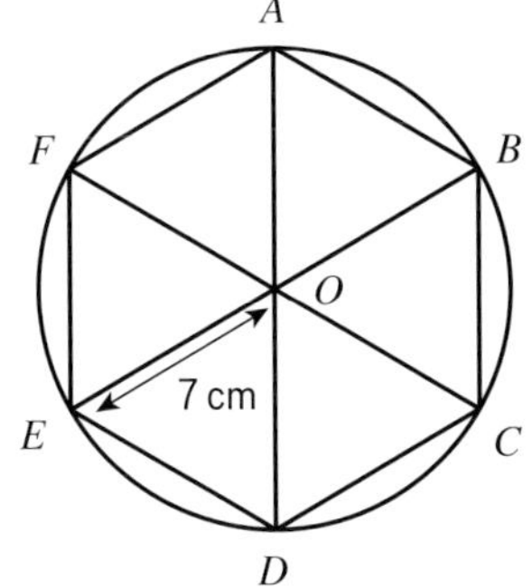

11

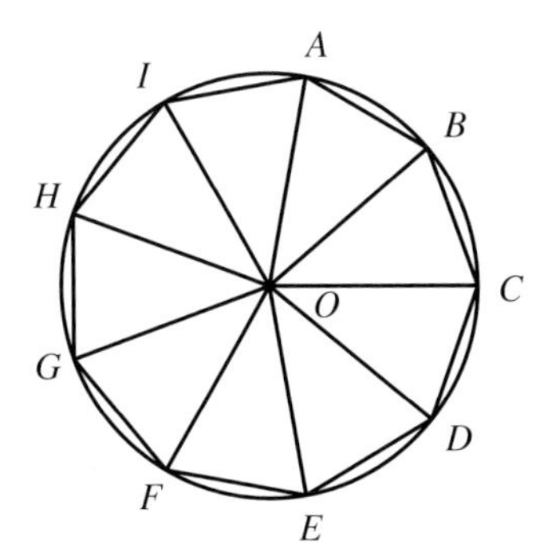

12 a

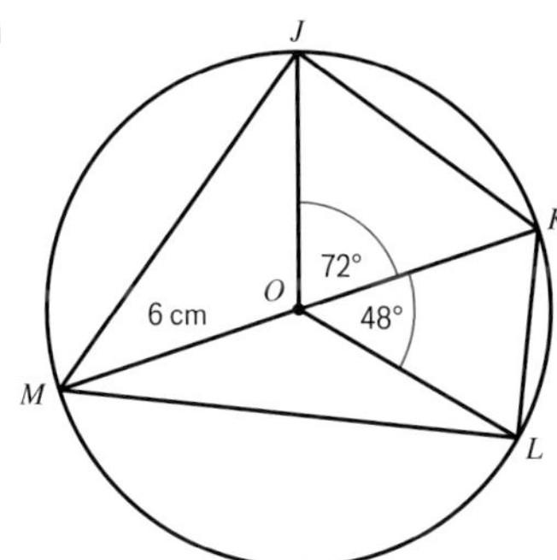

b $JOM = 108°$; $KLM = 90°$

15.6 Constructions

1 b

2

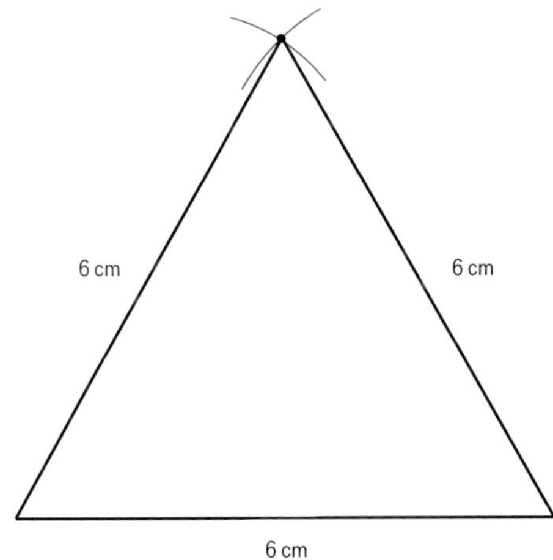

3 **a** Perpendicular bisector of 7 cm line accurately constructed

4 Perpendicular line accurately constructed through a point 4 cm from a line 9 cm long

5 Students' own answers, but should note in part **e** that the shortest distance from Q to the horizontal line AB is perpendicular to the line.

6 **a**

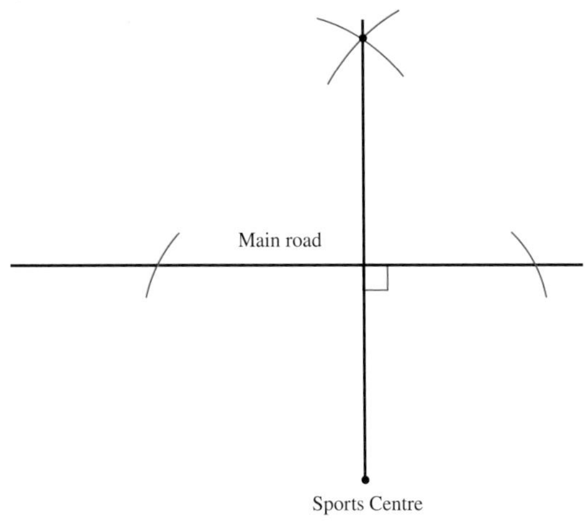

b 1.25 km

7

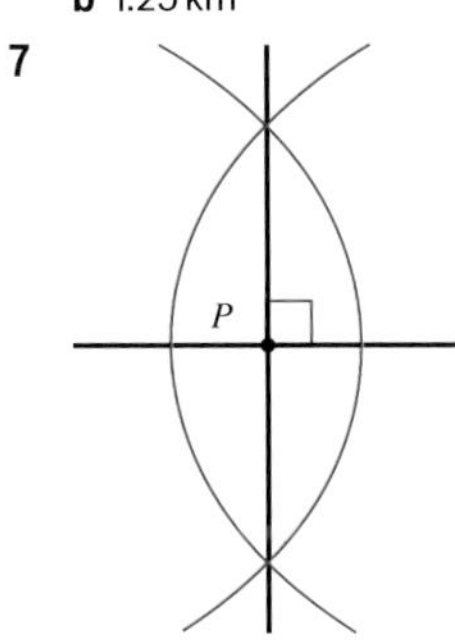

8 Accurate 70° angle drawn and bisected, showing construction arcs

9 Accurate copies of angle with angle bisected, showing construction arcs

10

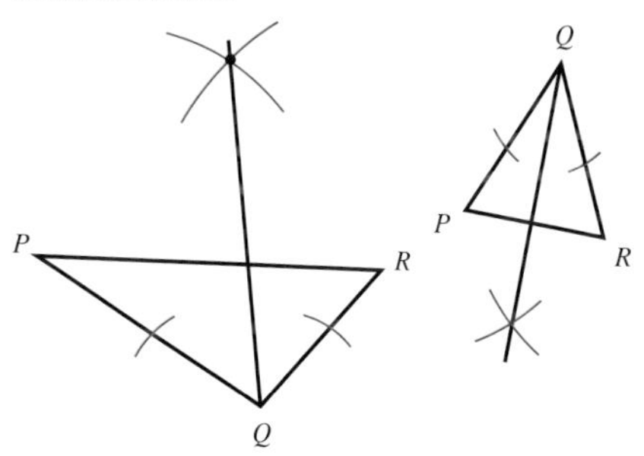

11 Accurate 85° angle drawn and bisected, showing construction arcs

12 Construct 90° by drawing a straight line and constructing the perpendicular bisector at an end point.
Construct 45° by first constructing 90° and then constructing its bisection.

13 **a, b**

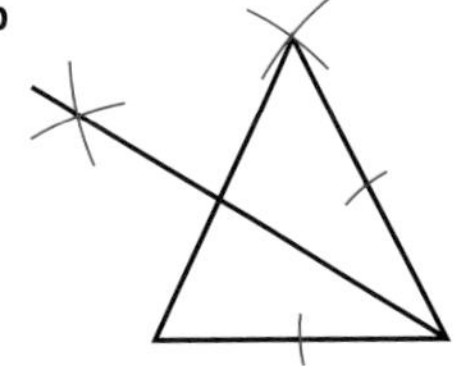

c Students' own answers, e.g. By constructing two adjoining equilateral triangles, or by using the external angle at one of the vertices of the triangle.

15.7 Loci and regions

1 **a** 20 km **b** 1.5 cm

2 Accurate line 76 mm drawn with accurate construction of its perpendicular bisector

3 **a** Circle

b A circle with centre P and radius 5 cm.

4

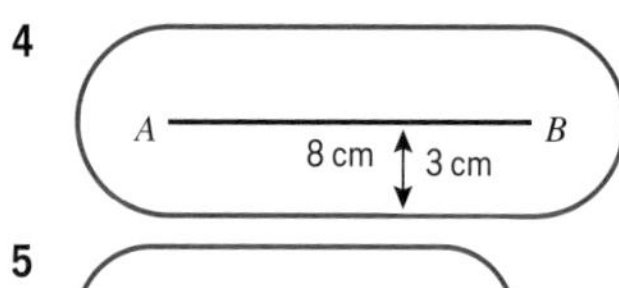

5

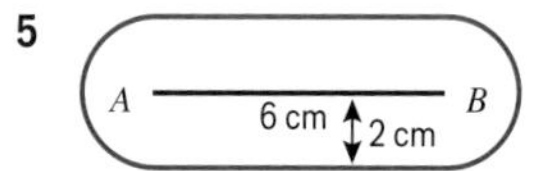

6 Students' own answers.

7

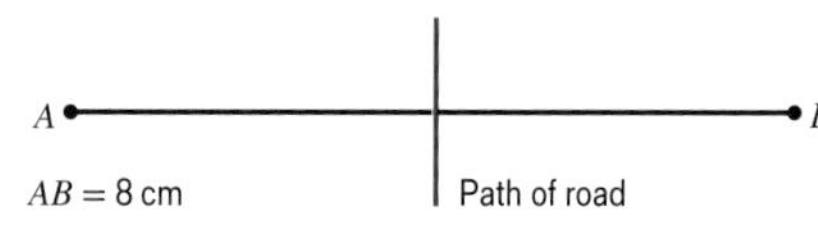

8

9

10 **a, c**

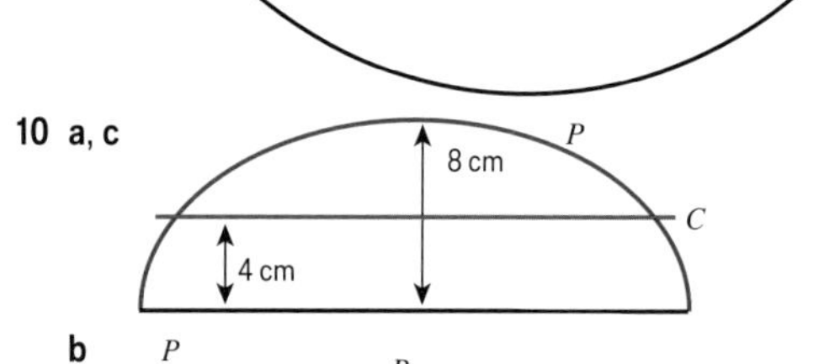

b

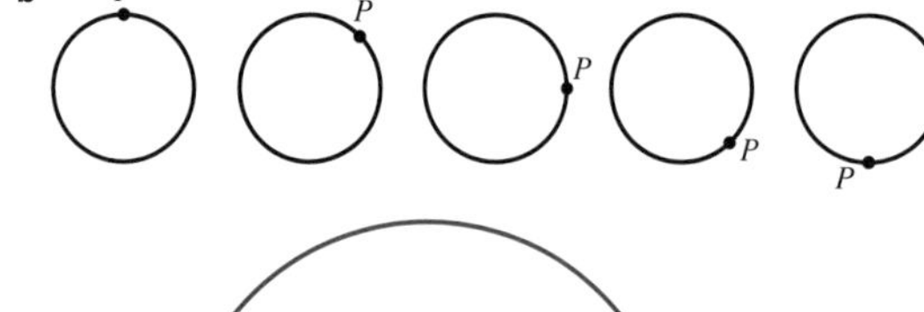

11

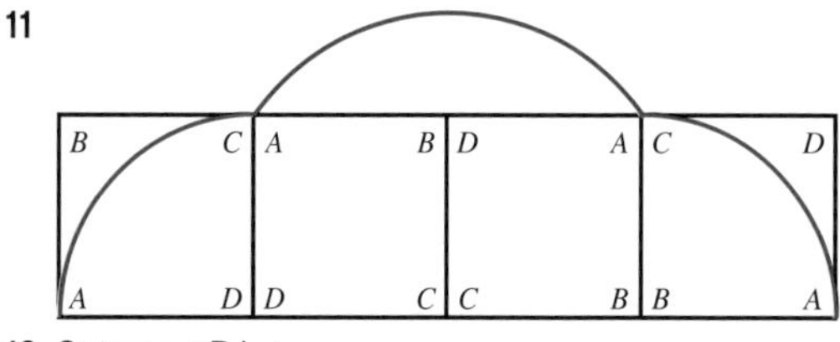

12 Statement B is true.

13

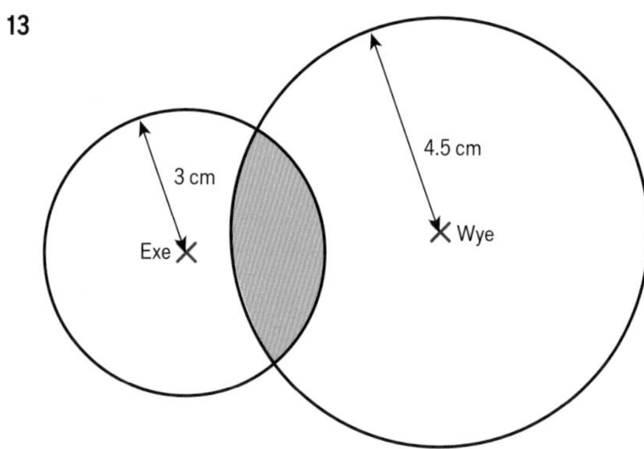

14

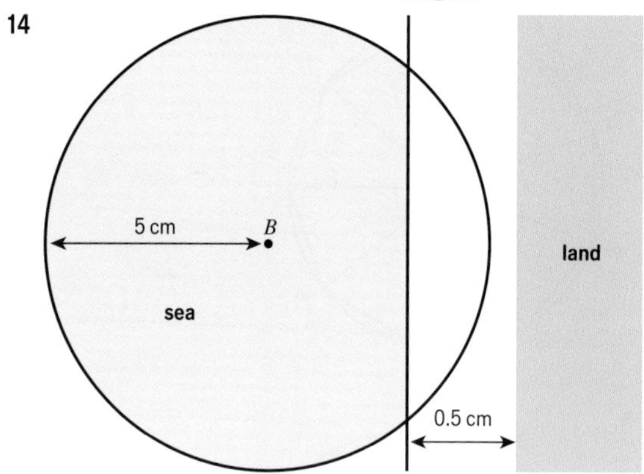

15

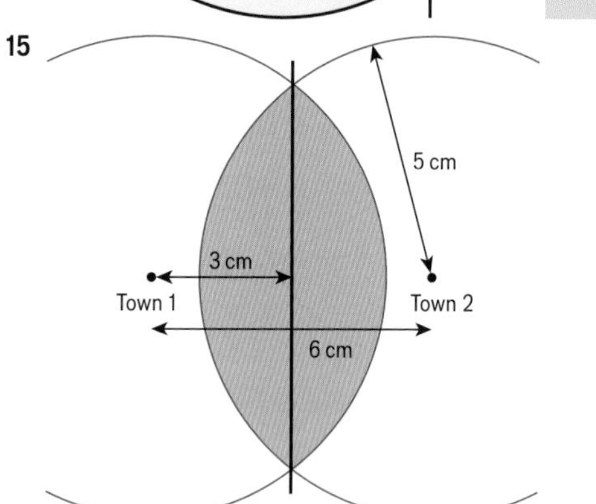

16 **a** A, B **b** QR, PQ

17

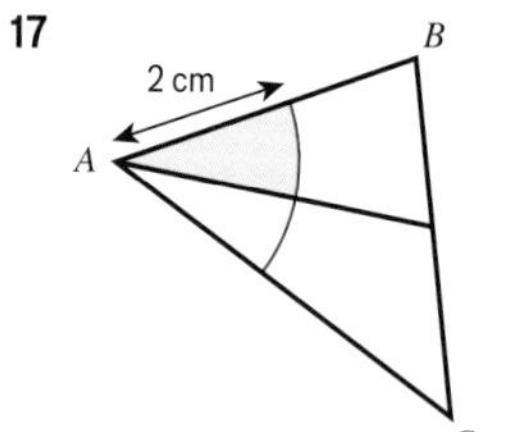

15.8 Bearings

1 **a** 100° **b** 80° **c** 110° **d** 250°

2 **a** 045° **b** 305° **c** 206° **d** 020°

3 **a** 2.25 km **b** 060°

4 **a** 166° **b** 289°

5 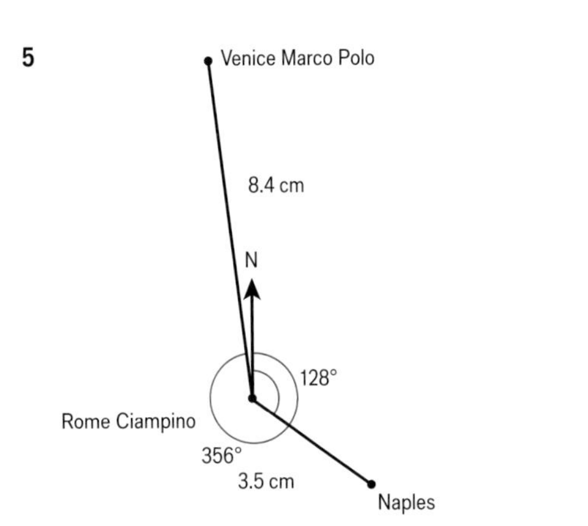

6 a Sketch showing approximate positions of P, Q and L.

b 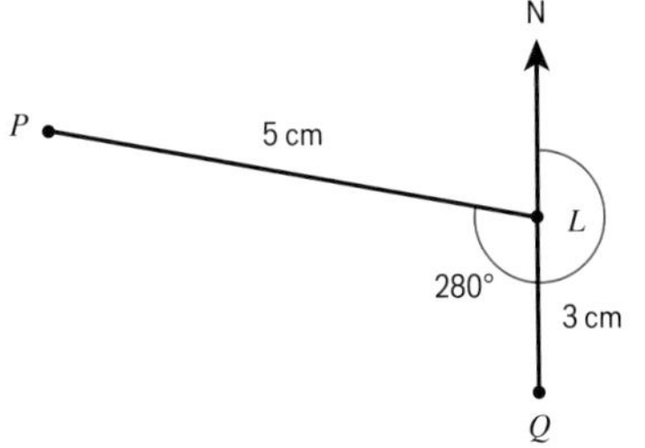

c 6.3 km

d i 128° ii 308°

e Students' own answers

7 a 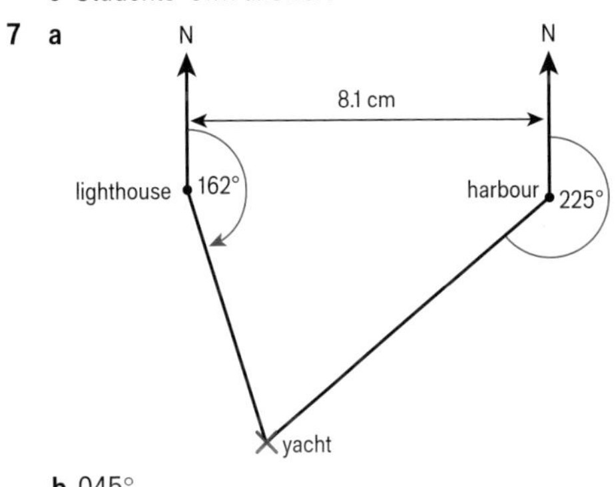

b 045°

8 a 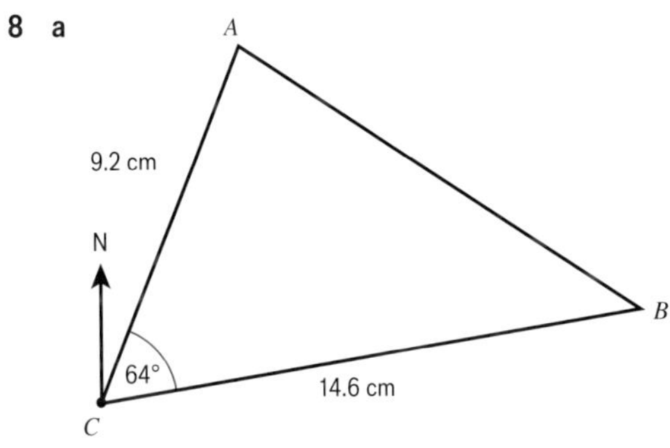

b 308° c 13.4 miles

9 a 260°

b i 290° ii 110°

10 a 240° b 310° c 200° d 040°

11 a 220° b 245°

15 Check up

1 a 6 b 9 c 5

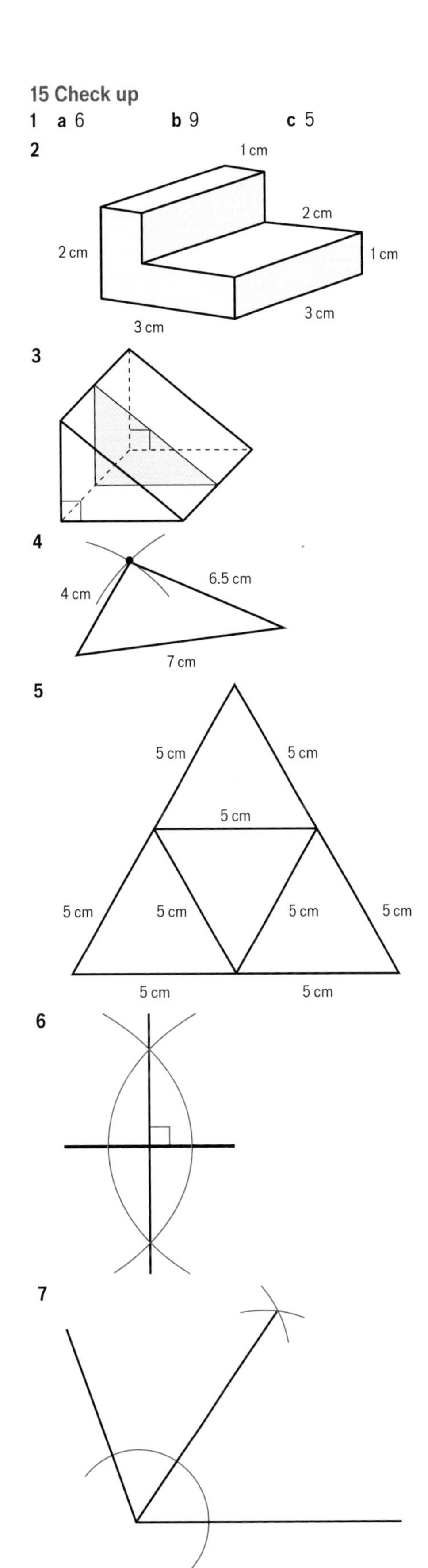

8 a, b

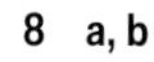

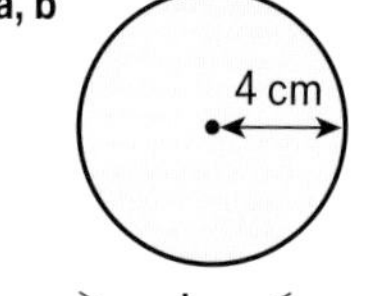

9

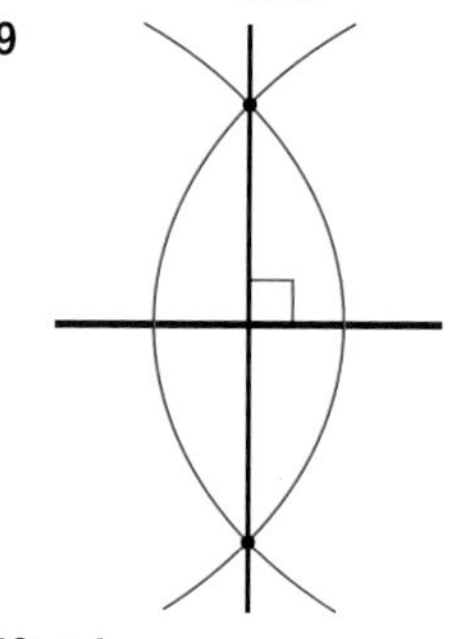

10 a, b

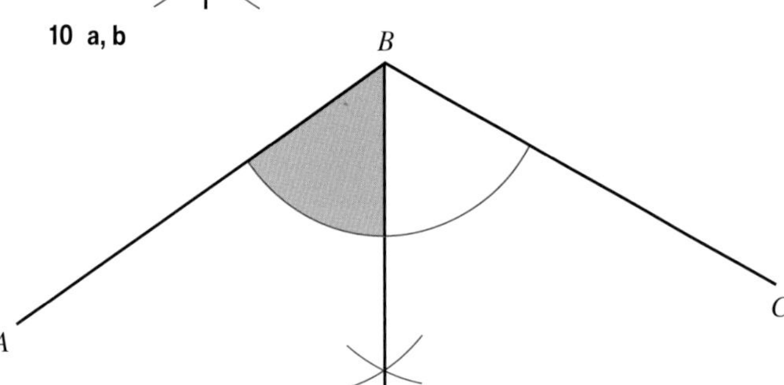

11 a 10 cm, 5 cm **b** 12 m

12 a 3 km **b** 16 cm

13 a, b

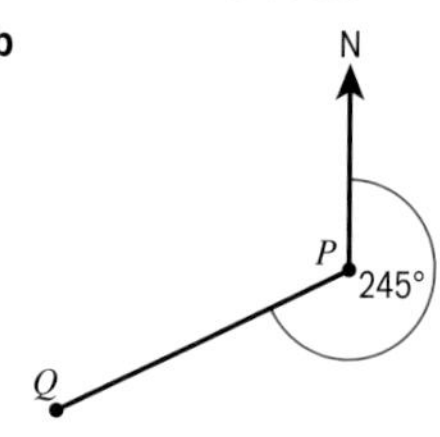

c 065°

14 a 075° **b** 255°

16 a $3 \times 4 \times 1$, $2 \times 6 \times 1$, $12 \times 1 \times 1$

b 3 **c** $2 \times 2 \times 3$

d

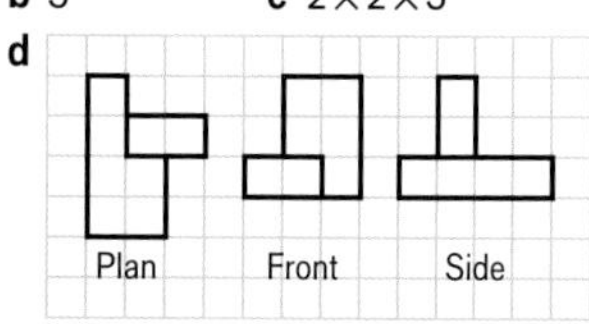

15 Strengthen

3D solids

1 A = vertex, B = face, C = edge

2 a, b, c

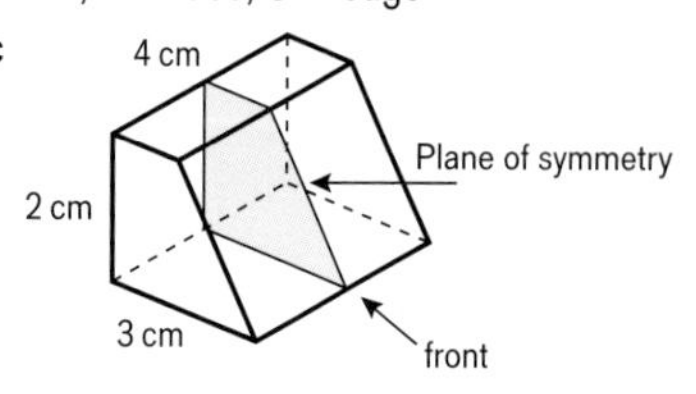

Constructions

1

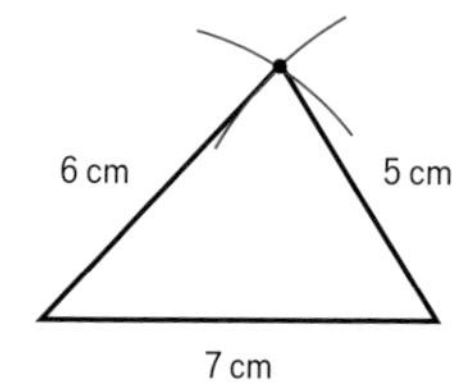

2 Accurate drawing of line 10 cm long, with its perpendicular bisector accurately constructed

3 Accurate drawing of 50° angle, with its bisector accurately constructed

Loci and regions

1 a

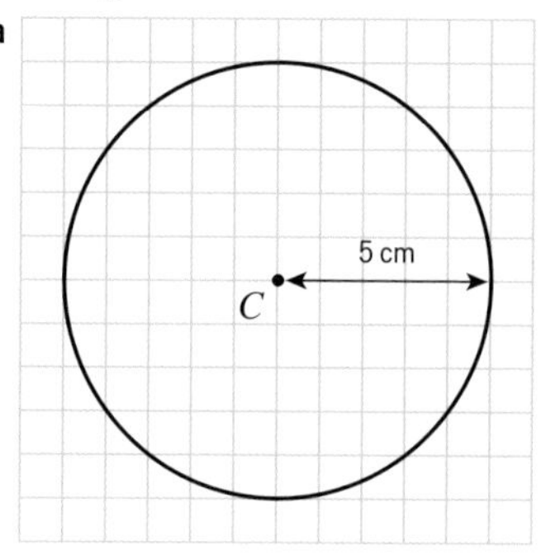

b Less

2

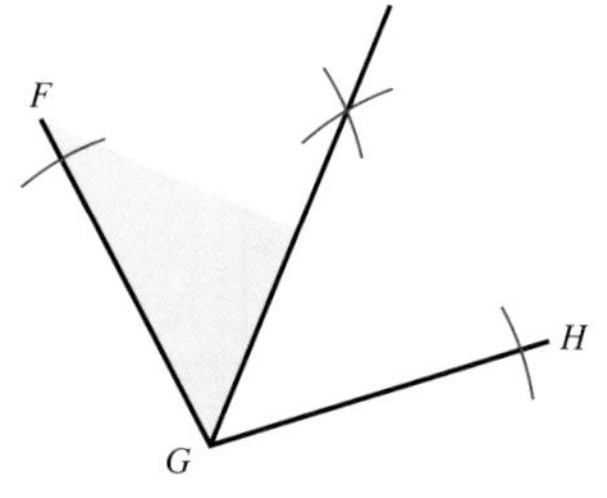

Scale drawings and bearings

1 i 4 cm **ii** 2.5 m

2 a 500, 0.5

b

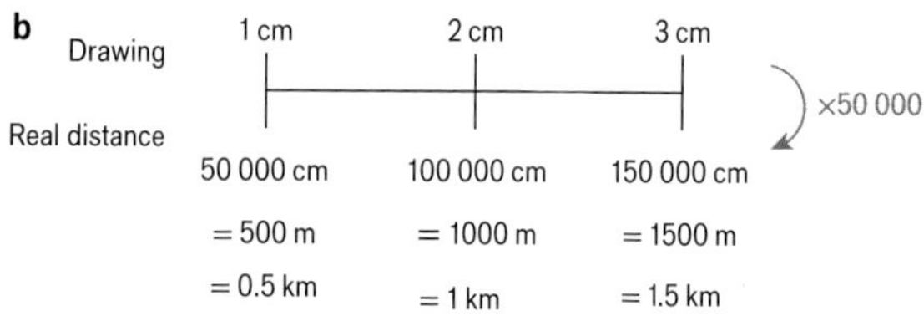

c 10 km **d** 9 cm

3 a 027° **b** 230° **c** 134° **d** 318°

4 a–e

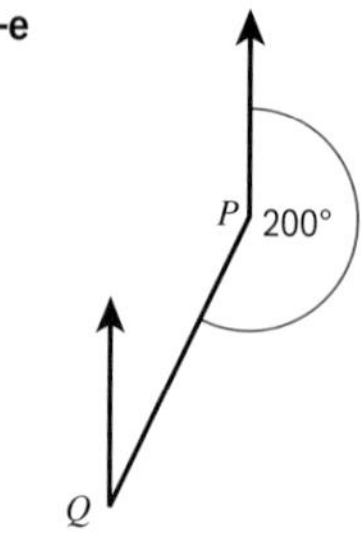

f 020°

5 a 40° **b** 220°

15 Extend

1 B. The length and width on the plan will be 21.5 cm and 12.5 cm respectively, so are closest to the measurements of a sheet of A4 paper.

2 No. 300 km journey needs 25 litres of fuel and she only has 22.5 litres.

3 a

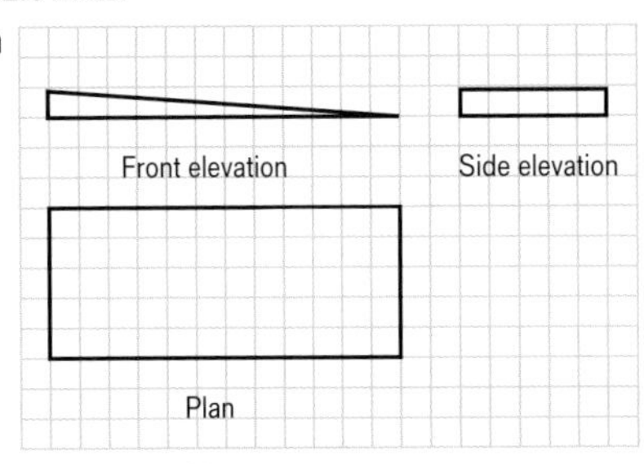

b 6.02 m or 602 cm

4 a The locus of points reached by each transmitter

b The region which is reached by both transmitters

c 35 miles

5

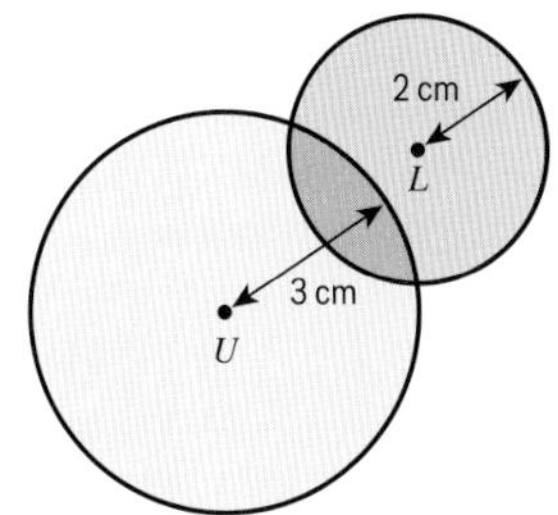

6 a i 000° ii 270°

b 180° c East

7 Accurate construction of triangle DEF drawn, with e.g. scale 1 cm represents 10 km, $DE = 8.6$ cm, $EF = 3.7$ cm

a 337° b 94 km

8 1 cm to 15 km or 1 : 1 500 000

9 a Accurate construction of equilateral triangle with side 9 cm

b 150°, 210°

10 a

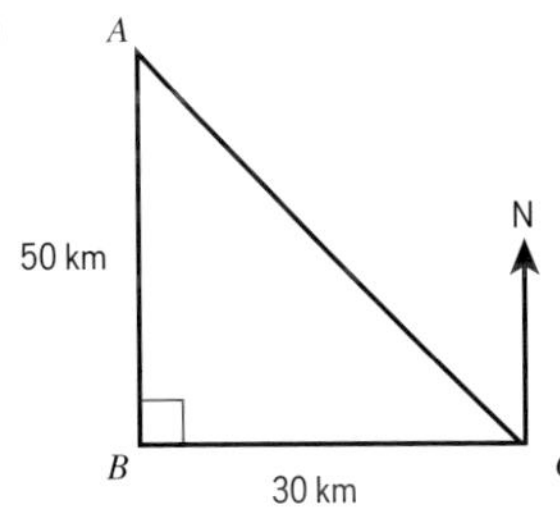

b Correct substitution into formula, angle $ACB = 59°$

c Bearing = 329°

d $AC = 58.3$ km

15 Test ready

Sample student answers

a Students' own answers, e.g. Draw accurately using a ruler and compasses.

b Students' own answers, e.g. No construction arcs, calculation of half angle shown.

15 Unit test

1 a 10 vertices

b Two planes of symmetry shown:

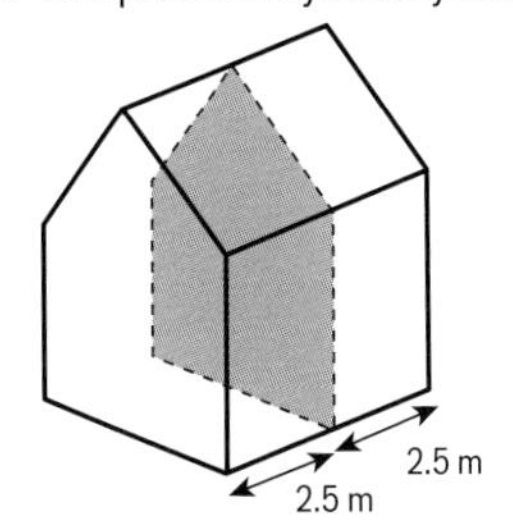

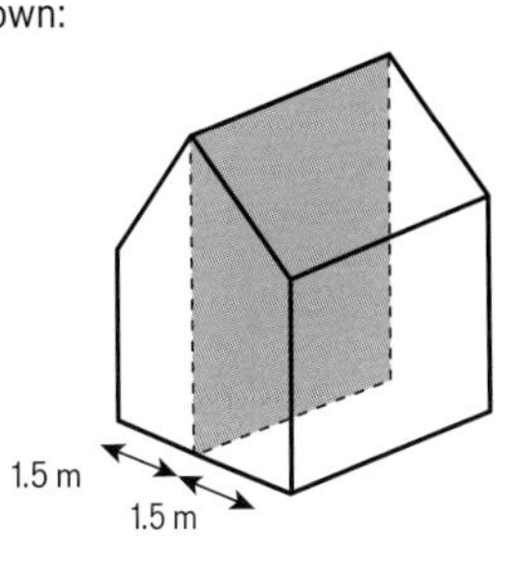

2 a Answer in the range 1.5–2 m

b 4 × answer to part **a**

3

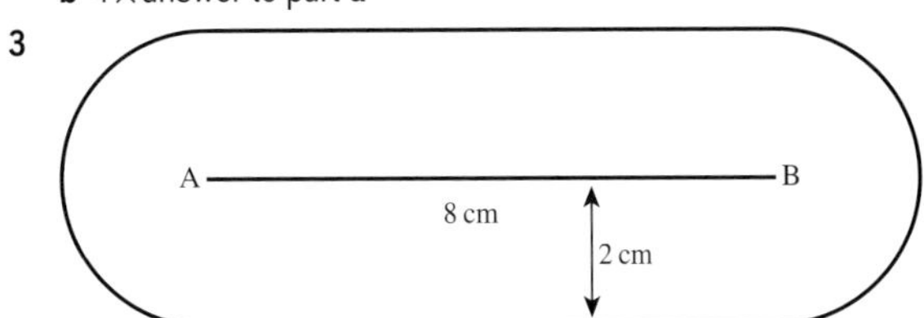

4

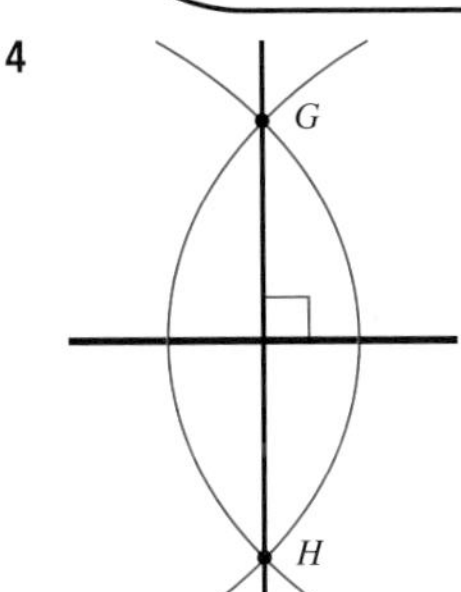

5

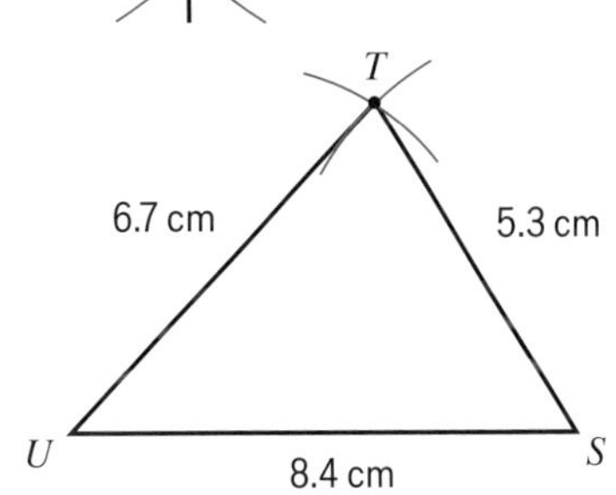

6

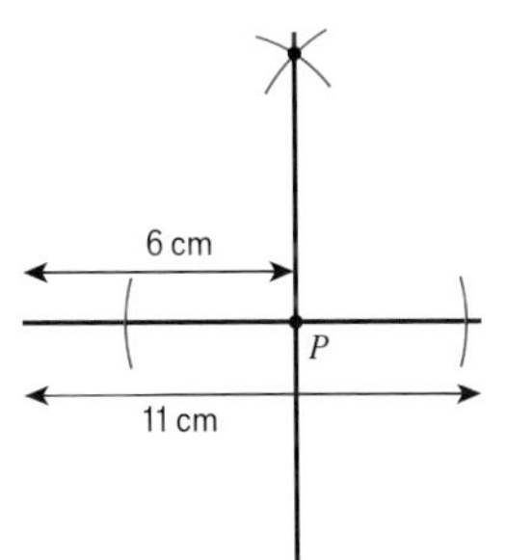

7

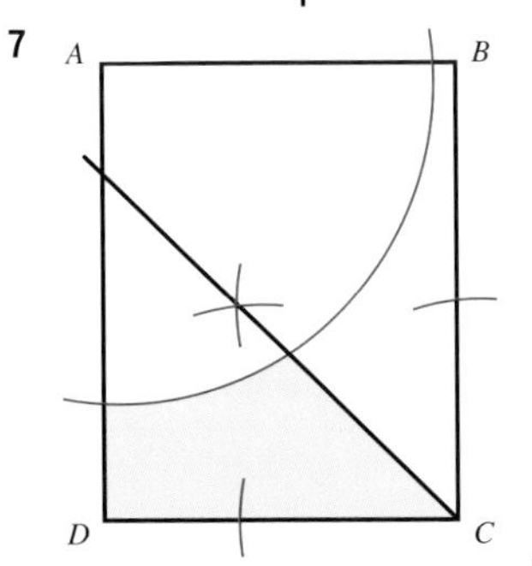

8 **a** 200° **b** 020°

9 3.5 cm

10 **a**

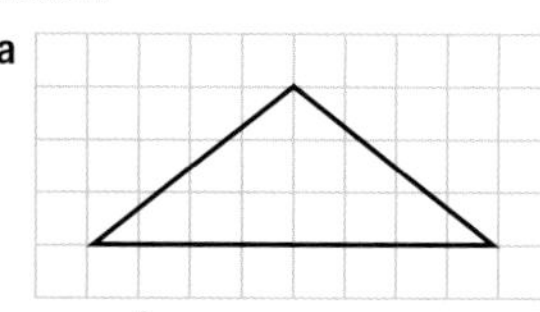

b $20\,\text{cm}^2$

11 **a, b**

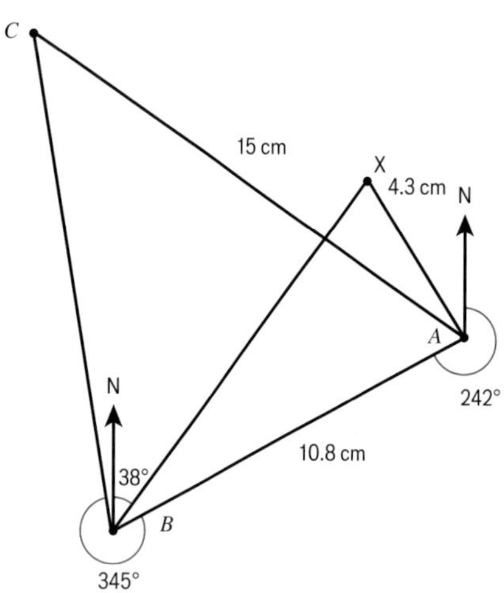

c Y is on a reciprocal course with C, so they may collide unless one of them diverts.

12

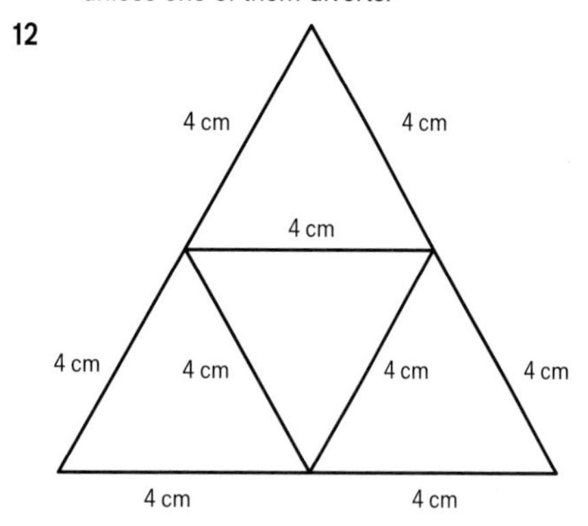

13 Students' own answers.

UNIT 16 Quadratic equations and graphs

16.1 Expanding double brackets

1 **a** $-9a$ **b** $-12b$ **c** $-3z^2$ **d** $6x^2$
e $8y$ **f** $-3t$ **g** m^2+9m+4
h $y^2-20y-3$

2 **a** $2a+4$ **b** $3x+12$ **c** m^2+8m **d** $2b^2-5b$

3 **a** $(n+4)(n+2)=n^2+4n+2n+8=n^2+6n+8$
b $(z+6)(z+3)=z^2+6z+3z+18=z^2+9z+18$
c $(a+1)(a+7)=a^2+a+7a+7=a^2+8a+7$

4 **a** x^2+3x+2 **b** $t^2+7t+12$
c $q^2+15q+54$ **d** $z^2+13z+12$
e $m^2+19m+88$ **f** $y^2+17y+70$
g j^2+6j+5 **h** $r^2+15r+54$
i Students' own answers.

5 **a** z^2-z-2 **b** m^2+m-30
c $a^2-5a-36$ **d** $n^2-3n-70$
e x^2-5x+6 **f** y^2-7y+6
g $b^2-7b-18$ **h** $k^2-10k+16$

6 5

7 Multiplying both ways gives $x^2-4x-21$ so Isabella is correct.

8 No, the answer is $a^2-11a+28$. Rex has made mistakes with the negative terms.

9 **a** $6+3t$ **b** $6x^2+15x$ **c** $m^2+13m+30$

10 **a**, **c**, **e**, **g** and **h**. A quadratic expression always contains a squared term as its highest power.

11 **a** x^2+4x+4 **b** $a^2+10a+25$
c $y^2-18y+81$ **d** $m^2-8m+16$
e t^2-6a+9

12 **a** $x^2+12x+36$ **b** $n^2+24n+144$
c $q^2-8q+16$ **d** $t^2-20t+100$
e y^2+6y+9

13 **A** and **H**, **B** and **G**, **C** and **E**, **D** and **F**

14 **a** $x^2+8x+16$
b $10(x^2+8x+16)=10x^2+80x+160$

15 $(n+6)(n+3)=n^2+3n+6n+18=n^2+9n+18$

16 $(x+4)(x-4)=x^2-16$. Answer is a quadratic expression as x^2 is the highest power. Both x^2 and 16 are squares.

17 $3x^2+5x+2$

18 **a** $2x^2+5x+2$ **b** $2x^2-3x-2$ **c** $2x^2+5x+3$
d $2x^2-x-3$ **e** $6x^2+7x+2$ **f** $6x^2-x-2$
g $8x^2+2x-3$ **h** $10x^2-9x+2$

19 $15x^2+x-2$

16.2 Plotting quadratic graphs

1 **a** 4 **b** 2 **c** 35 **d** 15

2 $x=3$

3 **a, b**

x	−4	−3	−2	−1	0	1	2	3	4
y	16	9	4	1	0	1	4	9	16

c

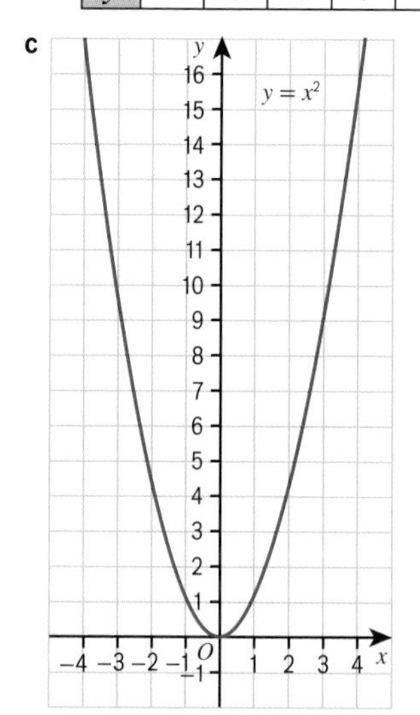

4 a

x	-4	-3	-2	-1	0	1	2
x^2	**16**	**9**	**4**	**1**	**0**	**1**	**4**
-3	**-3**	**-3**	**-3**	**-3**	**-3**	**-3**	**-3**
y	**13**	**6**	**1**	**-2**	**-3**	**-2**	**1**

b

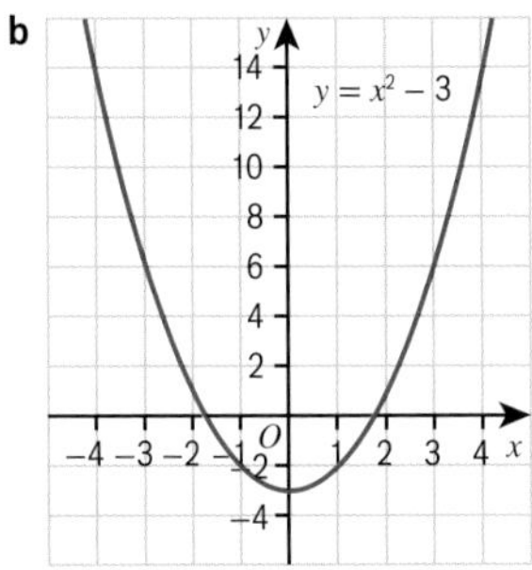

c

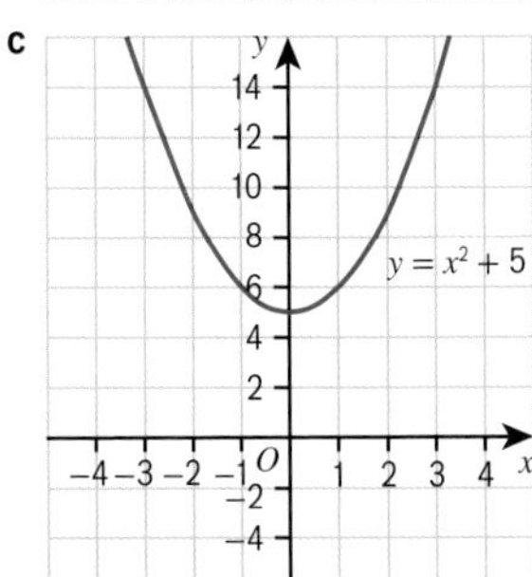

5 a

x	-3	-2	-1	0	1	2	3
y	10	4	0	-2	-2	0	4

b $y = x^2 - x - 2$

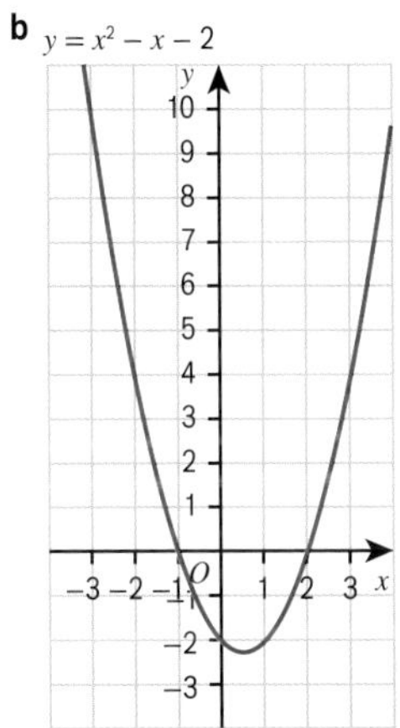

6 a

x	-4	-3	-2	-1	0	1	2
x^2	16	9	4	1	0	1	4
$+2x$	-8	-6	-4	-2	0	$+2$	$+4$
-4	-4	-4	-4	-4	-4	-4	-4
y	4	-1	-4	-5	-4	-1	4

b

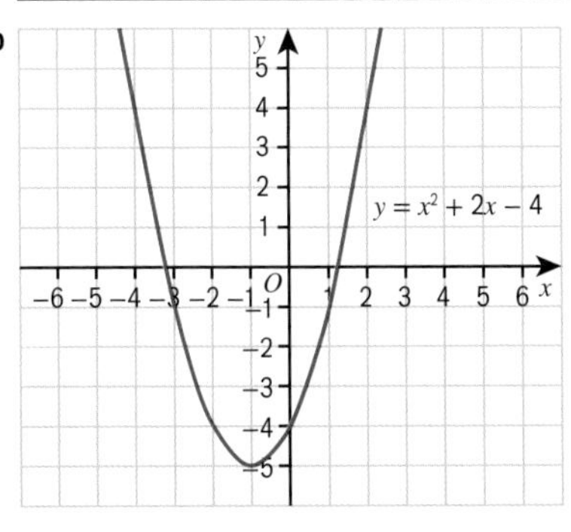

7 a i $(0, 4)$ **ii** $x = 1$ **iii** $(1, 3)$

b i $(0, -3)$ **ii** $x = -1$ **iii** $(-1, -4)$

8 a

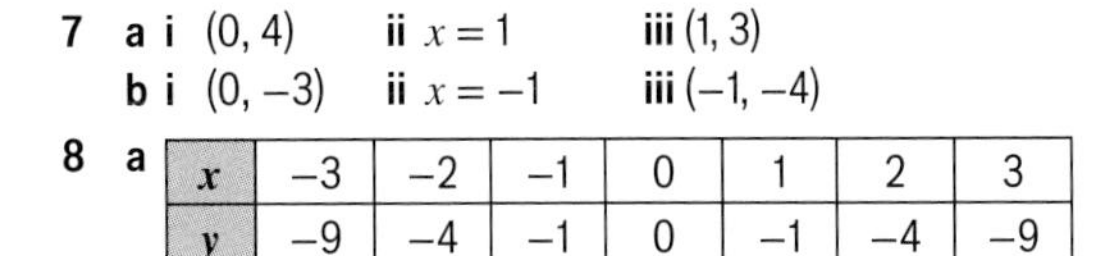

x	-3	-2	-1	0	1	2	3
y	-9	-4	-1	0	-1	-4	-9

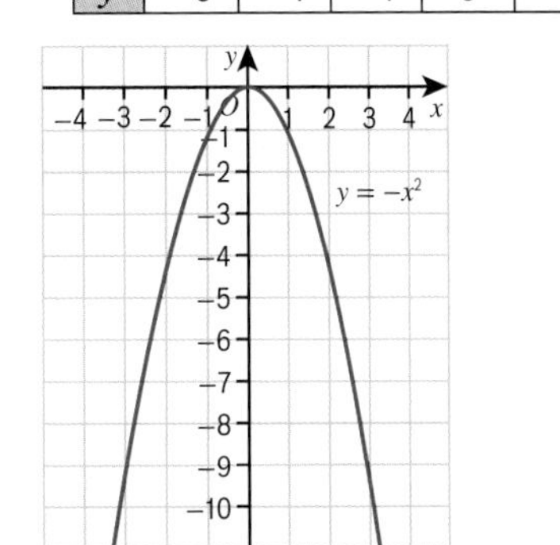

b i $(0, 0)$ **ii** $x = 0$ or the y-axis

c Same shape, opposite way up. Reflection of each other in x-axis. Same line of symmetry. Same y-intercept.

9 **b** and **d** are quadratic functions.

10 a i 13.3 m **ii** 1.6 seconds

b 1.2 seconds, 2.0 seconds

c 3.2 seconds

11 a i $1\,\text{cm}^2$ **ii** $16\,\text{cm}^2$ **iii** $6.25\,\text{cm}^2$

b Square. Area is always the square of the length.

12 a

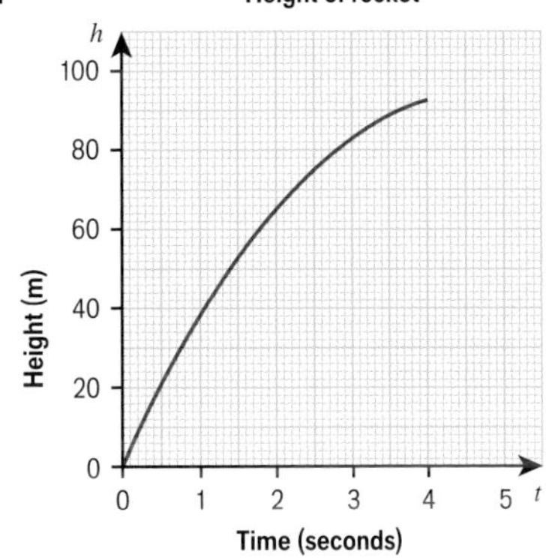

b 0.5 seconds **c** $76\,\text{m} \pm 2\,\text{m}$

16.3 Using quadratic graphs

1 a The point $(0, 0)$ **b** The x-axis

2 a

x	-3	-2	-1	0	1	2
y	10	4	0	-2	-2	0

b

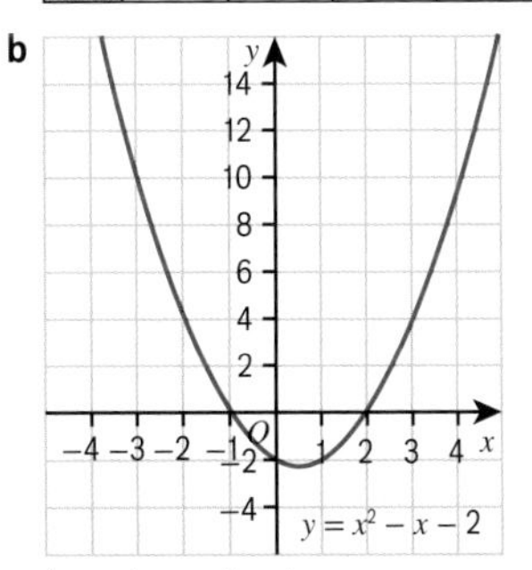

3 a $(-1, 0)$ and $(2, 0)$ **b** $x = -1$ and $x = 2$

4 a $x = -2$ or $x = 3$ **b** $x = -5$ or $x = 2$

c $x = 2$ **d** $x = -1.5$ or $x = 0.5$

5 **a** 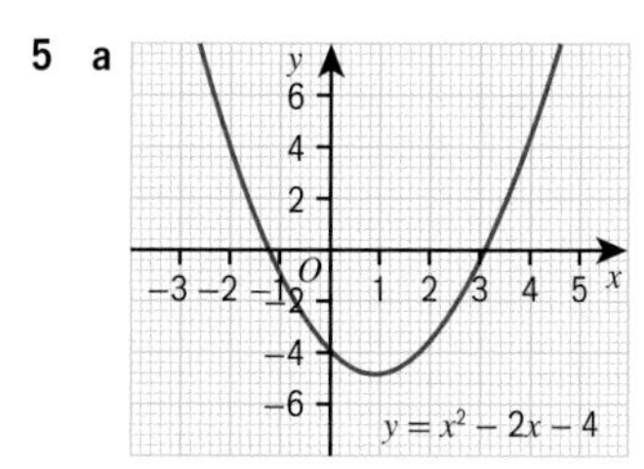

b $x = -1.2$ or $x = 3.2$

6 **a** 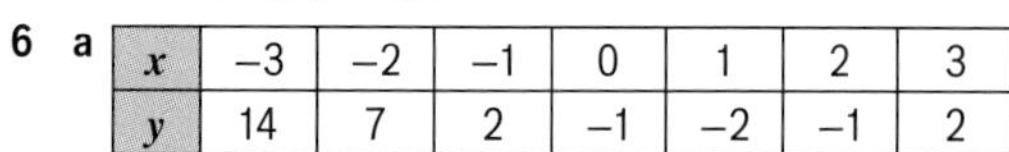

x	-3	-2	-1	0	1	2	3
y	14	7	2	-1	-2	-1	2

b 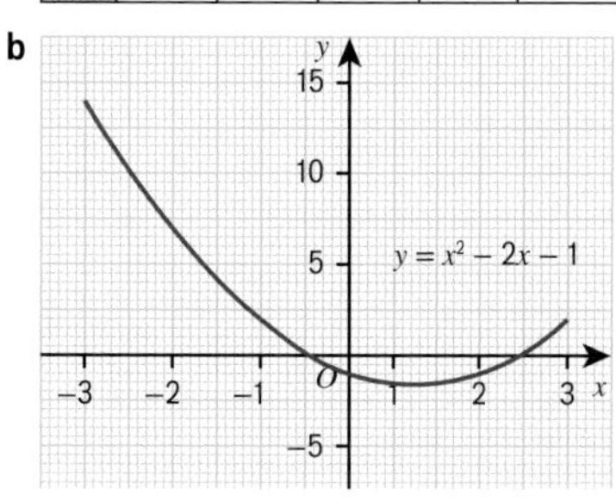

c -0.4 or 2.4 (or both)

7 **a** $x = 1$ or $x = -1$ **b** $x = 1$ or $x = 3$

8 **a** 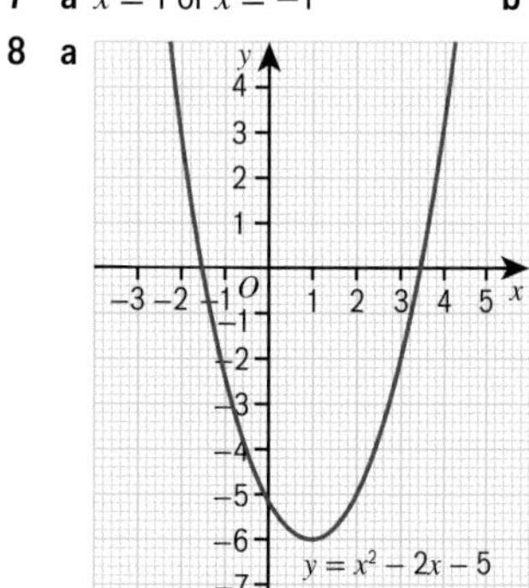

b $x = -1.4$ or $x = 3.4$ **c** $x = -2$ or $x = 4$

9 **a** $x = 0$ or $x = 3$ **b** $x = -0.6$ or $x = 3.6$

10 **a** 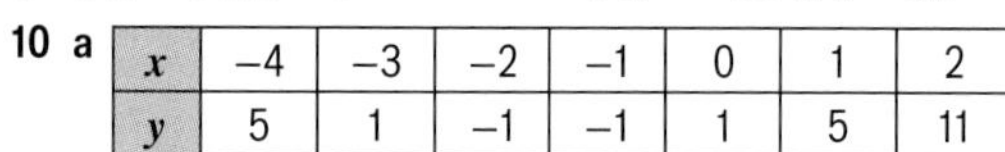

x	-4	-3	-2	-1	0	1	2
y	5	1	-1	-1	1	5	11

b 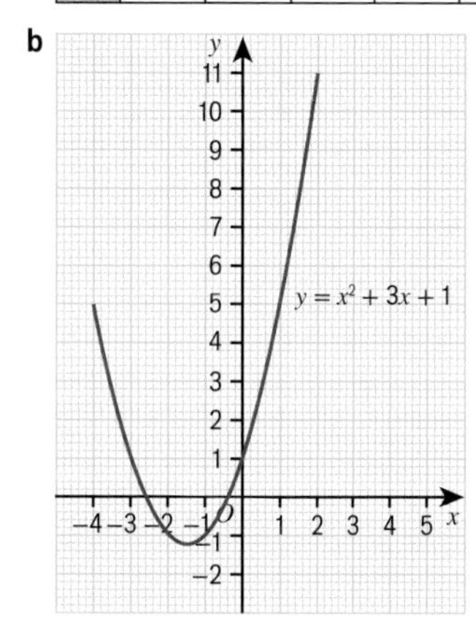

c $x = -3.3$ or $x = 0.3$

11 **a** $x = -1$ or $x = 3$

b The curve and the line do not intersect.

16.4 Factorising quadratic expressions

1 **a** x^2+7x+6 **b** $x^2+3x-40$ **c** x^2-3x+2 **d** $x^2+10x+25$

2 **a** -1 and -2 **b** -1 and -4, -2 and -2 **c** -1 and -6, -2 and -3 **d** -1 and -5 **e** -1 and -10, -2 and -5

3 **a** 2 and 4 **b** 3 and 4 **c** 3 and -2 **d** 1 and -10

4 **a** x **b** 7 **c** 5 **d** x, 4

5 **a** $(x+4)(x+3)$ **b** $(x+2)(x+4)$ **c** $(x+1)(x+3)$ **d** $(x+5)(x+2)$

6 $12x^2+5x-3$ **b** $(x+1)(x+2)$

7 **a** 1 and -6, -1 and 6, 2 and -3, -2 and 3
b -2 and 3
c $(x+3)(x-2)$

8 **a** $(x+5)(x-2)$ **b** $(x-5)(x+2)$ **c** $(x-3)(x+1)$ **d** $(x-1)(x+3)$ **e** $(x+4)(x-5)$ **f** $(x-4)(x+5)$ **g** $(x-2)(x+10)$ **h** $(x-10)(x+2)$ **i** $(x+5)(x+6)$ **j** $(x-2)(x+7)$ **k** $(x-4)(x+1)$ **l** $(x-2)(x-4)$

9 **a** $y+9$ **b** $144\,\text{cm}^2$

10 **a** $(x+4)^2$ **b** $(x+3)^2$ **c** $(x-1)^2$ **d** $(x-3)^2$

11 $(x+12)$ and $(x-2)$

12 **a** x^2-4 **b** x^2-9 **c** x^2-16 **d** x^2-100
e There is no x term, just an x^2 and a number term.

13 x^2-4; x^2-36; x^2-y^2

14 **a** z^2-9
b Difference of 2 squares

15 **a** $(x+5)(x-5)$ **b** $(x-6)(x+6)$ **c** $(x-8)(x+8)$ **d** $(x+y)(x-y)$

16 **a** n^2-n-12 **b** $(p-9)(p+9)$

16.5 Solving quadratic equations algebraically

1 **a** 3, -3 **b** 7, -7 **c** 10, -10 **d** 12, -12

2 **a** $(x+2)(x+5)$ **b** $(x-3)(x+4)$ **c** $(x+3)(x-3)$ **d** $(x+6)^2$

3 **a** $d = 18$ **b** $x = -1$ **c** $p = 4$ **d** $t = 11$

4 **a** $x = 6$ or $x = -6$ **b** $y = 9$ or $y = -9$ **c** $z = 10$ or $z = -10$ **d** $t = 4$ or $t = -4$

5 $x^2 = 9$, $x = 3$ or $x = -3$

6 **a** $x = 5$ or $x = -5$ **b** $y = 1$ or $y = -1$ **c** $n = 7$ or $n = -7$ **d** $m = 11$ or $m = -11$

7 **a** $x = 6$ or $x = -6$ **b** $x = 7$ or $x = -7$ **c** $x = 9$ or $x = -9$ **d** $x = 8$ or $x = -8$

8 **a** $x = 4$ or $x = -4$ **b** $x = 2$ or $x = -2$

9 **a** $x = 5$ or $x = -5$ **b** $x = 5$ or $x = -5$ **c** $x = 4$ or $x = -4$ **d** $x = 1$ or $x = -1$ **e** $x = 3$ or $x = -3$ **f** $x = 6$ or $x = -6$ **g** $x = 1$ or $x = -1$ **h** $x = 2$ or $x = -2$

10 **a** $x = -10$ or $x = 2$ **b** $x = -5$ or $x = 3$ **c** $x = 1$ or $x = 5$ **d** $x = -7$ or $x = 3$

11 **a** $x = -1$ or $x = -4$ **b** $x = 1$ or $x = 4$ **c** $x = 1$ or $x = -4$ **d** $x = 4$ or $x = -1$ **e** $x = -3$ or $x = -6$ **f** $x = -3$ or $x = 2$ **g** $x = 5$ or $x = 1$ **h** $x = 2$ or $x = 8$

12 A and E, B and G, C and F, D and H

13 $x = -5$ or $x = 6$

14 **a** $x(x-2)=0$, $x=0$ or $x-2=0$, $x=0$ or $x=2$
b $y(y+4)=0$, $y=0$ or $y+4=0$, $y=0$ or $y=-4$

15 **a** $n=0$ or $n=-7$ **b** $t=0$ or $t=6$
c $x=0$ or $x=9$ **d** $y=0$ or $y=-2$

16 $x^2-7x+12=0$

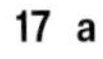

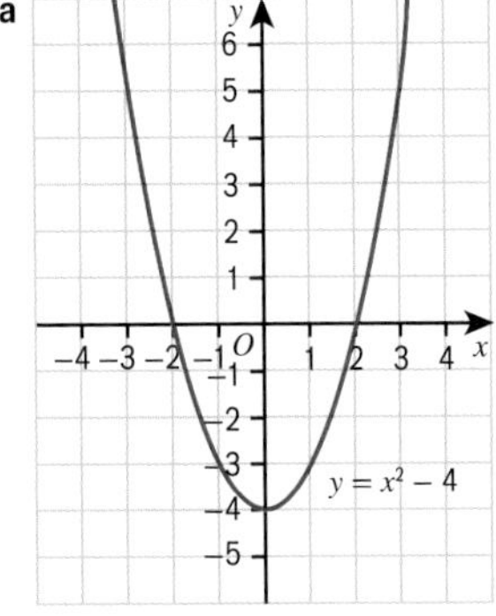

b $x=2$ or $x=-2$
c $x^2-4=0$
$(x+2)(x-2)=0$
$(x+2)=0$ or $(x-2)=0$
$x=2$ or $x=-2$

18 **a** x^2+x-6
b $x^2+x-6=24$
c $x^2+x-30=0$, $(x-5)(x+6)$, $x=5$ or $x=-6$
d $x=-6$
e $x=5$ cm

16 Check up

1 Graphs **a** and **c**

2 **a**

x	−1	0	1	2	3	4	5	6	7
y	7	0	−5	−8	−9	−8	−5	0	7

b

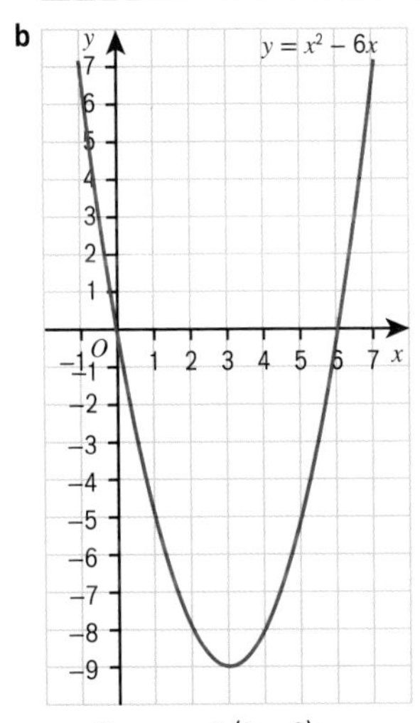

c $x=3$ **d** $(3, -9)$
e $(0, 0)$ **f** $x=0$ or $x=6$

3 **a** $x=-4.3$ or $x=-0.7$ **b** $x=-5.7$ or $x=0.7$

4 **a** 7.25 m **b** 1.5 seconds
c 0 seconds, 3 seconds **d** 3.1 seconds

5 **a** $t^2+10t+24$ **b** $f^2+2f-35$
c $n^2-14n+49$ **d** $4x^2+11x-3$

6 $(x+7)(x-7)$

7 **a** $x=-3$ or $x=3$ **b** $x=-6$ or $x=6$
c $y=3$ or $y=0$

8 **a** $(x+4)(x+5)$ **b** $(x-2)(x+4)$

9 **a** $(x+10)(x+3)$, $x=-10$ or $x=-3$
b $(x-6)(x-1)$, $x=6$ or $x=1$

11 **a** $(x+3)^2=0$ $(x-3)^2=0$ $(x+2)^2=0$ $(x-2)^2=0$
$(x+3)(x-3)=0$ $(x+2)(x-2)=0$
$(x+3)(x+2)=0$ $(x+3)(x-2)=0$
$(x+2)(x-3)=0$ $(x-3)(x-2)=0$

16 Strengthen

Quadratic graphs

1 **a**

x	−3	−2	−1	0	1	2	3
x^2	9	4	1	0	1	4	9
$+1$	+1	+1	+1	+1	+1	+1	+1
y	10	5	2	1	2	5	10

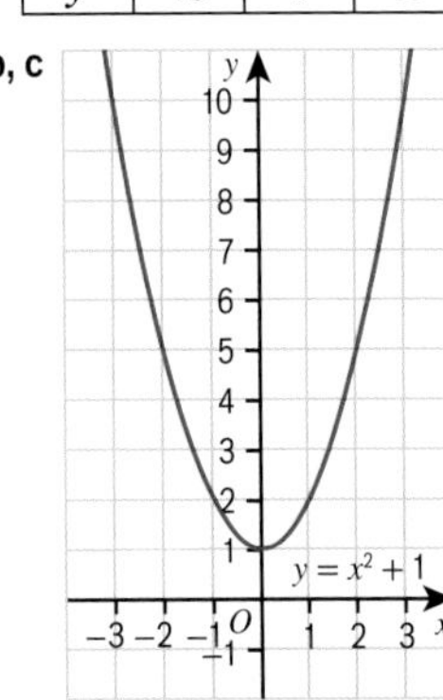

2 **a**

x	−3	−2	−1	0	1	2
x^2	9	4	1	0	1	4
$+x$	−3	−2	−1	0	1	2
y	6	2	0	0	2	6

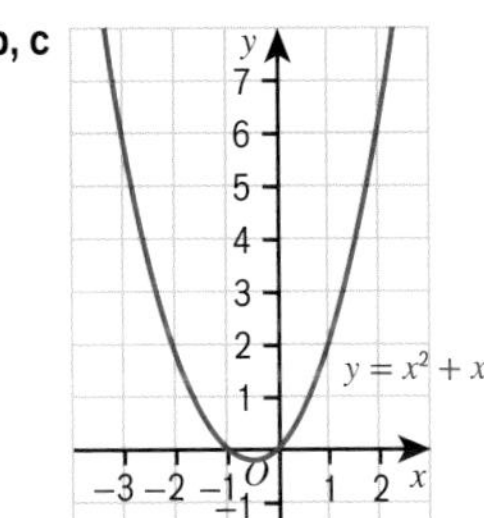

3 **a** Roots **b** y-intercept
c Turning point **d** Line of symmetry

4 **a** $(0, 8)$ **b** $(3, -1)$
c $x=3$ **d** $x=2$ or $x=4$

5 a

x	-5	-4	-3	-2	-1	0
x^2	25	16	9	4	1	0
$+5x$	-25	-20	-15	-10	-5	0
$+5$	$+5$	$+5$	$+5$	$+5$	$+5$	$+5$
y	5	1	-1	-1	1	5

b, c i

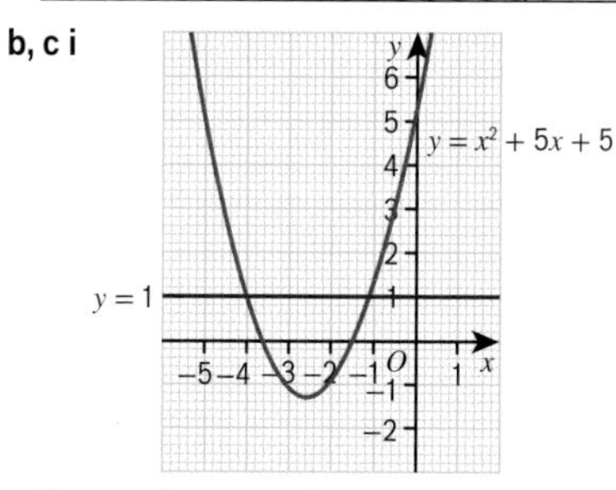

c ii $x = -4$ or $x = -1$

d $x = -3.6$ and $x = -1.4$

6 Graphs **a** and **d**

Quadratic equations

1 $a^2, 6, 2a, 3a, a^2 + 6 + 5a$

2 a $a^2 + 5a + 6$ **b** $t^2 + 5t + 4$
c $x^2 + 8x + 7$ **d** $y^2 + 15y + 50$
e $z^2 - z - 2$ **f** $f^2 + 2f - 63$
g $x^2 - 1$ **h** $x^2 - 16$

3 a $m^2 + 8m + 16$ **b** $x^2 + 16x + 64$
c $g^2 - 20g + 100$ **d** $y^2 - 4y + 4$

4 a $3x^2 + 7x + 2$ **b** $3x^2 - 7x + 2$
c $2x^2 - 9x - 5$ **d** $2x^2 + 9x + 4$
e $6x^2 + 5x + 1$ **f** $6x^2 - x - 1$
g $12x^2 + 25x + 12$ **h** $12x^2 - x - 6$

5 a $x^2 = 4, x = \sqrt{4} = \pm 2$ **b** $t^2 = 49, t = \sqrt{49}, t = \pm 7$

6 $x^2 - 9 = (x - 3)(x + 3)$
$x^2 - 49 = (x + 7)(x - 7)$
$x^2 - 16 = (x + 4)(x - 4)$
$x^2 - 4 = (x - 2)(x + 2)$

7 a $(x + 4)(x - 4)$ **b** $(p + 1)(p - 1)$
c $(y + 9)(y - 9)$ **d** $(k + 10)(k - 10)$

8 a 3 and 4, 2 and 6, 1 and 12 **b** 7, 8, 13
c i $(x + 3)(x + 4)$ **ii** $(x + 2)(x + 6)$
iii $(x + 1)(x + 12)$

9 a 15, 15, 5, 5 **b** 14, -14, 2, -2
c i $(x + 1)(x - 15)$ **ii** $(x - 5)(x + 3)$
iii $(x + 5)(x - 3)$ **iv** $(x - 1)(x + 15)$

10 a 0 **b** 0 **c** 0
d 0 **e** 0

11 a $x = -2$ or $x = -4$ **b** $x = 3$ or $x = -6$

16 Extend

1 a

x	-2	-1	0	1	2	3	4	5
$-x^2$	-4	-1	0	-1	-4	-9	-16	-25
$+3x$	-6	-3	0	3	6	9	12	15
$+4$	$+4$	$+4$	$+4$	$+4$	$+4$	$+4$	$+4$	$+4$
y	-6	0	4	6	6	4	0	-6

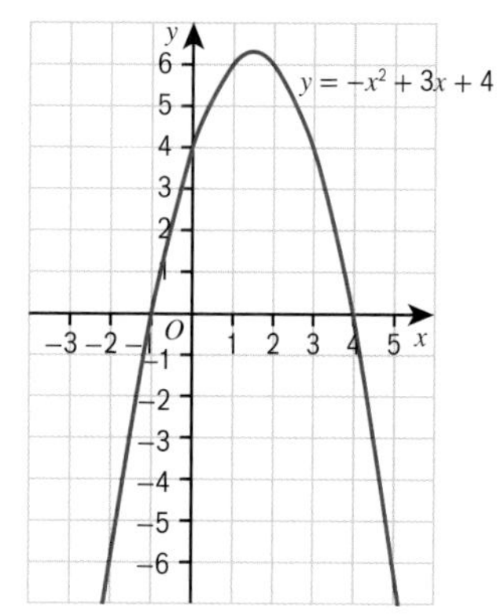

b i y-intercept = (0, 4) **ii** $x = 1.5$
iii (1.5, 6.25) **iv** $x = -1$ or $x = 4$

2 a $4(2x + 3)$ **b** $(y - 4)(y + 4)$

3 a

x	0	1	2	3	4	5
y	10	6	4	4	6	10

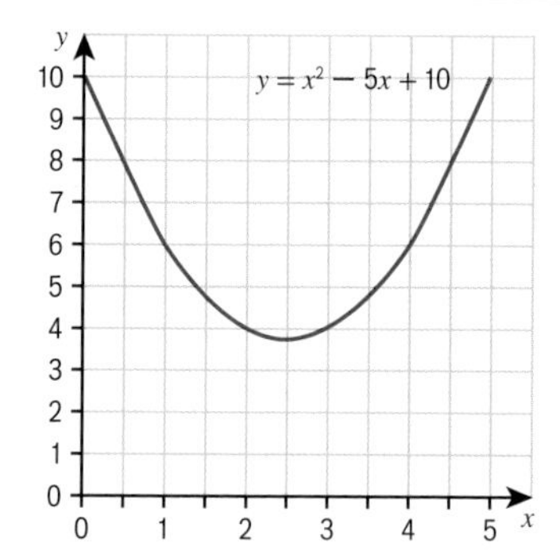

b i 10 m **ii** 3.75 m **iii** 5 m

4 a

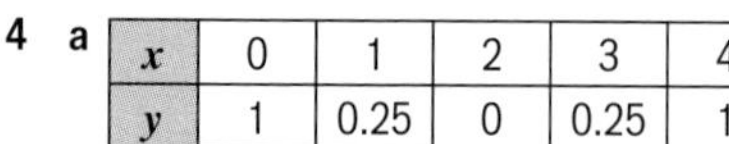

x	0	1	2	3	4
y	1	0.25	0	0.25	1

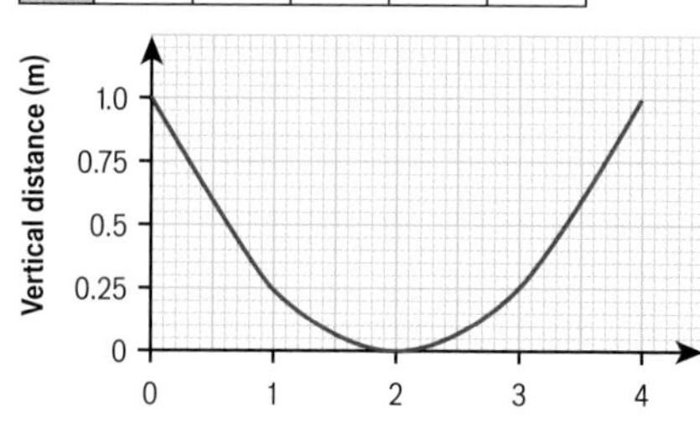

b i 1 m **ii** 4 m **iii** 0 m

5 a 5 m **b** 0 m
c It has landed on the ground.

6 $(x + 4)^2 = x^2 + 8x + 16, x^2 + 8x + 16 = 18, x^2 + 8x = 2$

7 a $x^2 - 4x + 4$ **b** $12(x + 1)$

8 a $x^2 = 3x$ **b** $x = 3$ **c** $9\,\text{cm}^2$

9 a $x = -1.3$ or $x = 2.8$ **b** $x = -0.7$ or $x = 2.2$

16 Test ready

Sample student answers

1 **a** Because the graph is not a smooth U-shaped curve.
b (–1, –2) is incorrect.
c Student B should draw a smooth curve with a turning point, not a flat (straight line) bottom.

2 It should be the product of two brackets, i.e. $(x+3)(x+4)$.

16 Unit test

1 **a** i, iii, **b** i

2 **a** (0, 0) **b** $x = 2.5$ **c** (2.5, –6.25)
d $x = 0$ or $x = 5$

3 **a**

x	–7	–6	–5	–4	–3	–2	–1	0
y	6	0	–4	–6	–6	–4	0	6

b

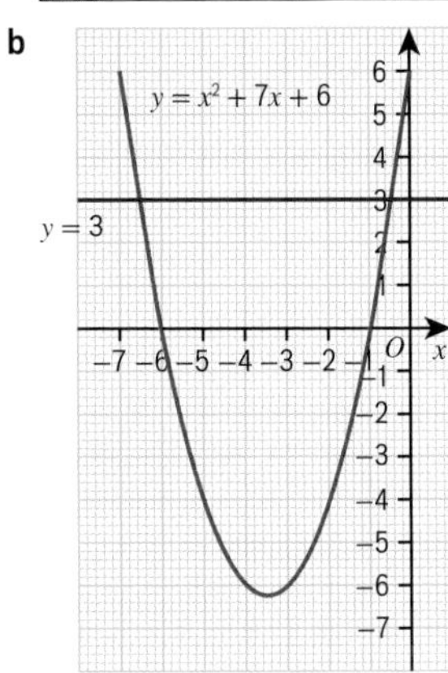

c $x = -6.5$ or $x = -0.5$

4 **a** $y^2 + 11y + 10$ **b** $z^2 - z - 56$
c $x^2 - 1$ **d** $n^2 - 18n + 81$

5 $a = 5$ or $a = -5$

6 **a** $x = 4$ or $x = -4$ **b** $6x^2 - 7x + 3$
c $(x+2)(x+2)$ or $(x+2)^2$

7 $x = 4$ or $x = -10$

8 $x - 6, x - 4$

9 $y = 7$ or $y = -7$

10 **a** 975 ± 10 m **b** 1120 ± 10 m
c 5 seconds and 25.5 seconds
d 15 seconds

11 **a**

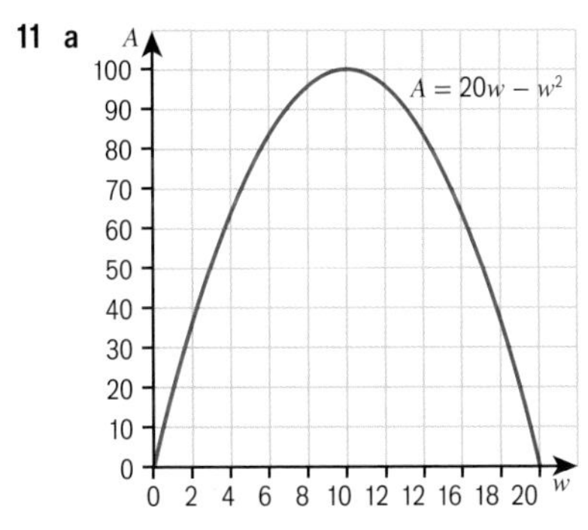

b 10 m **c** 10 m, 10 m

12 Students' own answers.

UNIT 17 Perimeter, area and volume 2

17.1 Circumference of a circle 1

1 **a** 3.8, 3.78 **b** 3490

2 $m = 45$

3 4.19

4 **a** 120 m **b** 5 inches **c** $r = \frac{d}{2}$ **d** $d = 2r$

5 2.6 inches

6 **a** All 3.14 **b** All 3.14 **c** $C = 3.14d$ **e** 57.5 m

7 **a** 16.7 cm **b** 1947.8 mm **c** 10.0 m
d 42.1 cm **e** 2450.4 mm **f** 15.1 m

8 34.6 cm

9 **a** 64.8 cm **b** 38

10 6 lengths

11 **a** 2136.283… mm (or 2.136 m) **b** 234

12 180π m

13 Madhu is correct; Paul forgot to double the radius.

17.2 Circumference of a circle 2

1 **a** 3 m **b** 3 m

2 **a** $l = \frac{A}{w}$ **b** 7.3

3 **a i** 7.5 cm **ii** 19.5 km **iii** 37.5 mm
b i Any number smaller than 8.5 but bigger than 8
ii Any number smaller than 20.5 but bigger than 20
iii Any number smaller than 38.5 mm but bigger than 38 mm

4 14.5 cm

5 **a** 1.625 m **b** 1.635 m **c** $1.625 \leqslant h < 1.635$

6 $25\,250 \leqslant n < 25\,350$

7 **a** $4.355 \leqslant n < 4.365$ **b** $715 \leqslant n < 725$
c $15.65 \leqslant n < 15.75$ **d** $445 \leqslant n < 455$

8 **a i** 4 **ii** 4.0
b i 4 **ii** 4.5
c i 4 **ii** 4.8
d i 4 **ii** 4.9
e i 4 **ii** 4.9
f i 5 **ii** 5.0
g i 5 **ii** 5.0
h i 5 **ii** 5.1

9 **a** 4 **b** 5 **c** $4 \leqslant n < 5$

10 $7.2 \leqslant N < 7.3$

11 Height of book could be up to 265 mm, minimum height of shelf is 255 mm (25.5 cm).

12 **a** 3.14159 **b** 3.142 **c** 15.7 cm
d 15.70796… **e** Using the π button

13 36 m

14 **a** Estimate $C = 3 \times 12 = 36$ m, $36 \div 4 = 9$ rolls
b Underestimate, because $C = 12\pi$ and $\pi > 3$, so true circumference > 36 m.

15 **a** 15.1 m **b** 19 mm **c** 4715.5 cm

16 **a** 111.4 cm **b** 0.9 m **c** 7.6 cm

17 **a** 107 m **b** 477 m **c** 8.28 cm

18 **a** 27.5 cm ⩽ circumference < 28.5 cm
b 4.4 cm **c** 4.4 ⩽ radius ⩽ 4.5 cm

17.3 Area of a circle

1 **a** 48 **b** 12 **c** 9 **d** 4

2 **a** $p = 40$ **b** $t = 108$

3 **a** 3.2 **b** 1.3 **c** 8 **d** 16.6

4 **a** 201 cm² (nearest cm²)
b 132.7 m² (1 d.p.)
c 138 540 cm² (nearest 10 cm²)
d 234 826.4 m² (1 d.p.)
e He has calculated the area as $\pi^2 r^2$ instead of πr^2.

5 a i $27\,mm^2$ ii 147 square feet iii $18.75\,km^2$
b i $9\pi\,mm^2$ ii 49π square feet iii $6.25\pi\,km^2$
c i $28\,mm^2$ ii 150 square feet iii $20\,km^2$

6 Rita is correct; Jack forgot to halve the diameter.

7 a $63.6\,m^2$ b 3180 g

8 a $7.1\,m^2$ b $\approx 20\,000$ slugs

9 a i 6.3 m ii 12.6 m
b i 2.1 cm ii 4.2 cm
c i 10 m ii 20 m
d i 1.2 cm ii 2.4 cm

10 11 m or 11.0 m

11 120 cm

12 a $10 \times 15 = 150$, $3 \times 3^2 = 27$, $150 - 27 = 123$, $123 \div 1.5 = 82$ packs
b Overestimate because $\pi > 3$ and $3.1 > 3$, so estimate of $27\,m^2$ for pond area is too small.
Therefore estimate for area for planting bulbs is larger than actual area.

13 $867\,cm^2$

14 a $49\pi\,cm^2$ b $4\pi\,cm^2$ c $45\pi\,cm^2$

15 Malia is correct. Area of large circle $= 36\pi$.
Area of shaded circle $= 9\pi$, $\frac{9\pi}{36\pi} = \frac{1}{4}$
(or using accurate calculations).

16 Students' own answers.

17.4 Semicircles and sectors

1 a 360° b 180° c 90°

2 a $\frac{1}{2}$ b $\frac{1}{4}$ c $\frac{1}{4}$ d $\frac{3}{4}$

3 a 90 b 36 c 72 d 4

4 a Chord b Tangent c Segment
d Arc e Sector

5 a i $32\pi\,cm^2$ ii $100.5\,cm^2$
b i $18\pi\,cm^2$ ii $56.5\,cm^2$
c i $2.56\pi\,cm^2$ ii $8.0\,cm^2$
d i $90.25\pi\,mm^2$ ii $283.5\,mm^2$

6 a 37.7 cm b 18.8 cm c 30.8 cm

7 a $9\pi + 18$ cm b 16.4 cm

8 a 44 cm b 176 cm c 22 cm
d 11 cm e 77 cm

9 a 144π cm b 6 c $\frac{1}{6}$
d i 12 and $\frac{1}{12}$ ii 10 and $\frac{1}{10}$
iii 3 and $\frac{1}{3}$ iv 8 and $\frac{1}{8}$

10 £3092.47

11 a i 1.05 cm ii 13.0 cm iii $3.14\,cm^2$
b i 101 mm ii 137 mm iii $905\,mm^2$
c i 18.8 mm ii 105 mm iii $403\,mm^2$
d i 22.3 m ii 38.3 m iii $89.4\,m^2$
e i 13.3 cm ii 32.3 cm iii $63.0\,cm^2$
f i 7.59 m ii 19.2 m iii $22.0\,m^2$

17.5 Composite 2D shapes and cylinders

1 A rectangle with one circle at each end

2 $60\,cm^3$

3 a $24\,cm^2$ b $4.5\pi\,cm^2$ c $4\pi\,cm^2$

4 a 22 cm b $3\pi + 6$ cm c $2\pi + 8$ cm

5 a Perimeter = 47.8 cm; area = $70.5\,cm^2$
b Perimeter = 28.9 cm, area = $43.6\,cm^2$

6 a 60 m b $7627\,m^2$ c 348 m d 3

7 $8.79\,cm^2$

8 $2.6\,cm^2$

9 a Circle b $A = \pi r^2$ c $V = \pi r^2 h$

10 a $113\,cm^3$ b $628\,cm^3$ c $1130\,cm^3$

11 a $21.2\,m^3$ b 1060

12 Cylinder B (volume of A = $277\,cm^3$; volume of B = $362\,cm^3$)

13 134.8 litres

14 a $1\,256\,637\,cm^3$ b 5403.5 kg

15 a $242.8\,cm^2$ b $294.1\,cm^2$

16 $243.1\,cm^2$

17 a $750\pi\,ft^3$ b $300\pi\,ft^2$

18 Students' own answers.

17.6 Pyramids and cones

1 5: a square and 4 triangles

2 a $60\,cm^2$ b $63.6\,cm^2$

3 a 6π b 45π c 12π d 80π

4 a $60.5\,cm^3$ b $30\,cm^3$ c $96\,cm^3$

5 21 509 583 cubic yards

6 a $144\,cm^2$ b $800\,cm^2$

7 a $64\,cm^3$ b $1280\,cm^3$

8 a $49\,000\,cm^3$ b $10\,500\,cm^2$

9 a Circle b $A = \pi r^2$ c $V = \frac{1}{3}\pi r^2 h$

10 a $288\,cm^3$ b $1018\,cm^3$ c $1571\,mm^3$

11 a 4 cm b $168\,cm^3$ c 29

12 a i $144\pi\,m^2$ ii $452\,m^2$
b i $224\pi\,cm^2$ ii $704\,cm^2$
c i $50\pi\,cm^2$ ii $157\,cm^2$

13 a i $128\pi\,m^3$ ii $402\,m^3$
b i $392\pi\,cm^3$ ii $1230\,cm^3$
c i $40\pi\,cm^3$ ii $126\,cm^3$

14 3.4 litres

15 a $3^2 + 4^2 = 25$; slant height, $l = \sqrt{25} = 5$ cm
b $15\pi\,cm^2$ c $9\pi\,cm^2$ d $24\pi\,cm^2$ e $75.4\,cm^2$

16 $2490\,cm^2$

17.7 Spheres and composite solids

1 a Volume of cuboid $= w \times d \times h$
Volume of pyramid $= \frac{1}{3} \times$ area of base $\times$ vertical height
Volume of cylinder $= \pi r^2 h$
Volume of cone $= \frac{1}{3} \times$ area of base $\times$ vertical height
b Area of circle $= \pi r^2$
Area of curved surface of cone =
$\pi \times$ base radius $\times$ slant height

2 a 8 b 32 c 8 d 144

3 a 8 b $10\frac{2}{3}$ c 4.19 d 84.95

4 a $2.304\pi\,cm^3$ b $7\,cm^3$

5 $75\,766\,mm^3$

6 Moira; Jill has used the formula πr^2.

7 $18\,cm^2$

8 $300\,000\,000\,km^2$

9 a $108\pi\,m^2$ b $339.29\,m^2$

10 $84\,m^3$

11 Volume of hemisphere $\frac{16}{3}\pi$ + volume of cylinder $80\pi =$ $\frac{256}{3}\pi = 268\,m^3$ (3 s.f.)

12 a $56.25\pi\,m^3$ b $36\,000\pi\,cm^3$ c $9\pi\,m^3$

13 **a** $297\pi\,cm^2$ **b** $933.1\,cm^2$

14 **a i** $235\pi\,cm^2$ **ii** $738\,cm^2$
b i $64\pi\,cm^2$ **ii** $201\,cm^2$
c i $297\pi\,cm^2$ **ii** $933\,cm^2$

15 Students' own answers.

17 Check up

1 **a** $10.5 \leqslant$ length < 11.5 **b** $5.315 \leqslant$ length < 5.325

2 The pen could be up to 145 mm but minimum length of pencil case is 141.5 mm.

3 **a** 12π cm **b** 37.7 cm

4 14 cm or 14.0 cm

5 $78.5\,cm^2$

6 7 m

7 **a** 82.3 cm **b** $402\,cm^2$

8 $30\,954\,mm^2$

9 **a** $19.2\,cm^2$ **b** 5.5 cm **c** 19.5 cm

10 **a i** $54\pi\,cm^3$ **ii** $170\,cm^3$
b i $40\,500\pi\,cm^3$ **ii** $127\,000\,cm^3$
c i $2304\pi\,cm^3$ **ii** $7240\,cm^3$

11 **a** $170\,cm^2$ **b** $17\,000\,cm^2$ **c** $1810\,cm^2$

12 Volume $= 23\,cm^3$; surface area $= 60\,cm^2$

14 **a** Cylinder 45π, cone 15π
b Volume of cone $= \frac{1}{3}$ volume of cylinder
c $\frac{1}{3}\pi r^2 h$ is $\frac{1}{3}$ the volume of cylinder formula

15 $x = 3$

17 Strengthen

Accuracy

1 **a** 3.5 cm
b i $3.5 \leqslant n < 4.5$
ii $2.45 \leqslant n < 2.55$
iii $5.455 \leqslant n < 5.465$

Circles and sectors

1 **a** 12π cm **b** 37.7 cm

2 **a** 8π cm **b** 25.1 cm

3 **a** $C = \pi d$ **b** 21 cm

4 **a** 10 cm **b** Area $= \pi \times 10^2\,cm^2$
c $314.2\,cm^2$

5 **a** 5 m **b** 10 m

6 **a** $\frac{1}{2}$ **b** $\frac{1}{4}$ **c** $\frac{1}{6}$ **d** $\frac{13}{36}$

7 **a** Area $= 314.2\,cm^2$; circumference $= 62.8$ cm
b i $50\pi\,cm^2$ **ii** 10π cm **iii** 20 cm
iv $(10\pi + 20)$ cm

8 **a** $630\,cm^2$ **b** $177\,cm^2$ **c** $453\,cm^2$

9 **a** 18.8 cm **b** $\frac{1}{8}$ **c** 2.4 cm

10 **a** $144\pi\,cm^2$ **b** $24\pi\,cm^2$ **c** 24π cm
d 4π cm **e** $(4\pi + 24)$ cm

Volumes and surface areas

1 **a**

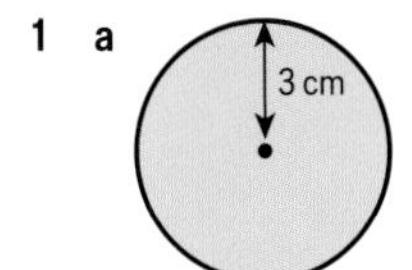

b $28.3\,cm^2$ **c** $339\,cm^3$

2 **a**

10π cm
5 cm
5 cm
6 cm

b i $78.5\,cm^2$ **ii** 31.4 cm **iii** $188.5\,cm^2$ **iv** $345.6\,cm^2$

3 **a** $48\,cm^2$ **b** 9 cm **c** $144\,cm^3$

4 **a** $324\pi\,cm^2$ **b** 24 cm **c** $8140\,cm^3$
d $540\pi\,cm^2$

5 **a** $33.5\,cm^3$ **b** $50.3\,cm^2$

17 Extend

1 **a** Maximum mass 187 g; minimum mass 153 g
b Maximum number 105; minimum number 95

2 £104.70

3 50.3 square units

4 $3.43\,cm^2$

5 **a** 31.4 m
b No, total distance is the same, number of points is the same, so mean $= \frac{\text{total distance}}{\text{number of points}}$ has not changed.

6 Area semicircle $= \frac{1}{2}\pi \times 4^2 = 8\pi$
Area quarter circle $= \frac{1}{4}\pi \times 8^2 = 16\pi$
Area rectangle $= 16 \times 8 = 128$
$\frac{24\pi}{128} = \frac{3\pi}{16}$

7 13.3 cm

8 No. Volume of tank $= 35^2 \times 140 \times \pi = 538\,783.14...\,cm^3$
$= 538.78...$ litres
Takes $538.78 \div 0.4 = 1346.95...$ seconds to fill
$= 22.4$ minutes, which is less than 1 hour.

9 **a** $216\pi\,cm^2$ **b** $448\pi\,cm^3$

10 1 : 6

11 Yes, density of $0.96...\,g/cm^3$ which is less than $1\,g/cm^3$

12 10

13 25 cm

17 Test ready

Sample student answers

a Student A is correct.

b Student B has used the circumference formula instead of the area formula for the quarter circle.

17 Unit test

Accept the use of $\pi = 3.14...$ throughout the test.

1 $75\,m \leqslant l < 85\,m$

2 $88 \div \pi = 28$ m (to nearest metre)

3 $254.5\,cm^2$

4 11 m

5 **a** $3980\,cm^2$ (to 3 s.f.) **b** 249 cm (3 s.f.)

6 **a** $10 + \frac{10\pi}{3}$ cm **b** $\frac{25\pi}{3}\,cm^2$

7 Total surface area of 2 tanks $= 34.683...\,m^3$. 8.7 tins.
He does not have enough paint.

8 $470\,000\,cm^3$ or 470 litres

9 $108\pi\,cm^2$

10 $3600\,cm^3$

11 **a** $4.7\,cm^3$ **b** 6π or $18.8\,cm^2$

12 Volume A (volume $A = 33\,432\,478\,\text{mm}^3$, volume $B = 23\,613\,427\,\text{mm}^3$)

13 Students' own answers.

Mixed exercise 5

1

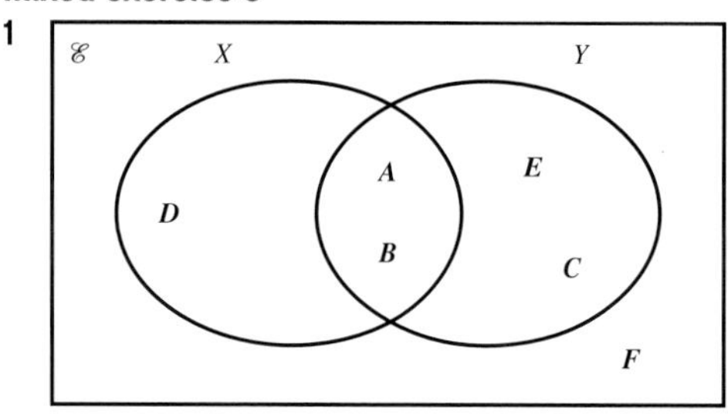

→ **13.4, 15.1**

2 27.5 cm → **15.4**

3 Cube → **15.2**

4

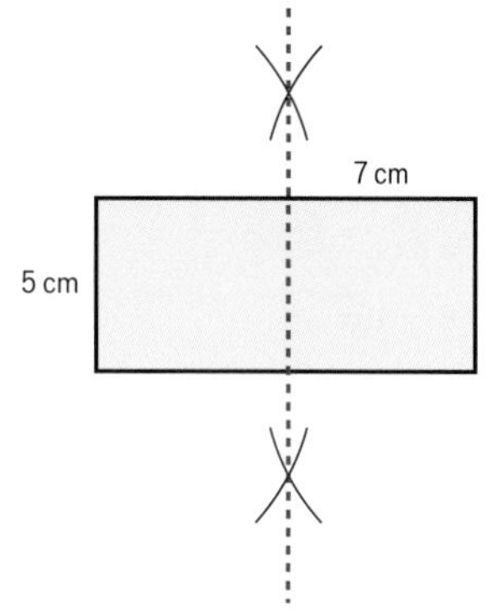

7 cm

5 cm

→ **15.5, 15.6**

5

5 cm

3 cm

6 cm

2 cm

Front elevation

Side elevation

→ **15.2**

6 $134.5\,\text{mm} \leqslant \text{circumference} < 135.5\,\text{mm}$ → **17.2**

7 a

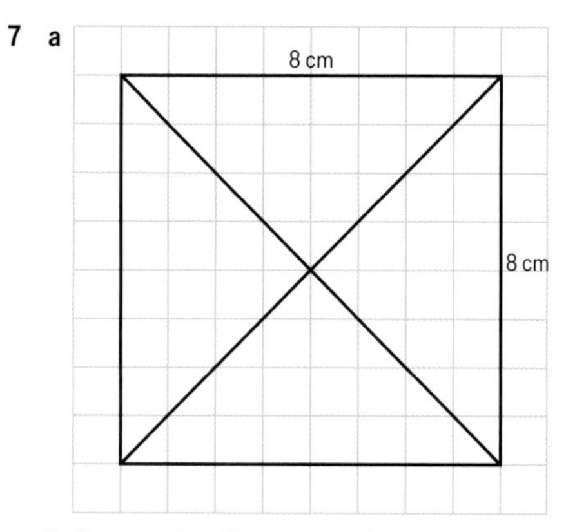

→ **15.2**

b Construction diagram not drawn to scale.

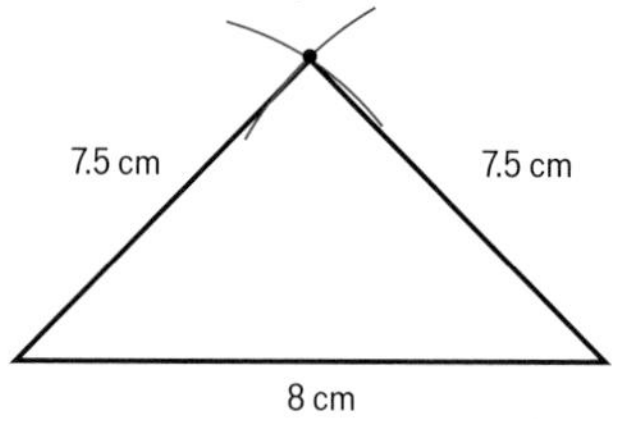

→ **15.3**

8 **a** iv, **b** ii, **c** v, **d** i, **e** vi, **f** iii → **16.1**

9 **a** She should have used a pair of compasses to construct the line, not a protractor. → **15.6**

b Construction diagram not drawn to scale.

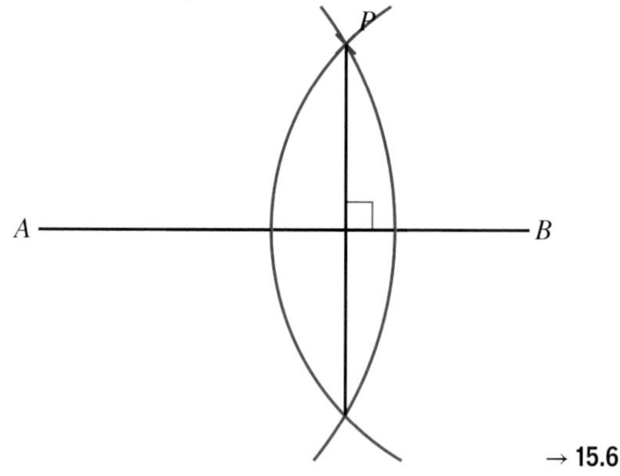

→ **15.6**

10 **a** **i** B **ii** A → **16.1**

b $x^2 + 10x + 24$ → **16.1**

c $10x + 48$ → **16.1**

11 Students' own answers, e.g. $(x+3)(x-2) = x^2 + x - 6$, not $x^2 - x - 6$. → **16.4**

12

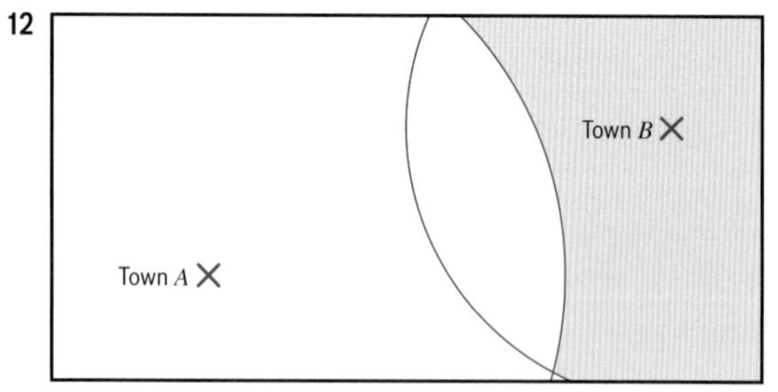

Scale: 1 cm represents 4 km

→ **15.7**

13 Distance from C to B is 1640 km, with C on a bearing of 010° from B → **15.8**

14 400 cm → **17.1**

15 a 9, 4, 1, 0, 1, 4, 9 → 16.2

b

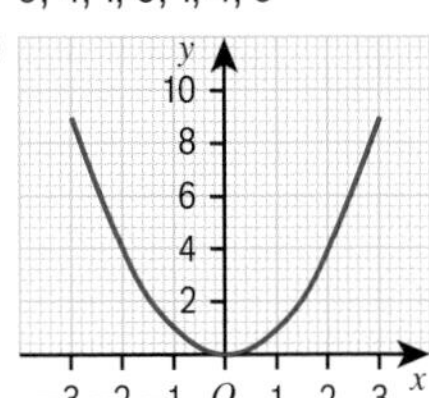

→ 16.2

c 2.1 to 2.3 or −2.3 to −2.1 → 16.2

16 $x^2+3x-28=(x+7)(x-4)=0$,
$(x+7)=0$ or $(x-4)=0$
$x=-7$ or $x=4$ → 16.5

17 Area $P=\frac{121}{4}\pi$, Area $Q=\frac{121}{16}\pi=\pi r^2$
$r=\sqrt{\frac{121}{16}}=\frac{11}{4}=2.75\,\text{cm}$ → 17.3, 17.4

18 $21.5\,\text{cm}^2$ → 17.5

19 a Volume $A=\frac{1}{3}\pi\times4^2\times9=48\pi$
Volume $B=\frac{1}{3}\pi\times6^2\times4=48\pi$ → 17.6

b Cone A: $l=\sqrt{97}\,\text{cm}$, Cone B: $l=2\sqrt{13}\,\text{cm}$ → 12.1

c No, surface area of cone $A=123.8\,\text{cm}^2$, surface area of cone $B=135.9\,\text{cm}^2$ → 17.6

20 Yes, Karis is right.
Volume $=16\,755.16\,\text{cm}^3=16.755\ldots$ litres,
$16.755\ldots$ litres $\div$ 2 litres/min $=8.38$ minutes,
which is less than a quarter of an hour. → 17.7

UNIT 18 Fractions, indices and standard form

18.1 Multiplying and dividing fractions

1 $1\frac{1}{4}$

2 $\frac{19}{5}$

3 a $7\frac{1}{2}$ **b** 16 **c** $\frac{12}{35}$
d $\frac{5}{12}$ **e** $26\frac{2}{3}$ **f** $3\frac{1}{2}$

4 a B **b** C **c** D **d** A

5 a $\frac{7}{10}$ **b** $\frac{7}{22}$ **c** $\frac{7}{23}$ **d** $\frac{3}{23}$

6 a $\frac{10}{3}$ **b** $\frac{10}{13}$ **c** $\frac{100}{33}$ **d** $\frac{100}{233}$

7 0.8

8 a $\frac{1}{4}$ **b** $\frac{1}{9}$ **c** $\frac{4}{9}$ **d** $\frac{8}{27}$

9 a $\frac{1}{2}$ **b** $\frac{3}{4}$ **c** $\frac{9}{10}$

10 a $1\frac{7}{20}$ **b** $1\frac{13}{20}$ **c** $1\frac{1}{2}$ **d** 3

11 a $\frac{3}{14}$ **b** $1\frac{3}{11}$ **c** $\frac{1}{2}$
d $2\frac{14}{15}$ **e** $\frac{54}{77}$

12 a $3\frac{21}{25}$ **b** $4\frac{4}{9}$ **c** $1\frac{7}{9}$
d 18 **e** $1\frac{11}{25}$

13 a $3\frac{1}{5}$ **b** $1\frac{11}{16}$ **c** $3\frac{5}{7}$
d 18 **e** $\frac{4}{7}$ **f** $\frac{25}{39}$
g $1\frac{1}{9}$ **h** $1\frac{2}{3}$ **i** $\frac{9}{22}$

14 a $3\frac{33}{35}$ **b** $267\frac{2}{9}$ **c** $7\frac{4}{13}$ **d** $56\frac{2}{5}$

15 26

16 $3\frac{3}{8}$ square inches

17 a 1 **b** 3 **c** 3 **d** 2

18 £7.40

18.2 The laws of indices

1 a 1 **b** −5 **c** 5 **d** −1

2 a 9 **b** 27 **c** 81
d 10 000 **e** 1 **f** 125

3 a 2^6 **b** 10^6 **c** 7^4
d 4^4 **e** 6^6

4 $11^2=121$

5 a $\frac{1}{36}$ **b** $\frac{1}{16}$ **c** $\frac{8}{125}$ **d** $\frac{1}{10\,000}$

6 a 3^6 **b** 3^5 **c** 3^{12} **d** 3^6
e 3^5 **f** 3^4 **g** 3^{12} **h** 3^6

7 a 3^8 **b** 3^{10} **c** 3^6 **d** 3^{18}

8 He should multiply the powers, not add them.

9

Power of 2	2^3	2^2	2^1	2^0	2^{-1}	2^{-2}	2^{-3}
Value	8	4	2	1	$\frac{1}{2^1}=\frac{1}{2}$	$\frac{1}{2^2}=\frac{1}{4}$	$\frac{1}{2^3}=\frac{1}{8}$

10 a i 1 **ii** 1 **iii** 1
b Students' own answers, e.g. They are equal to 1.

11 a i 0.25 **ii** 0.2 **iii** 0.1
b i $\frac{1}{4}$ **ii** $\frac{1}{5}$ **iii** $\frac{1}{10}$
c Students' own answers, e.g. A number raised to the power −1 is equal to one divided by the number.

12 a $\frac{1}{8}$ **b** $\frac{1}{7}$ **c** $\frac{1}{6}$ **d** $\frac{1}{x}$

13 a i 0.01 **ii** 0.001 **iii** 0.0001
b i $\frac{1}{100}$ **ii** $\frac{1}{1000}$ **iii** $\frac{1}{10\,000}$
c Students' own answers, e.g. A number x raised to the power $-n$ is equal to $\frac{1}{x^n}$.

14 a $\frac{1}{8}$ **b** $\frac{1}{49}$ **c** $\frac{1}{64}$ **d** $\frac{1}{125}$

15 a 8 **b** 25 **c** $\frac{81}{4}$ **d** $\frac{27}{125}$

16 $\frac{16}{25}$

17 a 2^3 **b** 2^{-3} **c** 8^0 **d** 4^{10}
e 12^7 **f** 11^{14} **g** 3^{-8}

18 a 81 **b** 1 **c** 400

19 a a^3 **b** $\frac{1}{b^3}$ **c** $\frac{1}{m^5}$
d n^5 **e** $\frac{1}{x^3}$ **f** c^0 or 1

20 a 0.25 **b** 16 **c** 1

21 100^{-2}, 10^{-3}, 1^4 and 1^{-5}, 10^3, 100^2

22 5^{-4}, 5^{-1}, 0.5^2, 0.5

23 a 2 **b** 30 **c** 0.18 **d** 18

24 a 0.004 **b** 9.57 **c** 9.85
d 762 **e** 7480 **f** 1.93

18.3 Writing large numbers in standard form

1 a 100 **b** 1000 **c** 10 000 **d** 100 000

2 a 500 **b** 3500 **c** 87.5 **d** 903 000

3 a 1 000 000 **b** 1 000 000 000

4 a 10 **b** 10 **c** 100

5 a 300 **b** 5 000 000 **c** 70 000
d 900 000 000 000

6 a 5×10^2 **b** 3×10^5 **c** 9×10^9 **d** 7×10^6

7 a 47 000 b 9 210 000
c 830 000 000 000 d 923 000
e 630 000 000 000 f 9 050 000
g 6 702 000 000 h 407 000 000

8 a C b A c F
d E e D f B

9 a 4.5×10^{5} b 3.2×10^{4} c 1.5×10^{5}
d 7.25×10^{6} e 6.291×10^{6} f 1.5×10^{6}
g 7.03×10^{8} h 7.6×10^{10}

10 4.503×10^{9}

11 a C; it is the only one written as a number between 1 and 10 multiplied by a power of 10.
b 4.2×10^{3}, 3.34×10^{4}, 4×10^{2}

12 3.2×10^{9}, 3.2×10^{8}, 3.2×10^{5}, 3.2×10^{4}, 320

13 a 7.05×10^{8} b 34 500 000

14 a 8×10^{12} bytes b 4.6×10^{6} metres
c 1.77×10^{10} litres d 9.5×10^{2} grams

15 a $<$ b $<$ c $>$

18.4 Writing small numbers in standard form

1 a 300 b 9.2 c 0.891 d 0.03245

2 a 0.001 b 0.000 001
c 0.1 d 0.01

3 a $\times\frac{1}{10}, \times0.1, \times10^{-1}$ b $\times10^{-2}, \times\frac{1}{100}, \times0.01$

4 a 0.000 03 b 0.08
c 0.4 d 0.000 000 000 07

5 a 3×10^{-2} b 5×10^{-3} c 1×10^{-4} d 3×10^{-7}

6 a 0.000 045 b 0.000 003 8
c 0.000 000 008 34 d 0.1401

7 a 5.2×10^{-2} b 7.1×10^{-4} c 5.69×10^{-4}
d 2.41×10^{-3} e 1.4×10^{-5} f 1.09×10^{-3}
g 3.04×10^{-5} h 6.102×10^{-1}

8 a D b A c F d B
e H f C g E h G

9 a Lucy: The first number should be between 1 and 10; she must divide by a further power of 10.
Ali: The zero between 9 and 7 is a place holder and should be included.
Sam: The power of 10 should be '–4'. otherwise you are making the number larger.
Tom: The number should be multiplied by a power of 10.
b 9.07×10^{-4}

10 a

Element	Radius (m)
lithium	1.45×10^{-10}
sodium	1.8×10^{-10}
phosphorus	1×10^{-10}
nitrogen	6.5×10^{-11}
tin	1.45×10^{-10}

b Nitrogen, phosphorus, lithium/tin, sodium

11 a 7.05×10^{-7} b 0.000 32

12 a 7×10^{-12} grams b 1.4×10^{-6} seconds
c 5.93×10^{1} metres d 1.05×10^{-8} volts
e 3.8×10^{-4} amps f 9.9×10^{-2} litres

13 Students' own answers according to calculator display. For example, some may give 5×10^{-11} where the power is displayed above a reduced-size 10.

18.5 Calculating with standard form

1 a 0.8 b 10 c $\frac{7}{11}$ d 12

2 a 3.2×10^{5} b 1.8×10^{9} c 9×10^{3} d 5.96×10^{-5}

3 a 10^{9} b 10 c 10^{4} d 10

4 a i 195 ii 1.95
b i 7 ii 0.7
c 159 000 d 0.00174

5 a 6×10^{5} b 7.5×10^{-7} c 6×10^{4}
d 1.8×10^{-5} e 4×10^{-3} f 2×10^{3}
g 5×10^{7} h 2×10^{-12}

6 a 2.1×10^{11} b 2×10^{14} c 3×10^{6}
d 1.2×10^{5} e 9.6×10^{6} f 7.77×10^{-11}

7 a 4×10^{-10} b 3×10^{-12} c 5×10^{-6}
d 2.2×10^{5} e 9.01×10^{3} f 8×10^{5}

8 a 6800 b 30 000 c 80 000

9 6.1×10^{11}

10 a 3.82×10^{11} b 1.50×10^{6} c 3.46×10^{-9}
d 1.65×10^{-5} e 6.37×10^{8} f 1.41×10^{18}

11 72 m

12 5×10^{4}

13 a i 32 000 000 ii 190 000 000
b 222 000 000
c 2.22×10^{8}

14 a, b i 4.9×10^{11} ii 5.07×10^{8}
iii $6.130\,72\times10^{7}$ iv 4.12×10^{-2}
v 7.654×10^{-5}

15 a i 0.000 19 ii 0.000 034
b 0.000 156 c 1.56×10^{-4}

16 a, b i 2.3×10^{9} ii 5.92×10^{-7}
iii 5.219×10^{5} iv 3.153×10^{-4}
v 4.6491×10^{-10} vi 9.9856×10^{-15}

17 a $5.199\,958\,9\times10^{-7}$ b 4.5882×10^{14}

18 a 5.95×10^{5} b 1.53×10^{-11}

19 5.8×10^{7}

20 a 57 600 000 000 m b 5.76×10^{10} m
c 192 seconds

18 Check up

1 $\frac{5}{7}$

2 a $1\frac{13}{35}$ b $2\frac{5}{6}$ c 4

3 a 6 b $1\frac{9}{11}$

4 a 5^{7} b 2^{-12} c x^{2} d 3^{-13}

5 a $\frac{1}{4}$ b $\frac{1}{8}$ c 1
d $2\frac{7}{9}$ e 1

6 a 400 000 000 b 5260
c 0.35 d 0.000 000 809 9

7 a 1.9×10^{5} b 1.05×10^{9}
c 7×10^{-6} d 4.52×10^{-5}

8 a 8×10^{9} b 6.6×10^{5} c 1.5×10^{8}

9 a 3×10^{6} b 5×10^{6} c 3.5×10^{-6}

10 a 3.24×10^{12} b 9.065×10^{-5}

11 a 4.14×10^{-8} b 7.86×10^{6}

13 $6\frac{1}{2}\times5\frac{3}{4}=37\frac{3}{8}$

18 Strengthen

Reciprocals and fractions

1 a 2, 2, 5 b $\frac{5}{10}, \frac{10}{5}, 2$ c 25, 25, 4
d $\frac{5}{100}, \frac{100}{5}, 20$ e $3, \frac{10}{13}$ f $\frac{1}{10}, \frac{21}{10}, \frac{10}{21}$

2 a $\frac{17}{5}$ b $\frac{17}{5}\times\frac{2}{3}=\frac{17\times2}{5\times3}=\frac{34}{15}=2\frac{4}{15}$

3 a $\frac{5}{6}$ b $1\frac{7}{25}$ c $1\frac{13}{14}$

4 $\frac{10}{3}\times\frac{7}{4}=\frac{5}{3}\times\frac{7}{2}=\frac{35}{6}=5\frac{5}{6}$

5 a $\frac{1}{2}$ b 2 c $3\frac{1}{2}$

6 $\frac{5}{4}\times\frac{7}{3}=\frac{5\times7}{4\times3}=\frac{35}{12}=2\frac{11}{12}$

7 a $2\frac{2}{15}$ b $10\frac{1}{2}$ c $1\frac{2}{3}$

8 $\frac{7}{3}\times\frac{3}{2}=\frac{7}{2}=3\frac{1}{2}$

9 a $\frac{16}{5}\div\frac{1}{4}=\frac{16}{5}\times\frac{4}{1}=\frac{64}{5}=12\frac{4}{5}$ b $\frac{8}{25}$ c $2\frac{1}{4}$

Indices

1 a i $\frac{64}{64}=1$ ii $8^{2-2}=8^0$ iii 1
b 1 c 1

2 a i 7^{-3} ii $\frac{1}{7^3}$ iii $7^{-3}=\frac{1}{7^3}$
b i 4^{-2} ii $\frac{1}{4^2}$ iii $4^{-2}=\frac{1}{4^2}$

3 a $\frac{1}{8}$ b $\frac{1}{5}$ c 1
d 1 e 2

4 a 2^{-1} b 4^3 c 8^{-7} d 19^{-10}

5 a 7^{-5} b 12^{10} c 6^7 d 2^{-3}

6 a 3^8 b 5^{14} c 19^{30} d 6^6
e 5^{-6} f 5^{-12} g 9^6 h 6^8

7 a $\frac{1}{3}\times\frac{1}{3}=\frac{1}{9}$; $\frac{1}{3}\times\frac{1}{3}\times\frac{1}{3}=\frac{1}{27}$ b $\frac{25}{4}$; $\frac{125}{8}$

8 a $\frac{9}{4}$ b 64 c $\frac{49}{16}$ d $\frac{1000}{27}$

Standard form

1

10^{-6}	0.000 001
10^{-5}	0.000 01
10^{-4}	0.0001
10^{-3}	0.001
10^{-2}	0.01
10^{-1}	0.1
10^{0}	1
10^{1}	10
10^{2}	100
10^{3}	1000
10^{4}	10 000
10^{5}	100 000
10^{6}	1 000 000

2 a 0.0003 b 500 c 0.0009 d 1200
e 0.000 057 f 112 g 0.000 903 h 1 010 000

3 a 1.8×10^4 b 9.6×10^5 c 4×10^4
d 9×10^6 e 7.51×10^5 f 1.08×10^6

4 a 6×10^{-2} b 4×10^{-5} c 3.6×10^{-3}
d 1.2×10^{-4} e 2.34×10^{-1} f 5.08×10^{-6}

5 $3\times2\times10^7\times10^3=6\times10^{10}$

6 a 5×10^{11} b 4×10^{-6} c 6.8×10^{-9}

7 a 1.5×10^6 b 2.9×10^4 c 3.61×10^5
d 4.28×10^7 e 1.72×10^{-2}

8 a 1.2×10^4 b 7.2×10^3 c 3.5×10^{14} d 1.26×10^{-3}

9 3×10^4

10 a 4×10^5 b 2.3×10^7 c 4×10^{-7} d 5×10^6

11 a i 52 000 ii 3500
b i 55 500 ii 48 500
c 5.55×10^4, 4.85×10^4

12 a i 0.0015 ii 0.000 022 3
b i 0.001 522 3 ii 0.001 477 7
c 1.5223×10^{-3}, 1.4777×10^{-3}

13 a i 0.000 068 71 ii 0.010 650 05
iii $4.432\,903\,226\times10^{-7}$
b i 6.87×10^{-5} ii 1.07×10^{-2}
iii 4.43×10^{-7}

14 a 7.90×10^{11} b 7.95×10^7 c 2.94×10^{-11}

18 Extend

1 $\left(\frac{2}{3}\div3\frac{4}{5}\right)$, 57%, $\frac{7}{9}$, 0.98, $\left(3\frac{1}{3}\div1\frac{2}{3}\right)$

2 4.57

3 a $n=4$ b $a=6$

4 a 1 000 000 b $-\frac{1}{2}$ c $\frac{1}{64}$ d 4
e 4 f 12 g 1 h $\frac{1}{144}$

5 a i 2^2 ii 2^6 iii 2^{48}
b i 5^3 ii 5^{-9} iii 5^{-81}
c 3^{-80}

6 a $\frac{2}{x}$ b $\frac{1}{2x}$ c $\frac{2}{x^2}$ d $\frac{1}{4x^2}$

7 a 125 million b 1.25×10^8

8 2.88×10^{-4}

9 a $9\times10^{-6}\,\text{m}^3$ b $3\times10^{-3}\,\text{m}^2$

10 a 10^6 b 10^3 c 10^{15}

11 400 000

12 $2.99\times10^{-23}\,\text{cm}^3$

13 $3.9182\times10^9\,\text{km}$

14 $2.67\times10^4\,\text{km/h}$

15 $5.6\times10^4-3.2\times10^3$, 6.5×10^2, $(3^2)^2$, $2\frac{1}{3}\times3\frac{3}{5}$

16 a $5.3\times10^8\,\text{km}^2$ (2 s.f.) b $1.2\times10^{12}\,\text{km}^3$ (2 s.f.)
c $1.6\times10^8\,\text{km}^2$ (2 s.f.)

17 a 3.2×10^{-2} b 3×10^{-8} c 0.12 d 2.5×10^{-7}

18 a $4\times10^{-5}\,\text{cm}^2$ b $8\times10^{-5}\,\text{cm}^2$
c $2.048\times10^{-2}\,\text{cm}^2$

19 $p=2.0\times10^2$

18 Test ready

Sample student answers

1 Students' own answers, e.g. Student A gives the better answer. There are 600 sheets with a thickness of 5×10^{-3} each. To find the total thickness multiply the thickness of each sheet by the number of sheets.
Student B has made a mistake, as $\frac{1}{0.002}$ is 500, not 5000.

2 a Students' own answers.
b Students' own answers, e.g. Student A should have written 8×10^{-3}; Student B has not written the answer in standard form.

18 Unit test

1 a $\frac{12}{5}$ or $2\frac{2}{5}$ b 10

2 a $\frac{15}{4}$ or $3\frac{3}{4}$ b $\frac{25}{12}$ or $2\frac{1}{12}$

3 a 3^{-5} b 10^{11} c 11^{-10}

4 a 30 400 000 b 0.0021

5 a 9.07×10^9 b 3.14×10^{-5}

6 a 1 b 2 c 64 d 0.25

7 a $\frac{1}{64}$ b $\frac{64}{27}$

8 6×10^{-6}

9 a 500 000 000 b 5.00×10^{8}

10 $1\div0.001$, 1×10^{3}

11 $(0.05)^{-2}=400$, $4.5\times10^{2}=450$, $\frac{(5^{3})^{2}}{5^{2}}=5^{4}=625$

12 9.9×10^{-7}

13 a i 6×10^{7} ii 60 000 000
b i 2×10^{-13} ii 0.000 000 000 000 2
c i 5.6×10^{6} ii 5 600 000
d i 6.2×10^{-4} ii 0.000 62

14 a 8.12×10^{7} b 5.65×10^{-8}

15 3.14×10^{21}

16 Mean mass of a feather = 0.01325 g

17 Students' own answers, e.g. $4^{-6}\times4^{1}$, $4^{-8}\div4^{-3}$, $\frac{1}{4^{5}}$
$4\times10^{6}\div10^{11}$

18 Students' own answers.

Unit 19 Congruence, similarity and vectors

19.1 Similarity and enlargement

1 a More b Less

2 $\frac{6}{8}$, $\frac{15}{20}$

3 a i 2 ii $\frac{1}{2}$
b i 3 ii $\frac{1}{3}$
c i $\frac{1}{2}$ ii 2

4 *B* and *D*. Rectangle *D* is an enlargement of rectangle *B*, because its sides are twice the length of the sides in rectangle *B*.

5 7.5 cm

6 No. $1\times3=3$, whereas $5\times3=15$ so the large rectangle would need to be 15 cm long for them to be similar.

7 *A* and *C*. Triangle *A* is an enlargement of triangle *C*.

8 a 37°, 53°, 90°. Both triangles have the same angles.
b 45°, 45°, 90°.
c Similar triangles have the same angles.

9 a i *XY* ii *AC* iii *YZ*
b i 2 ii $\frac{1}{2}$
c i 18 cm ii 6 cm
d i Angle *ZXY* ii Angle *ACB*
iii Angle *XYZ*
e $\frac{4}{9}$, $\frac{8}{18}$. The fractions are equivalent.

10 a *QR* b 7.5 cm
c i 30° ii 62°

11 a Yes. They have the same angles.
b *UW*
c i 3 ii *QR* iii $\frac{1}{3}$, *UW*

19.2 More similarity

1 a *P* b *Q* c *R*

2 Students' own answers, e.g.

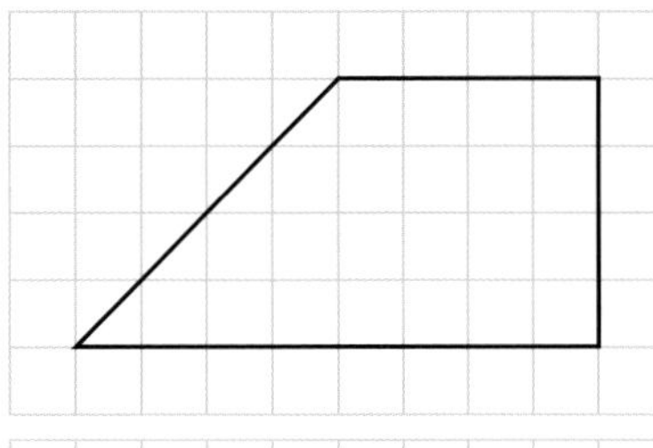

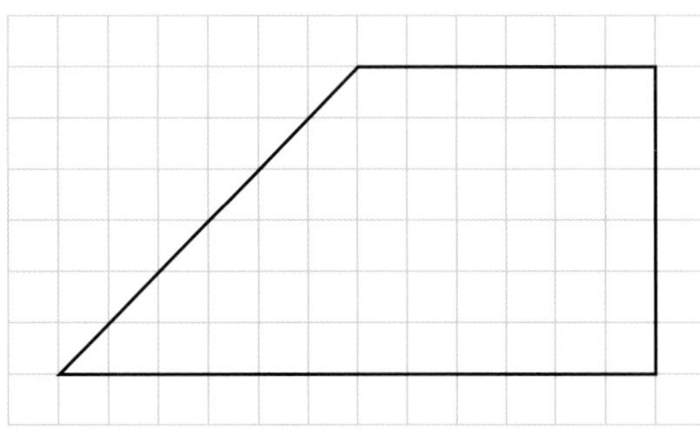

3 $x=4$

4

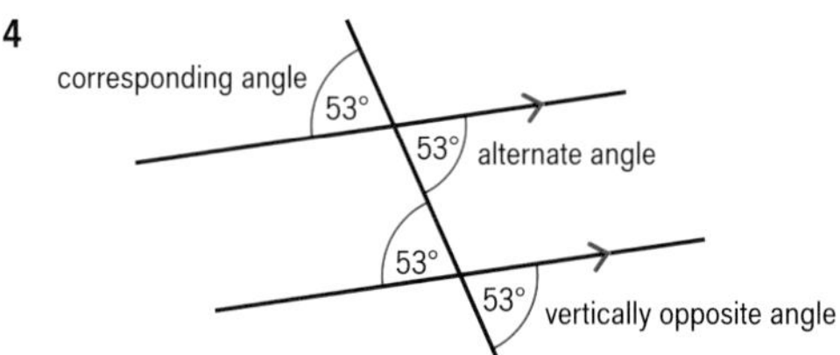

5 a They have the same angles.
b 2 c 10 cm

6 a 10 b $x=10$

7 a It has the same angles. b $\frac{1}{2}$

8 a They have the same angles.
b 1.4 or $\frac{5}{7}$
c $x=5.1$ cm to 1 d.p., $y=16.1$ cm to 1 d.p.

9 $a=6$ cm, $b=7$ cm

10 a 8.7 cm b 5.9 cm to 1 d.p.

11 a *TSR* b 106° c *TR*
d $\frac{3}{4}$ e 9 cm

12 a Angle *PQR* = angle *TSR*
Angle *QPR* = angle *STR*
Angle *PRQ* = angle *TRS*
b i 4.8 cm ii 3.2 cm

13 a 70° b 13.5 cm c 4.5 cm

19.3 Using similarity

1 No. They do not have the same angles (missing angles are 106° and 21°).

2 28 cm

3 a 10 cm b 5 cm

4 a $AB=4$ cm, $DF=11.7$ to 1 d.p.
b No
c No. Corresponding sides are not in the same ratio.
d *AC* and *EF* are not corresponding sides. Nor are *BC* and *DE* (the hypotenuse of one triangle can't correspond to a non-hypotenuse side of another triangle).

5 a Need more information
b Yes
c Need more information
d Need more information

6 No. Corresponding sides are not in the same ratio.

7 No. They don't have the same angles.

8 a Need more information
b Need more information
c Yes

9 Yes. They have the same angles and corresponding sides are in the same ratio.

10 a, b Yes. They have the same angles and corresponding sides are in the same ratio.

11 a 3 b 18 cm, 24 cm, 24 cm c 32 cm
d 96 cm e 3
f The scale factor of the perimeters is the same as the scale factor of the lengths.

12 20 m

13 32 m

19.4 Congruence 1

1 *GHI*

2 a *PQ* b *RPQ*

3 a Accurate constructions of triangles *ABC* and *YZX*.
b Yes
c Yes

4 a $x = 120°, y = 35°$ b $x = 32°, y = 28°$

5 SSS

6 a SAS b RHS

7 No. They have the same angles but are not the same size.

8 No. They only share one angle.

9 **C** and **D** (ASA)

19.5 Congruence 2

1 a *AB* and *AC*; *ABC* and *ACB*
b *DE* and *GF*; *DG* and *EF*; all angles
c all sides; *KHI* and *KJI*; *JKH* and *HIJ*
d *LM* and *ON*; *LO* and *MN*; *LMN* and *NOL*; *OLM* and *MNO*

2 a $x = 148°$, vertically opposite are equal
$y = 148°$, corresponding (or alternate) angles are equal
$z = 32°$, angles on a straight line add up to 180°
b $x = 37°$, alternate angles are equal
$y = 73°$, angles in a triangle add up to 180°
$z = 73°$, alternate angles are equal

3 a *MB* b *MC* c Angle *BMD*

4 *C*

5 a 5.4 cm
b i 51° ii 83°

6 a They are congruent (SSS or SSA).
b Angle *BDC* (alternate angles are equal)

7 a 110°, opposite angles in a rhombus are equal
b Yes (SAS or SSS)

8 a *AC* b Yes (SSS) c i 20° ii 25°

9 a $AM = MC$, $BM = MD$, $AB = CD$,
angle $AMB = 47°$ = angle CMD,
angle BAM = angle DCM, angle ABM = angle CDM
b Yes (SSS or SAS)

10 a Yes, triangle *PQR* is congruent to triangle *TSR* (ASA).
b $RT = RP = 14$ cm

11 a 50° (vertically opposite)
b All 6 cm (radii) c Yes; SAS

12 a 43° (alternate angles are equal)
b 54° (alternate angles are equal)
c Yes (ASA)
d 8 cm (corresponding positions in congruent triangles)

13 a Angle *NLO* (alternate angles)
b Yes (SAS)

14 a 65° (angles in a triangle sum to 180°)
b Yes (ASA) c *RQ* (ASA)

19.6 Vectors 1

1 a 2 b −2 c 10 d −7

2 a, b

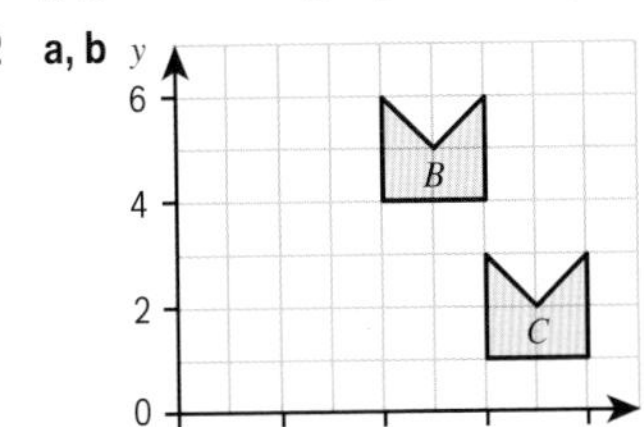

c A translation of 5 units to the right, 1 unit up

3 Add them together

4 a $\begin{pmatrix}5\\6\end{pmatrix}$ b $\begin{pmatrix}7\\3\end{pmatrix}$ c $\begin{pmatrix}7\\4\end{pmatrix}$
d $\begin{pmatrix}7\\-1\end{pmatrix}$ e $\begin{pmatrix}0\\0\end{pmatrix}$

5 $\begin{pmatrix}2\\5\end{pmatrix}$

6 a $\begin{pmatrix}4\\-2\end{pmatrix}$ b $\begin{pmatrix}-5\\1\end{pmatrix}$ c $\begin{pmatrix}0\\-2\end{pmatrix}$

7 a i $\begin{pmatrix}3\\3\end{pmatrix}$ ii $\begin{pmatrix}4\\-2\end{pmatrix}$ iii $\begin{pmatrix}7\\1\end{pmatrix}$
b $\begin{pmatrix}3\\3\end{pmatrix}+\begin{pmatrix}4\\-2\end{pmatrix}=\begin{pmatrix}7\\1\end{pmatrix}$

8 a $\overrightarrow{NT}$ b $\overrightarrow{PR}$ c $\overrightarrow{VT}$

9 a–c

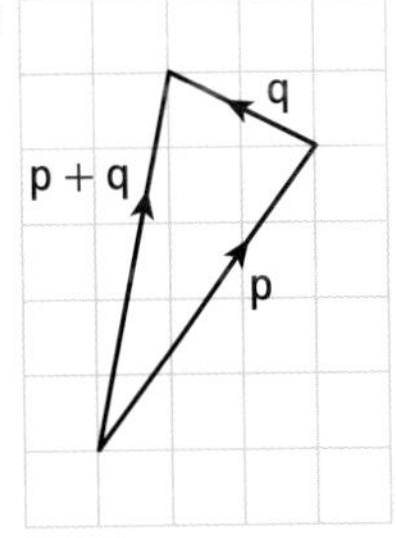

d

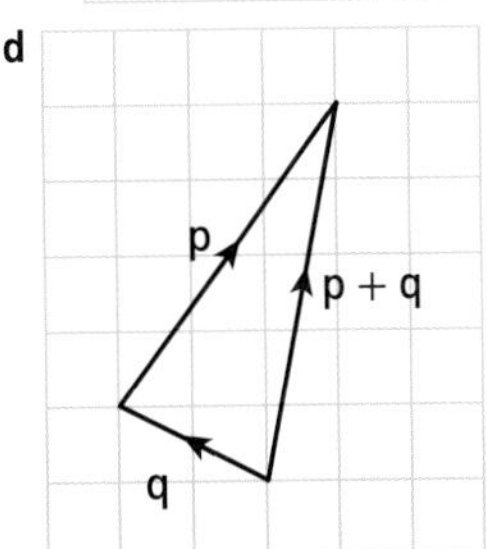

e No. The vectors are the same.

10 a $\begin{pmatrix}-1\\-14\end{pmatrix}$ b $\begin{pmatrix}-4\\4\end{pmatrix}$

19.7 Vectors 2

1 a −6 b 4 c −10 d 12

2 a i DC ii AD
b i 2 ii 2 iii 2 iv 2

3

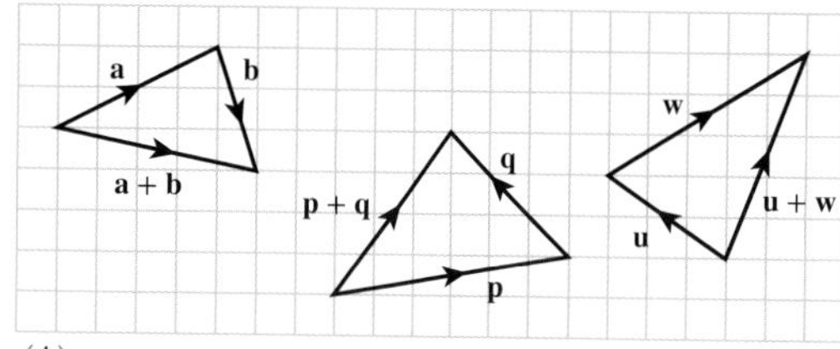

4 $\begin{pmatrix} 1 \\ 4 \end{pmatrix}$

5

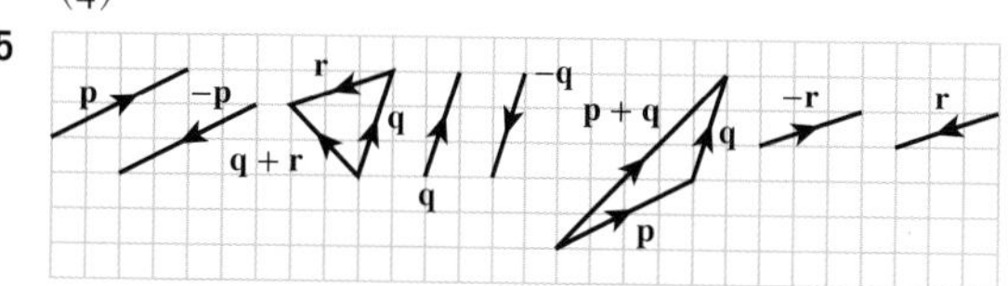

6

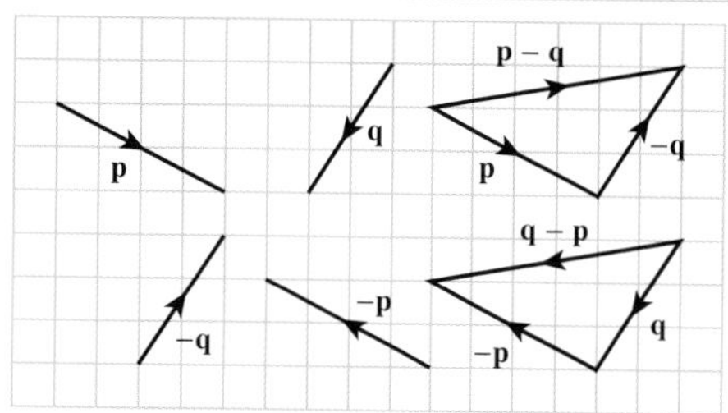

7 a $\begin{pmatrix} 5 \\ 1 \end{pmatrix}$ b $\begin{pmatrix} -3 \\ 0 \end{pmatrix}$ c $\begin{pmatrix} 5 \\ -5 \end{pmatrix}$

8 a $\mathbf{r} = \begin{pmatrix} 5 \\ -3 \end{pmatrix}$ b

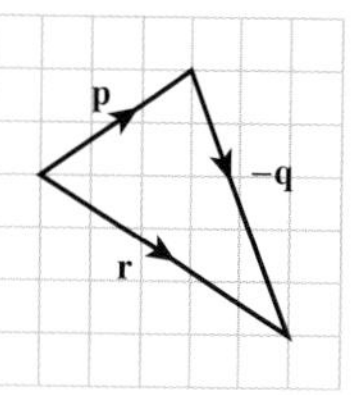

9 a $\mathbf{z} = \begin{pmatrix} 4 \\ 2 \end{pmatrix}$ b

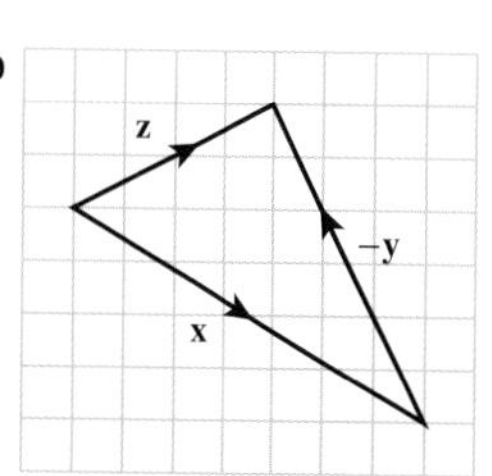

10 a $\begin{pmatrix} 4 \\ 10 \end{pmatrix}$ b $\begin{pmatrix} 9 \\ -6 \end{pmatrix}$ c $\begin{pmatrix} 13 \\ 4 \end{pmatrix}$ d $\begin{pmatrix} -8 \\ -20 \end{pmatrix}$
e $\begin{pmatrix} -15 \\ 10 \end{pmatrix}$ f $\begin{pmatrix} -23 \\ -10 \end{pmatrix}$ g $\begin{pmatrix} 5 \\ 3 \end{pmatrix}$ h $\begin{pmatrix} 0 \\ 19 \end{pmatrix}$

11 $\begin{pmatrix} 7 \\ -6 \end{pmatrix}$

12 a, b, c, d

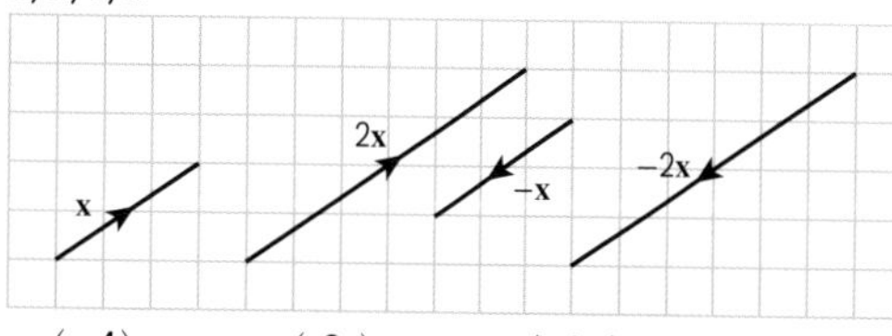

13 a $\begin{pmatrix} -4 \\ 8 \end{pmatrix}$ b $\begin{pmatrix} 2 \\ -4 \end{pmatrix}$ c $\begin{pmatrix} 6 \\ -12 \end{pmatrix}$
d, e $\begin{pmatrix} -1 \\ 2 \end{pmatrix}$ or $\begin{pmatrix} 1 \\ -2 \end{pmatrix}$

14 a a + b b −a c −b
d −a − b e 2b

15 a, c, e

16 a i, ii

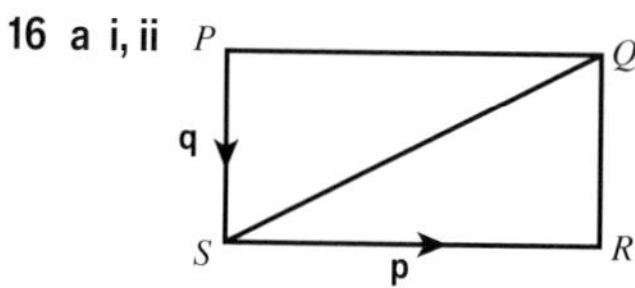

b p − q

c

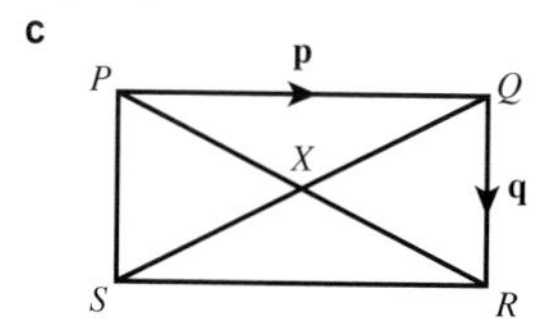

d Yes. $\overrightarrow{SX}$ is $\frac{1}{2}\overrightarrow{SQ}$

17 a 2c b d − c c −c − d

19 Check up

1 a $\frac{1}{3}$ b 16 cm

2 a 2.5 b 4 cm

3 a PQ b 9 cm
c i 40° ii 64°

4 a Yes. ASA (work out the missing angles)
b No. The 17 cm side is not between the 25° and 100° angles in the second triangle.

5 a **a** = 35° alternate angle; **b** = 106° interior or supplementary angle or angles in a triangle sum to 180°.
b Yes (ASA)

6 $\begin{pmatrix} 1 \\ 8 \end{pmatrix}$

7 a $\overrightarrow{UW}$ b $\begin{pmatrix} 6 \\ 0 \end{pmatrix}$

8 a $\begin{pmatrix} 9 \\ -3 \end{pmatrix}$ b $\begin{pmatrix} 7 \\ 9 \end{pmatrix}$

9 a, b, c, d

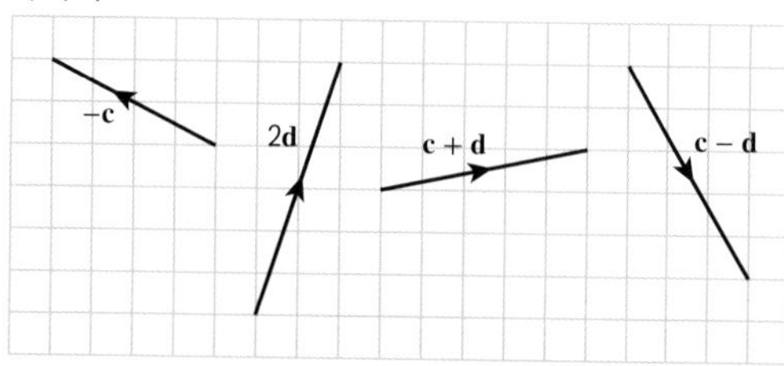

11 a a b −b c −c
d a + b e −a − b − c

19 Strengthen

Similarity and enlargement

1 a, b, e

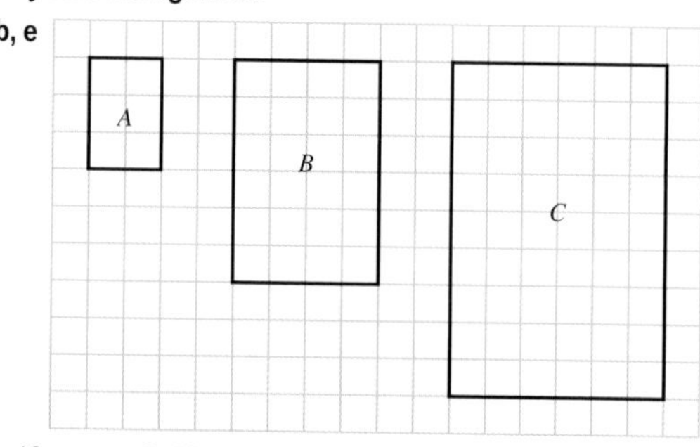

c i 10 ii 20
d 2 f 30

2 **a** 4
  b **i** 2.5 **ii** $\frac{1}{2}$

3 Triangle B 48 cm; triangle D 30 cm

4 **a** **i** Alternate angles are equal.
  ii Alternate angles are equal.
  iii Vertically opposite angles are equal.
  b **i, ii, iv**

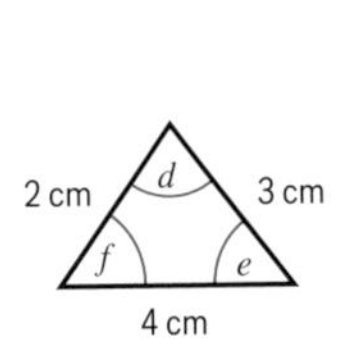

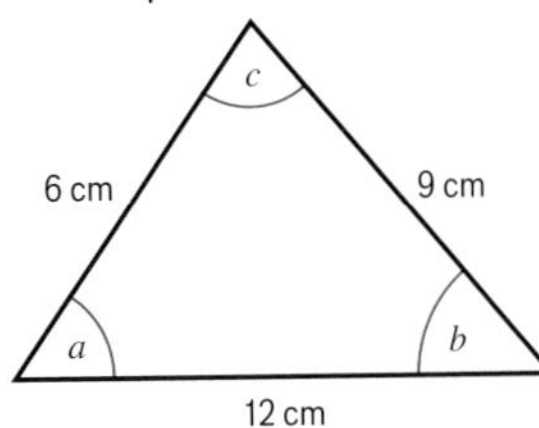

  iii 3

5 **a** XY **b** 2 **c** 14 cm
  d **i** 46° **ii** 60°

Congruence

1 Students' own answers.

2 **a**

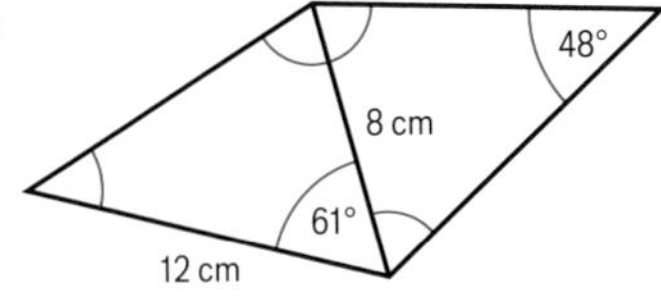

  b Accurate drawings of triangles STU and SVU. They are congruent (ASA).
  c UV

Vectors and translations

1 **a** $\begin{pmatrix}6\\5\end{pmatrix}$ **b** $\begin{pmatrix}-1\\2\end{pmatrix}$ **c** $\begin{pmatrix}-2\\-8\end{pmatrix}$
  d $\begin{pmatrix}3\\-2\end{pmatrix}$ **e** $\begin{pmatrix}9\\2\end{pmatrix}$

2 **a** $\begin{pmatrix}1\\4\end{pmatrix}$ **b** $\begin{pmatrix}1\\4\end{pmatrix}$ They are the same.

3 $\begin{pmatrix}-1\\7\end{pmatrix}$

4 $\overrightarrow{PX}$

5 **a** $\begin{pmatrix}12\\3\end{pmatrix}$ **b** $\begin{pmatrix}-4\\12\end{pmatrix}$ **c** $\begin{pmatrix}8\\15\end{pmatrix}$ **d** $\begin{pmatrix}-16\\9\end{pmatrix}$

6 **b** **i** D
  ii B
  c **i**

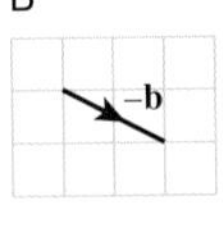

  ii

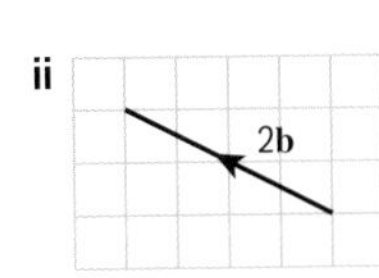

  d A
  e **a** + **b**

7 **a** $\begin{pmatrix}1\\9\end{pmatrix}$ **b** $\begin{pmatrix}-5\\1\end{pmatrix}$ **c** $\begin{pmatrix}5\\-1\end{pmatrix}$ **d** $\begin{pmatrix}4\\13\end{pmatrix}$

19 Extend

1 **a** (0, 20) **b** (80, 0)

2 **a** 3.5
  b **i** 10.5 m **ii** 7.5 m

3 **a** Angle A is common to both triangles.
  Angle AED = angle ACB (corresponding angles)
  Since two angles are the same, the third angle in each triangle must be the same, so the triangles are similar.
  b 1.25 **c** 1.75 cm

4 **a** 5.4 cm **b** 3.6 cm

5 **i** $a = 10$ cm, $b = 8.7$ cm
  ii $c = 8.7$ cm **iii** $d = 8.7$ cm
  iv $e = 5$ cm, $f = 8.7$ cm **v** $g = 17.3$ cm
  a i, ii, iii, iv
  b v is similar to each of the other triangles.

6 **a** 13 cm **b** 13 cm **c** Yes

7 $\overrightarrow{FV}$

8 **a** 2**a** **b** **b** − **a** **c** **a** + 2**b** **d** 2**b** − 3**a**

9 **a** $\frac{1}{2}$**p** **b** $\begin{pmatrix}4\\0\end{pmatrix}$

10 **a** −**a** + **b** **b** 2**b** **c** −**a** + 2**b**

19 Test ready

Sample student answers

1 The student has drawn the direction lines from A to B to help show that the directions are positive.
The student has labelled the axes with the coordinates known, to more clearly see how far they are counting, which is more difficult without grid lines.

2 The student applied the scale factor to one side only.
The scale factors for the corresponding sides are $\frac{15}{3} = 5$, $\frac{8}{1.6} = 5$, $\frac{17}{3.4} = 5$. They are all the same so the triangles are similar.

19 Unit test

1 **a** 1.5 **b** 48°; corresponding angle with angle ADE
  c 4 cm

2 54 cm

3 $x = 16$ cm, $y = 3$ cm

4 **a** They are congruent (SAS): angle ABC = angle PQR, $BC = QR$, $AB = PQ$
  b 12.9 cm
  c 38.9°

5 $\begin{pmatrix}5\\-2\end{pmatrix}$

6 **a** Yes: SSS. $CD = AB$, $AD = BC$, AC is common.
  b Angle ADC is the same as angle ABC (corresponding angles in congruent triangles)

7 $\begin{pmatrix}4\\-7\end{pmatrix}$

8 **a** $\begin{pmatrix}6\\-8\end{pmatrix}$ **b** $\begin{pmatrix}6\\-5\end{pmatrix}$ **c** $\begin{pmatrix}2\\-4\end{pmatrix}$

9 $\begin{pmatrix}1\\5\end{pmatrix}$

10 **a** −**a** **b** −**a** + **b** **c** $\overrightarrow{AC}$ = **a** + **b** ≠ $\overrightarrow{BD}$

11 **a** Similar: same angles, may not be the same size
  b Congruent: right-angled triangles with same hypotenuse and one side
  c Neither: different angles
  d Congruent: same angles, same size

12 Students' own answers.

UNIT 20 More algebra

20.1 Graphs of cubic and reciprocal functions

1 **a** $\frac{1}{4}$ **b** 3 **c** 2 **d** $-\frac{1}{5}$ **e** −4

2 $x = -3$ or $x = 1$

3

x	−3	−2	−1	0	1	2	3
y	−27	**−8**	**−1**	**0**	**1**	8	**27**

4 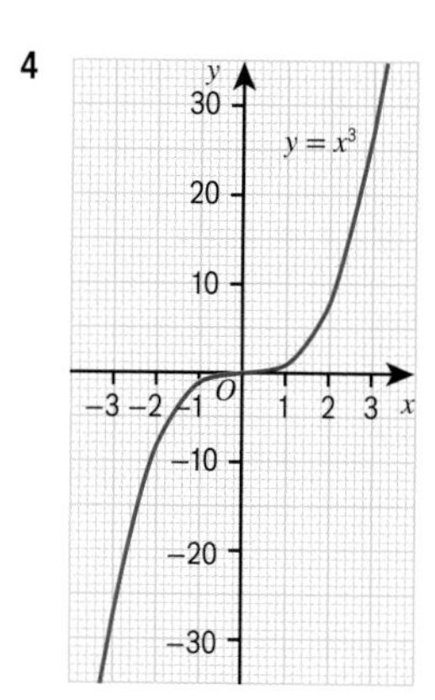

5 **a** Students' estimates from their graphs

i Approx. 12 **ii** Approx. 2.4

b **i** 12.167 **ii** 2.47 (2 d.p.)

c Part **b**, because in part **a** you cannot read very accurately from a graph.

6 **a**

x	−3	−2	−1	0	1	2	3
y	27	8	1	0	−1	−8	−27

b 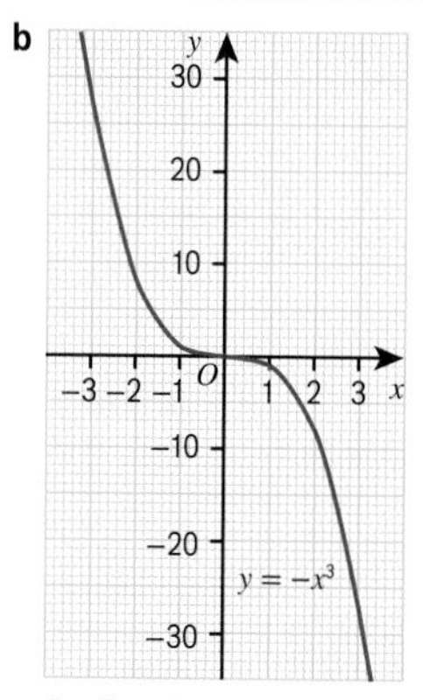

c Students' own answers, e.g. The graph of $y = -x^3$ is a reflection of $y = x^3$ in the y-axis.

7 **a**

x	−3	−2	−1	0	1	2	3
x^3	**−27**	−8	**−1**	**0**	**1**	**8**	27
+2	+2	+2	+2	+2	+2	+2	+2
y	**−25**	−6	**1**	**2**	**3**	**10**	29

b, d 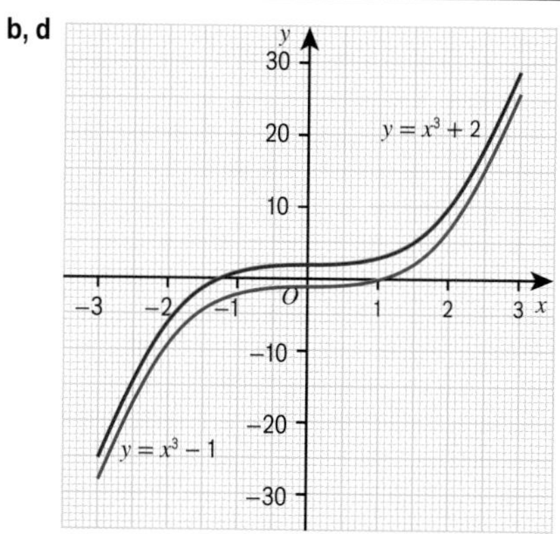

c

x	−3	−2	−1	0	1	2	3
x^3	−27	−8	−1	0	1	8	27
−1	−1	−1	−1	−1	−1	−1	−1
y	−28	−9	−2	−1	0	7	26

e Same shape; cross the x- and y-axes at different places.

f Students' own answers, for example, They would have the same shape as $y = x^3$ but y-intercepts of (0, 5) and $(0, -\frac{1}{2})$.

8 Answers between −1.2 and −1.4

9 **a**

x	−3	−2	−1	0	1	2	3
y	**−12**	**2**	4	0	**−4**	**−2**	12

b 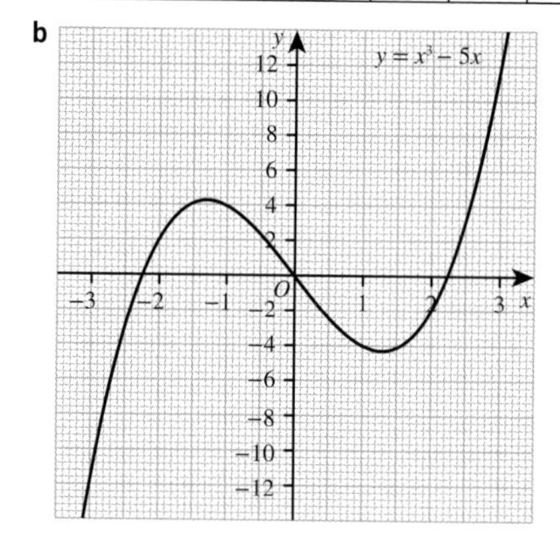

10 a

x	−4	−3	−2	−1	$-\frac{1}{2}$	$-\frac{1}{4}$	$\frac{1}{4}$	$\frac{1}{2}$	1	2	3	4
y	$-\frac{1}{4}$	$-\frac{1}{3}$	$-\frac{1}{2}$	−1	−2	−4	4	2	1	$\frac{1}{2}$	$\frac{1}{3}$	$\frac{1}{4}$

b, c

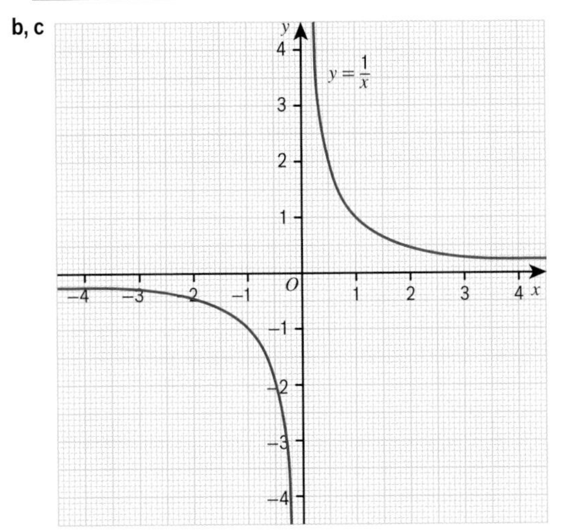

d $x \approx 0.3$

11 Students' own answers, for example, The x-axis is an asymptote of $y = \frac{1}{x}$.

12 a

x	0.5	1	1.5	2	2.5	4	5
y	**8**	4	**2.7**	2	**1.6**	**1**	0.8

b

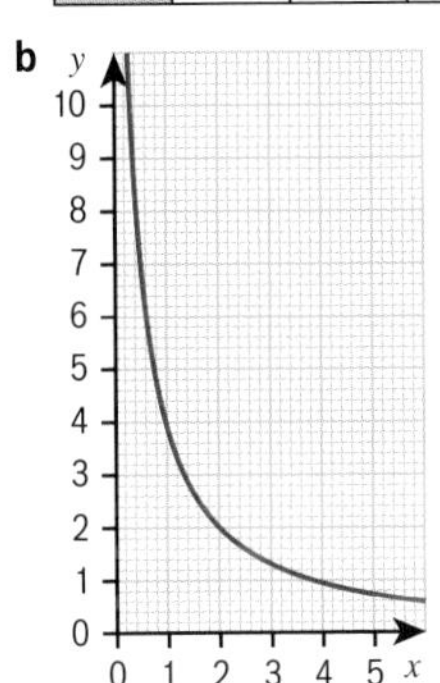

13 a

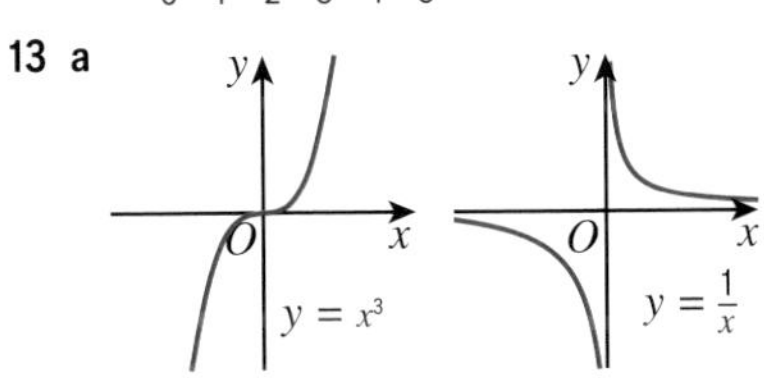

b Reflect the graph in the x-axis or y-axis.

14 a ii **b** iv **c** v
d iii **e** vi **f** i

20.2 Non-linear graphs

1 a $y = \frac{1}{x}$ **b** $y = x$

2 a $8\,\text{cm}^3$ **b** $125\,\text{cm}^3$ **c** $x^3\,\text{cm}^3$

3 1 hour 18 minutes

4 a $y = x^3 + 8$

b Students' own answers, e.g.

x	0	1	2	3	4
y	8	9	16	35	72

c

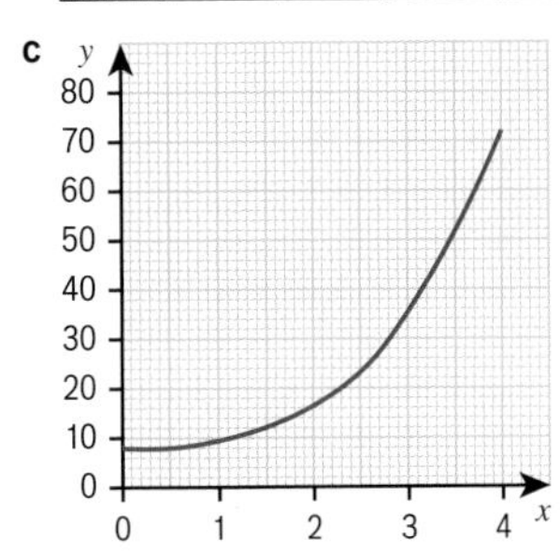

d Answers between 26 and 29

e Answers between 3.1 and 3.3

5 a 20 counts per second

b 6 weeks

c 2 weeks

d Mathematically, the answer is no, because you can keep halving; in real-life, however, the answer is yes because eventually there are no more atoms left to decay (i.e. you can't halve an atom).

6 a Inverse proportion

b Answers between 6.6 and 6.8 hours

c 14

7 a 100 litres

b 20 litres

c i 4 minutes

ii Answers between 8 and 9 minutes

d Between 0 and 2 minutes, 30 litres emptied out. Between 8 and 10 minutes, about 6 litres emptied out. This shows that the tank emptied less quickly as time passed and more water emptied out.

8 a Speed is inversely proportional to time.

b Estimates between 3 hours 20 minutes and 3 hours 30 minutes

c Estimates between 47 km/h and 49 km/h

9 a

Time (hours)	0	1	2	3	4
Number of cells	2	4	8	16	32

b First term is 2; multiply by 2 to get next term

c 18 or 19

d Answers between 3 hours 45 minutes and 3 hours 50 minutes

10 a £1000 **b** £1220
c £40 **d** 4%

20.3 Solving simultaneous equations graphically

1 **a** **i** $3x$ **ii** $2y$ **iii** $3x + 2y$

b $3x + 2y = 9$

2 **a**

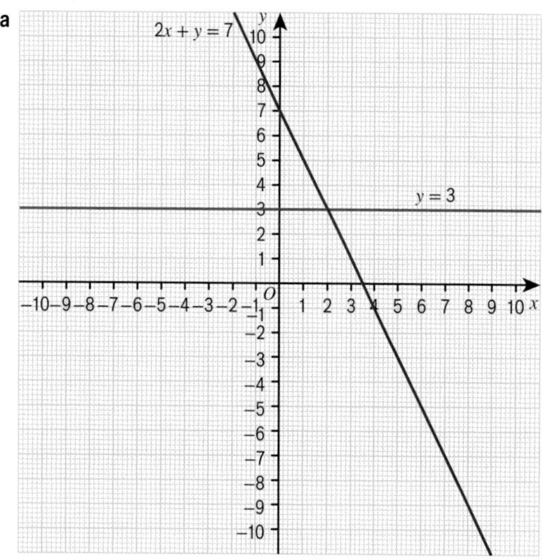

b (2, 3)

3 **a, b**

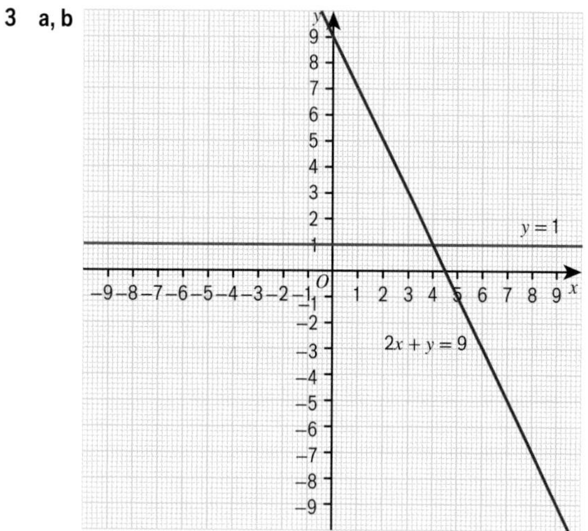

c $x = 4,\ y = 1$

4

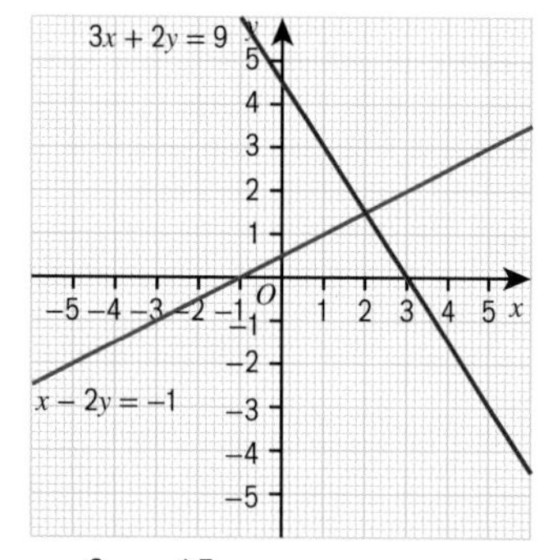

$x = 2,\ y = 1.5$

5

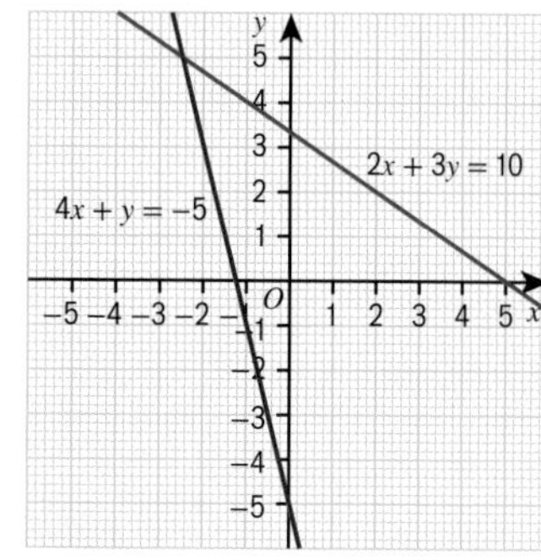

$x = -2.5,\ y = 5$

6 **a** $x = 5,\ y = 3$

b $x = 3.8,\ y = 1.8$

c $x = 7,\ y = 1$

d No, as the graphs do not all cross at one single point.

7 **a**

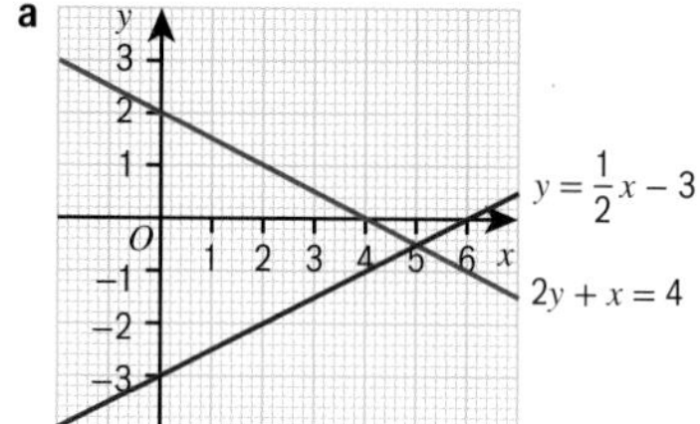

b $x = 5,\ y = -0.5$

8

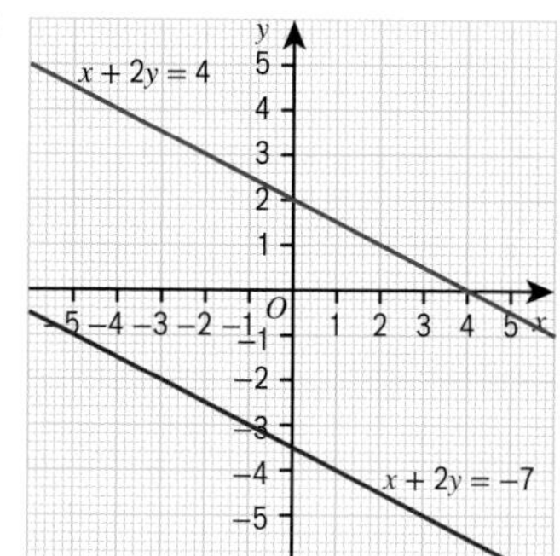

They do not have a solution because the graphs are parallel lines, so do not intersect.

9

$y = x + 13$

$y = 23 - x$

5, 18

10 a $x+y=18$ **b** $x-y=4$

c

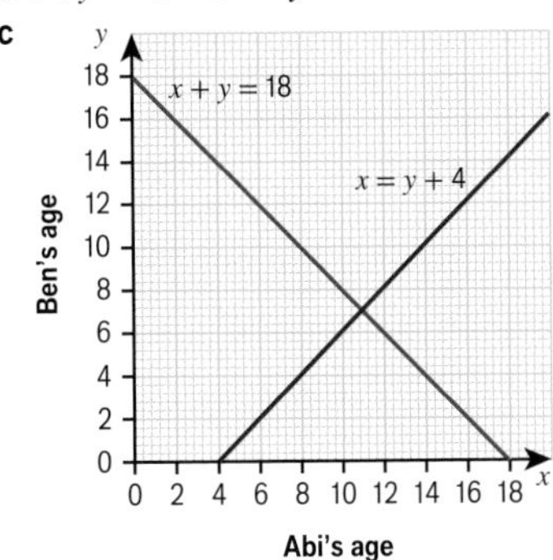

d Abi is 11 and Ben is 7.

11 a $x+y=16$ **b** $x+4y=34$

c

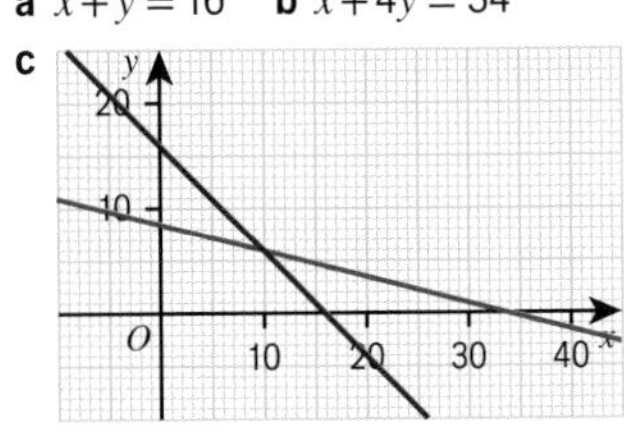

d £10 **e** £6

12 a x cost of adult ticket; y cost of child ticket

b $4x+9y=112$

c $x=£10$, $y=£8$

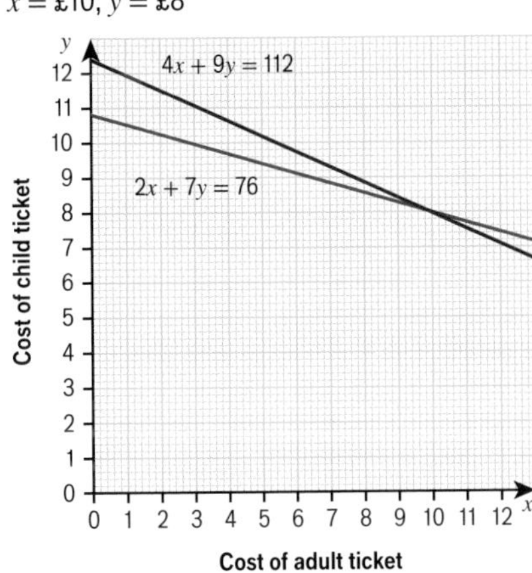

13 a $2x+y=60$; $x+2y=84$

b

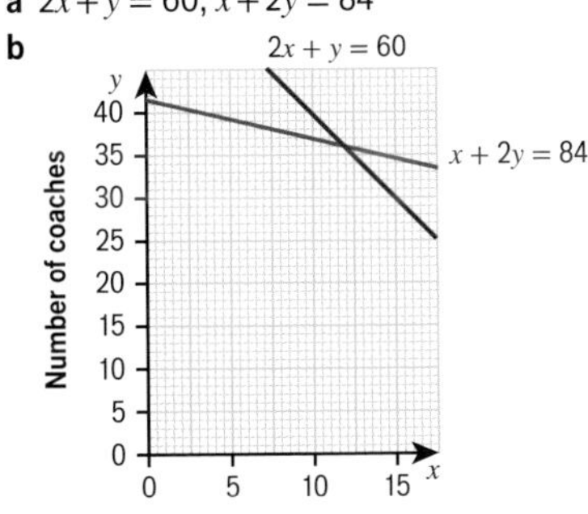

Minibus 12, coach 36

14 $2x+y=6.8$
$3x+2y=11.1$

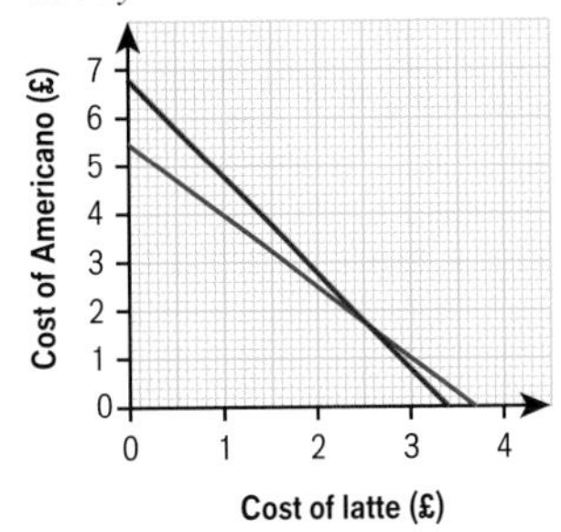

a £2.50 **b** £1.80

15 Set B would be harder to solve from a graph because to fit 115 and 5 on a graph you would need a very small scale, which can be difficult to read accurately.

20.4 Solving simultaneous equations algebraically

1 a 0 **b** $4x$ **c** 0 **d** 0

2 a + **b** − **c** + **d** −

3 a $y=6$ **b** $x=-1.5$ or $-\frac{3}{2}$
c $y=-4$ **d** $x=2$

4 a $2x+1=9$ **b** $x=4$
c $x=4, y=1$ **d** Yes

5 a $x=5, y=3$ **b** Yes

6 a $x=1, y=3$ **b** $x=-2, y=5$
c $x=2.75, y=3$ **d** $x=-\frac{2}{3}, y=1$

7 a $x=5, y=-2$ **b** $x=\frac{1}{4}, y=1$
c $x=-\frac{4}{5}, y=4\frac{3}{5}$
d If the signs are the same, subtract; if the signs are different, add.

8 $x=\frac{1}{6}, y=-\frac{3}{2}$ or -1.5

9 $x+y=24$ and $x-y=14$
The two numbers are 5 and 19.

10 a 4, 2 **b** 4, 2, 6 **c** 6 **d** $y=13$

11 a $x=2, y=5$ **b** $x=4, y=-1$
c $x=0.5, y=1.5$ **d** $x=2.4, y=0.8$

12 a $6x+9y=33$ **b** $6x+8y=30$
c $x=1, y=3$

13 a $8x+12y=44$ **b** $9x+12y=45$
c $x=1, y=3$ **d** Students' own answers.

14 a £7.50 **b** £10

15 a At point A: $4=2m+c$
At point B: $10=m+c$
b $m=-6, c=16$ **c** $y=-6x+16$

16 a $y=x+3$ **b** $y=\frac{3}{2}x+5$

20.5 Rearranging formulae

1 a $-x$ **b** $+z$ **c** $\div y$ **d** $\div -s$
e $\times m$ **f** $\sqrt{\ }$ **g** $\square^2$

2 a $x=3$ **b** $x=4$ **c** $x=2$

3 a i and iv **b** ii and iii

4 a C **b** d **c** V **d** A

5 a $\frac{A}{5}=w$ **b** $2M=V$ **c** $d=\frac{C}{3}$
d $R=\frac{V}{I}$ **e** $d=st$ **f** $r=pq$
g $m=dv$ **h** $F=PA$

6 a $q=\frac{4}{p}$ **b** $A=\frac{F}{P}$ **c** $v=\frac{m}{d}$ **d** $I=\frac{V}{R}$

7 **a** $D = ST$ **b** $T = \frac{D}{S}$ **c** Students' own answers.
d **i** $M = DV$ **ii** $V = \frac{M}{D}$

8 **a** $t = \frac{v-u}{a}$ **b** $z = \frac{x+y}{m}$ **c** $p = \frac{q-l}{r}$ **d** $h = \frac{e-f}{g}$

9 **a** $y = -3x + 2.5$
b Gradient $= -3$; y-intercept $= 2.5$

10 **a** and **d**

11 $3y - 12x + 9 = 0$
$3y = 12x - 9$
$y = 4x - 3$, gradient $L_2 = 4 =$ gradient L_1

12 **a** $r = \frac{C}{2\pi}$ **b** $l = \frac{V}{wh}$ **c** $w = \frac{V}{lh}$ **d** $h = \frac{A}{2\pi r}$

13 2.4 cm 15 cm circumference, radius 2.4 cm
3.2 cm 20 cm circumference, radius 3.2 cm
4.0 cm 25 cm circumference, radius 4.0 cm

14 **a** $T = \frac{100I}{PR}$ **b** $T = \frac{PV}{k}$ **c** $b = \frac{2A}{h}$
d $w = \frac{P-2l}{2}$ or $w = \frac{P}{2} - l$
e $n = \frac{X-m^2}{m}$ or $n = \frac{X}{m} - m$
f $n = 5M - 2$
g $Q = \frac{2P}{3} + t$
h $h = \frac{A}{2\pi r} - r$

15 **a** $y - 2 = \frac{x+1}{a}$
$a(y-2) = x + 1$
$a(y-2) - 1 = x$
b $x = b(z-5) + 1$
c $x = n(m-p) - 3$

16 **a** $4y - y = g + 6$
$3y = g + 6$
b $-3y = -7 - q$ or $3y = q + 7$
c $-2y = 2t + 3$ or $2y = -2t - 3$

17 **a** $x = \sqrt{y}$ **b** $z = \sqrt{\frac{y}{5}}$ **c** $x = \sqrt{2y}$
d $r = \sqrt{\frac{A}{\pi}}$ **e** $x = y^2$ **f** $s = \frac{t^2}{3}$
g $t = P^2 - r$ **h** $t = 2m^2$ **i** $r = \sqrt{\frac{A}{4\pi}}$
j $r = \sqrt{\frac{V}{\pi h}}$ **k** $h = \frac{3V}{\pi r^2}$ **l** $r = \sqrt[3]{\frac{3V}{4\pi}}$

18 **a** $u = \sqrt{v^2 - 2as}$ **b** $a = \frac{v^2 - u^2}{2s}$
c $u = \frac{s}{t} - \frac{at}{2}$ or $u = \frac{s - \frac{1}{2}at^2}{t}$
d $a = \frac{2(s-ut)}{t^2}$ or $a = \frac{s-ut}{\frac{1}{2}t^2}$

19 $b = \sqrt{c^2 - a^2}$

20 $x = 5y^2 - 2$

20.6 Proof

1 **a** Expression **b** Equation **c** Formula
d Expression **e** Equation **f** Equation

2 **a** $x^2 - 2x$ **b** $x^2 - 3x - 4$ **c** $x^2 + 4x + 4$

3 **a** $3(x+1)$ **b** $x(x-3)$ **c** $2(2x-1)$

4 $30\,\text{cm}^2$

5 **a** $x = 7$
b No, only true for $x = 7$
c Equation

6 **a** Equation **b** Formula
c Equation **d** Expression

7 **a** LHS $\equiv x \times x^2 - x \times 4 \equiv x^3 - 4x \equiv$ RHS
b LHS $\equiv x^2 - 2x - 2x + 4 + 4x \equiv x^2 + 4 \equiv$ RHS
c RHS $\equiv x^2 + 5x + 5x + 25 - 4x \equiv x^2 + 6x + 25 \equiv$ LHS
d RHS $\equiv x^2 \times 2x - x^2 \times 5 \equiv 2x^3 - 5x^2 \equiv$ LHS
e LHS $\equiv x^2 - 4x + 4x - 16 \equiv x^2 - 16 \equiv$ RHS
f RHS $\equiv x^2 - ax + ax - a^2 \equiv x^2 - a^2 \equiv$ LHS

8 **a** $x^2 + 5x + 6 + x + 3$
b $x^2 + 6x + 9$
c $x^2 + 6x + 9$
d Statement in part **b** = statement in part **c**

9 **a** $(x+3)(x+1) = x^2 + 4x + 3$
b $x^2 - x$
c $x^2 + 4x + 3 - (x^2 - x) = 5x + 3$

10 Area of rectangle $= (x-1)(x+1) = x^2 - 1$
Height of triangle $= 3x - 1 - (x-1) = 2x$
Area of triangle $= \frac{1}{2} \times 2x(x+1) = x^2 + x$
Total area $= x^2 - 1 + x^2 + x = 2x^2 + x - 1$

11 **a** $n, n+1, n+2$
b $n-1, n, n+1$

12 **a** $n-2, n-1, n, n+1, n+2$
b Range $= n + 2 - (n-2) = 4$
c Mean $= \frac{n-2+n-1+n+n+1+n+2}{5} = \frac{5n}{5} = n$

13 **a** $n, n+1, n+2$ **b** $3n + 3$
c $3(n+1)$ **d** $3(n+1)$, which is $(n+1) \times 3$

14 Four consecutive numbers: $n, n+1, n+2, n+3$
Sum $= 4n + 6 = 2(2n+3)$
2 is a factor of $2(2n+3)$ so it is even.

15 **a** **i** 2, 4, 6, 8, 10 **ii** 1, 3, 5, 7, 9
b **i** $2n$ **ii** $2n - 1$ or $2n + 1$
c Odd

16 $2m + 2n = 2(m+n)$
This is a multiple of 2, so it is an even number.

17 $2m + 1 + 2n + 1 = 2m + 2n + 2 = 2(m + n + 1)$.
This is a multiple of 2, so it is an even number.

18 **a** $(2m \times 2n) = 4mn =$ multiple of 4 = even
b $(2n)^2 = 2n \times 2n =$ even number, because product of 2 even numbers is even, from part **a**
or $(2n)^2 = 4n^2 =$ multiple of 4 = even

19 $2(x-a) = x + 4$
$2x - 2a = x + 4$
$x - 2a = 4$
$x = 2a + 4 = 2(a+2)$
Since a is an integer, $a + 2$ is an integer, and $2 \times$ an integer is a multiple of 2, or an even number.

20 **a** Example: $2(2+4) = 2 \times 6 = 12 =$ multiple of 4
b $2(2m + 2n) = 4(m+n) =$ multiple of 4

20 Check up

1 a

x	−3	−2	−1	0	1	2	3
y	−29	−10	−3	−2	−1	6	25

b,c

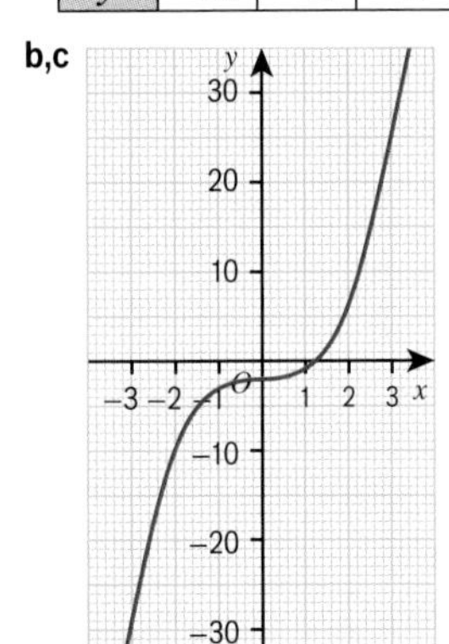

2 a

x	−3	−2	−1	−0.5	0.5	1	2	3
y	$-\frac{1}{3}$	$-\frac{1}{2}$	−1	−2	2	1	$\frac{1}{2}$	$\frac{1}{3}$

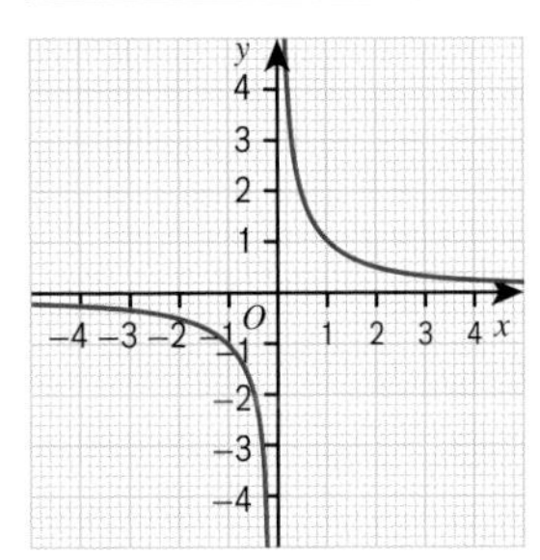

b i Answers between $y = -0.6$ and -0.8
ii Answers between $x = 0.2$ and 0.4

3 a Answers between 6 and 6.4 litres
b 2.5 kilopascals
c Answers between 7.5 and 9 litres

4 $3y + 2x = 9$

5 $x = 2, y = 3$

6 a $x + y = 24$; $x - y = 8$ b 16 and 8

7 $x = 4, y = 2$

8 a $t = m - pr$ b $c = \frac{z-y}{a}$
c $m = \frac{p}{3} - n$ d $r = \frac{pq}{v}$

9 a $t = \sqrt{R}$ b $p = \frac{nx}{t}$
c $x = 8 - 2y + 2z$ d $x = \frac{y-4}{5}$

10 Expanding the brackets: $xy(x-y) \equiv x^2y - xy^2$

12 Star = 3, triangle = 5, square = 1

20 Strengthen

Graphs

1 a Negative   b Positive
c A is $y = -x^3$, B is $y = x^3$
d Graph A with the label $y = -x^3$ and Graph B with the label $y = x^3$

2 a (−2, 7) and (3, −26)   b (−2, −9) and (3, 26)

3 a

x	−3	−2	−1	0	1	2	3
y	−26	−7	0	1	2	9	28

b

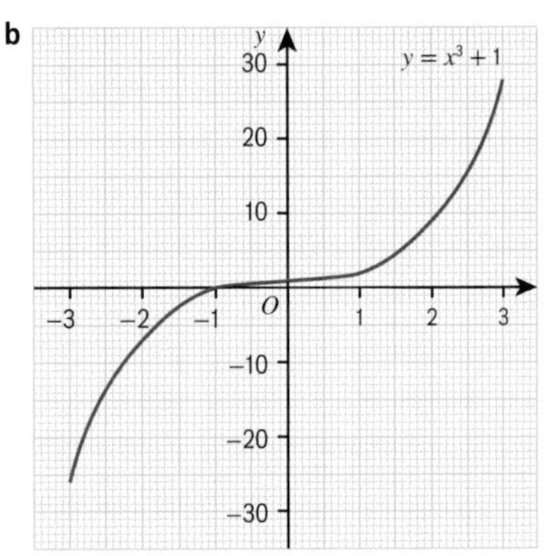

4 a $y = x^2$ b $y = \frac{1}{x}$ c $y = x$

5 a (2, 2) and (−2, 0.5)
b (2, 0.5) and (−2, −0.5)
c

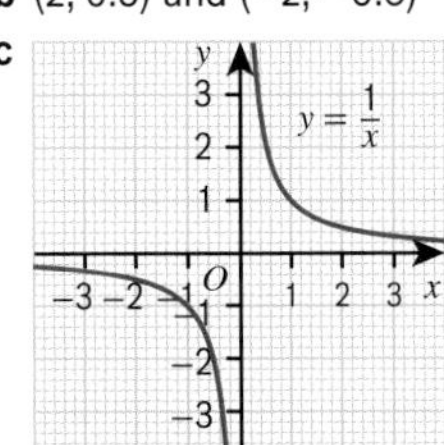

6 a Time in weeks
b Number of rats
c i 14   ii 32
d The number of rats increased by 18.

Simultaneous equations

1 a $2x + 3y = 14$ b $3x + 5y = 10$
c $6x + 10y = 70$ d $2x + 4y = 18$
e $5x + y = 65$

2 a $x = 6, y = 2$
b i $x = 3, y = 5$ ii $x = 1, y = 3$

3 a $x = 10, y = 2$ b $x = 14, y = 6$
c $x = 7, y = 3$ d $x = 5, y = 8$

4 b $2x$ is a multiple of x c $2x + 4y = 10$
d $x = 3, y = 1$

Using algebra

1 a $a = \frac{s-3}{t}$ b $a = qy - p$
c $a = v - 4x$ d $a = \frac{y-b}{4}$

2 a $d = \frac{x}{2} - 3$ or $d = \frac{x-6}{2}$
b $d = \frac{s}{2} + t$ or $d = \frac{s+2t}{2}$
c $d = \frac{2y}{t}$ d $d = \frac{kr}{t}$

3 a $n = 9$ b $n = c$ c $n = \sqrt{r}$
d $n = \sqrt{T}$ e $n = \sqrt{6y}$

4 a $x = 18$
b i $m = \frac{np}{3}$ ii $m = \frac{2k}{t}$ iii $m = \frac{qr}{p}$ iv $m = \frac{vz}{b}$

5 a $x = 2$
b i $x = y + 6$ ii $x = 2y$
iii $x = 7 - y$

6 a $x = 4$
b i $x = 3y - 15$ or $3(y-5)$ ii $x = a(z-2)$
iii $x = b(t - y - 4)$

7 **a** LHS simplifies to RHS $8a+b$

b LHS simplifies to RHS $5x^2+x-5$

c LHS simplifies to RHS $2x+8$

d Both sides simplify to x^2y+2x^2-4y

20 Extend

1 **a**

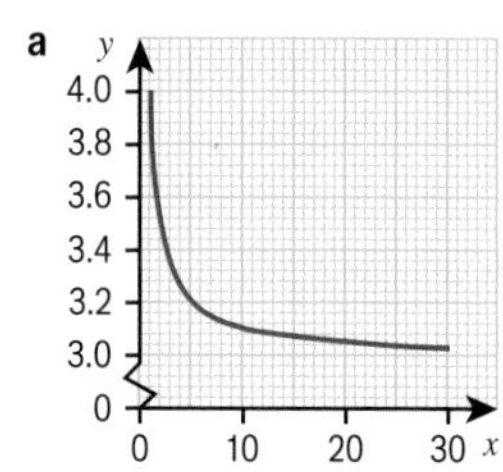

b Neither, as xy is not a constant.

2 **a** Estimates between 90 and 100

b Estimates between 70 and 80

c Between week 6 and week 7

d Week 8

3  **a**

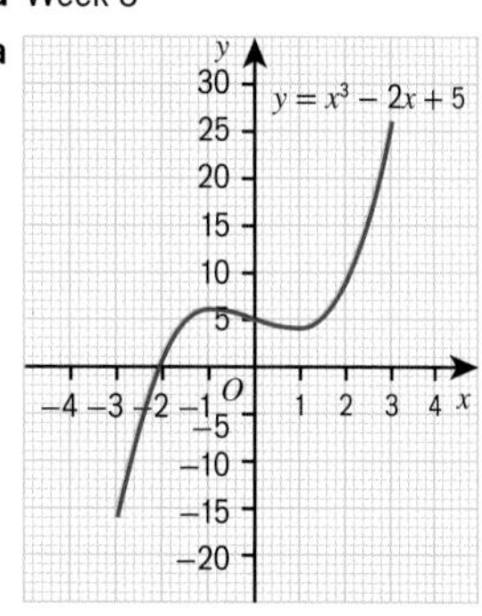

b –2.1

4

Sample	Mass (g)	Volume (cm^3)
a	25	2.8
b	60	6.7
c	90	10.1
d	125	14.0
e	240	26.9
f	350	39.2

5 4%

6 **a** $x=\frac{p-m-n}{r}$ **b** $x=\frac{c+ab}{a}$

c $x=\frac{de}{f-d}$ **d** $x=\frac{gh+n}{m-g}$

e $x=\frac{rs+pq}{p-r}$ **f** $x=\frac{cd-a^2}{a}$

7 **a** $8+14=9+13=22$

b Students' own answers.

c 8, n; 9, $n+1$; 13, $n+5$; 14, $n+6$

d $n+n+6=n+1+n+5=2n+6$

8 **a** $x^2+3x+12=35$, so $x^2+3x=23$

b When $x=3$, $x^2+3x=18$
When $x=4$, $x^2+3x=28$
So a value of x between 3 and 4 gives value of $x^2+3x=23$.

9 Volume of cube $=x^3$; volume of cuboid $=8x$.
So $x^3=8x+100$.
Rearrange to give $x^3-8x=100$.

10 **a** Example: $3^2=9$

b $(2n+1)^2=(2n+1)(2n+1)=4n^2+4n+1$
$=4(n^2+n)+1$
= multiple of 4 (even) + 1

20 Test ready

Sample student answers

1 The student should multiply equation (2) by 2 first, so that adding will eliminate the y term.

2 The student has subtracted $2s$ instead of dividing both sides by $2s$: $\frac{v^2-u^2}{2s}=a$

3 After $c^2-b^2=a^2$, the next line of working should be $a=\sqrt{c^2-b^2}=\sqrt{169-144}=5$ cm.

20 Unit test

1 **a** $x=\frac{z-4y}{4}$ **b** $x=\frac{rs}{t}$

2 $j=e+10$ and $j+e=58$

a 34 **b** 24

3 **a** **i** Answers between 32 000 and 34 000 Pa

ii Answers between 16 000 and 20 000 Pa

b Answers between 6400 and 6600 m

c From 0 m to 1000 m, actual change = 12 000 Pa
Percentage change $=\frac{12\,000}{100\,000}\times100=12\%$

4 **a** $r=n^2+s$ **b** $r=F(x+3)-2$

5 Inverse proportion

6 **a** $x=3, y=\frac{1}{2}$

b $M(4, 1)$

7 **a**

x	–2	–1	0	1	2	3
y	–21	–11	–7	–3	7	29

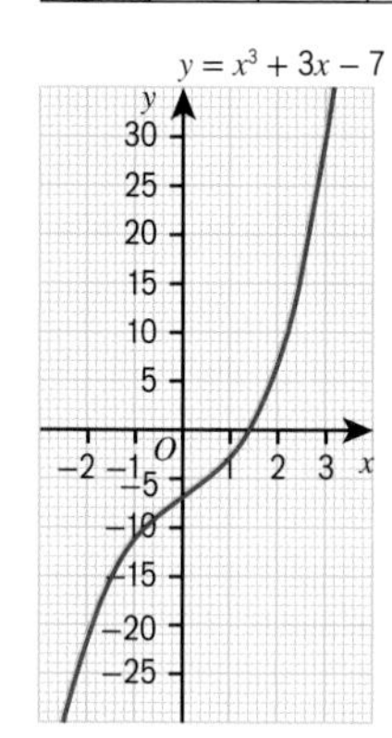

b Answer between 1.4 and 1.5

8 $x=7.5, y=-2.5$

9 LHS $\equiv(n+2)^2\equiv n^2+4n+4$
RHS $\equiv(n+1)^2+2n+3\equiv n^2+2n+1+2n+3\equiv n^2+4n+4$
LHS ≡ RHS

10 £16

11 **a** $x=\frac{2s}{T^2}$ **b** $x=-\frac{n^2}{27}+\frac{5}{3}$

12 a Students' own answers, e.g.

5	6
15	16

b Students' own answers, e.g. $5 \times 16 = 80$, $6 \times 15 = 90$, difference 10.

c

n	$n+1$
$n+10$	$n+11$

$n(n+11) = n^2 + 11n$
$(n+1)(n+10) = n^2 + 11n + 10$
$(n+1)(n+10) - n(n+11) = 10$

13 Students' own answers.

Mixed exercise 6

1 $1\frac{19}{30}$ → **18.1**

2 Equations: $14 = 3x + 2$, $3(x+2) = 21$, $r^2 = 4$
Expressions: $3(x+2)$, $3x+2$
Formulae: $S = \frac{D}{T}$, $A = \pi r^2$
Identities: $3x + 6 = 3(x+2)$ → **20.6**

3 a E → **20.1**
b F → **20.1**

4 $15\frac{1}{8}$ litres → **18.1**

5 B, C, A, D → **18.3**

6 $5(2n+1) + 3(2n+5) = 10n + 5 + 6n + 15 = 16n + 20$
$= 4(4n+5)$
As this is $4 \times$... this is always a multiple of 4. → **20.6**

7 $F = \frac{9C}{5} + 32$ or $F = \frac{9C+160}{5}$ → **20.5**

8 a 6 cm → **19.2**
b 6 cm → **19.2**
c $x = 35°$ → **19.1**

9 7×10^{-6}, 6×10^{-5}, 0.002, 9×10^{-3} → **18.4**

10 No, as the scale factors for the length and width are not the same. The scale factor for the length is $\frac{2}{3}$, but the scale factor for the width is $\frac{1}{2}$. → **19.2**

11 Area $= \frac{1}{2}(x + 2 + x + 6)(x+1)$
$= \frac{1}{2}(2x + 8)(x+1)$
$= (x+4)(x+1)$
$= x^2 + 5x + 4$ → **20.6**

12 a 1.56×10^7 → **18.3**
b 0.000 375 → **18.4**
c No, as 41 020 000 is greater than 5 345 000. → **18.3**

13 a A regular polygon has angles that are all the same size, so the angles are the same size in both pentagons. → **19.3**
b 144° → **19.3**

14 a Inverse proportion → **20.2**
b 5 cm → **20.2**
c 40 cm^3 → **20.2**

15

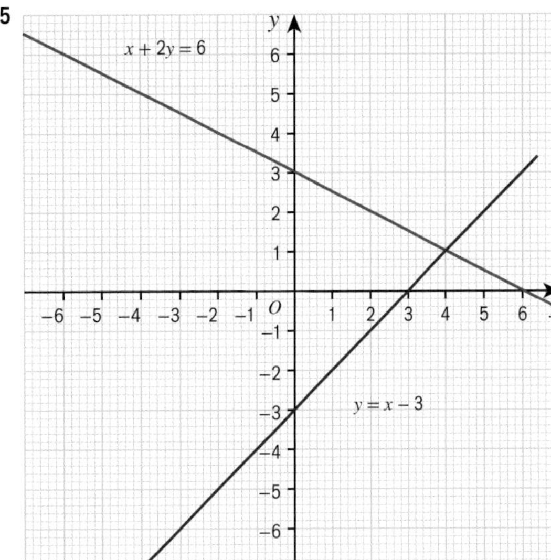

$x = 4$ and $y = 1$ → **20.3**

16 1.881×10^{-2} → **18.5**

17 a $\begin{pmatrix} 6 \\ 7 \end{pmatrix}$ → **19.6**
b $x = 3$ → **19.7**

18 a 6 cm → **19.2**
b 2 cm → **19.5**
c 54.5° → **19.4**

19 Tea £1.90, coffee £2.60 → **20.4**

20

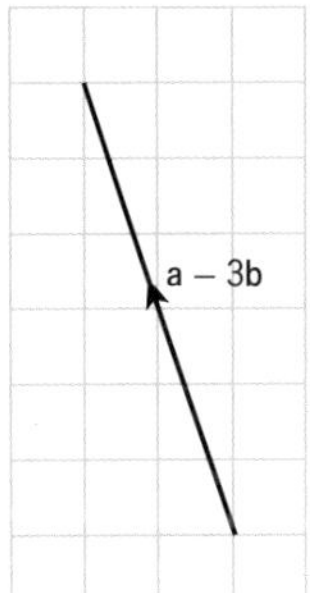

→ **19.6, 19.7**

21 0.0002, 2^{-3}, 0.2, 2^{-1}, 2^0 → **18.2**

22 24 cm → **19.5**

23 Gradient of $L_2 = \frac{6 - -2}{-1 - 7} = -1$
Equation of L_2 is $y + 2 = -(x - 7)$ so $x + y = 5$
Equation of L_1 can be rewritten as $3x - y = 7$
Adding the 2 equations gives $4x = 12$, $x = 3$
Substituting in either equation gives $y = 2$
So point of intersection is (3, 2) → **20.4**

Index